EMC Simulation of Electronic Systems

Jan Hansen • Andreas Barchanski

EMC Simulation of Electronic Systems

Jan Hansen
Institute of Electronics
Graz University of Technology
Graz, Austria

Andreas Barchanski
Simulia Electromagnetics
Dassault Systemes
Darmstadt, Germany

ISBN 978-3-032-09236-6 ISBN 978-3-032-09237-3 (eBook)
https://doi.org/10.1007/978-3-032-09237-3

This Springer imprint is published by the registered company Springer Nature Switzerland AG
The registered company address is: Gewerbestrasse 11, 6330 Cham, Switzerland

Foreword

This is an excellent book on the simulation of EMC problems. It teaches you how to select and use the appropriate tool at the right time to analyze and optimize EMC designs. It also provides insight into the underlying methodology of the tools. Each chapter covers different situations that you may encounter when learning to apply simulation to EMC. The examples are implemented using CST, but the authors have organized the book such that they can be solved with other simulation tools of the same functionality.

However, please do not expect this to be an engineering book that tells you how to solve EMC problems; this "translation" of the methodology is highly specific to particular problems. All examples are tailored to situations and tasks that occur in the EMC engineers' life. They are also made to help the readers to develop their own modeling strategies.

From EMC research and teaching, I can say that good simulation skills come from more than just knowing how to use software. You also need to understand the physics behind what you are simulating and build models carefully. As with measurements, you need to formulate your expectations before measuring or starting a simulation so that you can improve your understanding of both the system behavior and the test setup or simulation model. This book captures those principles with clarity and depth. Even experienced EMC engineers will be surprised at how much they can learn from the material presented here.

I am confident that readers—from students to professionals—will find this book a useful guide and a resourceful reference.

Institute of Electronics, Graz, Austria
May 2025

David Pommerenke

Acknowledgments This book would not have been possible without a helping hand and mind from many contributors.

We like to thank the Dassault Systèmes Simulia senior management for supporting this project and providing Andreas with a license of the simulation software package. Particularly the division leader of electromagnetic simulation Peter Hammes for approving this publication.

Furthermore, we thank Andreas' direct manager Frank Scharf who supported this project from day one and gave him the opportunity to reduce his daily work time to concentrate on writing this book.

We thank Bernd Deutschmann, Head of the Institute of Electronics of Graz University of Technology, for the encouragement and support in writing this book, without which it may have been published much later or never.

We thank David Pommerenke for his detailed advice and all discussions, as well as his guidance when writing Chap. 14 on ESD.

We highly appreciate the contributions of Marcel Plonka who designed all the printed circuit boards shown in this book. Without these examples, it would not have been possible to showcase any application treated in this book.

We thank Frank Demming-Janssen from the company Simuserv for supporting the license at the Technical University Graz.

We thank our colleagues from industry, Sayantan Dhar, Sergey Kochetov, Bibhu Nayak, Deepak Pai, and Patrick DeRoy for providing their insights into EMC simulation carried out in companies.

We thank Jan Henninger for testing, running, and investigating many of the electromagnetic models used in this book. We thank Patrick Gsöls for contributing the simulations and evaluations made in the machine learning chapter. We appreciate the production of many of the figures by Christoph Maier and Julian Tischler, and the final check of the book made by Christian Riener, Jan Henninger, Simon Profanter, and Julian Tischler.

We like to thank all reviewers of the individual sections or chapters, as there are Michelangelo Bandinu, Thomas Bauerfeind, Egor Dranischnikow, Patrick Gsöls, Herbert Hackl, Matthias Hampe, Martin Heinisch, Jan Henninger, Christoph Keller, Christian Kremers, Clemens Pechstein, Scott Piper, Mustafa Raya, Christian Rieckmann, Christian Riener, Daniel Sanchez, Yvonne Späck-Leigsnering, and Radu Voina.

We appreciate all the detailed discussions with Djamel Guezgouz, Stephan Koch, Tobias Leitgeb, Sascha Ranftl, Richard Sjiariel, and Martin Stoiber.

Jan Hansen thanks his former colleagues at Robert Bosch GmbH, who taught him how much it takes to design an EM compatible industrial product.

We appreciate the use of the Inkscape Electrical Symbols by Oliver Wallscheid. We thank the Center for Collision Safety and Analysis at George Mason University for their vehicle models. We like to mention the high professionality of the collaborative, cloud-based LaTeX editor overleaf, with which this book was written independently at two different places.

And both authors like to thank their families for their patience and support during the writing of this book.

Competing Interests The authors have no competing interests to declare that are relevant to the content of this manuscript.

Darmstadt, Germany
Graz, Austria
May 2025

Andreas Barchanski
Jan Hansen

How It All Started…

"…one should write a book about all this stuff…," "I always wanted to," "Hey, me too!"

This short conversation in November 2022 in a call in which we were discussing the content of a joint paper is the starting point of a long process that led to the book you are now reading.

Contents

Chapter 1
Introduction

This is a book about computer simulations of the EMC of electronic systems. EMC is concerned with the emission of electromagnetic disturbances and the immunity against them. Designing for EMC requires addressing multiple scales in geometry, signal amplitudes, and frequencies, making EMC simulation a challenging task.

Despite these challenges, EMC simulation has become a success story. For many leading companies, it has evolved into an indispensable asset in their design processes. This success began when engineers recognized the impracticality of computing entire models as a single entity using a single method. Instead, EMC models of electronic systems, much like the systems themselves, are assembled from component models. Each component model has its own specific description—it could be a circuit model, measurement data, or an EM model derived from geometric data, with multiple approaches available to compute its properties.

There are countless ways to construct an EMC simulation model, a process we refer to as modeling. In the fast-paced environment of competitive industrial design, identifying the most efficient modeling approach is crucial. However, modeling through simulations is not a simple push-button solution. Achieving reliable and fast simulations requires a deep understanding of the physical effects being modeled and the numerical methods used to solve them.

This book combines information about the domains that a simulation engineer should master, in order to successfully model and simulate EMC phenomena. These domains are summarized in Fig. 1.1.

Each domain in Fig. 1.1 represents a key domain of knowledge covered in this book:

Electromagnetic Compatibility (EMC) The EMC domain defines the scenarios simulated in this book, focusing on challenges EMC engineers face daily. Topics include emission, susceptibility, crosstalk, and signal and power integrity. Additionally, we explore the modeling of various EMC test methods through simulation.

J. Hansen, A. Barchanski, *EMC Simulation of Electronic Systems*,
https://doi.org/10.1007/978-3-032-09237-3_1

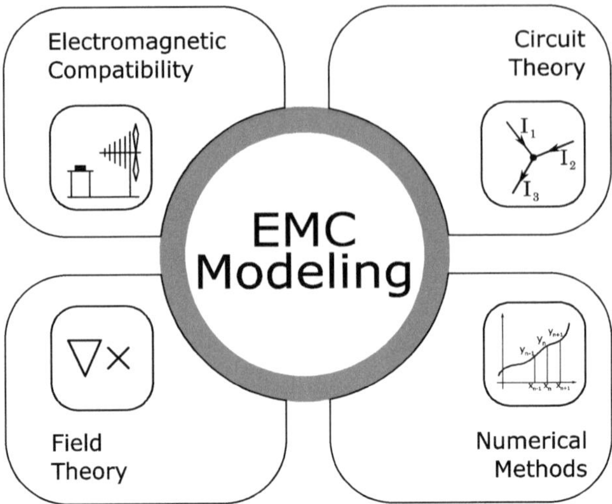

Fig. 1.1 EMC modeling requires knowledge of EMC, of circuit and field theory, and of numerical methods. These topics are treated in this book

Circuit Theory This domain covers the representation of electronic systems through circuit simulation. It examines the use and limitations of different circuit models and explores how circuit models can be derived from electromagnetic field computations.

Field Theory Field theory focuses on the use of electromagnetic field computation to model electronic systems or their components. We introduce a number of numerical methods for the solution of Maxwell's equations and explain their application in the simulation of electronic systems. Emphasis is placed on interpreting results from field computations and understanding their relevance in EMC simulation.

Numerical Methods This domain focuses on the techniques used to simulate and solve circuit and field computation models. We discuss the trade-offs in simulation speed and usage of computational resources among various methods and provide guidance on selecting the most efficient approach for specific problems.

The goal of EMC simulation is to support design decisions whenever it provides more appropriate results than any other available method. Additionally, EMC simulation enables the investigation of EMC properties of an electronic system even in the absence of physical hardware or measurement equipment.

In this book, we demonstrate—using the classical topics of EMC, as described, e.g., by Ott (2009) and Paul (2005)—how to generate EMC predictions and design decisions by simulation. Each chapter focuses on a specific topic: we define a problem to be solved, explain its background, and describe the most suitable numerical methods and modeling approaches. We give one or a few practical examples to illustrate the selected approach and discuss its limit. With each chapter, more numerical methods become available, and their combination allows to solve more advanced EMC engineering tasks.

In later chapters, we examine the simulation of setups defined in EMC standards, which result in computationally intensive models. To enhance simulation performance, we introduce simplifications and analyze the deviations they introduce. The primary focus remains on industrial EMC design and efficient workflows.

If explanations of numerical methods or theoretical background information become more detailed than necessary for solving a task, we present them in dedicated advanced sections at the end of each chapter. We employ a commercially available state-of-the-art software, with which (or any equivalent software of similar capability) trained EMC engineers can solve the discussed problems.

This book is not an introduction to EMC phenomena. The reader needs to have EMC knowledge and familiarity with electronic systems. The focus is on modeling—that is, identifying and integrating methods to efficiently solve specific EMC engineering problems through simulation. The book does not cover general EMC design principles. Neither does it discuss the validation of simulations through measurements. For EMC design principles, we refer the reader to EMC design books such as those listed in Keller et al. (2025). For validation methods, we refer to technical and scientific literature or encourage the reader to compile a dedicated reference.

Readers may choose to follow the chapters sequentially, gradually learning new methods. Alternatively, they can select chapters of interest independently. If methods from previous chapters are necessary for understanding a topic, references are provided.

1.1 EMC Modeling

The physical mechanisms that determine the EMC of electronic systems are contained in the equations provided by electromagnetic theory. EMC simulation challenges their solution.

EMC is monitored over a wide frequency range, from some Hertz in case of magnetic fields to several GHz in case of radiated emission and immunity. Amplitudes are also monitored over several orders of magnitudes, such as several 100 V supply voltage for an automotive DC DC converter down to some dBμV at FM frequencies. Coupling paths are subtle, changes in capacitances of some fF may or may not alter the emission or immunity of an electronic system.

The simulation model must reproduce the EMC properties of the electronic system. Figure 1.2 shows two such setups: On the left, we see the conducted emission test of a the traction inverter of an electric vehicle, displaying the inverter in black, and shielded high-voltage supply wires in orange and the load/break on the right. In the right-hand figure, the inverter is shown as mounted in an electric vehicle.

Fig. 1.2 Two scenarios relevant for EMC design. Left: Conducted emission test of a traction inverter. Right: Traction inverter mounted in an electric vehicle

In both scenarios, we encounter a multi-scale problem both in the physical dimensions and in time: the smallest dimension within the traction inverter is in the order of μm, which is the thickness of a trace of a PCB, and the geometrical details of the transistors are even smaller. The dimension of the measurement table of the conducted emission voltage setup, or that of the vehicle in which the inverter is mounted, is in the order of meters. The ratio of both is a measure for the complexity of the resulting simulation model, and it is 1:1,000,000. Likewise, the smallest time step in the operation of the inverter is in the range of nanoseconds, referring to the switching transistors. The machine rotates about 1000 times per second, again yielding a ratio of 1:1,000,000 in timescale. Current control and driving cycles have even longer time ranges.

With commonly available resources in industry, it is impossible to model such a system in a single piece. One solution is to stock up to computer farms as, e.g., used for crash simulations. With this book, we demonstrate that relevant EMC problems and tasks can be tackled without high-end hardware. All presented modeling strategies work on a single, well-equipped, desktop computer (details are given in Sect. 1.5). We guide the reader to find the most efficient modeling approach to solve her or his EMC problem and to raise awareness for its potentials and limits.

Whereas computing resources are available for money, model data are not. The product design process determines at which point of time which data are provided by whom. A typical design process is shown in the form of a V-model in Fig. 1.3. It starts with the requirements analysis of the product, followed by the design phase, the integration of the design into a larger system, and it is finalized with the qualification of the system by testing.

In the beginning of the design process, there are only little data about the prospective product available. No computer farm is needed to solve complex models, because there is neither layout nor mechanical data, nor a detailed specification of electronic components. The further we go in the design process, the more

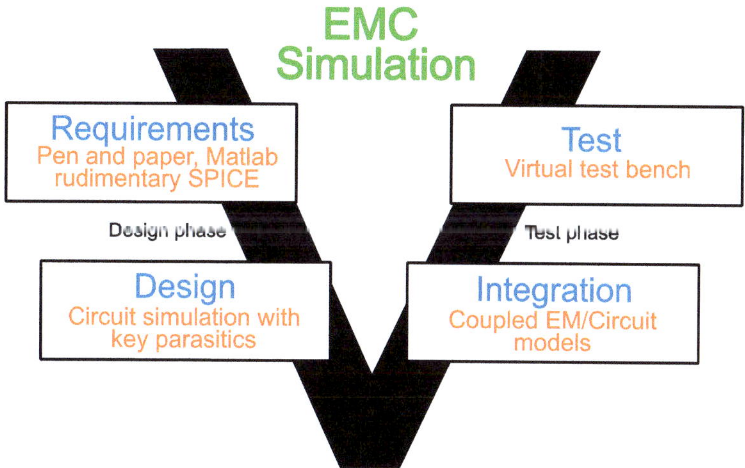

Fig. 1.3 The V-model of EMC simulation

information is released. The more information there is, the more precise a model can be, and the more elaborate is its generation and computation.

For EMC simulation, the design and the integration phase offer the best trade-off between model complexity and model effectiveness. In the design phase, models are rough, often on circuit level, but there are still many degrees of freedom to change in the system under test. In the integration phase, mechanical CAD and hardware prototypes are available. Model results match measurement results more closely, but the modeling effort and computation time are higher. Design improvements are harder to realize, so that the actual impact of simulation on the present design is lower. At the end of the design process, the EMC qualification, simulation is possible but hardly used. Simulation models are complex. At this point in time, the computer farm could speed up the simulation time significantly. However, C-sample hardware is also available, and the measurement performed on this hardware is, at least up to now, the reference. The simulation model does not contribute much.[1]

Modeling proceeds quicker and more accurate if engineering knowledge is considered. In EMC engineering, the common approach decomposes an EMC problem into the three contributions of source—coupling path—sink (Fig. 1.4). This decomposition is particularly useful to compose a simulation model, because it helps to select the appropriate modeling level for each contribution. At the beginning of each section of this book, the icons shown in Fig. 1.4 indicate the section's focus on one or more of these contributions. Often one contribution is dominant, and its modeling requires the highest attention.

[1] An exception is fire-fighting close to product release. We claim that if simulation is applied appropriately in the previous steps of the V-model, fire-fighting is unlikely.

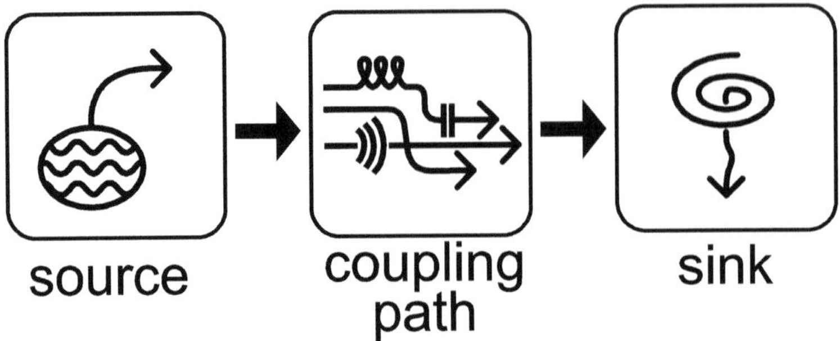

Fig. 1.4 We can decompose any EMC problem into the chain source—coupling path—sink

Before composing a simulation model, there should be answers to the following questions:

- Which questions shall the model answer?
- Which information is available?
- How much time do I have?

With little time and information, a modeling level of low complexity must be chosen. On such a low level, some questions can be answered, but some others may not. The medium and high modeling levels may answer these other questions, but could consume more time than necessary to answer simple questions.

The different modeling levels apply for each individual contribution of source, coupling or sink. Examples are given in the following:

Low (Design phase) Uses broadband sources and simplified geometries for fast simulations. Suitable for comparing design variants; provides quick but approximate insights.

Medium (Integration phase) Incorporates improved source modeling and is capable of predicting absolute values. Often represents the optimal balance between accuracy and effort.

High (Test phase) Enables comparison with measurements. Requires detailed modeling of the measurement setup and high-accuracy component models. Time-consuming to build and validate.

We demonstrate these levels using the simulation of conducted emission from a DC-DC converter:

A simple simulation model contains, e.g., the S-parameters of the input filter of the converter that characterizes its insertion loss. For this simulation, only the layout of the filter is required and no information about the actual switching of the converter. Another model for the same converter might include only the switching loop of the converter. We can calculate the inductance of the switching loop and optimize it as reducing the loop inductance typically reduces the emission.

A medium complexity model includes the complete converter with the power supply, input filter, the switching stage, the output filter, and a load. We assume the load to be fixed and located close to the output of the converter. The switching of the converter is modeled with the frequency that is used by the converter. If available, we use a vendor model of the transistor that is used as the switch of the converter. The result of such a simulation is an emission spectrum at the LISN. We can analyze how the emission spectrum changes if modifications to the design are applied.

A high complexity model that is comparable to measurements needs to include further details: If, for example, the converter uses a spread spectrum technique for the switching, it is required to model the spread accurately. Furthermore, if the converter is driving a non-constant load like charging a battery over a longer cable, it is required to include the behavior of the load and the cable in the simulation model. Otherwise, the comparison to the measurement might not be valid.

The V-model of the product development process shown in Fig. 1.3 serves as a guideline for EMC model development. When building simulation models, we start simple and increase the detail level step by step. Component models from vendors should be checked upfront in simple test simulations in order to make sure that they behave as expected. For the complete model to work properly, all components need to be tested individually. It might be tempting to build a complex model from scratch. If, however, there are some problems in one of the individual building blocks, debugging such a complex model is highly time-consuming. The complex model needs to be dissected into simple model parts until the problem is identified. In consequence, this comes down to testing all components individually, with a lot of time lost beforehand.

1.2 Simulation Methodology

One key aspect of the simulation methods used throughout this book is the coupling of circuit simulation to EM simulation methods. In nearly all presented examples, we employ coupled EM and circuit simulations.

We begin with a definition of the terms circuit simulation and EM simulation: A circuit simulation model operates on voltages and currents. A circuit simulation calculates voltages and currents in a circuit assembled from circuit component models. We use three types of circuit simulation: transient simulation, AC analysis, and S/Y/Z parameter-based tasks.

An EM simulation model, on the other hand, calculates electric and magnetic fields by applying numerical methods to a geometry defined by the simulation engineer. Various EM methods are introduced in this book, including:

- Static and quasi-static methods in Chap. 5.
- Full-wave methods, introduced in Chap. 6 and used extensively in later chapters.
- The MTL method in Chaps. 8 and 11.

The results of an EM simulation can be converted into a circuit component model in different ways:

Manual Approach The simplest, manual approach involves calculating R, L, or C values of geometric entities using an EM method and placing these values as a simple RLC model on a circuit schematic. This method is discussed in Chap. 5.

Semiautomatic Approach A semiautomatic approach generates models containing all RLC values in a single circuit component for use on a circuit schematic. This is also covered in Chap. 5.

Automatic Approach An automatic approach involves generating a (black-box) circuit component model based on S-parameters directly from the EM simulation of the geometry. The S-parameter data are then transferred to the circuit schematic, maintaining a link between the schematic and the geometry. This method is discussed in Chap. 6.

In Chaps. 13 and 14, we discuss co-simulations, where the circuit and EM simulation models are simulated simultaneously.

We use the term "coupled simulation" to refer to a sequential simulation chain, where the EM simulation and circuit simulation are run one after the other. The term "co-simulation" is reserved for cases where the EM and circuit simulations are performed simultaneously.

An important aspect of coupling circuit simulation and EM methods is the format used to transfer EM simulation results to the circuit model. Engineers familiar with SPICE circuit simulations understand coupled schematics where RLC models generated by the EM solver are used on a circuit schematic. However, EM models based on S-parameters require circuit schematics that differ from those using RLC models. We explain why the schematic for port-based S-parameters differs from that of RLC models. A general introduction to S-parameter ports is provided in Sect. 2.4.3.1, with further explanations on using S-parameter-based circuit component models obtained from EM simulations in Sect. 6.7.

Since S-parameters represent frequency domain data, they cannot be directly used in transient circuit simulations. A conversion process is required to generate a macromodel compatible with transient circuit simulations. This process, known as vector fitting, is introduced in Sect. 10.3, with advanced details covered in Sect. 10.9. We use "vector fitting" as a synonym for macromodel generation.

The link between circuit and EM simulation methods described in the automated model transfer is referred to as "combine results." This name originates in the terminology used in the simulation package, and we have taken over this naming into this book. Combined results are detailed in Sect. 6.7.5.

An overview of the applied simulation and coupling methods is shown in Fig. 1.5.

1.3 Simulation Reliability

The cardinal question often asked about EMC simulations is, "How does the simulation compare to reality?" Here, reality is a synonym for "the measurement that was performed," overlooking the fact that measurement results also exhibit

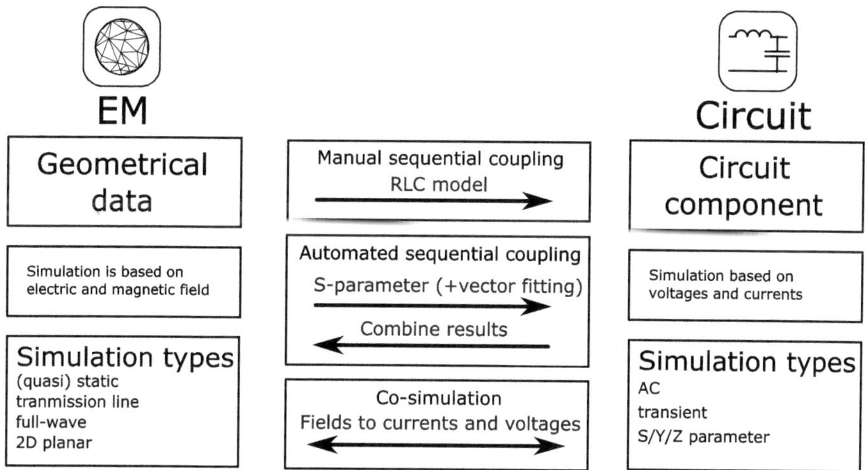

Fig. 1.5 An overview of the simulation methods and coupling approaches used in this book

a variance. The minor part of the variance comes from the measurement device uncertainty itself. The major differences usually come from the difficulty to specify all measurement details as exactly as needed to reproduce the results by a model. This includes the properties of the mounted active and passive components, the routing and grouding of wires, and the software used to operate the system.

1.3.1 Simulation Accuracy

The simple answer to the above-stated question is, “Simulation results accurately predict whatever has been modeled.” However, such a simplistic question and answer are not meaningful. Instead, we break down the term accuracy into two aspects.

Numerical Accuracy Software vendors of commercial simulation tools have relied on various examples to test their solvers for more than 20 years. The results of well-defined simulation models are accurate. The task of a simulation engineer is to create such well-defined simulation models. This includes the appropriate definition of mesh, materials, boundary conditions, and solver settings. We discuss the common pitfalls in circuit simulations in Sect. 3.7 and those in EM simulations in Sect. 4.8.

Modeling Accuracy Modeling accuracy refers to the quality and applicability of the models used to represent the components of an electronic system in a simulation. The simulation models presented in later chapters of this book are built from multiple component models. Consequently, the overall accuracy of a simulation is inherently limited by the least accurate component model. In simple terms: "a chain is only as strong as its weakest link."

As an illustrative example, we refer to the conducted emission simulation discussed in Chap. 10. This simulation requires a circuit model of the transistors and analyzes the emission spectrum up to the 1000th harmonic. However, the transistor models provided by vendors are often not validated at such high frequencies. Similarly, we assume the capacitance of the capacitors to be voltage-independent and the common mode choke to operate in its linear (unsaturated) region. The motor is modeled as an RLC circuit, which is valid only at low frequencies and near its first resonance, but not at higher resonances. The cables connecting the motor and the PCB are represented using generic CAD wire models.

Under these assumptions, the simulation yields accurate results within a limited frequency range and for specific operating conditions. At higher frequencies or under different conditions, more sophisticated models may be necessary. Whether such enhanced models are required to match simulation results to measurements often cannot be determined in advance; it becomes evident only through detailed analysis during the model development process. Although this process can be time-consuming, it frequently provides valuable insights into system behavior—particularly in identifying the components that dominate the EMC performance of a device under test (DUT). This understanding is critical to an effective design process.

Throughout this book, several examples illustrate the limitations of simulation models in terms of frequency range or geometric detail. For instance, the shielding effectiveness simulation of the enclosure in Sect. 7.4 treats the enclosure as a perfect electric conductor (PEC), thereby neglecting low-frequency field penetration. The bulk current injection clamp (BCI) clamp in Sect. 13.1, the antenna in Sect. 12.2.3, and the electrostatic discharge (ESD) generator in Sect. 14.2.2 are represented by simplified, generic models that may deviate from the actual laboratory equipment used in specific measurements.

Understanding the assumptions and limitations of the individual component models used to assemble the simulation model of an electronic system is crucial for the simulation engineer. This knowledge enables a well-founded assessment of the accuracy that can be expected from a given system-level simulation model.

1.3.2 Simulations vs. Measurements

The goal of this book is to explain the modeling process, the simulation setup, and the selection of the most efficient numerical solvers for a variety of scenarios relevant to EMC design. In Riener et al. (2022), we demonstrate an example in which the conducted emission workflow introduced in Chap. 10 is successfully compared to a measurement.

Table 1.1 Properties of EMC simulations and measurements

Simulation	Measurement
Accuracy depends on knowledge about the problem	Knowledge about the problem is unnecessary
Non-ideal simulations are challenging to achieve	Ideal measurements are challenging to perform
Applicable at any stage of development	Restricted to the final stages of development
Models take time to build and validate	Measurements can be performed quickly
Issues can be examined in isolation	Requires the entire system to operate
Provides unlimited information	Information is limited by measurement constraints

When a measurement is performed to investigate the same scenario, the simulation and measurement results will match, provided that all relevant aspects of the measurement setup are accurately replicated in the simulation model. This replication may, but does not necessarily need to, include certain hardware components used solely for measurement purposes.

For relatively simple electronic systems or in test scenarios in which all three, the source—coupling path—sink are well known and modeled accurately, a good correlation between measurement and simulation can often be achieved even by less experienced engineers. However, many challenges in EMC design arise from subtle root causes and even small parasitic couplings can make a significant difference (Benz et al. 2019).

Understanding the influence of such effects in a measurement setup requires a high level of expertise. Without this experience, achieving accurate correlation between measurement and simulation is challenging. Numerous publications address the agreement between simulation results and experimental measurements. Their scope and diversity are too broad to be comprehensively listed here, and we regret that the book that reports on them still needs to be written.

Rather than focusing on the direct comparison of simulation results to measurements, we explore how simulations and measurements can complement each other in a successful EMC design process. Table 1.1 summarizes the key differences between simulations and measurements, which are discussed in detail below. Table 1.1 describes the approach to EMC design within a company with a typical measurement and development budget. Leading companies that have heavily invested into measurement and prototyping capabilities are able to circumvent some of the measurement limitations given in Table 1.1.[2] However, such large investments are limited to a small number of companies and are not available to most EMC engineers.

---

[2] With permission from Scott Piper. Taken from: S. Piper, What I wish I knew about EMC simulation when I first started, in *2024 IEEE International Symposium on Electromagnetic Compatibility, Signal & Power Integrity*.

The value of a simulation lies in its ability to predict the EMC performance of an electronic system. However, using simulation models exclusively to predict the outcome of a measurement limits their potential. A successful simulation engineer leverages simulation capabilities beyond what measurements can offer by creating models that deliver insights relevant to specific stages of the design process, as discussed in Sect. 1.1.

One of the key strengths of simulation is the ability to study parts of an electronic system in isolation and to obtain field values, voltages, and currents at arbitrary positions within the system. These capabilities provide information that is often inaccessible through measurements. However, the simulation engineer must learn to utilize these tools efficiently. Chapter 6 discusses the use of field probes and field monitors in detail, illustrating how they can be applied to gain deeper insights into the electromagnetic behavior of an electronic system.

For EMC simulations of electronic systems, accuracy is often linked to simulation time. Electronic systems are geometrically complex, resulting in high computational demands. Moreover, EMC standards often require large surrounding spaces to be included in the measurement setup, further increasing simulation times and potentially hindering efficient design processes. Later chapters of this book discuss several examples where standardized EMC measurement setups are compared to simplified setups. These simplifications are quantified, and parameter studies are performed using the simplified models. Simplifications are explored for radiated emissions in Sects. 12.2.2 and 12.5.2, for radiated immunity in Sect. 13.4.3, and for ESD in Sect. 14.5.

1.4 Application Examples in This Book

In this book, we describe and examine more than 150 simulation models. An overview of the various application scenarios is presented in Fig. 1.6.

Fundamental properties of simulation are explained using simple models with low runtime, enabling a better understanding of how simulation settings affect the obtained results. They are complemented by application examples, which simulate industry-relevant scenarios and highlight the additional considerations required when incorporating more complex geometries into the simulation setup.

The complexity of the application examples presented in this book remains below that of simulation models encountered in industrial scenarios. Our examples provide an understanding of what is achievable through simulation without excessive details. In order to shed light on the use of EMC simulation in the electronics industry, we gratefully devote Chap. 17 to testimonies written by EMC simulation experts from major companies across the globe.

In several application examples, we use two PCB designs that were created for this book as realistic examples that cover a wide range of topics.

The first PCB is a power electronics example, featuring a motor inverter designed as a B6 bridge. This PCB is shown in Fig. 1.7. The input voltage at the power

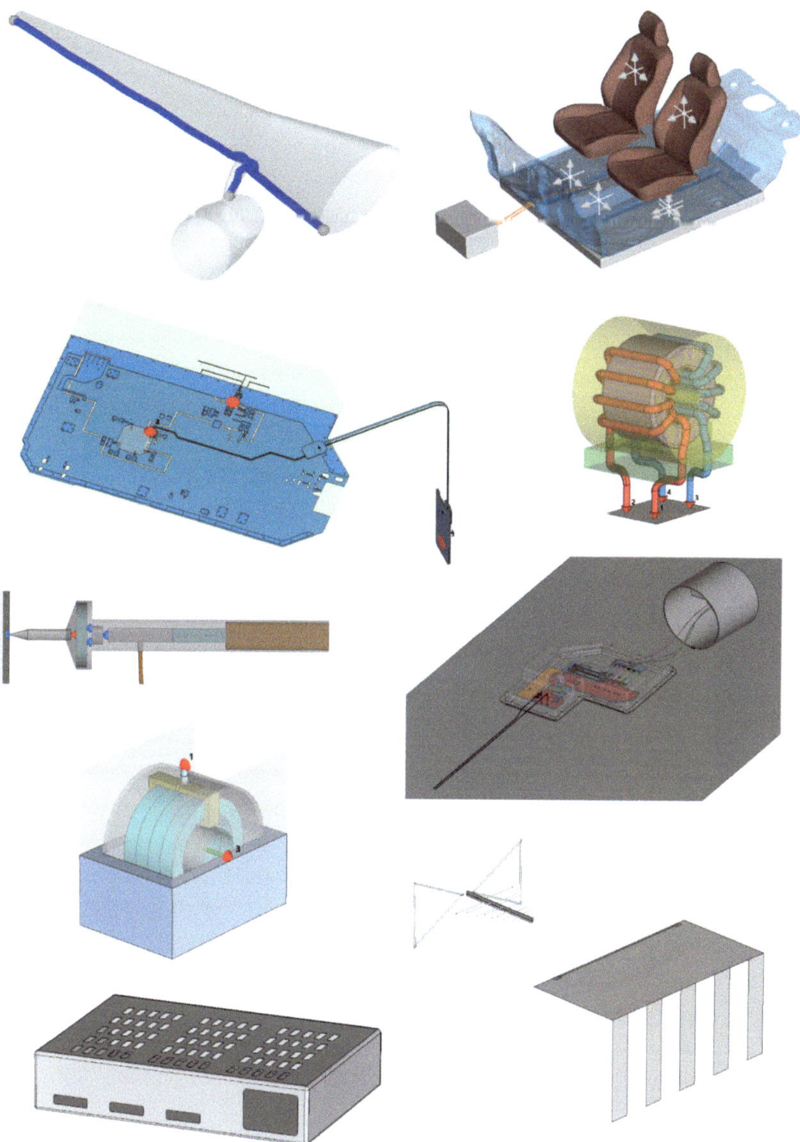

Fig. 1.6 A selection of simulation models found in this book

connector, located on the left-hand side of the PCB, is designed for 48 V. Behind the connector, an input filter with a common mode choke and C_x and C_y capacitors is placed. This filter is discussed in detail in Chap. 9. The layout also includes three DC-link capacitors and a B6 bridge constructed with six MOSFET transistors. The conducted emission from this board is simulated in Chap. 10. A shunt resistor

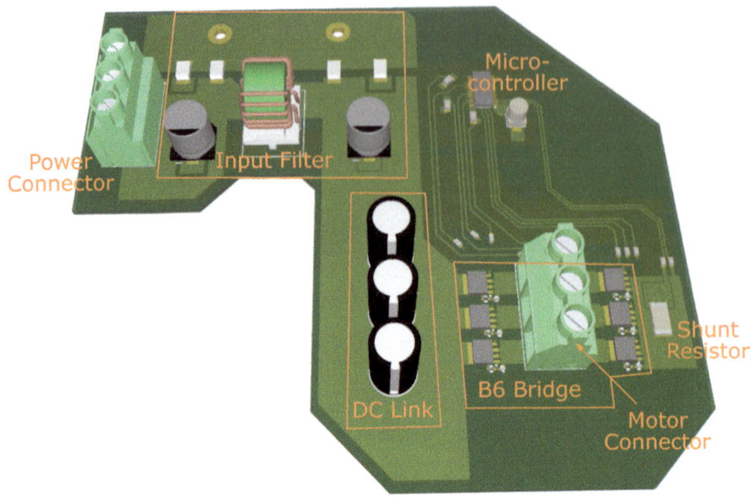

Fig. 1.7 PCB layout of an inverter board used in application examples in this book

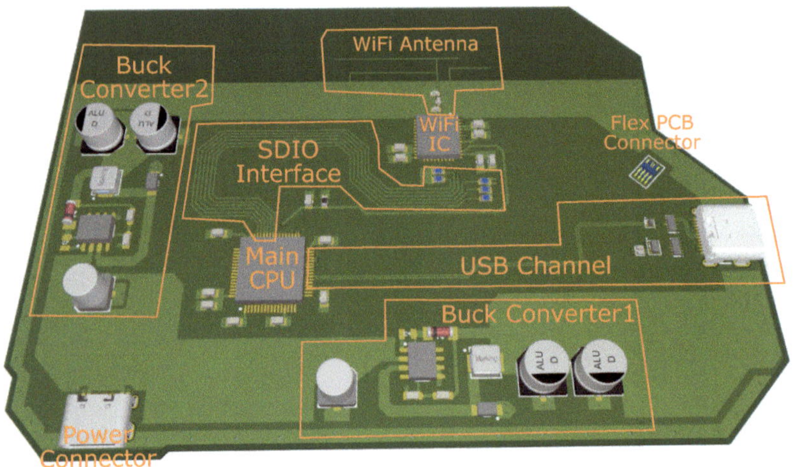

Fig. 1.8 PCB layout of a digital board used in application examples in this book

for current measurement is located on the low side of the bridge, and the effect of galvanic coupling on the shunt current measurement is examined in Sect. 5.3. Radiated emissions from the complete power electronics system, including the PCB and motor, are analyzed in Sect. 12.3.

The second PCB is a digital communication board, depicted in Fig. 1.8. The board is designed for a 12 V input, with the connector located in the lower-left corner. Two buck converters step down the input voltage. Buck Converter 1 reduces the voltage to 5 V, which is routed to the output USB connector on the left-hand side. A circuit-based simulation of this buck converter is discussed in Chap. 3. Buck

Converter 2 generates a 3.3 V output, used as the supply voltage for two digital ICs: a main CPU and a WiFi controller. The second buck converter and the ICs are connected by a power delivery network on the PCB, simulated in Sect. 11.2. The CPU and the WiFi IC communicate through a single-ended interface using the SDIO bus. The conducted emissions from the clock trace of the SDIO bus are analyzed in Sect. 5.4. Additionally, a USB 3.0 channel is routed from the main CPU to the USB connector on the left hand side. This channel is discussed in terms of signal integrity in Sect. 11.1 and radiated emissions in Sect. 12.5. The PCB also features a connector for a flexible PCB, which carries a simplified Low-Voltage Differential Signaling (LVDS) interface. The coupling from the LVDS interface into the WiFi antenna is analyzed in Sect. 15.2.

1.5 Hardware and Simulation Times

The hardware available for the simulation of circuits and EM models is diverse. It ranges from hand-held devices up to large computer clusters. Obviously, the choice of the hardware has a huge impact on the simulation time. In this book, we run the simulations on a single desktop computer of moderate capacity. The machine is equipped with an AMD Ryzen 5900X processor with 12 cores and out of the box overclocking. A total of 128 GB of RAM is available. The GPU is an Nvidia RTX 4080 Super graphics card, a model used for video games. The systems run a Windows 11 Professional OS.

All reported simulation times are obtained on this system. The simulation time can be reduced using stronger hardware. The reported simulation times should not be understood as an absolute measure. They serve to compare the time required to simulate a particular setup with a particular method. With further improvements of the computing power to come in future, the reported times will decrease.

1.6 Software Package

All simulations in this book are run using the CST STUDIO SUITE(R) 2024 (CST) package. This is a commercial software package offered by Dassault Systèmes. We have taken care that the examples and explanations are as generic as possible and can be replicated with other software packages. If a certain feature or simulation setup is related to a special functionality of this simulation package, we note this in the corresponding chapter. Some of the terminologies we use are aligned to the terminology of the CST, but can often be found under a different name in other packages.

Further Reading

Introduction to EMC and EMC Simulation Tools

Altair, *Electronic System Design*. https://altair.com/product-showcase#filter=.application-electronic-system-design (visited on 2025)

Ansys, *Best-In-Class 3D High Frequency Structure Simulation Software*. https://www.ansys.com/products/electronics/ansys-hfss (visited on 2025)

L. Ashok Kumar, Y. Uma Maheswari, *Electromagnetic Interference and Electromagnetic Compatibility Principles, Design, Simulation, and Applications*, 1st edn. (CRC Press, Boca Raton, 2023)

J. Benz et al., Analysis of immunity failures and optimization measures in automotive sensors, in *2019 International Symposium on Electromagnetic Compatibility (EMC Europe)*, Barcelona (2019), pp. 1014–1019

C. Brench, B.R. Archambeault, O.M. Ramahi, *EMI/EMC Computational Modeling Handbook*, 2nd edn. The Springer International Series in Engineering and Computer Science (Kluwer Academic Publishers, Boston, 2001)

H.-D. Brüns et al., Modeling challenging EMC problems. IEEE Electromagn. Compat. Mag. **6**(3), 45–54 (2017). https://doi.org/10.1109/MEMC.0.8093837

Cadence, *System Design and Analysis Platforms*. https://www.cadence.com/en_US/home/tools/system-design-and-analysis.html

S. Caniggia, F. Maradei, *Signal Integrity and Radiated Emission of High-Speed Digital Systems* (Wiley, Chichester, 2008)

COMSOL, *COMSOL Multiphysics: Multiphysics Software for Simulating Real-World Designs, Devices, and Processes*

Dassault Systèmes Simulia, *CST Studio Suite: Electromagetic Simulation Software*. https://www.3ds.com/products/simulia/cst-studio-suite (visited on 2025)

EMCoS, *Consluting and Software*. https://emcos.com/ (visited on 2025)

N.V. Kantartzis, T.D. Tsiboukis, *Modern EMC Analysis Techniques* (Springer, Cham, 2008)

R. Keller, S. Landolt, I. de Mendizábal, *Academy of EMC*. https://www.academyofemc.com/ (visited on 2025)

Keysight, *EDA Software*. https://www.keysight.com/us/en/products/software/pathwave-design-software.html (visited on 2025)

S. Mee, B. Adamczyk, In compliance magazine. Internet (2018). https://incompliancemag.com/bulk-current-injection-bci-substitution-method-and-closed-loop-method-with-power-limitation/

H. Ott, *Electromagnetic Compatibility Engineering* (Wiley, Hoboken, 2009)

C.R. Paul, *Introduction to Electromagnetic Compatibility* (Wiley, Hoboken, 2005). ISBN: 9780471758150. https://doi.org/10.1002/0471758159

S. Piper, Modeling fundamentals the importance of validation for all simulations, in *2017 IEEE International Symposium on Electromagnetic Compatibility & Signal/Power Integrity (EMCSI)* (2017), pp. 1–48. https://doi.org/10.1109/ISEMC.2017.8078053

C. Riener et al., Broadband modeling and simulation strategy for conducted emissions of power electronic systems up to 400 MHz. Electronics **11**(24) (2022). ISSN: 2079-9292. https://www.mdpi.com/2079-9292/11/24/4217

U. Rüde et al., Research and education in computational science and engineering (2017). arXiv:1610.02608v4 [cs.CE]

Siemens, *Electronic Design Automation*. https://eda.sw.siemens.com/en-US/pcb/hyperlynx/ (visited on 2025)

Simyog, *Agile for Hardware*. https://simyog.com/

Chapter 2
Circuit Component Models

Circuit component models serve as building blocks for EMC simulations at the circuit level. In this chapter, we discuss circuit component models, their properties, typical applications, and limitations. We interpret circuit models as impedances, which are used in SPICE simulations to solve circuit simulation tasks. The introduction of circuit components follows the decomposition of an EMC problem into three key elements: source, coupling path, and sink. Additionally, we introduce the concept of electrical size and explore network parameter models, which provide an efficient approach for representing electrically large coupling paths within a circuit simulation. In the advanced sections, we discuss the linearity of circuit components and explain how lumped models can be derived from network parameter models.

Some of the concepts introduced in this chapter are explored further in later chapters: An example illustrating the assembly of circuit models into a circuit-based EMC simulation is provided in the following Chap. 3. Further details on the decomposition of an EMC problem and its modeling are discussed in Sect. 5.2.

2.1 Introduction to Circuit Component Models

Domain: 

Modeling the interactions between components of an electronic system using only voltages and currents at their terminals significantly reduces computational time and eliminates the need for detailed knowledge of the components' internal structures. Many circuit component models are supplied by third-party vendors, allowing them to be incorporated into electronic system simulations without disclosing the internal design of the components. However, this approach neglects coupling effects between components that arise from field interactions.

J. Hansen, A. Barchanski, *EMC Simulation of Electronic Systems*,
https://doi.org/10.1007/978-3-032-09237-3_2

A circuit component interacts with other components exclusively through the current, I, and voltage, V, at its terminals, which are sometimes referred to as pins in circuit simulators. For a two-terminal component, the circuit model defines the relationship between the current flowing through the component and the voltage across it. A circuit model for a component with n terminals needs to relate all $n-1$ possible voltages and current combinations across and through the component. A circuit component model of a component can either be composed from other circuit models, allowing the evaluation of internal voltages and currents, or it can be a black-box model, which provides no information about V and I within the component. Throughout this book, we treat components as black boxes; thus, only the voltages and currents at the terminals are considered in the simulations presented.

For example, in a transistor, detailed semiconductor physics are not required to model the voltage and current at its terminals. The behavior can be represented using a current source with an appropriate parameter set, eliminating the need to model semiconductor physics in detail. A semiconductor vendor can publish a circuit component model of the transistor without disclosing the geometry of the semiconductor material, thereby protecting their intellectual property.

In simulation models of electronic systems, a circuit component model often aligns with the physical component mounted on a PCB. The simulation model is constructed from component models of capacitors, inductors, transistors, and other elements. The assembly of these components into an EMC simulation model is performed in a circuit simulator, as described in Chap. 3.

It is important to note that our definition of circuit component models is not restricted to these components. We also apply circuit component models to interconnects such as cable harnesses, connectors, or PCB traces, as long as describing these components using voltages and currents at their terminals is sufficient. The limitation of this approach is reached when electromagnetic field interaction between components becomes critical to the EMC performance of the product.

For certain cases, such as the coupling of two inductors, circuit simulation allows the inclusion of additional coupling elements, as shown in Sect. 2.4.2. However, scenarios involving radiative coupling, such as unintentional or intentional interactions between antennas, are extremely challenging to model using circuit models based solely on voltage and current interactions.

Whether parts of an electronic system can be treated as a circuit component depends on how it interacts with its surroundings. If it couples to other components only through voltages and currents, a circuit model suffices; if electromagnetic field interactions matter, the circuit model is not sufficient. This distinction is not always obvious beforehand. Heuristics—such as comparing a component's electrical size to the relevant wavelength or noting the spacing between elements—offer an initial guide, but the decisive answer comes from simulation.

Take the stray field of an inductor: to see whether it affects emissions, create two simulations—one that models the inductor as a circuit component and another that includes its physical geometry in an EM model—and compare the results.

2.1.1 *The Impedance Interpretation*

Different types of circuit models can use various mathematical approaches to describe the relationship between voltages and currents at the component's terminals. These models are discussed throughout this chapter. In order to allow us to better understand the components behavior, we will interpret the component as an impedance. The impedance relates voltage and current as.

$$Z = \frac{V}{I} \tag{2.1}$$

Impedance has the unit of Ohm, and electrical engineers typically gain significant insight into a component's behavior from its impedance value. Although impedance is formally a complex quantity that relates sinusoidal signals, we will also use the term impedance to describe component behavior in the time domain. At this stage, the defined impedance is not required to possess any specific properties, as long as it relates V and I. It may be time-dependent, vary with the operating conditions of the device, or depend on the voltage and current through the component. For the sake of readability, we refer to the magnitude of the impedance simply as the impedance throughout the book.

2.2 SPICE Simulation

All circuit simulations presented in this book were performed using a SPICE-like simulator. SPICE is widely used in EMC simulations due to its ability to model and simulate the electrical behavior of complex circuits. SPICE offers a large number of built-in component models for resistors, capacitors, inductors, transistors and diodes. It is well established and used across various industries for circuit simulation. Its widespread adoption leads to a broad availability of circuit component models provided by component vendors. Furthermore, SPICE offers time and frequency domain analysis. Throughout this book, we often switch between both a time and frequency domain analysis, depending on which one is more suitable at a given point of the modeling process.

SPICE simulators offer a convenient way to connect the individual circuit component models described in this chapter to a circuit that models the behavior of an electronic system. For the solution of the resulting circuit, SPICE uses a (modified) nodal analysis, which results in a set of equations based on the Kirchhoff's Current Law (KCL).

$$\sum_{i=0}^{n} I_i = 0 \tag{2.2}$$

The KCL is given in Eq. 2.2 and states that the sum of all currents at a node of the circuit is zero. The nodal equations, along with the equations of the individual circuit models, are solved using numerical methods. Details of the analysis are presented in Sect. 3.8. The results of a SPICE analysis are voltages at every circuit node and current through every component of the circuit.

At the time of writing this book, the freeware tool LTspice (Analog Devices 2025) is probably the most often used SPICE simulator. SPICE was introduced to the public in the 1970s with the latest official language definition being the Berkeley SPICE 3f5 version that was released in 1993 (Rabaey 2025). Since the official language reference for SPICE is quite outdated, a number of dialects have emerged that add functionality and syntax to the SPICE format. Besides LTspice, the commercial package PSpice from Cadence Design Systems (Cadence PSpice 2025) has also defined a number of additional keywords and model types to the SPICE format. Most other circuit simulators are able to use models written in the different dialects, but there is no guarantee that all proprietary extensions are supported. We recommend to the reader to get familiar with the SPICE language as a deeper understanding of the SPICE format opens a number of advanced options to the simulation engineer (Vladimirescu 1994).

Unfortunately, the abovementioned SPICE simulators do not support the usage of network parameter models like S-parameters. Also, the support for IBIS models is not included in these tools. A number of circuit simulation packages are available, which offer SPICE simulation functionality and extend the available circuit models to include network parameter and IBIS models. Throughout this book, we use the circuit simulator contained in CST for our simulations. Other popular tools on the market include Keysight's Advanced Design System (ADS) and Ansys Nexxim (part of the Ansys Electronics Desktop).

2.3 Source Models

Domains:  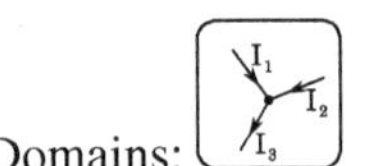

To build an EMC simulation model based on the EMC decomposition introduced in Sect. 1.1, an appropriate representation of the noise source is essential. This section focuses on source models used in the simulation of emissions. Laboratory sources for immunity testing are addressed separately in Chap. 13.

A noise source in an EMC simulation model differs from the power source in the functional description of an electronic system. Therefore, before discussing specific circuit models for noise sources, we examine the distinctions between functional and EMC modeling of sources. A detailed explanation of how a complete functional schematic is transformed into an EMC simulation model is treated in Sect. 3.1 using the example of a buck converter.

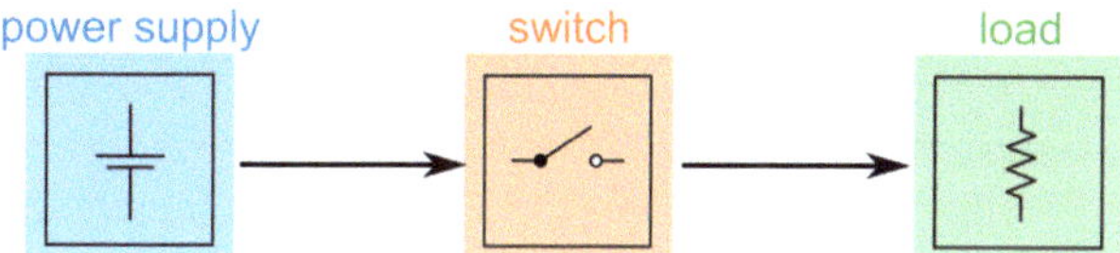

Fig. 2.1 Flow of power in an electronic system

In electronic systems, a power source delivers power to a load, as illustrated in Fig. 2.1. For EMC considerations, systems where power is switched between the load and source are of particular interest, as this switching process generates emissions. To accurately calculate these emissions, an EMC simulation must properly model the power flow through the electronic system. Depending on the type of electronic system being modeled, simplifications may be applied to the chain shown in Fig. 2.1. For the purposes of this discussion, we classify electronic systems into two main categories:

- **Power electronics systems**, such as inverters and switched-mode power converters.
- **Digital systems**, which transmit data using modulated signals.

In a **power electronics system**, switching is used to shape the properties of the voltage and current delivered to a load. For example, in a buck converter, which will be discussed in the next chapter, a switching pattern is used to convert a 12 V input from the power supply to a 5 V output to the load. The flow of power from the supply to the switch and from the switch to the load contributes to the emission, and it is necessary to include the complete power path in simulation models of power electronics. When modeling the complete path, the noise source and the power source are treated as separate entities. The power source is explicitly defined in the simulation model, while the noise source is represented by the switching element.

In a **digital system**, the switch is referred to as the transmitter or driver, while the load is the receiver or termination of the digital channel. A common approach in the simulation of digital systems is to segment the power flow into two segments. The segment from the power supply to the transmitter is analyzed separately from the segment from the transmitter to the receiver. The analysis of the segment from the power supply to the transmitter is called power integrity (PI), whereas the analysis of the segment from the transmitter to the receiver is called signal integrity (SI). We discuss these two fields in Chap. 11.

The motivation for this segmentation lies in the distinct implementation of these segments within a digital electronic system. The segment from the power supply to the switch, also referred to as the Power Delivery Network (PDN), is implemented on the PCB. A well-designed PDN provides a low-impedance path for high-frequency signals through the use of decoupling capacitors.

The segment from the transmitter to the receiver, on the other hand, often spans large distances and, in many systems, is routed from the PCB through a connector to a cable. This routing introduces impedance discontinuities and imbalance, making this segment the dominant contributor to emissions in a digital system.

In simulation models where it is assumed that the segment from the transmitter to the receiver dominates the emissions, the segment from the power supply to the switch is omitted. In such models, the noise and the power source are both located at the transmitter, which differs from their representation in the functional schematic.

2.3.1 Voltage and Current Sources

SPICE offers built-in voltages and current sources that we use throughout the book. They are simple to define and allow to model both power and noise sources. In order to model an electronic system, it is essential to understand how the properties of these simple sources are chosen, ensuring the correct representation of the power flow within the system.

An ideal voltage source creates a user-defined voltage across its two terminals. The impedance across the source is zero. The source behaves like a short, allowing any current to flow through it. An ideal current source creates a user-defined current from one terminal to the other. The impedance across the source is infinite, any current that flows through the source results in a voltage between its two terminals. With this definition, connecting a low impedance load to a voltage source leads to a high current while connecting a high impedance load to a current source leads to a high voltage. Ideal sources might significantly overestimate the actual power rating of the system we are modeling, as can be seen in the following example:

Let us consider a simple wire loop made from copper, for which we estimate the resistance and inductance using Eq. 3.13. For a loop radius of $r_{\text{loop}} = 10\,\text{cm}$, a wire radius of $r_{\text{wire}} = 0.5\,\text{mm}$, and the conductivity of copper $\sigma = 5.8 \cdot 10^7\,\frac{\text{S}}{\text{m}}$, we obtain an inductance of $L = 676\,\text{nH}$ and a resistance of $R = 13\,\text{m}\Omega$. Connecting an ideal 1 V source to the loop leads to a current of 77 A at DC, requiring the source to deliver 77 W of power.

To check the power requirement when a current source is connected, we choose a frequency of 10 MHz. Using $Z = j\omega L$, we calculate the magnitude of the impedance to be $42.5\,\Omega$ at 10 MHz. To drive a current of 1 A into the loop at 10 MHz, 42.5 W of power is required.

In order to limit the power delivered by the source, we add internal resistances to the source model as depicted in Fig. 2.2. A resistor R_{s} is added in series to the voltage source or in parallel to the current source resulting in sources with a limited power rating. By applying the Thévenin-Theorem, we are able to convert one source type to the other. To find the appropriate value for R_{s}, we use the voltage rating of the DUT's source. Using $R_{\text{s}} = \frac{V^2}{P}$ with P being the maximal power the source is able to deliver, we estimate the series resistance for the voltage source. SPICE simulators allow to define the internal resistance directly in the definition of the source, but we can also explicitly add the resistors to our schematic.

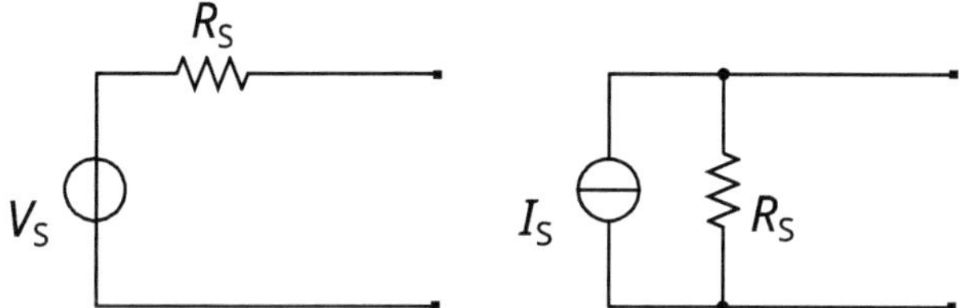

Fig. 2.2 Schematic of non-ideal voltage (left) and current (right) sources

Because the power delivered by a voltage source decreases when the load impedance is increased, it is the more suitable source type for most simulation models. It is a more realistic model of the behavior of common power supplies, like batteries. We use current sources only in a few applications, such as the calculation of shielding of magnetic fields in the low kHz range in Sect. 7.3.

The series resistance of the voltage source plays an additional role when used to model a digital transmitter. If the switching frequency of the digital signal is high enough to require consideration of reflections in the signal path, the series resistance acts as a termination impedance on the transmitter side. Thus, it governs the reflection level at the transmitter.

This is best illustrated by an example: We model a microstrip using a circuit model provided by the circuit simulator. Such models are commonly available in advanced circuit simulators and utilize transmission line theory to model the microstrip. The modeling details are not relevant for this example; we discuss them in Chap. 8. The microstrip is 5 cm long, 2 mm wide, and situated above a substrate of 1 mm height with a relative permittivity $\varepsilon_r = 4.3$. This configuration results in a characteristic impedance of $50\,\Omega$.

The line is driven by a 1 V signal at a frequency of 20 MHz with a rise/fall time of $t_r = 10\,\text{ns}$, terminated by a $10\,\text{k}\Omega$ resistor. Figure 2.3 shows the voltage at the driver (green curve) and the voltage at the termination for a driver impedance of $R_s = 0\,\Omega$ (blue curve) and $R_s = 10\,\Omega$ (orange curve).

We observe greater ringing of the voltage at the receiver when a source impedance of $R_s = 0\,\Omega$ is used.

2.3.2 *Spectrum of Trapezoidal Signals*

After covering source power and impedance ratings, we shift our focus to signal properties. In electronic systems, power switching typically follows a trapezoidal waveform. Understanding the relationship between the time domain and frequency domain representations of a trapezoidal signal is helpful in estimating the bounds of the expected emission limits from the system. This understanding also aids in designing filters to effectively control and reduce these emissions.

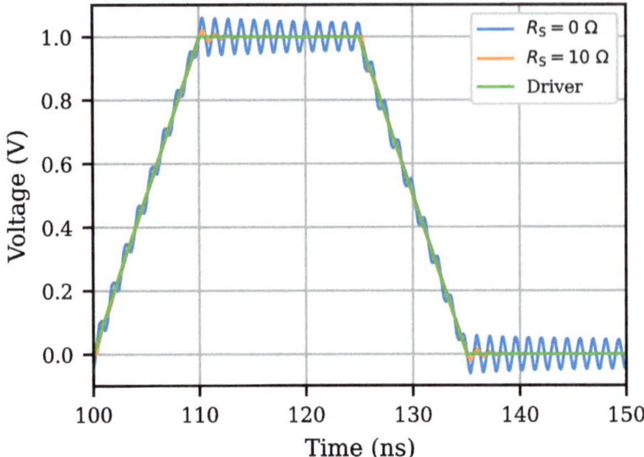

Fig. 2.3 Receiver voltage for different driver impedances

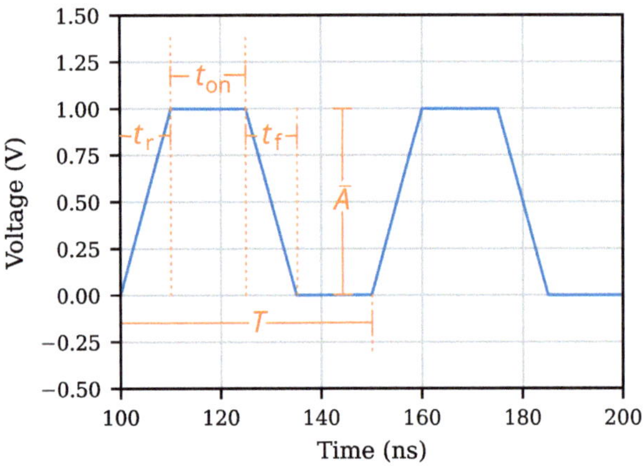

Fig. 2.4 Trapezoidal signal definition in SPICE simulators

A trapezoidal signal is characterized by its amplitude, frequency, and rise/fall time of the signal. SPICE simulators offer a convenient way to define a trapezoidal signal by a number of parameters: on amplitude, off amplitude, delay time, rise time, hold time, fall time, and total time, that we use in the following discussion. An overview of a signal defined in this way is plotted in Fig. 2.4.

Clayton Paul derives the spectral bounds of trapezoidal signals in his book (Paul 2005).For a trapezoidal signal with an Amplitude $\bar{A}$, an on time t_{on}, a total time T, and a fall/rise time t_{r}, three distinct regions that bound the spectral content of the pulse are present:

- in the frequency range from 0 to $\frac{1}{\pi t_{\mathrm{on}}}$ the bound is constant with an amplitude of $2\bar{A}\frac{t_{\mathrm{on}}}{T}$

- in the frequency range from $\frac{1}{\pi t_{\text{on}}}$ to $\frac{1}{\pi t_{\text{r}}}$ the bound decays with a slope of -20 dB per decade
- in the frequency range above $\frac{1}{\pi t_{\text{r}}}$, the bound decays with a slope of -40 dB per decade

We deduce that t_{r} has a significant influence on the spectral content of a trapezoidal pulse at high frequencies. When t_{r} is decreased, e.g., the signal rises faster, the second bound is moved toward higher frequencies. This increases the range in which the signal decays by -20 dB/dec, thus increasing the emission at higher frequencies. When t_{on} approaches t_{r}, the flat region with -20 dB/dec is wider, and the spectrum looks flat in the lower frequency region. Thus, choosing a realistic t_{r} is crucial for the simulation model. This is sometimes difficult, as the value of t_{r} might not be explicitly stated in a data sheet of the component we model. In such cases, an educated guess based on the driver technology is required. We have collected typical values of signal rise times for digital communication buses in Table 2.1 and for power electronics switches in Table 2.2.

In Fig. 2.5, we have overlaid the theoretical limits onto a spectrum calculated from a 1 MHz signal. The following parameters are used: $\bar{A} = 1$ V, $t_{\text{on}} = 0.5\,\mu$s, $T = 1\,\mu$s, and $t_{\text{r}} = 10$ ns. A total of 500 pulses are simulated to obtain the low-

Table 2.1 Rise times of signals for digital communication buses

Digital bus	Rise time
I²C	20–300 ns
SPI	5–20 ns
UART	5–100 ns
CAN	5–25 ns
Ethernet (10 Mbps)	10–25 ns
Ethernet (100 Mbps)	3–10 ns
Ethernet (1 Gbps)	Sub-nanosecond
USB 2.0 (High-Speed)	0.5–2 ns
USB 3.0 (SuperSpeed)	Sub-nanosecond
PCIe	100–300 ps
DDR (DDR3/DDR4/DDR5)	100–300 ps
HDMI	100–300 ps
LVDS	500 ps–2 ns

Table 2.2 Rise times of signals in power electronic systems

Semiconductor technology	Rise time
Silicon (Si) Power MOSFET	10–100 ns
Si IGBT	50–500 ns
Gallium Nitride (GaN) FET	1–10 ns
Silicon Carbide (SiC) MOSFET	10–50 ns
GaN HEMT	<10 ns
Wide Bandgap Diodes (SiC, GaN)	10–20 ns
Si Schottky Diode	10–50 ns
Si Fast Recovery Diode	50–200 ns

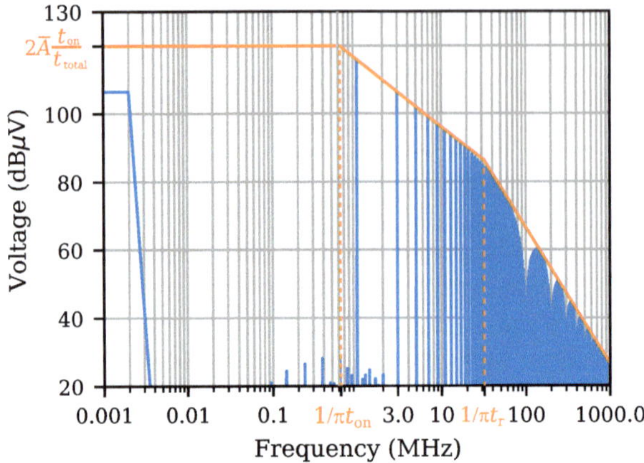

Fig. 2.5 Spectrum of a trapezoidal pulse with characteristic limits

frequency response with a resolution bandwidth of 2 kHz. A Hamming window is applied to suppress side lobes.

Most circuit simulators can compute the Fourier transform of any time-domain waveform. Keep in mind, though, that a numerical FFT can offer several user-selectable options like normalization conventions and window functions.

The spectral bounds presented in this section are efficient for performing worst-case estimations of emissions using frequency-domain circuit simulations. We use an envelope of this type for the emission simulation of a buck converter in Sect. 3.5.2 and discuss further details in the advanced Sect. 3.8.2.

In the simulation performed in this book, we also use a number of special signal types typically employed in electronic systems:

Pulse Width Modulation—PWM Signals are used in power electronics systems to control the voltage or power delivered to a load. In PWM, a constant-frequency signal is used and the width of pulses is varied in order to adjust the duty cycle.

Pseudo-Random Binary Sequence—PRBS A deterministic binary signal that appears random but is generated by a defined algorithm. PRBS signals alternate between on and off states in a pattern that repeats after a specific period.

Spread Spectrum Signals spread a signal's frequency over a wider bandwidth by modulating the signal frequency with another signal of a typically much lower frequency.

For the above-described signal types, it is possible to import signal definitions generated by external data processing tools. When generating these signals, we need to properly sample the rise/fall time; otherwise, the spectral content of the source might be inaccurate. This also holds for the import of measured signals. Fast measurement equipment may generate signals with a very small time step that slows down the circuit simulator.

2.3.3 *Analog Noise Sources*

In simulations where the entire power flow through the system is modeled (Fig. 2.1), an accurate representation of the switching element is crucial. The parasitic properties of the switch, particularly its capacitance, significantly influence the system's emission, as they generate resonances.

The switching element can be passive or active. In the power electronics systems covered in this book, passive switches are implemented by using diodes, while active switches use transistors. A diode is a two terminal component, which allows the current to only flow in one direction. A transistor is a three terminal component, which allows to control the impedance in one terminal pair by the voltage applied to another terminal pair.

For diodes and transistors, semiconductor manufacturers provide SPICE models of their products on their web pages. In the following section, we use vendor models to demonstrate the properties of these components. At the end of this section, we discuss the possibility of creating a simple switch model, which is necessary in cases where no vendor model of the active switching element is available.

2.3.3.1 Diodes

In our simulation models, diodes are passive switches, as their blocking properties depend solely on the applied signal, and there is no possibility of actively control their switching. The switching leads to the generation of harmonics if a signal with both polarities is applied to the diode. For an emission simulation, the circuit model of a diode requires to model this switching behavior and the capacitance across the diode. As we assume that the diode is operated in its intended operating conditions, the modeling of the diodes behavior in the event of a breakthrough is not required.[1]

For the demonstration of the harmonic generation, we use a simple schematic: an ideal sinusoidal voltage source with the amplitude of 1 V driving a diode in series with a 1 kΩ load resistor. We have chosen the 1N4148 diode type, modeled by a SPICE circuit obtained from a vendors web page (Nexperia).

The right-hand side of Fig. 2.6 depicts the voltage spectrum of the driver and the voltage spectrum at the load with a 1 MHz excitation. The driver spectrum, plotted in orange, consists of a peak at the fundamental frequency of 1 MHz (with some numerical noise due to FFT sampling). The voltage spectrum at the load, plotted in blue, exhibits a significant level of harmonics originating from the switching behavior of the diode.

Every diode has a characteristic recovery time required to block current after the current polarity changes. If the signal applied to the diode's terminals has a duration much shorter than this recovery time, the diode cannot block the reverse current

[1] Transient voltage suppressor diodes are an exception and only used in Chap. 14, where we discuss ESD simulations.

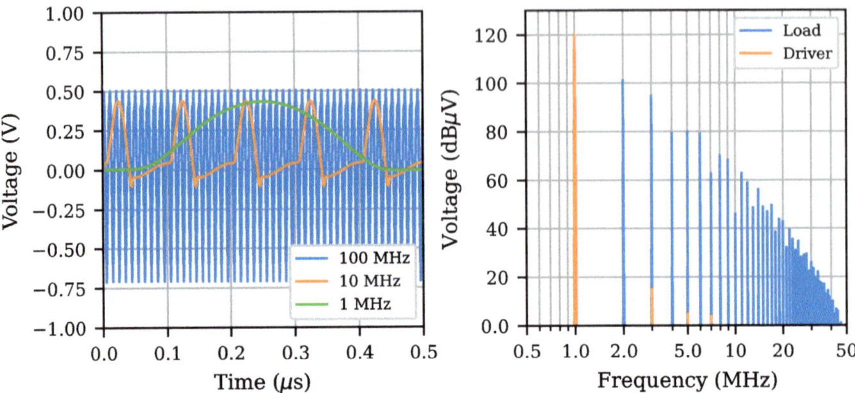

Fig. 2.6 Left: Time-domain voltage at the load for a sinusoidal signal with different frequencies. Right: Spectrum of the signal excitation and the signal at the load for a sinusoidal driving signal of 1 MHz

and becomes transparent to these fast signals, as illustrated on the left-hand side of Fig. 2.6.

In our example model, the voltage at the load is rectified for a signal frequency of 1 MHz (green line). For a signal frequency of 10 MHz (orange line), overshoots to the negative polarity occur, while a signal frequency of 100 MHz can no longer be rectified (blue line). Under such operating conditions, the diode behaves like a capacitance. This capacitance originates from the PN junction of the diode. In our example, the capacitance of the 1N4148 diode is directly obtainable from the SPICE model and amounts to 0.7 pF.

2.3.3.2 Transistors

Transistors are components that allow the control of the impedance of a terminal pair by applying a voltage to another terminal pair. Transistors are used in numerous electronic circuits, such as amplifiers, feedback loops, or control circuits. For an EMC simulation, the primary focus is on the properties of the transistor when it is used as a switch, as depicted in Fig. 2.1.

The circuit model of the transistor in an EMC simulation must represent the switching characteristics of the transistor, particularly the rise and fall times of the switching event, as these have a significant impact on the emission spectrum (cf. Sect. 2.3.2). Additionally, the transistor model must include the parasitics of the device to effectively model potential ringing, overshoots, and undershoots. In the next paragraph, we discuss how transistors are modeled in SPICE simulations. Subsequently, we present a number of simple circuits utilizing a vendor-provided transistor model, which allow us to test the transistor's behavior. The final part of this section is dedicated to creating a simplified switch model that can approximate the switching behavior of a transistor in EMC simulations.

Already Berkeley SPICE offers a number of models (Rabaey 2025) for different transistor types, including models for the BJT, JFET, MOSFET, and MESFET (but not IGBT). The idea behind these models is to obtain a behavioral model of the transistor based on properties of the actual semiconductor design. Examples of these properties include the width and length of the channel, carrier mobility, and gate oxide thickness, to name a few.

For the MOSFET type, different levels are available, varying in the number of parameters and the physical effects inside the semiconductor material that are considered (BSIM Group).[2] With the introduction of new semiconductor designs, some of these models became obsolete, and semiconductor manufacturers began implementing their own descriptions of transistors into SPICE models. In these models, the original SPICE transistor types are no longer used; instead, the device behavior is modeled by equations provided by the manufacturer.

This trend has been accelerated by the introduction of transistors based on novel semiconductor materials like SiC and GaN. For these materials, no native models are available in SPICE; they need to be completely defined by equations, effectively transforming the SPICE language into a set of mathematical expressions. The numerical performance and stability of circuit simulations vary depending on the equations used to describe the transistor.

Some manufacturer models are provided with a thermal circuit. The thermal circuit is used to model the thermal dependence of the transistor's properties. However, thermal modeling adds to the numerical complexity and can sometimes lead to convergence problems in circuit simulations. If thermal modeling is not explicitly required, we do not recommend including it in EMC simulations.

The predominant transistor type in circuits where transistors are used as switches is likely the MOSFET. In the following example, we study the behavior of a SPICE model of a MOSFET. For this task, we use the circuit depicted on the left-hand side of Fig. 2.7. Such testing circuits are very useful for verifying models downloaded from vendor web pages before including them in more complex schematics. They provide a good understanding of how the transistor model behaves in an isolated environment.

As a test case, we have selected the IRFZ44 MOSFET, with the SPICE file downloaded from Infineon. This is a typical general-purpose MOSFET with a voltage rating of 55 V, a maximum supported current of 49 A, and a drain-source on-resistance of 17.5 mΩ.

In the first simulation, we determine the static properties of the transistor. We remove R_{GS} and L_{DS} and set $R_{\mathrm{DS}} = 1\,\Omega$ to limit the maximum current in the drain-source branch according to the transistor specifications. The drain-source voltage V_{DS} is stepped from -10 V to 50 V, and the gate-source voltage V_{GS} is swept from 0 V to 10 V. (In the simulation model, we define the sweep over V_{GS} using a time-

[2] Various BSIM manuals are also accessible in the internet.

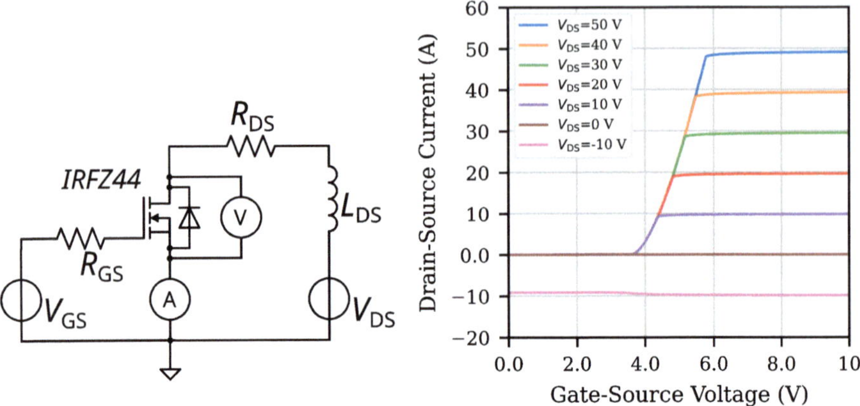

Fig. 2.7 Left: Test circuit for MOSFET characteristic. Right: Drain-source current vs gate-source voltage for different drain-source voltages

domain signal with a rise time of 1 $\frac{\text{V}}{\text{s}}$, as this is simpler to define for a continuous output.)

The results are depicted on the right-hand side of Fig. 2.7. From these results, we deduce the following properties of the transistor model:

- the model includes the MOSFET's body diode, as we see a constant negative current for $V_{\text{DS}} = -10\,\text{V}$.
- the threshold voltage at which the linear region starts is 3.6 V.
- the linear region extends up to 5.7 V for the maximal current rating of approximately 50 A. The width of the linear region depends on the total drain source current through the transistor. It is wider for higher current ratings.

Dynamic testing of the switching of a MOSFET is performed using a double-pulse setup. With this setup, the switching properties and losses of a MOSFET for a given inductive load are estimated. The simulation of a double pulse test is presented in Sect. 5.6. At this point, we use a simple setup to generate a ringing of voltage and current at the transistor. The ringing is generated with the simple test schematic we have already used for the estimation of the static properties of the MOSFET.

For the dynamic test, we keep $R_{\text{DS}} = 1\,\Omega$ to limit the maximum current and use $V_{\text{DS}} = 25\,\text{V}$. We set $R_{\text{GS}} = 10\,\Omega$ and use a 100 kHz driving pulse at V_{GS} ($t_{\text{r}} = 100\,\text{ns}$, $t_{\text{total}} = 10\,\mu\text{s}$, $t_{\text{on}} = 5\,\mu\text{s}$, $V_{\text{peak}} = 7\,\text{V}$).

Ringing at the drain source of the transistor is generated by a resonance of the transistor's drain-source capacitance with the inductance of the drain-source loop, which is the inductance of the commutation cell. We include this inductance in our test circuit by setting $L_{\text{DS}} = 20\,\text{nH}$.

The ringing voltage across the transistor and the ringing current are depicted in Fig. 2.8. When the transistor is switched off, the inductance is discharged, and the current direction in the drain-source loop reverses. Due to the body diode of

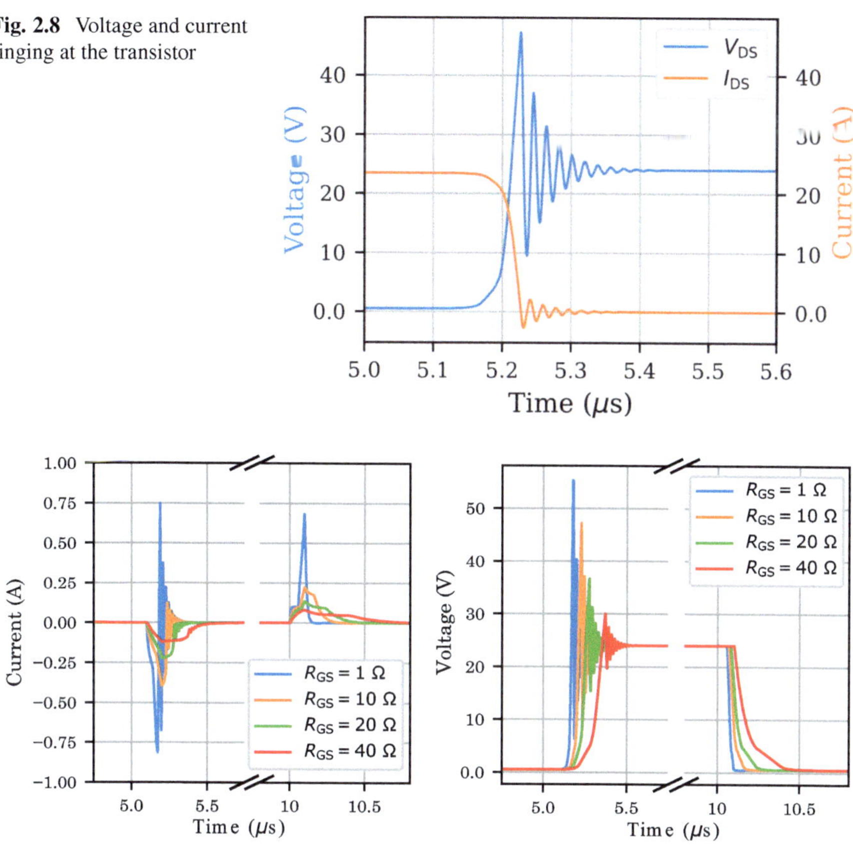

Fig. 2.8 Voltage and current ringing at the transistor

Fig. 2.9 Impact of the gate resistor. Left: Gate-source driving current. Right: Drain-source voltage

the MOSFET, the current flows in the opposite direction. The commutation cell inductance and the drain-source capacitance (parallel to the body diode) form a resonant circuit, resulting in the ringing. The effect of the ringing on the emission is further studied in the next chapter, Chap. 3, using a buck converter as an example.

Not only does the inductance in the drain-source path affect the ringing, but so does the current driving the gate. To analyze this effect, we vary the gate resistor R_{GS} from 1 Ω to 40 Ω. The results of this study are plotted in Fig. 2.9.

On the right-hand side of the figure, we observe that as the value of R_{GS} increases, the amplitude of the ringing decreases. This reduction in ringing is due to the fact that increasing the resistor value decreases the driving current at the gate, as shown on the left-hand side of the figure. The reduction in driving current results in a reduced slew rate at the transistor, which in turn reduces the spectral content at higher frequencies by increasing the rise time of the switching event. This finding underscores the need to define the proper driving current at the gate resistor to accurately represent the actual ringing at the drain source.

2.3.3.3 Simple Switch Model

When building a simulation model for an electronic system, we may encounter the issue of not being able to use the SPICE model of the actual transistor designed into the system. This can happen for various reasons: the manufacturer does not offer a SPICE model, the model is encrypted and can only be simulated in a circuit simulator to which we do not have access, the SPICE model suffers from convergence issues, or the switch is integrated into a more complex IC for which no SPICE model exists.

In such cases, we need to find a suitable replacement to continue with the model-building process. One option is to look for components with similar ratings from other manufacturers that provide a SPICE model. Another option is to construct a simplified switch model from basic circuit elements. Both approaches introduce some degree of uncertainty into the simulation model, which may cause deviations in the absolute values calculated by our EMC simulation. However, this approach still allows us to proceed with the simulation and obtain realistic estimations. In the following, we demonstrate the construction of a simple switch model of a MOSFET.

The schematic of the proposed switch is shown in Fig. 2.10. Our goal is to create a simplified switch model that delivers the same voltage spectrum across the switch (drain source) as the SPICE model of the IRFZ44. The switching element is a voltage-controlled switch (VCSW), which is a built-in SPICE component.

A VCSW sets the resistance between the switch terminals as a function of the voltage between the measurement terminals (indicated by the arrow). It is defined by a threshold voltage $V_{\text{threshold}}$, an on-resistance R_{on}, an off-resistance R_{off}, and a hysteresis voltage $V_{\text{hysteresis}}$. The hysteresis voltage allows for modeling a smooth transition from R_{on} to R_{off}. Without defining the hysteresis voltage, the switch instantaneously changes its resistance once the threshold voltage is

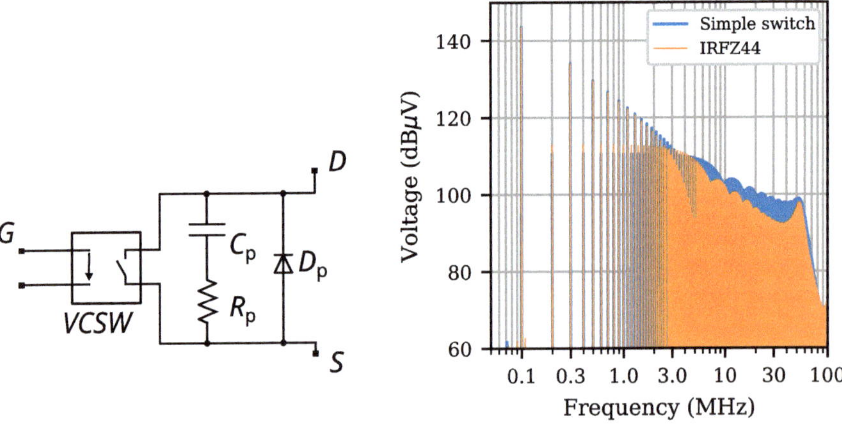

Fig. 2.10 Left: simple switch model. Right: comparison of the drain-source voltage spectrum of the simple switch with the vendor SPICE model

reached at the control terminals. Such an instantaneous change leads to a significant underestimation of the rise and fall times of the switching event.

The switch model does not include the modeling of the gate driver current. It perfectly isolates the switch from the control side. The drain-source capacitance is modeled by a parallel capacitor C_p with a damping resistor R_p in series with the capacitor. The body diode of the MOSFET is added in parallel to the switch and the capacitor. To keep the model as simple as possible, we use a diode with default SPICE parameters. For the same drain-source circuit and signal as in the previous paragraph ($R_{GS} = 10\,\Omega$), we manually tune the parameters of the simple switch to obtain a matching spectrum. For the switch, we use: $V_{threshold} = 3\,V$, $R_{off} = 1\,M\Omega$, $R_{on} = 0.01\,\Omega$, and $V_{hysteresis} = -4\,V$. The capacitor values are set to $C_p = 200\,pF$ and $R_p = 0.7\,\Omega$.

Using these values, we obtain a spectrum and compare it to the MOSFET spectrum in Fig. 2.10. The maximum deviation between the two spectra is approximately 6 dB in the frequency band from 100 kHz to 80 MHz. The deviation increases beyond 80 MHz, reaching about 20 dB at 100 MHz. At 100 MHz, we analyze the 1000th harmonic of the signal.

There is no guarantee that the vendor SPICE models accurately reproduce the spectrum at high frequencies. We illustrate these deviations by comparing SPICE models of the IRFZ44 transistor from different vendors. We compare models downloaded from the websites of Infineon and Vishay. Running the exact same simulation setup results in a difference of approximately 4 dB above 90 MHz.

The presented modeling might raise the question: is the effort to generate the simple switch model justified? We have matched the parameters to an existing SPICE model, but how can we tune the parameters if no SPICE model is available? This is precisely the use case for such simple switch models: they allow us to quickly build simulation models that can estimate realistic bounds of the emission within the parameter space of the switch properties. The simple model enables quick variation of the drain-source capacitance and the switch rise time (via $V_{threshold}$) to check the resulting spectra. Introducing such variations into a vendor SPICE model is far more complex.

Additionally, the simple switch model significantly reduces simulation time. In our example, the SPICE model for the IRFZ44 is relatively simple, yet the simulation time of the simple switch model is more than three times faster. For a simple schematic like our test case, the difference is 7 s versus 27 s simulation time for a total signal duration of 5 ms. For models including an EM block, as introduced in Chap. 10, the circuit simulation time increases significantly, and saving a factor of 3 in simulation time becomes substantial.

In Chap. 3, we use the simple switch model for the emission simulation of a buck converter.

2.3.4 Digital Noise Sources

Following the concept given in Fig. 2.1, a digital transmitter is considered a noise source. If no detailed information about the transmitter model is available, we model the transmitter using SPICE voltage sources, as introduced in Sect. 2.3.1. If a more realistic model of the signals generated by the transmitter is needed for the simulation model, a detailed description of the switching element is required.

The signal generation inside a modern transmitter IC consists of multiple stages, including multiple transistors. Using a transistor-level simulation to model the noise source is not practical: a transistor-level SPICE simulation would be very time-consuming without providing additional meaningful information to the EMC simulation model. Moreover, the network inside the IC is proprietary intellectual property of the IC vendor and is not disclosed.

For an EMC simulation, a behavioral model of the transmitter is required—one that defines the signal generated by the transmitter without modeling how this signal is created.

The same challenges were faced by digital designers in the 1990s. With increasing clock speeds, the importance of SI was rising, and SI simulations promised to become an integral part of the design flow for digital products. The requirements for models used in SI simulations at that time led to the development of the IBIS standard. Initially developed at Intel as paper-based information sheets, an industry alliance called the IBIS Open Forum was founded, and the first version of the IBIS standard was published in 1993.

Since then, the specifications for IBIS models have been extended and updated, with the latest release being IBISver7.2, ratified in January 2023 (IBIS). The documentation of the 7.2 standard spans more than 400 pages, offering a wealth of advanced methodologies for SI simulations.

In this book, we cover only the basics of IBIS models and their application in EMC simulations.

2.3.5 IBIS Models

IBIS models are behavioral models. They describe the behavior of signals at the output and input terminals of an IC. An IBIS file is text-based, containing a number of keywords along with associated tables of data. One IBIS file typically represents one IC, with data for each individual terminal of the IC. A terminal is associated to a buffer type. In the current specification, 17 buffer types are supported. In the following, we only describe an input and a three-state buffer, as these are the models we use throughout the book.

Let us first consider an input buffer. Figure 2.11 depicts the anatomy of an IBIS buffer, with the appropriate IBIS keywords placed in square brackets. The IBIS model includes a section describing the package parasitics, consisting of a series

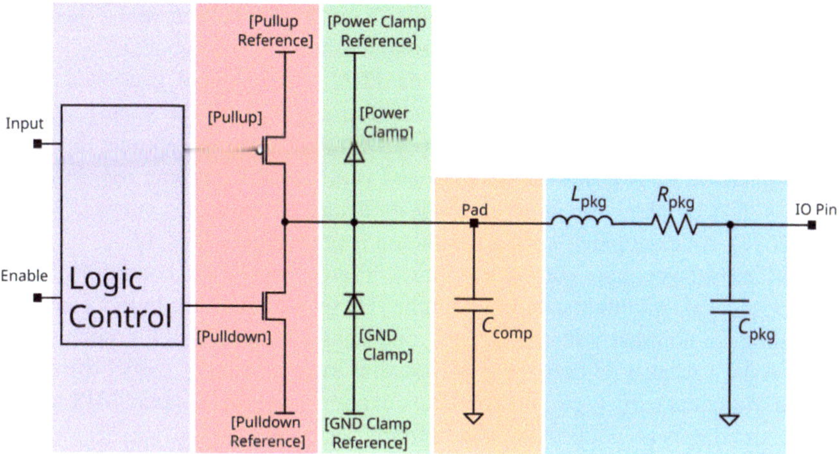

Fig. 2.11 Anatomy of an IBIS model

resistor R_{pkg}, a series inductor L_{pkg}, and a shunt capacitor C_{pkg}, as indicated by the blue box.

The capacitance C_{comp} in the orange box represents the buffer's input capacitance. The two diodes in the green box represent the ESD protection circuit at the buffer's input. In an IBIS model, these diodes are described by voltage–current $(V - I)$ curves with respect to their references, which differs from a SPICE description of a diode.

These three building blocks suffice to describe the input buffer, as they provide the termination impedance of the signal chain, corresponding to the load in Fig. 2.1.

A three-state buffer operated in high-impedance mode behaves analogously to the input buffer. For modeling the driving states, the red section with two drivers is added. The [Pullup] keyword provides a table with the $V - I$ behavior of the pull-up component with respect to the [Pullup Reference]. The same modeling is applied to the low side.

To include the dynamic properties of switching, rising and falling waveforms are provided as $V - t$ tables, describing the voltage behavior during the state change over time. Additionally, a ramp rate is provided, which describes the rate at which the voltage transitions from one state to another (measured at 20% and 80% of the transition edge).

Every output buffer also contains a logic section, indicated in purple, that allows control of the output of the buffer stage. This is achieved by connecting voltage sources to the Input and Enable terminals.

For the generation of a signal at the IO terminal of the IBIS model, a pull-up reference voltage source needs to be connected to the model. If we only model the coupling path from the transmitter to the receiver, the voltage source is directly connected to the IBIS model, neglecting the coupling path from the supply to the switch and providing an ideal reference voltage to the IBIS model. This is the modeling level chosen in the radiated emission examples in this book.

If we want to model the effect of the PDN on the signal generated by the transmitter, the IBIS model must include information about the supply voltage's effect on the signal. Since the release 5.0, IBIS models can incorporate this effect. This is referred to as power-aware IBIS and is achieved by adding new keywords along with data tables for more accurate modeling of the driver current (Mayder et al. 2013). An example of such a simulation is presented in Sect. 11.3.

IBIS models contain curves for three different operating conditions, called corners: typical, minimum, and maximum. These represent the buffer's behavior depending on nominal voltage, process, and temperature. The IO terminals of an IBIS model can also be defined as differential to conveniently model differential signals. As noted earlier, circuit simulators explicitly need to support IBIS models. SPICE simulators do not provide this support.

To demonstrate an IBIS simulation, we create a simple schematic consisting of a differential driver and a differential termination using a demonstrator IBIS model that does not represent a physical device. We monitor the single-ended and differential voltages at the output of the driver for three different driving schemes:

- 333 MHz clock signal with a rise time of 100 ps and a duty cycle of 0.5.
- PRBS with the same frequency and rise time as the clock signal.
- the same clock as in the first simulation but with a spread spectrum modulation of 2% center spread and a modulation frequency of 100 kHz

The PRBS pattern is used to model the data flow in a digital system. The spread spectrum modulation is created by a script that generates a PWL file that is imported into the simulation task.

The results of the simulation are plotted in Fig. 2.12. On the left-hand side, the time-domain voltages of the PRBS sequence are shown. We see the different

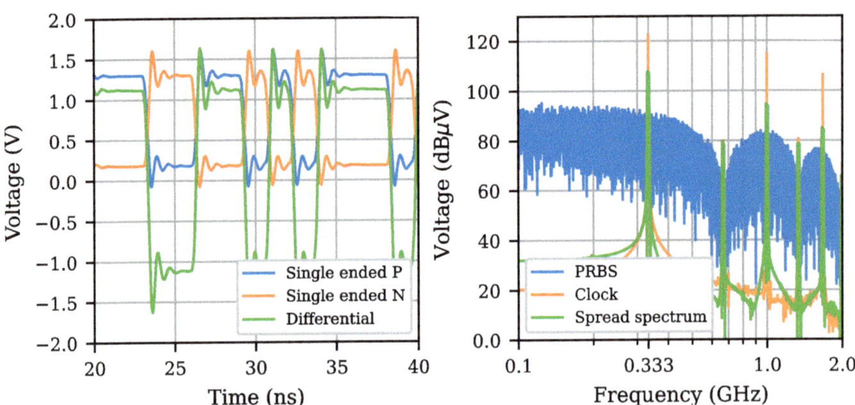

Fig. 2.12 Time- and frequency-domain voltages from a simple IBIS simulation

individual voltages corresponding to the bit sequence generated by the PRBS source (blue and orange curve) and the differential voltage (green curve), which is obtained by simply subtracting the N signal from the P signal.

The right-hand side of Fig. 2.12 compares the spectra of the differential voltage of the clock and the PRBS signals. It reveals an important property of PRBS sequences: PRBS sequences generate broadband noise, which is significantly different from the clock signal spectrum already studied in Fig. 2.5. The maximum voltage level of the PRBS spectrum is more than 30 dB below the clock's fundamental peak.

Because of this difference, clock signals are critical and should be the first signal types to be considered in a simulation. The spread spectrum clock results lie between the PRBS and the non-modulated clock. With the chosen modulation settings, there is a difference of about 15 dB compared to the original clock signal.

It is important to note that converting narrowband noise to (more) broadband noise reduces the peak values but does not reduce the overall energy of the noise. The energy of the noise signal is now distributed over a larger frequency band and can still lead to interference, for example, at RF receivers, as we explore in Chap. 15.

A correctly defined IBIS model is well suited for the modeling of digital sources in EMC simulation. Unfortunately, the vendor support for IBIS models is of mixed quality. Some vendors do not offer IBIS models at all, some vendors only on request. Full libraries with easy download and search ability are rare.

2.4 Circuit Models of the Coupling Path

Domains: 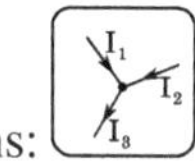

In the previous section, we introduced circuit models of noise sources. As described in Sect. 1.2, the modeling approach using coupled EM and circuit simulation is a key concept of this book. Thus, to model the noise propagation of an electronic system, it is necessary to connect the circuit model of the noise source to a circuit model of the coupling path.

For a clearer understanding of the modeling process, we segment the coupling path into discrete components and interconnects:

Discrete components are commonly resistors, capacitors, and inductors. These components are parts of the coupling path for which we use a lumped RLC model to describe their behavior.

The motivation for this modeling approach stems from the geometrical dimensions of these discrete components. Capacitors and resistors are composed of structures with geometrical dimensions much smaller than the electronic system itself. Including the interior structure of a capacitor in the EM simulation model is unnecessary, as the fields inside the capacitor do not interact with other components.

However, the physical dimensions of inductors are often comparable to the dimensions of the electronic system, and the stray fields of inductors can couple to other components. For this reason, we model inductors both as circuit components and as EM models in this book. In this section, we discuss inductor modeling using circuit models. The creation of EM models of inductors is addressed in Sect. 9.6.

In our segmentation of the coupling path, interconnects refer to all physical elements that electrically connect individual components. These can include traces or planes on a PCB, connectors, or cables. Throughout this book, we determine the electrical properties of the interconnects using EM simulations and employ two approaches to incorporate the results from the EM simulation into a circuit model.

The first approach is the use of RLC models to describe the properties of interconnects. This approach is presented in Chaps. 3 and 5. If the interconnect geometry is electrically long, the description of the interconnect using a lumped RLC circuit is no longer valid. We discuss the concept of electrical size in the next section and examine the limits of lumped RLC models in Sect. 2.4.2.5.

For generating circuit models of electrically large interconnects, a description using network parameters is more suitable. We begin using network parameters to characterize the coupling path in Chap. 6 and continue their application in all subsequent chapters. The fundamental properties of network parameter models used in EMC simulations are presented in Sect. 2.4.3, and the process of generating network parameter models from EM simulations is discussed in Sect. 6.7.

The segmentation into a discrete component and an interconnect is not always straightforward, as it requires precise knowledge of the method used to obtain the circuit model of the component. When using a circuit model obtained from measurements, the model must include only the properties of the component, and the effect of the mounting fixture during the measurement must be removed from the circuit model. This is necessary because the mounting of the component in the system being modeled is accounted for in the EM simulation model of the system.

If the effect of the mounting during the measurement is not properly removed from the circuit model of the component, using such circuit models in the simulation of the system will result in deviations from the modeling intent. We discuss this effect in detail using a measured model of a common mode choke in Sect. 9.4.2.

2.4.1 Electrical Size

A lumped RLC description is only valid as long as the component is electrically short. To explain this in detail, we introduce the electrical length l_λ of a component as:

$$l_\lambda = \frac{l}{\lambda} \tag{2.3}$$

In Eq. 2.3, l refers to the physical length of the component and λ to the wavelength of interest. l_λ essentially measures the length of the component in fractions of a wavelength. As the frequency increases, the electrical length of a component also increases.

A component is considered electrically short if $l_\lambda < \frac{\lambda}{10}$. This criterion is based on the assumption that one period of a sine wave can be reasonably accurately sampled with 10 samples, and that the voltage and current in each $\frac{\lambda}{10}$ short section remain constant. Throughout this book, the concept of electrical size is a key principle used to determine the required mesh resolution in an EM simulation or to select the most efficient EM simulation method.

For components that are not electrically short, a lumped RLC model does not sufficiently describe the component. This limitation is demonstrated in Sect. 2.4.2.5. To accurately model components that are not electrically short, two options are available:

- A cascade of lumped RLC models, in which every RLC section is electrically short.
- A black box description using network parameters.

The cascading of RLC blocks leads to the concept of the Multiconductor Transmission Line Method (MTL), as described in Chaps. 8 and 11. This method works well for geometries that are homogeneous in length, such as cables or traces on a PCB, but it is not applicable to arbitrary geometries.

Network parameters are black-box models that can be applied to arbitrary geometries to describe their electrical characteristics. They efficiently characterize the coupling path formed by interconnects and can be easily obtained from an EM simulation, even when the geometry of the coupling path is complex.

In the following, we first present lumped RLC models for resistors, capacitors, and inductors. Then, we introduce the use of network parameters. In Sect. 2.4.3, we discuss important properties of network parameters that must be understood to set up coupled EMC simulations that describe the properties of interconnects using network parameters.

2.4.2 *Lumped Models of Discrete Components*

2.4.2.1 Ideal Components

Before discussing lumped models of real components with parasitic values, we first introduce the basic behavior of ideal RLC elements. The impedance of an ideal resistor remains constant across all frequencies. The impedance of an ideal inductor increases with frequency at a rate of $+20\,\mathrm{dB}$ per decade, while the impedance of an ideal capacitor decreases with frequency at the same rate. The frequency-dependent magnitude of the impedance of ideal RLC elements is shown in Fig. 2.13.

For the plot in Fig. 2.13, we used a resistance value of 1 Ω, a capacitance of $\frac{1}{2\pi \cdot 10^6} \approx 0.159\,\mu\mathrm{F}$, and an inductance of $\frac{1}{2\pi \cdot 10^6} \approx 0.159\,\mu\mathrm{H}$, each resulting in an impedance of $1\,\Omega$ at 1 MHz.

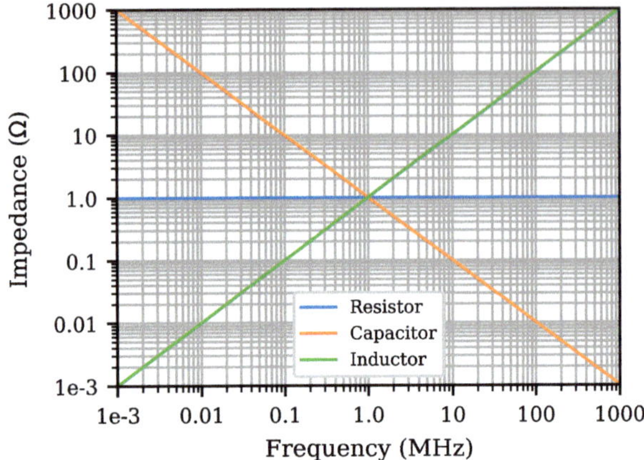

Fig. 2.13 Frequency dependence of the magnitude of the impedance of ideal RLC elements

The impedance of an ideal capacitor approaches zero as $f \rightarrow \infty$, while the impedance of an ideal inductor approaches zero as $f \rightarrow 0$. However, real components deviate from this ideal behavior, always exhibiting some finite impedance. To accurately model their realistic behavior, parasitic values must be added to the ideal components.

2.4.2.2 Resistors

Resistors are characterized by the nominal value of the resistance, which is constant for all frequencies in an ideal resistor. The parasitic model of a resistor is plotted on the left-hand side of Fig. 2.14. A series inductance L_{series} is added in order to account for the inductance of the connecting leads. Across the resistance, a parallel capacitance C_{parallel} is placed to include the capacitance effect of the leads and electrodes.

Depending on the nominal value R of the resistor, either the parasitic capacitance or the inductance dominates the frequency behavior of the impedance. Let us consider two nominal values, $R_1 = 100\,\text{m}\Omega$ and $R_2 = 1\,\text{k}\Omega$. Using $L_{\text{series}} = 1\,\text{nH}$ and $C_{\text{parallel}} = 1\,\text{pF}$ for both resistors, we obtain the impedance curves plotted on the right-hand side of Fig. 2.14.

For the low-value resistor R_1, the inductance L_{series} significantly affects the impedance, causing an increase in the low MHz region. This effect is critical for shunt resistors used to measure current. For the high-value resistor R_2, the capacitance C_{parallel} causes the impedance to decrease starting around 10 MHz. A resonance occurs at 5 GHz, where the impedance R_2 drops to 1 Ω. It is important to note that the value of $C_{\text{parallel}} = 1\,\text{pF}$ is relatively high and is chosen to illustrate the resonance effect.

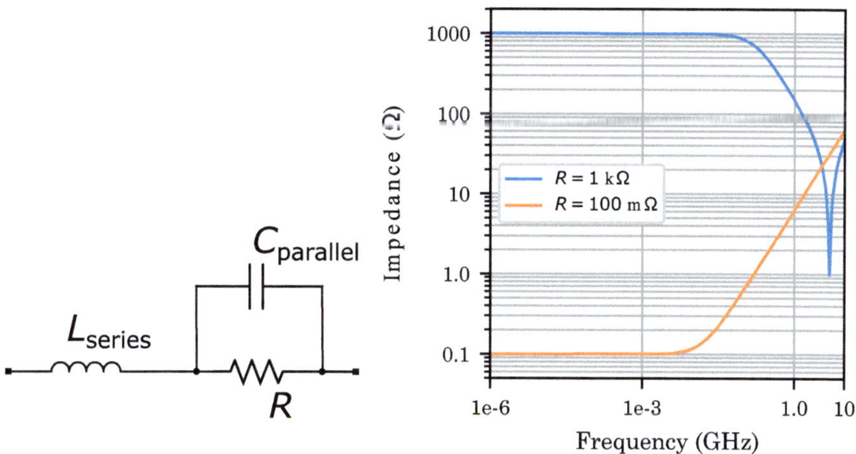

Fig. 2.14 Left: A high-frequency parasitic model of a resistor. Right: Impedance of two example resistors

Two circular leads with a diameter of 0.4 mm, a separation of 2 mm, and a length of 1 cm result in $C_{\text{parallel}} = 0.1\,\text{pF}$, pushing the resonant frequency as high as 16 GHz. For SMD resistors with values above a few ohms, the parallel capacitance is even lower. In such cases, the parasitic values can be omitted, and only the nominal resistance value is used in simulation models where the frequency of interest is in the low GHz range.

2.4.2.3 Capacitors

Capacitors are probably the predominant component on PCBs today. The impedance of an ideal capacitor decreases monotonically with frequency. When placed between a power and GND net, capacitors are referred to as “decoupling” capacitors. They act as energy storage for switching elements and provide a path to close high-frequency current loops. Capacitors are also used as high-pass filter elements, diverting high-frequency current to GND.

In the high-frequency parasitic model of a capacitor, the resistance R_{series} and the inductance L_{series} of the connections are added in series to the nominal capacitance value, as shown on the right-hand side of Fig. 2.15. These values are often referred to as equivalent series resistance (ESR) and equivalent series inductance (ESL). As in the case of resistors, the parasitics are governed by the leads and electrodes of the capacitor. However, unlike resistors, the parasitic inductance becomes significant in the MHz range, even for small form-factor SMD capacitors.

On the right-hand side of Fig. 2.15, the impedance of a 47 nF capacitor is plotted. For this plot, ESR is fixed at 100 mΩ, and ESL is varied. For an ESL of 0.5 nH, the resonant frequency is approximately 33 MHz; for 1 nH, it is around 10 MHz. At

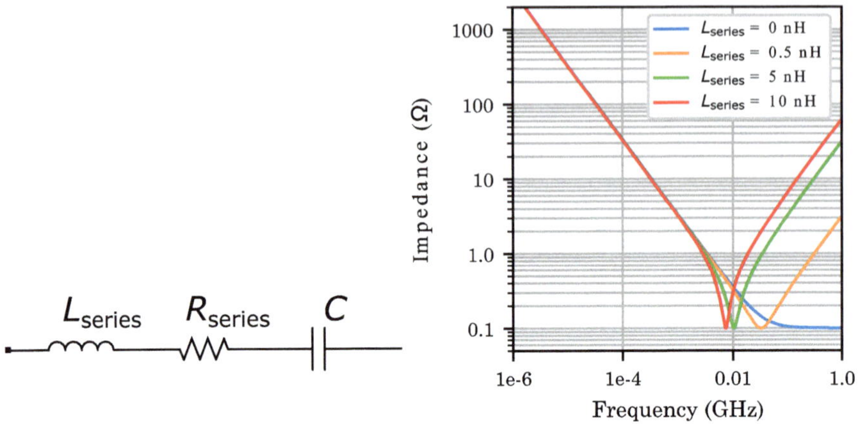

Fig. 2.15 Left: A high-frequency parasitic model of a capacitor. Right: Impedance of a 47 nF capacitor with varying ESL

Table 2.3 ESR and ESL by capacitor size code

Size code	Approx ESR (mΩ)	Approx ESL (nH)
0402 (1005 M)	20–100	0.5–1.0
0603 (1608 M)	10–50	0.5–1.2
0805 (2012 M)	5–30	0.5–1.5
1206 (3216 M)	3–20	0.5–2.0
1210/1812	1–10	0.5–2.5

the resonant frequency, the impedance value equals ESR. Beyond this frequency, the impedance starts to increase, and the capacitor behaves like an inductor. This behavior is critical for every EMC simulation model, and we include ESL and ESR in all our simulation models.

The estimation of ESL and ESR can be performed with varying levels of fidelity. For quick results, we assign typical values for ESL and ESR to capacitors in our simulation models based on the size of the capacitor's footprint. Table 2.3 summarizes approximate ESL and ESR values of capacitors in terms of their size code.

This approach is not highly accurate, as the values of ESL and ESR also depend on the capacitor type. However, such a simplified choice allows for the quick construction of simulation models without the overhead of searching for actual component models on vendor homepages.

In 2025, the availability of accurate capacitor models provided by vendors for download is very good. All the major capacitor manufacturers, such as Murata, Kemet, AVX, Panasonic, and TDK (see the references for a selection of internet addresses), offer simulation models of their physical components. These models are typically created from measurements and are available in different formats, including raw measured S-parameter files or SPICE models fitted to match the measurements.

Table 2.4 Capacitor size codes and part numbers for the impedance plot in Fig. 2.16

Size code	Part number
0402	GCM155R71H103KA55
0603	GCM188R71H103KA37
0805	GCJ216R71H103MA01
1206	GRM3195C1H103JA01

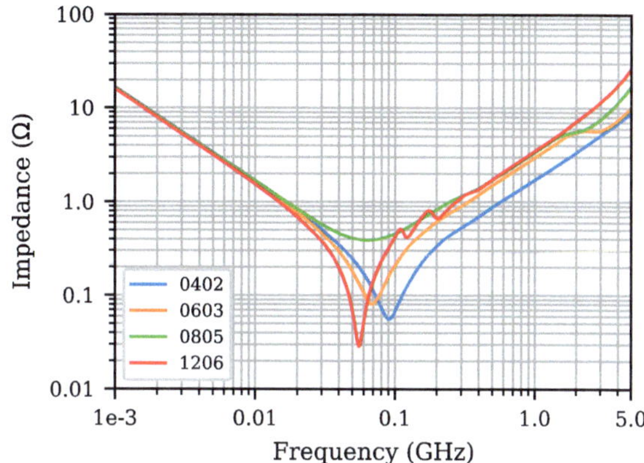

Fig. 2.16 Impedance curves of capacitors as downloaded from a vendor's webpage

To demonstrate typical values, four SPICE models were downloaded from Murata's SimSurfing application, a web-based resource for component information. All capacitors are SMD ceramic with a nominal capacitance of $C = 10\,\text{nF}$, a voltage rating of 50 V, and different size codes, see Table 2.4. Their impedance is plotted in Fig. 2.16. We observe that the resonant frequency decreases as the footprint size increases.

The nominal capacitance of a capacitor varies within the tolerance range defined by the vendor. Additionally, it is affected by several other parameters, such as temperature, age, and, for certain capacitor types, the applied voltage. Circuit models of capacitors that include dependencies on these parameters are sometimes available, depending on individual vendors. In cases in which the simulation engineer wants to study these effects, but cannot obtain a model from his vendor, a good possibility is to use an RLC model instead of the vendor model and vary the value of the capacitance within the expected range.

2.4.2.4 Inductors

Inductors are used to store magnetic energy and to block high-frequency currents. In the high-frequency model of an inductor, a series resistor is added to account for ohmic losses in the inductor and its leads. The inductance of the leads is omitted because, being in series with the larger nominal inductance, it is negligible. Throughout this book, we use a simple parasitic model that neglects the core losses in the inductor.

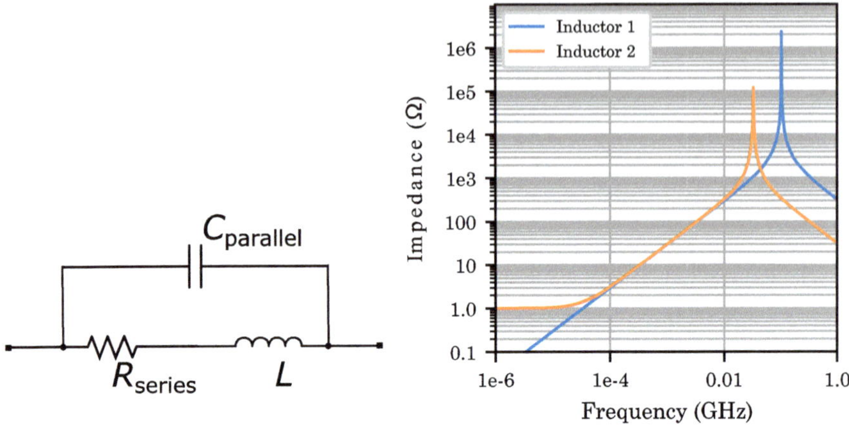

Fig. 2.17 Left: A high-frequency parasitic model of an inductor. Right: Impedance of two 4.7 nH inductors with different parasitic values

A capacitance C_{parallel} is placed in parallel with the inductance and resistance, as shown in Fig. 2.17. This capacitance is sometimes referred to as the self-capacitance of the inductor because it leads to the inductor's self-resonance. However, in theory, the term self-capacitance originates from electrostatics and should not be used for C_{parallel}. Since most inductors consist of a conductor wound in turns around a shape, the main contribution to the parallel capacitance is the capacitance formed between the turns.

Let us analyze the impedance of an inductor with parasitic elements. We select two inductors with a nominal inductance of 4.7 µH. The parasitics of inductor 1 are set to $R_{\text{series}} = 10\,\text{m}\Omega$ and $C_{\text{parallel}} = 0.5\,\text{pF}$, while the parasitics of inductor 2 are $R_{\text{series}} = 1\,\Omega$ and $C_{\text{parallel}} = 5\,\text{pF}$. The impedances of both inductors are plotted on the right-hand side of Fig. 2.17.

We observe three distinct regions in the impedance curve. For inductor 1, at the lowest frequency of 1 Hz, the impedance equals R_{series}. For inductor 2, the parasitic resistance is chosen so low that the first region is not visible in the plot.

In the second region, as the frequency increases, the impedance is governed by the inductance. Since the nominal inductance is the same for both inductors, their impedance curves overlap in this region.

In the third region, above the self-resonance frequency, the behavior becomes capacitive. The self-resonance frequency is determined by the value of the parallel capacitance, assuming a fixed nominal inductance. As the capacitance increases, the resonance frequency shifts downward.

If a permeable material is used in the inductor, the behavior of the component becomes more complex than the parasitic model we have proposed. Due to dispersion, the permeability of the material decreases at higher frequencies, altering the slope of the impedance curve. Additionally, due to saturation effects, the inductance may decrease with the applied current through the component.

SPICE models can incorporate these effects into simulations (see Sect. 9.7 and the corresponding references). For a quick estimation of the impact of these effects, we use the same approach as proposed for capacitors: employ the simple parasitic model of the inductor and vary the nominal values of the component to determine whether the reduction in inductance significantly affects the simulation results.

For more detailed studies, we construct EM models of inductors that can account for the effects of dispersion and saturation. The creation of a dispersive inductor model is described in Sect. 9.6, while the simulation of saturated inductors is discussed in Sect. 9.7.

In many electronic circuits, inductors are designed to efficiently couple energy from one inductor to another. A transformer is the most prominent example of such a design, and a common mode choke is also based on coupled inductors. Inductive coupling can be modeled using the SPICE element coupling factor, k.

The coupling factor k relates the current flowing through one inductor to the current flowing through another inductor. It is connected to the mutual inductance of the coupled inductors by the following simple formula:

$$k_{1,2} = M_{1,2}\sqrt{L_1 \cdot L_2} \tag{2.4}$$

For transformers, the coupling factor is close to 1 (or -1, for opposite polarity). We use the coupling factor also to model unintended coupling such as that of two closely spaced wires. However, it is not straightforward to estimate the appropriate value of the coupling factor for arbitrary conductor arrangements using analytical formulas (Margueron and Keradec 2007a,b, see also Sect. 9.6). We see in Sect. 5.5 how coupling factors are estimated using EM simulation.

2.4.2.5 Limits of Lumped RLC Models

At the beginning of this section, we stated that lumped RLC models are not suitable for modeling the electrical characteristics of a coupling path that is electrically large. In this section, we present a detailed example to demonstrate the limitations of lumped RLC models. This example serves as a motivation for the introduction of network parameters, which are discussed in the next section.

For the demonstration, we analyze the input impedance Z_{in} of a wire pair. At one end of the wire pair, we calculate the input impedance Z_{in}. At the other end, we connect a loading resistor R_{L}.

We use two straight and parallel copper wires with a length of $l = 50\,\text{cm}$, a wire diameter of $d = 1\,\text{mm}$, and a center-to-center separation of $s = 1.5\,\text{mm}$. This configuration results in a characteristic impedance of $Z_0 = 120\,\Omega$ for this wire pair.

An accurate circuit model of the cable is created using the MTL method, which is discussed in Chap. 8. The length of 50 cm corresponds to an electrical length of 1λ at 599.585 MHz. That is, an electromagnetic wave with a frequency of 599.585 MHz has, in vacuum, a wavelength of 50 cm. For the sake of readability, we round this value to 600 MHz.

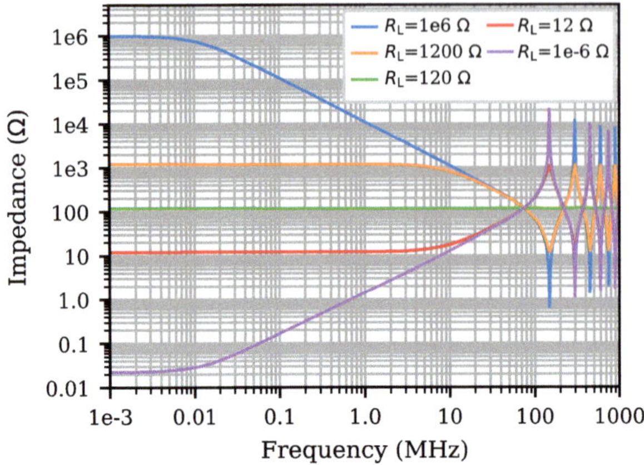

Fig. 2.18 Input impedance of a wire pair of 50 cm length with different loads

The input impedance Z_{in} of the wire pair for a varying load R_{L} is plotted in Fig. 2.18. We have used five loading conditions:

Case	Load	R_{L}
1	Open	$R_{\text{L}} = 10^6\,\Omega$
2	$10 \cdot Z_0$	$R_{\text{L}} = 1200\,\Omega$
3	Z_0	$R_{\text{L}} = 120\,\Omega$
4	$\frac{1}{10} \cdot Z_0$	$R_{\text{L}} = 12\,\Omega$
5	Short	$R_{\text{L}} = 10^{-6}\,\Omega$

Below 1 kHz and down to DC, the input impedance of the wire pair is the sum of the wires' resistance R_{loop} in series with the load Z_{L}. For cases 1–4, where $Z_{\text{L}} \gg R_{\text{loop}}$, the impedance value below 1 kHz essentially equals Z_{L}. In case 5, where the wire pair is terminated with a short, we determine $R_{\text{loop}} = 21.9\,\text{m}\Omega$ from the plot.

In case 3, where the wire pair is terminated with the characteristic impedance $Z_{\text{L}} = Z_0 = 120\,\Omega$, the impedance curve remains flat across the entire frequency band. With this termination value, the wire pair is matched to the load, resulting in no resonances. When the load resistor value deviates from Z_0, resonances (low impedance) and anti-resonances (high impedance) begin to appear. The amplitude of these resonances increases further as Z_{L} deviates more from Z_0.

The first resonance, located at 150 MHz, corresponds to an electrical length of $\frac{\lambda}{4}$, and the second resonance at 300 MHz corresponds to $\frac{\lambda}{2}$, and so on. At a resonance frequency, the input impedance of the wire pair is low. For the open termination at 150 MHz, Z_{in} is less than $1\,\Omega$. Even a small voltage can drive a significant current into the wire pair, leading to radiation from the cable and potentially causing emission problems. For an EMC simulation, it is crucial that circuit models account for this effect.

We create a lumped RLC model of the cable by calculating the RLC values using analytical formulas provided in Sect. 3.3. For the resistance, assuming a copper conductivity of $\sigma = 5.8 \cdot 10^7 \frac{\text{S}}{\text{m}}$, we obtain $R_{\text{analytical}} = 21.9\,\text{m}\Omega$, which matches with the value obtained in case 5 in Fig. 2.18.

For the inductance, a value of $L_{\text{analytical}} = 192\,\text{nH}$ is calculated using Eq. 3.10. For the capacitance, Eq. 3.7 yields $C_{\text{analytical}} = 14.4\,\text{pF}$. The resulting schematic of the lumped RLC model is shown in Fig. 2.19.

A comparison of the input impedance calculated using the MTL-based cable model and the lumped RLC model is shown in Fig. 2.20. The frequency of 60 MHz is highlighted as it corresponds to an electrical length of $\frac{\lambda}{10}$. Below this frequency, the wire pair is considered electrically short.

We observe that the lumped RLC circuit models the input impedance of the wire pair very well below this frequency, as postulated at the beginning of this section. The difference around 0.1 MHz, in the case where the pair is terminated with a short (orange and red curves), arises because the MTL cable model accurately calculates the inductance of the wire, which comprises both inner and outer inductances (cf. Sect. 5.6.1). In contrast, the RLC model considers only the outer inductance. A detailed discussion of inner and outer inductances is provided in Sect. 5.6.1. This difference results in a deviation of approximately 30% in Z_{in} at 0.1 MHz. We

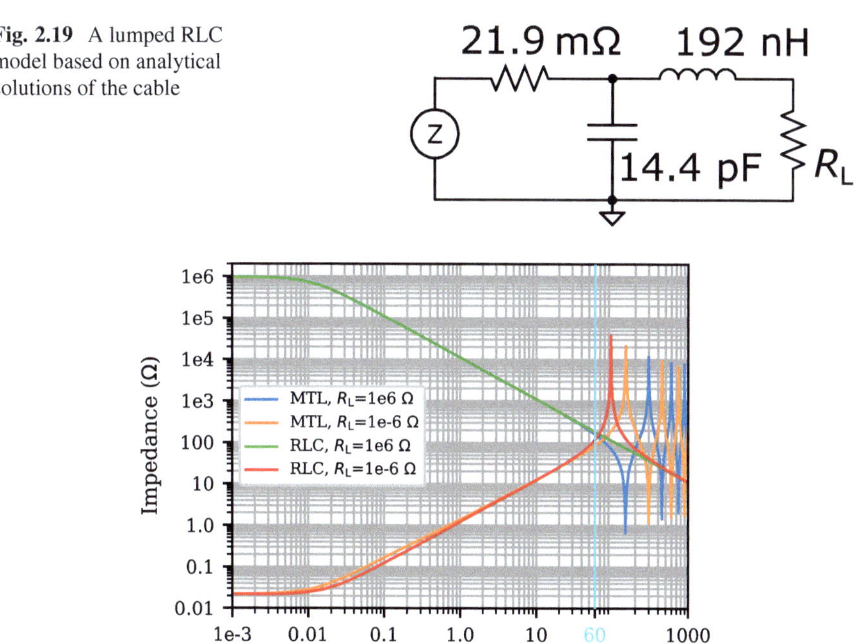

Fig. 2.19 A lumped RLC model based on analytical solutions of the cable

Fig. 2.20 Input impedance of the transmission line based cable model compared to the impedance of the lumped RLC model for two different loads

also examine the deviation between the RLC and the cable model at 60 MHz. The deviation is 29% for the short case and 22% for the open case.

This example gives us a good idea about the limits of lumped RLC models. We could now take the RLC values as a building block and cascade them to a ladder RLC model. As already mentioned, this leads to the MTL model of the cable pair and is discussed in Sect. 8.1.

However, for arbitrary distributed conductor arrangements, the generation of a lumped description is difficult as we would need to build a geometry-based mesh of RLC elements. Such a process leads us to EM computation and is actually the idea of the TLM solver that we introduce in Chap. 4.

Please note that the models of cables obtained from the (Multiconductor) Transmission Line Theory are labeled as MTL throughout this book, while the method of using a three-dimensional impedance matrix is called Transmission Line Matrix Method and abbreviated by TLM. These two methods should not be confused.

2.4.3 Network Parameters

Network parameters provide a means to characterize the coupling path irrespective of its electrical size. This property makes them the preferred choice for characterizing coupling paths, and starting in Chap. 6, we use network parameters to convert the results of full-wave EM simulations into a circuit model.

Network parameters describe the relationship between the voltages and currents at the input and output of the network parameter model. The inputs and outputs of a network parameter model are called ports. A port consists of two terminals, and the network parameter model defines the voltage across these two terminals and the current flowing through them. The concept of a network parameter port is elaborated in the next section.

Network parameters for n-port components are typically represented in matrix form. For an n-port, a complex-valued matrix of size $n \times n$ is used to define the relationship between the n currents and n voltages at all ports. Depending on which current or voltage variables are considered independent, several representations for the matrix entries are available: Z-parameters, Y-parameters, H-parameters, G-parameters, and ABCD-parameters (Pozar 2012). Instead of discussing each definition in detail, we focus on an impedance-based interpretation to understand the component's behavior. Unlike other network parameters, S-parameters use power waves, rather than voltages and currents, to describe the coupling between ports. The benefits of using power waves are discussed in detail in Sect. 2.4.3.2.

In the way we use network parameters, the network parameter matrix serves as a look-up table that characterizes the coupling path in the frequency domain. We limit the network parameter characterization in this book to linear networks (see Sect. 2.6). Due to linearity, the value of each matrix element at a given frequency does not depend on other frequencies. This allows us to define a frequency-dependent table of all matrix entries. Such tables are typically written into text files, for example, in the Touchstone format.

Standard SPICE simulators do not directly support the Touchstone format, but simulators listed at the end of Sect. 2.2 do. These simulators load network parameter files and calculate voltages and currents at the ports within the frequency band provided in the file. Between the defined frequency samples, interpolation of the tabulated data is used. Apart from the requirement of linearity, there is no inherent frequency limit for such a description. This enables modeling of the resonant behavior of a component.

In a network parameter-based characterization, the origin of the data used to generate the network parameter model does not matter. Data can be sourced from measurements, simulations, or other sources, as long as it can be written in the form of network parameters. This approach allows the characterization of even highly complex coupling paths in a circuit model.

A tabulated network parameter dataset, as described above, works well for frequency-domain circuit simulations. However, for transient circuit simulations, including switching, these models cannot be used in a straightforward way. A circuit component model in a transient circuit simulation must be passive and causal, but network parameter models do not inherently guarantee these properties.

To use network parameter models in a transient circuit simulation, an additional step is required to convert the frequency-domain data into a representation that is both causal and passive. This conversion process is known as macromodel generation and is based on vector fitting. The basics of this step are described in Sect. 10.3.

The methodology for generating macromodels is relatively new compared to circuit simulation, with its first usage dating back only to the late 1990s. At the time of writing this book, macromodel generation has matured into an essential component of our simulation strategy, enabling the coupling of circuit and EM simulations. Nevertheless, the process requires the simulation engineer to understand the underlying principles and may also require control over the accuracy of the macromodel generation. Details of the process are discussed in Sect. 10.9.

2.4.3.1 Network Parameter Ports

In this section, the basics of ports are explained, as they need to be understood when building EMC simulation models in which the coupling path is described using network parameter models.

In circuit theory, a port is the interface of the network parameter model. It is formed by a pair of terminals, allowing us to connect the network parameter model to an external circuit. The terminal pair must satisfy the port condition: the sum of the currents entering and leaving the port must be zero. This is indicated by the blue arrows in Fig. 2.21.

The network parameter matrix provides information only about the voltage across the port and the current through the port. The voltage across the port is a local

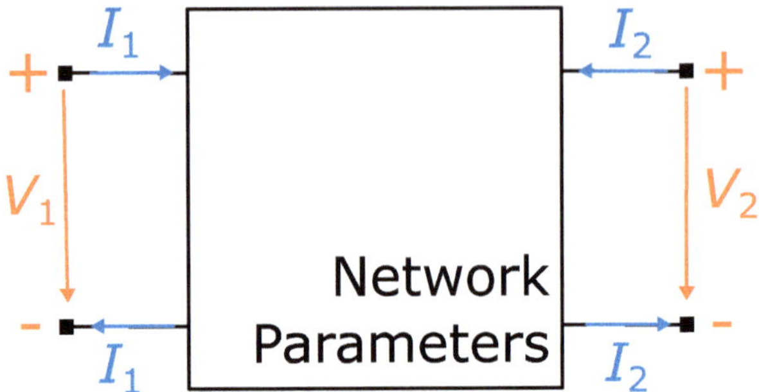

Fig. 2.21 A two-port network parameter model

voltage between two conductors, such as the voltage across an SMA connector. A network parameter model does not include information about a global reference. The concept of a global reference is applicable only to electrically short structures, where we can assume the reference conductor to be an equipotential.

Consider, for example, a network parameter model characterizing the coupling between two antennas. Antenna 1 is a monopole, and the voltage is measured between the antenna arm and its ground plane. Antenna 2 is a dipole, and the voltage is measured between the two arms of the antenna. Both voltages are local to each antenna and do not require a global reference.

The concept of a local voltage unfortunately collides with the concept of a ground reference in a circuit simulation and often causes confusion. In a circuit simulation, the voltage at every circuit node is referenced to a global (circuit) reference, indicated by the ground symbol. This global reference is necessary for the circuit simulator to solve the corresponding system of equations. Without defining a zero-potential node (the ground symbol), the system of equations becomes underdetermined. This issue is discussed in detail in Sect. 3.8.

A network parameter model does not contain information about global potentials. This discrepancy is resolved by employing a proper methodology when connecting network parameter models to an external circuit. In the following, we illustrate this methodology using an π-filter example. The resulting circuit notation may be unfamiliar to the SPICE community but is an efficient representation when network parameter models are included in a circuit schematic.

Let us discuss the π-filter depicted on the left-hand side of Fig. 2.22. We have added parasitics to the nominal values and aim to convert the RLC based schematic into a black-box network parameter model. First, we need to define the external ports in the RLC model. Throughout this book, we use hexagonal boxes to indicate a port.

In the RLC model, both ports are defined between the input and output of the filter and the circuit ground. For port 2, we follow the common convention that the second terminal of a port is automatically connected to the circuit ground and not explicitly displayed. Thus, electrically, both ports are referenced to the circuit ground. In our example, we assume that the reference for both ports is equal. In

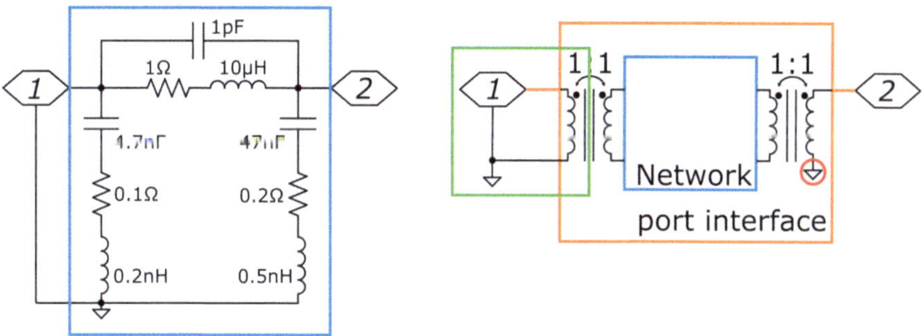

Fig. 2.22 Left: An π-filter modeled as a RLC network with external ports. Right: Connection of external ports to a network parameter block

general, this is not required, and the reference of port 1 can be different from the reference of port 2.

We now calculate the 2 port S-parameters of the schematic on the left-hand side and export them into a Touchstone file. While we could choose another representation for network parameters and a different file format, S-parameters and Touchstone files are the most practical choice. For the export, we need to select the appropriate frequency band and sampling points for calculating the S-parameters.

When the Touchstone file is imported into another schematic, the typical representation appears as the orange box (with the orange terminals), as depicted on the right-hand side of Fig. 2.22. This representation is a block with two terminals, which we refer to as a network parameter block. By default, the reference terminals of a network parameter block are not displayed, but circuit simulators often provide the option to display them.

We could now connect our circuit between the two terminals of the block. However, this approach leads to potential issues. To connect the network parameter model to an external circuit, the circuit simulator inserts a port interface with ideal 1:1 transformers. These transformers are required to ensure that the port condition (sum of currents at a port = 0) is satisfied at every port. Without this, combinations of network parameter blocks could violate the port condition.

These transformers are not explicitly shown on the schematic, but they are always present. As the block connects to the external circuit through the transformers, the circuit inside the green box becomes floating (i.e., its voltage with respect to the circuit reference is not defined). This is intuitive because network parameters provide only information about the voltage between terminal 1 and terminal 2, not the voltage between a terminal and the circuit reference.

However, the circuit simulator requires a reference for the floating circuit section; otherwise, the network equations cannot be solved. The common solution is to connect one terminal of the port internally within the block to the circuit ground, as shown for port 2 in Fig. 2.22. This connection is indicated by the red circle.

Recall that the external port 2 is also referenced to the circuit ground, although this reference is not explicitly displayed. As a result, port 1 and port 2 on the right-hand side of Fig. 2.22 are connected to the network parameter block in exactly the same way, with a more compact notation at port 2.

To further become familiar with the notation of network parameters, we modify our model so that only the two capacitors are included in the S-parameter block, while the inductor is connected externally to the block. The required port definition is explained in Fig. 2.23. We have also included a voltage and current probe in the setup to illustrate how circuit probing works in the case of network parameter blocks.

If we probe the network parameter block as shown on the right-hand side, we measure the voltage across the inductor and the current through the inductor. However, we cannot measure the input or output voltage of the filter at port 3. The most intuitive way to understand the notation is to imagine that the entire circuit located between the n-terminal of the block and the ground symbol of the block is effectively placed between the two terminals of the port. This is indicated by the orange boxes in Fig. 2.23.

From this, it should become clear that understanding the port definition (e.g., series and shunt) is essential for every network parameter block we use. In measurements, we are constrained by the connectors and cables required for the measurement. In EM simulation models, we have more flexibility in defining ports.

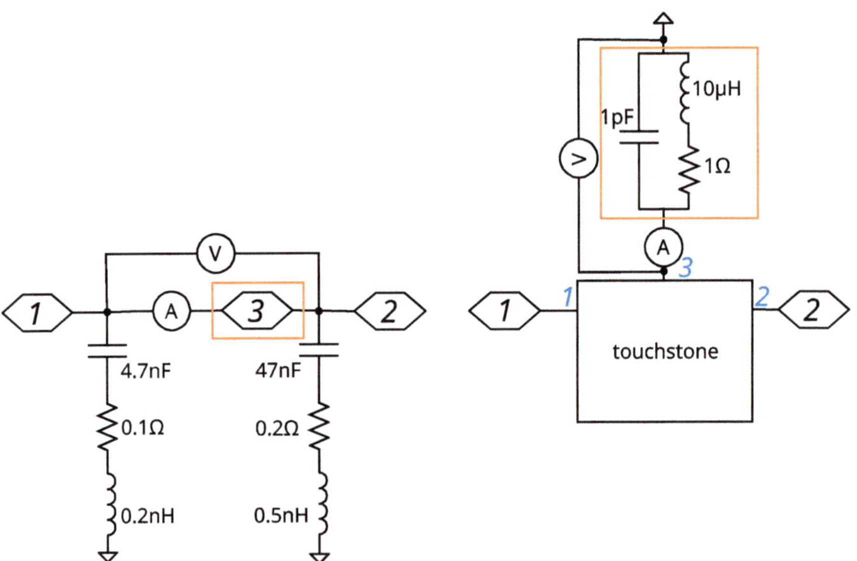

Fig. 2.23 Left: A π-filter modeled as an RLC network with three external ports. Right: Connection of external ports and an inductor to a network parameter block

In this section, we have discussed the basics of the port definition. The discussion continues in Sect. 6.7, where we focus of network parameter models obtained from an EM simulations.

2.4.3.2 S-Parameter

S-parameters are a form of network parameters, where instead of using voltages or currents to characterize the components behavior (see Fig. 2.21), the so-called power waves are used. At every port, we define the incident a_i and reflected b_i power wave as Pozar (2012),

$$a_i = \frac{1}{2}\left(\frac{V_i}{\sqrt{Z_i}} + \sqrt{Z_i} I_i\right), \; b_i = \frac{1}{2}\left(\frac{V_i}{\sqrt{Z_i}} - \sqrt{Z_i} I_i\right). \tag{2.5}$$

The S-parameter relates the incident and reflected power wave by the matrix equation

$$\mathbf{b} = \mathbf{S}\mathbf{a}. \tag{2.6}$$

In component notation, this is

$$\begin{pmatrix} b_1 \\ \vdots \\ b_n \end{pmatrix} = \begin{pmatrix} S_{11} & \dots & S_{1n} \\ \vdots & \ddots & \vdots \\ S_{n1} & \dots & S_{nn} \end{pmatrix} \begin{pmatrix} a_1 \\ \vdots \\ a_n \end{pmatrix}. \tag{2.7}$$

Following the definition in Eq. 2.5, a reference impedance Z_i is required at every port. With the definition of Z_i, we obtain a description that allows us to calculate port voltages and currents from the power waves and vice versa.

One key advantage of S-parameters is their flexibility in measurement and analysis. Unlike Z- or Y-parameters, S-parameters allow for measurements with arbitrary terminations. For instance: In Z-parameter measurements, the current at the non-excited port must be zero. In Y-parameter measurements, the voltage at the non-excited port must be zero. At high frequencies, meeting these requirements is often impractical. In contrast, S-parameters are compatible with realistic terminations as long as the exact termination impedance is known. This flexibility has made S-parameters the standard for network analysis in high-frequency systems, with network analyzers being the most accurate tools for measuring linear networks.

In EMC simulation, we benefit from these properties of S-parameters and use them to generate circuit models from EM simulations. A complete simulation flow is demonstrated in Chap. 9.

The notation of S-parameters is S_{ij}, where j is the excited port and i the receiving port. For example, S_{21} refers to the coupling from port 1 to port 2.

Throughout this book, we use S-parameters to characterize filters and antennas.

2.5 Sink Models

Domains: 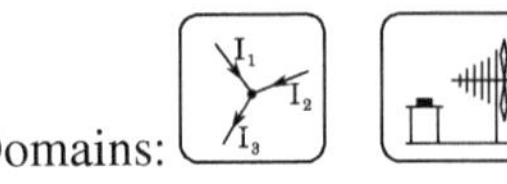

In EMC engineering, the sink refers to the part of the setup on which the source is acting. In emission measurements, the sink is typically a piece of laboratory equipment that measures the magnitude of the noise emitted by the DUT.

In a simulation, we can define generic sinks, as we can calculate quantities such as voltages, currents, or field values directly, without needing to include the measurement equipment in the simulation model. We begin this section by discussing such generic sink models in circuits.

After introducing generic sinks, we discuss models of Line Impedance Stabilization Networks (LISN), as LISNs are present in many measurement setups. Antenna measurements are briefly introduced in this section, as a detailed discussion about antennas in radiated emission testing is provided in Chap. 12.

The last part of this section examines the measurement principles of EMI receivers and how they can be included in a simulation.

2.5.1 Generic Circuit Sink Models

A simulation model allows us to obtain voltages and currents without needing to include specific measurement devices in the model. Voltages and currents can be calculated directly at ports and lumped elements in the EM simulation model or across circuit elements in a coupled EM-circuit simulation. This type of sink is used in many examples throughout the book, where we calculate voltages and currents at circuit components.

In simulation models, a resistor can be used to measure voltages and currents at arbitrary positions. For example, placing a low-ohmic resistor into a signal conductor allows us to measure the current in this conductor without modeling the current measurement device required in a physical setup. Similarly, placing a high-ohmic resistor between two conductors allows measurement of the voltage between these conductors.

2.5.2 LISN

In electronic systems where the section from the noise source to the power source is an essential coupling path (Fig. 2.1), a well-defined termination at the supply side is required. For this purpose, the Line Impedance Stabilization Network (LISN), also known as the Artificial Mains Network (AMN), is introduced to standardize the coupling path to the power supply.

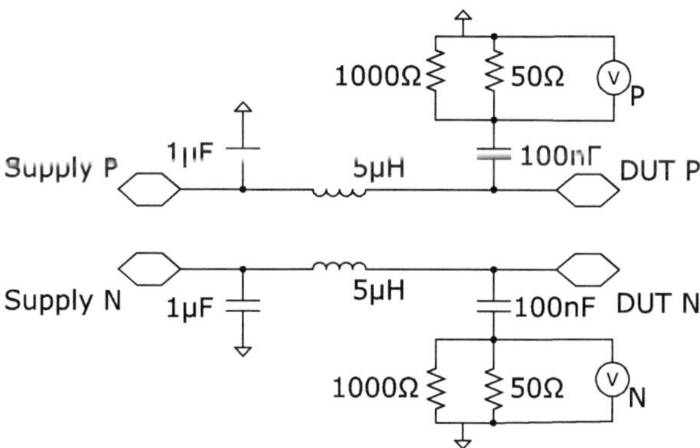

Fig. 2.24 Schematic representation of the LISNs of the CISPR 25 conducted emission test

The LISN serves multiple purposes:

- To ensure repeatability of experiments.
- To decouple noise on the grid itself from the experiment.
- To provide a connection for the measurement equipment.

Since they are standardized, LISNs are generally convenient to model. Datasheet information is typically available in terms of circuit schematics. The circuit schematic of the Schwarzbeck NNBM 8124 LISN, designed for CISPR 25 emission testing (e.g., Schwarzbeck), is shown in Fig. 2.24. We use P for the battery supply and N for the neutral line.

The network is designed to represent the power supply path toward the battery within a vehicle. This path is assumed to consist of several meters of wire within a metallic chassis. The 5 μH inductance and the 1 μF capacitor are used to decouple external noise from the supply to the measurement equipment. The output to the EMI receiver is decoupled using the second 100 nF capacitor. The 1 kΩ resistor in parallel with the 50 Ω resistor protects the measurement equipment from discharge events during connector removal. The 50 Ω resistor is actually not part of the LISN, but characterizes the impedance of the measurement equipment, mostly an EMI receiver (see Sect. 2.5.4). Sometimes, the 1 kΩ resistor is omitted when modeling the LISN. The addition of this resistor results in a slightly lower impedance of about 48 Ω when viewed from the DUT side, as shown in Fig. 2.25.

For different testing standards, different LISNs exist. CISPR 16, FCC, and MIL 461/462 specify a 50 μH inductor instead of the 5 μH used in CISPR 25. All standards define a tolerance for the impedance seen from the DUT side, typically as high as 20%.

The impedance seen from the DUT side for both 50 μH and 5 μH LISNs, calculated from the schematic in Fig. 2.24, is plotted in Fig. 2.25. We observe that the different inductance values shift the frequency at which the 50 Ω impedance is

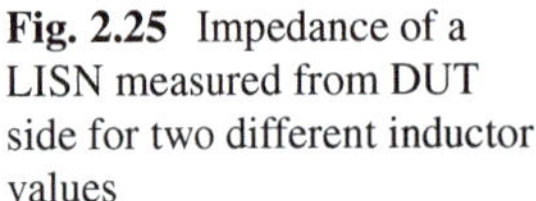

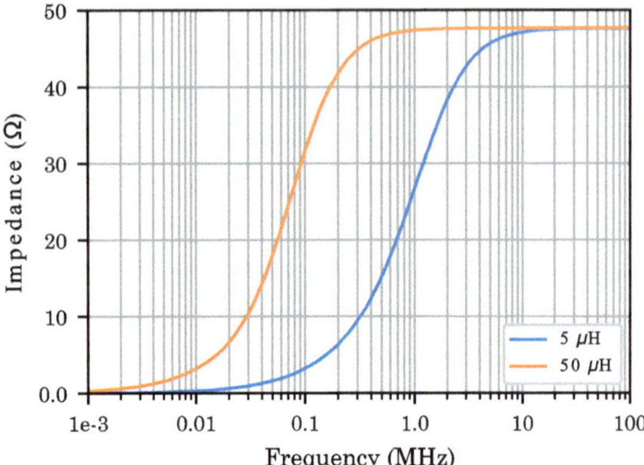

Fig. 2.25 Impedance of a LISN measured from DUT side for two different inductor values

reached. Using the simple schematic in EMC simulation models allows us to quickly modify the inductor value if required.

The conducted emission test is specified up to 108 MHz, and so is the required impedance of the LISN. Above this frequency, LISNs exhibit arbitrary behavior. The authors of Schertlen et al. (2018) report resonant behavior above 200 MHz, caused by the parasitic capacitance of the 5 μH air-coil within the LISN.

The simple circuit model shown in Fig. 2.24 does not accurately describe the behavior of any hardware used in the lab above 108 MHz, as each piece of hardware exhibits its own unique impedance. In simulation, however, the impedance of the LISN above 108 MHz can be easily modified (Schertlen et al. 2018) to determine whether such modifications impact the measurement results.

2.5.3 *Antennas*

Radiated emission testing defines EMI limits in terms of field strengths. In a measurement, antennas are required to measure the fields. In a simulation, however, including the antenna is not necessarily required, as the EM simulation calculates field strength within the computational domain. These field strengths are evaluated using field probes, which are introduced in Sect. 6.2.

A detailed discussion on the differences between the calculated fields and the values obtained when including the antenna in the simulation model is provided in Chap. 12.

2.5.4 EMI Receiver

The EMI receiver processes emission voltages such that they can be evaluated with respect to the limit lines defined in standards. Limits are defined in frequency domain. The EMI receiver generates the frequency-domain response of a measured time-domain signal, based on definitions made in the CISPR 16-1-1 standard.

In this section, we explain how the frequency spectrum visualized by an EMI receiver is obtained from simulation data. We highlight the differences between the three detector types, peak (PK), average (AV), and quasipeak (QP), and their relation to a DFT. There are many codes available to produce the readings of an EMI receiver, and their application may take some effort and time. For simulation engineers, it is useful to understand under which circumstances an EMI receiver code must be applied, and when the application of a DFT, which runs much faster, is sufficient.

The EMI receiver is used in standardized emission setups and connected, e.g., to a LISN, a current clamp or an antenna output. For the EMI receiver, no extra geometric component is introduced into the simulation. The only interaction with the measurement setup is its input impedance of 50 Ω. If, e.g., attached to a LISN, the EMI receiver is modeled by a 50 Ω resistor in parallel to the LISN's 1 kΩ resistor (see Fig. 2.24).

The EMI receiver measures a voltage in time domain. Heterodyne detectors measure the spectrum frequency by frequency. Time-domain receivers work faster and generate the spectrum using a short-time Fourier transform (STFT) of the time-domain data. In simulation, we use the latter approach. An STFT with window length T_{capt} is applied to the recorded time-domain voltage of length T_{rec}, which is measured across its 50 Ω input impedance. The function of the EMI receiver is reproduced in a postprocessing step, and for each detector type, one evaluation is needed. In this section, we give an overview only, a precise description of a time-domain measurement receiver is given in Keller and Feser (2007).

According to CSIPR, a resolution bandwidth (RBW) of Δf_{RBW} must be specified. Table 2.4 specifies, according to the four frequency bands of CISPR 16 ranging from 9 kHz to 1 GHz, RBWs between 200 Hz and 120 kHz. To achieve a resolution bandwidth of Δf_{RBW}, the STFT needs to be evaluated over a capture time of about $T_{\mathrm{capt}} = \frac{1}{\Delta f_{\mathrm{RBW}}}$. The bandwidth increases with increasing frequency in order to save evaluation time. The postprocessing of the time-domain waveforms must consider the RBW when computing the spectral amplitude.

Table 2.4 CISPR 16 band frequency ranges and resolution bandwidths

CISPR 16 Band	A	B	C	D
Frequency Range	9–150 kHz	150 kHz–30 MHz	30– 300 MHz	300 MHz–1 GHz
Resolution Bandwidth	200 Hz	9 kHz	120 kHz	120 kHz

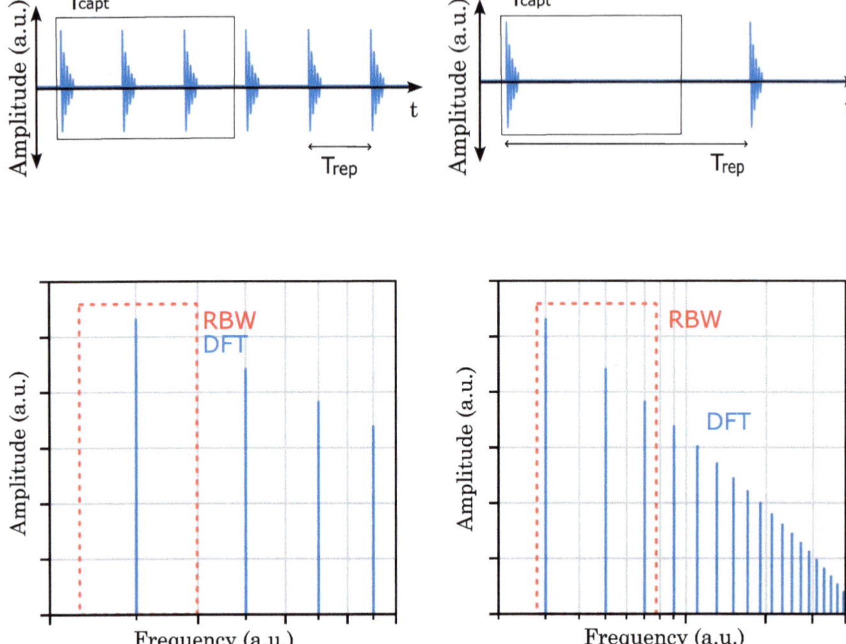

Fig. 2.26 Time-domain (top) and sketch of frequency-domain representation (bottom) of pulsed signals. Left: narrowband. Right: wideband (cf. Keller and Feser 2007; Hackl and Deutschmann 2017)

The capture time T_{capt} has impact if the measured signal is pulsed with a repetition frequency of f_{rep}, or, respectively, with a pulse repetition time $T_{\text{rep}} = \frac{1}{f_{\text{rep}}}$. The two possible situations are shown in Fig. 2.26 (cf. Keller and Feser 2007). On the upper left-hand side, $T_{\text{capt}} > T_{\text{rep}}$, the capture time is longer than the pulse repetition time. On the upper right-hand side, we have the reverse situation $T_{\text{capt}} < T_{\text{rep}}$.

On the lower side of Fig. 2.26, we sketch the equivalent situation in the spectral domain. The frequency content of the signal appears as multiples of the pulse repetition frequency, displayed by the blue line. Since $T_{\text{rep}} = \frac{1}{f_{\text{rep}}}$, $T_{\text{capt}} = \frac{1}{\Delta f_{\text{RBW}}}$, the time-domain signal on the left hand side has in frequency domain the property $f_{\text{rep}} > \Delta f_{\text{RBW}}$. The pulse repetition frequency is larger than the receiver bandwidth. Such a signal is called a narrowband signal. Sinusoidal signals and superpositions of them are typical narrowband signals, as long as their period does not exceed the capture time.

In case of $T_{\text{capt}} < T_{\text{rep}}$ we obtain $f_{\text{rep}} < \Delta f_{\text{RBW}}$, which is displayed on the lower right-hand side. Such a signal is called wideband signal. White noise and pulsed signals with long pulse repetition rate (namely $T_{\text{capt}} < T_{\text{rep}}$) are typical wideband signals.

The three detectors PK, AV, and QP differ as follows. For narrowband signals, all detectors result in the same amplitude, and they are identical to a DFT. This is particularly true for sinusoidal and superpositions of sinusoidal signals, i.e., periodic signals.

For wideband signals, the received signal energy at one measurement instant can be different from that at another one. In the time frame shown in the upper right of Fig. 2.26, there is one signal detected within T_{capt}. If the time window is slightly shifted, there is no signal detected.

Ideally, a wideband signal must be recorded sufficiently long to capture all its details, i.e., at least one signal period. Given such a recorded signal of length T_{rec}, the PK detector monitors the maximum amplitude of this signal. It is not sensitive to the pulse repetition frequency f_{rep}. The AV detector averages the signal detected within each capture time T_{capt} over the entire length T_{rec} of the recorded signal. It is sensitive to f_{rep}: the lower f_{rep}, the less is the average energy received during the measurement time and the less records the AV detector compared to the PK. The reading of the QP detector is in between the other two, originally defined to describe the subjectively perceived disturbance of a pulsed audio signal by a human.

The readings of the different detectors over the pulse repetition frequency f_{rep} are shown in Fig. 2.27, normalized to that of the PK detector for wideband signals. The plot is shown for a receiver bandwidth of $\Delta f_{\mathrm{RBW}} = 120\,\mathrm{kHz}$, which corresponds to the CISPR 16 bands C and D. In case of wideband signals, i.e., $f_{\mathrm{rep}} < \Delta f_{\mathrm{RBW}}$, the PK detector shows the maximum received signal strength during the recorded signal time T_{rec}. This maximum is not a function of the pulse repetition frequency f_{rep}. In relation to the PK detector, the AV detector decreases with 20 dB per decade with decreasing f_{rep}. The less the number of signals recorded within T_{rec}, the less is its reading. The QP detector is in between the two and in detail defined in the CISPR16-1-1 standard.

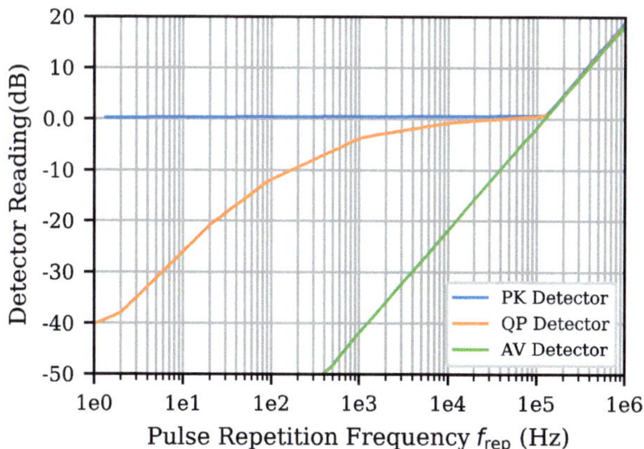

Fig. 2.27 Reading of the PK, AV, and QP detectors, relative to the PK detector for wideband signals. The curve is shown for band C and D, with RBW 120 kHz (Keller and Feser 2007)

Once the pulse repetition frequency f_{rep} is larger than Δf_{RBW}, all detectors have the same reading. The more pulses are captured within T_{capt}, the larger is the received signal energy, and the readings of all detectors increase proportional to f_{rep}.

A plot such as Fig. 2.27 exists for each Δf_{RBW}. For each, the PK detector is constant up to Δf_{RBW} and increases with 20 dB per decade for larger f_{rep}, the AV detector increases with 20 dB per decade for all f_{rep}, and the QP detector is in between the two. For $f_{\text{rep}} > \Delta f_{\text{RBW}}$, all detectors are equal, irrespective of Δf_{RBW}.

The different detectors of the EMI receiver allow to investigate the number of occurrences of an interference signal within T_{rec}. The difference between the reading of the PK and the AV detectors indicates whether the noise sources of the DUT emit few or many pulses, provided that T_{rec} is sufficiently long to record these pulses.

In the simulation, there is no need to characterize unknown noise sources. In a model, all sources are known. We need to emulate the different detectors only to compare simulation results using wideband signal sources with the limits. For qualitative studies and for narrowband signals, the application of an EMI receiver script is not required, a DFT is sufficient.

Two spectra produced by an EMI receiver are plotted in Fig. 2.28. Additionally, a DFT with a bandwidth of 1 kHz is shown. The DFT is displayed in blue, the PK detector in orange, and the AV detector in green. The detector readings are produced with the EMI receiver script provided by CST.

On the left-hand side, we evaluate the spectrum of a periodic sequence of short pulses. The period, i.e., the repetition frequency, is 50 kHz, and the duty cycle $\tau = 0.01$. The DFT shows the spectrum of a periodic signal with a fundamental harmonic of 50 kHz. We evaluate the EMI receiver spectrum according to CISPR 25 in band B and C above 150 kHz. In band B, $\Delta f_{\text{RBW}} = 9$ kHz, the pulse sequence is narrowband. The results of the PK and the AV receiver are identical. The spectrum of the DFT is 3 dB larger, because, according to CISPR-16-1-1 Annex E.7, the

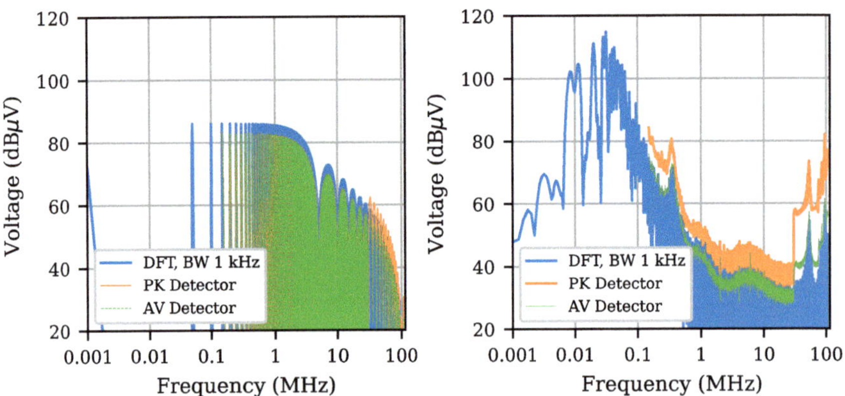

Fig. 2.28 Spectra computed using an EMI receiver postprocessing. Left: Short periodic pulse train with $f_{\text{rep}} = 50$ kHz. Right: Inverter with PWM sequence and switching frequency of 10 kHz

detectors refer to root-mean-square values, whereas the DFT is computed with respect to the peak amplitudes.

At 30 MHz, Δf_{RBW} increases from 9 kHz to 120 kHz. The pulse sequence with a fundamental harmonic of 50 kHz is now weighted different between AV and PK detector. The spectrum generated by the PK detector shows an increase in the amplitude, because for the signal with $f_{\text{rep}} = 50\,\text{kHz}$, more than one harmonic is captured within the receiver bandwidth. The captured energy increases as a matter of the position of the sliding window of the STFT, and the PK detector records the maximum. The AV receiver does not detect this increase, because on average, the energy content of the pulsed signal is unchanged. The difference between the readings of the AV and the PK detectors is $20\log_{10}(50/120) = -7.5\,\text{dB}$. This is the ratio between the pulse repetition frequency and the resolution bandwidth for band C. It corresponds to the difference between the AV and the PK detectors as it is indicated in Fig. 2.27.

On the right-hand side, we see the spectrum of the motor drive discussed in Sect. 10.6. The motor drive uses a PWM sequence with a modulation frequency of 500 Hz and a switching frequency of 10 kHz. The signal is wideband. Due to the modulation frequency, one period of the signal is 2 ms. Its spectrum is richer than that of a periodic sequence of short pulses.

There is no fundamental harmonic at 10 kHz. We see two side lobes, one to the left and one to the right of the fundamental harmonic. The spectrum computed by the AV detector is close to that of the DFT, the spectrum of the PK detector is 7 dB dB higher. At the transition from band B to band C, the peak detector increases by 17 dB, that of the AV detector by 7 dB. The spectrum of the DFT remains unchanged, because it uses a bandwidth of 1 kHz irrespective of the frequency being in band B or band C.

The large difference between the AV and the PK detectors for the PWM signal shown on the right-hand side indicates that the signal contains low-frequency components. The signals captured during T_{capt} do not represent the characteristics of the entire recorded signal. Slow-varying or rarely occurring events average out, so that the readings of the PK and the AV receiver are different. In measurement, this information helps to identify the dominant noise sources. In simulation, the noise sources are known. The employment of the EMI receiver is only needed if, in case of broadband signals with $f_{\text{rep}} < \Delta f_{\text{RBW}}$, simulation results are to be compared to limit lines or to EMI receiver measurements.

2.6 Linearity and Time Invariance of Components

Domains: 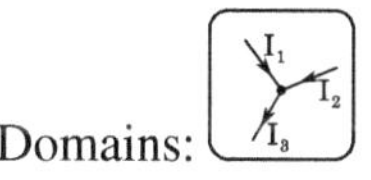

In Sect. 2.4.3, we briefly introduce the concept of component linearity, as network parameter models require linearity. The concept of linearity is important

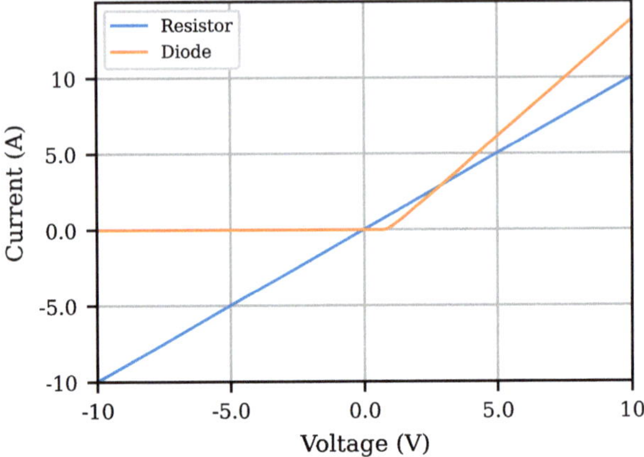

Fig. 2.29 VI characteristic of a resistor and a diode.

because it can be exploited to speed up circuit simulations. Linearity allows for the application of the principle of superposition. In combination with time invariance, it allows to analyze the system in frequency domain.

In the following, we first introduce the concept of linearity and time invariance for the circuit component models discussed in this chapter and then explore the properties of systems composed entirely of such components.

A circuit component is linear if its circuit component model (interpreted as an impedance) does not depend on the magnitude of the voltage and current at its terminals. We have plotted the VI characteristics of a 1 Ω resistor and the 1N4148 diode in Fig. 2.29. The characteristic curve for the resistor is linear, but for the diode it is not, as the current through the diode is a function of the voltage applied across it. The resistor is a linear component, whereas the diode is not.

The impedance of a capacitor or inductor depends on the rate of change of the current or voltage but not on its magnitude. At each frequency (rate of change), the impedance remains linear. Therefore, all components created from RLC circuits are linear. A transistor, however, is not linear, as its drain-source impedance depends on both the gate-source voltage and the drain-source voltage.

Time invariance is another important property. A component is time-invariant, if its response in time domain is not a function of when a signal is applied. RLC elements, and models of transistors and diodes are time-invariant. However, a model of a trace on a PCB that considers the change of the trace's resistivity over time due to a change in temperature is time-variant.

An IBIS model, as introduced in Sect. 2.3.4, is defined by $V - t$ data. It is time-dependent but time-invariant. The IBIS model is nonlinear. Twice the voltage of the input at the gate of the transistors does not generate twice the voltage at the output of the model.

The entire concept of network parameter models, as introduced in Sect. 2.4.3, refers only to linear, time-invariant components. A circuit composed of multiple elements is linear and time-invariant, if all of its components are. The complete system is then classified as a Linear Time-Invariant (LTI) system. An LTI system can be described by a set of linear differential equations of the system's voltages and currents with coefficients that are neither a function of time, nor of current or voltage. As in case of the components, time invariance implies that the response of the system to an input, be it current or voltage, is not a function of when the signal is applied. Linearity implies that the superposition principle holds. The response of any linear combination of input signals equals the linear combination of the system's responses to each individual signal. In particular, if a current or voltage signal at the input is scaled by a given factor, the output signal is scaled by the same.

We do not discuss LTI systems in detail, as they have been thoroughly described in numerous books and lectures (e.g. Hayt et al. 2023). Instead, we highlight some properties of LTI systems that allow us to speed up simulations and calculate transfer functions.

A linear system can be characterized in the frequency domain because its behavior at any particular frequency ω_0 does not depend on other frequencies. This property allows the system to be simulated frequency by frequency. In SPICE, this approach is commonly known as an AC analysis.

For an AC analysis, an AC source must be defined, as the sources introduced in Sect. 2.3.1 are focused on time-domain definitions. An AC source is defined analogously to the voltage and current sources in Sect. 2.3.1, specifying the magnitude and phase of the driving signal at every frequency to be simulated.

In the following, we refer only to voltage sources, but all properties discussed are equally valid for current sources. If we use a magnitude of 1 V and a phase of 0° for all source frequencies and observe the voltage at an arbitrary position in our system, we obtain the so-called transfer function $\mathcal{H}(\omega)$. The transfer function describes the transfer of the voltage from the source to the observation point.

In an LTI system, the voltage at the observation point, $V_{\text{out}}(\omega)$, for any other source $V_{\text{in}}(\omega)$, can be simply calculated as a product of the input voltage and the transfer function:

$$V_{\text{out}}(\omega) = \mathcal{H}(\omega) \cdot V_{\text{in}}(\omega) \tag{2.8}$$

This concept is used for the characterization of filters, as we see in Chap. 9. In EMC, the transfer function is interpreted as the response of the coupling path, independent of the source and sink. This interpretation is particularly helpful when comparing different geometries and sources.

To calculate $\mathcal{H}(\omega)$, an AC source is placed at the position of the noise source, and the transfer function is determined from the noise source to the sink. This provides information about the coupling path in the frequency domain, which is much easier to interpret compared to the narrow peak spectrum of a noise source with rectangular pulses, as the former has a continuous spectrum.

When comparing changes in the coupling path geometry, this approach gives a clear picture of which frequency bands are affected by the change. We employ this method in Chap. 10. We can also use a nonuniform spectrum at the source, similar to the PWL definition of time-domain sources. For example, we can use the spectrum of the bounds of a rectangular pulse, as shown in Fig. 2.5, as a source and directly calculate the voltage V_{out} from the AC simulation. The advantage of an AC simulation over a transient simulation lies in its simulation speed. We introduce the usage of AC simulations in Sect. 3.5.2.

SPICE simulators also allow the usage of nonlinear circuit elements, such as transistors and diodes, in AC analysis. When a nonlinear component is included in the system and an AC simulation is performed, the SPICE solvers linearize the element. However, a linearized element may behave quite differently from its actual nonlinear description. The process of linearization is discussed in detail for a diode in Sect. 3.8.2.

Linearity has another important consequence. For two different sources in the circuit, the transfer function of each source can be calculated separately, and the two output voltages at the observation point can then be added. Linear superposition is central to the combine results method, which we use to obtain field values from excitations defined on the circuit. This method is introduced in Sect. 6.7.5.

In EMC simulation, there are hardly any time-variant models; linearity is a much stronger assumption. For this reason, when we refer to linearity in this book, we implicitly also mean time invariance.

2.7 Advanced: From Network Parameters Back to Lumped Elements

Domains: 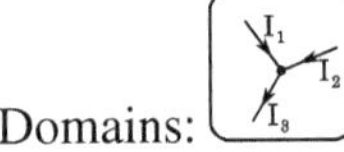

In this section, we discuss the possibility of calculating RLC values from a network parameter block. Since network parameters are black-box models, simply reading the table contained in a file does not reveal the actual characteristics described by the model. To interpret the behavior of the black-box model, we present a technique to extract impedance and RLC values from the model.

This technique is useful for network parameter models obtained from both measurements and simulations. Extracting and analyzing the RLC values from a black-box model allows us to understand its nominal and parasitic characteristics.

In the following, we describe the extraction technique using the π-filter from Sect. 2.4.3.1. The extraction of RLC values for models generated by an EM simulation is discussed in Sect. 6.7.4.

Network parameter models can be converted from one representation to another. For the RLC extraction described in this section, we use the impedance representation provided by Z-parameters. In a two-port S-parameter block, where both reference impedances are equal to Z_0, $S_{1,1}$ can be converted into $Z_{1,1}$ using

$$Z_{1,1} = Z_0 \frac{((1 + S_{1,1})(1 - S_{2,2}) + S_{1,2}S_{2,1}}{((1 - S_{1,1})(1 + S_{2,2}) - S_{1,2}S_{2,1}}, \tag{2.9}$$

and in matrix notation as

$$Z_{1,1} = Z_0 \left(\frac{\mathbf{I}_2 + \mathbf{S}}{\mathbf{I}_2 - \mathbf{S}} \right)_{1,1}. \tag{2.10}$$

Circuit simulators offer this conversion in automated postprocessing steps. We use this functionality throughout this book. All impedance plots are created by using this technique instead of manually calculating the ratio of voltage and current.

From the complex impedance in the form $Z = R + jX$ we can obtain the frequency dependent RLC values using:

$$R = R \ , \ L = \tfrac{X}{\omega} \ , \ C = -\tfrac{1}{\omega X} \tag{2.11}$$

This is a generalized definition of RLC values, which can result in negative values. For example, the capacitance calculated from the impedance of an ideal inductor is a negative number across the entire frequency band, and we do not aim to interpret these numbers.

Instead, we focus on a method to extract RLC values from the black box that are straightforward and meaningful to understand. Similar to the S-parameter to Z-parameter conversion, the operations described in Eq. 2.11 are typically offered by circuit simulators as postprocessing steps.

In the following, we use the same π-filter as in Sect. 2.4.3.1 to demonstrate the technique. The main idea is to reduce the n-port model to a one-port model, where we measure the input impedance and convert this impedance to an RLC model.

With the appropriate choice of termination impedances in the reduction, we can select a particular branch for the current inside the network parameter block. When analyzing the π-filter model from port 1, we identify three possible branches for the current, as depicted in Fig. 2.30. The path through the parallel capacitance of the inductor is not highlighted, as it is not relevant to the extraction.

At sufficiently low frequencies, by choosing the termination impedance R_L at port 2, we control the current in the green branch. If R_L is chosen to be high-ohmic, no current flows through the green branch; if R_L is low-ohmic, the green current branch dominates all current paths. The schematic, now utilizing the S-parameter block, is simplified as shown in Fig. 2.31.

The impedance at port 1 for high and low ohmic R_L is plotted in Fig. 2.32. Above 1 MHz, the impedance is no longer a function of R_L. This region above the resonance is not of interest, as per the limitations of RLC models described in Sect. 2.4.2.5, we must use a frequency well below the first resonance to extract RLC values.

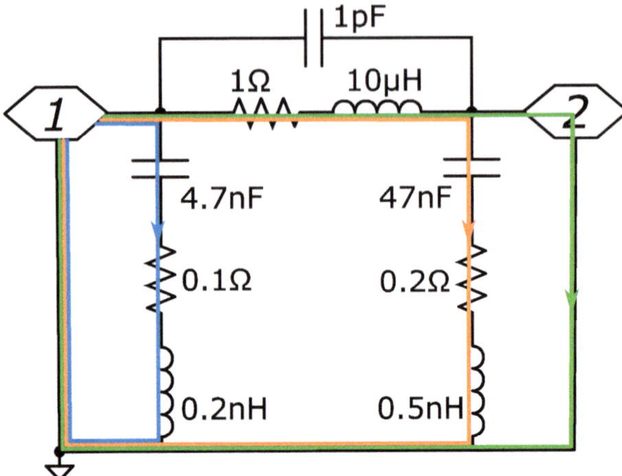

Fig. 2.30 Possible current branches in the π-filter when excited from port 1

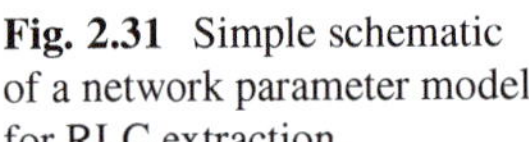
Fig. 2.31 Simple schematic of a network parameter model for RLC extraction

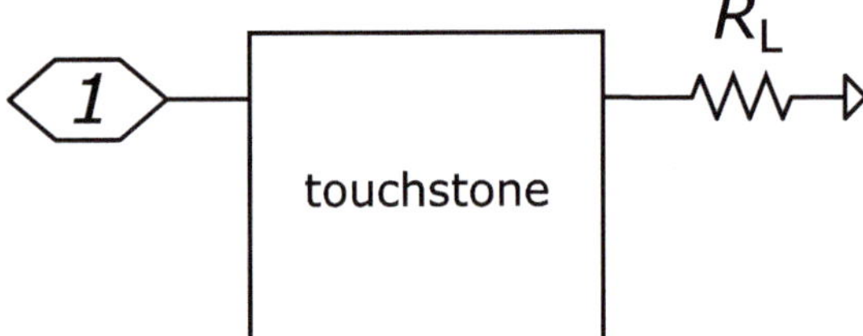

At 1 Hz, in the case where the termination is a short, the impedance is 1 Ω. This represents the series impedance of the inductor, as the current flows exclusively through the green branch, and the contributions from the other two branches to the total current are negligible. Above 0.01 MHz, the contribution from the inductance becomes noticeable as the impedance begins to increase.

In the case of an open termination, we observe a high impedance, as there is no current flow through any branch. From the decreasing impedance in the blue curve, we conclude that the behavior is capacitive.

For the extraction, we use postprocessing to calculate the RLC values from the Z-parameters using Eq. 2.11. The results of the extraction are plotted in Fig. 2.33. Let us discuss these results in detail, focusing on the values at 1 Hz:

For the resistance, plotted in part (a), we obtain 1 Ω for the case where the termination is a short. This corresponds to the series resistance of the inductor. In the case of an open termination, we observe an increase in resistance starting around 100 Hz. This is due to all three current paths being high-ohmic and connected in parallel.

For the extracted inductance in the shorted case, we obtain 9.995 μH at 1 Hz, very close to the actual value of 10 μH. In the case of an open termination, the inductance at 1 Hz is negative and therefore not visible in the logarithmic plot.

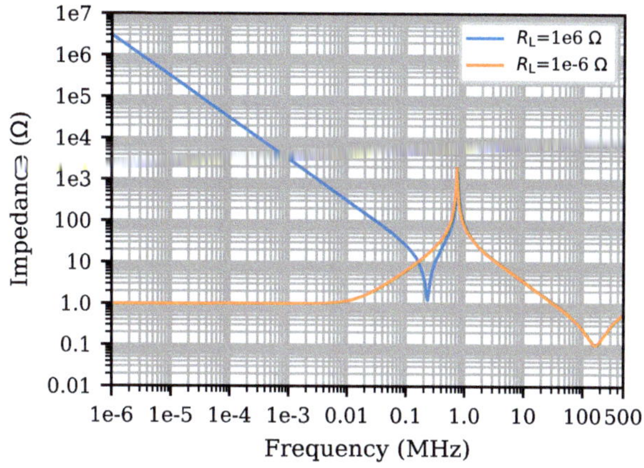

Fig. 2.32 Impedance at port 1 for open and short termination at port 2

For the capacitance extraction, the situation is reversed: the shorted case delivers a negative value, which is not visible in the plot. The extraction from the case where $R_L = 10^6\,\Omega$ yields a capacitance of 51.7 nF. This represents the value of both capacitors in parallel.

If we terminate at port 1 and measure the impedance at port 2, we obtain the same capacitance for the open case but a slightly different inductance of 9.95 μH. This difference arises because the capacitor at port 2 has a lower impedance at 1 Hz compared to the capacitor at port 1. Consequently, the ratio of the impedances for the different current branches is slightly different at port 1 compared to port 2.

From the above study, we conclude that for the extraction of resistance and inductance, a low-ohmic termination is the appropriate choice. For the extraction of capacitance, a high-ohmic termination is required. The extraction is not perfect because it is not possible to completely isolate the individual branches of the current.

This extraction technique is very useful for true black-box models that we intend to use in our simulations. For example, when using a measured or downloaded Touchstone file of a component, we can run the extraction to gain insight into its electrical characteristics.

Another application is the extraction of RLC values from simulation models. In later chapters of this book, we generate large n-port circuit models of PCBs. The extraction technique presented here enables us to select a specific current loop by connecting the appropriate terminations to the circuit model and extracting the RLC values of the selected loop. This process is demonstrated in Sect. 6.7.4.

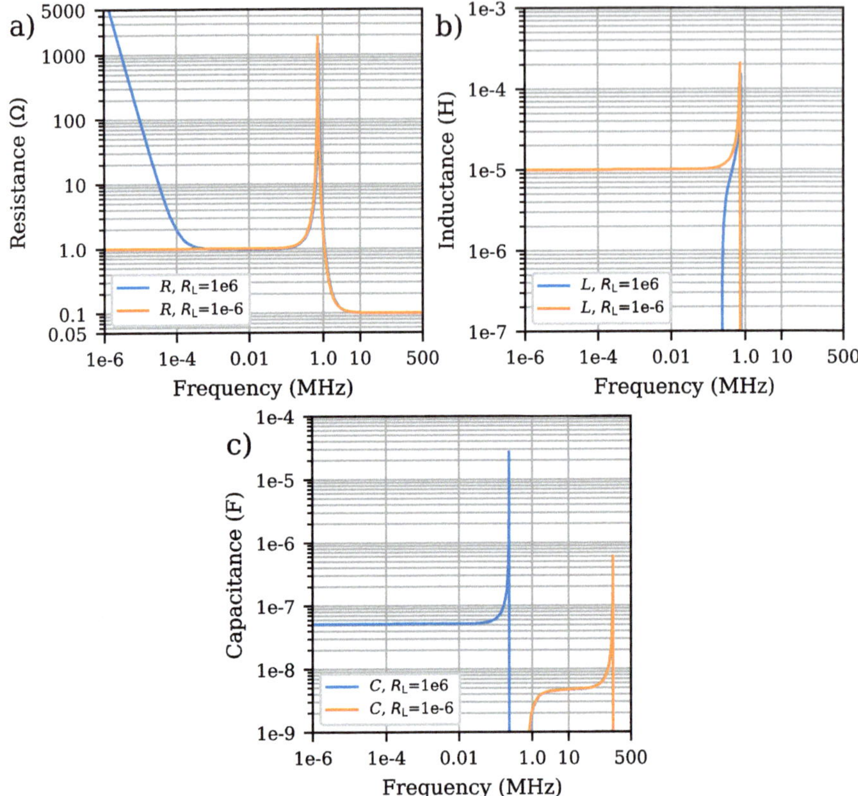

Fig. 2.33 RLC values extracted from the network parameter model of the π-filter

Further Reading

Source Models

BSIM Group, *The BSIM Family*. https://www.bsim.berkeley.edu/models/ (visited on 2025)

D. Cittanti et al., Role of parasitic capacitances in power MOSFET turn-on switching speed limits: a SiC case study, in *2017 IEEE Energy Conversion Congress and Exposition (ECCE)* (2017), pp. 1387–1394. https://doi.org/10.1109/ECCE.2017.8095952

Y. Ding et al., IBIS model simulation accuracy improvement by including nonlinear power supply induced Jitter effect. IEEE Trans. Signal Power Integr. **4**, 55–64 (2025). https://doi.org/10.1109/TSIPI.2025.3551671

M. Fuchs, L. Spielberger, K. Odreitz, Fast and simple model generation for superjunction power MOSFETs. Elektrotech. Inftech. **138**, 48–52 (2021). https://doi.org/10.1007/s00502-020-00860-3

M. Gholizadeh et al., Modeling a GaN transistor and its impact on conducted emission up to 300 MHz, in *International Symposium on Electromagnetic Compatibility, EMC Europe*, Krakow, Poland (2023)

M. Hervé et al., A multi-physics model of the VJFET with a lateral channel, in *Proceedings of the 2011 14th European Conference on Power Electronics and Applications* (2011), pp. 1–10

P. Hillenbrand et al., Sensitivity analysis of behavioral MOSFET models in transient EMC simulation, in *2017 International Symposium on Electromagnetic Compatibility – EMC EUROPE* (2017), pp. 1–6. https://doi.org/10.1109/EMCEurope.2017.8094762

C. Hu, *Modern Semiconductor Devices for Integrated Circuits* (2010). https://www.chu.berkeley.edu/modern-semiconductor-devices-for-integrated-circuits-chenming-calvin-hu-2010/

IBIS, *IBIS Open Forum*. https://ibis.org/ (visited on 2025)

Infineon, *IRFZ44 Power MOSFET*. https://www.infineon.com/dgdl/Infineon-IRFZ44N-DataSheet-v01_01-EN.pdf?fileId=5546d462533600a40153563b3a9f220d (visited on 2025)

K. Iokibe et al., EMC macro-modeling of CMOS inverter using LECCS-I/O model with additional current source, in *2008 IEEE International Symposium on Electromagnetic Compatibility* (2008), pp. 1–4. https://doi.org/10.1109/ISEMC.2008.4652105

IRFZ44NPbF, *HEXFET Power MOSFET*. Ed. by Infineon. https://www.infineon.com/dgdl/Infineon-IRFZ44N-DataSheet-v01_01-EN.pdf?fileId=5546d462533600a40153563b3a9f220d (visited on 2025)

L. Jing et al., An improved behavior model for IGBT modules driven by datasheet and measurement. IEEE Trans. Electron. Dev. **67**(1), 230–236 (2020). https://doi.org/10.1109/TED.2019.2954329

K. Li, S. Sen, A fast and accurate GaN power transistor model and its application for electric vehicle. IEEE Trans. Veh. Technol. **73**(4), 4541–4553 (2024). https://doi.org/10.1109/TVT.2023.3340297

H. Li et al., Improved double pulse test for accurate dynamic characterization of medium voltage SiC devices. IEEE Trans. Power Electron. **38**(2), 1779–1790 (2023). https://doi.org/10.1109/TPEL.2022.3210749

Y. Mabuchi et al., LECCS (linear equivalent circuit and current source) modeling technique for ICs, in *16th International Zurich Symposium and Technical Exposition on Electromagnetic Compatibility* (2005), pp. 659–677. https://doi.org/10.23919/EMC.2005.10806455

X. Margueron, J.P. Keradec, Identifying the magnetic part of the equivalent circuit of *n*-winding transformers. IEEE Trans. Instrum. Meas. **56**(1), 146–152 (2007a). https://doi.org/10.1109/TIM.2006.887402

X. Margueron, J.-P. Keradec, Design of equivalent circuits and characterization strategy for *n*-input coupled inductors. IEEE Trans. Ind. Appl. **43**(1), 14–22 (2007b). https://doi.org/10.1109/TIA.2006.887001

R. Mayder et al., Using power aware IBIS v5.0 behavioral IO models to simulate simultaneous switching noise, in *DesignCon2013* (2013)

Nexperia, *1N4148 High-speed Diodes*. https://assets.nexperia.com/documents/data-sheet/1N4148_1N4448.pdf (visited on 2025)

NXP, *NXP Semiconductors*. Internet. https://www.nxp.com/store:STORE (visited on 2025)

C.R. Paul, *Introduction to Electromagnetic Compatibility* (Wiley, 2005). ISBN: 9780471758150. https://doi.org/10.1002/0471758159

P. Pulici et al., A modified IBIS model aimed at signal integrity analysis of systems in package. IEEE Trans. Circuits Syst. I Regul. Pap. **55**(7), 1921–1928 (2008). https://doi.org/10.1109/TCSI.2008.918203

S.R. Rao et al., Black-box behavioral DC–DC converter IC emission model, in *2022 IEEE International Symposium on Electromagnetic Compatibility & Signal/Power Integrity (EMCSI)* (2022), pp. 570–574. https://doi.org/10.1109/EMCSI39492.2022.9889598

I. Scott et al., Integration of behavioral models in the full-field TLM method. IEEE Trans. Electromag. Compat. **54**(2), 359–366 (2012). https://doi.org/10.1109/TEMC.2012.2183133

M. Souilem et al., Signal and power integrity IO buffer modeling under separate power and ground supply voltage variation of the input and output stages. IEEE Trans. Very Large Scale Integr. (VLSI) Syst. **31**(6), 874–886 (2023). https://doi.org/10.1109/TVLSI.2023.3244144

M. Turzynski, W.J. Kulesza, A simplified behavioral MOSFET model based on parameters extraction for circuit simulations. IEEE Trans. Power Electron. **31**(4), 3096–3105 (2016). https://doi.org/10.1109/TPEL.2015.2445375

Vishay, *IRFZ44 Power MOSFET*. https://www.vishay.com/docs/91291/irfz44.pdf (visited on 2025)

O. Wada et al., High-speed simulation of PCB emission and immunity with frequency-domain IC/LSI source models, in *2003 IEEE Symposium on Electromagnetic Compatibility. Symposium Record (Cat. No.03CH37446)*, vol. 1 (2003), pp. 4–9. https://doi.org/10.1109/ISEMC.2003.1236554

J. Wang, H. Shu-hung Chung, R. Tin-ho Li, Characterization and experimental assessment of the effects of parasitic elements on the MOSFET switching performance. IEEE Trans. Power Electron. **28**(1), 573–590 (2013). https://doi.org/10.1109/TPEL.2012.2195332

L. Wang et al., Modelling and optimization of SiC MOSFET switching voltage and current overshoots in a half-bridge configuration. IET Power Electron. **14**(9), 1684–1699 (2021). https://doi.org/10.1049/pel2.12146. eprint: https://ietresearch.onlinelibrary.wiley.com/doi/pdf/10.1049/pel2.12146, https://ietresearch.onlinelibrary.wiley.com/doi/abs/10.1049/pel2.12146

Z. Zhang et al., Evaluation of switching performance of SiC devices in PWM inverter-fed induction motor drives. IEEE Trans. Power Electron. **30**(10), 5701–5711 (2015). https://doi.org/10.1109/TPEL.2014.2375827

Z. Zhang et al., Methodology for wide band-gap device dynamic characterization. IEEE Trans. Power Electron. **32**(12), 9307–9318 (2017). https://doi.org/10.1109/TPEL.2017.2655491

Component Models

A. Asmanis et al., 3D modeling of surface-mount capacitors and mutual couplings between them, in *2017 International Symposium on Electromagnetic Compatibility – EMC EUROPE* (2017), pp. 1–6. https://doi.org/10.1109/EMCEurope.2017.8094717

A. Ayachit, M.K. Kazimierczuk, Self-capacitance of single-layer inductors with separation between conductor turns. IEEE Trans. Electromag. Compat. **59**(5), 1642–1645 (2017). https://doi.org/10.1109/TEMC.2017.2681578

J. Biela, J.W. Kolar, Using transformer parasitics for resonant converters-A review of the calculation of the stray capacitance of transformers. IEEE Trans. Ind. Appl. **44**(1), 223–233 (2008). https://doi.org/10.1109/TIA.2007.912722

G.R. Choi et al., Near-field coupling effect analysis of SMD inductor using 3D-EM model. Electronics **12**(13) (2023). ISSN: 2079–9292. https://doi.org/10.3390/electronics12132845, https://www.mdpi.com/2079-9292/12/13/2845

P. Chowdhuri, Calculation of series capacitance Eor transient analysis of windings. IEEE Trans. Power Del. **2**(1), 133–139 (1987). https://doi.org/10.1109/TPWRD.1987.4308083

H. De Gersem, K. Hameyer, A finite element model for foil winding simulation. IEEE Trans. Mag. **37**(5), 3427–3432 (2001). https://doi.org/10.1109/20.952629

M. Fuchs, M. Sievers, B. Deutschmann, Voltage dependence and characterization of ceramic capacitors under electrical stress, in *2020 IEEE Applied Power Electronics Conference and Exposition (APEC)* (2020), pp. 2815–2821. https://doi.org/10.1109/APEC39645.2020.9124365

T. Makharashvili et al., Circuit models for the inductance of eight-terminal decoupling capacitors. IEEE Trans. Comp. Packag. Manuf. Technol. **10**(1), 142–150 (2020). https://doi.org/10.1109/TCPMT.2019.2952138

T. Makharashvili et al., Accurate inductance models of mounted two-terminal decoupling capacitors. IEEE Trans. Electromag. Compat. **63**(1), 237–245 (2021). https://doi.org/10.1109/TEMC.2020.2987995

A. El-S. Mostafa, M.K. Gohar, Determination of voltage, current, and mlagnetic field distributions together with the self-capacitance, inductance and HF resistance of single-layer coils. Proc. IRE **41**(4), 537–547 (1953). https://doi.org/10.1109/JRPROC.1953.274240

N. Nagaoka, D.J. Wilcox, Impedance calculation of a single-layer air-core coil by means of resonant modal theory. IET Gener. Transmission Distrib. **1**, 133–139 (2007). https://doi.org/10.1049/iet-gtd:20050392. eprint: https://digital-library.theiet.org/doi/pdf/10.1049/iet-gtd%3A20050392. https://digital-library.theiet.org/doi/abs/10.1049/iet-gtd%3A20050392

A.J. Palermo, Distributed capacity of single-layer coils. Proc. Inst. Radio Eng. **22**(7), 897–905 (1934). https://doi.org/10.1109/JRPROC.1934.226741

C. Riener et al., 3D modeling of inductive and capacitive coupling between surface-mounted multilayer-capacitors. Int. J. Numer. Model. Electron. Netw. Dev. Fields **36**(3), e3046 (2023). https://doi.org/10.1002/jnm.3046. eprint: https://onlinelibrary.wiley.com/doi/pdf/10.1002/jnm.3046, https://onlinelibrary.wiley.com/doi/abs/10.1002/jnm.3046

H. Zuhrt, Eine quasistationäre Berechnung der Eigenwellen einlagiger Flachund Zylinderspulen. Teil I. Archiv für Elektrotechnik **27**(9), 613–636 (1933). ISSN: 1432-0487. https://doi.org/10.1007/BF01656916

Sink Models

H. Hackl, B. Deutschmann, An EMI receiver model to minimize simulation time of long data transmissions, in *2016 Austrochip Workshop on Microelectronics (Austrochip)* (2016), pp. 57–62. https://doi.org/10.1109/Austrochip.2016.022

H. Hackl, B. Deutschmann, Computing the electromagnetic emission spectrum of pulses by convolution in frequency domain, in *2017 International Symposium on Electromagnetic Compatibility – EMC EUROPE* (2017), pp. 1–6. https://doi.org/10.1109/EMCEurope.2017.8094699

K. Ichikawa et al., Simulation of integrated circuit immunity with LECCS model, in *2006 17th International Zurich Symposium on Electromagnetic Compatibility* (2006), pp. 308–311. https://doi.org/10.1109/EMCZUR.2006.214932

T. Karaca, B. Deutschmann, G. Winkler, EMI-receiver simulation model with quasi-peak detector, in *2015 IEEE International Symposium on Electromagnetic Compatibility (EMC)* (2015), pp. 891–896. https://doi.org/10.1109/ISEMC.2015.7256283

C. Keller, K. Feser, Fast emission measurement in time domain. IEEE Trans. Electromag. Compat. **49**(4), 816–824 (2007). https://doi.org/10.1109/TEMC.2007.908282

C. Rauscher, *Fundamentals of Spectrum Analysis* (Rohde & Schwarz GmbH & Co. KG., München, 2001)

R. Schertlen et al., Einfluss der Bordnetznachbildung auf Störfestigkeitsmessverfahren (zB BCI) oberhalb 100 MHz, in *EMV: Internationale Fachmesse und Kongress für Elektromagnetische Verträglichkeit, Düsseldorf, 2018* (mesago, Frankfurt/Main, 2018), pp. 465–472

D. Schwarzbeck, *The EMI-Receiver According to CISPR 16-1-1*. Application Note. Schwarzbeck Mess-Elektronik. https://www.schwarzbeck.de/en/application-notes.html (visited on 2025)

E. Takahashi et al., Evaluation of LSI immunity to noise using an equivalent internal impedance model, in *2002 EMC-Europe Sorrento* (2002), pp. 1–6. https://doi.org/10.23919/EMC.2002.10879982

Networks, Spice Modeling

Analog Devices, *LTspice*. https://www.analog.com/en/resources/design-tools-and-calculators/ltspice-simulator.html (visited on 2025)

S. Bauer et al., Investigation of SPICE models for overvoltage protection devices with respect to fast transients. IEEE Lett. Electromag. Compat. Pract. Appl. **1**(1), 20–25 (2019). https://doi.org/10.1109/LEMCPA.2019.2912587

Cadence PSpice, *Advanced Circuit Simulation*. https://www.cadence.com/en_US/home/tools/pcb-design-and-analysis/analog-mixed-signal-simulation/pspice.html (visited on 2025)

W. Hayt et al., *Engineering Circuit Analysis*, 10th edn. (McGraw Hill, 2023). ISBN: 9781264149919

D.M. Pozar, *Microwave Engineering*, 4th edn. (Wiley, 2012). ISBN: 978–1-118-21363-6

J.M. Rabaey, *The Spice Page*. https://bwrcs.eecs.berkeley.edu/ (visited on 2025)

A. Vladimirescu, *The SPICE Book* (Wiley, 1994). ISBN: 978-0-471-60926-1

J. Zhang et al., Causal RLGC(f) models for transmission lines from measured S-parameters. IEEE Trans. Electromag. Compat. **52**(1), 189–198 (2010). https://doi.org/10.1109/TEMC.2009.2035055

Passive Component Vendor Model Repositories and Design Tools

Bournes, *Resources*. https://bourns.com/resources

Coilcraft, *Transformer and Inductor Tools*. https://www.coilcraft.com/en-us/tools/

Kyocera AVX, *Design Tools*. https://www.kyocera-avx.com/design-tools/

Murata, *Design Support Software Simsurfing*. https://www.murata.com/en-eu/tool/simsurfing

Panasonic Industry, *Design and Simulation*. https://industrial.panasonic.com/ww/design-support/ds-tools

Sumida, *Design Support*. https://products.sumida.com/ProductsInfo/Support/

Taiyo Yuden, *Design Support Tools*. https://www.yuden.co.jp/en/product/#anc-03

TDK Electronics, *Design Tools*. https://www.tdk-electronics.tdk.com/en/180524/design-support/design-tools

Vishay, *Design Tools*. https://www.vishay.com/en/how/design-support-tools/

Würth Electronic *Component Library*. https://www.we-online.com/en/support/design-tools/libraries

Yageo Kemet, *Featured Products*. https://www.kemet.com/en/us.html

Chapter 3
EMC Assessment Using Circuit Simulation

Circuit simulation offers a straightforward and accessible approach for analyzing and estimating the EMC of an electronic system. Its ease of use makes it a practical tool for engineers. To model the EMC of an electronic system, it is crucial to understand how to combine the models of the noise source, the coupling path, and the sink, as introduced in Chap. 2.

We use a buck converter as an example. A buck converter with its attached load and supply incorporates the essential components of an electronic system while remaining straightforward to analyze. The switching transistor and diode act as the sources of EMI. The commutation cell, along with the attached wires, load, and ground, defines the coupling path. The voltage is measured at the LISNs, which serve as the sink.

We analyze the conducted EMI generated by an SMPS for a USB application. The scenario is illustrated in Fig. 3.1. The buck converter steps down the 12 V battery voltage of a road vehicle, $V_{\text{in}} = 12$ V, from the power supply on the left-hand side to $V_{\text{out}} = 5$ V for the USB on the right-hand side, where a load is connected. Our focus is on the conducted emissions of the buck converter toward the 12 V power supply.

We start with the functional model, which consists of the switching transistor and the diode, the DC link capacitor, the load, and the power supply. We extend the models of these components by parasitic elements to make them valid for EMC simulation, as described in Chap. 2. Additionally, we study how to account for the parasitic elements of the interconnects. We analyze the EMI behavior of the resulting model using time-domain and frequency-domain simulations. In doing so, we encounter the limitations of circuit simulation and discuss common pitfalls.

J. Hansen, A. Barchanski, *EMC Simulation of Electronic Systems*,
https://doi.org/10.1007/978-3-032-09237-3_3

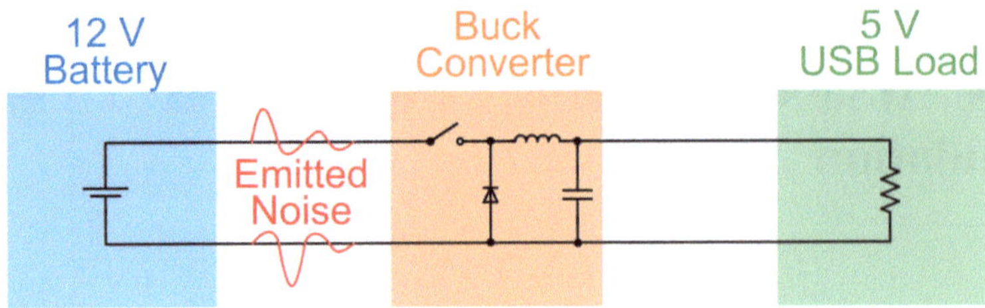

Fig. 3.1 Schematic picture of the switching noise of the buck converter propagating to the 12 V battery

3.1 From Functional Circuits to EMI

Domains: 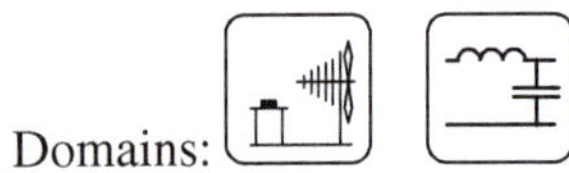

Any hardware development begins with a functional model that hardware engineers use to determine the bill of materials required to make the hardware operational. This functional model can be efficiently extended to support EMI prediction.

Figure 3.2 illustrates the functional schematic of a buck converter. This model includes all essential circuit elements required to simulate the functional behavior of the circuit, specifically to step down the DC input voltage, $V_{\text{in}} = 12\,\text{V}$, to the desired output voltage, $V_{\text{out}} = 5\,\text{V}$, suitable for the USB.

On the input side, the buck converter is equipped with a DC link capacitor of $C_{\text{DCL}} = 10\,\mu\text{F}$. The circuit includes a switching transistor controlled by a pulsed gate signal. The transistor features a drain-source voltage V_{DS}, a drain current I_{DS},

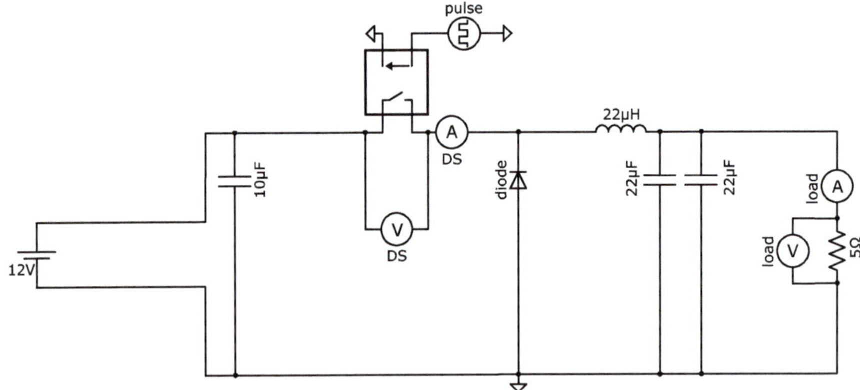

Fig. 3.2 Functional schematic of the buck converter

and an on-resistance R_{on}. Its duty cycle τ is determined by the voltage down-conversion ratio between the input and output.

For a buck converter, the duty cycle is given by $\tau = \frac{V_{out}}{V_{in}}$. In this chapter, we use $\tau = 0.475$, resulting in a load voltage of $V_L = 5.06\,V \approx 5\,V$. The circuit includes a diode, an output inductance of $L_{out} = 22\,\mu H$, and two parallel output capacitors, each with a capacitance of $C_{out} = 22\,\mu F$.

All circuit elements are modeled using their nominal values without considering parasitic elements. It is important to note that the power source providing the voltage V_{out} is not the noise source of the electronic system; in other words, it is not the source of the conducted emissions (cf. Sect. 2.3).

Our primary concern is the impact of the high-frequency harmonics generated by the switching transistors of the buck converter on the 12 V supply network. Since other electronic devices may be connected to this supply, we need to consider EMI regulations to ensure that these harmonics do not cause interference.

There are two nonlinear elements in this circuit that produce high-frequency harmonics: the transistor and the diode. We analyze both as sources of interference.

The impact of the USB controller is not explicitly modeled; instead, we only represent its impedance. The load impedance is chosen to yield $I_L = 1$ A and is calculated as $R_L = \frac{V_L}{I_L} = 5\ \Omega$. This corresponds to a load power of $P = RI_L^2 = 5$ W, which is a realistic value for a USB drive. The impact of the load power on the high-frequency harmonics is low and is not further investigated.

A functional model could also include the gate circuitry and output current control. The gate circuit determines the switching behavior, which influences frequency components above $f_{tr} = \frac{1}{\pi t_r}$, where t_r is the rise time. For buck converters in integrated SMPS with rise and fall times around 10 ns, $f_{tr} \simeq 30$ MHz. This means the model becomes less accurate at frequencies above this threshold and serves more as a fingerprint than a precise prediction.

The output current control depends on the control algorithm of the SMPS, which is typically not detailed in data sheets. The typical settling time of control algorithms is larger than some milliseconds; thus, they only affect the spectrum in a range below 1 kHz. This is much lower than the switching frequency f_{sw} of a SMPS, that is, typically in the MHz range. Any oscillations from the control algorithm would appear well below the harmonics of f_{sw}. For this reason, we neglect the control algorithm in our model.

We also disregard any impact from nonlinear elements in the load attached to the USB. The load is modeled solely by the current flowing through R_L. Any additional nonlinear elements would require a separate model, which could eventually be connected to the converter model.

Such a concatenated approach provides a more accurate representation of the total system's emissions but comes at the cost of increased computational effort and reduced clarity regarding the specific physical properties of the buck converter. Therefore, concatenated models should only be considered after thoroughly understanding the properties and impacts of each individual model.

Figure 3.3 presents the results of a circuit simulation based on the model shown in Fig. 3.2. The transistor is modeled as an ideal switch operating at a switching

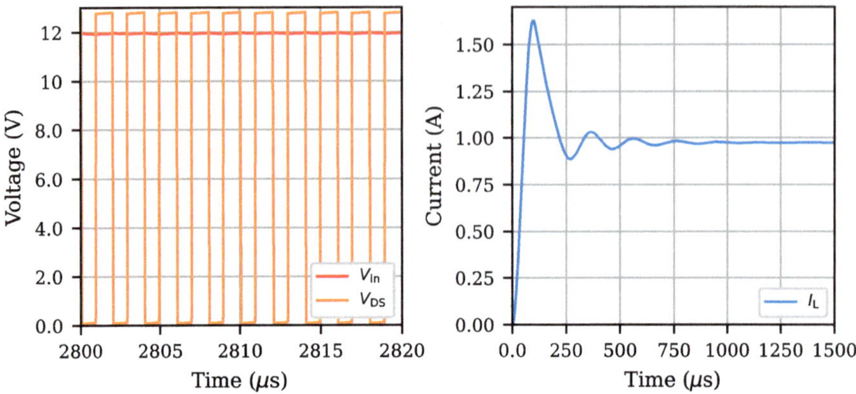

Fig. 3.3 Supply voltage V_{DC}, drain source voltage V_{DS}, and load current I_L of the buck converter of Fig. 3.2

frequency of 500 kHz, and an ideal diode model is used. All circuit components are assigned their nominal values, as indicated in the figure. The simulation time is 7 s.

On the left side of Fig. 3.3, the supply voltage V_{in} and the drain-source voltage V_{DS} of the transistor are shown over a time frame of 20 μs. The supply voltage remains constant at $V_{in} = 12$ V because it is modeled as an ideal voltage source. The drain-source voltage appears as a rectangular pulse with approximately 50% duty cycle (actual value: 45.8%). When the transistor is open, $V_{DS} = 12.8$ V due to the threshold voltage of 0.8 V of the diode model.

On the right side, the load current I_L is depicted over a longer time window, from 0 μs to 1500 μs. The load current requires approximately 750 μs to reach its steady-state value of 1 A. This shows that the load voltage stabilizes at $V_{out} = 5$ V, as intended.

Our focus is on the conducted switching noise generated by the buck converter on the supply lines connected to the power supply V_{in}. As described in Sect. 2.5.2, conducted emissions in the power supply path are measured using LISNs with a 50 Ω measurement receiver (see Sect. 2.5.4). One LISN is placed in the positive supply path, and another in the negative supply path. The grounds of both LISNs are connected to a metal table.

To simulate the conducted voltage at the LISNs, the LISN models and the metal table are incorporated into the circuit model, as illustrated in blue on the left-hand side of Fig. 3.4. The simulation time of this model is 10 s.

We now evaluate the voltage $V_{LISN,P}$ at the 50 Ω measurement resistor of the LISN in the positive supply path. On the left side of Fig. 3.5, the voltage $V_{LISN,P}$ is compared with the drain-source voltage V_{DS} of the transistor over the time range from 2800 μs to 2820 μs. The upper limit of the voltage is clipped at 0.03 V.

The shape of $V_{LISN,P}$ is (nearly) triangular, which results from the charging and discharging of the LISN capacitance (100 nF) during the on-state and off-state transitions of the transistor.

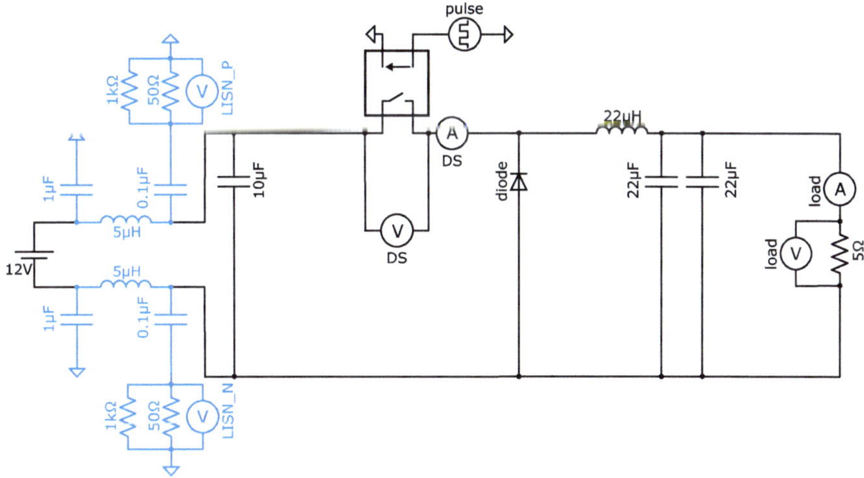

Fig. 3.4 Functional schematic extended by two LISNs (blue) and the metallic plate of the measurement table (circuit ground)

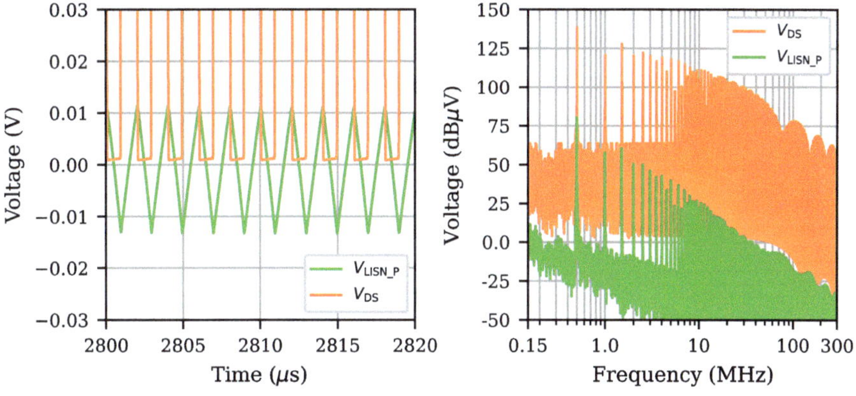

Fig. 3.5 Drain-source voltage V_{DS} and voltage $V_{LISN,P}$ of the model extended by table and LISNs. Plots are in time and frequency domains

The DFT of V_{DS} and $V_{LISN,P}$ is shown on the right side of Fig. 3.5. Both spectra exhibit harmonics at multiples of the switching frequency, $f_{sw} = 500\,\text{kHz}$. The spectrum of V_{DS} follows a decay of $-20\,\text{dB}$ per decade, as expected for a rectangular waveform, which exhibits a $\frac{1}{n}$ decay, where n represents the harmonic number. In contrast, the spectrum of $V_{LISN,P}$ decays at $-40\,\text{dB}$ per decade, reflecting the triangular shape of the waveform, which results in a $\frac{1}{n^2}$ decay. Additionally, the even harmonics are slightly lower in amplitude than the odd harmonics due to the duty cycle τ being close to 50%.

By incorporating the LISNs, we are able to simulate and analyze the conducted EMI. However, the circuit model currently lacks parasitic elements, which are essential for a more realistic representation of the components (cf. Sect. 2.4) and the interconnects. Consequently, the model does not accurately predict the EMI at frequencies where these parasitics begin to have impact.

3.2 Impact of Component Parasitics

Domains: 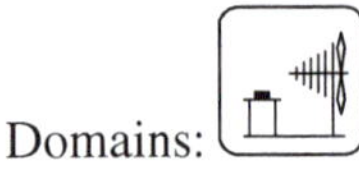

In Sect. 3.1, we extended the functional model to simulate the EMI on the supply lines. However, this model remains incomplete. As discussed in Sect. 2.4, the impedance of discrete components deviates from their nominal values above a certain frequency.

In this section, we examine the impact of parasitic elements added to the nominal component values. While we may only have approximate values for these parasitics, studying their influence allows us to better understand their effect on the EMI behavior of the buck converter.

Figure 3.6 presents the circuit schematic from Fig. 3.4, now extended to include the parasitic elements described in Sect. 2.3 for the noise sources (i.e., the transistor and the diode) as well as the parasitic elements of the discrete components (cf. Sect. 2.4).

In particular, a parasitic inductance of $L_{\mathrm{DCL}} = 2.2\,\mathrm{nH}$ and a parasitic resistance of $R_{\mathrm{DCL}} = 20\,\mathrm{m}\Omega$ are added in series with the DC link and output capacitors. These

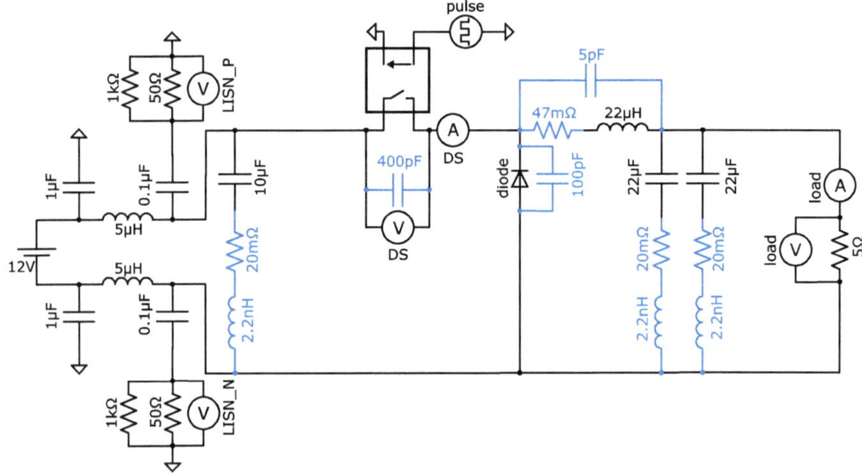

Fig. 3.6 Functional schematic showing two LISNs and the metallic plate of the measurement table. The parasitic elements of the DC link capacitor, the transistor, and the diode are included (highlighted in blue)

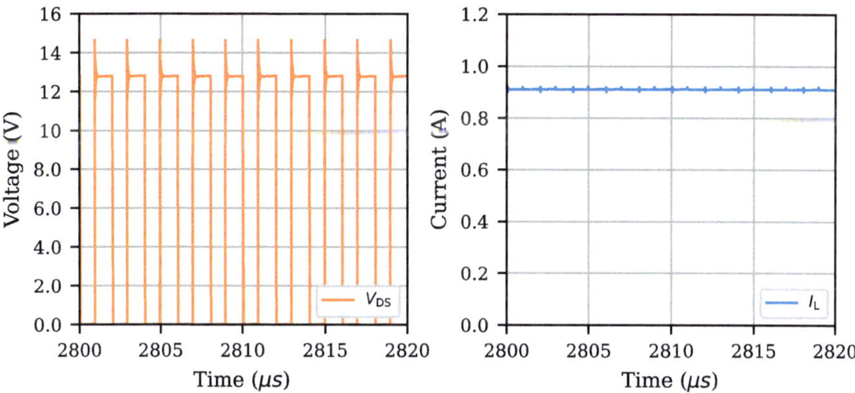

Fig. 3.7 Time-domain waveforms of the extended functional model, incorporating parasitic elements for the capacitors, the output inductor, the transistor, and the diode

values are typical for SMD components and can be obtained from data sheets. The parasitic values of the interconnects are not yet considered.

For the transistor and diode, typical parasitic capacitances are $C_{DS} = 400\,\text{pF}$ and $C_D = 100\,\text{pF}$ (Schottky diode), respectively, as specified in data sheets. These capacitances vary with the applied voltage; however, for simplicity, we model C_{DS} and C_D as constants, using their values at the open-loop voltage. The impact of this simplification can be assessed by varying the capacitance values in simulations. The inductor on the output side is modeled with an additional series resistance of $R_{L,out} = 47\,\text{m}\Omega$ and a parallel parasitic capacitance of $C_{L,out} = 5\,\text{pF}$, which are also taken from data sheets. The load resistance remains unchanged. The simulation time of this model is 1 min and 14 s.

The resulting drain-source voltage at the transistor, V_{DS}, and the load current, I_L, are shown in Fig. 3.7 for the time frame from 2800 μs to 2820 μs. The load current has decreased by 10% to approximately $I_L = 0.9\,\text{A}$ due to the parasitic resistance of the inductor, $R_{L,out} = 0.47\,\Omega$, which increases the total load resistance, $R_L = 5\,\Omega$, by 10%. A voltage overshoot of 2 V is observed whenever the transistor turns off. Additionally, each switching event induces a small ripple in the load current, I_L.

The observed ringing indicates an L-C resonance, and understanding the contributing inductances L and capacitances C is important for EMI analysis. In Fig. 3.8, we examine the current through the transistor. The ringing is evident on both the rising and falling edges of the current waveform. Closer inspection reveals that the ringing on the falling edge has a period that is twice as long as the period of the ringing observed on the rising edge of the current. This is an important observation, because we see this ringing as spectral peaks at the LISNs.

The voltage at the positive LISN, $V_{LISN,P}$, exhibits similar behavior, as shown in Fig. 3.9. The time-domain waveform reveals a ringing event occurring once per microsecond. The ringing with a shorter decay, observed at around 2802 μs, is

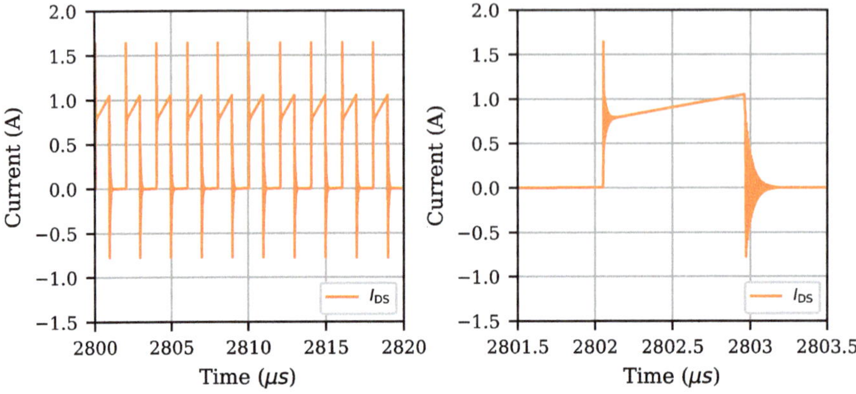

Fig. 3.8 The current I_{ds} through the transistor exhibits ringing during both the turn-on and turn-off transitions

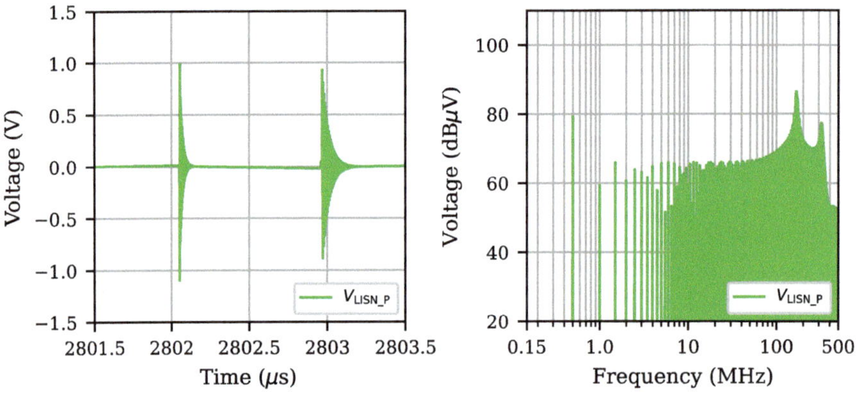

Fig. 3.9 Voltage $V_{LISN,P}$ at the positive LISN, represented in both the time and frequency domains

generated when the transistor switches on and V_{DS} drops to a low value. In contrast, the ringing with a longer decay occurs when the transistor switches off.

The spectral content of these waveforms shows the clock frequency of 500 kHz as the fundamental harmonic. The two ringing events manifest as distinct resonant peaks at the frequencies $f_{r1} = 167$ MHz and $f_{r2} = 324$ MHz.

Next, we investigate the origin of these resonances. The resonances are linked to the two capacitances, $C_D = 100$ pF and $C_{DS} = 400$ pF, which interact with one or more inductances. When the transistor switches from a closed (short) to an open state, the coil drives a current through the diode. This current excites a resonance in the loop formed by $C_{DS} = 400$ pF and the inductance of the DC link capacitor, $L_{DCL} = 2.2$ nH. The DC link capacitance, $C_{DCL} = 10\,\mu$F, effectively acts as a short circuit in this scenario.

The resonance frequency of this circuit is given by

$$\frac{1}{2\pi\sqrt{L_{\mathrm{C,DCL}}\,C_{\mathrm{DS}}}} = 170\,\mathrm{MHz} \approx f_{\mathrm{r1}} = 167\,\mathrm{MHz}. \tag{3.1}$$

This value is close to the resonance frequency f_{r1} observed in Fig. 3.8. The slight mismatch of 3 MHz suggests that the effective capacitance or inductance in the simulation model is marginally higher. Additional parasitic elements, such as those from the output choke or the output capacitors, contribute to the shift in f_{r1} in the simulation model.

When the transistor switches on, its parasitic capacitance is no longer effective. The current through the transistor flows into a series resonant circuit composed of C_{D}, $L_{\mathrm{C,DCL}}$, and $R_{\mathrm{C,DCL}}$. Since $C_{\mathrm{D}} = \frac{1}{4}C_{\mathrm{DS}}$, the resonance frequency f_{r2} of this circuit is approximately twice the first resonance frequency, calculated as $2 \cdot 170\,\mathrm{MHz} = 340\,\mathrm{MHz}$. The observed resonance $f_{\mathrm{r2}} = 324\,\mathrm{MHz}$ is about 5% lower, suggesting that in the simulation model, the effective inductances and capacitances are slightly larger than assumed.

In the voltage across the transistor, only the resonance frequency $f_{\mathrm{r1}} = 167\,\mathrm{MHz}$ is observed. When the transistor is in the on-state, its low on-resistance, $R_{\mathrm{on}} = 0.1\,\Omega$, results in a voltage drop of only a few millivolts.

Compared to the results in Sect. 3.1, accounting for the parasitic elements of the components introduces resonances in the supply line voltage, which are evident in the voltage measured at the LISN. The root cause of these resonances can be traced to L-C resonances. Understanding these root causes enables insight into how changes in component parasitics affect the system. For instance, if the parasitic capacitance C_{DS} of the transistor increases, the resonance f_{r1} shifts to a lower frequency. This understanding is particularly valuable in the early stages of design for selecting components. If a new component has larger parasitic values, corresponding resonances shift to lower frequencies; if it has smaller parasitic values, the resonances shift to higher frequencies in the EMI spectrum.

The quality factor of the resonances is very high. For a more realistic assessment, we need to include the resistance of the circuit at the resonance frequencies, which is a value not available in data sheets.

We gain these insights from the model because only a few L-C pairs contribute to each resonance. The example in this chapter demonstrates how a specific L-C pair explains the resonances f_{r1} and f_{r2}. However, for circuits with a higher number of parasitic elements—required, for example, to describe their behavior accurately at higher frequencies where more resonances occur—this type of root-cause analysis becomes significantly more complex.

3.3 Resistance, Capacitance, and Inductance of Interconnects

Domains: 

As with components, interconnects also influence the electromagnetic properties of an electronic system. Additional modifications to the functional buck converter model are required beyond those shown in Fig. 3.6. In Chap. 2, we demonstrate how to enhance component models with parasitic elements to capture their electrical properties over a wider frequency range. The same principle applies to geometric structures, such as interconnects.

When interconnects conduct currents, these currents generate magnetic fields, causing the structure to behave like an inductor. Similarly, if interconnects store charges, their electrical behavior is described by a capacitance. In industrially manufactured electronic systems, all conductors exhibit losses, imparting resistive characteristics to the conductors.

In this section, we present analytical formulas to estimate the parasitic properties of geometrically simple conductors. These estimates are then applied to the buck converter discussed in this chapter to analyze the impact of interconnects and other mechanical components on EMI.

Resistance The resistance of a conductor is defined as the real part of the ratio between voltage, V, and current, I,

$$R = \Re\left(\frac{V}{I}\right), \tag{3.2}$$

i.e., it describes the voltage drop across a conductor when a current of $I = 1\,\mathrm{A}$ flows through it. The resistance of a conductor with a length l and a cross-sectional area A is given by

$$R_{\mathrm{cond}} = \frac{1}{\sigma_{\mathrm{c}}}\frac{l}{A}, \tag{3.3}$$

where σ_{c} represents the conductivity of the material. This formula assumes a constant current density across A, and the skin effect is neglected.

The conductivity of most conductors used in electronics lies between that of iron $\left(\sigma_{\mathrm{iron}} \approx 1 \cdot 10^7\,\frac{\mathrm{S}}{\mathrm{m}}\right)$ and silver $\left(\sigma_{\mathrm{Ag}} \approx 6.3 \cdot 10^7\,\frac{\mathrm{S}}{\mathrm{m}}\right)$, with copper being a common choice at $\sigma_{\mathrm{Cu}} \approx 5.8 \cdot 10^7\,\frac{\mathrm{S}}{\mathrm{m}}$.

For example, a $50\,\mu\mathrm{m}$ thick copper trace with a width of $0.15\,\mathrm{mm}$ and a length of $1\,\mathrm{cm}$ has a resistance of approximately $23\,\mathrm{m}\Omega$. Similarly, a $10\,\mathrm{cm}$ piece of copper wire with a cross-sectional area of $1.5\,\mathrm{mm}^2$ has a resistance of about $0.8\,\mathrm{m}\Omega$.

Whether the resistance of conductors is significant depends on other resistances in the DUT. For instance, the on-resistances of transistors typically range between

10 mΩ and 100 mΩ, which are of the same order of magnitude or larger than the resistance due to the finite conductivity of conductors.

Carefully inspecting the model or testing it with and without the conductor resistances determines whether the finite conductivity of conductors has a meaningful impact on the system's behavior.

Capacitance The capacitance, C, is defined as the ratio of the charge, Q, accumulated on two conductors to the voltage, V, between them,

$$C = \frac{Q}{V}, \tag{3.4}$$

where V, according to the electrostatic definition, is expressed as the line integral of the electric field between the conductors, $\mathbf{E}$, along a path curve l,

$$V = \int_l \mathbf{E} \cdot \mathbf{ds}. \tag{3.5}$$

A simple yet effective method to estimate the capacitance between two flat conducting bodies is to use the formula for a parallel plate capacitor,

$$C_{\text{plate}} = \varepsilon_{\text{r}}\varepsilon_0 \frac{A_{\text{plate}}}{d}, \tag{3.6}$$

where A_{plate} is the surface area of the smaller plate, and d is the distance between the two plates.

Equation 3.6 is exact for two thin, parallel plates of equal size separated by a distance that is small compared to the diameter of the plates, as it neglects the effects of fringing fields. The influence of fringing fields is minimal when the plates are close together but becomes more significant as the distance and thickness of the plates increase.

The capacitance between two parallel wires of length l, separated by a distance s, and having radii r_{w1} , r_{w2}, is given by

$$C_{\text{w,para}} = \frac{2\pi\varepsilon_0\epsilon_r l}{\operatorname{arccosh}\left(\frac{s^2 - r_{\text{w1}}^2 - r_{\text{w2}}^2}{2 r_{\text{w1}} r_{\text{w2}}}\right)}, \tag{3.7}$$

where the wires are embedded in a homogeneous dielectric material with a dielectric constant ε_{r}.

Various scenarios involve a smaller, flat metallic conductor near a larger one. Examples include a trace or plane above a ground plane in a layout or the bottom of a housing above a metallic table during a CISPR 25 emission test. If the assumption that both conductors are of equal size and close to each other is not met, Eq. 3.6 serves as a rough and simple estimator.

Table 3.1 Accuracy of the plate capacitance formula given a 15 cm by 15 cm plate with varying distance and thickness

Distance d (mm)	Thickness η (mm)	EM model result	Equation 3.6	Deviation
0.1	0.01	2.17 nF	1.99 nF	9%
1	0.01	217 pF	199 pF	9%
20	0.01	15.9 pF	9.96 pF	60%
50	0.01	8.28 pF	3.98 pF	108%
0.1	5	2.21 nF	1.99 nF	11%
20	5	16.8 pF	9.96 pF	68%

To assess the accuracy of this estimation, we compare results from Eq. 3.6 with those obtained from a numerical model (Table 3.1). We consider a square conductor with a side length of 15 cm and finite thickness η, positioned at varying distances d from a larger square conductor with a side length of 30 cm. The smaller plate resembles a printed circuit board, while the larger plate represents, for instance, a table. The larger plate is assumed sufficiently large to capture fringing fields.

The analytical results are compared with those obtained from an EM model, which accurately accounts for fringing fields and the finite thickness η of the plate. These computations are performed using a static electric solver (see Sect. 5.7), with each calculation requiring a few seconds. After mesh refinement, the model consists of approximately 12k mesh cells.

For small distances ($d = 0.1$ mm or 1 mm), the results of Eq. 3.6 and the EM model differ by less than 10%. However, for larger distances, the deviations become significant. At $d = 20$ mm, the capacitance computed by the EM model is 60% larger, and at $d = 50$ mm, the deviation reaches 108%. This increase occurs because fringing fields become more dominant at larger distances.

The thickness η of the plates has a minor impact. As η increases from 0.01 mm to 5 mm, the deviation between Eq. 3.6 and the EM model rises from 9% to 11% for $d = 0.1$ mm and from 60% to 68% for $d = 20$ mm.

Recognizing that Eq. 3.6 underestimates capacitance as the distance between the plates increases, this equation remains a practical tool for evaluating the potential impact of parasitic capacitances during the early design phase. By using circuit simulations, the influence of worst-case, average-case, and best-case estimations of the capacitance can be systematically analyzed.

(Self-Partial) Inductance Inductance arises from Faraday's law of induction, which states that a time-varying current, $\frac{\mathrm{d}I}{\mathrm{d}t}$, flowing in a conductor generates a magnetic field that induces a voltage, V_{ind}. This induced voltage opposes the change in current.

The self-inductance, L, is the proportionality constant relating the induced voltage, V_{ind}, to the rate of change of current,

$$V_{\text{ind}} = L\frac{\mathrm{d}I}{\mathrm{d}t}. \tag{3.8}$$

In addition to self-inductance, mutual inductance describes the voltage induced in a second conductor due to a changing current in a first conductor.

If the conductor carrying the current does not form a closed loop, the inductance is referred to as a partial inductance. The concepts of self-inductance, partial inductance, and methods for their computation are discussed in detail in Sect. 5.6.

Numerous analytical formulas exist for calculating inductances, including mutual and partial inductances (Chapter 3 and Appendix A in Caniggia and Maradei 2008). Many inductance calculators are available online. It is wise to double-check the results from such calculators against the literature in the references section at the end of this chapter.

Below, we list the self-inductances for a few geometries that are commonly used in EMC engineering.

The partial self-inductance of a long, thin wire with a length l and radius r, $r \ll l$, is given by

$$L_{\text{p,wire}} = \begin{cases} \frac{\mu_0}{2\pi} l \left[\ln\left(\frac{2l}{r}\right) - \frac{3}{4} \right], & \text{low frequencies,} \\ \frac{\mu_0}{2\pi} l \left[\ln\left(\frac{2l}{r}\right) - 1 \right], & \text{high frequencies.} \end{cases} \tag{3.9}$$

The low-frequency case considers an increase of inductance by $\frac{\mu_0}{8\pi} l$ due to the uniform distribution of the current inside the conductor that generates an additional magnetic field (internal inductance Paul 2010). This inductance is not present at high frequencies where the current is confined to the surface of the conductor and no current flows through its interior (cf. Sect. 5.6.1 and Fig. 5.28).

The self-inductance of two parallel wires in a homogeneous medium with length l, distance s, and radii r_{w1}, r_{w2}, with the second wire being the return current wire of the first, is given by

$$L_{\text{p,w,para}} = \frac{\mu_0}{2\pi} l \operatorname{arccosh}\left(\frac{s^2 - r_{\text{w1}}^2 - r_{\text{w2}}^2}{2 r_{\text{w1}} r_{\text{w2}}} \right). \tag{3.10}$$

Equation 3.10 does not consider fields inside the conductor, i.e., it assumes the high-frequency case. It is proportional to the inverse of Eq. 3.7, because, in homogeneous media, the product of the per-unit-length inductance and the per-unit-length-capacitance equals $\mu\varepsilon$ (cf. Sect. 8.7 and Paul 2007).

The self-inductance of a trace with length l, width w, thickness η, and height h above the ground on a printed circuit board is (at low-frequencies)

$$L_{\text{p,trace}} = \frac{\mu_0}{2\pi} l \left[\ln\left(\frac{2h}{w + \eta} \right) + \frac{3}{2} \right], \quad w, h \ll l. \tag{3.11}$$

The self-inductance of a coaxial wire with length l, outer conductor radius r_o, and inner conductor radius r_i is (at high frequencies) (Paul 2007, 2010)

$$L_{p,coax} = \frac{\mu_0}{2\pi} l \ln \frac{r_o}{r_i}. \tag{3.12}$$

The self-inductance of a circular loop with loop radius r_{loop} and wire radius r_w is given by (high frequencies) (Paul 2010)

$$L_{loop} \approx \mu_0 r_{loop} \left(\ln \left(\frac{8 r_{loop}}{r_w} \right) - 2 \right), \quad r_{loop} \gg r_w. \tag{3.13}$$

The formulas for self-inductance are valid only for structures where the cross-sectional dimensions are small compared to the length. For example, Eq. 3.11 can yield a negative inductance if $2h < w \exp^{\frac{3}{2}}$ for small η.

For a layer thickness of $h = 0.13$ mm, the formula produces a negative inductance for traces wider than 6 mm. This situation arises in power electronics, where traces designed to carry supply currents can exceed this width.

For straight structures of length l, the equations above share the common factor $\frac{\mu_0}{2\pi} l = 2 \cdot 10^{-7} l$ H. When l is expressed in millimeters, this corresponds to 0.2 nH per mm. A practical approximation for geometries with a length much larger than their height or width (e.g., traces or bond wires) is to use 1 nH per mm as an initial estimate (Ruehli 1972). Overestimating the inductance in such cases introduces a safety margin, as larger inductance often exacerbates EMI issues.

For wide structures with large w, where Eq. 3.11 is not valid, the inductance is substantially lower. In such cases, we recommend using a numerical method to compute the inductance directly from CAD models. The appropriate method is detailed in Sect. 5.6.

Using the equations for resistance, capacitance, and self-inductance described above, we can qualitatively evaluate the impact of mechanical structures in an electronic system. Parasitic capacitances and inductances contribute to resonances, which occur within a frequency range dictated by their values. For typical parasitic capacitances and inductances (usually in the range of pF and nH), these resonances are not without bounds. The resonance frequency f_r can be approximated using the formula $f_r = \frac{1}{2\pi\sqrt{LC}}$. For example, if $L = 10$ nH and $C = 10$ pF, the resonance frequency is approximately $f_r \approx 500$ MHz. If either L or C increases, f_r decreases anti-proportionally to $\sqrt{LC}$, shifting the resonance to lower frequencies. Due to uncertainties in the values of parasitic elements, the exact resonance frequency f_r is not precisely known. However, if resonances are observed, the EMC engineer can use the estimated values of parasitic capacitance C or inductance L to determine whether a specific part of the geometry could be responsible for an EMI issue in the corresponding frequency range. The study of these resonances with the aid of EM simulation is treated in several later sections of this book, e.g., in Chaps. 9 or 10.

3.4 Lumped Model of the DC-DC Converter

PWM switching of a power transistor

LISN

RLC models from analytical equations

None

AC, Transient

0–300 MHz

In this section, we extend the circuit model of the buck converter to include the parasitic elements that characterize the influence of interconnects on the EMI of the converter. To achieve this, we outline the geometry of the PCB and apply the formulas from Sect. 3.3 to calculate the parasitic element values.

The buck converter is a part of the digital board introduced in Sect. 1.4 and is shown in Fig. 3.10. It is located on the lower side of the PCB, enclosed by the orange outline. The components of the converter are positioned between the 12 V and 5 V USB-C connectors, located to the left and right on the board, respectively. The voltage input, $V_{\text{in}} = 12\,\text{V}$, is connected to a wide power plane that serves as the positive supply path. At the end of this power plane, there is a DC link capacitor,

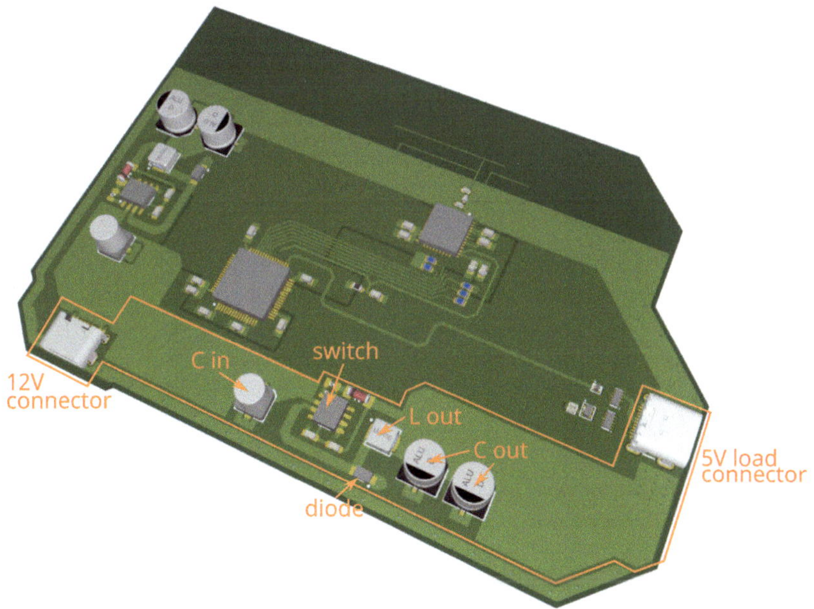

Fig. 3.10 Buck converter position on the digital PCB

$C_{\text{in}} = 10\,\mu\text{F}$, and a buck-mode switching regulator in an SOIC-8 package, labeled as "switch." This regulator contains the switching transistor. On the output side, the circuit includes an inductor, $L_{\text{out}} = 22\,\mu\text{H}$, and two parallel capacitors, $C_{\text{out},i} = 22\,\mu\text{F}$ $(i = 1, 2)$. A diode is located adjacent to L_{out}. The output path, leading to the 5 V USB-C load connector, is also implemented as a power plane. The layer below the top layer serves as the ground plane. While the PCB has four layers in total, only the top layer and the ground plane directly beneath it are relevant to this analysis of the buck converter.

We use the equations from Sect. 3.3 to estimate the parasitic element values of the layout. However, even for a layout as simple as this, these equations serve only as approximations. The layout does not adhere to the geometry of a plate capacitor, nor does it consist of long, narrow traces—key assumptions required for the equations to hold accurately. Despite these limitations, the estimated parasitic element values provide useful insight into the frequency ranges where L-C combinations induce resonances. These resonances influence the EMI of the buck converter.

We focus on two key geometrical features: the inductance of traces and the dominant capacitance of structures with large surface areas to the underlying ground plane. To analyze these effects, we estimate the inductances and parasitic capacitances of the power planes in three regions: between the supply connector and the SOIC-8 package, between the SOIC-8 package and the load, and from the diode to the ground.

In particular, we introduce eight parasitic elements into the circuit:

1. The inductances $L_{\text{p,in}}$, $L_{\text{p,load}}$, $L_{\text{p,FrW}}$ for the traces and planes, namely
   - $L_{\text{p,in}}$ for the power plane on the supply side (left of the SOIC-8 package),
   - $L_{\text{p,load}}$ the power plane towards the load (right of the SOIC-8 package),
   - $L_{\text{p,FrW}}$ the trace to the freewheeling diode.

   Each inductance is accompanied by a parasitic resistance connected in series, to model the DC resistance.
2. The parasitic capacitances for the power planes, which are
   - $C_{\text{p,in}}$ for the power plane on the supply side (left of the SOIC-8 package),
   - $C_{\text{p,load}}$ for the switching node toward the load (right of the SOIC-8 package).

The resulting circuit is shown in Fig. 3.11, with the already parameterized parasitic values. The estimation of the parameters is shown in the following.

Using Eq. 3.3, we estimate the resistances of the power planes to be about $1.5\,\text{m}\Omega$, and the resistance of the trace with the free-wheeling diode as $R_{\text{p,FrW}} = 2.4\,\text{m}\Omega$. Note that the value of the parasitic resistance of the power plane on the load side is irrelevant, because it is in series with the much larger load resistor R_{L}.

To compute the self-inductances of the power planes, we must not use Eq. 3.11, as it applies to a trace rather than a plane. The power plane on the supply side has a length of approximately 30 mm. The estimate of 1 nH per mm is too conservative for a geometry of this width. As a rough estimate, we use $\frac{1}{10}$ of that value, i.e., $L_{\text{p,in}} =$

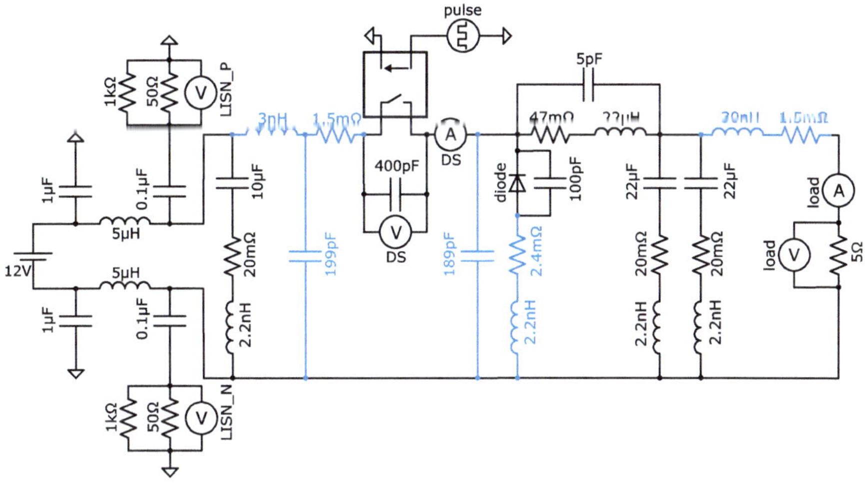

Fig. 3.11 Functional schematic extended by two LISNs and eight parasitic elements (blue color) to represent the interconnects. The GND symbol references the metallic table

$\frac{1}{10}30\,\text{mm} \cdot 1\,\frac{\text{nH}}{\text{mm}} = 3\,\text{nH}$. In Sect. 5.6.1, we address how to compute inductances numerically. In the remainder of the book, we learn how to avoid lumped elements at all.

The estimation of the value of $L_{\text{p,FrW}}$ cannot be done using Eq. 3.11. Its width is more than 10 times larger than its distance from the ground plane, so that the equation becomes negative. The trace has a length of 7 mm. Since the rule of 1 nH per mm greatly overestimates the inductance for a trace of this width, we set $L_{\text{p,FrW}} = 5\,\text{nH}$ as a rough estimate.

The estimation of the inductance $L_{\text{p,load}}$ is even more uncertain. The length of the power plane toward the connector is about 40 mm. After the connector, a coaxial wire is connected, the length of which is unknown. In the model, we use $L_{\text{p,load}} = 30\,\text{nH}$, which must be regarded as a placeholder. Once the impact of the value of this inductance is understood, conclusions can be drawn about the effect of changing this value. In the later discussion, it turns out that the value of $L_{\text{p,load}}$ has no impact on the resonances observed within the frequency range of interest.

We use Eq. 3.6 to estimate the parasitic capacitances of the power planes. We have $\varepsilon_{\text{r}} = 4.3$ and $d = 0.127\,\text{mm}$. For the two planes having an area of $665\,\text{mm}^2$ and $632\,\text{mm}^2$, respectively, we obtain $C_{\text{p,in}} = 199\,\text{pF}$ and $C_{\text{p,load}} = 189\,\text{pF}$. The capacitance between the trace to which the diode is attached and the ground layer is $C_{\text{FrW}} = 14\,\text{pF}$. We neglect it. Being more than a factor of 10 lower than $C_{\text{p,in}}$ and $C_{\text{p,load}}$, this other capacitance may generate resonances that are at frequencies a factor of more than $\sqrt{10}$ larger than those caused by $C_{\text{p,1}}$ and $C_{\text{p,load}}$.

In Fig. 3.12, the voltage $V_{\text{LISN,P}}$ is presented in both the time and frequency domains. Ringing is observed during the switching events of the transistor, both when it turns on and off. Compared to Fig. 3.9, the decay of the ringing is

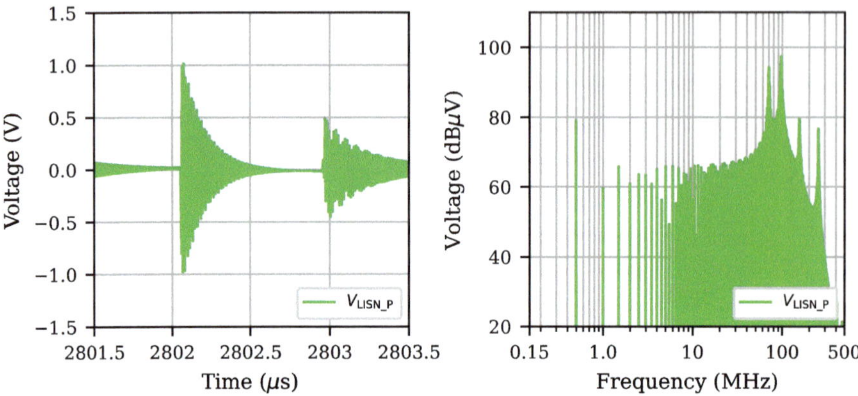

Fig. 3.12 Time-domain signal and spectrum of $V_{\mathrm{LISN,P}}$ of the model using parasitic elements for both the components and the interconnects

noticeably slower, and the amplitude of the second ringing has decreased. In the frequency domain, we see that there are four resonances instead of two. The larger number of resonances is because of the higher number of parasitic capacitances and inductances introduced into the circuit. Two of the resonance frequencies are located at or below 100 MHz. The addition of parasitic capacitance due to the power planes, along with an increase in total inductance, has shifted the frequency of the first resonance to a lower value compared to the case without layout parasitics. This analysis highlights how the layout of this buck converter affects the EMI spectrum, with the impact becoming evident in the FM frequency range. The simulation time of this model is 2 min and 26 s.

3.5 Circuit Simulation of a Buck Converter

Domains: 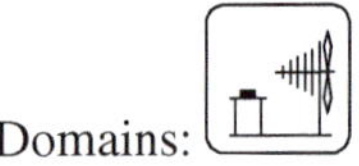

Circuit simulations with estimated parasitic elements provide a convenient method to study resonance frequencies and their root causes. They help identify scenarios where combinations of (parasitic) circuit elements may negatively impact the circuit's EMI.

The model derived in Sect. 3.4 suggests that the combination of parasitic elements representing the layout and the components causes two resonances that strongly dominate the emission between 70 MHz and 100 MHz. This is concerning because, as defined by the ITU, the FM radio band spans frequencies from 85MHz to 108 MHz and is subject to particularly strict EMI limits.

In previous sections, we discussed the uncertainty of the parasitic element values. The values applied in the model are estimated. Even if these values were computed, in industrial design, the layout could be adapted to meet new requirements, leading to changes in these values. Consequently, the two resonances observed in Fig. 3.12 are unlikely to appear at exactly the same frequencies in the final product. However, a resonance frequency is inversely proportional to the square root of the inductance and capacitance. A 30% error in estimating these values would result in only about a 15% shift in the resonance frequency. For example, a resonance at 100 MHz would shift to 85 MHz, which still lies within the FM range. The conclusion is that, although the exact resonance frequencies may vary due to the parameter estimation uncertainty discussed in Sect. 3.3, these resonances pose a significant risk because they dominate the FM, irrespective of the exact values of the involved capacitances and inductances.

In this section, we investigate the root causes of the two resonances observed in Fig. 3.12 at 70 MHz and 97 MHz, which we denote as $f_{70\,\mathrm{MHz}}$ and $f_{97\,\mathrm{MHz}}$, respectively. Our objective is to identify the conditions required to shift both $f_{70\,\mathrm{MHz}}$ and $f_{97\,\mathrm{MHz}}$ out of the FM band. These investigations can be conducted using simulations in either the time domain or the frequency domain.

3.5.1 Time-Domain Analysis

We analyze the spectrum of $V_{\mathrm{LISN,P}}$ based on transient simulations in the time domain using the circuit shown in Fig. 3.11. To compute the spectrum, we rely on the DFT. The EMI receiver is not applied because, in the frequency range of interest, the circuit model provides only qualitative accuracy. While the model is suitable for investigating the root causes of resonances, it is not intended for quantitative predictions. Implementing the EMI receiver script requires additional effort but does not contribute to a deeper understanding of the system.

Due to the self-inductance $L_{\mathrm{p,in}} = 3\,\mathrm{nH}$ of the power plane in the input side and $L_{\mathrm{p,FrW}} = 5\,\mathrm{nH}$ at the diode, the resonance f_{r1} of Eq. 3.1 shifts from 170 MHz (Eq. 3.1) to

$$\frac{1}{2\pi\sqrt{(L_{\mathrm{C,DCL}} + L_{\mathrm{p,in}} + L_{\mathrm{p,FrW}})\,C_{\mathrm{DS}}}} = \frac{1}{2\pi\sqrt{(2.2\,\mathrm{nH} + 3\,\mathrm{nH} + 5\,\mathrm{nH})\,C_{\mathrm{DS}}}} = 78.8\,\mathrm{MHz}. \tag{3.14}$$

This calculated value is 12% higher than the simulated resonance at $f_{70\,\mathrm{MHz}}$. The discrepancy arises because Eq. 3.14 does not account for the parasitic layout capacitances $C_{\mathrm{p,in}}$ and $C_{\mathrm{p,load}}$, which increase the total capacitance and, consequently, lower the resonance frequency. This effect is confirmed by removing these capacitances, which shifts the resonance to 78 MHz in accordance to Eq. 3.14.

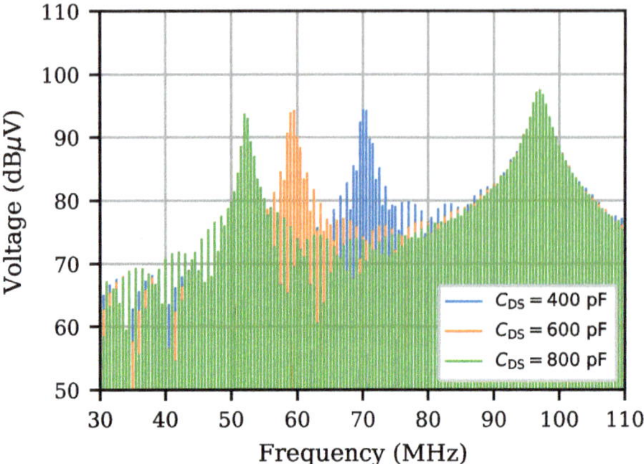

Fig. 3.13 The resonance frequency $f_{70\,\text{MHz}}$ shifts to nearly 50 MHz as C_{DS} is increased from 400 pF to 600 pF and 800 pF

We aim to shift this frequency away from the FM band. According to Eq. 3.14, this can be achieved by increasing the transistor's drain-source capacitance C_{DS}. The simulation results are shown in Fig. 3.13, covering the frequency range from 30 MHz to 110 MHz. With $C_{\text{DS}} = 400\,\text{pF}$, the resonance occurs at 70 MHz (blue), but it shifts below 60 MHz when $C_{\text{DS}} = 600\,\text{pF}$ (orange) and further to 52 MHz with $C_{\text{DS}} = 800\,\text{pF}$ (green). In the latter case, the EMI in the FM band is dominated by the resonance $f_{97\,\text{MHz}}$, making further increase in C_{DS} unnecessary.

The simulation suggests that using a transistor with twice the drain-source capacitance of the originally selected one effectively eliminates the impact of the resonance caused by C_{DS} and the inductance of the circuit loop, which includes the diode and the DC link capacitance, on the FM band. However, increasing the drain-source capacitance slows down the switching speed, which leads to higher power losses—a trade-off that is unpopular in hardware design. This simulation helps quantify the benefits of increasing C_{DS} to reduce EMI, enabling discussions about transistor selection to be based on concrete data rather than intuition.

Secondly, we address shifting the resonance $f_{97\,\text{MHz}}$. This resonance is caused by the combination of the capacitances $C_{\text{p,in}}$ and $C_{\text{p,load}}$ with the parasitic inductances of the commutation cell, which comprises the layout loop formed by the DC link capacitor, the transistor, and the switch. Reducing these capacitances shifts the resonance frequency to a higher value, as demonstrated in Fig. 3.14. The switching node capacitance $C_{\text{p,load}}$ is reduced from 189 pF (blue) to 90 pF (orange) and further to 45 pF (green). Consequently, the resonance frequency increases by 10 MHz and 15 MHz, respectively. The maximum EMI within the FM band decreases from 97 dBμV at 97 MHz to 85 dBμV at 108 MHz, the upper frequency limit of the FM band. This represents a reduction of 12 dB.

As with the transistor capacitance C_{DS}, such a layout change incurs certain costs. Reducing the area of the switching node increases its parasitic inductance

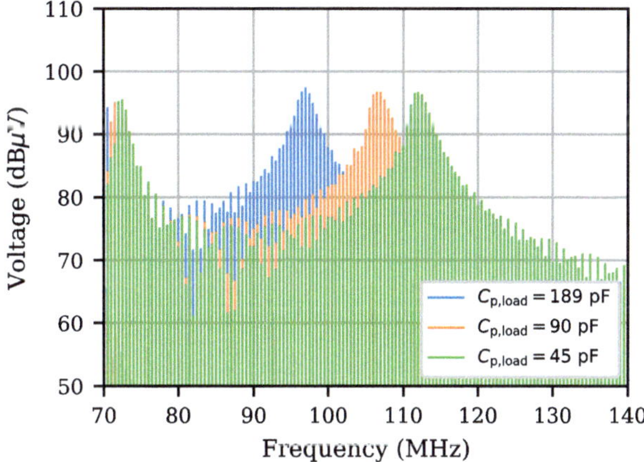

Fig. 3.14 The resonance frequency $f_{97\,\text{MHz}}$ is shifted to 112 MHz by reducing the switching node capacitance $C_{\text{p,load}}$

and may also lead to higher ohmic losses. The simulation quantifies the impact of this layout change on the EMI, but it does not address other design requirements. Handling conflicting requirements through multi-objective optimization is discussed in Chap. 16.

In Sect. 10.7, we compare the results obtained from estimations of the parasitic values presented in this chapter to a full-wave coupled simulation.

3.5.2 *Frequency-Domain Analysis*

In the previous Sect. 3.5.1, we used time-domain circuit simulations to determine the values of circuit elements that contribute to resonances within or close to the FM band, which is subject to strict EMI limits. When the amplitude of the EMI spectrum is not the primary concern, the shift in resonance frequencies can be analyzed using frequency-domain circuit simulations. Frequency-domain simulations are faster, making them more suitable for running multiple simulations with varying circuit component parameters. The type of the frequency-domain source, either current or voltage, is important and affects the simulation results.

To investigate the resonances observed in Fig. 3.12 using a frequency-domain simulation, we replace the EMI noise sources—the transistor and the diode—with ideal unit-amplitude AC sources (cf. Sect. 2.3.1). This simplifies the simulation by focusing solely on the coupling path while abstracting the complexity of the sources. This approach is particularly beneficial during the early stages of design, where detailed transistor models are either unavailable or the selected transistor may change due to pending vendor contracts. By analyzing the coupling path with an emphasis on resonance frequencies, this method offers insights into the underlying

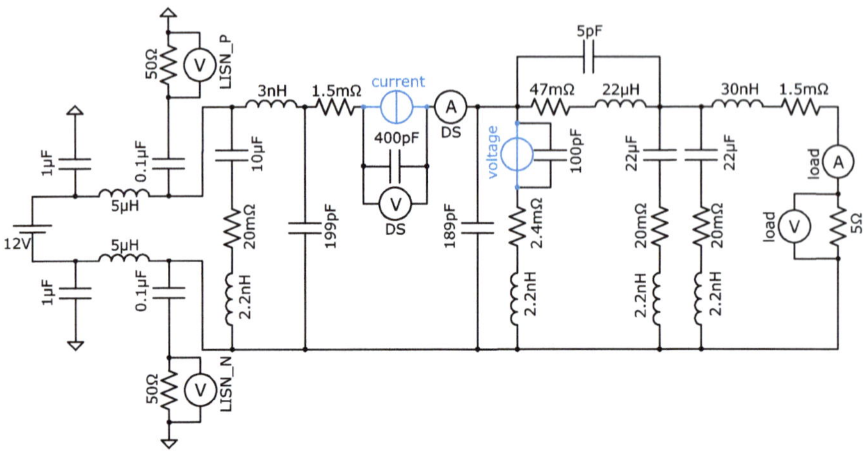

Fig. 3.15 Circuit diagram with linearized AC sources

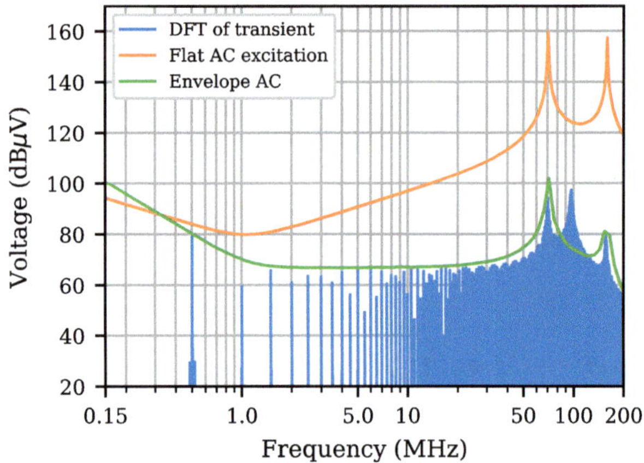

Fig. 3.16 Comparison of the EMI spectra of the simulations in time domain, with frequency-flat sources, and with spectrally weighted sources

coupling mechanisms. The simulation time of this model is 1 s, thus much faster than the 2 min and 26 s of the corresponding time-domain model of Fig. 3.12.

In Fig. 3.15, the transistor is replaced with a current source, and the diode with a voltage source. The current source has infinite source impedance, while the voltage source has zero source impedance (cf. Sect. 2.3.1).

The resulting spectrum is shown in Fig. 3.16 (orange) and compared to the spectrum obtained from the time-domain simulation (blue, same spectrum as in Fig. 3.12). Two resonances are observed: the lower frequency resonance, $f_{70\,\mathrm{MHz}}$, corresponds to the ringing when the transistor switches off and is examined in Eq. 3.14. The higher frequency resonance, at 160 MHz, is caused by the two layout capacitances, $C_{\mathrm{p,in}}$ and $C_{\mathrm{p,load}}$.

The resonance $f_{97\,\mathrm{MHz}}$, observed in the spectrum from the time-domain simulation, is absent in the AC simulation. This resonance results from the ringing when the transistor switches on, involving the diode's capacitance C_{D}. In the AC simulation, the diode is replaced by a voltage source, effectively acting as a short circuit. This replacement eliminates any voltage across C_{D}, thereby suppressing the resonance $f_{97\,\mathrm{MHz}}$. To excite $f_{97\,\mathrm{MHz}}$ in the AC simulation, a current source could be used in place of the diode. However, this would prevent the excitation of the resonance $f_{70\,\mathrm{MHz}}$, as the infinite impedance of the current source would open the loop required for this resonance to occur.

Since the amplitudes of the chosen sources are constant over frequency, the spectrum exhibits extremely large amplitudes at frequencies above 2 MHz. Referring to Fig. 2.5, the envelope of the spectrum of a trapezoidal signal with amplitude $\bar{A}$, period T, and equal rise and fall times $t_{\mathrm{r}} = t_{\mathrm{f}}$ decays at rates of -20 dB per decade and -40 dB per decade as functions of T and t_{r}, respectively. For the buck converter, $T = \frac{1}{500\,\mathrm{kHz}}$. Above $f_1 = \frac{500\,\mathrm{kHz}}{\pi} \approx 170\,\mathrm{kHz}$, the decay rate of -20 dB per decade takes effect. Below this frequency, the amplitude is given by $\frac{2\bar{A}t_{\mathrm{r}}}{T}$. From the time-domain simulation models, the rise and fall times of the voltages at the diode and the transistor are approximately $t_{\mathrm{r}} = t_{\mathrm{f}} \approx 3$ ns. This results in $f_2 = \frac{1}{\pi t_{\mathrm{r}}} \approx 100\,\mathrm{MHz}$.

Instead of using constant-current or constant-voltage excitation, we employ excitation sources with appropriately weighted amplitudes. Both the current and voltage sources have an amplitude of $\frac{2\bar{A}t_{\mathrm{r}}}{T}$, where the current source is assigned an amplitude of 1 A, and the voltage source 5 V. These values are derived from the functional simulation in Sect. 3.1. We have $T = \frac{1}{500\,\mathrm{kHz}}$, which is the inverse of the switching frequency. Above $f_1 \approx 170\,\mathrm{kHz}$, the amplitude is set to decay at a rate of -20 dB per decade on a logarithmic scale, and above $f_2 \approx 100\,\mathrm{MHz}$, the decay increases to -40 dB per decade. The resulting spectrum is shown as the green curve in Fig. 3.16. The deviation between the green curve and the orange curve, which corresponds to flat-spectrum sources, becomes evident above the frequency f_1. The steep decay of the spectrum above 100 MHz in the green curve is attributed to the spectral decay of the noise sources. The missing resonance at $f_{100\,\mathrm{MHz}}$ remains absent in the spectrum because the green curve is simply a scaled version of the orange curve.

Replacing the switching transistor and diode with AC sources for frequency-domain simulations of the DC–DC converter reduces simulation time, from several minutes to 1 s. Such time savings are valuable when examining the root causes of resonances through parameter variations. However, when setting up the simulation, it is crucial to carefully select the excitation sources. The impact of the selection must be studied by comparison with time-domain simulations.

3.6 Limits of Circuit Simulation

EMI spectra depend on both the hardware implementation and the parasitic elements of the circuit. Considering the tight design cycles in industry, circuit simulation offers a convenient and efficient entry point to EMC analysis. Simulations can be conducted before the design freeze, enabling designers to make adjustments for improved EMC performance prior to hardware fabrication. Even when interconnects are not yet finalized or vendor components are not available, the interactions between (parasitic) capacitances and inductances can still be evaluated based on worst-case and best-case assumptions. Furthermore, if new requirements emerge later in the design process, necessitating changes to a previously fixed design, a new EMC assessment can be quickly carried out.

However, there are also limits of circuit simulation. In the following, we list the advantages and disadvantages.

Advantages—EMI assessment using circuit simulation is beneficial for:

Early Design In early design or for concept studies, there are no layout, mechanical data, or selected components available. Any system assessment is a best guess. Circuit simulation is an efficient method to produce qualitative insight into EMC-relevant product properties.

Quick System Analysis Data collection is a time-consuming aspect of EMC modeling. Experienced simulation engineers distill complex systems into their essential coupling elements. A quick system analysis using a simplified circuit can reveal EMC-relevant properties without requiring the construction of complex simulation models.

Prediction in LW, MW, and SW Ranges Circuit simulation is accurate for hardware design at frequencies where parasitic inductances and capacitances do not produce resonances or inductive coupling. The applicable frequency range depends on the size of the system under consideration. For ECUs with dimensions of approximately 10 cm, this range extends from the LW band (a few hundred kHz) potentially into the MW or SW range (1–6 MHz). For smaller devices, such as sensors with dimensions of a few centimeters, circuit simulation can remain accurate well beyond the FM range (i.e., above 100 MHz).

Disadvantages—Circuit simulation, as described in this chapter, has limitations in the following scenarios:

Electrically Large Models Parasitic capacitances and inductances are often spatially distributed. In circuit simulations, these are typically represented as lumped elements. While this approach is accurate at the first system resonance, it becomes less reliable at higher frequencies, where the model becomes electrically large and resonance frequencies are not well reproduced.

Accurate Prediction For complex electronics, estimating parasitic elements is often imprecise. In particular, mutual coupling is challenging to quantify and may influence behavior even in the MW and SW ranges (a few MHz). Circuit simulations based on estimated lumped parameters carry the risk of overlooking coupling effects.

Modeling the Impact of Control Current or voltage control is essential for the functional assessment of a power electronic system. While it is possible to include control algorithms in SPICE simulations, doing so is cumbersome. Simulators such as PLECS (PLECS Plexim), which simplify control implementation, often rely on fixed time steps, making them slow at high frequencies. Additionally, their performance tends to degrade when incorporating many parasitic elements or imported S-parameters. However, at low frequencies, where control algorithms have the most significant impact, modeling parasitic elements is unnecessary. Consequently, for the implementation of control algorithms, simulators other than SPICE are more suitable.

3.7 Avoiding Common Pitfalls in Circuit Simulation

In this chapter, we utilize analytically estimated values for the parasitics of the layout, while in the following chapters, EM simulations are employed for this task. Circuit simulation always is a cornerstone of the simulation and modeling strategy, making the correct setup of circuit simulation models essential.

In this section, we discuss common pitfalls that simulation engineers may encounter when creating circuit simulation models. To provide a comprehensive reference for readers, we bundle these topics into one section, including particularities of coupled simulations where layout parasitics are obtained from EM simulations—even though this simulation flow has not yet been introduced.

We categorize the common pitfalls into two groups: modeling errors and simulation setting errors and begin with the most frequent issues.

For circuit-level EMC simulations, it is critical to include the parasitic values of the components mounted on the board. Capacitors, in particular, must include ESL and ESR values; otherwise, the results at higher frequencies become invalid. The modeling of component parasitics is discussed in detail in Sect. 2.4.2.

In transient circuit simulations, the correct choice of rise and fall times (t_r and t_f) for signal sources is crucial. Section 2.3.2 discusses the spectrum of trapezoidal signals and analyzes the effect of t_r and t_f. For functional simulations of power electronics systems, t_r and t_f are sometimes set to zero for simplicity. However, this is not permissible in EMC simulations, which consider frequencies above $1/(\pi t_r)$ or $1/(\pi t_f)$ (cf. Fig. 2.5). When exact rise and fall times are unknown, simulation engineers must make an educated guess, as outlined in the table in Sect. 2.3.2.

For a thorough analysis, once the circuit is correctly defined, simulation engineers should review component models in detail to understand the effects of tolerances, temperature dependence, or potential nonlinear behavior. These effects vary based on the specific components used in the design and may or may not play a significant role.

When setting up coupled EM-circuit simulations using S-parameter models, special attention must be paid to the orientation of nonlinear components such as diodes or transistors. In such simulations, the port orientation in the EM model affects the schematic, unlike standalone circuit simulations. Guidance on selecting the correct orientation is provided in Sect. 6.7.1.

When using component models based on measurements in coupled simulations, proper port definitions in the EM model are also necessary. This topic is discussed in Sect. 9.4.

We summarize the common modeling pitfalls as:

- Ideal circuit models without parasitics.
- Inaccurate signal settings (t_r and t_f).
- Lack of consideration for tolerances, temperature dependence, and nonlinearities in circuit components.
- Coupled simulations: Incorrect pin connections or component orientation.
- Coupled simulations: 1-port vs. 2-port configuration.

For transient circuit simulations, it is essential to ensure that the adaptive time-stepping scheme used by SPICE simulators does not select a time step larger than the maximum frequency intended to be resolved (cf. Sect. 3.8). Therefore, the simulation engineer should estimate the maximum allowable time step for the frequency band of interest and define it in the simulation settings.

For frequency-domain (AC) circuit simulations, appropriate sampling is required. Choosing a too coarse sampling rate may result in missing peaks in the spectrum, particularly for narrowband signals.

When nonlinear elements are included in AC simulations, it is necessary to ensure that these elements are linearized at the correct operating point. The linearization process is discussed in Sect. 3.8.2. Linearizing certain schematics, such as those involving a diode rectifier, is challenging. In such cases, it can be advantageous to approximate the operating points of the nonlinear element using a resistor.

For transient simulations involving S-parameter data, vector fitting is required to generate a passive and causal macromodel. The process of generating these macromodels is detailed in Sect. 10.3, while advanced techniques for controlling the accuracy of this process are discussed in Sect. 10.9.

Below, we summarize the common pitfalls related to simulation settings:

- Transient: Too coarse time step (exceeding the maximum frequency of interest).
- AC: Insufficient frequency sampling.
- AC: Incorrectly linearized schematics.
- Inaccurate vector fitting in schematic including S-parameter data.

3.8 Advanced: Circuit Simulation Algorithms

Domains: 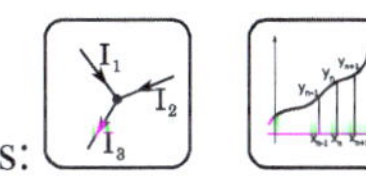

In this section, we discuss the details of circuit simulation. Understanding how SPICE simulation operates enables users to select the most suitable and efficient analysis type. It also aids in debugging simulation models and managing the simulation process when issues arise. However, this knowledge is not strictly required to build an EMC simulation model.

SPICE is based on the method of nodal analysis (Hayt et al. 2023), which calculates the voltages V_n at the connections of circuit components, known as nodes. Nodal analysis applies the Kirchhoff Current Law (KCL), which states that the sum of all currents at a node is zero (Eq. 2.2). The branch currents I_b are expressed in terms of the circuit node voltages V_n, requiring an admittance representation for each component, $I_b = Y_{bn} \cdot V_n$. While circuit simulators use admittances, this book primarily characterizes components by their impedances, as these are more intuitive and easier for engineers to interpret. The admittance representation facilitates the formulation of all KCL-defined equations into a system of linear equations,

$$\mathbf{I} = \mathbf{Y} \cdot \mathbf{V}. \tag{3.15}$$

The vector **I** represents the source currents, and solving the system yields the vector **V**, which contains the unknown voltages at all nodes. An advantage of using the admittance representation is that it eliminates the need for division by resistance. The equation $I_b = Y_{bn} \cdot V_n$ is, for very small values such as the DC resistance of short conductor stubs, numerically more stable compared to $I_b = \frac{1}{R_{bn}} \cdot V_n$. Using this definition, a parallel connection of components can be represented as a simple addition in the admittance form. This simplifies the composition of the elements Y_{bn}, particularly when multiple parallel components are involved. For a series connections, no disadvantage arises from this definition, as the series connection of two components requires two entries in the current vector **I**.

A limitation of the nodal analysis as described above is that only currents are allowed as sources. If we want to use a voltage source, this concept needs to be extended to the so-called modified nodal analysis (Hayt et al. 2023; Ho et al. 1975). The modified nodal analysis uses not only voltages as unknowns and currents as sources. In case of the modified nodal analysis, the entries in the vectors in Eq. 3.15 are composed of voltages, currents, or even charges, if this description is better suited for the components in the circuit.

Let us set up the equations for a simple system with a current source and a resistor, as depicted in Fig. 3.17. The system of equations consists of just one equation, $\frac{1}{R} \cdot V = I_S$. However, there are two unknowns V_i, $i = 0, 1$. To solve the system, we need to define a fixed value for one of the voltages.

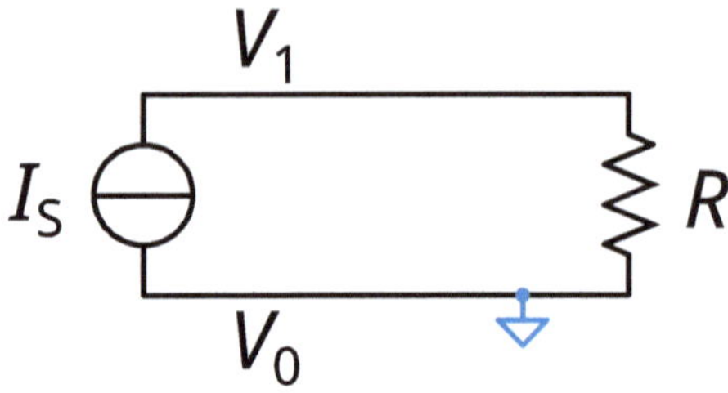

Fig. 3.17 A simple circuit for a nodal analysis. The ground node is indicated in blue

This is done by defining a node at zero voltage, called the ground node, as indicated by the blue ground symbol. From this simple exercise, we conclude that every SPICE simulation requires at least one ground node. If there are multiple instances of ground nodes defined in a circuit, they are all connected to each other. Probing a voltage calculated by the circuit simulator is equivalent to calculating the voltage of the particular node relative to the ground node.

If one of the components in the circuit is nonlinear, the admittance matrix **Y** of the system depends on the voltage across or the current through the component, and the system of equations is no longer linear. For a nonlinear system of equations, no direct solutions can be derived. Instead, the equation must be solved approximately using a numerical method, as discussed in the next section.

3.8.1 *Transient Circuit Simulations*

The flow of a transient SPICE simulation running from an initial time $t = 0$ to a time t_{end} is illustrated in Fig. 3.18. The simulation starts with the calculation of the DC operating point. This step determines the state of the circuit at the time $t = 0$. For this calculation, capacitors are removed from the circuit, as no DC current flows through them. Inductors are replaced by a short, as they pose no resistance to DC current. All sources are set to their value at $t = 0$, and the operating point of any nonlinear components is calculated iteratively. The details of this iterative solution are explained later in this chapter when discussing the transient solution, as it operates similarly to the computation of an operating point at every given time step t.

Here, we focus on the convergence of the DC operating point analysis.
Consider the case of two capacitors in series. During the operating point calculation, the capacitors are removed from the circuit, leaving the node between them disconnected and floating. As discussed in Fig. 3.17, floating nodes create a system of equations that cannot be solved. If there are two floating regions, the admittance matrix contains two separate regions that are not connected, rendering the matrix rank-deficient and non-invertible. To address this issue, the circuit simulator employs regularization methods to improve convergence. SPICE employs several

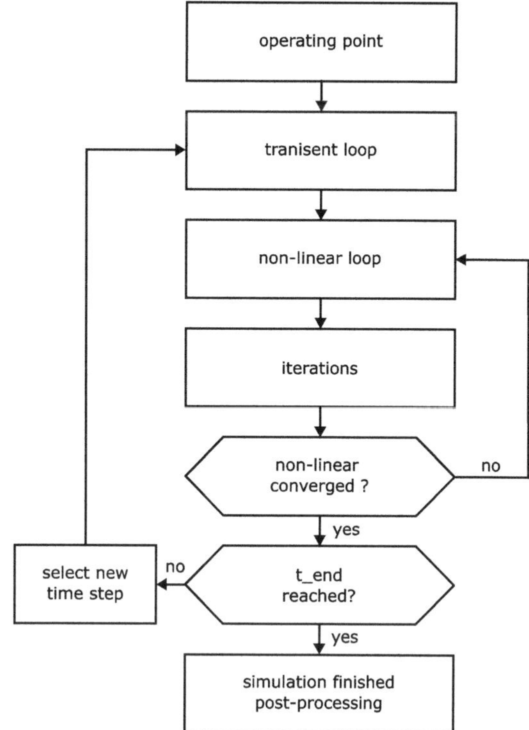

Fig. 3.18 A flow chart of the steps in a transient SPICE simulation

such convergence aids, which are outlined in this section. A detailed description can be found in Vladimirescu (1994).

To resolve the problem of floating nodes, the concept of GMIN is used. At a floating node, a connection to ground is added through a large resistor, $R = \frac{1}{\text{GMIN}}$. This connection fixes the voltage at the node without significantly affecting the current sum in the KCL, as the resistor is chosen to be sufficiently large. The default value in SPICE is $\text{GMIN} = 10^{-12}$, introducing an acceptable error in the current sum of 1 pA for a source of 1 V.

Another issue arises with the current magnitude through very small resistors. For instance, a voltage drop of $1\,\mu$V across a $1\,\text{n}\Omega$ resistor results in a current of 1 kA. Such a high current can lead to incorrect guesses in the nonlinear solver (as discussed in the next paragraph) and hinder convergence. To prevent this, SPICE uses the concept of RMIN. All resistance values smaller than RMIN are replaced with RMIN, which typically has a default value of $\text{RMIN} = 1\,\mu\Omega$. This highlights the importance of choosing realistic resistance values for the voltage source and switch in the buck converter example discussed in this chapter.

After the calculation of the operating point, the first time step is performed. Before discussing the time-stepping process, let us first discuss the solution of the nonlinear system where the admittance matrix depends on the voltage. This system needs to be recalculated at every time step. As mentioned earlier, there is no direct solution for a nonlinear system. Instead, we solve it approximately using an iterative method, the so-called Newton–Rhapson method.

The Newton–Rhapson method begins with an initial guess for the voltages, denoted as $\mathbf{V}_{k=0}$. In the next step k, every nonlinear element is linearized around the initial guess. This is done by numerically calculating the derivative of the V-I characteristic $f(\mathbf{V})$ of the component at the voltage value of the initial guess.

When performed for all circuit nodes, this process results in the Jacobian matrix $\mathcal{J}_f(\mathbf{V})$. To refine the voltage estimate, the next step involves solving the linear system

$$\mathbf{V}^{(k+1)} = \mathbf{V}^{(k)} - \mathcal{J}_f^{-1}\big(\mathbf{V}^{(k)}\big) \cdot f\big(\mathbf{V}^{(k)}\big). \tag{3.16}$$

This step requires not only solving the linear system but also inverting the Jacobian matrix. The iteration process continues until predefined convergence criteria are met. Thus, it is essential to define suitable convergence criteria. SPICE employs two convergence criteria: the residual criterion and the update criterion.

The residual criterion ensures that the sum of the currents at a node falls below a certain threshold, indicating how well KCL is satisfied. This criterion is defined based on the relative current at the node and the absolute current in the circuit. For every node n, we require the sum of all branch currents I_b flowing into that node to obey

$$\sum_b I_b < \mathrm{RELTOL} \cdot |I_{\max}| + \mathrm{IABSTOL} \tag{3.17}$$

where $I_{\max}$ is the maximum current in the entire circuit. The default is RELTOL $=$ 0.001, and IABSTOL $= 1$ pA.

The update criterion ensures that the rate of change in the calculated voltage from one iteration to the next is below a defined threshold. At every node n, the condition

$$|V_n^{(k)} - V_n^{(k-1)}| < \mathrm{RELTOL} \cdot \max\big(V_n^{(k)}, V_n^{(k-1)}\big) + \mathrm{VABSTOL} \tag{3.18}$$

is checked.

If either of the convergence criteria—the residual criterion or the update criterion—is satisfied, the nonlinear iteration loop is terminated, and the operating point is considered successfully calculated.

This iterative process is applied during the initial operating point calculation and for the determination of the operating point at every time step. If the iteration does not converge, it is terminated after a predefined maximum number of iterations. For a well-behaved circuit with a good initial guess during transient time stepping, typically no more than 10 iterations are required. The DC operating point, however, requires more iterations due to the absence of an initial guess. The maximum number of iterations is typically set to around 150. Some simulators are equipped to apply alternative methods if the Newton–Raphson iteration fails, but this capability depends heavily on the specific SPICE implementation being used.

After successfully estimating the operating point, the solver checks whether the maximum time t_{end} has been reached. If not, it proceeds to select the next time step. The operating point is only calculated at discrete time samples; there is no method to generate a continuous solution over time.

The choice of the time step significantly affects the simulation time since the nonlinear iteration is performed at every time step. By default, SPICE employs adaptive time stepping, dynamically adjusting the time step to the circuit's state. During events such as switching, smaller time steps are chosen to accurately resolve the behavior, whereas larger time steps are used during intervals of minimal signal variation to accelerate the simulation.

The adaptive time stepping process involves:

1. Selecting the next time step based on previous time steps,
2. Predicting the voltages at the next time step using a Taylor expansion with voltage values from previous steps,
3. Calculating the actual operating point at the next time step.

This process is known as a predictor-corrector scheme, one of several methods available to estimate new time steps. The predicted values are compared with the actual values, and their difference is called the Local Truncation Error (LTE). If the LTE exceeds a defined threshold

$$|\mathrm{LTE}| > \left|\mathrm{LTERATIO} \cdot \left(\mathrm{RTOL} \cdot V_n^{(k+1)} + \mathrm{VABSTOL}\right)\right| \tag{3.19}$$

the time step is rejected, and a smaller time step is chosen. If the LTE is within the acceptable range, a larger time step may be selected in the next iteration.

Additionally, the simulator explicitly places time samples at points where the source signal is known to exhibit discontinuities, such as the start of a ramp or at specified points in a PWL file.

In EMC simulations, we are often interested in harmonics at frequencies much higher than the system's switching frequency. In such cases, it is critical to limit the maximum allowed time step according to the Nyquist Theorem to ensure that all frequencies of interest are adequately resolved.

After a time step is completed, the solver integrates voltages and currents over the last time period. These integral values are necessary for modeling capacitors and inductors. Various time integration methods are available for this step, but the standard choice is a second-order Gear method, which uses three time samples for integration. Once t_{end} is reached, the SPICE simulation is finished.

We have added a post-processing step to the flow chart in Fig. 3.18, as EMC simulations often focus on the spectrum of the signals. Post-processing can involve a simple FFT of the time-domain signals calculated by the solver or a more elaborate analysis based on an EMI receiver, as described in Sect. 2.5.4.

The transient SPICE simulation represents a highly sophisticated algorithm that has revolutionized the daily work of electrical engineers by enabling the analysis of complex electrical networks. In EMC simulations, transient analysis is used to examine switched circuits in detail. For calculating a circuit's frequency response, the AC analysis offers an efficient alternative for linear circuits, as already introduced in Sect. 2.6.

3.8.2 Frequency-Domain Circuit Simulation

An AC analysis is computationally much simpler than a transient analysis because it avoids solving a nonlinear system. Instead, it operates on a linearized version of the circuit, using phasors to represent steady-state sinusoidal signals.

The phasor represents a sinusoidal function with Amplitude $\bar{A}$ and initial phase θ. Both are time-invariant, and the function has a fixed angular frequency ω. For a harmonic function $a(t) = \Re\left(\bar{A} \exp j\left(\omega t + \theta\right)\right)$, we introduce the phasor $\underline{A}$ as

$$\underline{A} = \bar{A} e^{j(\omega t + \theta)}. \tag{3.20}$$

Correspondingly, the phasor notations for the AC voltage and AC current are $\underline{V} = \bar{V} \exp j\left(\omega t + \theta\right)$ and $\underline{I} = \bar{I} \exp j\left(\omega t + \theta\right)$ with $\bar{V}$ and $\bar{I}$ the peak voltage and peak current.

As in transient analysis, the first step of AC analysis is the calculation of the operating point. The admittance matrix is linearized at the operating point, and the resulting linear system is solved for the desired frequency samples. Since these frequency samples are independent, they can be processed efficiently in parallel.

To understand the impact of linearization, we examine the behavior of a 1N4148 diode (Nexperia). In AC analysis, the operating point calculation considers only DC sources, while AC sources are replaced by their impedances and remain inactive. This differs from transient simulations, where sources are initialized at their $t = 0$ values.

The simple circuit depicted on the left side of Fig. 3.19 demonstrates this concept. A DC source V_{DC} biases the diode, and the impedance is monitored at the position marked by the Z monitor by calculating the impedance Z at this position. The values of Z for three biasing voltages $V_{\mathrm{DC}} = -5, 0, 5\,\mathrm{V}$ are plotted on the right side of Fig. 3.19.

For $V_{\mathrm{DC}} = -5\,\mathrm{V}$ and $V_{\mathrm{DC}} = 0\,\mathrm{V}$, the diode is in the blocking state during the operating point calculation. After linearization, it remains in the blocking state across all frequencies analyzed in the AC simulation. The observed decrease in impedance at higher frequencies arises from the parallel capacitance included in the diode's model. For $V_{\mathrm{DC}} = 5\,\mathrm{V}$, the diode operates in the forward direction, resulting in a low impedance across the frequency band. This example illustrates the effect of linearization: in AC simulations, the diode cannot alternate between on and off states but remains in a single state determined by the operating point.

Circuit simulators supporting network parameters can also perform S-parameter calculations. An S-parameter analysis is analogous to AC analysis, with the addition of defining a reference impedance for each port. S-parameters are then derived

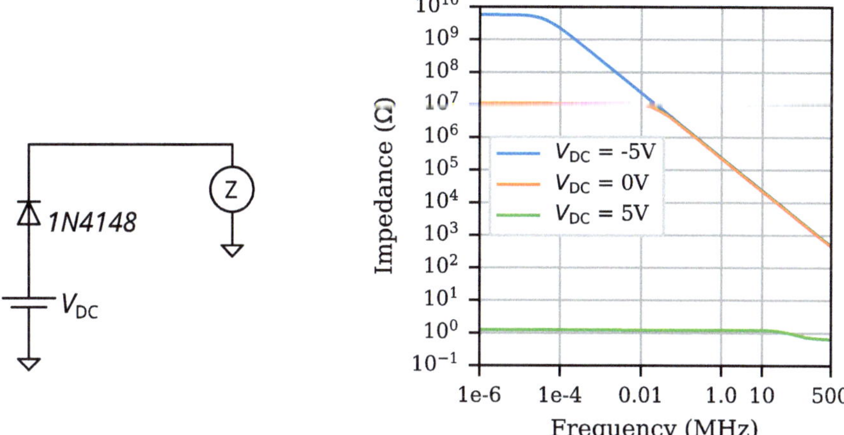

Fig. 3.19 Left: Schematic for the impedance calculation of a linearized diode. Right: Impedance of a diode for different biasing conditions

from the AC simulation using Eqs. 2.5 and 2.6. Following the methodology in Sect. 2.7, this enables the extraction of impedances and RLC values by converting S-parameters to Z-parameters in a post-processing step. This approach is used in the diode example to calculate impedance, rather than directly computing $\frac{\underline{V}}{\underline{I}}$.

Further Reading

Circuit Simulation

H. Bishnoi et al., EMI modeling of half-bridge inverter using a generalized terminal model, in *2011 Twenty-Sixth Annual IEEE Applied Power Electronics Conference and Exposition (APEC)* (2011), pp. 468–474. https://doi.org/10.1109/APEC.2011.5744638

H. Bishnoi et al., Analysis of EMI terminal modeling of switched power converters. IEEE Trans. Power Electron. **27**(9) 3924–3933 (2012)

H. Bishnoi et al., EMI behavioral model of DC-fed three-phase motor drive systems. IEEE Trans. Power Electron. **29**(9), 4633–4645 (2014)

A. Domurat Linde, E. Hoene, Investigation and PEEC based simulation of radiated emissions produced by power electronics converters, in *International Conference on Integrated Power Electronics Systems* (2010)

J. Hansen, B. Deutschmann, Efficient electromagnetic interference modeling of a gallium nitride half bridge. Elektrotechnik und Informationstechnik **140**(1), 95–102 (2023). ISSN: 0932-383X, https://doi.org/10.1007/s00502-022-01119-9

W. Hayt et al., *Engineering Circuit Analysis*, 10th edn. (McGraw Hill, 2023). ISBN: 9781264149919

P. Hillenbrand et al., Understanding conducted emissions from an automotive inverter using a common-mode model, in *2015 IEEE International Symposium on Electromagnetic Compatibility (EMC)* (2015), pp. 685–690. https://doi.org/10.1109/ISEMC.2015.7256246

C.-W. Ho, A. Ruehli, P. Brennan, The modified nodal approach to network analysis. IEEE Trans. Circuits Syst. **22**(6), 504–509 (1975). https://doi.org/10.1109/TCS.1975.1084079

Y. Ishii et al., Accurate conducted EMI simulation of a buck converter with a compact model for an SiC-MOSFET, in *2020 IEEE Applied Power Electronics Conference and Exposition (APEC)* (2020), pp. 2800–2805. https://doi.org/10.1109/APEC39645.2020.9124318

T. Langbauer et al., Closing the loop between circuit and thermal simulation: a system level co-simulation for loss related electro-thermal interactions, in *2019 25th International Workshop on Thermal Investigations of ICs and Systems (THERMINIC)* (2019), pp. 1–6. https://doi.org/10.1109/THERMINIC.2019.8923595

N. Moonen, F. Buesink, F. Leferink, Enhanced circuit simulation using mutual coupling parameters obtained via 3D field extraction, in *2016 Asia-Pacific International Symposium on Electromagnetic Compatibility (APEMC)*, vol. 1 (2016), pp. 181–183. https://doi.org/10.1109/APEMC.2016.7523002

Nexperia, *1N4148 High-speed Diodes*. https://assets.nexperia.com/documents/data-sheet/1N4148_1N4448.pdf (visited on 2025)

K. Odreitz, M. Musil, B. Deutschmann, Resonance mode analysis on conducted emissions of an inverting buck-boost converter, in *2024 International Symposium on Electromagnetic Compatibility—EMC Europe* (2024), pp. 973–978. https://doi.org/10.1109/EMCEurope59828.2024.10722672

PLECS Plexim, *The Standard Simulation Software for Power Electronics*. https://www.plexim.com/products/plecs (visited on 2025)

E. Rondon-Pinilla et al., Modeling of a buck converter with a SiC JFET to predict EMC conducted emissions. IEEE Trans. Power Electron. **29**(5), 2246–2260 (2014). https://doi.org/10.1109/TPEL.2013.2295053

J.-L. Schanen, J. Roudet, Built-in EMC for integrated power electronics systems, in *IEEE 5th International Conference on Integrated Power Systems (CIPS)* (2008)

G. Shen et al., Terminal modeling of DC–DC converters with stochastic behavior. IEEE Trans. Electromagn. Compat. **60**(6), 2011–2018 (2018). https://doi.org/10.1109/TEMC.2018.2805283

K. Takahashi, T. Ibuchi, T. Funaki, Noise-source model for frequency-domain EMI simulation of a single-phased power circuit. IEEE Trans. Electromagn. Compat. **63**(3), 772–782 (2021). https://doi.org/10.1109/TEMC.2020.3022887

A. Vladimirescu, *The SPICE Book* (Wiley, 1994). ISBN: 978-0-471-60926-1

L. Zhai et al., Mitigation emission strategy based on resonances from a power inverter system in electric vehicles. Energies **9**(6) (2016). ISSN: 1996–1073. https://doi.org/10.3390/en9060419, https://www.mdpi.com/1996-1073/9/6/419

Inductance Calculation

S. Caniggia, F. Maradei, *Signal Integrity and Radiated Emission of High-Speed Digital Systems* (Wiley, Chichester, 2008)

F.W. Grover, *Inductance Calculations: Working Formulas and Tables* (Dover Publications Inc., Mineola, 2004). ISBN: 978-0486495774

C.R. Paul, *Analysis of Multiconductor Transmission Lines*, 2nd edn. (Wiley-IEEE Press, 2007). ISBN: 978-0-470-13154-1

C.R. Paul, *Inductance Loop and Partial* (Wiley, New York, 2010)

A.E. Ruehli, “Inductance calculations in a complex integrated circuit environment. IBM J. Res. Dev. **16**(5), 470–481 (1972). https://doi.org/10.1147/rd.165.0470

Chapter 4
Electromagnetic Field Computation

We have seen in the previous chapter that with the addition of parasitic values for components and interconnects to the functional system schematic, we can predict the EMC behavior (of a part) of an electronic system.

While we have used analytical estimations, analytical equations exist only for simple geometries. For realistic geometries of electronic systems, analytical estimation becomes impractical due to their geometrical complexity. By using numerical field computation instead of the analytic approach, we can extract component models for arbitrary interconnect geometries.

Additionally, numerical field computation enables the calculation of field quantities, as required for radiated emission analysis. Thus, our next step is to incorporate numerical field computational methods into EMC simulations.

In the next sections, we introduce the ingredients required to perform a field computation: the geometrical data used to define the structure to be simulated, the computational mesh, the material models, and the boundary conditions. In the advanced sections, we provide additional background information about the numerical methods and the origins of their properties. These sections are not mandatory but are recommended for engineers interested in understanding why we recommend particular methods for specific problem groups. For readers who choose to skip the advanced sections, practical recommendations are also provided in Chaps. 5 and 6. Our goal is not to present an exhaustive list of all available methods but rather to introduce the methods we use for the EMC simulation of electronic systems.

We focus on three numerical methods: The Finite Integration Theory (FIT), the Transmission Line Matrix Method (TLM), and the Finite Element Method (FEM), as these are the primary methods used throughout this book. We also briefly introduce the Method of Moments (MoM), though it is not used in our simulation models.

J. Hansen, A. Barchanski, *EMC Simulation of Electronic Systems*,
https://doi.org/10.1007/978-3-032-09237-3_4

In electronics design, field computation is still less common than pure circuit simulation. However, in recent years, there has been a noticeable increase in the use of EM simulators for EMC problems, as observed in numerous publications.

A variety of commercial and free packages are available on the market, each offering different numerical methods for calculating electromagnetic fields. Well-known free packages include Elmer FEM, a Finite Element-based open-source code, and Concept-II, a package based on the Method of Moments. Commercial packages include Ansys Electronics Desktop, which offers several field solvers, with the Finite Element method tool HFSS being the most recognized. Altair's tool suite also offers multiple solvers, including the Method of Moments-based tool FEKO. EMCoS Studio is another primary Method of Moments–based simulation package. As described in Sect. 1.6, we use the CST STUDIO SUITE for field computation throughout this book.

Before discussing the field computation methods, we provide an introduction to Maxwell's Equations, as they form the foundation of every field computation method.

4.1 Electromagnetic Fields and Maxwell's Equations

Domains: 

For the functional part of an electronic design, a circuit description based on voltages and currents is sufficient and simpler to understand. However, the situation changes dramatically when wireless functionality is included in the design, such as in the case of antenna design. In such scenarios, the description using electromagnetic fields becomes essential. Antenna designers have heavily relied on field computation tools for the last 20 years, and these software packages incorporate many concepts from RF engineering.

These RF concepts are not a limitation to EMC simulation but rather a foundation upon which we build, as they allow for a thorough analysis of resonant behavior at high frequencies. The S-parameters introduced in Sect. 2.4.3.2 are an excellent example of such RF concepts.

Since we are dealing with the computation of electromagnetic fields, some understanding of electromagnetic field theory is essential. In this book, we do not provide a comprehensive introduction to electromagnetic field theory, as there are numerous detailed textbooks available on the subject. Instead, we offer a brief, illustrative introduction to the concepts of curl and divergence and cite Maxwell's Equations, from which we derive the numerical methods used for field calculations throughout this book.

This brief introduction is not exhaustive, and readers who are completely unfamiliar with Maxwell's Equations may find it insufficient to gain a full understanding. However, for interpreting the results of field computations, it is not necessary to follow all the mathematics and derivations from a field theory textbook. What is truly helpful is an understanding of the fundamental properties of electromagnetic fields, enabling effective interpretation of the field patterns obtained from field computations.

- What is an electric and a magnetic field.
- Sources of electromagnetic fields: charges and currents.
- How do electromagnetic fields behave at different frequencies.
- Divergence and Curl of a vector field.
- Which properties describe the interaction of electromagnetic fields with materials: conductivity, permittivity, permeability.
- Boundary conditions, interface conditions

The concept of divergence and curl is helpful in understanding the behavior of fields. However, it is often associated with complicated analytical calculations involving the nabla operator, while the graphical explanation is much more straightforward.

To understand the divergence of the electric field, it is simplest to think of a charged sphere. The electric field is perpendicular to the sphere and points outward. If we integrate the electric field over a volume that completely encompasses the charged sphere, we always obtain the same value, regardless of the size of the integration volume. This is because the charge is the source of the electric field, and divergence can be understood as quantifying the magnitude of the source.

For the curl of a field, we use the right-hand rule as illustrated by Ampère's Law (see Table 4.1). Assuming a static DC current, the time derivative of the electric field is zero, and the curl of the magnetic field is equal to the current density. Imagine a single wire and point your thumb in the direction of the current flow—then your fingers of your right hand indicate the direction of the magnetic field. The field resulting from the curl operator does not point from a source to a destination. Instead, the field lines of a curl field are always closed around the wire.

We illustrate both types of fields in Fig. 4.1. You can also recall your first lesson about electromagnetism in high school, where electric and magnetic fields were often visualized using iron powder. For the curl in Faraday's Law, imagine a time-varying magnetic field pointing in one direction and the corresponding electric field circulating around it, similar to the wire case.

Table 4.1 Macroscopic Maxwell's equations in differential and integral form

Equation	Differential form	Integral form
Gauss's law	$\nabla \cdot \mathbf{D} = \rho$	$\oiint_S \mathbf{D} \cdot d\mathbf{S} = \iiint_V \rho dV$
Gauss's Law for Magnetism	$\nabla \cdot \mathbf{B} = 0$	$\oiint_S \mathbf{B} \cdot d\mathbf{S} = 0$
Faraday's Law	$\nabla \times \mathbf{E} = -\frac{\partial \mathbf{B}}{\partial t}$	$\oint_l \mathbf{E} \cdot d\mathbf{l} = -\frac{d}{dt} \iint_S \mathbf{B} \cdot d\mathbf{S}$
Ampère's Law	$\nabla \times \mathbf{H} = \mathbf{J} + \frac{\partial \mathbf{D}}{\partial t}$	$\oint_l \mathbf{H} \cdot d\mathbf{l} = \iint_S \mathbf{J} \cdot d\mathbf{S} + \frac{d}{dt} \iint_S \mathbf{D} \cdot d\mathbf{S}$

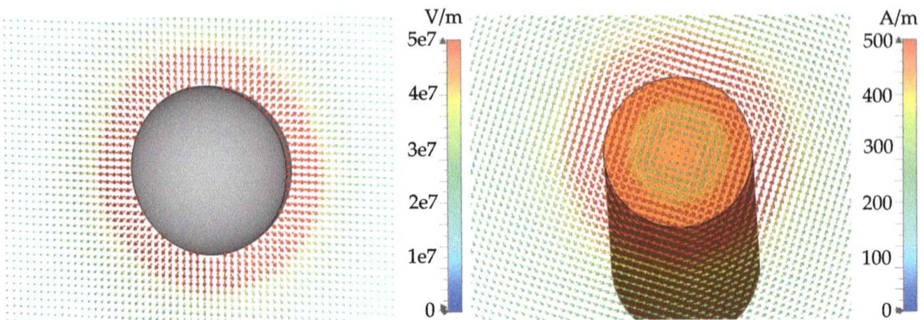

Fig. 4.1 Left: Electric field of a sphere illustrating charge as the source of the electric field. A charge of 1 μC was used, which corresponds to a 1 μF capacitance at 1 V. Right: Magnetic field around a wire carrying 1 A of current. The field arrows illustrate the curl of the magnetic field

4.1.1 Introduction to Numerical Solutions of Maxwell Equations

In order to calculate electromagnetic fields, it is necessary to solve Maxwell's Equations. Maxwell's Equations can be written in either differential or integral form, and it is a matter of personal preference which formulation is used to explain electromagnetic effects.

Throughout this book, we primarily use the differential form when explaining electromagnetic phenomena. Both forms are presented in Table 4.1. In the table, V denotes an integration volume, S represents the integration surface of the volume, and l refers to the boundary of the surface S.

Depending on the desired results of a field calculation, it may be sufficient to solve only a subset or approximation of Maxwell's Equations. For example, the computation of the static electric field, as shown in Sect. 5.4, requires solving only one equation: Gauss's law. In static and quasi-static problems (i.e., electrically short structures), the electric and magnetic fields are solved separately, as demonstrated in Chap. 5. Radiation problems, such as calculating the radiation pattern of an antenna, require solving the complete set of Maxwell's Equations. Methods that solve the coupled electric and magnetic field equations are referred to as full-wave methods. Full-wave methods are applied to EMC problems starting from Chap. 6.

The description using voltages and currents, as introduced in the two previous chapters, is inherently included in Maxwell's Equations. Kirchhoff's laws can be derived from Maxwell's Equations: Kirchhoff's voltage law is obtained from Faraday's law, and Kirchhoff's current law from Ampère's law together with Gauss's law. The derivation assumes that fields do not change significantly over the integration paths—a condition fully satisfied only for static fields. However, when the integration path is electrically small, the error introduced by this assumption remains acceptable, as shown in Sect. 2.4.2.5.

Thus, Kirchhoff's laws can be interpreted as a quasi-static formulation of Maxwell's Equations. The relationship between the electric and magnetic fields and voltage and current is further discussed in Sect. 4.7.

4.2 Geometrical Input

For the numerical calculation of electromagnetic fields in a given geometry, a computer-based geometrical description is required. All simulation tools mentioned at the beginning of this chapter use a CAD kernel to display and interact with the geometry to be simulated. A CAD kernel provides the functionality to describe a geometry based on geometrical entities such as vertices, edges, and faces that form geometrical shapes. It also defines how intersections of geometrical shapes are handled and visualized.

The 3D geometry modeling functionality in commercial tools allows for the interactive construction of a CAD model using primitives such as cuboids, cylinders, spheres, or tori. These tools also offer advanced features for modeling wires or wire bonds. This functionality is sufficient to create simple structures, including traces, vias, enclosures, or basic cable cross sections. Such simple models are particularly useful for what-if analyses or studying design concepts. Examples of these applications include the enclosure model in Chap. 7 and the filter model in Chap. 9.

In industrial product development, the design of a product is handled by dedicated teams that must address a wide range of requirements, including thermal and mechanical performance as well as aesthetic considerations. For EM simulations, we aim to import these designs into the simulation tool from popular CAD software such as Dassault Systèmes CATIA and Solidworks, PTC Creo, Siemens NX, or Autodesk Inventor, to name the most common ones. If the simulation tool does not support native imports, generic CAD formats can be used.

The STEP format is formally defined in the ISO-Standard 10303 and serves as a robust exchange format for mechanical CAD data. Real-world product data often include numerous details that are not required for accurate simulation, such as stamped logos on plastic enclosures or fully detailed screw threads. Geometrical details smaller than approximately $\frac{\lambda}{10}$ can typically be removed without losing result accuracy, provided they do not alter the conductive connections between objects. Simulation packages offer tools to modify and simplify imported data to reduce computational load.

If simplification is performed manually, it can be time-consuming, requiring the simulation engineer to find an appropriate trade-off between the time spent simplifying the model and the increased simulation time caused by unnecessary details. This balance strongly depends on the quality and complexity of the CAD data and the specific simulation being run. Unfortunately, we cannot provide simple guidelines for this process in this book.

One important CAD concept is the distinction between solid and shell objects (referred to as sheets in CST). A solid CAD shape is a volumetric shape extending in three dimensions, while a shell is a 2D body with zero thickness. This distinction is crucial when applying material properties to different object types, as certain material models can only be applied to one or the other CAD entity. More about this difference is explained in Sect. 4.4.

Up to this point, we have only discussed mechanical CAD data (MCAD). For the design of PCBs and packages, the so-called EDA tools are used. In the following, we use the term ECAD (Electronic CAD) data to refer to the output of these design tools. For the design of semiconductor components, another category of tools exists under the TCAD (Technology CAD) label. Throughout this book, we do not simulate at the silicon level; therefore, the TCAD landscape is not discussed.

There are several commercial software packages used for the design of PCBs, with four major vendors: Altium, Cadence, Siemens (formerly Mentor), and Zuken. The PCBs used throughout this book are designed using the free-of-charge version of Altium, called Circuit Maker. It is also worth mentioning Eagle (now owned by Autodesk) and the open-source tool Ki-Cad, as they are both popular and widely used.

The native import of ECAD data into simulation tools is currently very limited. However, as in the case of MCAD, there are exchange formats that typically rely on the fabrication output generated by an EDA tool. The simplest format is the so-called Gerber format, which dates back to the 1980s. Gerber files provide a 2D geometry description for each PCB layer, with each layer saved in a separate file. They are accompanied by drill files that specify the positions of the vias. While it is possible to import a set of Gerber files into a simulation tool, the process is highly manual and time-consuming, requiring the definition of each layer individually. Additionally, the Gerber format does not include any information about the components mounted on the PCB, necessitating manual re-definition of all components and their positions—an extensive effort for modern PCBs.

There are two widely used exchange formats that enable seamless ECAD data transfer into simulation tools: ODB++ and IPC-2581. The ODB++ format is proprietary and owned by Siemens, whereas IPC-2581 is defined by a consortium of EDA tool vendors. For the purpose of creating EMC simulation models, both formats provide equivalent functionality. They include data about all net names (which Gerber files do not), enabling users to navigate and select specific nets for import into the simulation model. This ability to reduce the number of imported nets helps in simplifying the simulation model.

Additionally, these formats contain information about the positions of assembly components, their reference designators, and part numbers. When properly defined, they can also include component values and details of the stack-up. This information significantly reduces the time and effort required to construct a simulation model. Using the above-described procedures, we can combine MCAD and ECAD inputs, make manual modifications as necessary, and generate the input required for field computation.

4.3 Spatial Discretization: Mesh

Domains: 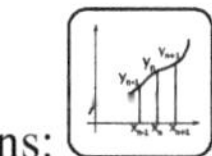

Following the previous section, we are now able to create a geometry in which we aim to calculate the electromagnetic field distribution. Unlike analytical methods, where the solution is provided in continuous space, a numerical calculation is performed for a limited number of discrete positions. The volume to be discretized is referred to as the computational domain, while the computational mesh is essentially a list of all locations at which the fields are calculated.

The process of generating the mesh is called meshing. A mesh is a critical component of the simulation, as its primary purpose is to resolve the spatial distribution of the electromagnetic field in such a way that the results obtained from the field computation achieve the desired accuracy. This concept is illustrated with an example later in this section.

First, we turn our attention to the interface conditions of electromagnetic fields, as they impose requirements on the proper choice of the mesh. The electromagnetic field at the interface of two materials is defined by the interface conditions. With $\hat{\mathbf{n}}$ being the unit normal vector of the interface surface, the interface conditions for the electric field **E** and the electric flux density **D** are given by:

$$\hat{\mathbf{n}} \times (\mathbf{E}_2 - \mathbf{E}_1) = 0 \ ; \ (\mathbf{D}_2 - \mathbf{D}_1) \cdot \hat{\mathbf{n}} = \rho \tag{4.1}$$

where ρ is the surface charge density of unbound charges, and **D** is sometimes referred to as the electric displacement field. The subscripts denote the sides of the interface.

In simpler terms, the tangential component of the electric field at the interface of two dielectrics is continuous, while the normal component of the electric flux density changes according to the surface charge. At the interface of a dielectric and a Perfect Electric Conductor (PEC: $\sigma \to \infty$), the tangential electric field is zero. In the dielectric, the electric field is perpendicular to the conductor and zero inside the conductor.

For the magnetic field **H** and the magnetic flux density **B**, the corresponding conditions are

$$\hat{\mathbf{n}} \times (\mathbf{H}_2 - \mathbf{H}_1) = \mathbf{J}_\mathrm{S} \ ; \ (\mathbf{B}_2 - \mathbf{B}_1) \cdot \hat{\mathbf{n}} = 0 \tag{4.2}$$

where $\mathbf{J}_\mathrm{S}$ is the surface current. At a material interface with a non-permeable material ($\mu_\mathrm{r} = 1$) and a Perfect Magnetic Conductor (PMC: $\mu_\mathrm{r} \to \infty$), the magnetic field **H** is perpendicular to the magnetic material and zero inside.

From the above explanation, we conclude that the interface conditions significantly affect the distribution of the electromagnetic field. This is why the mesh must

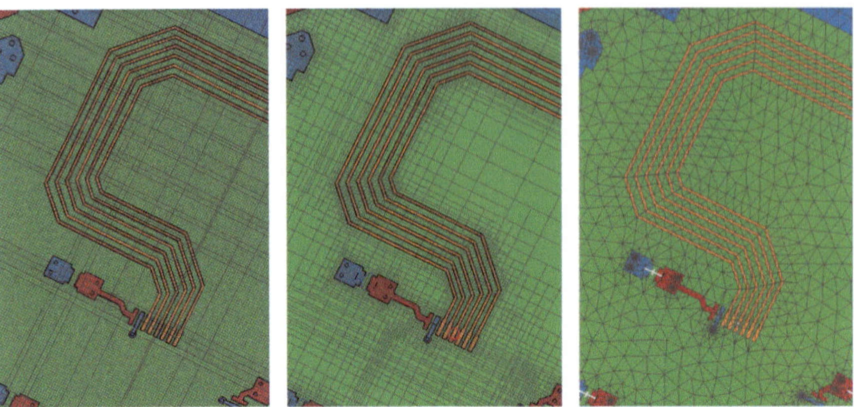

Fig. 4.2 Different mesh types for the discretization of singal traces on a PCB. Left: hexahedral mesh of the FIT. Center: Octree mesh of the TLM. Right: Tetrahedral mesh of the FEM

resolve the interfaces of different materials and why mesh generation in simulation tools places mesh lines at these interfaces.

In the simulation of electronic systems, the interface between dielectrics and metals is particularly important. For example, consider a signal trace in a PCB, routed above a reference layer. The mesh must resolve the fields between the trace and the reference layer. Additionally, it must resolve the gaps between neighboring signal traces; otherwise, there is a risk of short circuits between neighboring conductors due to overly coarse meshing.

There are numerous possible choices for structuring individual mesh elements: hexahedrons, tetrahedrons, prisms, wedges, pyramids, and quadrilaterals, to name a few. Electromagnetic field computation methods have been developed for most of these shapes, but many are not widely used. The most common volumetric mesh types in commercial simulation tools are hexahedral or hexahedral octree meshes for time-domain methods, like FIT and TLM, and tetrahedral meshes for FEM. These three mesh types are illustrated in Fig. 4.2, and their properties are discussed throughout this book. For the MoM, surface meshes based on tets or quads are used. The differences between volume- and surface-based meshes are explained in more detail in Sect. 4.11.

In addition to resolving material interfaces, the mesh must resolve the wave behavior of the electromagnetic field. During mesh generation, a wavelength criterion is applied to ensure that the sinusoidal distribution of the field in space is properly sampled. The mesh depends on the maximum frequency of interest, which must be specified before mesh generation. This wavelength criterion is based on the concept of electrical size introduced in Sect. 2.4.

For first-order methods (see Sect. 4.11), where the field along one mesh edge is constant, at least 10 steps per wavelength are required. For improved spatial resolution, 12–15 steps per wavelength are good default settings for time-domain solvers without mesh adaptation. Mesh adaptation will be discussed in Sect. 4.3.1.

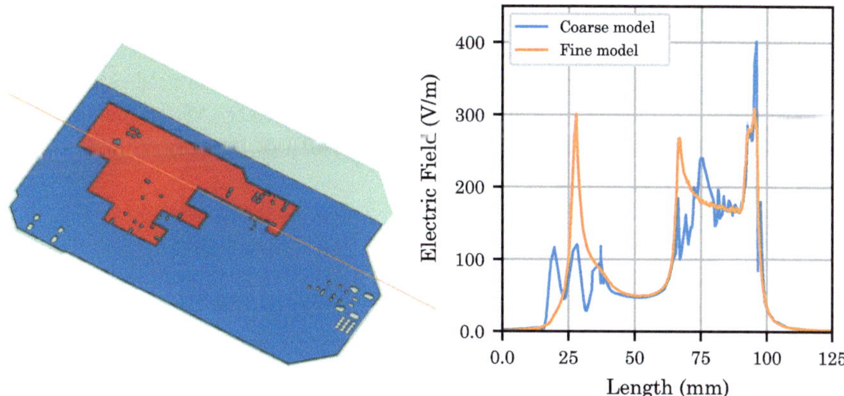

Fig. 4.3 Left: Model for demonstrating the effect of the mesh on the calculation of the capacitances. Right: Absolute value of the electric field along the orange line

Using a mesh with fewer than 10 steps per wavelength is not recommended for first-order methods. As the wavelength in materials is shorter than in vacuum, the geometrical distance between mesh lines must be chosen correspondingly shorter in materials.

In addition to the wavelength, the skin depth is an important metric that governs the spatial distribution of electromagnetic fields. In most electromagnetic simulations, the skin depth is not explicitly resolved by the mesh but is handled by material models. Skin depth is discussed further in Sect. 4.4.

Let us return to the requirements on the mesh and their dependence on the results of interest. To explain this in detail, we consider an example: We aim to calculate the capacitance between a power plane and a ground plane in a PCB, as illustrated on the left-hand side of Fig. 4.3. We use the electrostatic solver described in Sect. 5.4. At this point, we focus solely on the effect of the mesh.

In the model shown in Fig. 4.3, the red plane represents the power plane, the blue plane represents the ground plane, and the transparent green shape represents the substrate between the metal layers, set to $\varepsilon_\mathrm{r} = 4.2$. The orange line is located 1 mm above the PCB and is used to plot the magnitude of the calculated electric field. For this demonstration, we created a model with both a coarse and a fine mesh.

The coarse model uses the default starting mesh without any manual adjustments, resulting in 83k tetrahedrons. The fine model was configured to use a mesh resolution of 0.625 mm on all solids and in a volume around the evaluation curve, resulting in 833k tetrahedrons, 10 times as many as in the coarse model. The capacitance value calculated from the coarse model is 122.3 pF, while the fine model delivers 113.9 pF. This corresponds to a relative difference of 7.4%.

The absolute value of the electric field along the orange line is plotted on the right-hand side of Fig. 4.3. We observe that the local field values at the position of 30 mm differ by more than 100%. For accurate local field values, a fine mesh in the evaluation region is required. However, derived results, like the capacitance in this case, are often obtained from numerical integration of fields and can be calculated with satisfactory accuracy even with relatively coarse meshes.

The simulation time, including meshing and the electrostatic field calculation, is 9 s (240 MB RAM) for the coarse model and 53 s (1.18 GB RAM) for the fine model. This example illustrates that, depending on the results of interest, the requirements on the mesh can vary significantly.

We have now defined a number of properties that the mesh needs to satisfy. In the following, we discuss the practical implications of these properties. For engineers starting with field computation, the choice of a proper mesh might appear to be a chicken-and-egg problem: we need a mesh that resolves the fields accurately, but the field distribution is not known beforehand because the distribution is precisely what we want to calculate.

In the next section, we introduce the concept of mesh adaptation. Mesh adaptation is a very powerful method that iteratively determines a mesh satisfying certain conditions. However, mesh adaptation is not a magic tool that solves everything. Spending some time on defining the initial mesh before using mesh adaptation is highly recommended, as it greatly speeds up the entire simulation process.

Building a simulation model for a complete system, such as the inverter in Chap. 10, typically requires several iterations, even for simulation experts. In the early stages, we need to check the basic functionality of the model without requiring highly accurate results. At this level, a well-defined initial mesh is sufficient, and mesh adaptation is not necessary. Once we have a model delivering the expected functional results, we activate mesh adaptation to ensure that the quality of the EM simulation results is not negatively affected by an overly coarse mesh.

There are a number of rules we use to manually define a suitable initial mesh:

- Avoid conductor shorts or conductor breaks caused by a too coarse mesh.
- Ensure that the mesh is fine in regions where the field is strong, such as near sources and excited traces.
- Use a coarser mesh in regions far from the excitation, such as in large areas of a PCB away from excited traces.
- Apply a fine mesh at positions where field values are evaluated.
- Always respect the wavelength criterion.

Due to their nature, hexahedral meshes allow for more flexible manual definition, as they define local mesh steps separately for each coordinate axis. Simple finite-difference time-domain (FDTD) codes using hexahedral meshes struggle with the approximation of curved shapes due to the so-called staircase effect: a mesh cell in FDTD can only be fully filled with a single material, leading to a coarse approximation of curved objects.

Both solvers in the CST Studio Suite employ the so-called Perfect Boundary Approximation (PBA), a conformal meshing technique. PBA allows multiple materials in one mesh cell: two materials for TLM and three materials for FIT. PBA greatly improves discretization accuracy and enables the use of a coarser mesh compared to simple FDTD. It is most efficient for conductor-to-dielectric interfaces. Details of the PBA technique are discussed in Sect. 4.11.

Tetrahedral mesh generation typically only allows for the definition of a minimum mesh step in a volume, rather than in individual coordinate directions. A simple and popular way to control the mesh is the use of so-called meshing dummies. These are CAD bodies defined as vacuum, used to manually define the mesh within their volume. Since they are defined as vacuum, they do not alter the field distribution. While this may not be the most elegant solution, it is a highly practical and effective procedure.

4.3.1 Mesh Adaptation

Mesh adaptation is a method used to iteratively determine a mesh based on predefined criteria. In the first step, an initial solution is calculated. Then, the mesh is refined in regions selected by the software, for example, at positions where the absolute field values are high, as these have a significant impact on the results.

In the next step, the field is calculated on the refined mesh, and the results of both simulations are compared. If the difference between the results obtained from the two meshes exceeds the defined threshold criterion, the adaptation is repeated. This time, the positions for refinement are determined from the local differences in the field values obtained from the starting and refined meshes. The process continues until the difference between consecutive steps falls below the requested threshold or the maximum number of allowed adaptation passes is reached.

The user controls this process by defining the minimum and maximum number of allowed adaptation steps, the percentage of mesh positions to refine at each step, and the requested criterion and threshold.

The time required to run a mesh adaptation differs significantly between time-domain solvers and frequency-domain FEM solvers. In FEM, the adaptation is typically performed at single frequency samples. Instead of calculating the complete solution over all frequencies, the mesh is adapted for the field distribution at a single frequency sample. By default, the maximum requested frequency of the simulation model is used for this adaptation. This default choice assumes that the highest frequency requires the finest mesh, but it is not always optimal.

For instance, consider the simulation of a filter. If the adaptation frequency is placed in the stop band, the field inside the computational domain is very low, making the mesh adaptation inefficient. In such cases, it may be beneficial to manually select adaptation frequencies that align with the anticipated field distribution.

In time-domain solvers, the solution is calculated for the entire requested frequency band (more details are provided in Chap. 6 and Sect. 4.12). During each adaptation step, the full solution is computed, making automatic mesh adaptation for time-domain solvers more time-consuming compared to FEM.

Table 4.2 Comparison of meshing and mesh adaptation properties of hexahedral and tetrahedral meshing

Tetrahedral mesh generation	Hexahedral mesh generation
Can fail for complex and low-quality CAD input	Robust
Automated meshing process	Manual fine-tuning possible
Efficient adaptation at selected frequencies	Adaptation across the complete frequency band

Additionally, FEM mesh generation is less prone to discretization errors, such as unintended conductor shorts. These advantages might be some of the reasons why FEM is often considered easier to use and labeled as a "push-button solution." On the other hand, tetrahedral mesh generation is less robust than hexahedral meshing and can even fail to generate a mesh for complex or low-quality CAD input data.

For broadband problems and electrically large structures, time-domain methods outperform the finite element frequency-domain method, as discussed in Sect. 4.10. With the ability to control the hexahedral mesh along each axis separately, a few manual steps upfront to define a good initial mesh can save a significant amount of computational time.

We have compiled a simple overview of the differences between hexahedral and tetrahedral meshes in Table 4.2.

Simulation tools often come with a pre-defined set of conditions to use out of the box to define the goals of the mesh adaptation. Full-wave solvers, which deliver S-parameters as their main output, typically adapt the mesh based on S-parameter criteria. This is a good choice for most application scenarios, but as in the case of the filter in the stop band described above, the user should verify if the chosen criterion is meaningful for the problem at hand.

Static solvers, which will be described in Chap. 5, typically use the energy in the computational domain as the adaptation criterion. In cases where the user wishes to define a different criterion, most simulation packages offer the possibility to specify adaptation criteria based on results obtained from post-processing, such as the extracted inductance or the local field value at a particular position.

4.3.2 Mesh Quality

A selected mesh has a strong impact on the simulation time. It is quite obvious that a mesh consisting of a large number of mesh elements requires higher computational time than a mesh with a lower element count. Thus, the mesh should be as fine as required, but not finer. However, there is also further influence on the simulation speed that goes beyond the mere number of mesh elements. These factors are referred to as mesh quality. The metrics that determine the quality of the mesh differ between time-domain and frequency-domain solvers.

Let us focus on time-domain solvers first. A (full-wave) time-domain simulation is run at discrete time steps. However, the stepping method differs from the SPICE time stepping discussed in Sect. 3.8.1. The time stepping in time-domain solvers is not adaptive; instead, fixed time steps are used. These fixed steps are subject to the so-called stability criterion, which is discussed in detail in Sect. 4.12. In simple terms, the stability criterion implies that there exists a maximum stable time step that depends on the minimum cell size. The time-domain simulation cannot use a larger time step than this maximum, so the smallest mesh cell has a direct effect on the simulation time. Smaller mesh cells require reduced time steps, increasing the overall simulation time.

Hexahedral mesh generators aim to prevent the creation of excessively small mesh steps, but there are limits because material interfaces (mainly conductor interfaces) must be resolved even if they are close. Before running a time-domain simulation, it is worth checking if the smallest mesh cell is truly required to resolve conductors or if it can be increased. In the simulation of electronics, the smallest mesh step required to resolve conductors on a PCB typically corresponds to the thickness of a trace, which is around 50 µm, but might become even smaller at positions where traces are routed close to each other.

For tetrahedral meshes, the aspect ratio of the individual tetrahedra is an important metric that determines the convergence speed and the solution accuracy of the solver. We have plotted two tetrahedra in Fig. 4.4. The ideal case is when all edges of the tetrahedron are of equal length. In this scenario, the aspect ratio is 1. If a tetrahedron is flat, such as the tetrahedron on the right-hand side of Fig. 4.4, the aspect ratio is lower than 1.

For FEM solvers, aspect ratios down to 10^{-4} are acceptable. However, elements with ratios below this threshold might negatively affect the solution accuracy; we refer to these as low-quality elements. In a geometrically complex model consisting of PCBs, housing, and connectors, the effort to ensure the absence of low-quality elements might be substantial. In such cases, it is essential to check the location of these elements.

If low-quality elements are situated in regions where significant fields are expected, for example, near excited traces, the mesh must be manually improved

Fig. 4.4 Aspect ratio of tetrahedrons. Left: tetrahedron with a aspect ration close to 1. Right: Flat tetrahedron with poorer aspect ratio

at these positions, such as by using a finer local mesh. However, if the low-quality elements are in regions where the field is expected to be low, they can be accepted without impacting the simulation accuracy.

4.4 Material Modeling

Domains: 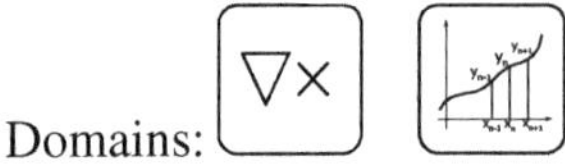

We have seen in the previous section that the interface conditions govern the behavior of electromagnetic fields at material interfaces. In electromagnetic field theory, the interaction of electromagnetic fields with materials is derived from microscopic models and can be extremely complex. For EMC simulations, we limit all interactions to macroscopic scales, as the smallest scales that we resolve are seldom smaller than 1 μm. Furthermore, we limit the frequency of interest to be well below optical frequencies and the material models to linear cases. Nonlinearities due to the saturation of ferrites are discussed separately in Sects. 4.4.3 and 9.7.

With these restrictions, we define a number of material types that accurately and efficiently describe the interaction of the fields with materials in an EMC simulation. Materials that are not commonly used in electronics are not discussed. In frequency domain, the relationship between the electric field and the electric flux density is

$$\mathbf{D}(\mathbf{r}, \omega) = \varepsilon_0 \varepsilon_\mathrm{r}(\mathbf{r}, \omega) \cdot \mathbf{E}(\mathbf{r}, \omega). \tag{4.3}$$

And similar for the magnetic field and the magnetic flux density,

$$\mathbf{B}(\mathbf{r}, \omega) = \mu_0 \mu_\mathrm{r}(\mathbf{r}, \omega) \cdot \mathbf{H}(\mathbf{r}, \omega). \tag{4.4}$$

The macroscopic material relations[1] $\varepsilon_\mathrm{r}(\mathbf{r}, \omega)$ and $\mu_\mathrm{r}(\mathbf{r}, \omega)$ are frequency- and spatially dependent. In this book, we also use the notations $\varepsilon = \varepsilon_0 \varepsilon_\mathrm{r}$ and $\mu = \mu_0 \mu_\mathrm{r}$. The spatial dependency is resolved by the mesh, the frequency dependence needs to be defined by the material model. Both material relationships are complex-valued numbers, in order to include losses.

They are defined as

$$\varepsilon_\mathrm{r}(\omega) = \varepsilon_\mathrm{r}'(\omega) - j\varepsilon_\mathrm{r}''(\omega) \tag{4.5}$$

[1] For simplicity, we have denoted the permittivity and the permeability as scalars. In general, they are tensors.

and

$$\mu_{\mathrm{r}}(\omega) = \mu_{\mathrm{r}}'(\omega) - j\mu_{\mathrm{r}}''(\omega). \tag{4.6}$$

The imaginary part of the complex permittivity is directly connected to the conductivity σ_{c} by the relation:

$$\sigma_{\mathrm{c}} = \varepsilon_{\mathrm{r}}'' \cdot \omega \cdot \varepsilon_0. \tag{4.7}$$

For a metal, where the conductivity is constant below optical frequencies (Jackson 2021), $\varepsilon_{\mathrm{r}}''$ is still frequency-dependent. The material relations are defined in every mesh element, ensuring that the interface conditions are satisfied. Depending on the actual material properties, they can be simplified, resulting in the different material models described in the following:

Perfect Electric Conductor (PEC) We start with the PEC material model. The PEC model is used for modeling metals. It assumes that the material has an infinite conductivity, which is equal to zero resistivity. Even though the resistivity is zero, the impedance is not, due to the inductance of the shape modeled by a PEC. Because the PEC material does not model the resistance of the conductor, it is only suitable at frequencies where the impedance of the structure is dominated by the inductance and losses are not relevant. Using PEC material in the simulation model can lead to sharp resonance peaks with an exaggerated quality factor at the structure's resonances, as the resonances are not damped.

We compare the different material models later in this section and demonstrate in an example the frequency bands at which PEC models are applicable. The main advantage of the PEC material is its simplicity. We use it, for example, to model the conductive table in the later chapters dealing with emissions. The tangential electric field on the surface of a PEC material is zero, and no field penetrates into the material. Both the electric and magnetic fields inside the material are also zero. Even a sheet shape of zero thickness completely decouples the fields on each side of the sheet. In terms of EMC, this is a perfect shield.

Perfect Magnetic Conductor (PMC) The PMC material model is the magnetic equivalent of the PEC material. The permeability is assumed to be infinite, and no fields can penetrate into the material. This type of material is not commonly used in EMC simulations. CST does not offer this material type, but there are simulation packages that do.

Surface Impedance Material The surface impedance material is an extension of the PEC model that efficiently allows for the inclusion of losses of a metallic body. As the name implies, the modeling of the loss mechanism is simplified by assuming that all losses occur on the surface of the metal. The fields still do not penetrate the material, and interior fields are set to zero. On the surface of the shape, a one-dimensional relationship between the tangential magnetic and tangential electric field is imposed:

$$\mathbf{E}_t(\mathbf{r}, \omega) = Z(\mathbf{r}, \omega) \cdot \mathbf{H}_t(\mathbf{r}, \omega) \times \hat{\mathbf{n}}(\mathbf{r}) \tag{4.8}$$

where $\hat{\mathbf{n}}(\mathbf{r})$ is the normal vector of the conductor surface. The term $\mathbf{H}_t(\mathbf{r}, \omega) \times \hat{\mathbf{n}}(\mathbf{r})$ is simply the surface current on the conductor. Thus, this equation is essentially Ohm's law. For standard metals, the impedance is obtained from analytical equations. These models are called "Lossy Metal" in CST.

More elaborate models can be provided for the definition of $Z(\mathbf{r}, \omega)$: One example is metals coated with thin dielectrics, as this material model allows the effects of the coating layer to be included without requiring the mesh to resolve it. More important for electronics is the ability to model the surface roughness of conductors with an appropriate choice of the surface impedance (Huray 2010). In high-speed design, the surface roughness of copper conductors becomes an important factor for Signal Integrity.

The Lossy Metal material model is the most commonly used material type for metals throughout this book. Even though it is not highly accurate toward DC, it is sufficiently accurate in the frequency ranges defined by emission standards, and it has the advantage that only the surface of the material needs to be meshed. In cases where an accurate DC resistance value is required, we must use a volumetric material model, which is described in the following section.

Volumetric Materials The two material models described so far are used for metals. For non-metallic structures, such as dielectrics or ferrites, the field distribution inside the material's volume must be calculated. In CST, this material type is referred to as "Normal." Since volumetric material models resolve fields within a volume, the reader should note that applying these models to a sheet body has no effect on the fields.

This material type is the obvious choice for PCB substrates, plastic enclosures, rubber insulators, and similar materials. Volumetric materials can be defined as anisotropic, meaning their properties differ along different axes. Unlike metals, the material properties of dielectrics and ferrites are often frequency-dependent within the frequency bands relevant to EMC standards. These dispersive properties are addressed in the next Sect. 4.4.2.

At this point, we return to the modeling of metals to explore the differences between material models and the treatment of skin depth and resistance, as these are often discussed topics among simulation engineers.

The skin depth δ_{sk} is a measure of the ability of a magnetic field to penetrate a conductor. For metals, it is given by the formula:

$$\delta_{\mathrm{sk}} = \sqrt{\frac{2}{\omega \sigma_{\mathrm{c}} \mu}}. \tag{4.9}$$

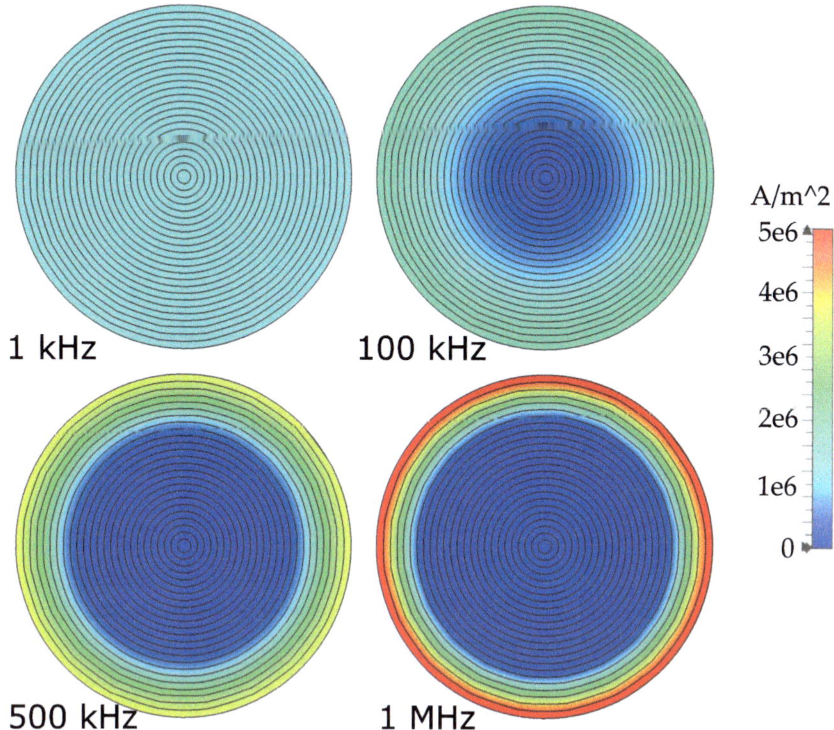

Fig. 4.5 Current density for different frequencies in a thin copper wire

At a distance of δ_{sk}, the magnetic field has decayed to $\frac{1}{e}$ (approximately 0.37) of its value at the surface. The penetration depth of the magnetic field corresponds to the effective thickness that carries the current in the conductor, as the magnetic field and current density are related through Ampère's law (see Table 4.1). Due to the reduced conductive region in the material, the AC resistance R_{AC} increases with the square root of the frequency.

For example, in copper, the skin depth is:

- 652 μm at 10 kHz,
- 65.2 μm at 1 MHz, and
- 6.52 μm at 100 MHz.

While it is possible to account for skin depth in simulations, care must be taken to ensure that the computational mesh resolves the field penetration accurately. As an example, we performed a simulation of a thin copper wire with a diameter of $D = 1\,\text{mm}$ and a length of $l = 1\,\text{cm}$, driven by a current of 1 A. The resulting current density in the wire's cross section is plotted in Fig. 4.5.

The circular lines in Fig. 4.5 indicate the positions of the mesh lines in the cross section of the wire. For this simulation, we used an extremely fine mesh with a

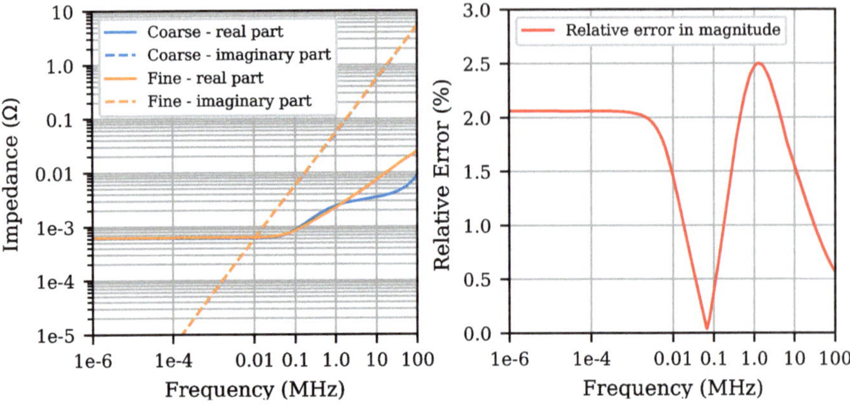

Fig. 4.6 Left: Impedance of a 1 cm long wire pair for two different mesh resolutions. The dashed lines are perfectly overlapping. Right: relative error in the magnitude of the impedance among coarse and fine model

radial resolution of 20 μm. For a wire length of 1 cm, this model already contains 220k tetrahedrons. Extending the wire length to 1 m with the same fine mesh would result in a mesh count that exceeds the capabilities of our available hardware.

The AC resistance R_{AC} is a critical parameter for power electronics designers, as minimizing losses is of paramount importance. Thus, it is essential to evaluate the impact of R_{AC} on the results of interest in an EMC simulation.

To illustrate this, we compare the impedance of a wire loop constructed from two wires, each 1 cm in length, using two different meshing approaches:

- Fine mesh: This model uses the same fine radial resolution of 20 μm as depicted in Fig. 4.5. The resulting mesh contains 395k tetrahedrons, and the simulation time is 41 minutes.
- Coarse mesh: This model resolves the wire diameter with just two mesh steps, resulting in a radial resolution of 500 μm. The total mesh count is significantly reduced to 11k tetrahedrons, with a simulation time of only 8 seconds.

The impedance of the wire pair for both models is plotted on the left-hand side of Fig. 4.6, with the real and imaginary parts shown separately. Above 2 MHz, we observe that the real part of the coarse model no longer follows the theoretically predicted dependency on the square root of the frequency. This discrepancy arises because the mesh in the coarse model does not resolve R_{AC} correctly. However, the imaginary part and the DC resistance of both models match perfectly.

At 2 MHz, where the real part of the impedance begins to deviate between the two models, the real part in the fine model is 3.1 mΩ, while the imaginary part is 109.6 mΩ. Thus, even for a wire loop length of only 2 mm, the imaginary part is nearly two orders of magnitude larger than the real part.

The relative error in the magnitude of the impedances between the two models is plotted on the right-hand side of Fig. 4.6. Over the complete frequency band, the error remains below 2.5%, while the simulation time of the fine model is 300 times longer.

From the above study, we conclude that:

- For an accurate calculation of R_{AC}, the mesh must resolve the skin depth.
- For the magnitude of the impedance in this example, the error is relatively small if the skin depth is not accurately resolved.

Resolving the skin depth in the mesh requires significant computational effort, which is not justified in this case. However, there are scenarios in EMC simulation where accurate resolution of the skin depth is critical. For example, in Chap. 7, we discuss the magnetic penetration of metallic layers in the kHz range. To accurately model this penetration, a more advanced material model is necessary, one that accounts for the skin depth without requiring spatial resolution of the penetration by the mesh.

Thin Panel The term thin panel is a material-type terminology used in CST, originally introduced in the TLM solver for RF applications. With its recent integration into the full-wave FEM solver, the thin-panel material has gained significance in simulations at kHz and low MHz frequencies.

To understand the thin-panel material, envision a sheet body where the fields on one side are coupled to the other side by a two-port admittance matrix. This allocation is illustrated in Fig. 4.7. A reciprocal admittance matrix can be represented as a π-network, and its individual entries are associated with the sheet body as shown in the figure.

In the illustration, the thickness of the material is finite to explain the concept; however, in simulation models, the corresponding CAD body must have zero thickness to apply the thin-panel material model. The coupling term $Y_{1,2}$ represents the penetration of the field through the material, making this model particularly suited for applications such as shielding, as we discuss in Chap. 7.

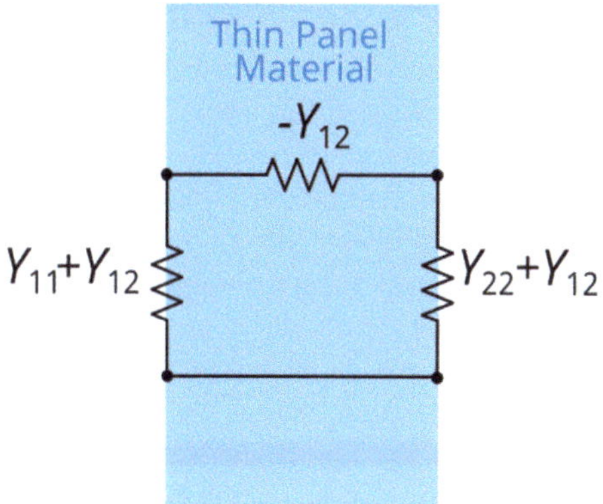

Fig. 4.7 Illustration of the allocation of a two-port admittance matrix to a sheet body modeled as a thin-panel material

Before running the actual EM simulation, the admittance matrix of the thin-panel material must be determined:

Simple Materials: For straightforward cases, such as thin metallic layers, analytical equations provide the admittance matrix.

Complex Materials: For materials consisting of multiple layers with dispersive properties, the matrix is obtained from a transmission line model.

Custom Definitions: The generality of this concept allows importing a two-port S-parameter file. By converting the S-parameters into an admittance representation, the relationship between the fields on either side of the material can be established.

The thin-panel material model efficiently handles skin depth without requiring its resolution by the mesh. While accurate results for fields at the material surfaces are achievable, internal field distributions cannot be calculated. For example, generating a plot like Fig. 4.5 is not feasible with this material model.

A crucial limitation of this approach lies in the name: thin panel. The model is valid only for layers that are electrically short, meaning their thickness η must satisfy the condition $\eta < \frac{\lambda}{10}$.

4.4.1 Material Model Comparison

To evaluate the differences between the four material models introduced earlier, two simulation setups are employed, and the calculated loop impedances are analyzed. Both setups are simulated using the full-wave FEM solver, with all models incorporating default mesh settings and mesh adaptation.

The first setup involves a 10 cm long wire pair with a cylindrical wire diameter of 1 mm and a separation of 3 mm. These wires are assigned the conductivity of copper, $\sigma_{\mathrm{Cu}} = 5.8 \cdot 10^7 \, \frac{\mathrm{S}}{\mathrm{m}}$. At one end of the wire pair, a discrete port is used for excitation, while the other end is shorted to estimate the loop impedance. When applying the thin-panel material model to this setup, the wires are represented as hollow cylinders where the surface represents the conductive region. The thin-panel model assumes a material thickness of 0.5 mm.

The second setup is the DC current path of the buck converter discussed in Chap. 3. The simulation model includes an excitation at the input connector. The switch and inductor are shorted, and the output voltage pin is connected to the reference conductor at the output connector. For the thin-panel model, all conductors of the PCB are modeled as planar sheets. This allows for a straightforward definition, given the planar geometry of the PCB.

Both simulation setups are illustrated in Fig. 4.8. The red cones indicate the positions of excitation and termination, while the blue cones in the PCB model mark the shorts at the switch and inductor.

Let us first analyze the results of the wire pair, as shown on the left hand side of Fig. 4.9. Below 10 kHz, the four curves deviate, whereas above 10 kHz, the

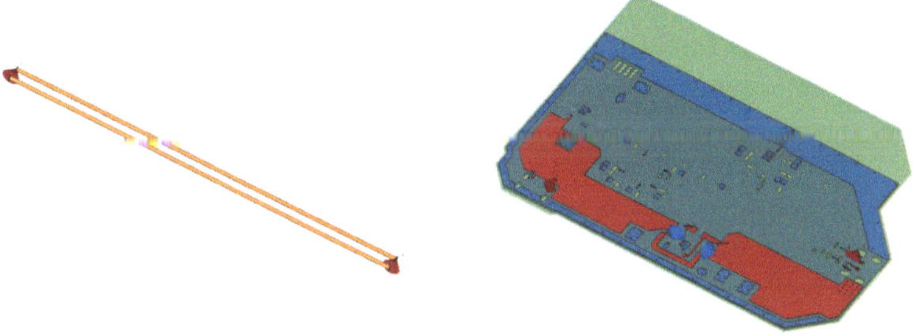

Fig. 4.8 Simulation models for the comparison of different material types

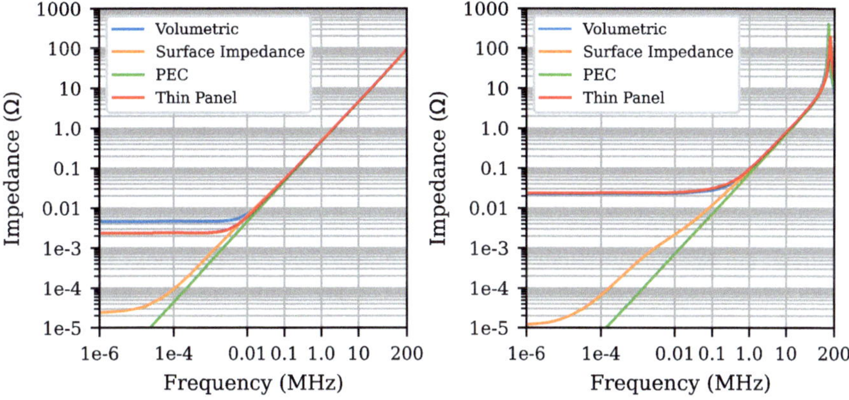

Fig. 4.9 Comparison of the four material models for the two examples. Left: Two wires. Right: PCB planes

Table 4.3 Impedance at 150 kHz calculated for the two examples with the different material models

	Volumetric	Lossy metal	PEC	Thin panel
Wires	74.5 mΩ	74.8 mΩ	67.5 mΩ	74.6 mΩ
PCB Planes	30.0 mΩ	16.9 mΩ	10.5 mΩ	33.8 mΩ

surface impedance and the volumetric conductor match. The impedance values obtained at 150 kHz are summarized in Table 4.3. This frequency represents the lower bound in conducted emission measurements. For the wires, the impedance calculated with the PEC model is slightly below the values obtained using the other three material models. Notably, the impedance deviation due to different material models occurs below 150 kHz, ensuring that all material models are suitable for conducted emission analysis according to CISPR 25.

For the current path in the buck converter, the frequency below which the curves start to deviate is higher, around 1 MHz (Fig. 4.9, right-hand side). The current loop in the buck converter exhibits a higher resistance and a lower inductance compared to the wire pair, as the copper traces are shorter and wider than the wires. The impedance values at 150 kHz, listed in Table 4.3, show a difference of up to a factor of 3 among the models. This discrepancy could potentially affect the conducted emission result. However, in the case study presented in Chap. 3, where a load resistor of 5 Ω is used, the impact is negligible, as the impedance difference is two orders of magnitude smaller than the load resistance. Similarly, when connecting the 10 cm wire pair to the PCB and recalculating the impedance, the frequency at which the material models begin to diverge shifts toward lower frequencies, rendering the differences in the PCB model insignificant.

The significance of material model differences depends on the specific problem. For low-frequency excitations in the kHz range, or when the load impedance is in the mΩ region, the volumetric conductor or thin-panel models should be employed. Conversely, for results above a few hundred kHz, the lossy metal model provides sufficient accuracy. In cases of uncertainty, it is advisable to compare the complete model using different material types.

The thin-panel material model offers notable accuracy at lower frequencies compared to the lossy metal model, despite both not discretizing the conductor volume. This advantage arises from the explicit definition of the metal layer's thickness in the thin-panel material, enabling precise resistance calculations. In contrast, the lossy metal model, as a surface impedance model, lacks information about the geometrical dimensions of the CAD body. While not yet widely adopted in the EMC simulation community, the thin-panel material model's accuracy and efficiency make it a good choice for modeling of the resistance at lower frequencies. However, the absence of the physical thickness of the metal layers can lead to an underestimation of the capacitance between them. Table 4.4 summarizes the simulation time and mesh count required for all material models.

Furthermore, using the thin panel material requires additional preparation steps for the CAD model. Volumetric shapes must be converted into sheets, and the appropriate thickness must be assigned to the material model. This process may take extra time during model setup. Therefore, the simulation engineer should evaluate in advance whether high accuracy in the low kHz range is necessary, or if it can be considered irrelevant based on the termination impedance of the functional circuit being simulated. Furthermore, the usage of thin layers might underestimate the capacitance between the conductors.

Table 4.4 Simulation time and mesh count for the two examples used to compare the different material models

	Volumetric		Lossy metal		PEC		Thin panel	
	Time	#Tets	Time	#Tets	Time	#Tets	Time	#Tets
Wires	1 m 13 s	50,809	0 m 44 s	32,117	0 m 36 s	32,136	0 m 40 s	44,642
PCB Planes	4 m 56 s	406,697	2 m 20 s	241,747	2 m 38 s	254,942	2 m 02 s	214,202

4.4.2 *Dispersive Materials*

As already hinted in the previous section, the normal material model allows the definition of dispersive materials in a full-wave simulation. A dispersive material is characterized by the frequency dependence of its material properties in the frequency band of interest. Materials have distinct dispersion properties, and a complete discussion of all possibilities is well beyond the scope of this book. We discuss two dispersion models that are relevant in the modeling of electronics: the electric $\tan(\delta)$ model for losses in dielectrics and a Debye first-order model for modeling the dispersive characteristics of ferrites.

We have introduced the complex permittivity in Eq. 4.5. From this equation, the electric $\tan(\delta)$ model (in the following, we simply use $\tan(\delta)$) is defined as:

$$\tan(\delta(\omega)) = \frac{\varepsilon_{\mathrm{r}}''(\omega)}{\varepsilon_{\mathrm{r}}'(\omega)} \tag{4.10}$$

where $\varepsilon_{\mathrm{r}}''(\omega)$ represents the imaginary part and $\varepsilon_{\mathrm{r}}'(\omega)$ the real part of the complex relative permittivity. The parameter $\tan(\delta)$ is commonly used to quantify losses, particularly the polarization losses in dielectric substrates within electronics. For instance, the value $\tan(\delta) = 0.02$ is widely recognized for the commonly used FR-4 material. As full-wave solvers inherently model material properties as complex-valued, it is straightforward to incorporate these losses into the material model: For a material with user-defined ε_{r} and $\tan(\delta)$, Eq. 4.10 can be applied to determine $\varepsilon_{\mathrm{r}}''$. While the frequency-domain FEM method can directly utilize these complex material properties, the same is not true for time-domain methods.

In time-domain simulations, material models must be both passive and causal, whereas a constant $\tan(\delta)$ down to 0 Hz does not meet these criteria. The real and imaginary components of a passive and causal material model are required to satisfy the Kramers–Kroning relations (Jackson 2021). To create a dispersive material suitable for time-domain simulations, it is necessary to compute an $\varepsilon_{\mathrm{r}}''(\omega)$ for a given $\varepsilon_{\mathrm{r}}'(\omega)$ and $\tan(\delta)$ that adheres to the Kramers–Kroning relations within the frequency band of the simulation model. This results in a frequency-dependent material model where $\tan(\delta)$ gradually approaches zero as the frequency approaches 0 Hz. Consequently, a constant $\tan(\delta)$ cannot be used down to 0 Hz in time-domain simulations. When comparing results from frequency-domain methods to those from time-domain methods, it is essential to ensure that the frequency-domain solver also employs a passive and causal material model.

An alternative to the $\tan(\delta)$ model is to define the real and imaginary parts of the material properties using a table of frequency-dependent values, often obtained from measurements. This approach allows for a more accurate representation of material behavior compared to a constant-over-frequency $\tan(\delta)$ value. However, such detailed data are often unavailable. When this type of table is provided, it can be directly applied in frequency-domain methods. For time-domain methods, simulation software typically fits the data to generate a passive and causal material model.

Significant deviations between the input data and the generated material model may indicate low-quality input, often stemming from measurement inaccuracies.

For standard materials commonly used in electronic systems, such as FR-4 substrates for PCBs and cable insulators, dispersive material properties generally have negligible impact on system behavior below approximately 500 MHz. Consequently, these effects are usually insignificant in power electronics simulations. However, for signal integrity simulations involving multi-GHz signals, dispersive effects become relevant and must be appropriately accounted for.

The above-described process of defining a frequency-dependent material also applies to magnetic dispersion, characterized by the complex-valued $\mu(\omega)$. Similar to $\varepsilon(\omega)$, ensuring the causality and passivity of the material is necessary for time-domain simulations. An alternative to providing a table of frequency-dependent values is using physical dispersion models, which offer explicit representations in the time domain. While numerous dispersion models describe the interaction of matter with electromagnetic fields, covering all of them is beyond the scope of this book.

We focus on the simplest dispersion model, the first-order Debye model, which describes a relaxation process. This model is particularly effective for characterizing the dispersive properties of ferrite materials, such as those used in simulations of inductors and common mode chokes. The choice of the Debye model is not based on a detailed physical description, as ferrite dispersion is governed by a gyrotropic dispersion relation (Tsutaoka 2003). Instead, the Debye model is chosen for its simplicity. In cases where material data for ferrites are unavailable, fitting material parameters to measured impedance curves is often necessary, as demonstrated in Sect. 9.6. The Debye model requires the estimation of only two parameters while providing reasonable accuracy. The first-order Debye model for permeability is defined by the equation:

$$\mu_{\mathrm{r}}(\omega) = \mu_{\infty} + \frac{(\mu_{\mathrm{s}} - \mu_{\infty})}{1 + j\omega\tau_{\mathrm{r}}} \tag{4.11}$$

It uses three parameters that are obtained from material data sheets. The parameter μ_{s} represents the static permeability, μ_{∞} corresponds to the relative permeability at infinite frequency (typically set to 1). For example, with $\mu_{\mathrm{s}} = 5000$, $\mu_{\infty} = 1$, and a relaxation time $\tau_{\mathrm{r}} = 1\,\mu\mathrm{s}$, we obtain the frequency-dependent curve for $\mu_{\mathrm{r}}(\omega)$ plotted in Fig. 4.10.

The relaxation time τ_{r} determines the relaxation frequency f_{relax}, at which $\mu_{\mathrm{r}}'(\omega)$ reaches exactly half the value of μ_{s}, and $\mu_{\mathrm{r}}''(\omega)$ achieves its maximum. The relationship $f_{\mathrm{relax}} = \frac{1}{2\pi\tau_{\mathrm{r}}}$ simplifies adjusting the relaxation frequency by modifying only one parameter, allowing the curves to be shifted to lower or higher frequencies to match data sheet values.

The material properties of common ferrites on the market, such as MnZn or NiZn, can be approximated for EMC simulations using the first-order Debye dispersion model.

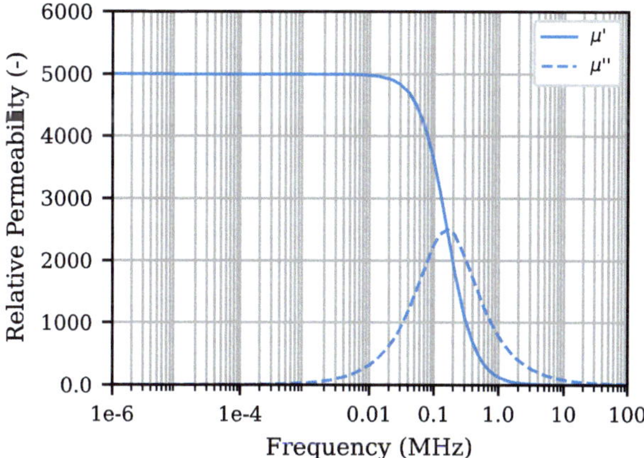

Fig. 4.10 Permeability of a Debye first-order dispersion model with the parameters given in the text

4.4.3 Nonlinear Materials

Similar to a nonlinear component discussed in Sect. 2.6, the properties of a nonlinear material depend on the field strength interacting with the material. A variety of nonlinear material models are available for time-domain solvers, but most of these models have little relevance in EMC simulation. Notably, the full-wave FEM solver in CST does not support nonlinear material types, which is also true for most commercial full-wave FEM implementations.

At this point, it is important to distinguish between nonlinear materials and nonlinear circuit elements in the simulation model. In the time-domain solver, it is possible to include a lumped element model—such as a diode—within the EM model and perform the field calculation. Coupled simulations that integrate nonlinear circuit components with results from a (linear) full-wave simulation are discussed in Chap. 10. Additionally, the combination of EM and circuit co-simulations is explored in Chap. 14. This section focuses on nonlinear materials.

In EMC applications, the saturation of ferrite materials plays a significant role. When a ferrite material becomes saturated, its permeability value decreases, leading to a corresponding reduction in inductance. For instance, consider a coil with a ferrite core used as a filter inductor. If the inductance of the coil decreases due to saturation, the filter response may change and fail to meet its specifications.

The nonlinearity of magnetic materials is typically described by a BH-curve. These curves are derived from measurements, under static conditions or at low kHz frequencies, and are used in the magnetostatic solver to evaluate saturation effects. Dynamic saturation caused by pulsed fields is not covered in this book.

We use the magnetostatic solver to calculate coupling inductances in Sect. 5.5. At this point, we only present a short introduction on how a nonlinear solution is calculated. We treat an example in the advanced Sect. 9.7.

In order to calculate saturation effects, an iterative scheme is applied. Before the iteration, a $\mu_r(H)$ curve is calculated from the BH-curve. In the first step of the iteration, the solver calculates the magnetic field for a non-saturated material. After the calculation, the magnetic field in every mesh cell is used to calculate a new value for $\mu_r(\mathbf{r})$ from the $\mu_r(H)$ curve. This results in a spatially dependent distribution of $\mu_r(\mathbf{r})$ that is used in the next iteration to obtain a new magnetic field, which is again used to calculate a new permeability distribution (Kaltenbacher 2015).

This process is repeated until a convergence criterion is met. The numerical method behind the nonlinear calculation is the Newton algorithm, which we also used for the nonlinear operating point calculation in Sect. 3.8.1. To apply this method, a BH-curve is required. This curve is either imported as a table from measurement data or defined using analytical equations.

4.5 Boundary Conditions

Domains: 

The time-domain methods and the finite element method are volume-based methods. They discretize the volume of the computational domain using a mesh. At the boundaries of the computational domain, a so-called boundary condition needs to be applied, which defines how the fields behave at the boundary.

We can interpret the boundary condition as a material at the outer extent of the computational domain. Depending on the choice of the material type, the field at the boundary exhibits different properties. Boundary conditions are typically set individually for the six planes encompassing the cuboidal computational domain. As in the material section, we begin with the simplest choice.

Electric Boundary The electric boundary is equivalent to a PEC material. The tangential electric field $\mathbf{E}_t$ at the boundary is set to zero, $\mathbf{E}_t = \hat{\mathbf{n}} \times \mathbf{E} = \mathbf{0}$. All fields generated inside the computational domain are reflected back into the domain.

If all boundaries are set to electric, the simulation model is effectively enclosed within a metallic box. By applying this boundary only to one side of the computational domain, while setting all other boundaries to open boundaries (explained later), we can model an infinite metallic plane. This is particularly useful for simulating the surface of a metallic measurement table without explicitly including it as a CAD shape. Instead, the boundary itself defines the conductive plane.

Conductive Wall A conductive wall is a surface impedance material that behaves like an electric boundary but additionally includes surface losses on the boundary. The losses are calculated using Eq. 4.8

Magnetic Boundary The magnetic boundary is a PMC. It forces the tangential magnetic field $\mathbf{H}_t$ to zero, $\mathbf{H}_t = \hat{\mathbf{n}} \times \mathbf{H} = \mathbf{0}$. This type of boundary is not relevant for EMC simulation.

Open Boundary An open boundary is the typical choice for many EMC simulations. The purpose of an open boundary is to model the field at the boundary such that the exterior of the computational domain behaves like open space. In the following, we discuss open boundaries specifically for full-wave solvers.

A straightforward approach to model open space is by applying a Surface Impedance Boundary Condition (SIBC) at the boundaries of the computational domain. Using the plane wave impedance of 377 Ω, the boundary impedance is matched to the field impedance of an electromagnetic wave in free space, introducing losses to the fields at the boundary. These losses prevent the fields from being reflected back into the computational domain, effectively simulating an open space. For the SIBC to perform efficiently, the impedance of the fields striking the boundary must be close to the plane wave impedance. As a result, there needs to be a sufficient distance between the structure and the boundary.

A more advanced type of open boundary is the Perfectly Matched Layer (PML). The PML boundary introduces additional layers of absorbing material around the computational domain, significantly reducing reflections at the boundary compared to the SIBC model. However, this also increases the computational effort, as the added layers are included in the simulation. The PML boundary allows the distance from the structure to the boundary to be smaller than what is required for the SIBC.

We discuss the impact of open boundaries on field results in Sect. 6.3. For simulation models where accurate field evaluations are necessary, we recommend using PML boundaries.

Open Boundary (Add Space) For the open boundary to efficiently absorb impinging fields, a certain distance from the field source to the boundary is required. This distance can be chosen automatically by the simulation tool. In CST, as a standard setting, $\frac{\lambda}{4}$ at the center frequency of the requested frequency band is used. At kHz frequencies, this added space is large and increases the computational effort significantly. In these cases, we have to manually define the extent of the computational domain, use the standard open boundary, and accept some reflections at the boundary.

4.6 Sources

Domains: 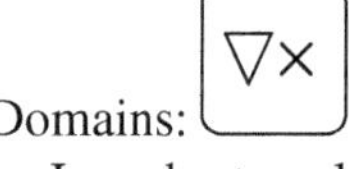 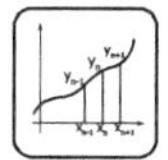

In order to calculate a field distribution, at least one source needs to be defined in the model. This source is part of the simulation setup and should not be confused with the EMC concept of a noise source introduced in Chap. 2. Depending on the solver type, different sources are used. In the following, we introduce the source types used throughout this book.

There is one important property of sources that we will discuss upfront, and it concerns sources with zero thickness. One of the simplest ways to define a source is to directly apply a voltage or current to an edge of the mesh. For the low-frequency solvers in CST, this is called a voltage or current path, while for the full-wave solvers, the equivalent is the discrete edge port. Unfortunately, the simplicity of a source defined on a mesh edge is physically problematic. Such a source can be interpreted as a wire of zero thickness. However, the analytically calculated inductance of a zero-thickness wire diverges, because the wire radius appears in the denominator (Eq. 3.9).

These sources can still be used, but the results depend heavily on the local mesh around the wire, leading to slow convergence during adaptive meshing. For this reason, we recommend avoiding simple wire sources and instead choosing more well-behaved sources, such as a face port or a coil segment.

Potentials In the electrostatic solver setup, we can define a potential on a PEC shape. This definition is used in Sect. 5.4.

Coil Segments We use this source type for the stationary current solver in Sect. 5.3 and for the magnetostatic solver in Sect. 5.5. A coil segment defines a current source by specifying an entry and an exit face on a CAD body designated as the source. These entry and exit faces are connected to other conductors to form a resulting current loop. Since the body of the coil segment has a finite thickness, the inductance is well behaved. We simply define the magnitude of the source current. The current loop must be closed for the magnetostatic solver but does not need to be closed for the stationary current calculation.

RLC Nodes RLC nodes are used in the pRLC solver in Sect. 5.6. RLC nodes are current sources and sinks that are used to define a current path through the structure. An RLC node is defined on the face of the CAD shape we would like to excite (different from the coil segment that requires the definition of a CAD body as the source). In the pRLC solver, the current loop does not need to be closed as it calculates partial results. The concept of partial inductances is also treated in Sect. 5.6

Plane Wave The plane wave is a source type available for the full-wave solvers. It is characterized by its incidence direction, field magnitude, and polarization. We will use a plane wave source when discussing shielding simulations in Chap. 7 and radiated immunity in Chap. 13.

Ports Ports are the most commonly used excitation type for full-wave solvers. They are categorized into two types: lumped or discrete ports and wave or waveguide ports.

We begin the discussion with discrete ports. A discrete port applies the voltage and current source definitions discussed in Sect. 2.3.1 to a 3D model, driving an excitation signal into the computational domain. While it is possible to use voltage and current ports, they have the drawback of not supporting power normalization, making S-parameter calculations impossible. As outlined in Sect. 2.4.3.2, S-parameters require a reference impedance at each port. Therefore, we exclusively use S-parameter ports. When voltage or current values need to be excited, we extract the S-parameter block from the EM simulation and apply it in a subsequent circuit simulation. This approach is more efficient and versatile than directly using current or voltage sources in the EM model.

The choice of reference impedance can impact simulation accuracy. A significant mismatch between the port impedance and the structure it drives results in most of the energy being reflected at the port, with only a small portion entering the computational domain. Sophisticated impedance estimation is unnecessary; by default, we use a reference impedance of 50 Ω for digital systems and 1 Ω for power electronics applications. The rationale behind the 1 Ω choice is elaborated in Chap. 10.

A discrete port can be defined as either an edge type or a face type. The use of edge-type ports is discouraged, as noted earlier in this section. Face-type discrete ports are the most frequently employed port type throughout this book.

Waveguide ports are a special form of ports that provide a source matched to the structure they drive. They are widely used in HF design because they minimize reflections at the port. A waveguide port is defined by an area within the computational domain, placed in front of a conductor arrangement that supports a (Quasi)TEM mode, such as coaxial connectors, cables, microstrips, and striplines. In this area, the port computes the corresponding 2D field distribution, representing the mode supported by the given conductor arrangement. From this 2D solution, the characteristic impedance of the mode is calculated, enabling the structure to be driven with minimal reflection. More details are provided in Sect. 8.2.1.

Waveguide ports allow for high-accuracy analysis of specific sections of a PCB. For instance, when simulating a planar RF filter, waveguide ports can be placed at both ends of the filter to optimize the filter section without simulating the entire PCB. Similarly, when analyzing a connector mounted to a PCB, signal traces can be cut shortly after the transition from the connector to the PCB, enabling a TDR analysis of the connector independently of the rest of the PCB. While waveguide ports are an excellent choice for component-level simulations, they are less commonly used in system-level simulations like conducted and radiated emission analysis.

It is essential to note that ports are physically present within the computational domain. A discrete port includes a connecting section that links the port position to the rest of the structure. In CST, the port's actual position is marked by a red cone, while the connecting sections are PEC elements: wires for edge ports and faces for face ports. The physical presence of these connecting sections introduces inductance and capacitance to the signal path and should be kept as short as possible. This physical connection ensures continuity of the fields, as required to satisfy Maxwell's equations.

4.7 From Fields to Voltages and Currents

Domains: 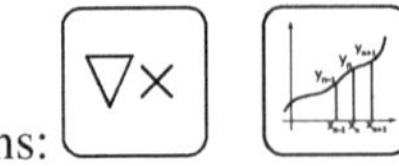

As we have seen in the previous sections, the EM simulation enables the calculation of the electric and magnetic fields within the computational domain. For radiated emission simulations, these fields are often the primary results of interest. However, in many scenarios, we aim to extract circuit models of the coupling path from the EM simulation. In this context, we focus on the results obtained from full-wave solvers, as they are the primary solver type utilized throughout this book.

Circuit models extracted from full-wave solvers are based on S-parameters. According to Eq. 2.5, obtaining the S-parameters requires the calculation of the voltage and current at a port. In the following, we explain the methodology for extracting voltages, currents, and S-parameters from the electric and magnetic fields computed in the simulation.

The voltage between two points $\mathbf{r}_\mathrm{a}$ and $\mathbf{r}_\mathrm{b}$ in space is defined as

$$V_\mathrm{ab} = \int_l \mathbf{E} \cdot d\mathbf{l}. \tag{4.12}$$

where l is a path between the points $\mathbf{r}_\mathrm{a}$, $\mathbf{r}_\mathrm{b}$. The scalar product between the integration path l and the electric field $\mathbf{E}$ ensures that only the electric field component tangential to the path is integrated. In a discretized model, this corresponds to summing the tangential electric field along mesh edges. This concept is used at discrete ports to calculate the voltage at the port.

When calculating voltage from the electric field, we must be aware that the voltage is not always independent of the integration path between two points in space:

In Maxwell's Equations, there are two contributors to the electric field: Gauss' Law and Faraday's Law (Table 4.1). The contribution from Gauss' Law is of divergence type, representing the electric field generated between charges. This is the voltage definition familiar from circuit simulators and is independent of the integration path.

The contribution from Faraday's Law is of curl type, depending on the time derivative of the magnetic field. The rotational field from Faraday's Law is not confined to conductors and exists even in vacuum. If the electric field along the integration path in Eq. 4.12 has significant contributions from the rotational part, the voltage depends on the integration path's choice.

This concept might seem unfamiliar to electronics designers, as rotational contributions are not represented in circuit models. To avoid dependence on the integration path, voltages must be measured only along electrically short paths. In such paths, $\frac{\partial \mathbf{B}}{\partial t}$ can be assumed to be small, and consequently, $\nabla \times \mathbf{E}$ is also small. As a rule, the length of a port should be shorter than $\frac{\lambda}{10}$.

It is also important to note that an electrically long conductor is not an equipotential. For example, estimating the voltage between two electrically long wires depends on the estimation position along their length. This is valid, as the voltage between the wires varies along their length. However, using a port to measure the voltage from one end of the wire to the other end is not correct.

The current I through a closed surface S with the surface contour l is calculated using:

$$\oint_l \mathbf{H} \cdot d\mathbf{l} = \oiint_S \left(\mathbf{J} + \frac{\partial}{\partial t}\mathbf{D} \right) \mathrm{d}\mathbf{S} = I. \tag{4.13}$$

The two contributions to the current are the conductive current and the displacement current. The displacement current contribution results in a non-zero current even when integrating over a loop in free space. Typically, the loop encloses some conductors and provides the resulting current on all enclosed conductors, analogous to using a current clamp in a measurement. For discrete ports, this is achieved by integrating the magnetic field along a loop defined around the port.

Using Eqs. 4.12 and 4.13, we calculate the voltage across and the current through a discrete port. Along with the port's reference impedance, the S-parameter can then be determined using Eq. 2.5. However, when multiple ports are present in the simulation model, they cannot be excited simultaneously. This is because the solver calculates the resulting field, which satisfies Maxwell's Equations as a whole and cannot be separated into individual contributions from each port.

To calculate the full S-matrix, we must excite the ports sequentially. In the first step, port 1 is excited while all other ports are set as passive terminations. From this calculation, we obtain all S-matrix entries $S_{x,1}$. Next, port 2 is excited, yielding $S_{x,2}$, and so forth, until all ports have been individually excited. For a system with n ports, this procedure requires n field calculations.

The sequential excitation of sources significantly affects simulation time, as discussed in Sect. 4.9. This procedure is not limited to S-matrix calculations. In scenarios where multiple sources are present and a matrix-type result, such as a capacitance, inductance, or resistance matrix, is required (as in Chap. 5), a stepping scheme of the sources is applied.

4.8 Avoiding Common Pitfalls in EM Simulations

Domains: 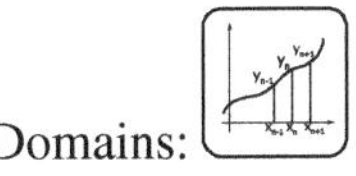 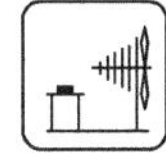

Building correct EM simulation models can be a challenging task, particularly for beginners. For complex geometries in modern electronic systems, the challenges increase significantly, requiring experienced simulation engineers to apply great care. Often, the first iteration of a simulation model will require modifications.

In this section, we discuss common pitfalls and best practices for building EM simulation models.

During the development of EMC simulation models, simulation engineers should assess the feasibility of results at every step. If the simulation results deviate significantly from expectations, it often indicates a fundamental modeling error. In such cases, the setup should be checked for common issues, including:

- disconnected ports,
- broken CAD geometry causing conductor breaks or shorts,
- overly coarse mesh resulting in conductor breaks or shorts,
- incorrect material models for the targeted frequency band, and
- inappropriate boundary conditions.

Once the general response of the model appears correct, mesh adaptation can be activated to evaluate the effect of the mesh resolution on the results. However, simply examining S-parameters may not suffice, as their values strongly depend on the reference impedance and are not always intuitive. For extracted circuit models, analyzing the impedance of current loops in the simulation model is an effective approach to evaluate the impact of mesh adaptation. As a rule of thumb, achieving an accuracy of 5% in the extracted impedance is often sufficient. The method for calculating impedance from an S-parameter matrix is detailed in Sect. 2.7, with a focus on simulated S-parameters in Sect. 6.7.4.

For models where results are within the expected range but still raise concerns, potential discretization errors should be considered, such as:

- coarse mesh or poor mesh quality,
- insufficient frequency-domain sampling,
- inadequate energy decay in time-domain simulations (see Sect. 4.10.1.1),
- insufficient distance to boundaries, and
- undersized waveguide ports.

To understand how specific settings affect the final results, the simplest approach is to perform a direct comparison between two models with differing settings. Run the simulation twice, altering only one parameter at a time, to clearly identify its impact. If doubts about the results persist, running the simulation with an alternative solver and comparing the outcomes is a viable option.

For transient circuit simulations using the extracted circuit model, special attention must be paid to the accuracy of the vector fitting process. This topic is addressed in Chap. 10.

4.9 Time- vs. Frequency-Domain Methods: The Basics

Domains: 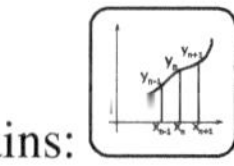

An important first step in building a simulation model is selecting the appropriate method (or solver) for electromagnetic field calculations. In the following, we limit the discussion to full-wave methods, as they are predominantly used throughout this book. Due to the differences in the way the time- and frequency-domain methods operate, the simulation times of a particular setup can be quite different. In this section, we provide an overview of the factors that determine the efficiency of each method for a given model. These descriptions use simple categories based on the simulation application. More detailed explanations, including mathematical background and examples, are provided in Sects. 4.10–4.11. We begin with the concept of the so-called system matrix.

After the meshing process, all field computation methods discussed in this book generate a system [2] $\mathcal{A}$. The system matrix is derived from the discretization of fields in the computational domain and relates the discretized field values in such a way that they satisfy Maxwell's equations. The structure of $\mathcal{A}$ depends on the problem type we aim to solve and the method used for field discretization. Section 4.11 provides examples of how various methods allocate field values within the mesh. In general, the entries of $\mathcal{A}$ encapsulate information about the magnitude of mesh entities, such as edge lengths and face areas, as well as material properties within each mesh cell.

At this stage, the significance of mesh quality in the FEM method, discussed in Sect. 4.3, becomes evident: the dimensions of mesh entities are directly included in the system matrix, influencing the magnitude of its entries. Since computer simulations process these entries with finite precision, rounding errors may occur, for example, when subtracting a very small number from a very large number. While simulation tools employ techniques to manage these challenges, simulation engineers may still encounter situations where rounding errors affect result quality.

In addition to the system matrix, which describes field behavior, we require an excitation vector to represent sources in the computational domain. Together, these components form a system of equations:

$$\mathcal{A}\mathbf{x} = \mathbf{b}_s \tag{4.14}$$

which must be solved to obtain the values of the field quantities in the computational domain. In Eq. 4.14, $\mathbf{b}_s$ represents the vector containing source information, $\mathcal{A}$ is the system matrix, and $\mathbf{x}$ is the vector of unknown field values we aim to calculate. In the following, we describe how time- and frequency-domain methods compute the solution to Eq. 4.14.

[2] Mathematically, $\mathcal{A}$ is an operator, but we use the term system matrix for simplicity.

4.9.1 Frequency Domain

In a frequency-domain method, the system matrix is assembled for each frequency at which we want to calculate the electromagnetic field. All matrix entries must be independent of the entries of the field vector **x**, making the system in Eq. 4.14 a linear system of equations. There are two main approaches to solving such a system: direct and iterative methods.

The direct method essentially computes the inverse of the system matrix, $\mathcal{A}^{-1}$. After calculating this inverse,[3] the solution vector **x** can be obtained as:

$$\mathcal{A}^{-1}\mathcal{A} \cdot \mathbf{x} = \mathbf{I} \cdot \mathbf{x} = \mathcal{A}^{-1}\mathbf{b}_s \tag{4.15}$$

where **I** is the identity matrix, having ones on the main diagonal and zeros elsewhere.

The direct solution offers a significant advantage: once the inverse of the system matrix is computed, the fields for any source vector $\mathbf{b}_s$ can be determined simply by multiplying the inverse with $\mathbf{b}_s$. Recall from the section on simulation results that calculating the full S-parameter matrix requires stepping through individual ports one at a time. The direct solution method proves highly efficient for models with a large number of ports, making it our preferred choice for frequency-domain simulations. However, this efficiency comes at the cost of increased memory consumption.

The iterative method is an alternative approach to solving the system of equations, particularly useful when the memory requirements of the direct method exceed the available hardware. This method employs an iterative scheme to calculate the field vector **x**. While it consumes less memory than the direct method, it requires solving the complete system for each port, which can increase simulation time significantly if many ports are present. An example of memory requirements for iterative and direct methods is provided in Sect. 4.10.

The solution obtained by the iterative solver is not exact. Since the exact solution is unknown, we cannot measure the true error directly. Instead, we measure the residual $\tilde{\mathbf{r}}$ as:

$$\tilde{\mathbf{r}} = \mathbf{b}_s - \mathcal{A}\mathbf{x} \tag{4.16}$$

The residual represents "what is left" when subtracting the obtained solution from the source vector. The norm of this vector is used to control the accuracy of the iterative scheme. The iterative process continues until the residual norm falls below a predefined threshold. For full-wave frequency domain solvers, a relative residual between 10^{-4} and 10^{-6} is typically used.

[3] Modern simulation tools typically use factorization methods instead of explicitly storing the inverse matrix. This simplified explanation is provided for better conceptual understanding.

Due to finite computational precision, the solution obtained by the direct solver is also not exact and has a non-zero residual. As a rule of thumb, if the direct solver fails to achieve a residual smaller than 10^{-5}, this might be an indication of an ill-conditioned model. In such cases, the mesh quality and material settings should be thoroughly checked. A common issue is, for example, the use of PEC materials at frequencies close to DC, where the impedance of PEC tends toward zero, potentially leading to stability problems in the solution of the system.

An ill-conditioned matrix can even affect simulation time. For instance, solving a matrix in double precision may result in faster computations compared to single precision, despite the higher memory requirements for double precision.

To obtain results across the frequency band of interest, adaptive interpolation schemes are employed. The process begins by calculating the first frequency sample, typically at the highest frequency within the specified band. A second sample is then calculated at the lowest frequency of the band, and the S-parameter values between these samples are interpolated. The interpolation error is assessed, and the next sample is selected at the frequency where the interpolation error is largest. This process continues iteratively until the interpolation error falls below a user-defined threshold, which is typically set to a default value of 2%.

For scenarios requiring finer resolution within the frequency band, additional frequency samples can be manually specified. Since each frequency sample necessitates a complete field computation, the simulation time increases proportionally with the frequency band's width. The broader the band, the more samples are required. For electrically long structures with many resonances, the efficiency of the frequency-domain method diminishes.

4.9.2 Time Domain

The system matrix of a time-domain method is assembled for the complete frequency band of interest. Unlike the frequency-domain method, where a separate system matrix is assembled for each frequency sample, the time-domain approach uses a single system matrix that describes the field coupling throughout the computational domain for the entire band. Consequently, all dispersive material properties must be defined in their time-domain representation, as detailed in Sect. 4.4.2.

Different from the frequency-domain method, the matrix entries in time-domain methods can be field-dependent or time-dependent, allowing for the inclusion of nonlinear materials or nonlinear component models when necessary. However, when the goal is to extract a circuit model based on S-parameters, it is essential to avoid introducing nonlinearities and time dependencies into the EM simulation model.

In the Finite Integration Technique (FIT) and the Transmission Line Matrix Method (TLM), the system Eq. 4.14 is solved using a so-called explicit scheme by stepping through time. To achieve this, the relationship of the field values between consecutive time steps—referred to as the update equations—must be defined. In the most common approach, often employed to ensure stability, the time step is halved: during one half-step, the electric field is updated, and during the other half-step, the magnetic field is updated. This method is known as the leapfrog integration scheme, and the corresponding equations are detailed in Sect. 4.12.

At $t = 0$, all field vector values $\mathbf{x}$ are initialized to zero. At each subsequent time step, the source vector $\mathbf{b}_s$ is updated according to the source's time-dependent value, and the fields are recalculated. This stepping process continues as the solver calculates the propagation of a finite pulse injected into the computational domain. The choice of the pulse is discussed later in this section. Before proceeding, we must establish a criterion to determine how long the field propagation needs to be tracked. A fixed time length can be defined, or the solver can terminate the calculation based on an energy criterion.

If the excitation is a finite energy pulse, the solver can monitor the energy in the computational domain by calculating it from the field values. When the energy decays below a specified threshold, time-stepping halts, as further changes in the field values become negligible. Several mechanisms contribute to energy decay: energy is absorbed by ports, radiated through open boundaries, or dissipated due to material losses. Default values for the energy criterion typically range from $-30\,\mathrm{dB}$ to $-40\,\mathrm{dB}$. For a threshold of $-30\,\mathrm{dB}$, the solver continues until only 0.001 of the injected energy remains in the computational domain.

Two key conclusions can be drawn about the performance of time-domain solvers:

- Accurately resolving low frequencies requires long time domain signals, leading to extended simulation times. A common rule of thumb is that resolving a response at 1 kHz requires a simulation duration of approximately 1 ms, which is significant for time-domain solvers applied to electronic systems.
- For resonant structures, energy decay is slow, resulting in longer simulation times.

The calculation of S-parameters from time-domain data involves performing a Fourier transform on the time-domain signals at the ports. When defining the frequency band of interest, the solver generates a time-domain signal that ensures all frequencies in the band are excited during the simulation. By default, FIT employs Gaussian pulses, while TLM uses impulse signals. To calculate S-parameters at a port using a Gaussian pulse, the Fourier transform of the calculated output signal is divided by the Fourier transform of the input signal.

For calculating the full S-matrix, the transient simulation must be run for each port. Consequently, the simulation time increases with the number of ports.

4.9.3 Model Properties Determining Solver Choice

We now outline the most important properties of an EM model that determine whether a time- or frequency-domain method will compute faster. The model properties discussed in this section are presented in simple terms and serve as rules of thumb for selecting the appropriate simulation method.

For certain applications, time- and frequency-domain methods yield comparable performance. This creates an opportunity to use both methods on the same model to validate simulation settings. If both methods produce matching results, it indicates that the mesh choice is correct. This approach is especially beneficial for beginners in simulation who may be uncertain about the appropriate mesh configuration. To choose the most suitable simulation method, consider the following factors:

- The bandwidth of the problem
- The number of ports
- The electrical size of the problem
- The quality factor of the structure

As the time-domain simulation calculates broadband results from a single solver run, it is faster than frequency-domain methods when a broad frequency band is of interest. To provide a numerical estimation, the geometrical properties of the structure being simulated must be included. Since the time step of the transient solver depends on the smallest mesh cell, as described in Sect. 4.3, the frequency band where the time-domain method outperforms the frequency-domain method can only be defined with knowledge of the smallest dimensions required to be resolved by the hexahedral mesh of the time-domain solver.

Assume that we simulate a PCB, resolving structures of approximately 50 μm, which corresponds to the thickness of a trace in the signal layer. Using this assumption, the frequency-domain solver is faster for models below 2–3 GHz. Between 3 GHz and 6 GHz, the performance is similar, and above 6 GHz, time-domain methods become faster. This estimation assumes a low port count and minimal resonance in the structure. When the port count is high (e.g., >10), these frequency thresholds shift in favor of the frequency-domain method, assuming that sufficient memory is available for a direct solver. Conversely, if the smallest mesh cell exceeds 50 μm, the thresholds shift in favor of the time-domain method.

The electrical size of the computational domain is another critical factor for solver choice. In some models, the bandwidth and the electrical size are not necessarily the same, such as an antenna mounted on a vehicle. The bandwidth is determined by the antenna, while the electrical size is governed by the vehicle. As a rule of thumb, for computational domains smaller than 2 λ, the frequency domain method is faster. Between 2 λ and 5 λ, the methods perform similarly, and for domains larger than 5 λ, the time-domain method is faster. These estimations also assume a low port count. Time-domain methods require less memory and can solve models up to 50 λ, and in extreme cases, 100 λ, on a single machine. Actual examples of simulation time and memory consumption are provided in Sect. 4.10.

Table 4.5 Preferred solver choice for different EMC-related applications

Application	Preferred solver choice
Emission: power electronics	Frequency Domain
Emission: fast digital electronics	Time and Frequency Domain
Susceptibility: BCI	Frequency Domain
Susceptibility: ESD	Time and Frequency Domain
Susceptibility: radiated	Time Domain
Antennas	Time and Frequency Domain
Filters, Inductors	Frequency Domain

When the simulated structure is resonant (e.g., with a high quality factor), the frequency-domain method is typically the better choice. However, precise numbers are not provided since an EMC engineer may not know the quality factor upfront. To simplify the selection process, common EMC-related applications and their preferred solver choices are summarized in Table 4.5. This table serves as an initial recommendation rather than an exhaustive guide, as performance comparisons may vary in certain cases.

Throughout this chapter, we have introduced the fundamental concepts of electromagnetic field computation. While these topics may seem somehow distant from the daily responsibilities of an EMC engineer, the authors believe that a solid understanding of these principles is highly beneficial for simulation engineers.

The next Sect. 4.10, provides two solver comparisons, offering actual data to support the rules of thumb discussed earlier. Following this, two advanced sections delve deeper into the theoretical background of field computation. These sections are intended for readers who wish to explore the subject in greater depth. However, they are not essential for building simulation models and can be skipped if desired.

4.10 Solver Comparisons

Domains: 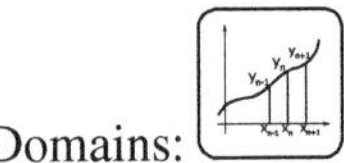

In this section, we analyze and compare the performance of different full-wave solvers using two examples. In the previous section, we provided only general rules of thumb for solver selection. The comparisons presented here offer concrete data in terms of simulation time and memory consumption.

The first study compares the FEM and FIT methods, highlighting how the electrical size of the computational domain influences solution time and memory usage. The second study examines the FIT and TLM solvers, focusing on the impact of local meshing on simulation time. As discussed earlier, simulation time depends heavily on the properties of the simulation model. When comparing solvers, it is crucial to ensure that any conclusions drawn are directly attributable to specific model properties.

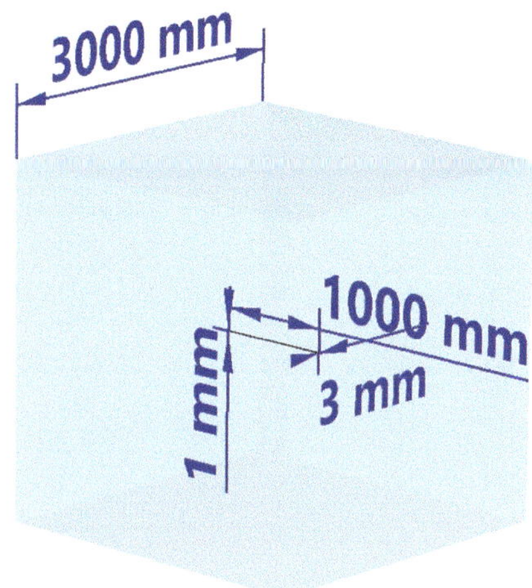

Fig. 4.11 Dimensions of the simulation model

There are publications and benchmarks comparing different simulation methods where either the problem is poorly defined or the simulation is conducted by non-experts, leading to incorrect conclusions about solver performance. Well-conducted studies are, e.g., Giacomini et al. (2022). However, keep in mind that the results of solver comparison studies are often specific to the analyzed problem.

4.10.1 Solver Performance in Terms of Electrical Size

In the first study, we compare the simulation times and memory requirements of the FIT and FEM solvers, analyzing their dependence on the electrical size of the computational domain. A very simple geometry was chosen: two straight wires, each with a length of 1 m, a rectangular cross section of 1×1 mm, and a separation distance of 3 mm. At both ends of the wire pair, 50 Ω ports are placed to excite and terminate the structure. The computational domain is fixed at a size of 3 m along each axis, and the boundary conditions at the domain edges are set to open. A sketch of the setup is provided in Fig. 4.11.

In the first step of the study design, we need to decide how the electrical size of the simulation model is varied. For a valid comparison, it is essential to ensure that all simulation models deliver results with the same accuracy. Therefore, we do not simply change the geometrical length of the wire pair to vary the electrical length, as this would make the comparison of results challenging. Instead, we use a fixed wire length and a fixed computational domain size, increasing the maximum frequency at which the model is simulated. With a computational domain length of 3 m, one side corresponds to 1 λ when the maximum frequency is 100 MHz. By adjusting the maximum frequency, we control the electrical size of the simulation model.

As discussed in Sect. 4.3, the wavelength criterion determines the number of mesh elements required to discretize the volume. For the FIT, we use 15 mesh cells

Table 4.6 Number of mesh elements for the FIT and FEM methods at varying maximum frequencies

Max frequency (MHz)	1	10	100	500	1000	1500	2000	3000	5000
Electrical Length (λ)	0.01	0.1	1.0	5.0	10	15	20	30	50
Electrical Volume (λ^3)	10^{-6}	10^{-3}	1	125	1000	3375	8000	27,000	125,000
Mesh Elements FEM	60 k	60 k	60 k	71 k	170 k	478 k	–	–	–
Mesh Elements FIT	58 k	58 k	58 k	716 k	4.2 M	13.3 M	29.9 M	96.9 M	436.6 M

per wavelength. For the FEM, four steps per wavelength are sufficient because the FEM method uses second-order elements. The concept of mesh element order is explained in Sect. 4.11. For now, it suffices to understand that higher-order elements allow the field along one mesh edge to vary, reducing the number of edges required to resolve the wavelength. The FIT uses an energy criterion of −50 dB, while the FEM iterative solution employs a residual of 10^{-4}.

The highest field values are located between the two wires, requiring finer mesh resolution in this region. To maintain consistency across models, we define a fixed mesh between the wires with a local mesh step of 1 mm, rather than using mesh adaptation. Mesh adaptation introduces additional steps in the mesh generation process, which may result in varying local meshes during the study. The manually selected mesh between the wires ensures a smallest mesh step of 1 mm for the FIT. Recall from Sect. 4.3.2 that this is a critical metric, as it determines the time step used by the time-domain method.

In our study, the maximum frequency is varied from 1 MHz up to 5000 MHz, with steps outlined in Table 4.6. At the lowest frequency of 1 MHz, the electrical size of one edge of the computational domain is 0.01 λ, while at the highest frequency of 5000 MHz, the electrical size amounts to 50 λ.

The required number of mesh elements is also provided in Table 4.6. At 100 MHz and below, the mesh count remains constant for both solvers due to the fixed local mesh defined between the wires, which does not vary with the electrical size. Beyond 100 MHz, the number of mesh elements needed to discretize the computational domain increases cubically. Since both the FIT and FEM methods are volume-based, the entire volume must be discretized. As a result, the number of mesh elements is governed by the electrical volume of the computational domain rather than its electrical length. This is why Table 4.6 also includes the electrical volume of the computational domain.

However, because the electrical length is a more intuitive parameter, the plots comparing computational times and memory requirements presented in the following sections use length and frequency as the primary metrics rather than volume.

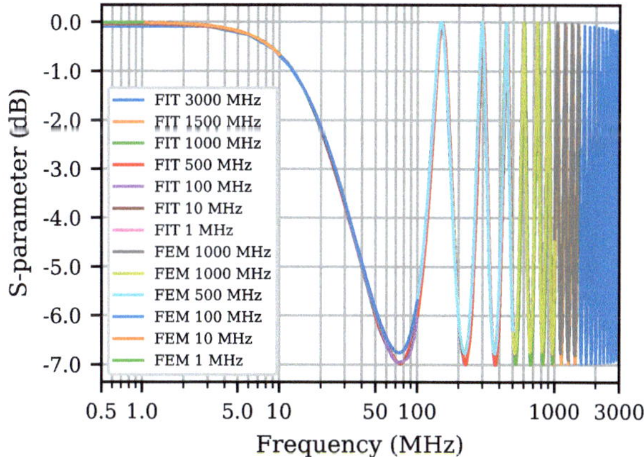

Fig. 4.12 Transmission S-parameter for all solver runs

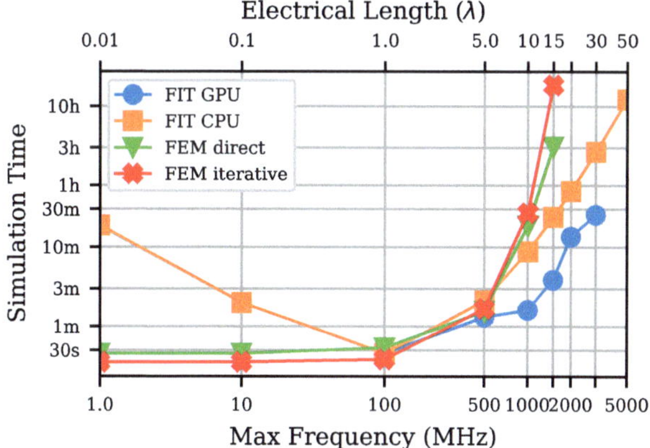

Fig. 4.13 Simulation times for the FEM direct, FEM iterative, FIT CPU and FIT GPU simulations

In order to check the accuracy of the results, we compare the transmission S-parameter S_{21} in Fig. 4.12. The difference of the S-parameters of all solver runs is less than 0.3 dB over the complete frequency band.

The simulation times for this study are plotted in Fig. 4.13. For the FEM, both the direct and iterative solvers are used. The FIT simulations are performed on a CPU and on a GPU. For the comparison of simulation times, the total time required to mesh the model and simulate both ports is considered. For this model, the meshing time is relatively small due to the simple geometry. This differs significantly for complex geometries involving numerous CAD shapes.

At 1 MHz and 10 MHz, the FIT method is slower than the FEM method. At these lower frequencies, the excitation signal used in the FIT method is long, leading

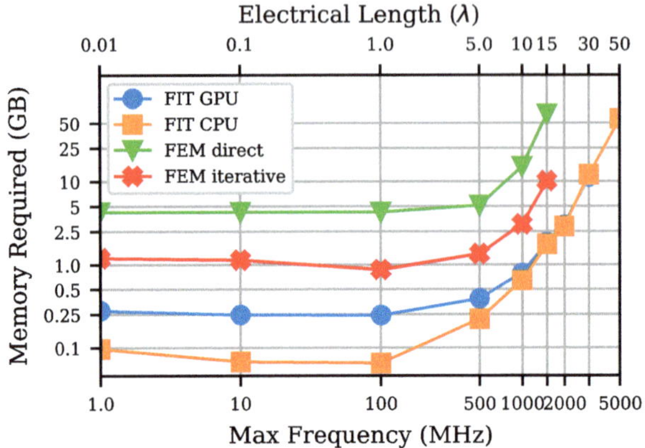

Fig. 4.14 Memory consumption for the FEM direct, FEM iterative, FIT CPU and FIT GPU simulations

to extended simulation times. This phenomenon is explored in greater detail in Sect. 4.10.1.1.

At 100 MHz, corresponding to an electrical length of 1 λ, both FIT and FEM achieve comparable simulation times, approximately 1 minute. At 5 λ (500 MHz), their performance remains similar, with simulation times ranging from 1 to 3 minutes. However, starting at 10 λ (1000 MHz), the FIT method becomes noticeably faster, and the benefits of GPU acceleration become apparent. Below 1000 MHz, the mesh count of the FIT simulation is too small to leverage GPU acceleration effectively.

At 15 λ, corresponding to a maximum frequency of 1500 MHz, the FEM direct solver requires approximately 3 hours to complete the simulation, while the GPU-accelerated FIT simulation is completed in under 4 minutes. The GPU results exhibit minor inconsistencies in the curve, which can be attributed to variations in the meshing time included in the study.

The memory consumption of the different models is plotted in Fig. 4.14. From the vertical offset of the four curves, we observe that the FEM direct solver requires the largest memory, while the FIT on CPU requires the least. At smaller mesh counts, the GPU implementation introduces some overhead, which becomes negligible for models above 1000 MHz, where the memory consumption of FIT on CPU and GPU becomes comparable.

The largest model that could be run on the available GPU was at 3000 MHz, requiring 12 GB of RAM. The memory consumption of the FIT solver depends on the number of lossy metal cells and field monitors defined. In this simulation, no monitors were used, and only PEC material was included.

The largest model simulated in this study was at 50 λ, using the CPU implementation of the FIT. This model required 436 million mesh cells and 59 GB of RAM. The FEM solver was executed up to an electrical size of 15 λ. At this size,

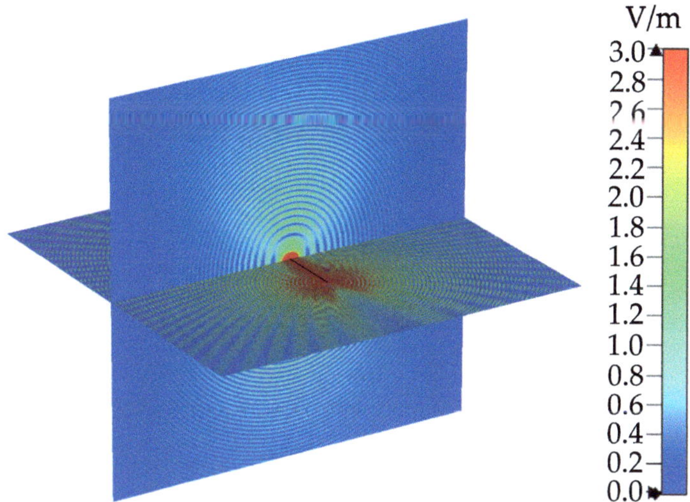

Fig. 4.15 Electric field of the wire pair at 5000 MHz

the direct FEM solver required 67 GB of RAM. It is worth noting that the memory requirements of the FEM solver depend on the element order used in the simulation (see Sect. 4.11).

The memory requirement of the FEM iterative solver lies between the direct solver and the FIT solver. At 15 λ, the iterative solver required 10 GB of RAM, enabling the simulation of larger models. However, the study was capped at this size due to computational time (see Fig. 4.13). While the direct solver required 3 hours for 15 λ, the iterative solver needed 18 hours.

As a reminder, all results were obtained using the hardware setup described in Sect. 1.5. Before continuing, we plot the electric field of the wire pair at 5000 MHz in Fig. 4.15. This plot provides an idea of the dimensions of 50 λ, the largest electrical size analyzed in this study. As the plot shows the absolute value of the electric field, we observe 100 maxima over one dimension of the computational domain.

4.10.1.1 Energy Criterion in Time-Domain Simulations

Let us return to the discussion of simulation times for the FIT at low frequencies. As discussed in Sect. 4.9, time-domain methods inject a pulse into the structure and track the energy within the computational domain. The Gaussian pulse used by the FIT is configured based on the frequency bounds defined by the model. Figure 4.16 shows the field energy plotted for different maximum frequencies used in Study 1. The x-axis of Fig. 4.16 is logarithmic. All energy curves exhibit a bell shape.

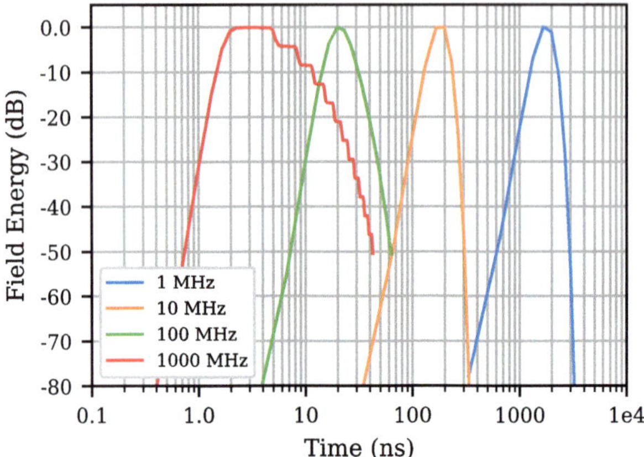

Fig. 4.16 Energy in the computational domain of the FIT simulation for different maximum frequencies

During the injection of the excitation signal into the computational domain, the energy rises. Once the excitation signal is completed, the energy begins to decay. For the lowest maximum frequency of 1 MHz, the complete signal duration is approximately 3000 ns. This duration ensures accurate resolution of the frequency band around 1 MHz.

As the maximum frequency increases, the duration of the time signals shortens, resulting in reduced simulation times. The shape of the energy curve provides insights into the structure's behavior: At 1 MHz and 10 MHz, the pulse essentially travels once through the structure and is fully absorbed by the second port at the end of the wire. By the end of the simulation, the energy has decayed below −80 dB. At 100 MHz, some reflections occur at the ports due to a slight mismatch between the termination and the characteristic impedance of the wire pair. However, these reflections have almost entirely decayed by the time the excitation signal is complete. This conclusion is supported by the smooth energy curve, which has decayed to −50 dB.

At 1000 MHz, the energy decays in discrete steps. These steps result from partial reflection and partial absorption of the signal at the port when it reaches the end of the wire pair. The energy curve reveals these steps, with each step width corresponding to the length of the wire pair.

4.10.2 Hexahedral and Octree Mesh Comparison

The second study compares the FIT and TLM methods applied to the simulation of a USB differential pair on a PCB. As both methods are time-domain solvers, they operate in a similar manner: tracking energy decay and determining the time step based on the smallest mesh step. The primary difference between these two methods lies in the way their meshes are defined.

The FIT method employs a simple hexahedral mesh, which is entirely regular. With PBA, it allows for three materials in a single mesh cell. This regularity results in mesh lines extending across the entire computational domain. Consequently, a fine mesh step defined in one region extends over the entire computational domain.

In contrast, the TLM method uses an octree mesh. The octree mesh enables local splitting of mesh cells, allowing for regions with fine discretization that do not extend across the entire computational domain. During octree mesh generation, the structure is initially meshed with a fine discretization. In a subsequent step, regions without material interfaces are coarsened, a process referred to as cell lumping. The octree mesh also supports PBA, but permits only two materials per mesh cell.

Due to these differences, the number of mesh cells in FIT and the lumped mesh cells in TLM can vary significantly, especially in models with extensive regions where a coarse mesh suffices. Because the mesh is completely regular in the FIT but mesh cells can split in the TLM, the simulation time scales different with the number of mesh cells.

We have plotted the model to be simulated in Fig. 4.17. The selected model is a differential pair designed to carry a USB signal. The length of the differential pair is 8 cm, with each trace having a width of 200 μm and a separation of 84 μm. As in the previous study, we have manually defined a mesh that accurately resolves the traces, ensuring consistent results from both solvers.

To evaluate performance, we systematically increase the size of the computational domain around the PCB, analyzing the ratio of the local mesh to the surrounding space. The octree mesh used by the TLM solver refines the traces locally, whereas the mesh of the FIT solver extends uniformly across the entire computational domain. This distinction is illustrated in the bottom portion of Fig. 4.17. For this study, the maximum frequency was set to 500 MHz. By incrementally adding space in all three spatial dimensions around the PCB, we obtained the mesh cell counts summarized in Table 4.7.

For surrounding spaces up to 500 mm, the mesh count for the TLM solver remains nearly constant, ranging between 1.3 and 1.5 million mesh cells. In contrast, the mesh count for the FIT solver doubles, increasing from 24.2 million to 56.2 million. At 1000 mm and beyond, the wavelength criterion begins to affect the mesh, resulting in an increase in mesh count also for the TLM solver. Despite this, the experimental setup is well designed, with realistic settings for the maximum frequency that align with typical application requirements. When examining the mesh counts at the smallest surrounding space of 50 mm, we observe that the FIT solver requires nearly 20 times more mesh cells compared to the TLM solver.

The simulation times of both solvers are plotted in Fig. 4.18. We simulate only one port of the set up, with an energy criterion of −50 dB. The meshing time is included. A GPU is used in both simulations.

For a surrounding space of 50 mm and 100 mm, the FIT solver is faster than TLM, even though the mesh cell count in the FIT is higher (Table 4.7). Due to the simplicity of the FIT mesh, the update of the fields at every time step is extremely

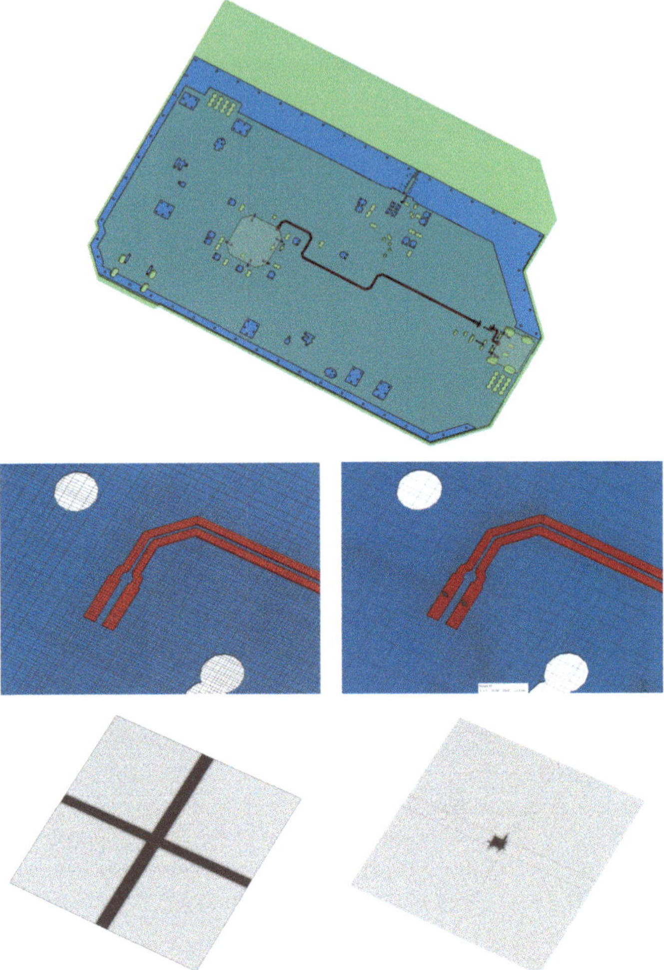

Fig. 4.17 Simulation model for the second study. Top: overview of the PCB and the routing of the differential trace pair. Bottom left: mesh for the FIT simulation. Bottom right: mesh of the TLM model

Table 4.7 Mesh cell count for FIT and TLM in study 2

Surrounding space (mm)	50	100	500	1000	1500	2000
Mesh Cells FIT	24.2 M	27.0 M	56.2 M	101.3 M	–	–
Lumped Cells TLM	1.3 M	1.3 M	1.5 M	2.8 M	5.8 M	11.4 M

fast compared to the TLM solver. The splitting of the cells in the octree mesh introduces the requirement of keeping an array of mesh connectivity information, which is not required for the simple, regular mesh of the FIT.

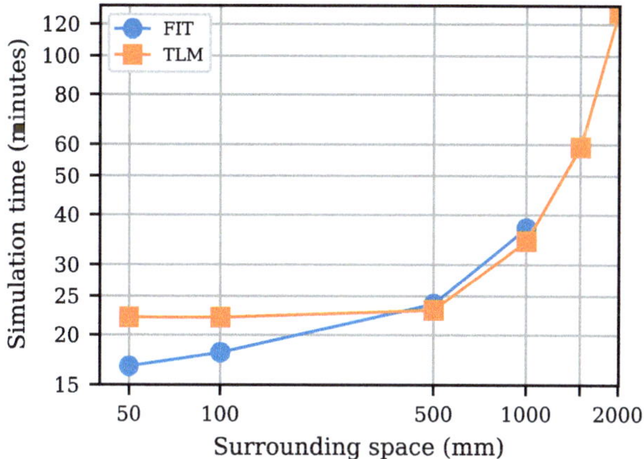

Fig. 4.18 Simulation time of the FIT and TLM solver

For a surrounding space of 500 mm, the TLM solver becomes faster, but the performance of the FIT method is not much different, even at a surrounding space of 1000 mm.

Although the speed is similar, the octree mesh of the TLM has the additional advantage of lower memory consumption. At 1000 mm surrounding space, the FIT requires 14 GB of RAM, close to the limit of the available GPU. Thus, the simulation of models with a larger surrounding space is no longer possible using the FIT on the available GPU, while they remain possible using the TLM.

Summarizing the second study, the TLM solver is more time- and memory-efficient in cases where a fine local mesh and a large surrounding space are required. Such cases are common when fine structures need to be resolved, which are not aligned with the principal axis of the mesh. In this example, the PCB is aligned in the z-plane of the mesh; thus, the fine discretization is only required along the x and y axes. If PCBs are tilted or bent, as is often the case for flex PCBs, the octree mesh of the TLM is the better choice when a time-domain method is used.

4.11 Advanced: Spatial Discretization

Domains: 

The fields described by Maxwell's Equations are a function of space and time. In this section, we discuss how the three field computation methods treated in this book discretize space. The FIT and the TLM are finite difference methods, while the FEM is a finite element method. Both method types discretize the complete volume of the computational domain and define the coupling of the fields locally. The field

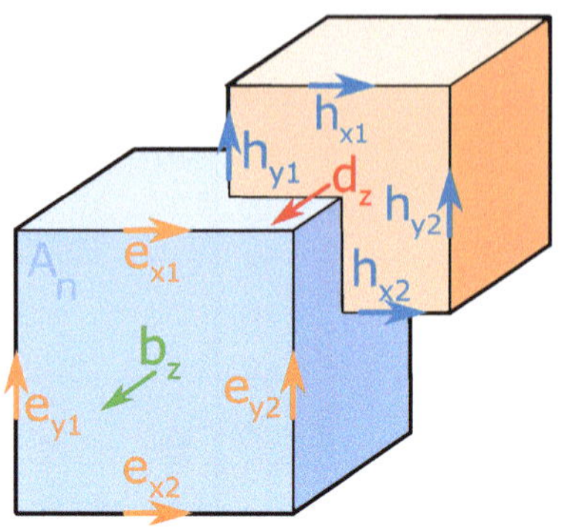

Fig. 4.19 Allocation of integral field quantities to the mesh used by the FIT

in a mesh element only couples to fields in neighboring elements. This is different from integral methods, such as the MoM, where all mesh elements couple to each other.

The local coupling results in system matrices that are sparse, meaning that even though the dimension of the system matrix is large, only a few entries in every row and column of the matrix are populated. The scaling of the memory consumption and solving time of sparse matrices with increasing element numbers is superior compared to methods requiring a full matrix.

Each method has a unique way to allocate field quantities in the mesh. We discuss the differences in this section. Most of these details are not needed to set up a simulation model; however, they help the interested reader form an understanding of how the field computation methods work. We start with the FIT, as the field allocation scheme of the FIT is probably the easiest to understand.

4.11.1 Field Allocation in the FIT

In the FIT, the discretization is performed on Maxwell's Equations in their integral form. The calculation is not carried out directly on the fields but rather on quantities integrated over the edges or faces of each mesh cell. These are referred to as grid voltages or grid fluxes. A staggered, dual orthogonal grid is employed, similar to FDTD methods. In Fig. 4.19, the primary grid is indicated in blue, while the secondary grid is shown in orange. It is worth noting that the mesh visualization in CST displays only the primary grid; the secondary grid remains hidden.

This directly allocated scheme simplifies the relationships of the fields as defined by Maxwell's Equations. For instance, Faraday's Law on a single grid facet can be calculated using straightforward operations:

$$e_{\mathrm{x2}} + e_{\mathrm{y2}} - e_{\mathrm{x1}} - e_{\mathrm{y1}} = -\frac{d}{dt} b_{\mathrm{z}} \tag{4.17}$$

In Eq. 4.17, e represents the electric grid voltages, which correspond to the electric field integrated over the edge. Similarly, b represents the magnetic flux, which is the magnetic flux density integrated over the corresponding mesh face.

To couple the fields and fluxes, the material relationships within each cell must be defined. This is achieved through material matrices that are generated by averaging schemes in every mesh cell, assuming that the cell is filled with a single material. This approach is commonly referred to as the staircase approximation. The FIT method employs a direct material description using material matrices for permittivity, permeability, and conductivity. This differs from the TLM method, as discussed in the following sections.

Particularly the interface between dielectric materials and metals is of interest in electronics simulation. The various material models have been introduced in Sect. 4.4. At the interface between a volumetric material and a PEC (or Surface Impedance), the PBA method offers a significant improvement over the staircase approximation. It enables the handling of multiple materials within a single mesh cell, enhancing the efficiency of the method as less mesh cells are required to approximate shapes not aligned with the mesh.

Three different metal and volumetric material interfaces over a mesh facet are depicted in Fig. 4.20. On the left-hand side, we observe one region with PEC and another region with the volumetric material. This represents the standard PBA case. During meshing, the position at which the material boundary intersects the individual mesh lines is calculated, and the grid voltage is computed for the shortened section, as indicated by the orange lines. Inside the PEC, the field is set to zero.

Recall from Sect. 4.3.2 that the time step of the time domain method depends on the shortest mesh cell. The position of the cut could be very close to the actual mesh line, creating a reduced mesh edge that is significantly shorter than the original mesh edge, which can lead to a very small time step. The achievement of PBA is to control this reduced time step in every mesh cell and deliver improved spatial resolution without suffering from unreasonably small time steps.

PBA has been extended to handle three materials in one mesh cell. This extension is called the Thin Sheet Technique (TST) and is illustrated in the center of Fig. 4.20. In the configuration on the right-hand side of Fig. 4.20, four material regions are present over the mesh face. This configuration is not supported and would result in

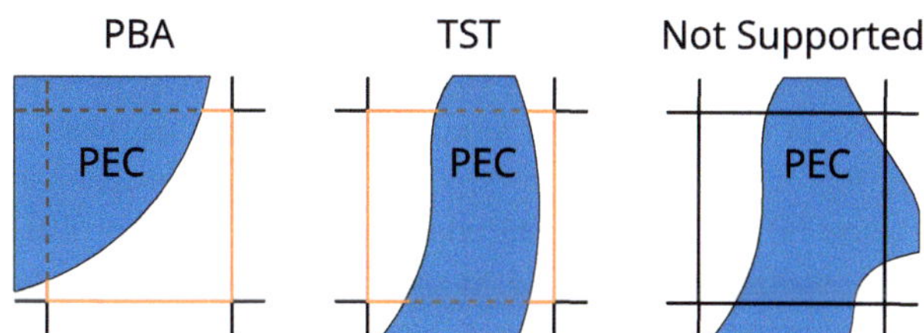

Fig. 4.20 Handling of PEC and volumetric material in a mesh cell with PBA

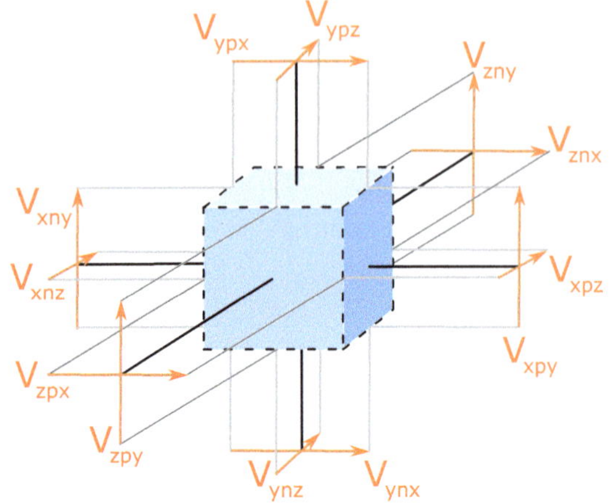

Fig. 4.21 Allocation of voltages to the mesh used by the TLM

a completely filled face. For such cases, it is necessary to add an additional mesh line.

The discretization performed by the FIT is not limited to the time-domain solver. The same discretization scheme can also be used in frequency-domain simulations, leading to a FDFD method. While the FDFD simulation based on FIT is still available in CST, it is less performant than the FEM method in the frequency domain and is therefore seldom used.

4.11.2 *Field Allocation in the TLM*

As we have already discussed in Sect. 4.10.2, the TLM solver uses a hexahedral octree mesh. The allocation of the mesh quantities is plotted in Fig. 4.21. Similar to the FIT, the TLM also assigns voltages to the grid, but these voltages are defined across the cell and positioned at the cell center.

Figure 4.21 illustrates the electric voltages indicated by V_{abc}, but the magnetic voltages (integrals of the magnetic field over a mesh edge) are allocated at exactly the same positions. Unlike the FIT, a dual grid is not used in TLM.

For the calculation of material properties, the TLM calculates an impedance description inside every mesh cell. This approach differs from the FIT, which uses material values directly. The outcome of the TLM meshing process is a grid of impedances obtained from the calculations in each cell. One can imagine this as a three-dimensional impedance network spanning the computational domain. The TLM also supports PBA with two materials but does not support TST.

The network resulting from the TLM discretization can also be solved in the frequency domain, as reported in Jin and Vahldieck (1992). However, the authors are not aware of any publicly available implementation of the TLM in the frequency domain.

4.11.3 Field Allocation in the FEM

The roots of the FEM lie in structural mechanics, as this method, in its general formulation, can be applied to various physical problems. Unlike the FIT and TLM, the FEM does not discretize Maxwell's Equations directly. Instead, the problem is reformulated as follows: Let **E** represent the solution to the differential equation we aim to solve. We approximate **E** with a function $\mathbf{E}_\mathrm{h}$, which is expressed as a linear combination of basis functions:

$$\mathbf{E} \approx \mathbf{E}_\mathrm{h} = \sum_i c_i \boldsymbol{\psi}_i \tag{4.18}$$

$\boldsymbol{\psi}_i$ are the so-called basis functions, and c_i are the coefficients that need to be calculated to approximate **E** with $\mathbf{E}_\mathrm{h}$.

The FEM provides a systematic way to convert continuous functions into a discrete space defined by the mesh. Instead of requiring the solution $\mathbf{E}_\mathrm{h}$ to solve the differential equation in continuous space, the so-called weak or variational formulation requires that $\mathbf{E}_\mathrm{h}$ solves the problem only in the discrete space. The coefficients c_i are then obtained from the solution of a linear system of equations, as described in Sect. 4.9. The mathematical foundation of the FEM is well developed, providing accurate methods for estimating error bounds. However, discussing these mathematical details is beyond the scope of this book. Instead, we focus on the differences between the FEM and methods like FIT and TLM, as well as the implications of these differences for electromagnetic field computation.

In the FEM, different basis functions can be chosen for Eq. 4.18. In the lowest-order scheme, a basis function is associated with each edge, where the line integral of the basis function equals one for its corresponding edge and zero for all other edges. These localized basis functions result in sparse matrices when applying the variational formulation. However, richer sets of basis functions can also be employed, leading to what is known as element order in the FEM. While FIT and TLM assume a constant voltage on the edge or across the cell, FEM allows the use of higher-order basis functions, resulting in multiple values of the electric field over a single edge.

The allocation of the electric field in the FEM method is depicted in Fig. 4.22. The arrows indicate the number of field values along one mesh edge. For a first-order element, as shown on the left-hand side of the figure, the electric field is constant over the edge, similar to the time-domain methods discussed previously. In the center, a second-order element is illustrated. In second-order elements, two electric field values are calculated along the edge. On the right-hand side, a third-order element is shown, where three electric field values are assigned to a single mesh edge.

The element order has a direct impact on the wavelength criterion, as mentioned in Sect. 4.10. Since higher-order elements calculate more field values along a single mesh edge, fewer mesh elements are required to satisfy the wavelength criterion.

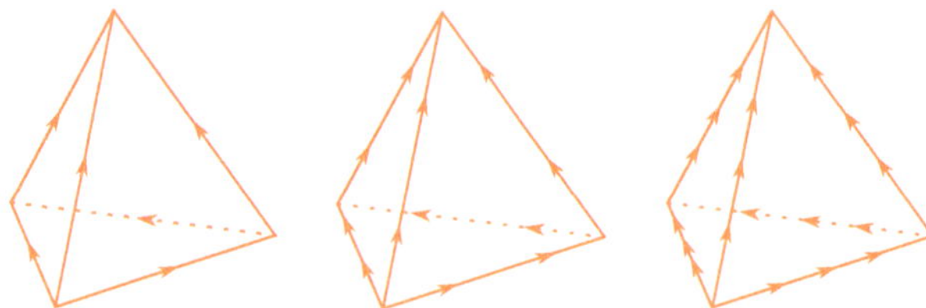

Fig. 4.22 Allocation of the electric field to the mesh in the FEM. Left first order, center second order, right third order

However, higher-order elements increase both memory requirements and simulation time. Their primary advantage lies in efficiently discretizing large domains, while their efficiency near material interfaces is comparable to that of a fine mesh with low-order elements.

When discretizing a simulation model, it is not necessary to fix the element order for all tetrahedrons. Instead, an order can be assigned to each individual tetrahedron, a technique known as variable order. This flexibility can be leveraged during mesh adaptation, where the mesh can not only be refined, as described in Sect. 4.3.1, but the appropriate element order for individual tetrahedrons can also be chosen. Refining both the mesh and the element order is referred to as h-p adaptation. Additionally, curved finite elements can be used to better approximate rounded structures. In curved finite elements, the mesh edges are not straight but curved to align more closely with material interfaces.

4.11.4 Method of Moments

The full-wave Method of Moments (MoM) is not used in this book; however, we would like to briefly introduce the method's strengths and weaknesses, as it is a well-known method in the EMC simulation community.

The MoM is an integral method and does not rely on finite differences, unlike the three methods we have described so far. The mesh in the MoM is surface-based, meaning it does not require the discretization of the entire volume. The core idea of the MoM is to compute the current distribution on the discretized surface elements and then calculate the fields in space (using analytical functions) with the current patches serving as sources.

Since every surface element contributes to the resulting field, the system matrix in the MoM is fully populated. As the number of surface elements increases, the memory requirements grow rapidly. To address this, the coupling between elements can be limited by dividing the computational domain into several regions, where surface elements only couple within their respective region, while regions themselves couple to each other. This technique is referred to as the Fast Multipole

Methods (FMM). An advanced version of this is the Multi-Level Fast Multipole Method (MLFMM), where the region definition is hierarchical.

Because the MoM does not discretize the volume of the computational domain, it is inherently an open method. The natural boundaries of the computational domain are open by default, eliminating the need for SIBC or PML. This makes the MoM advantageous for radiation pattern calculations. However, calculating the field distribution inside the computational domain requires an additional explicit step.

Due to its fully populated system matrix, the memory consumption of the MoM is higher than that of volume-based methods. The MoM is efficient for large volumes with few surfaces to discretize, but for structures with densely packed surfaces in a small volume, its memory consumption becomes excessive. Using MLFMM for models with tightly coupled surfaces in small volumes is generally not recommended. Most of the simulation models in this book, such as PCBs with dense packing, are poorly suited for the MoM, which is why this method is not used in this book.

Traditionally, the MoM is used for metal-dominated structures, such as car chassis or airplane bodies. For many years, it was the method of choice for calculating windscreen antennas, where the vehicle body must be included in the simulation.

Since the MoM discretizes only the surface, the growth of the number of mesh elements due to the wavelength criterion is slower compared to volume-based methods. The MoM is capable of simulating electrically large structures beyond what time-domain solvers can handle, often extending to several hundreds of wavelengths. This capability makes the MoM suitable for applications like Radar Cross Section (RCS) calculations, where entire airplanes are simulated at double-digit GHz frequencies.

Another notable application of the MoM is wire antenna simulation. The MoM can model the finite thickness of wires without explicitly meshing their cross-section. Using wire edge elements, the MoM avoids the issue of diverging inductance for zero-thickness wires, as described in Sect. 4.6. Not surprisingly, one of the earliest computer codes for electromagnetic field computation, called NEC, remains widely used for wire antennas.

The possibility to use wire elements in the MoM full-wave method has led to a number of commercial codes that couple the MTL with the MoM. As we discuss in Chap. 8, the 2D solution required for the MTL is calculated using a quasi-static MoM. In CST, the MTL is coupled only to time-domain methods.

4.12 Advanced—Time Discretization

Domains: 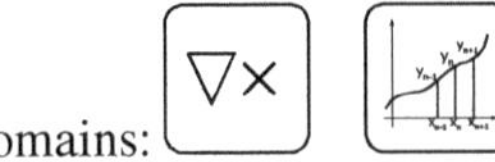

In the previous section, we have discussed how Maxwell's Equations are discretized in space. In this section, we discuss how the discretization of the time is performed. The differences between the time- and frequency-domain methods that we have listed in Sect. 4.9 are actually the results of the choice of the discretization in time.

4.12.1 *Explicit – Time-Domain Methods*

For the time-domain method, we directly discretize the time derivatives in Faraday's Law and in Ampère's Law using a central finite difference scheme. Additionally, we use the material relation $\mathbf{D} = \varepsilon\mathbf{E}$, and the time step Δt:

$$\nabla \times \mathbf{E} = -\frac{\partial \mathbf{B}}{\partial t} \Rightarrow \nabla \times \mathbf{E}|_t = -\frac{\mathbf{B}|_{t+\Delta t/2} - \mathbf{B}|_{t-\Delta t/2}}{\Delta t} \tag{4.19}$$

$$\nabla \times \mathbf{H} = \mathbf{J} + \varepsilon\frac{\partial \mathbf{E}}{\partial t} \Rightarrow \nabla \times \mathbf{H}|_{t+\Delta t/2} = \varepsilon\frac{\mathbf{E}|_{t+\Delta t} - \mathbf{E}|_t}{\Delta t} + \mathbf{J}|_{t+\Delta t/2} \tag{4.20}$$

Using $\mathbf{B} = \mu\mathbf{H}$, we rearrange these equations to derive an explicit scheme that computes the next (half) time step from the previous time steps:

$$\mathbf{H}|_{t+\Delta t/2} = \mathbf{H}|_{t-\Delta t/2} - \frac{\Delta t}{\mu} \cdot (\nabla \times \mathbf{E}|_t) \tag{4.21}$$

$$\mathbf{E}|_{t+\Delta t} = \mathbf{E}|_t + \frac{\Delta t}{\varepsilon} \cdot \left(\nabla \times \mathbf{H}|_{t+\Delta t/2}\right) - \frac{\Delta t}{\varepsilon} \cdot \left(\mathbf{J}|_{t+\Delta t/2}\right) \tag{4.22}$$

The Eqs. 4.22 and 4.21 are the update equations that we need to apply at every half time step. This time stepping is called a leap frog scheme because we jump over the time steps as indicated by Fig. 4.23.

We have mentioned the dependence of the time step on the smallest mesh step through this chapter. This is the CFL criterion:

$$\Delta t \leq \frac{\sqrt{\varepsilon\mu}}{\sqrt{(\frac{1}{\Delta x})^2 + (\frac{1}{\Delta y})^2 + (\frac{1}{\Delta z})^2}} \tag{4.23}$$

where Δx, Δy and Δz are the smallest mesh steps in the respective axis. Mathematically, the CFL is a requirement on the time step in order to ensure the stability of

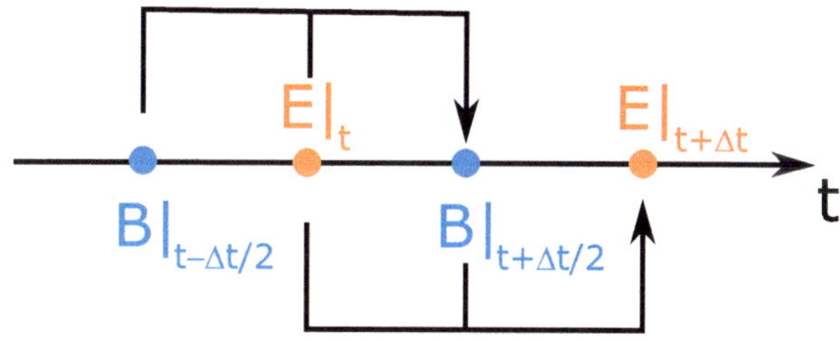

Fig. 4.23 Schematic of time stepping in the Leap frog algorithm

the explicit integration scheme. In case it is violated, the field values start to diverge, i.e., the magnitude grows toward infinity.

We can also interpret the stability criterion based on physical reasoning: In order to obtain meaningful results from the time stepping scheme, the (numerical) propagation distance of the electromagnetic field at every time step is not allowed to be larger that the propagation distance of light.

4.12.2 Implicit—Frequency-Domain Methods

For the discretization of the time derivatives in the Maxwell Equations in the frequency domain, we use time-harmonic solutions of the equations. A time-harmonic function is of the form:

$$\mathbf{E}(t) = \Re(\underline{\mathbf{E}} \cdot e^{j\omega t}) \tag{4.24}$$

The underlines indicate that the fields are described by phasors. The phasor notation is explained in more detail in Sect. 6.1.1, where we introduce the plotting of frequency-domain fields. The time derivative of a harmonic function as in Eq. 4.24 is the multiplication of the harmonic function by $j\omega$,

$$\frac{d}{dt}\mathbf{E}(t) = \Re(j\omega\underline{\mathbf{E}} \cdot e^{j\omega t}) \tag{4.25}$$

If we use this identity for Faraday's Law, we obtain

$$\nabla \times \mathbf{E} = -\frac{\partial \mathbf{B}}{\partial t} \Rightarrow \underline{\mathbf{B}} = \frac{j}{\omega}\nabla \times \underline{\mathbf{E}}. \tag{4.26}$$

For the derivation, we have used the identity $\frac{1}{j} = -j$. For Ampère's Law, using $\mathbf{D} = \varepsilon\mathbf{E}$ we obtain

$$\nabla \times \mathbf{H} = \mathbf{J} + \frac{\partial \mathbf{D}}{\partial t} \Rightarrow \nabla \times \underline{\mathbf{H}} - j\omega\varepsilon\underline{\mathbf{E}} = \underline{\mathbf{J}} \tag{4.27}$$

In the next step, we split the current $\underline{\mathbf{J}}$ into a conductive part $\underline{\mathbf{J}}_c$ and a source part $\underline{\mathbf{J}}_s$ and use Ohm's Law $\underline{\mathbf{J}}_c = \sigma_c \underline{\mathbf{E}}$ on the conductive part. Equation 4.27 becomes:

$$\nabla \times \underline{\mathbf{H}} - j\omega\varepsilon\underline{\mathbf{E}} - \sigma_c\underline{\mathbf{E}} = \underline{\mathbf{J}}_s \tag{4.28}$$

Now we use $\mathbf{H} = \mu^{-1}\mathbf{B}$, insert Faraday's Law Eq. 4.26 into Eq. 4.28 and obtain

$$\left[\nabla \times \mu^{-1}\nabla \times + j\omega\sigma_c - \omega^2\varepsilon\right]\underline{\mathbf{E}} = -j\omega\underline{\mathbf{J}}_s \tag{4.29}$$

Equation 4.29 is a full-wave equation formulated using only the electric field. There are three contributions to the electric field in Eq. 4.29: a rotational term from the magnetic field, a conductive term, and a displacement current term. The frequency-domain FEM method requires solving only for the electric field. The magnetic field is subsequently obtained from the electric field using Eq. 4.26. Solving this full-wave equation in the kHz range and below is numerically challenging, as the term $\omega^2\varepsilon$ becomes small. The full-wave method requires a modification through the so-called low-frequency stabilization, which we do not discuss in this book.

Further Reading

Numerical Methods for Electromagnetics

B. Anderson et al., Full-wave analysis in SPICE via model-order reduction. IEEE Trans. Microwave Theory Tech. **52**(9), 2314–2320 (2004). https://doi.org/10.1109/TMTT.2004.834569

C.A. Balanis, *Advanced Engineering Electromagnetics*, 3rd edn. (Wiley, 2024). ISBN: 978-1-394-18001-1

P.A. Brennan, N. Raver, A.E. Ruehli, Three-dimensional inductance computations with partial element equivalent circuits. IBM J. Res. Dev. **23**(6), 661–668 (1979). https://doi.org/10.1147/rd.236.0661

S. Burgard, O. Farle, R. Dyczij-Edlinger, An ***h***adaptive sub-domain framework for parametric order reduction. IEEE Trans. Magn. **51**(3), 1–4 (2015). https://doi.org/10.1109/TMAG.2014.2353138

M. Eller et al., A symmetric low-frequency stable broadband Maxwell formulation for industrial applications. SIAM J. Sci. Comput. **39**(4), B703–B731 (2017)

A. Giacomini et al., Antenna gain calibration with improved accuracy modeling of pyramidal standard gain horns, in *2022 Antenna Measurement Techniques Association Symposium (AMTA)* (2022), pp. 1–6. https://doi.org/10.23919/AMTA55213.2022.9954972

D.J. Griffiths, *Introduction to Electrodynamics* (Pearson Education, 2013). ISBN: 978-0-321-85656-2

H. Jin, R. Vahldieck, A frequency domain TLM method, in *1992 IEEE MTT-S Microwave Symposium Digest*, vol. 2 (1992), pp. 775–778. https://doi.org/10.1109/MWSYM.1992.188101

M. Jochum, O. Farle, R. Dyczij-Edlinger, A new low-frequency stable potential formulation for the finite-element simulation of electromagnetic fields. IEEE Trans. Magn. **51**(3), 1–4 (2015). https://doi.org/10.1109/TMAG.2014.2360080

M. Kaltenbacher, *Numerical Simulation of Mechatronic Sensors and Actuators* (Springer, Heidelberg/Wiley, 2015). https://doi.org/10.1007/978-3-642-40170-1

S. Kurz et al., Hybrid modeling: towards the next level of scientific computing in engineering. J. Math. Ind **12**(1), 8 (2022). ISSN: 2190-5983. https://doi.org/10.1186/s13362-022-00123-0

Y. Li, Z.J. Cendes, High-accuracy absorbing boundary condition. IEEE Trans. Magn. **31**(3), 1524–1529 (1995). https://doi.org/10.1109/20.376320

R. Mertens et al., An algebraic multigrid method for solving very large electromagnetic systems. IEEE Trans. Magn. **34**(5), 3327–3330 (1998). https://doi.org/10.1109/20.717782

O. Schenk, K. Gärtner, Solving unsymmetric sparse systems of linear equations with PARDISO. Future Gener. Comput. Syst. 475–487 (2004). https://doi.org/10.1016/j.future.2003.07.011

D. Schmitt, T. Weiland, 2D and 3D computations of eigenvalue problems. IEEE Trans. Magn. **28**(2), 1793–1796 (1992). https://doi.org/10.1109/20.124054

A. Schultschik, O. Farle, R. Dyczij-Edlinger, An adaptive multi-point fast frequency sweep for large-scale finite element models. IEEE Trans. Magn. **45**(3), 1108–1111 (2009). https://doi.org/10.1109/TMAG.2009.2012639

A. Sommer, O. Farle, R. Dyczij-Edlinger, A new method for accurate and efficient residual computation in adaptive model-order reduction. IEEE Trans. Magn. **51**(3), 1–4 (2015). https://doi.org/10.1109/TMAG.2014.2352812

P. Thoma, T. Weiland, Numerical stability of finite difference time domain methods. IEEE Trans. Magn. **34**(5), 2740–2743 (1998). https://doi.org/10.1109/20.717636

T. Weiland, A discretization method for the solution of Maxwell's equations for six-component fields. Electron. Commun. AEU **31**(3), 116–120 (1977)

Material Modeling

M. Bali, H. De Gersem, A. Muetze, Finite-element modeling of magnetic material degradation due to punching. IEEE Trans. Magn. **50**(2), 745–748 (2014). https://doi.org/10.1109/TMAG.2013.2283967

F. Bruckner et al., Macroscopic simulation of isotropic permanent magnets. J. Magn. Magn. Mater. **401**, 875–879 (2016). ISSN: 0304-8853, https://doi.org/10.1016/j.jmmm.2015.11.005, https://www.sciencedirect.com/science/article/pii/S0304885315307769

H. De Gersem, I. Munteanu, T. Weiland, Construction of differential material matrices for the orthogonal finite-integration technique with nonlinear materials. IEEE Trans. Magn. **44**(6), 710–713 (2008). https://doi.org/10.1109/TMAG.2007.915819

P.G. Huray, *The Foundations of Signal Integrity* (Wiley, Hoboken, 2010)

D. Jackson, *Classical Electrodynamics*, 3rd edn. (Wiley, Hoboken, 2021). ISBN: 978-1-119-77076-3

D. Jiles, D. Atherton, Ferromagnetic hysteresis. IEEE Trans. Magn. **19**(5), 2183–2185 (1983). https://doi.org/10.1109/TMAG.1983.1062594

M.Y. Koledintseva et al., Wide-band Lorentzian media in the FDTD algorithm. IEEE Trans. Electromagn. Compat. **47**(2), 392–399 (2005). https://doi.org/10.1109/TEMC.2005.847406

F. Maradei, H. Ke, T.H. Hubing, Full-wave model of frequency-dispersive media with Debye dispersion relation by circuit-oriented FEM. IEEE Trans. Elec-tromagn. Compat. **51**(2), 312–319 (2009). https://doi.org/10.1109/TEMC.2009.2014639

T. Tsutaoka, Frequency dispersion of complex permeability in Mn–Zn and Ni–Zn spinel ferrites and their composite materials. J. Appl. Phys. **5**(5), 2789–2796 (2003). https://doi.org/10.1063/1.1542651

Chapter 5
Coupling Paths: Near Field

In Chap. 3, we demonstrate how to predict the EMI of a buck converter using a schematic that accounts for the parasitic effects of components and the physical characteristics of interconnects. Parasitic values for components are estimated based on vendor models and data sheets, while the physical properties of interconnects are derived using analytical formulas or general rules of thumb. However, these analytical formulas are limited in their accuracy.

In this chapter, we use EM simulation to determine the physical properties of the coupling path formed by the interconnects. All coupling path models in this chapter are described using lumped RLC elements. Recall from Sect. 2.4.2.5 that a lumped RLC model accurately describes the coupling path only if it is electrically short. This constraint aligns with the EM methods used in this chapter, as we compute the properties of the coupling path using static and quasi-static solutions to Maxwell's equations, which are valid only for electrically short geometries.

This chapter focuses on methods that are both instructional and, when applicable, computationally efficient. We categorize the coupling path into galvanic, capacitive, and inductive contributions, a framework commonly employed in EMC literature. The examples provided illustrate how to decompose a model into its source, coupling path, and sink. They also introduce practical approximations to Maxwell's equations and clarify the relationship between capacitance and inductance with the electric and magnetic fields.

Radiative coupling, which cannot be conveniently described using RLC models, is treated in Chap. 6.

J. Hansen, A. Barchanski, *EMC Simulation of Electronic Systems*,
https://doi.org/10.1007/978-3-032-09237-3_5

5.1 Near-Field Coupling

Domains: 

The electromagnetic field generated by a source is divided into three regions: the near field, the transition zone, and the far field. The term "near field" refers to a coupling path that is electrically short, so that it can be modeled with a lumped RLC circuit. In this section, we examine how the electric and magnetic near fields vary with distance from a loop of electrically small radius, considering different termination impedances. Additionally, we revisit the static and quasi-static approximations of Maxwell's equations. These approximations are applied to various EMC tasks in later sections of this chapter.

5.1.1 Electric and Magnetic Near Field

To illustrate how the field decays with distance from the source, an infinitesimally short electric dipole and an infinitesimally small current loop are commonly used as examples.

In the near field of an electrically short dipole, the electric field decays proportionally to $\frac{1}{r^3}$, while the magnetic field decays with $\frac{1}{r^2}$. Conversely, in the near field of a loop, the magnetic field decays as $\frac{1}{r^3}$ and the electric field as $\frac{1}{r^2}$. In the far-field region, both fields decay proportionally to $\frac{1}{r}$. These decay relationships are helpful for understanding the behavior of the electric and magnetic fields in different regions. However, the assumptions of a small dipole and a small loop do not always apply in practical scenarios. To illustrate this, we compare the differences between an infinitesimal loop and a loop of finite size in the following section.

For this simulation, we use the full-wave FEM solver to obtain both the electric and magnetic fields from the same model. We define a loop constructed from a copper wire with a wire diameter of 1 mm and a loop diameter of 10 cm. The loop is excited on one side and terminated with a load resistor R_L on the opposite side. All fields are evaluated at 1 kHz. At this frequency, the wavelength is approximately 300 km, making the loop electrically small and ensuring that all presented results are within the near-field region.

Since the DC resistance of the copper wire is not of interest, we model the loop material as a PEC. The excitation is implemented using a 1 V source with an internal resistance of 1 Ω to limit the current. The electric and magnetic fields of the loop at 1 kHz are depicted in Fig. 5.1.

In the first step, we demonstrate the effect of the termination resistance on the fields by varying R_L to 1 mΩ, 1 Ω, and 1 kΩ. In Fig. 5.2, we plot the magnitude of the electric and magnetic fields along a line perpendicular to the center of the loop, as indicated by the black line in Fig. 5.1.

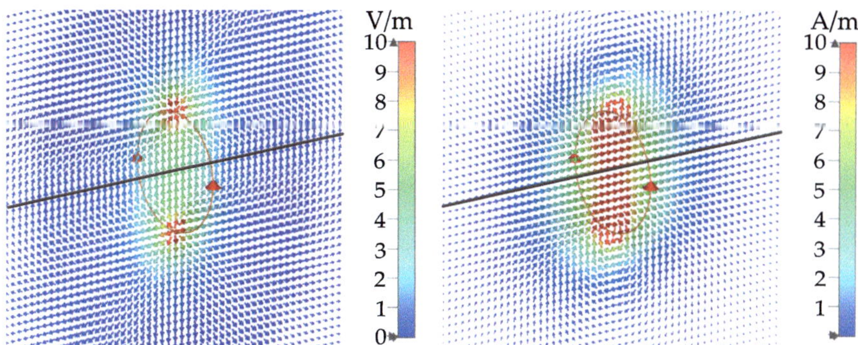

Fig. 5.1 Left: Electric field of the loop for a termination resistance of 1 kΩ. Right: Magnetic field of the loop for a termination resistance of 1 mΩ. The field is displayed in a cut-plane perpendicular to the loop's axis. The black line indicates the position at which we evaluate the fields

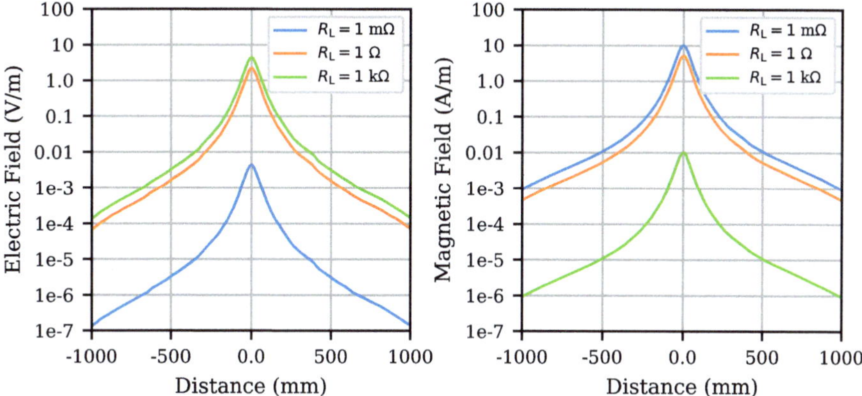

Fig. 5.2 Left: Electric field along the center of the loop for different termination impedances. Right: Magnetic field along the center of the loop for different termination impedances

For the termination with $R_L = 1\,\text{m}\Omega$, the current in the loop is 1 A, and the voltage at the termination is 1 mV. Thus, the magnetic field dominates the fields around the loop. For $R_L = 1\,\text{k}\Omega$, the current in the loop is 1 mA, and the voltage at the termination is 1 V. The electric field is a factor of 1000 higher than for the low-ohmic termination, while the magnetic field is a factor of 1000 lower. For the termination with $R_L = 1\,\Omega$, the current is 0.5 A, and the voltage is 0.5 V; both fields contribute significantly to the field of the loop.

This simple study illustrates that, depending on the termination resistance of the loop, either the electric or magnetic field predominates. Consequently, the coupling path from source to sink is influenced by the corresponding mechanism: capacitive coupling for a predominantly electric field at high termination impedances, and inductive coupling for a predominantly magnetic field at low termination

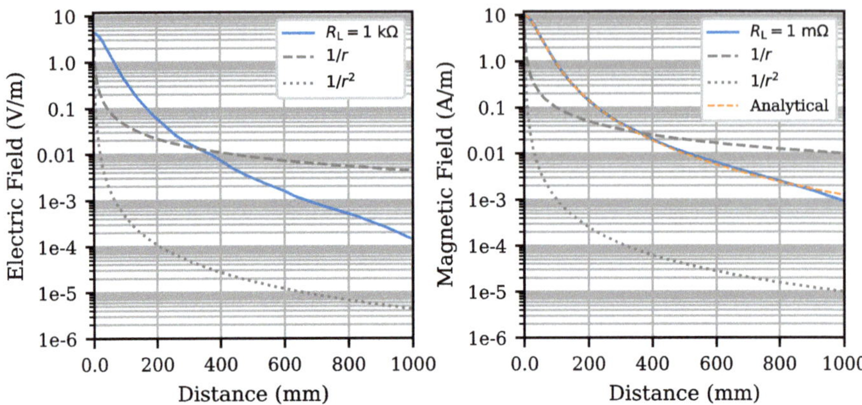

Fig. 5.3 Left: Decay of the electric field with distance to the loop. Right: Decay of the magnetic field with distance from the loop. For guidance, a $\frac{1}{r}$ and $\frac{1}{r^2}$ decay is plotted in gray dashed and dotted lines

impedances. In cases where both fields contribute significantly, it is not clear a priori which coupling mechanism is dominant.

In a second step, we compare the decay of the magnetic and electric field to the $\frac{1}{r}$ and $\frac{1}{r^2}$ laws, as these are often used to describe how fields decay with distance. For an infinitesimally small electric dipole, the electric field decays with $\frac{1}{r^3}$, and for an infinitesimally small loop, the magnetic field decays with $\frac{1}{r^3}$ in the near field.

The comparison of the calculated values with these decay laws, shown in Fig. 5.3, demonstrates that the decay laws for infinitesimally small sources do not hold in our example. Due to the finite size of the loop, the decay is significantly slower than the $\frac{1}{r^3}$ law. Below a distance of approximately 35 cm, the decay is even slower than $\frac{1}{r}$. This is consistent with the fact that the magnetic field along the central axis of a current loop with diameter d carrying a current I can be calculated analytically using:

$$H = \frac{I \cdot d^2}{2(r^2 + d^2)^{(3/2)}} \tag{5.1}$$

This equation can be derived from Biot–Savart's law and is widely available in many textbooks. In Fig. 5.3, we also plot the analytical results for the magnetic field, demonstrating excellent agreement with the results obtained from the field simulation. For the electric field, however, no straightforward analytical equation exists.

This study has illustrated how the different fields depend on the termination impedance and their respective decay with distance. A similar analysis is conducted for electrically long structures in the next chapter, in Sect. 6.4. In the following section, we explore simplifications to Maxwell's Equations applicable to electrically short models.

5.1.2 Static and Quasi-static Approximations

If an object is electrically short (e.g., $< \frac{\lambda}{10}$), we can assume that the surface of a conductor is an equipotential and that the current remains constant along the path. Under these assumptions, Maxwell's Equations become decoupled, allowing the magnetic and electric fields to be calculated separately.

In the previous section, we demonstrate that the magnetic or electric field dominates depending on the termination impedance of the loop. However, this observation does not yet prove that the fields are decoupled. A more rigorous demonstration of field decoupling is provided in Sect. 2.4.2.5. There, we calculate the inductance and capacitance of a wire pair using analytical equations that treat the magnetic and electric fields independently. We also show that the separately calculated inductance and capacitance can be combined into a lumped RLC model, which accurately describes the electrical properties of the wire pair, provided it remains electrically short.

When solving Maxwell's Equations in the static regime, the time derivatives $\frac{\partial \mathbf{B}}{\partial t}$ and $\frac{\partial \mathbf{E}}{\partial t}$ become zero, leading to the decoupling of the magnetic and electric fields. In the decoupled case, for the electric field, we solve Gauss's Law $\nabla \cdot \mathbf{D} = \rho$, while Faraday's Law reduces to $\nabla \times \mathbf{E} = \mathbf{0}$. The sources of the static electric field are charges, and there is no rotational component.

For the magnetic field, we solve Ampère's Law, which becomes $\nabla \times \mathbf{H} = \mathbf{J}$, along with the non-existence of magnetic monopoles, $\nabla \cdot \mathbf{B} = 0$. The latter condition implies that the current path must form a closed loop because the divergence of the magnetic flux is zero. These are the equations that are discussed in the following next three Sects. 5.3–5.5.

In Sect. 5.6, we use the Magnetoquasi-static (MQS) approximation. In the MQS approximation, we assume that the displacement current is negligible, setting $\frac{\partial \mathbf{E}}{\partial t} = 0$ in Faraday's Law, but we retain the induction term $\frac{\partial \mathbf{B}}{\partial t}$ in Ampère's Law. This allows to calculate induced currents in conductive media and the effects of skin depth, while excluding wave propagation. These induced currents are known as eddy currents. In MQS, the electric field in conductors, connected to the eddy currents via Ohm's Law, is purely rotational (a curl field).

In a curl field, the field lines form closed loops, as known from eddy currents. We illustrate such a curl field on the right-hand side of Fig. 4.1. Sources in the MQS regime are of current type, as no free-space electric field exists.

Finally, if the structure is no longer electrically short, the electric and magnetic fields do not decouple, and the full set of Maxwell's Equations must be solved, even in the near field of the structure. The complete solution of Maxwell's Equations is treated in Chap. 6. A summary of the different formulations is provided in Table 5.1.

Table 5.1 Different formulations for the solution of Maxwell Equations. PEC also includes Surface Impedance and Thin Panel materials

Formulation	Electric field	Magnetic field	Current density	Remarks	Approximation
Stationary current	Only inside conductors	–	Ohm's law	–	$\frac{\partial \mathbf{B}}{\partial t} = 0\,,\ \frac{\partial \mathbf{E}}{\partial t} = 0$
	Zero everywhere else	–	–	–	
Electrostatic	Only in non conductive regions	–	–	–	$\frac{\partial \mathbf{B}}{\partial t} = 0\,,\ \frac{\partial \mathbf{E}}{\partial t} = 0$
	Conductors are PEC	–	–	–	
Magnetostatic	–	Entire computational domain	As source current	For a distributed source current,	$\frac{\partial \mathbf{B}}{\partial t} = 0\,,\ \frac{\partial \mathbf{E}}{\partial t} = 0$
	–			the current distribution is calculated	
	–			before the magnetostatic solution	
Magnetoquasi-static	Only in conductors	Entire computational domain	Source and induced	Used by pRLC solver	$\frac{\partial \mathbf{E}}{\partial t} = 0$
Electroquasi-static	Entire computational domain	–	Ohm's law	Not used in this book	$\frac{\partial \mathbf{B}}{\partial t} = 0$
Full-wave	Entire computational domain	Entire computational domain	All volumetric conductors	Chapter 6	None
	But PEC	But PEC	Only surface current on PEC		

5.2 Circuit Model of the Decomposition of an EMC Problem

Domains: 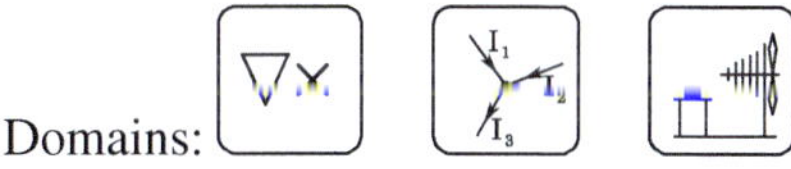

For the calculation of voltages and currents in an EMC simulation model, a circuit model of the decomposition of an EMC problem, as introduced in Sect. 1.1, is required. We define the circuit model of the decomposition for a generic coupling path in this section.

In the following sections, the schematic of the decomposition is explicitly plotted for application examples, in which the coupling path consists only of one coupling type: galvanic, capacitive, or inductive.

In later chapters, the coupling path is modeled using black-box S-parameter models, and the schematics no longer explicitly depict the decomposition. Nevertheless, the concept introduced here remains applicable to these models.

The circuit model of the decomposition for a generic coupling path is plotted in Fig. 5.4. For simplicity, the schematic uses a voltage source in the source and sink loops, but the concept is equally valid for current sources.

The schematic of the source is defined by a voltage source V_{source} and the impedance of the source loop Z_{source}. The voltage and current in the source path depend on Z_{source}, the internal impedance of V_{source}, and the source amplitude.

In the sink loop, in addition to the voltage source V_{sink} and loop impedance Z_{sink}, a current probe and a voltage probe are present. The magnitude of the voltage and current in the sink loop depends on the internal impedance of V_{sink}, the value of Z_{sink}, and the amplitude of V_{sink}.

The choice of the source and the sink is driven by the application and is not a property of the coupling path. The coupling path is symmetric, meaning that not

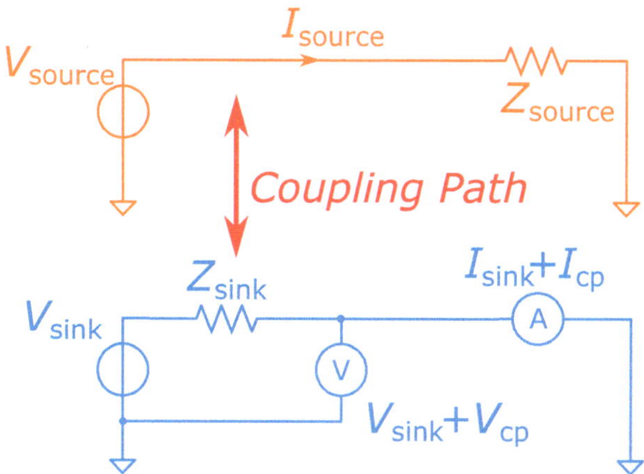

Fig. 5.4 Circuit schematic of the decomposition of an EMC problem

only does the source act on the sink, but the sink also acts on the source. To simplify the schematic, we can remove V_{sink}, leaving only the source acting on the sink and generating a voltage and a current in the sink loop.

For different measurements of the coupling into the sink loop, the following configurations are used:

Voltage measurement Replace V_{sink} with a short and set Z_{sink} to a high impedance. In this configuration, the current in the sink loop is negligible.

Current measurement Replace V_{sink} with a short and set Z_{sink} to a low impedance. In this configuration, the voltage in the sink loop is negligible.

Measurement at a LISN Replace V_{sink} with a short and set Z_{sink} to $50\,\Omega$. The coupling into the sink loop leads to a non-negligible current and voltage in the sink loop.

Effect on a signal in the sink loop Define the signal at V_{sink}. The termination Z_{sink} depends on the termination impedance of the system. Whether we are interested in the voltage or the current in the sink loop depends on the function of the system in the sink loop.

In electrically large models, the source and sink impedance can be frequency-dependent and distributed. Therefore, when probing voltage, the position of the probe in the geometry is crucial.

In the following application examples, we use EM simulations performed with static solvers to calculate the properties of the coupling path. The examples are chosen such that each focuses exclusively on one type of coupling path. This approach allows us to manually define the coupling path in the schematic of the decomposition using RLC values extracted from the EM simulation.

For the decomposition schematic to produce meaningful results, the coupling path model must align with the signal frequency band in both the source and sink loops. While the decomposition schematic can calculate voltages and currents even with an unsuitable coupling path model, it is the responsibility of the simulation engineer to ensure that the chosen coupling path model is appropriate for the specific electronic system.

The results of the examples in this chapter are valid only under the assumption that the model is electrically short and dominated by a single coupling type. These examples are deliberately selected to meet these conditions. In later chapters, when full-wave simulations are used to derive the coupling path circuit model, this process becomes simpler. The simulation engineer only needs to ensure that the selected frequency band in the EM simulation model matches the decomposition schematic.

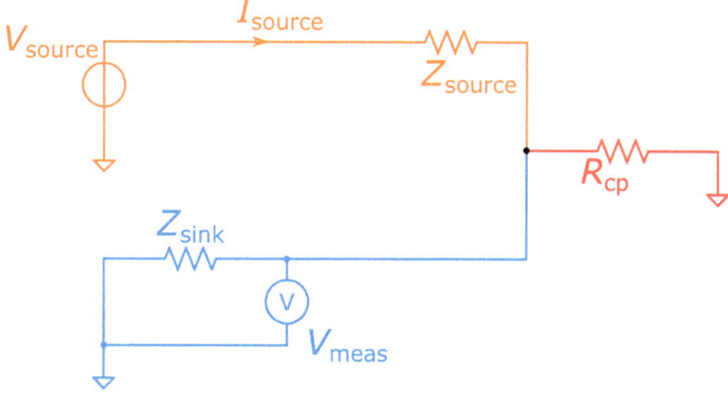

Fig. 5.5 Schematic of the galvanic coupling

5.3 Galvanic Coupling

Domains: 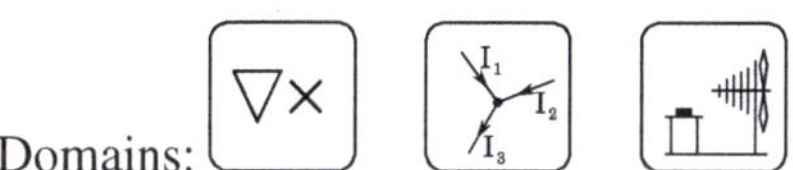

Galvanic coupling is described by the schematic plotted in Fig. 5.5. The source and sink current loops share a common edge containing a resistive element. The current flowing in one of the loops, represented here by the excited V_{source}, induces a voltage drop across the common element, R_{cp}. This voltage drop is measured in the second loop, as indicated by V_{meas}. For this voltage measurement, the impedance of the sink loop Z_{sink} is set to high ohmic. The magnitude of the sink voltage is determined by the value of the coupling resistor R_{cp} and the magnitude of the source current I_{source}.

In electronic systems, the galvanic coupling mechanism poses challenges when multiple signals share a common return current conductor. Consider a cable with a wire carrying a high current and an analog signal line, both sharing the same wire for the return current path. The current in the high-current wire generates a voltage drop in the analog signal loop, distorting the voltage of the analog signal.

5.3.1 Application Example Galvanic Coupling: Shunt Measurement

- Ideal circuit current source
- Voltage measurement across resistor
- Galvanic described by the resistance matrix
- Stationary current solver

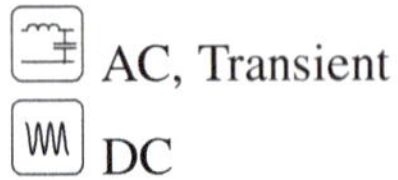

In a shunt current measurement, the galvanic coupling mechanism is intentionally employed to measure the current in the source loop. In this example, a shunt resistor R_{shunt} is used to measure the current flowing through an electric motor driven by a three-phase inverter. The measurement of the total current through the motor is essential for controlling the motor. The inverter is implemented on a PCB, and the shunt measurement is performed on the low side. Due to the galvanic coupling, the voltage measured across the shunt resistor allows the determination of the current I_{motor} flowing through the motor, which is the load of the inverter.

To minimize thermal losses, which are proportional to $R_{\mathrm{shunt}} \cdot I_{\mathrm{motor}}^2$, the value of R_{shunt} should be kept low. The principal schematic of a shunt measurement is depicted in Fig. 5.6. The voltage measurement V_{shunt} is evaluated inside a microcontroller, which is placed away from the shunt to avoid interference from the high-current path. The measured voltage is proportional to the current in the motor, with R_{shunt} as the proportionality constant. Since R_{shunt} is a design parameter, the current in the motor can be measured with this setup.

The schematic in Fig. 5.6 shows an ideal shunt measurement on the low side, without including any parasitic resistances such as those from the layout. If there is an additional resistance shared by both loops in the path, in the section indicated by the red lines, this resistance contributes to the voltage measured by the microcontroller. Consequently, the measured current becomes incorrect, as the proportionality constant now includes R_{shunt} plus the parasitic contribution.

In the following, we analyze the effect of the parasitic coupling resistance on the inverter layout depicted in Fig. 5.7. The DC connection to the battery is located in the lower-left corner. The current loop of the motor flows from the battery connector through a C-L-C filter with a cylindrical coil, three 100 µF DC link capacitors, and a three-phase bridge designed with BUK9Y72-type transistors to the output side, where the three phases of the electrical motor are connected. The current loop is indicated by orange arrows. The transistors have a maximum voltage of 80 V and a maximum current of 15 A.

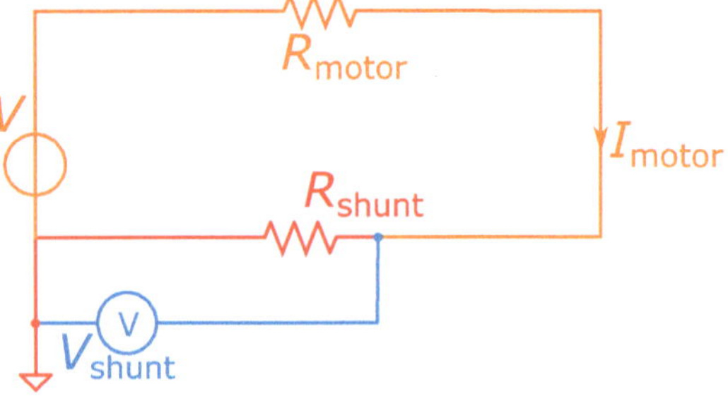

Fig. 5.6 Low-side shunt measurement of the current flowing in an electrical motor

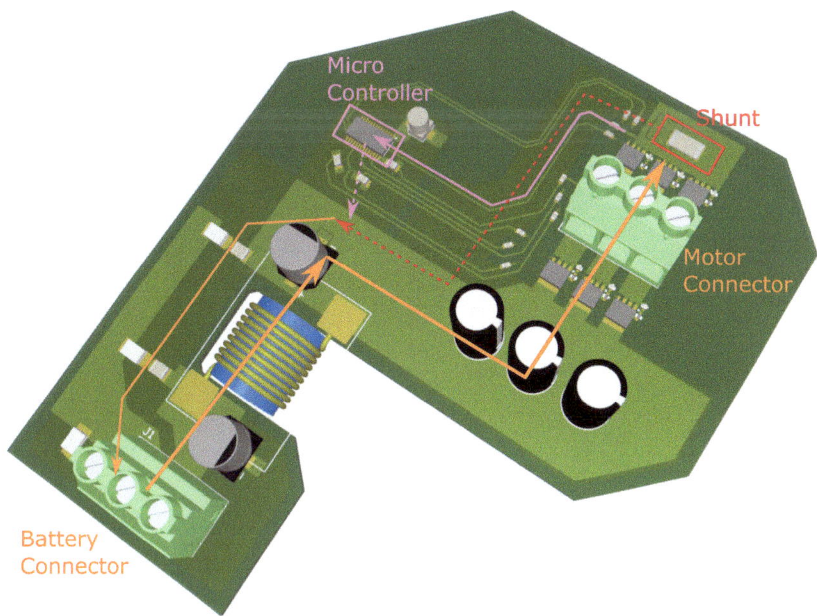

Fig. 5.7 Layout of the PCB of the three-phase inverter

The shunt is located on the low side, as indicated by the red box. At the position of the shunt, a (measurement) signal trace is connected and routed to the microcontroller on the left-hand side (for better visibility: pink line). This trace forms the measurement loop, depicted in blue in the schematic. The measurement loop and the current loop of the motor share not only the shunt resistance but also a section of the ground plane on the second layer, as indicated by the red dashed line. This shared section of the layout introduces the parasitic coupling resistance described above. In the following, we calculate the contribution of this shared section to the voltage measured by the microcontroller.

5.3.2 *Computation of the Resistive Coupling*

Resistive coupling is based on DC currents, allowing all time derivatives in the Maxwell Equations to be set to zero. In this regime, the electric field decouples from the magnetic field, enabling the calculation of the scalar electric potential ϕ in conductors. Details of the equations solved by the stationary current solver are presented in Sect. 5.7.

Due to the finite conductivity in the conductive region, a potential drop occurs along the path of the current. The difference in potential between two positions in the conductor is:

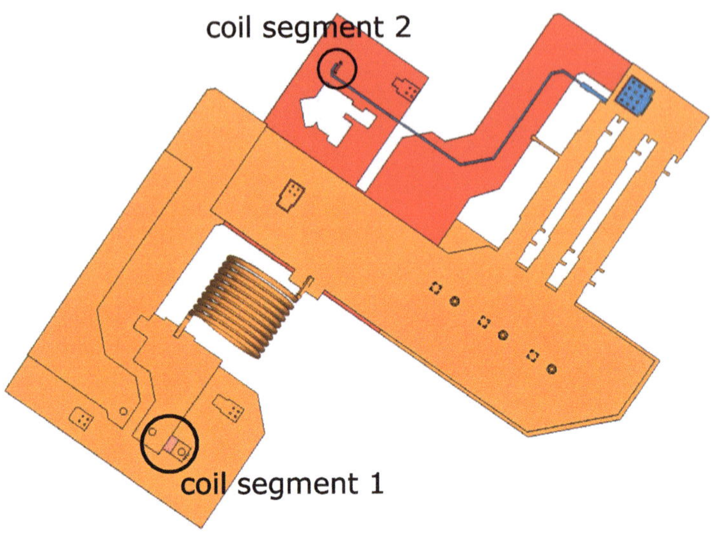

Fig. 5.8 Geometrical model used to compute the values of the elements of the resistive network, which characterizes galvanic coupling

$$V = \phi(\mathbf{r}_1) - \phi(\mathbf{r}_2). \tag{5.2}$$

From the scalar potential, the electric field is obtained as the gradient of the potential

$$\mathbf{E} = -\nabla\phi. \tag{5.3}$$

The current density is obtained from the electric field by multiplying it with the conductivity. This is the field-based formulation of Ohm's Law,

$$\mathbf{J} = \sigma_c \mathbf{E}. \tag{5.4}$$

In CST, these steps are implemented in the stationary current solver.

For the calculation of the resistance network, we prepare the PCB model as illustrated in Fig. 5.8. Since only the current loop through the motor and the current measurement loop toward the microcontroller are required, we import only the corresponding planes and traces of the PCB layout. The substrates and the ferrite core of the inductor are omitted, as they do not conduct any current.

The red section in Fig. 5.8 highlights the region where parasitic coupling is expected to occur. Since the conductor in this section is wide and the paths in both loops are not necessarily parallel, an analytical solution to the geometry is impractical. Field computation, however, enables us to determine the value of the coupling resistance, which can later be used in a circuit simulation.

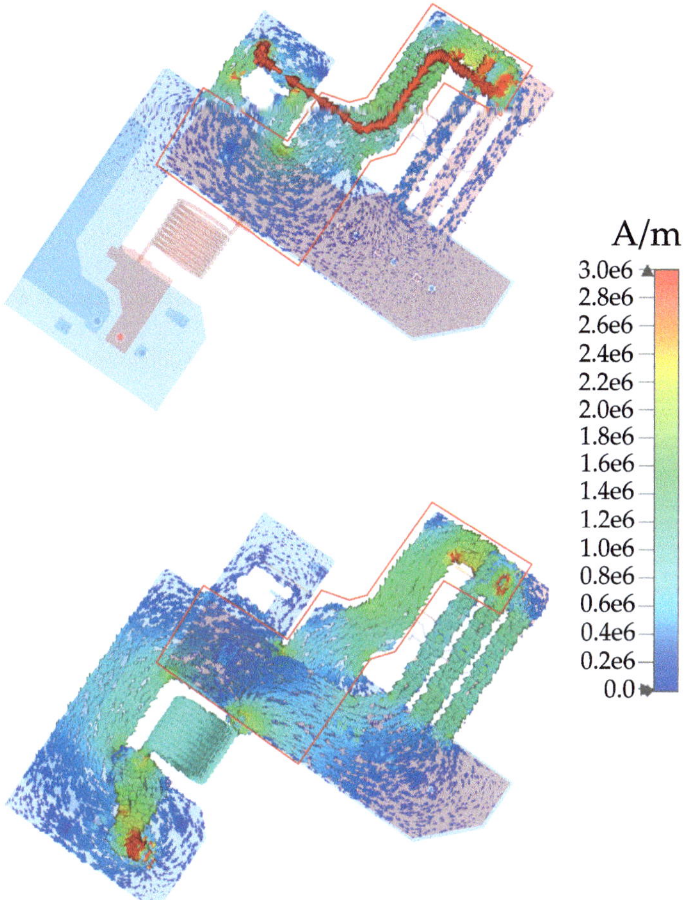

Fig. 5.9 Top: Current distribution in the measurement loop. Bottom: Current distribution in the motor loop. The red shape indicates the region where we expect the parasitic resistive coupling

For both current loops, we use a coil segment as described in Sect. 4.6. One coil segment is placed at the battery connector, and another is positioned between the measurement trace near the microcontroller and the ground plane on the second layer. The magnitude of the current in each coil segment is irrelevant for the extraction, as we aim to obtain a resistance value. For simplicity, we set the current to 1 A. The individual resistances of the source and sink loops, as well as the resistance of the shunt, do not affect the estimation of the coupling resistance. It is only necessary to ensure that the current loops are closed. To achieve this, the transistors of the B6 bridge are shorted, and at the position of the shunt, we add a short conductor section to short the shunt.

The current distribution in both loops is plotted in Fig. 5.9. The loop at the top of the figure is calculated when the measurement loop is excited, while the loop at the bottom is calculated when the motor loop is excited. As expected, in the region indicated by the red shape, both loops share the conductor, and the current in this region contributes to the parasitic coupling resistance.

We now extract the actual resistance values from the simulation model. For the calculation of the current distribution in Fig. 5.9, we individually excite the coil segments. Thus, one simulation model is excited by the coil segment at the battery connector, and the second simulation model by the coil segment at the microcontroller.

For the extraction of the resistances, both coil segments are included in the simulation model, and the solver option to calculate the conductance matrix is activated. Unfortunately, the stationary current solver in CST does not directly calculate the resistance matrix, which is why the conductance matrix is used as an intermediate step. The conductance matrix is defined as:

$$\mathbf{I} = \begin{pmatrix} G_{11} & G_{12} \\ G_{21} & G_{22} \end{pmatrix} \mathbf{V}. \tag{5.5}$$

where $\mathbf{G}$ is the real part of the admittance matrix. For the calculation of the conductance matrix, the solver steps through each source in the EM model, as described in Sect. 4.6. In our example, the first coil segment at the battery connector is excited, and the voltage drop in the first and second coil segments (at the microcontroller) is calculated. This results in G_{11} and G_{21}. Then, the second coil segment is excited, and we obtain G_{22} and G_{12}. To derive the resistance matrix $\mathbf{R}$, we use the relationship:

$$\mathbf{V} = \begin{pmatrix} R_{11} & R_{12} \\ R_{21} & R_{22} \end{pmatrix} \mathbf{I} \tag{5.6}$$

We invert the conductance matrix, $\mathbf{R} = \mathbf{G}^{-1}$. A matrix inversion is required, rather than a simple inversion of the individual matrix elements. This operation can be performed using an external tool, such as simple Python code.

The calculated values of the resistance matrix are used as a T-element in the schematic of Fig. 5.5. Specifically:

- $R_{11} - R_{12}$ equals Z_{source},
- $R_{22} - R_{12}$ equals Z_{sink},
- and R_{12} equals R_{cp}.

As the coupling is symmetric, $R_{12} = R_{21}$. The coupling resistance R_{12} can be positive or negative, depending on the current directions in both loops. It has a straightforward interpretation: for a current of 1 A in the source loop, a voltage V equal to the magnitude of R_{12} in Ohms is induced in the sink loop.

Even though the interpretation of R_{12} given above is used in a straightforward way to estimate the effect of the parasitic resistance, it is instructive to create a full circuit that demonstrates the effect of the resistive coupling on the shunt

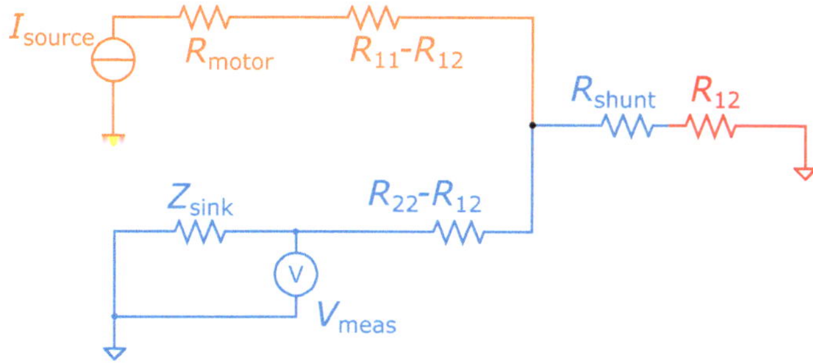

Fig. 5.10 Circuit used for the estimation of the parasitic coupling resistance on the shunt measurement

measurement. This circuit, shown in Fig. 5.10, is used in the next section for the analysis of different PCB layouts. In this circuit:

- The resistors in the motor loop (orange) do not influence the result, as we use an ideal current source.
- The resistor $R_{22} - R_{12}$ in the sink loop (blue) has no effect on the measurement because Z_{sink} is set to $1\,\text{M}\Omega$ to measure the voltage in the sink loop, making $(R_{22} - R_{12}) \ll Z_{\text{sink}}$.

5.3.3 *Comparison of Different Shunt Designs*

In this section, we study the effect of the coupling resistance on the shunt current measurement for three different PCB layouts. We select a shunt resistance of $R_{\text{shunt}} = 2\,\text{m}\Omega$, a typical value for shunt measurements. The three different layouts are depicted in Fig. 5.11.

The layout on the left is identical to the one shown in Fig. 5.8. The layout in the center aims to reduce the coupling resistance R_{12} by increasing the width of the

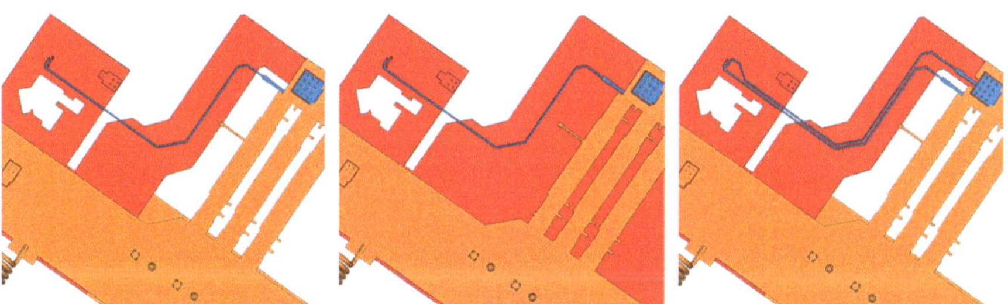

Fig. 5.11 Three PCB layout variants for the study of the parasitic coupling resistance

Table 5.2 Resistance values for different PCB layout variants

Layout variant	R_{11} (mΩ)	R_{22} (mΩ)	R_{12} (mΩ)
Original layout	13.3	50	2.1
Wide ground plane	12.1	50	0.9
Additional trace	13.3	100	0.6

ground plane. The third layout, shown on the right-hand side of Fig. 5.11, introduces an additional signal trace connected behind the shunt resistor. This additional signal trace closes the measurement current loop instead of the ground plane.

It is important to note that the second variant, in the center, represents a more-than-best case because it overwrites some signal traces originally routed on the bottom layer. In the early stages of design, such simplified cases are often evaluated. If they show no improvement, a more realistic layout does likely not improve performance either. Conversely, if they demonstrate a benefit, it may warrant a discussion about potential layout changes.

The computed resistances for the three layout variants are summarized in Table 5.2. All simulations are performed with the FEM stationary current solver, with a simulation time of less than 2 minutes for each design. As in the previous section, R_{11} represents the resistance of the motor current loop, R_{22} corresponds to the resistance of the measurement loop, and R_{12} is the parasitic coupling resistance.

For variants 1 and 3, R_{11} is 13.3 mΩ. In the layout with the wide ground plane, R_{11} decreases to 12.1 mΩ due to the increased width of the ground plane.

The value of R_{22} is significantly higher because of the smaller cross section of the trace leading to the measurement input of the microcontroller. For the return path being the ground plane, R_{22} is 50 mΩ, as the ground plane's wide size renders its resistance negligible compared to the trace. When the return path is also a trace, as in the third variant, R_{22} doubles to 100 mΩ.

Of most interest is the mutual resistance R_{12}. In the first layout, it is 2.1 mΩ, equivalent to the magnitude of the shunt resistance R_{shunt}. With the wide ground plane, the coupling resistance reduces to 0.9 mΩ, demonstrating that widening the ground plane reduces the parasitic resistive coupling. The most effective solution is routing with two signal traces, where the coupling resistance further decreases to $R_{12} = 0.6\,\text{m}\Omega$.

To study the effect of the layout resistances on the shunt measurement voltage, we drive the current source in Fig. 5.10 with a sinusoidal current at a frequency of 10 Hz and a peak value of $I_{\max} = 10\,\text{A}$, resembling a current through the electrical motor.

The measured shunt voltages for the three layout variants are shown in Fig. 5.12. Also, the hypothetical case of no coupling resistance, $R_{12} = 0\,\Omega$ is depicted. For the latter case, we see the expected voltage of $V_{\text{meas}} = R_{\text{shunt}} I_{\max}$ which is 20 mV for $I_{\max} = 10\,\text{A}$ and $R_{\text{shunt}} = 2\,\text{m}\Omega$.

In the case of two lines toward the voltage measurement at the microcontroller with $R_{12} = 0.6\,\text{m}\Omega$, we obtain a voltage offset of about $\frac{R_{12}}{R_{\text{shunt}}} \approx 30\%$. For the case

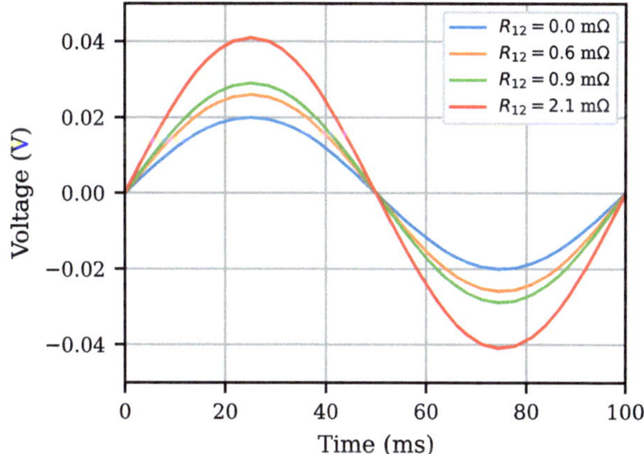

Fig. 5.12 Measured shunt voltage for a motor current of 10 A for the three different PCB layouts

with the increased ground plane, the error in the measurement is almost 50%. If $R_{12} \approx R_{\text{shunt}}$ as in the original layout, the error is about 100%.

The third variant, using two traces for the measurement, is the most suitable for the measurement of the motor current, but the error is still around 30% because both traces connect to the shunt at its side. To further improve the design, the connection of the measurement traces needs to be moved to the top of the shunt. That would, however, increase the required PCB area.

5.4 Capacitive Coupling

Domains: 

The schematic of the capacitive coupling is depicted in Fig. 5.13. The source and the sink loop are coupled by a capacitor C_{cp}. Different from the galvanic case, this schematic is frequency-dependent: The impedance of the coupling capacitor C_{cp} decreases with frequency, so that the coupling between sink and source increases with frequency and is zero for a DC voltage.

The magnitude of the coupling solely depends on the voltage and not on the current in the source loop. This is a result of the fact that the capacitance describes the coupling by the electric field, and the electric field is proportional to the voltage as we see in the introduction of this chapter. The time-varying electric field of the source induces a charge movement in the conductors of the sink, and this charge movement results in a current in the sink loop.

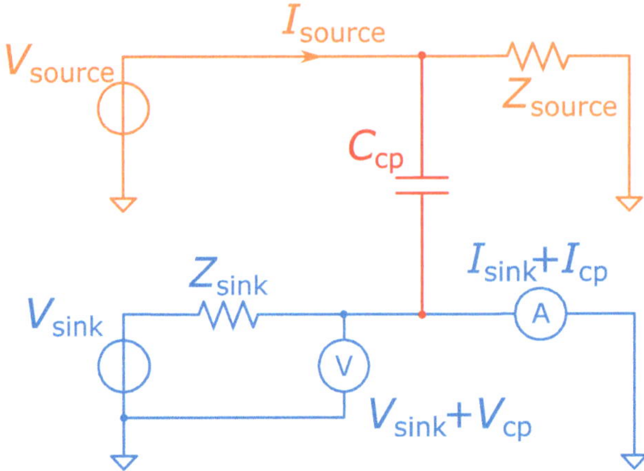

Fig. 5.13 Schematic of the capacitive coupling

Whether we measure a voltage or a current in the sink loop depends on the impedance of Z_{sink} and V_{sink}, as discussed in Sect. 5.2. We demonstrate the effect of the sink impedance in the application example.

5.4.1 *Application Example Capacitive Coupling: Common-Mode Voltage due to a Clock Signal*

- Ideal voltage source of a clock signal
- Voltage in a LISN circuit
- Capacitive, described by a capacitance matrix
- Electrostatic solver
- AC, Transient
- 0–1 GHz

In this application example, we demonstrate how the voltage on a clock signal trace can induce a common-mode voltage between the PCB ground and the measurement table. The setup models the clock trace of an SDIO bus.

The SDIO bus, originally designed for data transfer to SD cards, is also sometimes used as a communication bus between two ICs on a PCB. In our example, the SDIO clock trace connects the main microcontroller of the PCB to a WiFi chip.

Clock traces transmit voltage signals to synchronize the data stream between ICs. To minimize power consumption during communication, the current on clock traces is kept low. As a result, capacitive coupling, driven by the electric field, is the

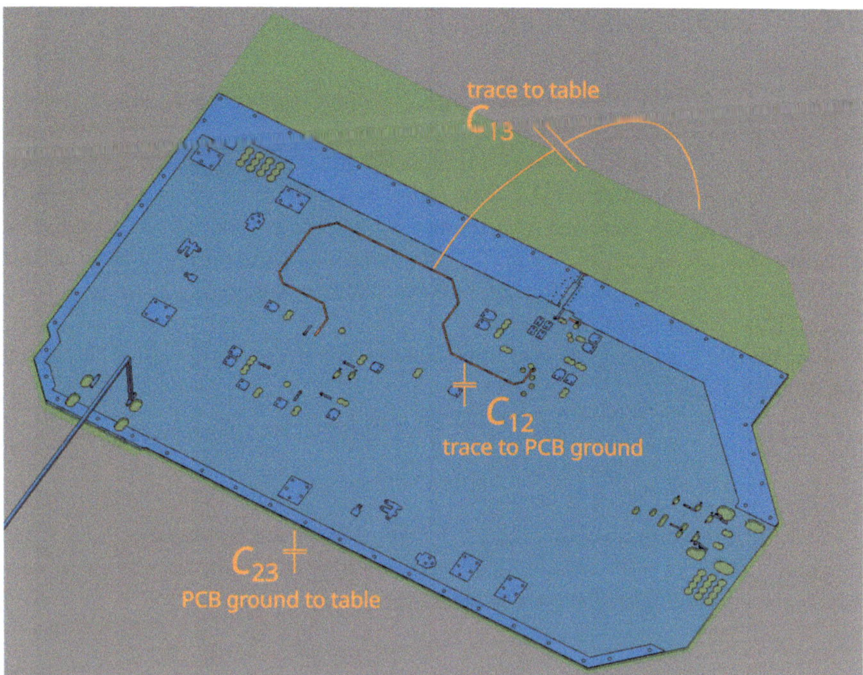

Fig. 5.14 Sketch of the PCB with a clock signal trace and the corresponding capacitances calaculated in this example

dominant coupling mechanism. Whenever the logic state of the clock trace changes, the rapid voltage transitions couple capacitively to the sink.

In this scenario, three capacitances influence the coupling: one between the clock trace and the PCB ground, another between the clock trace and the metallic table below the PCB, and a third between the PCB ground and the table. The PCB is located 5 cm above the table. An FR4 layer with a permittivity of $\varepsilon_r = 4.4$ is located between the trace and the PCB ground. Additionally, a short wire is connected to the PCB ground to model the negative supply wire.

To measure the common-mode emission, a 50 Ω resistor is placed between the ground wire and the measurement table. This simplified representation of an LISN suffices for this setup, as the supply path of the PCB is not included in the simulation. The complete setup is illustrated in Fig. 5.14. Although the capacitance between the trace and the table is only in the range of femtofarads, while the other capacitances are in the picofarad range, we observe that this small capacitance can still result in a significant voltage at the LISN for typical clock frequencies used in the SDIO bus.

5.4.2 Computation of the Capacitive Coupling

We use the electrostatic solver to calculate the capacitance matrix, which contains the capacitances indicated in Fig. 5.13. As with the stationary current solver, the electrostatic solver calculates the scalar electric potential ϕ. The key difference is that the source of the electric field in the electrostatic solver are charges or potentials on metals, and the potential distribution is determined outside of conductive domains. All conductors are set to PEC and treated as equipotentials, with no field distribution calculated inside the conductors. Since this calculation assumes a static case, this is a valid simplification.

Equations 5.2 and 5.3 are also applicable in the electrostatic case. The capacitance is defined as:

$$C = \frac{Q}{\phi} \tag{5.7}$$

Thus, it is a measure of how much charge is stored on a conductor for a given potential. Working with charge values is somehow not common for electronics engineers, so we employ the energy-based definition of capacitance,

$$W_\mathrm{e} = \frac{1}{2} C V^2. \tag{5.8}$$

Every conductor is characterized by its self-capacitance. The authors have observed that the concept of self-capacitance can be challenging for electronics engineers to understand, as they are more accustomed to capacitances defined between two metal conductors, such as capacitors mounted on a PCB or the capacitances shown in Fig. 5.13.

To grasp the concept of self-capacitance, one must consider the potential value of a conductor as described in Eq. 5.7. However, the absolute potential of a conductor cannot be directly measured; only potential differences, observed as voltages, are measurable. A common approach in the literature is to define the potential at infinity as zero, but this raises practical questions about how to measure the voltage relative to infinity. This reasoning often leads to the conclusion that self-capacitance is a theoretical construct. Nevertheless, self-capacitance can indeed be observed in practical experiments.

A Van de Graaff generator serves as an excellent example of self-capacitance. In a Van de Graaff generator, a rotating belt charges a metallic sphere, storing the energy from the rotation as electric field energy, as described by Eq. 5.8. In this case, only one conductor is present, and its potential increases because of the energy added. As a result, a voltage is generated between the sphere and its surroundings.

For the calculation of the capacitance matrix, the electrostatic solver requires a reference potential, that is, a position where the potential is set to zero. This reference can be a conductor within the computational domain, but it is typically chosen to be the boundary of the computational domain.

In our PCB example, we assign fixed potentials to the clock trace, the PCB ground, and the table. The self-capacitances of all three conductors are then calculated with reference to the boundary, resulting in a 3 × 3 capacitance matrix.

In the schematic of the clock trace and the PCB board that we set up later, the self-capacitance is omitted, and only the mutual capacitances are used. This approach aligns with the everyday definition of capacitance commonly used by electronics engineers. However, this is not the only possible method.

Alternatively, we could define the table as the reference and apply fixed potentials to the clock trace and the PCB ground. In this case, the self-capacitance of the clock trace and the PCB ground would be defined relative to the table, resulting in a 2 × 2 capacitance matrix. In this setup, the self-capacitance would need to be included in the schematic.

When calculating the capacitance matrix in the electrostatic solver, the three sources defined as fixed potentials are stepped sequentially. In the first step, source 1 is excited by a 1 V potential, the boundary and all other CAD shapes with a potential definition are set to 0 V, thus used as a reference. In the subsequent steps, source 2 and source 3 are excited individually in the same manner.

This process results in three potential distributions, from which the electric field is calculated using Eq. 5.3. To calculate the capacitance from these distributions, the energy definition in Eq. 5.8 is applied as follows: The energy in the electric field containing dielectric media can be calculated using the scalar product of the electric field and the electric flux density, integrated over the volume:

$$W_{\mathrm{e}} = \frac{1}{2} \iiint_V \mathbf{E} \cdot \mathbf{D} \, dV. \tag{5.9}$$

We can combine Eqs. 5.8 and 5.9. Since the solver uses $V = 1$ V when stepping through the sources, we obtain:

$$C_{ij} = \iiint_V \mathbf{E}_i \cdot \mathbf{D}_j \, dV. \tag{5.10}$$

This is how the electrostatic solver calculates the capacitance matrix. The result of the calculation is the so-called grounded capacitance matrix, alternatively referred to as the field-theory-based capacitance matrix or the Maxwell capacitance matrix. In the grounded capacitance matrix, the system is characterized by the potential relative to the selected reference. If the reference is the boundary, this definition can be rather impractical.

The grounded capacitance matrix can be converted to the so-called lumped capacitance matrix or the circuit-theory-based capacitance matrix. In the lumped capacitance matrix, the system is characterized by voltages between conductors (instead of voltages relative to the 3D reference), as used in circuit schematics. For this conversion, the mutual capacitances of the grounded capacitance matrix are simply multiplied by -1, and the self-capacitances of the lumped capacitance matrix are obtained by summing the columns of the grounded capacitance matrix.

While this might sound complicated, it is interpreted in a straightforward way if the self-capacitances are not used later on.

In the way we define the three fixed potentials—for the trace, the PCB ground, and the table—the self-capacitances do not play a role. In this setup, only the mutual capacitances are used, and the magnitudes of the mutual capacitances in both matrices are the same. The electrostatic solver in CST calculates both matrix types. In our schematic, we use the mutual capacitances from the lumped capacitance matrix.

Next, we consider the setting for conductors that do not carry any source definition. In the PCB example, potentials are assigned to all three conductors, and the boundary is used as the reference. If another conductor is added—for instance, another signal trace to the model—we need to define how it is treated by the solver.

The electrostatic solver, as well as the capacitance calculation of the pRLC solver in Sect. 5.6, provides a switch to treat conductors without any source definition as either grounded or floating. When set to grounded, all conductors without a source are used as the reference potential, even if they are not physically connected. This can be a desired option when we aim to assign the same potential to conductors that are not geometrically connected.

Setting non-connected conductors to the same potential is only feasible in a static solver. A full-wave solver inherently treats non-connected conductors as floating because, in a full-wave simulation, potentials cannot simply be set on conductors since conductors are no longer equipotentials.

From this discussion, it becomes evident that using the capacitance matrix for arbitrary conductor arrangements must be approached with great care. For two conductors, the mutual capacitance between them is straightforward to interpret. However, with multiple conductors, understanding the source setup and the definition of the reference potential in the EM model can become significantly more challenging.

Figure 5.15 depicts the schematic for calculating the coupling of the clock trace to the simplified LISN. The source loop, indicated in orange, is driven by a voltage

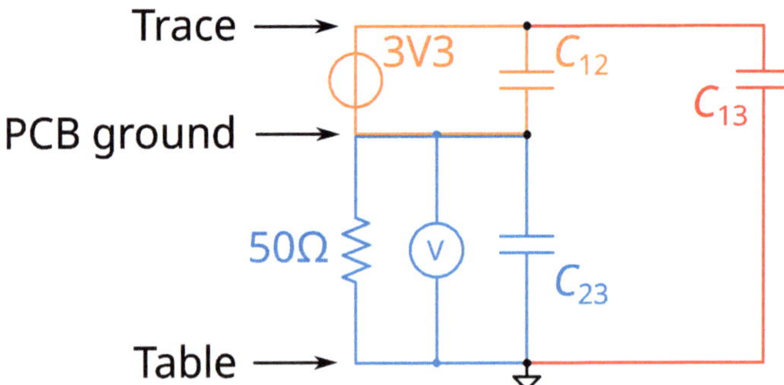

Fig. 5.15 Schematic for the calculation of the coupling of the clock trace to a LISN simplified as a 50 Ω resistor

source with a peak voltage of 3.3 V and terminated by C_{12}, which represents the capacitance of the signal trace to the PCB ground. This setup assumes that there is no termination of the clock trace on the receiver side.

The impedance in the sink loop consists of C_{23}, the capacitance between the PCB ground and the table, and a 50 Ω resistor, across which the coupling voltage is measured. The coupling path between the source and the sink is defined by the capacitance between the clock trace and the table, C_{13}.

In this setup, the table is connected to the circuit ground. Note that the circuit ground is not the same as the reference used by the electrostatic solver. The schematic only uses mutual capacitances, as self-capacitances are not required when the 3×3 capacitance matrix is defined with fixed potentials at all conductors and the boundary as the reference.

In the schematic, we have the freedom to place the circuit ground either at the trace or at the PCB ground. The measured voltage across the 50 Ω resistor remains unchanged regardless of whether the circuit ground is connected at the table or at the PCB ground.

5.4.3 Comparison of Different Clock Trace Layouts

In this section, we evaluate the voltage at the simplified LISN for two clock trace layouts shown in Fig. 5.16. In the original routing, the clock trace is routed on the top layer of the PCB. For a view of the complete PCB, please refer to Fig. 5.14. In the modified routing, shown on the right-hand side of Fig. 5.16, the clock trace is routed for nearly the entire trace length in an inner layer, with PCB ground planes both above and below the trace. Only the pins at the ends of the clock trace, required

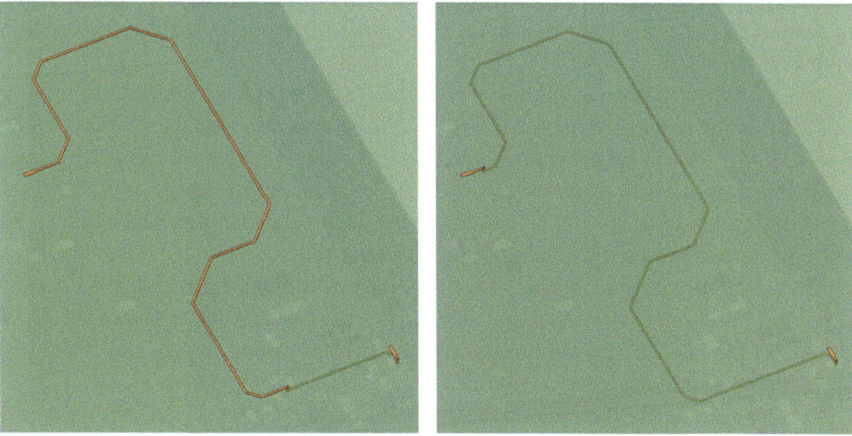

Fig. 5.16 Different routing of the clock trace. Left: routing of the clock trace on an outside layer. Right: routing of the clock trace in an inner layer

Table 5.3 Entries of the capacitance matrix for the two layout variants

Layout variant	C_{12} (pF)	C_{13} (fF)	C_{23} (pF)
Trace on top layer	13.1	4.88	5.5
Trace buried	18.9	0.36	5.5

to connect the trace to the IC, remain on the top layer. Directly behind the pins, a via routes the trace to the inner layer. With this modified routing, we expect the capacitance between the trace and the table to be minimized.

The capacitance obtained from the EM simulation of the two models is summarized in Table 5.3. The FEM-based electrostatic solver requires 2 minutes and 11 seconds for each simulation, including mesh adaptation. The capacitances between the clock trace and the PCB ground, as well as the capacitance between the PCB ground and the metallic table, are in the order of pF.

As expected, the capacitance C_{12} between the trace and the PCB ground increases when the trace is buried, as in this configuration, the PCB ground is located both above and below the trace for most of the routing length. The capacitance C_{23} between the PCB ground and the metallic table is not affected by the layout modification. The coupling capacitance C_{13} between the trace and the metallic table is of the order of fF. In the buried case, C_{13} is lower than in the top-layer routing case by a factor of 10.

The plots of the electric field in Fig. 5.17 illustrate the expected behavior of the electric field. For the calculation of the field distribution, the potential on the trace was set to 3.3 V, and the potential on the PCB ground and the table was set to 0 V. In the case where the clock trace is routed on the top layer, the electric field is visible above the trace, pointing toward the PCB ground and the table. In the case where the trace is routed on the inner layer, the field is confined to the inner part of the PCB. Because the distance between the PCB layers is in the order of 100 μm, the electric field is barely visible in the plot, with only two red arrows plotted. The region above the PCB, left of the clock trace, where some small blue arrows are visible, is due to the pin of the clock trace that is located on the top layer in the plane behind the visible plane.

We use the coupling capacitances C_{13} and C_{23}, computed from the EM simulation, within the network depicted in Fig. 5.15 to evaluate the voltage across the 50 Ω resistor. The voltage source driving the clock trace with respect to the PCB ground generates a signal with an amplitude of 3.3 V, a frequency of $f_{\text{clock}} = 25$ MHz, and a rise and fall time of $t_{\text{r,f}} = 5$ ns, modeling the SDIO clock signal. The clock signal waveform is shown on the left-hand side of Fig. 5.18.

On the right-hand side of Fig. 5.18, the voltage waveform across the 50 Ω resistor is plotted. The voltage is observed only during the rise and fall times of the clock signal, as the capacitive coupling increases with frequency and is zero when the clock signal amplitude remains constant.

For the case where the clock trace is routed on the top layer, the voltage across the 50 Ω resistor reaches an amplitude of 170 mV during the falling edge of the clock signal and − 170 mV during the rising edge, resulting in a peak-to-peak

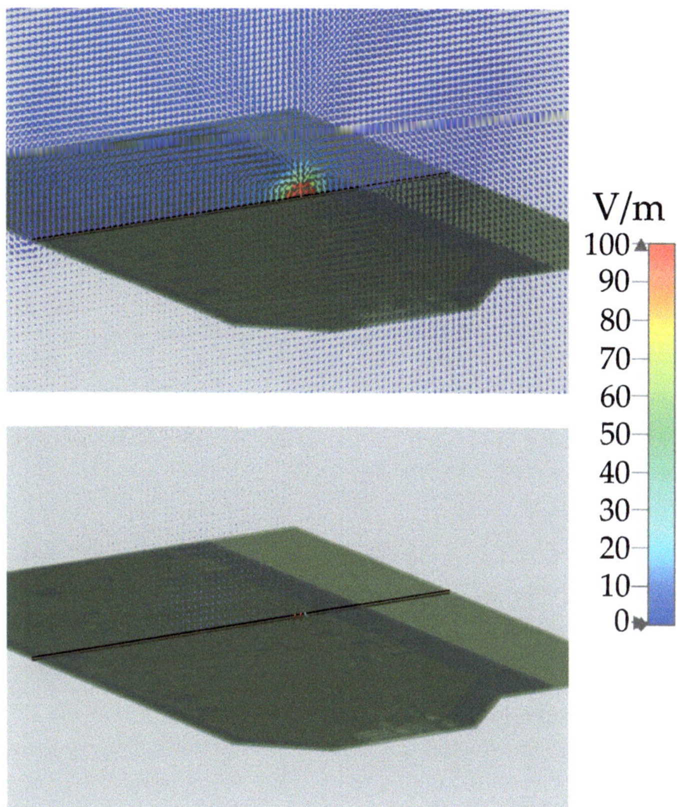

Fig. 5.17 Top: electric field around the clock trace routed on the top layer. Bottom: electric field around the clock trace routed on inner layer

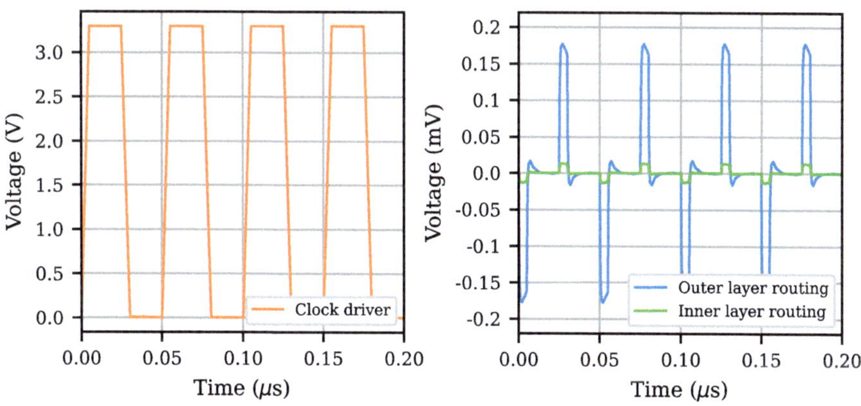

Fig. 5.18 Left: Simulated SDIO clock signal. Right: Voltage at the LISN resistor for the two board layouts

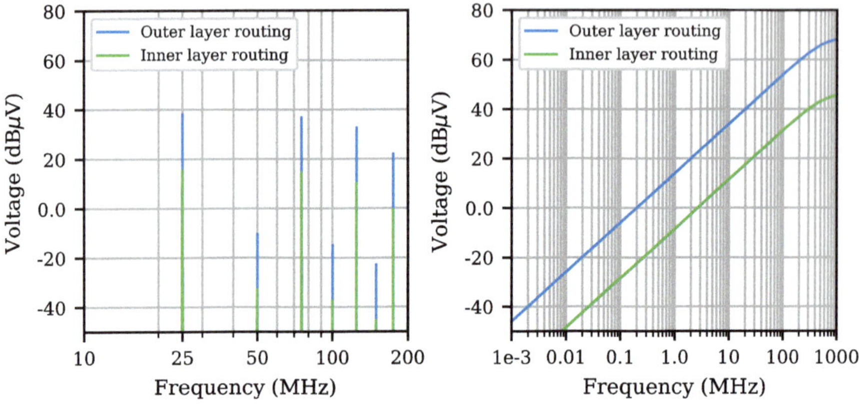

Fig. 5.19 Spectrum of the voltage at the LISN resistor due to capacitive coupling. Left: Spectrum when the trace is driven with a clock signal. Right: Circuit simulation with a 3.3 V AC source

voltage of 340 mV. In contrast, when the clock trace is routed on an inner layer, the voltage amplitude decreases to 13 mV, yielding a peak-to-peak voltage of 26 mV. This reduction in voltage directly corresponds to the difference in the coupling capacitance C_{13}, as shown in Table 5.3.

To further analyze the coupling to the 50 Ω resistor, we plot the spectrum of the voltage from the SDIO clock signal on the left-hand side of Fig. 5.19. At the clock frequency of 25 MHz, the voltage across the 50 Ω resistor is 38 dBμV for the case where the clock trace is routed on the outer layer, and 15 dBμV for the case with the routing on the inner layer. This indicates that routing the clock trace on the inner layer reduces the coupling by 23 dB.

A common EMC design recommendation is to route high-speed clock signals on inner layers. Using the simulation model, we can confirm that routing on inner layers significantly reduces emissions. Furthermore, the simulation provides a detailed evaluation of the potential improvement achieved with a specific layout change. Such an analysis would be highly challenging without the EM simulation model, as analytically calculating the capacitance of the clock trace to the table is difficult, due to the presence of the PCB ground layer between the clock trace and the table.

The results of the excitation with the clock signal allow us to estimate the voltage at the 50 Ω resistor for the SDIO clock signal. However, if we aim to analyze the frequency dependence of the coupling independently from the signal properties, a simulation using an AC source of fixed amplitude across the entire frequency band is more suitable.

On the right-hand side of Fig. 5.19, we plot the results of the AC simulation. The voltage source is set to 3.3 V at every frequency, and the maximum frequency is chosen to be 1 GHz. Since the distance between the PCB and the table is 5 cm, the model is electrically short only below 600 MHz (for a distance of $\frac{\lambda}{10}$). At 1 GHz, the model is no longer strictly valid for an RLC description of the coupling.

Nevertheless, we have decided to include frequencies up to 1 GHz in the plot, as it reveals the dependence of the coupled voltage on the impedance in the sink loop: up to a frequency of approximately 500 MHz, the capacitive coupling increases linearly with a slope of 20 dB/decade, as expected. Beyond this frequency, both curves begin to flatten while maintaining the 23 dB difference between them. This behavior is explained by analyzing the schematic in Fig. 5.15:

Since the voltage source is ideal, the voltage between the trace and the PCB ground is not influenced by the impedance of C_{12}. Therefore, C_{12} can be removed from the schematic when using an ideal voltage source. In the sink circuit, C_{23} and the 50 Ω resistor are in parallel, and the voltage is measured across this parallel combination. At approximately 570 MHz, the impedance of C_{23} equals 50 Ω and decreases further with increasing frequency. Thus, for higher frequencies, the 50 Ω resistor becomes negligible in the schematic. The resulting schematic simplifies to a capacitive voltage divider with a fixed ratio of C_{23} to C_{13}. This is an excellent example of how Z_{sink} affects the measurement of the coupling effect.

5.5 Inductive Coupling

Domains: 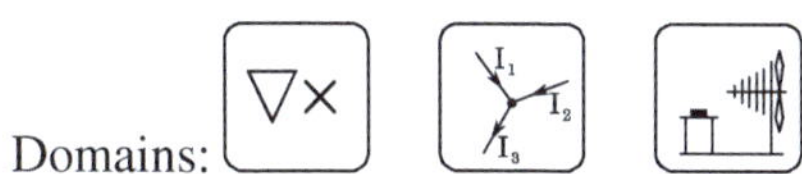

The schematic of the magnetic coupling is depicted in Fig. 5.20. The coupling path between the source and the sink loops is represented by a mutual inductance that couples the inductances of the two loops. Similar to capacitive coupling, the coupling from the source to the sink loop depends on the frequency. In magnetic coupling, a current in the source loop induces a voltage in the sink loop.

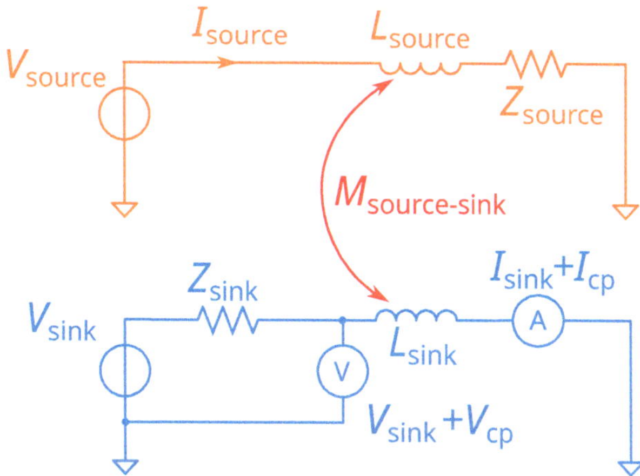

Fig. 5.20 Schematic of the inductive coupling

Consequently, the magnitude of the coupling is determined by the current in the source loop, unlike in capacitive coupling, where it depends on the voltage. As the frequency increases, the induced voltage and, thus, the coupling strength, also increase.

The physical origin of the mutual inductive coupling lies in the magnetic flux. The magnetic flux generated by the current in the source loop penetrates the surface of the sink loop, linking the magnetic fluxes of both loops. For linear materials, this flux linkage, which results in the mutual inductance, is independent of the current in the source loop. However, for nonlinear materials such as ferrites or other magnetic materials, the mutual inductance becomes a function of the current in both loops.

In the case of capacitive coupling, we do not typically discuss the dependence of the coupling path on the magnitude of the excitation because nonlinear dielectric materials are uncommon in electronic designs. However, nonlinear ferrite materials are frequently encountered in inductive coupling scenarios. As with the previous coupling mechanisms, whether a voltage or a current is measured in the sink loop depends on the impedance of Z_{sink} and V_{sink}, as outlined in Sect. 5.2.

In the following example, we also discuss how the frequency dependence of the coupling path can lead to an inaccurate schematic of the decomposition.

5.5.1 Application Example Inductive Coupling: Coupling Between Adjacent Power Inductors

Ideal current source
Voltage in a LISN circuit
Inductive, described by a inductance matrix
Magnetostatic solver
AC, Transient
0–100 MHz

In this application example, we study the coupling between two power inductors placed in close proximity. Let us assume a scenario with two switched buck converters on a PCB. One of the converters operates at 200 kHz and utilizes a larger inductor with an inductance of $L_{\text{large}} = 6.64\,\mu\text{H}$ and a 4040 footprint. The second converter operates at 2 MHz and uses a smaller inductor with an inductance of $L_{\text{small}} = 369\,\text{nH}$ and a 2525 footprint. These inductors are positioned with a center-to-center distance of 1 cm, as illustrated in Fig. 5.21.

The inductors are of a simple wire-wound type and are non-shielded, making them a very cost-effective choice. The wire is wound around a cylindrical ferrite core, and the windings are modeled in full 3D. Most inductor manufacturers do

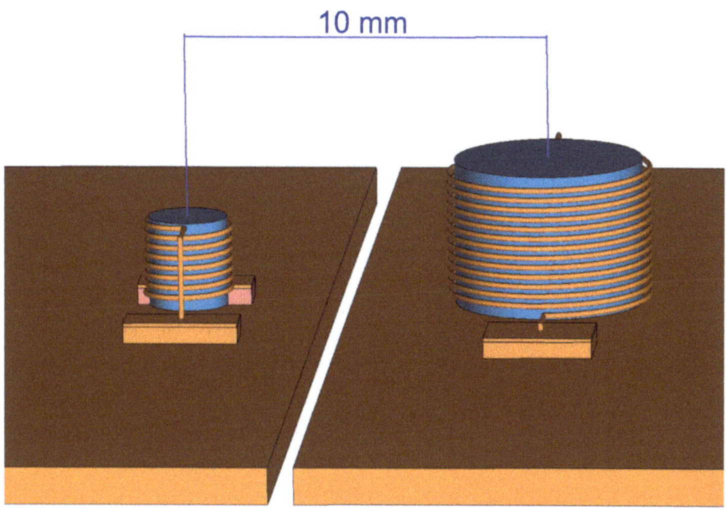

Fig. 5.21 Overview of the model for the calculation of inductive coupling

not provide ready-to-simulate EM models to protect their intellectual property. As a result, the 3D models used in simulations are often quite generic.

To create the models, we match the coil radius to the footprint dimensions and select an iron powder material with a permeability of $\mu_r = 120$, as such materials are commonly used for SMD inductors. We then calculate the number of windings required to achieve the specified inductance values. For the larger inductor, 13 windings are needed to achieve $6.6\,\mu$H. For the smaller inductor, 7 windings are required to reach 360 nH. This method allows reverse engineering of 3D component models to match the inductance values provided in datasheets.

While the generated model may not precisely resemble the real component, this reverse engineering approach is a useful practice when 3D models are not provided by the vendor. In Chap. 9, we explore how this method can be extended to build inductor models with accurate self-resonant frequency.

5.5.2 Computation of the Inductive Coupling

For the computation of the inductive coupling, we use the magnetostatic solver. The magnetostatic solver calculates the magnetic field from a given current distribution by solving Ampères law with the time derivative of the electric flux set to zero,

$$\nabla \times \mathbf{H} = \mathbf{J} \tag{5.11}$$

If the current density is distributed over volumetric conductors, as in our case, a stationary current solver (Sect. 5.3) is run in the first step to calculate the current distribution in the conductors before proceeding with the magnetostatic solution.

For the calculation of the inductance, we cannot use the circuit-based definition given in Eq. 3.8, which defines inductance via the induced voltage, because the magnetostatic solver computes a static solution where no induced voltage is present. Instead, we use the field-based inductance definition, which determines the inductance of a loop by the magnetic flux Φ through the closed surface of the loop S for a given current I in the loop:

$$L_{\text{loop}} = \frac{\oiint_S \mathbf{B} \mathrm{d}S}{I} = \frac{\Phi}{I}. \tag{5.12}$$

Using Eq. 5.12, we calculate the inductance matrix by applying a stepped approach similar to the two other coupling cases. In this method, the first current source is excited with a current of 1 A, while all other sources are set to zero. By calculating the magnetic flux through the surface of the first loop, we determine the self-inductance of the first loop. Similarly, by calculating the magnetic flux through the surface of the second loop, we determine the mutual inductance between the two loops.

Next, the second current source is set to 1 A, and the procedure is repeated for all sources. This iterative process yields the inductance matrix, which characterizes the inductive coupling between all loops in the model,

$$\begin{pmatrix} \Phi_1 \\ \Phi_2 \end{pmatrix} = \begin{pmatrix} L_{11} & -L_{12} \\ -L_{21} & L_{22} \end{pmatrix} \begin{pmatrix} I_1 \\ I_2 \end{pmatrix}, \tag{5.13}$$

where Φ_1, Φ_2 denote the magnetic fluxes through the loops 1 and 2, generated by the currents I_1, I_2 in either the loop 1 or the loop 2.

For the calculation of the field-based inductance, it is essential that the two loops are galvanically separated from each other. This is why a gap in the ground plane is included in our simulation model, as shown in Fig. 5.21. Without this gap, the loop areas cannot be separated, and the inductance cannot be calculated.

Unlike the magnetostatic solver, full-wave solvers compute the inductance from Z-parameters, as discussed in Sect. 2.7. When using Z-parameters, the separation of the loops is not required. The inductance from full-wave solver is the circuit-based definition.

To utilize the mutual inductance calculated by the EM solver in a schematic, it must be converted to a coupling factor, which is the input required for a SPICE simulation model. The equation for the coupling factor k is:

$$k = \frac{L_{12}}{\sqrt{L_{11} \cdot L_{22}}}. \tag{5.14}$$

In Fig. 5.22, we plot the magnetic flux density generated by the small coil. The flux penetrating the large coil is clearly visible, illustrating the inductive coupling between the two coils. Since the calculation was performed using the magnetostatic

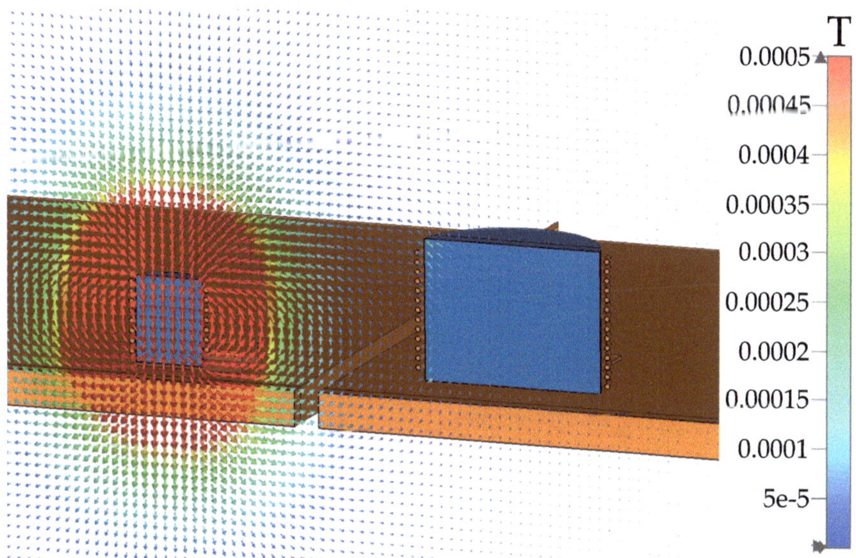

Fig. 5.22 Magnetic flux density **B** generated by the small coil

solver, the magnetic flux fully penetrates the ground plane. Induced currents in the ground plane are not accounted for in the magnetostatic simulation.

The induced currents in the ground plane and the skin effect in the coil wires result in a reduction of the inductance compared to its DC value. Also the coupling factor between the inductors is influenced by the presence of induced currents.

In the following analysis, we use the static inductance matrix up to frequencies of 50 MHz. This choice is justified by the purpose of this chapter, which is to demonstrate the coupling mechanism rather than provide accurate results. However, applying static inductance values at higher frequencies serves as a cautionary example of an incorrect coupling path model in the decomposition schematic, as discussed at the end of Sect. 5.2.

5.5.3 Inductive Coupling as a Function of the Distance Between the Coils

For the calculation of the inductive coupling between the two coils, we use the schematic shown in Fig. 5.23. The small coil is driven by a current source I_{source}. The inductance values of both coils are connected in series with a 1 Ω resistor, modeling the load in each loop. The coupling factor k couples the inductors. From the EM simulation, we obtain a coupling factor of $k = 0.001$ for the coils placed at a center-to-center distance of 1 cm, as depicted in Fig. 5.21.

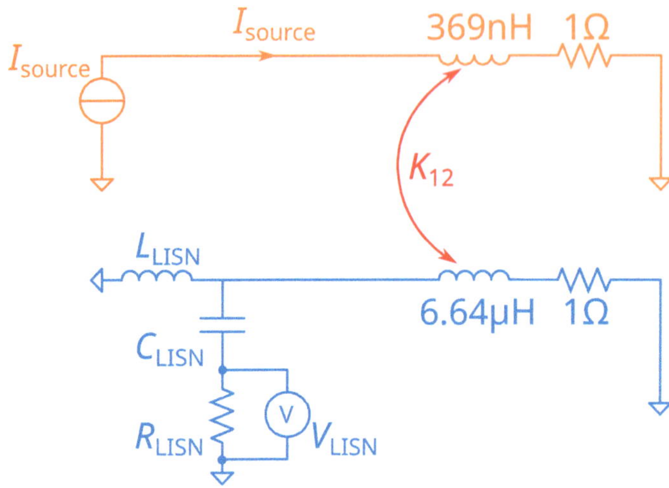

Fig. 5.23 Schematic for the calculation of the inductive coupling

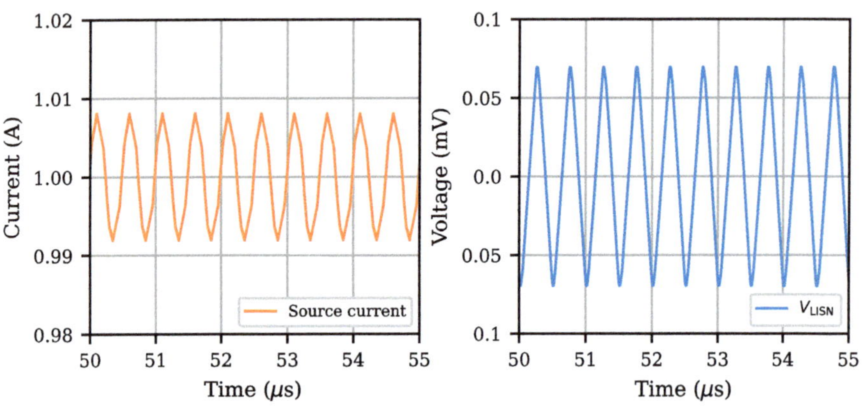

Fig. 5.24 Signals in the first simulation of inductive coupling. Left: Source current. Right: Voltage at the LISN

To measure the coupling to the sink loop, a LISN is added with $L_{LISN} = 5\,\mu H$, $C_{LISN} = 0.1\,\mu F$, and $R_{LISN} = 50\,\Omega$. The coupled voltage is measured across the LISN resistor.

In a first simulation, we drive the source loop with a 1 A current overlaid with a triangular ripple of 0.01 A. The frequency of the ripple is set to 2 MHz. The coupled signal, measured as a voltage at the LISN, is in the order of 0.1 mV, as shown in Fig. 5.24. The spectrum at the LISN resistor is plotted on the left-hand side of Fig. 5.25.

The current ripple in the source loop generates a LISN voltage of 35 dBμV at 2 MHz. In our design, the buck converter associated with the large coil was designed to operate at 200 kHz. Due to the coupling from the small coil, a signal at a much

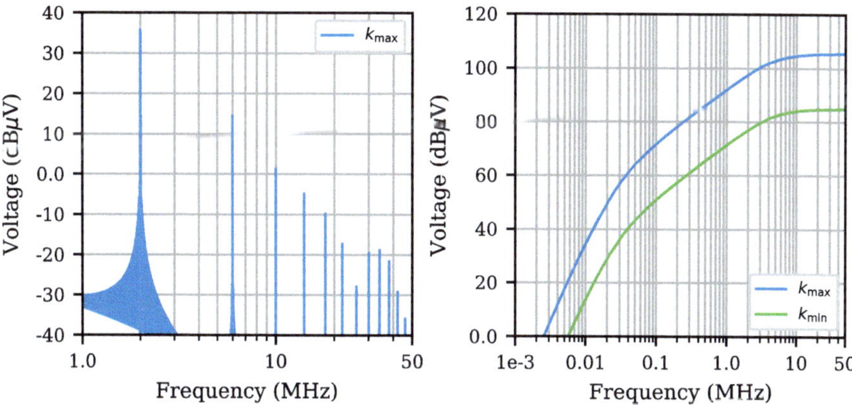

Fig. 5.25 LISN voltage in the simulation of inductive coupling. Left: Spectrum when driving the source loop with the current plotted in Fig. 5.24. Right: Voltage at the LISN, calculated by an AC simulation with a broadband 1 A excitation at the position with no offset and the position with maximal offset

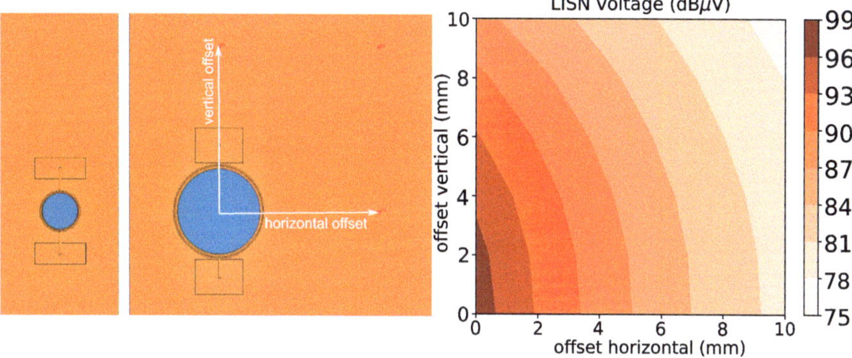

Fig. 5.26 The coupling coefficient decreases with the distance d between the coils. The LISN voltage is evaluated at 2 MHz

higher frequency is introduced into the buck converter with the large coil. If a filter is designed for the buck converter with the large coil, it is crucial to ensure that the coupled signal at the higher frequency is also attenuated effectively.

The inductive coupling is reduced by increasing the distance between the coils. In the next EM simulation, we parameterize the position of the large coil and introduce both a horizontal and a vertical offset from the original position. The coil is moved 1 cm horizontally and 1 cm vertically in 0.2 cm steps. A sketch of the offsets is shown on the left-hand side of Fig. 5.26, with the maximal extent of the offsets indicated by red points. The simulation of a single position takes slightly less than 1 minute. The parametric sweep includes 36 positions, requiring a total simulation time of approximately 30 minutes. To optimize the process, we manually set the mesh and deactivate the mesh adaptation during the sweep. Although manual mesh setup requires additional preparation time, it significantly reduces the simulation time for parametric studies involving many similar models, as in this case.

For this study, we use an AC simulation and excite the source loop with a current of 1 A. The coupling factor is calculated from the mutual inductance obtained from the parametric EM simulation (Eq. 5.14). On the right-hand side of Fig. 5.25, we plot the LISN voltage for the original position (no offset), where the coupling is the highest, and for the position with the largest offset, where the coupling is the lowest. The difference in the LISN voltage between these positions is 20 dB. Thus, moving the large coil 1 cm to the right and 1 cm upward reduces the coupling by 20 dB. As in the capacitive example, the flattening of the curves is due to the impedance relationship in the sink loop. Please also note that a current of 1 A at 50 MHz corresponds to a power of more than 100 W in the source loop.

To provide an overview of how the coil offset affects coupling, we create a contour plot that illustrates the LISN voltage as a function of the horizontal and vertical offsets. This plot is depicted on the right-hand side of Fig. 5.26. The LISN voltage is evaluated at 2 MHz. As expected, the horizontal offset is more effective in reducing coupling than the vertical offset because the distance between the centers of the inductors decreases more rapidly in the horizontal direction.

5.5.4 Frequency Dependence of the Inductive Coupling

As stated in the beginning of this section, the coupling factor between the inductors calculated in the previous section is based on the DC coupling factor. However, there are three potential contributions that can influence the actual value of the coupling factor and the inductance of the individual coils:

- The nonlinear behavior of the permeable material
- The dispersive properties of the permeable material
- The impact of induced currents

A power electronics designer typically selects an inductor that does not saturate during operation. Therefore, we did not analyze the effect of saturation in this study. Instead, we focus on comparing the results obtained using the DC coupling factor to a frequency-dependent coupling factor to understand the error introduced into the decomposition schematic.

For this comparison, the geometry with the minimal offset is simulated using the full-wave FEM solver. Two models are simulated: the first assumes a fixed permeability of $\mu_r = 120$, consistent with the magnetostatic simulation. The second uses a Debye first-order dispersive material model for the permeability, defined by $\mu_{static} = 120$ and a relaxation frequency of 10 MHz. The same permeable material is used in both coils.

The results of the AC simulation using different models for the coupling factor are plotted in Fig. 5.27. The AC simulation is performed with the same settings as

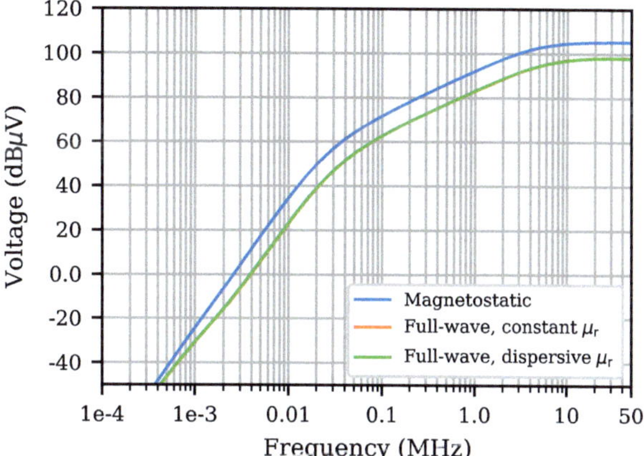

Fig. 5.27 Voltage at the LISN, calculated by an AC simulation with a broadband 1 A excitation at the position with no offset. Comparision of different coupling factor models

in the previous section. Comparing the voltages obtained using the coupling factor calculated by the magnetostatic solver to those calculated by the full-wave solver, we observe that, as expected, at very low frequencies, the curves are very close to each other, with a difference of about 2 dB.

At higher frequencies, the induced currents in the ground plane affect the coupling, resulting in a lower coupling factor in the full-wave solver results. At 50 MHz, the difference between the two results is approximately 8 dB. This demonstrates that using the DC simulation for the extraction of the coupling factor overestimates the coupling at higher frequencies. Nevertheless, the relative changes obtained from the study of the coil offset remain valid. It is ultimately up to the simulation engineer to decide whether the differences in the coupling factors are acceptable for the application.

The curves obtained from the full-wave model with and without dispersion are so similar that only one curve is visible. This indicates that the dispersion inside the permeable material does not significantly affect the coupling factor in this example.

5.6 Partial RLC Extraction

Domains:  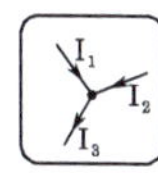 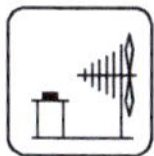

Using the various static solvers, we can extract RLC values through field computation and incorporate these values into a schematic to estimate the resulting voltages and currents. This process is manual: different source types are defined in the EM simulation models, and the RLC values are placed into the schematic manually.

It is desirable to improve this process by defining a single EM simulation that extracts an RLC circuit model directly usable in the schematic of the decomposition.

Because the extraction involves RLC models, the simulated geometry must remain electrically short, allowing the separation of electric and magnetic fields for individual computation. However, a unified set of sources is required for such a simulation model. Only by using the same sources for both the electric and magnetic field calculation, the extraction process can be automatized.

Several obstacles arise in unifying the sources. For the stationary current calculation, it is not strictly necessary to close the current loop. It is possible to calculate the resistance of a non-closed current path, by using current ports instead of coil segments. Conversely, the magnetic field calculation in Sect. 5.5 requires a closed current path. Lastly, the capacitance calculation in Sect. 5.4 allows only one source per conductor, which is incompatible with applying a voltage difference across the ends of a conductor.

5.6.1 Partial Inductance

We address the challenges described above by adopting a partial RLC extraction scheme. The concept of partial inductance has gained popularity in EMC modeling, and there is a wealth of literature available on this topic (Paul 2010; Caniggia and Maradei 2008). Partial inductance facilitates the extraction of inductance values that can be directly incorporated into circuit simulations without the need to close the current loop in the model used for the extraction. Since it is derived from the vector potential **A**, it depends on the choice of a gauge condition (e.g., Coulomb or Lorenz; Traub et al. 2012). This is different from the loop inductance, which is independent from the gauge condition.

In the following, we explain the basics of partial inductance, as understanding this concept is crucial for EMC modeling. In Sect. 5.5, we introduced inductance as the ratio of the magnetic flux through the area A enclosed by a current loop to the current I flowing in this loop (Eq. 5.12).

This definition inherently requires a closed loop. Consequently, it cannot be directly applied to partial inductance calculations, where the current loop is not closed. If we use the definition that inductance is a proportionality factor that measures the energy stored in the magnetic field W_m depending on the current:

$$W_\mathrm{m} = \frac{1}{2} L \cdot I^2, \tag{5.15}$$

We can think of a segmentation of the current path, where each segment stores a fraction of the energy and delivers a partial contribution to the loop inductance (this relates to the derivation of the partial inductance as shown in Caniggia and Maradei 2008). For a closed loop, the partial inductance equals the loop inductance.

We plot the partial inductance of an $l = 50\,\mathrm{cm}$ long copper wire with a radius of $r = 0.5\,\mathrm{mm}$ in two arrangements in Fig. 5.28. In the first arrangement, the wire is straight. The partial inductance of a straight wire is analytically calculated using

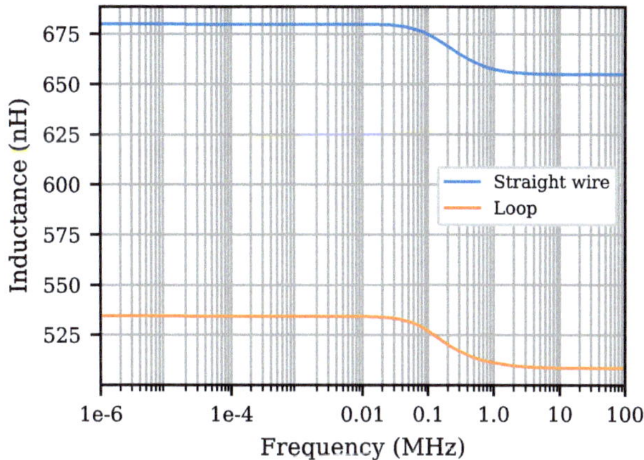

Fig. 5.28 Partial inductance of a 50 cm long wire

Eq. 3.9, resulting in 685 nH at low frequencies and 660 nH at high frequencies, while the EM simulation model delivers 680 nH and 655 nH.

In the second arrangement, we form a circular loop with the same wire dimensions and calculate the partial inductance by cutting out a 1 mm long section of the loop. Because the section we remove is very short, the so-calculated partial inductance is close to the loop inductance. The inductance of a loop is calculated using the approximate formula provided in Eq. 3.13.

The analytical equation gives 515 nH, a value that matches the result of the EM simulation model at 500 kHz. Both analytical formulas do not account for the frequency dependence of the inductance, as observed in Fig. 5.28.

The frequency dependence of the inductance of a conductor results from the skin effect: At low frequencies, the magnetic field exists both inside and outside the conductor. In the transition region, in this example between 0.1 MHz and 1 MHz, the magnetic field inside the wire decreases as the skin effect reduces the penetration depth of the magnetic field into the conductor. Above approximately 1 MHz, there is negligible field penetration into the conductor, and only the field outside the wire contributes to the inductance. This phenomenon produces the characteristic S-shape of the inductance curve, as shown in Fig. 5.28. The contribution of the field in the wire is sometimes referred to be the "inner inductance" and the contribution from the field outside of the wire is labeled "outer inductance."

The above example highlights an important property of the partial inductance: For a wire of the same length, the partial inductance of a straight wire is larger than the inductance of a loop formed by the same wire. This difference arises due to the absence of mutual inductance in the straight wire case.

To explain this, consider dividing the loop into small segments, as illustrated in Fig. 5.29. For each segment in the loop, there is a corresponding segment on the opposite side of the loop where the current flows in the opposite direction. The

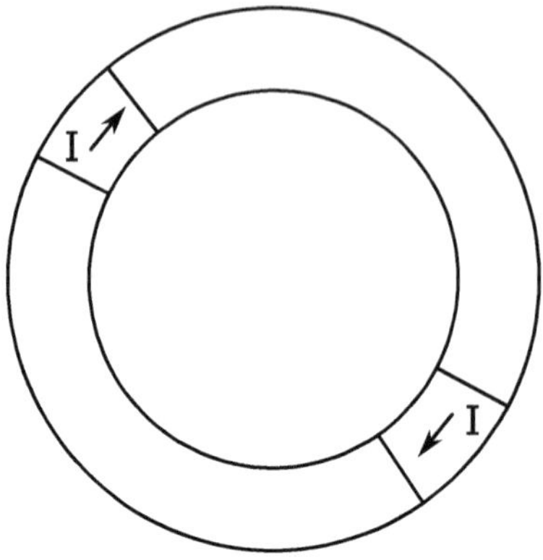

Fig. 5.29 Mutual inductance of segments in a current loop

mutual inductance between these two segments reduces the overall inductance of the loop.

This effect does not occur in the straight wire case: If we divide the current flow in the EM simulation model into sections and include only one current direction, the partial inductance calculated always overestimates the inductance of the current loop in the DUT (Caniggia and Maradei 2008, appendix).

In the calculation, we use the magneto quasi-static (MQS) approximation. The MQS approximation includes the induced current, which allows for accurate modeling of the skin effect. However, it does not include the displacement current, meaning the capacitance between individual segments of the loop, as shown in Fig. 5.29, is not accounted for in the calculation. As a result, the inductance value remains constant with increasing frequencies, and the effect of self-resonance of the loop is not calculated.

This might appear to be a limitation but aligns perfectly with the purpose of RLC extraction. For an RLC model to be valid, the structure must be electrically short, i.e., the frequency band in which the model is extracted must remain below the self-resonant frequency of the structure.

The calculation of partial inductance requires a specialized EM solver: Partial inductance cannot be calculated using the magneto(quasi-)static solver or a full-wave solver, as both require the current loop to be closed within the computational domain. Instead of directly calculating fields, the scalar potential ϕ and the vector potential $\mathbf{A}$ are computed. Details of this calculation are provided in the advanced section.

The PEEC method has been utilized for many years for partial inductance extraction (Brennan et al. 1979). The PEEC method is derived from the MoM by separating the solution into resistive, inductive, and capacitive contributions. Similar to the MoM, the PEEC method discretizes only the surfaces of conductors. On these surfaces, two meshes are generated—one for resistive and inductive contributions and another for capacitive contributions. Summing over these mesh elements provides the resulting RLC values.

In CST, a different method based on the FEM is used for partial RLC extraction. While both methods yield the same partial values, differences in their formulations (FEM versus MoM, see Sect. 4.11) lead to variations in performance. The system matrix of the PEEC method is fully populated, resulting in higher memory requirements as geometrical complexity increases compared to the FEM-based method. Additionally, the FEM-based method outperforms the PEEC for models incorporating magnetic materials, as these require a volumetric mesh. However, as a volume-based method, the FEM requires a computational domain large enough to allow the potentials to decay sufficiently. In contrast, the PEEC method's natural boundary conditions are open, enabling a smaller computational domain.

5.6.2 *Partial Resistance*

The introduction of partial resistances is straightforward, namely the resistance of a segment of a conductor of length l, cross section A, and electrical conductivity σ_c as described by Eq. 3.3.

5.6.3 *Partial Capacitance*

The introduction of partial capacitances often causes confusion, even among experienced engineers. Capacitance calculations are performed separately from inductance and resistance calculations, based on the assumption that electric and magnetic fields are decoupled, as explained in the first section of this chapter. If the permittivity does not vary with frequency, an electrostatic solution can be used to compute the capacitance between conductors.

However, when calculating partial inductances, a source and a sink of current must be placed on the same conductor. This approach conflicts with the definition used in the electrostatic solver, where each conductor is treated as an equipotential. To address this discrepancy, the capacitance calculation in the pRLC solver is performed in two steps. In the first step, the same capacitance matrix is calculated as in the electrostatic solver, by assigning a potential to each conductor.

In the second step, the capacitance of each conductor is distributed among the sources present on that conductor. This distribution creates a partial capacitance at each source on the conductor. While there is no physical interpretation for partial capacitance, this division is necessary for exporting the circuit model. The exported RLC model from the partial RLC EM simulation includes circuit terminals corresponding to the sources in the EM model. To create a valid circuit model, the capacitances must be distributed appropriately to the terminals.

5.6.4 *Application Example: Half-Bridge Module*

Switching of a power transistor, modeled by a SPICE circuit

Clamp current measured by a circuit probe

RLC matrix

Partial RLC solver

AC, Transient

0–100 MHz

Using a partial extraction allows access to the electromagnetic properties of specific parts of a DUT, while ignoring the impact of other parts. This approach is particularly useful during the early stages of design, where many details are not yet defined. The following example demonstrates this concept.

Consider a three-phase inverter driving a motor, as introduced in the galvanic coupling example in Sect. 5.3. This system consists of three half-bridges. By employing the partial extraction method, we can analyze a single half-bridge independently, without considering the complete inverter.

5.6.4.1 Double Pulse Test—DPT

The Double Pulse Test (DPT) is a measurement method used to analyze the switching behavior of semiconductors in a controlled environment. It provides insights into the switching waveforms and the impact of parasitic elements in the design. The DPT is commonly used to estimate switching losses and evaluate ringing on the input supply lines.

In this section, we analyze the switching current waveform in the negative supply path. Instead of performing a full conducted emission simulation, as described in Chap. 3, we focus on the performance of one half-bridge.

The schematic of a DPT is depicted on the left-hand side of Fig. 5.30. The test is configured as a half-bridge circuit in which the low-side transistor is investigated as the DUT. The high-side transistor in the bridge is the same device as the DUT.

During the test, the high-side transistor remains in the off state while the low-side transistor is switched. An inductive load is placed at the AC output of the half-bridge to limit the current flow during the test and mimic real system load conditions. The low-side transistor is switched by applying a pulsed voltage at V_{pulse}. A signal with a frequency of 20 kHz and rise/fall times of 1 ns is used to generate the current waveform plotted on the right-hand side of Fig. 5.30.

The current waveform exhibits three distinct regions:

Region 1 The low-side transistor is turned on, and current flows through the inductor and the low-side transistor.

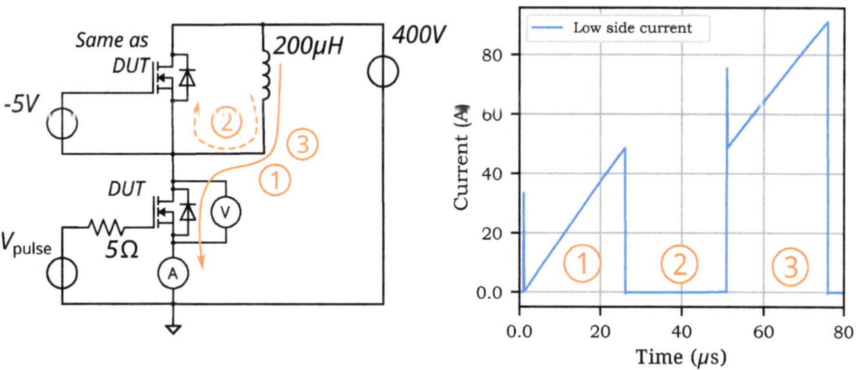

Fig. 5.30 Left: schematic of a DPT. Right: Current through low-side transistor

Region 2 The low-side switch is off, and the current flows through the high-side diode and the inductor.

Region 3 During the second pulse (hence the name DPT), the current flows along the same path as in Region 1.

An overshoot is present in the current waveform and originates from the reverse recovery of the diode on the high side.

In our example, we use a 1200 V SiC transistor model from Wolfspeed (CPM3-1200-0021A) and a SiC Schottky diode also from Wolfspeed (CPW4-1200-S020B). The SPICE models for both components are bare die models.

The current waveform on the right-hand side of Fig. 5.30 is calculated with only the parasitics inside the semiconductor material. Since we aim to compare different module layouts in the next step, these models need to be amended with the parasitic elements specific to each layout. SPICE models of packaged transistors cannot be used, as they already include packaging parasitics.

The two modules we analyze using the partial RLC extraction are plotted in Fig. 5.31. These designs do not represent any specific product but are created for demonstration purposes. Both modules are square-shaped with dimensions of 5 cm on each side.

Each module consists of two diodes and two transistors on both the high and low sides. To improve thermal performance, two transistors are used in parallel, distributing the current and reducing the load on each semiconductor. Each transistor and diode in both designs are bonded using four identical bond wires, each 8 mm long and 200 μm thick.

The copper layer in each module is 330 μm thick and is placed on an alumina substrate ($\varepsilon_r = 9.4$) of the same thickness. The baseplate below the substrate is also 330 μm thick.

In Fig. 5.31, the copper layer is color-coded as follows: red for the high-side input (labeled “P”), blue for the low-side input (labeled “N”), and orange for the

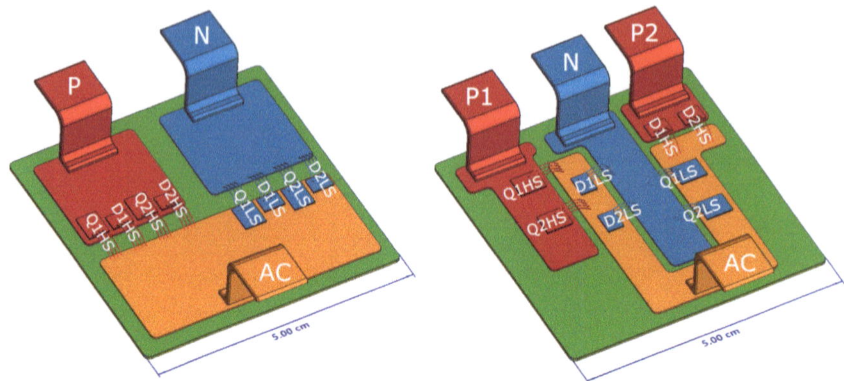

Fig. 5.31 Two example modules analyzed using partial RLC extraction. Left: layout 1 with reference semiconductor placement, Right: layout 2 with optimized semiconductor placement

commutation output (labeled "AC"). The module connections are realized using clamps.

The placement of the semiconductors in layout 1 is straightforward. All semiconductors on the high and low sides are arranged in a line, alternating between transistors and diodes. Layout 2 follows design concepts presented by Chen (Chen et al. 2013), which aim to reduce the commutation cell inductance by positioning the transistors of one side close to the diodes of the other side. Both models include only the copper traces of the power loop. Traces for connecting the gates of the transistors are excluded to keep the layouts simple. These models are designed for early-stage analyses to study different layout concepts.

For the partial RLC extraction, it is crucial to define the correct set of sources in the EM simulation model. In the partial RLC solver in the CST, the positions where the current enters or leaves the computational domain are referred to as RLC nodes. We place RLC nodes at the source and drain positions of each transistor, the anode and cathode of each diode, the connecting clamps, and one node at the baseplate. This results in 19 RLC nodes for module 1 and 20 RLC nodes for module 2, as module 2 features two high-side clamps.

To extract partial resistances and inductances, the RLC nodes are grouped into node pairs that form current paths. In each current path, one node acts as a source and the other as a sink. A total of 16 node pairs are required for a complete definition of all current paths.

In module 1, we set them as:

- 4 node pairs: P $\rightarrow$ HS drain and HS cathodes
- 4 node pairs: HS source and HS anodes $\rightarrow$ AC
- 4 node pairs: AC $\rightarrow$ LS drain and LS cathodes
- 4 node pairs: LS source and LS anodes $\rightarrow$ N

In module 2, the setup is

- 2 node pairs: P1 $\rightarrow$ HS drain
- 2 node pairs: HS source $\rightarrow$ LS cathodes
- 2 node pairs: LS cathodes $\rightarrow$ AC
- 2 node pairs: LS anodes $\rightarrow$ N
- 2 node pairs: AC $\rightarrow$ LS drain
- 2 node pairs: LS source $\rightarrow$ N
- 2 node pairs: LS drain $\rightarrow$ HS anodes
- 2 node pairs: HS cathodes $\rightarrow$ P2

This is a complete set of current paths because all other possible current paths can be derived as a linear combination of the 16 node pairs we have defined. It is also possible to generate the node pairs automatically by allowing the solver to select the current paths. For demonstration purposes, we define them manually.

On the baseplate, only one RLC node is placed. This setup allows for the calculation of the capacitance to the baseplate but does not include its partial resistance or inductance. It is not necessary to evaluate the currents in the baseplate, as the mutual inductance between the current in the top layer and the induced current in the baseplate is already included in the calculation.

The RLC extraction is performed using second-order elements with one mesh adaptation step. The simulation time is 61 minutes and 9 seconds for module 1 and 56 minutes and 14 seconds for module 2. The memory requirement is 36 GB for module 1 and 42 GB for module 2. The resulting RLC model consists of $16 \times 16 = 256$ (for the 16 node pairs) partial inductances and partial resistances and $20 \times 20 = 400$ (for 19 nodes plus self-capacitance) partial capacitances.

In order to use the RLC model calculated by the EM solver, we need to export it into a format that is compatible with a circuit simulator. Exporting the RLC model into a SPICE netlist is the most straightforward option, but there are several ways the netlist can be generated.

The partial inductance calculated by the partial RLC solver is frequency-dependent, as illustrated in Fig. 5.28. Similarly, the partial resistance also exhibits frequency dependence due to the skin effect, as discussed in Sect. 4.4. On the other hand, the frequency dependence of the partial capacitance is negligible, provided that the material properties remain constant over the frequency band of interest.

The simplest method for exporting the SPICE netlist is to neglect the frequency dependence and export the RLC values at a fixed frequency. This approach aligns closely with the general understanding of an RLC model. However, there are more sophisticated export options available.

The frequency dependence of R and L for each individual node pair can be represented using a ladder-type RLC circuit model, which incorporates the values of R and L at the lowest and highest frequencies and combines them into a circuit that approximates their frequency dependence. Alternatively, the results can be converted into an impedance or admittance form, and vector fitting can be applied, similar to the method used for fitting S-parameters.

In this study, we use the single-frequency export method for simplicity and comprehensibility. The extraction is performed at 1 MHz, where the inductance has already reached the lower part of the s-shaped curve (cf. Fig. 5.28). While this approach underestimates the inductance at the switching frequency of 20 kHz, it aligns with the simplified model chosen for this demonstration.

RLC models are human-readable, allowing us to inspect the content of the exported SPICE netlist. The netlist contains 374 lines, including the mutual coupling. Despite this relatively low node count, the model's complexity makes manual interpretation impractical. In practice, the model serves as a black-box representation, and we interact with it by connecting a circuit and calculating voltages and currents rather than interpreting the individual RLC values.

This highlights the advantage of the RLC extraction method for specific, isolated parts of the DUT. For circuits with a small number of RLC nodes, the RLC model offers easy human interpretability as well as fast and stable circuit simulation. However, as the number of RLC nodes increases, these benefits diminish: the model's complexity grows, potentially leading to convergence issues in simulations, and human interpretation becomes infeasible.

In contrast, for electronic systems requiring a large number of sources in the EM model, S-parameter black-box models are more effective. This approach enables the modeling of broader frequency ranges and provides an efficient representation of the frequency dependence of the coupling path. The advantages of using S-parameters are explored in greater detail in the conducted emissions chapter (Chap. 10).

Coming back to the extracted RLC model, we now create the schematic of the DPT using the extracted SPICE netlist, as depicted in Fig. 5.32. The blue labels on the pins of the SPICE netlist correspond to the RLC nodes defined in the EM simulation. To reduce clutter on the schematic, we omit individual labels for the semiconductors, as they are identified by the pin labels, with HS denoting the high side and LS the low side. The schematic in Fig. 5.32 represents module 2, which contains two pins on the high side input. For module 1, pin "P2" is absent.

Let us analyze the resulting model, which is constructed by combining the DPT schematic and the SPICE netlist. The SPICE netlist encapsulates all the couplings within the module, including those for all semiconductor components and the clamps.

At each transistor, we connect the RLC node from the EM model at the drain and source terminals. This is appropriate, as the distance between these nodes is only 330 μm. However, the inductor and the DC source are also connected to the RLC nodes located at the clamps, which are separated by several centimeters. This is necessary to construct the double pulse test schematic, even though the current flowing from one clamp to the other is not modeled in the EM simulation.

This modeling approach is not feasible in a circuit model derived from a full-wave solver, as the full-wave solver requires the current loops to be closed within the EM model. The partial extraction method uniquely allows for this flexibility, further demonstrating its suitability for investigating specific parts of a DUT. With this setup, we can analyze the module in detail while disregarding the impact of the physical size of the inductor and the connections to the DC source.

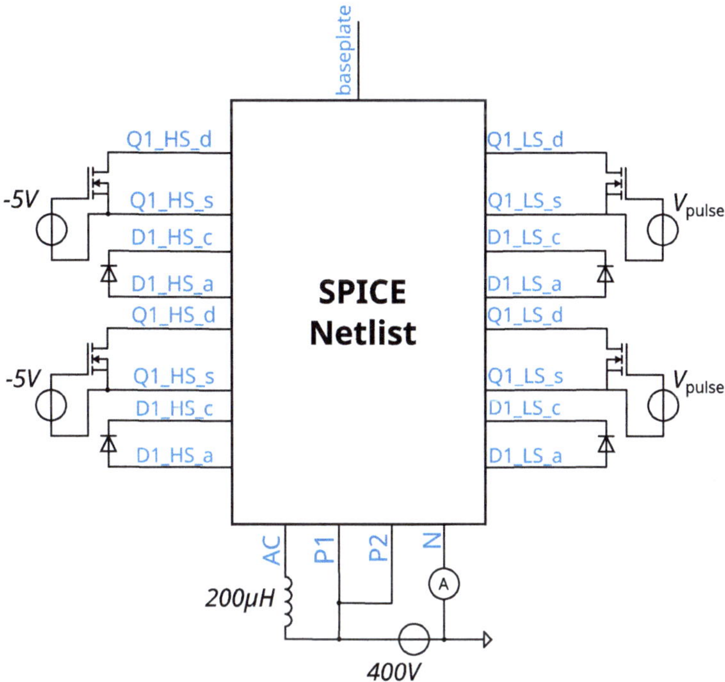

Fig. 5.32 Schematic for the double pulse test with the extracted SPICE netlist

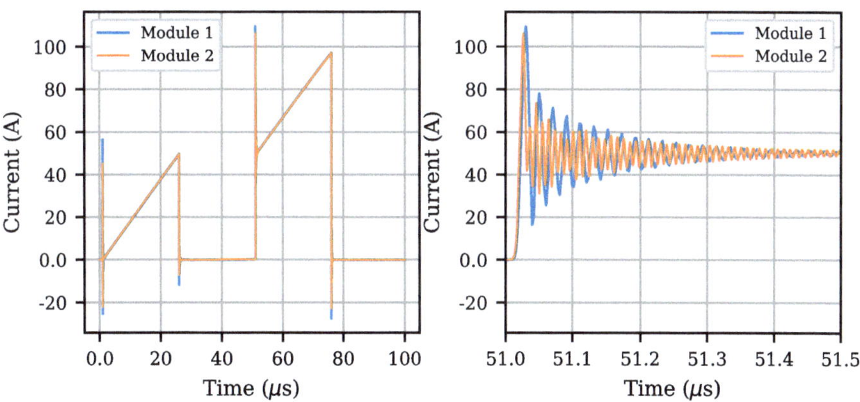

Fig. 5.33 Current on the N clamp during the DPT for both modules. On the right side, we zoom to see the details of the overshoot

The resulting current through the N clamp is plotted in Fig. 5.33. The left-hand side shows both pulses, while the right-hand side provides a zoomed-in view of the rising current during the second turn-on. The amplitude of the ringing is smaller for module 2, and the frequency of the ringing is higher compared to module 1.

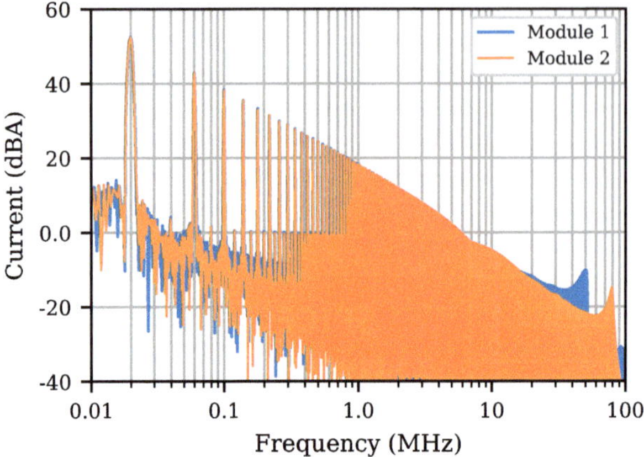

Fig. 5.34 Spectrum of the current at the N clamp during the DPT for both modules

The higher frequency confirms that the inductance of the commutation cell is smaller, as expected from the improved design. The time required for the circuit simulation of the 100 μs long signal is 15 minutes, while the DPT simulation without the SPICE netlist takes 4 minutes. The computationally intensive models of the SiC transistors contribute significantly to the simulation time, and the addition of the netlist from the RLC extraction further increases the complexity of the circuit, thereby extending the required simulation time.

To further evaluate the impact of the layout differences, we calculate the spectrum of the current at the N clamp. A reliable spectral analysis requires a longer simulation time, as only two pulses result in a coarse spectral resolution in the FFT. We extend the simulation time of the switching with a 50% duty cycle to 3 ms and then perform an FFT. In a laboratory measurement, generating such a long signal can be challenging due to thermal constraints, as the current increases with subsequent pulses until the inductor is fully charged. However, in simulation, this limitation is not present.

The current spectrum is plotted in Fig. 5.34. For module 1, the frequency of the ringing is approximately 50 MHz, while for module 2, the resonance peak is located at 79 MHz and is 6 dB lower than that of module 1. This example illustrates how partial RLC extraction can be utilized in component-level simulations to evaluate and quantify potential improvements in component layouts.

5.7 Advanced: Static Fields and Potentials

Domains: 

Both the electrostatic solver used in Sect. 5.4 and the stationary current solver used in Sect. 5.5 solve Maxwell's Equations under static conditions, for the scalar potential ϕ. From ϕ, the electric field **E** is calculated. Since the static condition implies $\frac{\partial \mathbf{B}}{\partial t} = 0$, we have $\nabla \times \mathbf{E} = \mathbf{0}$. The derivation of the equation for the calculation of the scalar potential is obtained straightforwardly from Gauss's Law. For the electrostatic case, we write Gauss's Law as

$$\nabla \cdot \mathbf{D} = \rho, \tag{5.16}$$

where **D** is the electric flux density and ρ is the charge density. By substituting $\mathbf{D} = \varepsilon \mathbf{E}$ and $\mathbf{E} = -\nabla \phi$, we obtain the governing equation for ϕ as

$$\nabla \cdot (\varepsilon \nabla \phi) = -\rho. \tag{5.17}$$

This equation is called the Poisson equation and is solved by the electrostatic solver to compute ϕ in non-conductive regions.

For the stationary current solver, we use Ohm's Law in its field form, $\mathbf{J} = \sigma_c \mathbf{E}$, where **J** is the current density, and σ_c is the conductivity. Since $\nabla \cdot \mathbf{J} = 0$ (continuity equation under static conditions), substituting $\mathbf{E} = -\nabla \phi$ gives

$$\nabla \cdot (\sigma_c \nabla \phi) = 0. \tag{5.18}$$

This equation is solved to calculate the potential distribution ϕ in conductive regions. Sources in the stationary current solver can be understood as boundary conditions in Eq. 5.18.

The similarity in the form of the equations for ϕ in the electrostatic and stationary current solvers highlights their shared foundation in Maxwell's Equations under static conditions. However, their physical interpretations differ: the electrostatic solver determines the potential due to charges, while the stationary current solver calculates the potential distribution due to steady currents.

Solving a potential equation, as applied in both electrostatics and the stationary current solver, offers the advantage of needing to calculate only a single potential value per mesh cell. In contrast, calculating fields with the full-wave FEM solver requires computing three field components (Sect. 4.12). Consequently, solving for ϕ requires less memory than field calculations. Additionally, the reduced matrix size contributes to shorter simulation times. Most examples in this chapter that utilize static solvers require less than 5 minutes of simulation time.

The potential also plays an important role in the computation of the partial inductance. Recall that it is not possible to calculate the partial inductance using the magnetic flux density, as the loop in the partial inductance calculation is not closed (Eq. 5.12). Instead of calculating the magnetic flux density, we calculate the vector potential **A**. The curl of the vector potential is the magnetic flux density, $\mathbf{B} = \nabla \times \mathbf{A}$. Using Stokes' theorem, we transform the surface integral for the calculation of the magnetic flux into a line integral of the vector potential. The equation for the calculation of inductance becomes:

$$L = \frac{\int_l \mathbf{A} d\mathbf{l}}{I}, \tag{5.19}$$

where l is the line along which the current flows. In case of partial inductance, l does not need to be closed (Caniggia and Maradei 2008).

The vector potential **A** is not unique but a matter of the chosen gauge condition. To match the partial inductance calculated by Eq. 5.19 to the partial inductance definition introduced by the PEEC method, the gauge that is used by PEEC must be applied. Different from the partial inductance, the inductance of a closed loop is not a function of the gauge condition.

Further Reading

Quasistatic Solvers

O. Biro, K. Preis, On the use of the magnetic vector potential in the finite-element analysis of three-dimensional eddy currents. IEEE Trans. Magn. **25**, 3145–3159 (1989). https://doi.org/10.1109/20.34388

O. Biro, K. Preis, K.R. Richter, Various FEM formulations for the calculation of transient 3D eddy currents in nonlinear media. IEEE Trans. Magn. **31**, 1307–1312 (1995). https://doi.org/10.1109/20.376269

M. Clemens, T. Weiland, Transient eddy-current calculation with the FI-method. IEEE Trans. Magn. **35**, 1163–1166 (1999). https://doi.org/10.1109/20.767155

M. Clemens et al., Transient electro-quasistatic adaptive simulation schemes. IEEE Trans. Magn. **40**, 1294–1297 (2004). https://doi.org/10.1109/TMAG.2004.824582

M. Kaltenbacher, S. Reitzinger, J. Schoberl, Algebraic multigrid method for solving 3D nonlinear electrostatic and magnetostatic field problems. IEEE Trans. Magn. **36**, 1561–1564 (2000). https://doi.org/10.1109/20.877737

S. Koch et al., Modeling thin conductive sheets using shell elements in magnetoquasistatic field simulations. IEEE Trans. Magn. **45**, 1292–1295 (2009). https://doi.org/10.1109/TMAG.2009.2012601

S. Koch, H. Schneider, T. Weiland, A low-frequency approximation to the Maxwell equations simultaneously considering inductive and capacitive phenomena. IEEE Trans. Magn. **48**, 511–514 (2012). https://doi.org/10.1109/TMAG.2011.2173163

J.F. Lee, D.K. Sun, Z.J. Cendes, Tangential vector finite elements for electromagnetic field computation. IEEE Trans. Magn. **27**, 4032–4035 (1991). https://doi.org/10.1109/20.104986

D. Vanoost et al., Two-dimensional magnetostatic finite-element simulation for devices with a radial symmetry. IEEE Trans. Magn. **50**, 1–4 (2014). https://doi.org/10.1109/TMAG.2013.2292672

S.H. Wong, Z.J. Cendes, Numerically stable finite element methods for the Galerkin solution of eddy current problems. IEEE Trans. Magn. **25**, 3019–3021 (1989). https://doi.org/10.1109/20.34356

RLC Extraction

G. Antonini et al., The partial elements equivalent circuit method: the state of the art. IEEE Trans. Electromagn. Compat. **65**, 1695–1714 (2023). https://doi.org/10.1109/TEMC.2023.3302700

R. Aranco, F. Maradei, Passive equivalent circuits of complex discontinuities: an improved extraction technique, in *2005 International Symposium on Electromagnetic Compatibility, 2005. EMC 2005*, Vol. 3 (IEEE, 2005), pp. 700–704. https://doi.org/10.1109/ISEMC.2005.1513614

J. Benz et al., Generation and EMC optimization of compact equivalent electrical circuits of industrial electronics. IEEE Trans. Electromagn. Compat. **65**, 1811–1819 (2023). https://doi.org/10.1109/TEMC.2023.3321078

N. Bondarenko et al., Development of simple physics-based circuit macromodel from PEEC. IEEE Trans. Electromagn. Compat. **58**, 1485–1493 (2016). https://doi.org/10.1109/TEMC.2016.2575847

P.A. Brennan, N. Raver, A.E. Ruehli, Three-dimensional inductance computations with partial element equivalent circuits. IBM J. Res. Dev. **23**(6), 661–668 (1979). https://doi.org/10.1147/rd.236.0661

S. Caniggia et al., 3-D numerical modeling and circuit extraction techniques for the analysis of unshielded twisted pairs. IEEE Trans. Magn. **43**, 1357–1360 (2007). https://doi.org/10.1109/TMAG.2007.892424

S. Caniggia, F. Maradei, *Signal Integrity and Radiated Emission of High-Speed Digital Systems* (Wiley, New York, 2008)

Z. Chen et al., A 1200 V, 60 A SiC MOSFET multi-chip phase-leg module for high-temperature, high-frequency applications, in *2013 Twenty-Eighth Annual IEEE Applied Power Electronics Conference and Exposition (APEC)* (IEEE, 2013), pp. 608–615. https://doi.org/10.1109/APEC.2013.6520273

C.R. Paul, *Inductance Loop and Partial* (Wiley, New York, 2010)

A. Schroeder, B. Wunsch, S. Kicin, Wideband macro-modelling of power modules for transient electromagnetic analysis, in *CIPS 2020; 11th International Conference on Integrated Power Electronics Systems* (2020), pp. 1–5.

F. Traub et al., Generation of physical equivalent circuits using 3D simulations, in *International Symposium on Electromagnetic Compatibility* (IEEE, Rome, 2012), pp. 486–491.

F. Traub et al., Automated construction of physical equivalent circuits for inductive components, in *International Symposium on Electromagnetic Compatibility (EMC Europe)* (IEEE, Brugge, 2013), pp. 67–72.

F. Traub et al., Eigenmodes of electrical components and their relation to equivalent electrical circuits, in *IEEE International Symposium on Electromagnetic Compatibility* (IEEE, Denver, 2013), pp. 287–293.

Chapter 6
Coupling Paths: Near and Far Field

In the previous chapter, we study near-field coupling and modeled the coupling path via RLC elements. We present in that chapter three examples where it is possible to separate the coupling path into galvanic, capacitive, and inductive contributions. This separation is only feasible as long as the electric and magnetic fields remain decoupled, which is the case for electrically short structures. However, as the size of the simulation model approaches the wavelength's dimension, this separation is no longer possible, and we must solve the fully coupled set of Maxwell's equations using a full-wave solver.

From this chapter onward, we no longer use static or quasistatic methods and instead rely on full-wave simulations, which provide a complete description of the coupling path.[1] A full-wave solution includes all types of coupling paths studied in the previous chapter and additionally considers radiative coupling, which cannot be addressed using the (quasi-)static methods previously discussed. To generate a circuit model encompassing all four coupling mechanisms from the full-wave simulation, we shift from RLC models to a description of the coupling path using S-parameters. In this chapter, we examine the behavior of fully coupled fields and introduce efficient techniques for analyzing them.

6.1 Evaluating Electromagnetic Fields

Domains: 

[1] For cables and SI simulations of PCBs, we introduce the MTL method in Chaps. 8 and 11.

J. Hansen, A. Barchanski, *EMC Simulation of Electronic Systems*,
https://doi.org/10.1007/978-3-032-09237-3_6

6.1.1 Electromagnetic Field Visualization in Frequency Domain

An EM simulation allows us to visualize the electromagnetic field within the computational domain and even beyond it. This capability is a powerful tool for simulation engineers, as it allows to examine fields where direct measurement is not possible.

Unlike physical measurements, which require specialized equipment to estimate field strength, simulations directly provide field values as part of the calculation. This eliminates the need for additional models of measurement devices, such as probes or antennas. A comparison between direct field calculations and simulations incorporating antenna models is presented in Chap. 12. In this chapter, we focus on methods for evaluating fields calculated by a full-wave solver.

A full-wave solver computes the coupled electric and magnetic field. As discussed in Sect. 5.1, one field type may dominate depending on the structure's excitation and termination. However, in the general case, both fields contribute significantly. Analyzing both fields is essential for a comprehensive understanding of the system.

The electromagnetic field is a vector quantity, meaning each field has three components at every position ($\mathbf{r}$) in space. These components are typically defined in a Cartesian coordinate system, giving us x, y, and z components at every point in the computational domain. To determine the absolute value of a field, we calculate the magnitude of the resulting vector. For the electric field, this is expressed as

$$E(\mathbf{r}) = \sqrt{|E_{\mathrm{x}}(\mathbf{r})|^2\mathbf{e}_{\mathrm{x}} + |E_{\mathrm{y}}(\mathbf{r})|^2\mathbf{e}_{\mathrm{y}} + |E_{\mathrm{z}}(\mathbf{r})|^2\mathbf{e}_{\mathrm{z}}}. \tag{6.1}$$

with E_i, $i =$ x,y,z the field strength of the field component in x,y,z direction. The same holds of the magnetic field.

In most EMC emission standards, a spectrum is compared to limit lines, and the results of interest are evaluated in the frequency domain. When using a frequency domain method like the FEM, the results are computed in the frequency domain. If a time domain solver is used, the frequency domain results are derived from the time domain simulation using a DFT. In this chapter, we focus on frequency domain results. Time domain results are explored in Chap. 14.

In the frequency domain, fields are represented as phasors. The phasor representation $\underline{\mathbf{E}}$ of the electric field is described as (cf. Sect. 3.8.2)

$$\bar{\mathbf{E}}(\mathbf{r})\cos(\omega t + \theta) + j \cdot \bar{\mathbf{E}}(\mathbf{r})\sin(\omega t + \theta) = \bar{\mathbf{E}}(\mathbf{r})e^{j(\omega t + \theta)} = \bar{\mathbf{E}}(\mathbf{r})e^{j(\omega t)} \cdot e^{j\theta}, \tag{6.2}$$

where $\bar{\mathbf{E}}(\mathbf{r})$ is the amplitude of the electric field at the position $(\mathbf{r})$ and θ is its phase. We can set ωt in Eq. 6.2 to zero and obtain an equation for the field that only depends on the amplitude and the phase,

$$\underline{\mathbf{E}}(\mathbf{r}, \theta) = \bar{\mathbf{E}}(\mathbf{r}) \cdot e^{j\theta}. \tag{6.3}$$

The same applies to the magnetic field $\mathbf{H}$.

For surface conductors modeled as PEC or as a surface impedance, full-wave solvers do not calculate a current density but rather a surface current. The surface current $\mathbf{J}_\text{S}$ is calculated as

$$\mathbf{J}_\text{S} = \hat{\mathbf{n}} \times \mathbf{H}. \tag{6.4}$$

The surface current is of the unit $\frac{\text{A}}{\text{m}}$. We plot the surface current in a number of examples in later chapters of the book.

Equation 6.3 calculates the so-called instantaneous field. When plotting fields in the computational domain, the plot of the instantaneous field enables us to rotate the phase and observe the propagation of the field. Such animations are often used to illustrate phenomena like the radiation of an electromagnetic field from a dipole. Since the plot corresponds to a specific frequency, the phase rotation applies solely to this frequency.

In the phasor description, the field can also be represented in terms of its real and imaginary components, as shown in Eq. 6.2. This representation occasionally leads to confusion: The physical fields are always real-valued and time-dependent, with no imaginary component in the time domain. The real and imaginary representation is exclusive to the frequency domain and serves as a means to express the phase dependency of time-harmonic fields.

6.2 Field Probes

Domains:  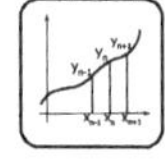

The first possibility to obtain field values from the simulation is by using a field probe inside the computational domain. In CST, these field probes are referred to as near-field probes, though this terminology can be misleading, and we avoid it. The term "near" solely reflects the fact that the probe is located within the computational domain.

The name distinguishes these probes from far-field probes, which are placed outside the computational domain. How fields outside of the computational domain are calculated is discussed in a later section of this chapter.

When a probe is placed inside the computational domain, it records the field at its location for the frequency band defined in the model. For field calculations, the concepts of near-field and far-field zones from antenna theory are irrelevant—there is simply the field, and the value recorded depends on the simulated structure and the probe's distance from it.

Field probes are typically used to analyze the frequency dependence of the field. They highlight the resonant frequencies of the structure at which the field exhibits peaks

However, a field probe provides information for only one specific position in space. If placed inappropriately, the results may not capture the full picture. For example, in simulations of symmetric structures, placing the probe in a symmetry plane where fields cancel out can lead to misleading results. To mitigate such effects, we recommend defining several probes rather than a single one and positioning them slightly away from symmetry planes. We discuss probe placement in detail in Sect. 12.2.3.

6.2.1 *Simple Radiated Emission Example*

Ideal AC voltage source
Field probes in the computational domain
Full-wave, S-parameters, combine results
FIT solver
AC, Transient
0–2 GHz

For the discussion of the behavior and the estimation of the coupled electromagnetic fields, we use a geometrically simple model depicted in Fig. 6.1. The model consists of two simple PCBs positioned 5 cm above a metallic table and connected by two wires, each 50 cm long. The table has dimensions of 1.5 m in length and 50 cm in width.

The PCBs measure 8 cm in length and 4 cm in width, and each consists of two layers: a full reference plane on the bottom layer and a signal trace on the top layer. The substrate between the layers is 200 μm thick with a relative permittivity of $\varepsilon_r = 4.4$. The signal trace is 5 cm long and 400 μm wide. At the edge of the board, a simple connector transitions from the PCB to the two wires. The wires have a rectangular cross section with a thickness of 400 μm and a spacing of 400 μm between them. One pin of the connector links to the signal trace, while the other pin connects to the reference plane.

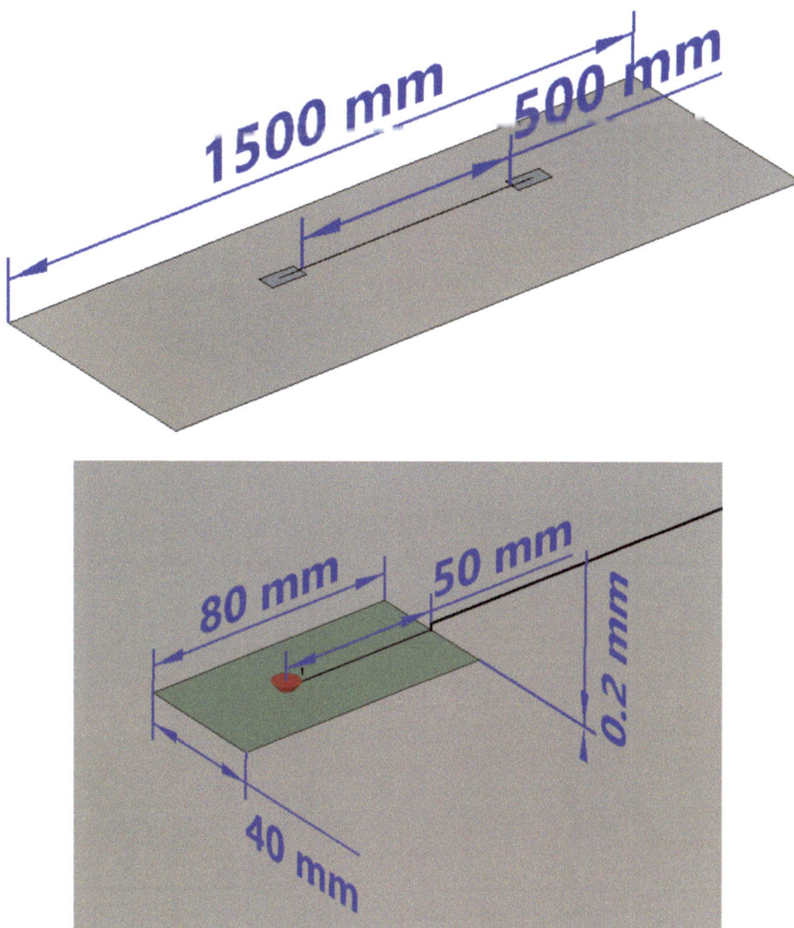

Fig. 6.1 Simple radiated emission example model consisting of two PCBs over a metallic table

In one of the PCBs, a port drives a signal between the trace and the reference layer, while in the other PCB, a 50 Ω resistor is connected between the trace and the reference layer.

The model is simulated up to a frequency of 2 GHz. Although the geometry is simple, the electromagnetic behavior is not. At 2 GHz, the cable is 3.3 λ long, and the table measures 10 λ, so multiple resonances are expected. Additionally, the transitions between the PCB, wires, and table significantly influence the electromagnetic field.

For the simulation, the FIT solver is employed. The results discussed in this section required 24 minutes of simulation time using a GPU. In the following section, we will explore which settings influence the simulation time in more detail.

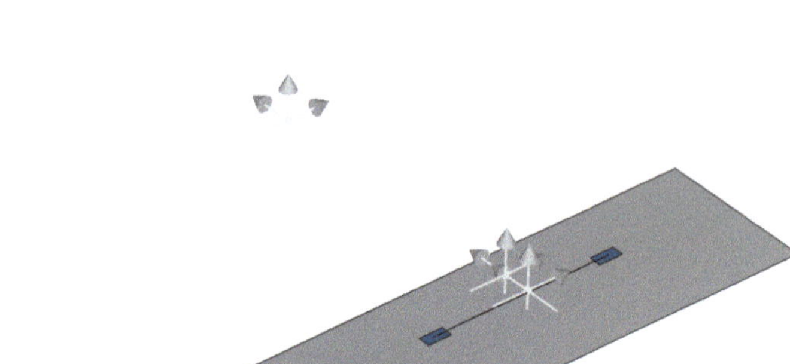

Fig. 6.2 Probe placement in the simple PCB model

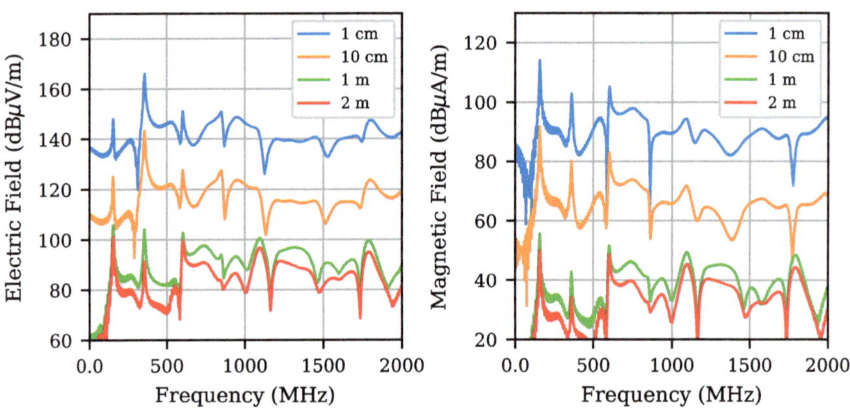

Fig. 6.3 Field values at the probe positions depicted in Fig. 6.2. Left: Electric field, Right: Magnetic field

In a first study, we place electric field probes and magnetic field probes at four positions within the simple PCB model. The probes are situated 5 cm above the plane of the table, 3 cm away from the symmetry plane orthogonal to the center of the wire, and at distances of 1 cm, 10 cm, 1 m, and 2 m from the PCB pair. The setup is illustrated in Fig. 6.2.

The port is excited using an AC simulation with a voltage of 1 V across all frequencies and a source impedance of 10 Ω, representing a typical impedance for a digital driver. All probes are defined within the computational domain. For the chosen maximum frequency of 2 GHz, this results in an electrically large computational domain.

The magnitude of the electric and magnetic field at the probe positions indicated in Fig. 6.2 are plotted in Fig. 6.3. Several observations can be made from these plots:

Field decay with distance At 10 cm distance, the field values are approximately 20 dB smaller than at 1 cm. Moving from 10 cm to 1 m, the field further decreases by 20 dB to 30 dB. Between 1 m and 2 m, the field magnitude reduces by an additional 5 dB to 10 dB. This gives a preliminary indication of how the field decays with distance. A more detailed analysis of field decay is presented later in this chapter using a field monitor.

Resonance peaks Below 1 GHz, three sharp peaks are observed at 150 MHz, 355 MHz, and 600 MHz. The sharpness results from the usage of PEC material in the simulation model. By comparing the relative magnitudes of these peaks, we gain insight into how the distance and position of the probes influence the field magnitudes:

- For the electric field, at 1 cm and 10 cm, the second peak is significantly higher than the first. At 1 m, the magnitudes of the first and second peaks are comparable, while at 2 m, the second peak becomes smaller than the first.
- For the magnetic field, the first peak is the highest at all distances.

Field similarity at larger distances At the distance of 1 cm and 10 cm, the ratios of peaks in the electric and magnetic fields differ significantly. However, at 2 m, the frequency dependence of the electric and magnetic fields becomes more similar, as observed by comparing the red curve on the left and right side of Fig. 6.3. This indicates that the behavior of the fields becomes similar at larger distances.

Resonant frequencies and signal path The total length of the signal path includes the trace on the PCB, the connectors on both sides, and the wire pair:

$$2 \cdot (50\,\text{mm} + 2.2\,\text{mm} + 5.2\,\text{mm}) + 500\,\text{mm} = 614.8\,\text{mm}.$$

The wavelength corresponding to 614.8 mm is approximately 487 MHz. However, no resonance is observed at this frequency. Instead, the resonant frequencies are shifted to lower values due to the size of the reference layer in the PCB. This can be confirmed by reducing the size of the reference layer.

We reduce the size of the reference layer to a minimum, retaining it only directly below the signal trace and at the side of the trace where the connector pin is located. The updated design is shown on the left-hand side of Fig. 6.4. On the right-hand side of Fig. 6.4, the electric field at the probe located 1 m away is plotted.

With this slim reference layer design, the resonances shift, and the second peak now occurs at 472 MHz, which is close to the value obtained analytically from the length of the signal path. The resonance frequency calculated by the EM simulation is slightly lower than the analytical prediction. This discrepancy arises because the analytical calculation does not account for the effect of the PCB substrate in the microstrip sections, which increases the electrical length of the complete signal path.

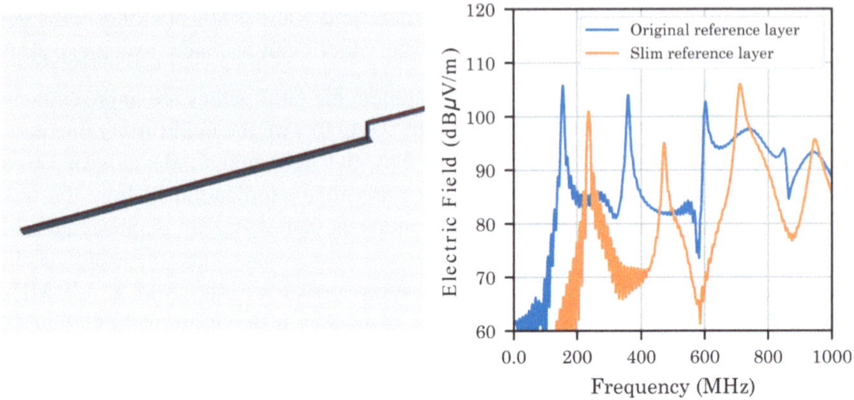

Fig. 6.4 Effect of the PCB reference layer size on the electric field. Left: The modified PCB with a reference layer as slim as the trace. Right: Electric field at the probe with 1 m distance for the original reference layer size and the slim reference layer

It is crucial to understand that while these plots show trends, they do not provide a complete picture of the emission. The evaluation is performed only at individual positions and does not capture the spatial variation of the field. A more comprehensive analysis of spatial variation is conducted later in this chapter.

Next, we calculate field values for a digital signal. Up to this point, we have used a broadband 1 V AC source with a 10 Ω inner impedance. This broadband source provides a comprehensive overview of the frequency behavior of the structure and helps identify resonant frequencies.

The simple PCB pair serves as a model for a digital communication channel between two PCBs. We model the digital channel as a clock signal resulting in a narrow-band noise source. To calculate the field generated by a digital signal, we redefine the source in the next step.

We use two different signals for the excitation: the first signal has a frequency of 50 MHz, while the second is set to 100 MHz. Both signals have an amplitude of 1 V, a duty cycle of 50%, and a rise and fall time of 1 ns. The source impedance is set to 10 Ω, as in the broadband simulation. In this simulation, we concentrate on the electric field at the probe located 1 m away.

The redefinition of the source does not require to resimulate the EM model. Since the EM simulation model is linear, we can modify the source in the schematic and scale the field values according to the new source. Exchanging the source and calculating the resulting fields requires only 30 seconds of simulation time. Details of this simulation flow, referred to as the transient combine results method, are provided in Sect. 6.7.5.

The voltage at the driving port for both signals is plotted on the left-hand side of Fig. 6.5. Due to the finite source impedance, the signal at the port is not ideally trapezoidal and includes some overshoots and ringing.

On the right-hand side, we plot the electric field calculated at the probe position located 1 m from the wire pair. We observe that the highest field value occurs at

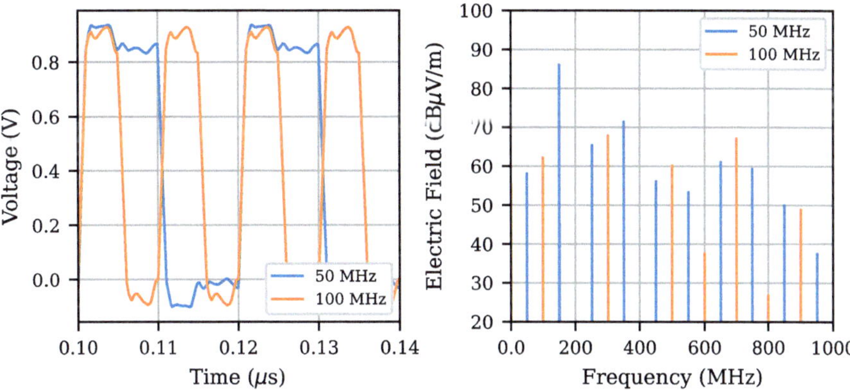

Fig. 6.5 Results obtained from the digital source in the PCB pair example. Left: Voltage at driving port. Right: Electric field at a distance of 1 m for

150 MHz for the simulation driven with the 50 MHz clock frequency. This peak is more than 15 dB higher than the highest peak generated by the 100 MHz signal. At first glance, this result might be surprising, but it can be explained by the resonance already observed when using the broadband source.

The third harmonic of the 50 MHz signal coincides with the first resonance at 150 MHz, generating the highest peak. In contrast, the harmonics of the 100 MHz signal do not align with this resonance. In the design of electronic systems, it is often impractical to modify functional signals to avoid resonances, as these depend on cable length. However, this example demonstrates how the interaction between the noise source and the coupling path can be analyzed through simulation. This enables the detection of potential emission issues at early design stages.

To ensure that the field results obtained from the simulation are reliable and that the analysis performed above is valid, we must understand the effects of the simulation settings on the fields. This topic is discussed in the next section.

6.3 Accuracy of Field Results

Domains: 

In this section, we study the effect of simulation settings on the field values calculated at the probe positions in the model described in the previous section (Sect. 6.2.1). For accurate field calculations, it is essential to understand which simulation settings influence the accuracy. Achieving accurate results often requires increased computational resources and time. Selecting the correct simulation settings ensures an optimal balance between the desired accuracy and the computational effort.

In the next subsection, we discuss how the signal duration of the time domain simulation affects the calculated field results. Subsequently, we analyze the impact of boundary conditions on the field results.

6.3.1 *Effect of Signal Duration in Time Domain Simulations*

Recall from Sect. 4.9 how time domain methods calculate fields: the propagation of the fields within the computational domain is tracked, and the time domain fields are recorded. Once a suitable termination criterion is met, the EM simulation stops, and frequency domain results are obtained via a Fourier transform.

To achieve accurate results at field probes, it is crucial to ensure that the time domain signal at the probes has sufficiently decayed before stopping the simulation. Termination criteria can be defined based on the probe results. For instance, the simulation can stop if the probe spectrum changes by less than 1% over a defined signal interval.

For the following comparison, we do not rely on this automatic criterion. Instead, we manually define the signal duration to control the effect of signal duration on the probe results more explicitly. Simulations are conducted with signal durations of 50 ns, 100 ns, and 200 ns, and the results are compared for the probe located 1 m away from the PCB pair.

The results of the simulations with different signal durations are plotted in Fig. 6.6. The left-hand side depicts the complete spectrum up to 2 GHz, while the right-hand side focuses on the low-frequency band up to 400 MHz.

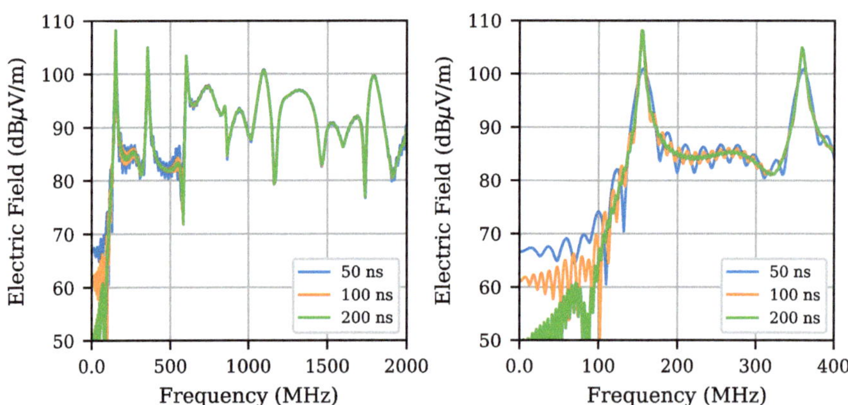

Fig. 6.6 Dependence on the simulation length of the electric field at the probe at 1 m distance. Left: up to 2 GHz, Right: up to 400 MHz

In the band above 1 GHz, the impact of the signal duration is negligible. The frequency components in this band have sufficiently decayed even for the shortest simulation signal duration of 50 ns.

In the band below 400 MHz, as shown on the right hand side, differences among the results of the three simulations become apparent. Between the two peaks, the shorter signal durations exhibit ringing. This ringing originates from the Fourier Transform performed on a truncated signal. For the simulation with a signal duration of 200 ns, the ringing is no longer present between the peaks, though it remains visible below 100 MHz.

Another effect of the signal duration is the peak magnitude. For a 50 ns duration, the lowest peak at 150 MHz is 10 dB below the value calculated with a 200 ns duration. For a duration of 100 ns, the difference compared to the 200 ns result is 3.3 dB.

A third effect is the magnitude of the field toward DC. As the signal duration increases, the field magnitude below 100 MHz decreases. This is expected because high-frequency components decay more quickly in time domain simulations, making the effect of signal duration more prominent at lower frequencies.

To quantify the impact of signal duration on simulation time, we perform a simple estimation: Since the meshing time remains constant for all models, doubling the signal duration approximately doubles the simulation time. In this example, the actual simulation times are as follows: 12 minutes for 50 ns, 21 minutes for 100 ns, and 41 minutes for 200 ns.

From this study, we gain insight into how the signal duration in a time-domain solver affects the calculated field values and can adjust the duration based on the required accuracy. It is recommended to use shorter signal durations while refining the simulation model. Once the setup is finalized, the signal duration can be increased to improve accuracy at low frequencies. For all following examples presented in this chapter, we use a signal length of 100 ns as a balanced trade-off between simulation duration and accuracy. In frequency domain FEM, setting the signal duration is not necessary.

6.3.2 *Effect of the Size of the Computational Domain*

The second effect we study is the influence of the size of the computational domain on the field values. As FIT, TLM, and FEM are volume-based methods, an open boundary condition is required at the boundaries of the computational domain to model the propagation of fields out of the computational domain. The basics of boundary conditions are introduced in Sect. 4.5.

In Sect. 4.5, we discuss how the distance between the boundary and the simulated structure can affect the calculated field values. In this section, we quantify this effect. The open boundary condition (add space) automatically sets the boundary distance to $\frac{\lambda}{4}$ at the center frequency of the chosen simulation frequency interval.

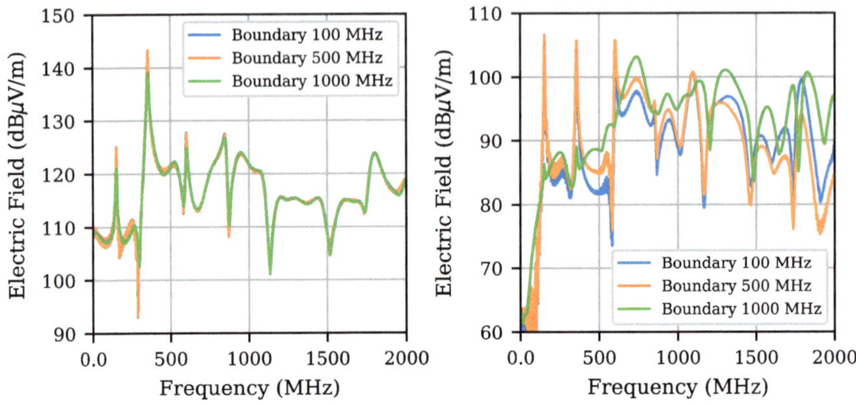

Fig. 6.7 Effect of the size of the computational domain on the electric field. Left: probe at 10 cm distance, Right: Probe at 1 m distance

In the simple PCB simulation model, the frequency band is defined from 0 to 2 GHz. The automatic setting selects $\frac{\lambda}{4} = 7.5$ cm at the center frequency of 1 GHz as the default boundary distance. We compare the results obtained with this default setting to those with distances of $\frac{\lambda}{4}$@500 MHz = 15 cm and $\frac{\lambda}{4}$@100 MHz = 75 cm. In the discussion below, we refer to the distances by their corresponding frequencies, as this is how the open (add space) boundary condition defines them. Additionally, we employ the PML boundary, as it offers superior performance compared to the SIBC boundary.

The magnitude of the electric field for the probe at distances of 10 cm and 1 m, with varying boundary distances, is plotted in Fig. 6.7. Note that the y-axis intervals differ in the two plots.

For the probe at 10 cm, the electric field shows only minor variations across the three boundary distances, differing by just a few dB. However, at 1000 MHz with the smallest boundary distance, the amplitude at the peaks in the low-frequency band is smaller compared to the results obtained with the other two boundary distances.

At the 1 m probe distance, we observe significant differences between the boundary settings. The 1000 MHz setting fails to resolve the first two peaks and exhibits large deviations in the frequency behavior across the entire band. For the 500 MHz setting, the frequency behavior aligns more closely with that of the 100 MHz setting, but the amplitude differs by approximately 5 dB, particularly at 1800 MHz. In this simulation model, using a larger boundary distance, such as $\frac{\lambda}{4}$@100 MHz = 75 cm, is necessary for accurate evaluation of field probes located at the larger distance of 1 m.

The reason for this effect lies in the shape of the computational domain. The absorption of fields at SIBC and PML boundary conditions depends on the incidence angle of the field relative to the boundary. Since the probes are placed only on one side of the table, the bounding box of the computational domain is elongated in the direction perpendicular to the long axis of the table, while the dimensions above and below the table are comparatively short. This results in a shallow incidence angle, particularly for the boundary above the table, leading to inaccuracies when the top boundary is too close to the simulated structure.

To improve the results, it would have been sufficient to move only the top boundary further away, instead of adjusting all boundaries with the open (add space) setting. However, we chose to use the global setting for a more generalized approach to boundary distance definition. Before the simulation, the simulation engineer should determine the lowest frequency of interest and set the boundary distance accordingly.

Increasing the boundary distance also increases the simulation time, as the size of the computational domain grows and more mesh cells are required. Using a signal duration of 100 ns as described in the previous section, we obtain the following simulation times: 8 minutes for the 1000 MHz boundary, 9 minutes for the 500 MHz boundary, and 21 minutes for the 100 MHz boundary. This observation should not be easily generalized, as the implementation of open boundaries is solver-specific and includes other options to optimize their performance.

Since we identified the shape of the computational domain as the primary cause of inaccuracies in this example, we introduce an alternative to using probes within the computational domain in Sect. 6.5. Before continuing the discussion on probes, we will introduce field monitors as a second method for evaluating field results in the next section.

6.4 Field Monitors

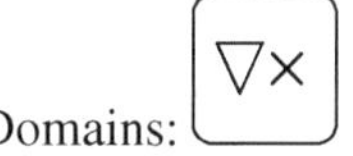

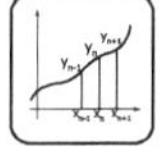

Field monitors allow the visualization of field quantities across the entire computational domain. As mentioned earlier in this chapter, the fields can be computed in either the time or frequency domain. In this section, we focus on frequency domain monitors. Visualizing the field distribution is a powerful tool, offering numerous options for plotting the field. This section provides an overview of these options, helping readers select the most appropriate visualization for their specific problem. We demonstrate these options using the PCB example, with an additional 2 m of space surrounding the PCB pair and the table.

A three-dimensional plot of the electric field surrounding the PCB pair is shown in Fig. 6.8. On the left-hand side, an arrow plot is displayed, with arrows distributed throughout the computational domain. This plot illustrates the field distribution across the entire domain, but the high density of arrows in three-dimensional space can make interpretation challenging. On the right-hand side, a three-dimensional contour plot is presented, visualizing the field on the surfaces of the CAD shapes. At 360 MHz, we observe three field maxima and higher field values at the edges of the table. While this type of plot effectively shows field behavior on geometry surfaces, it does not capture fields in the space between the CAD shapes.

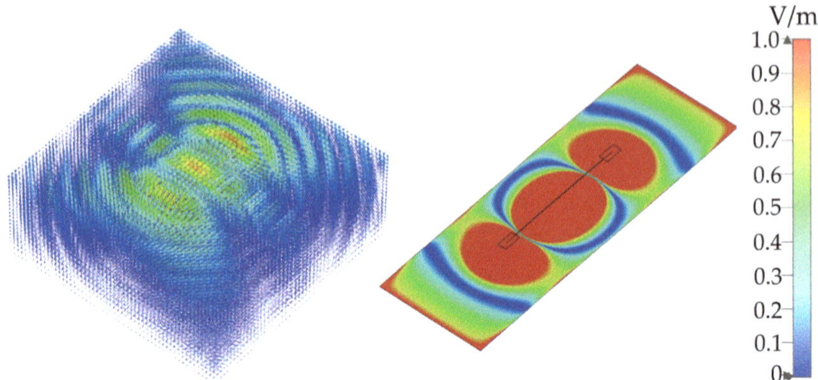

Fig. 6.8 Electric field of the PCB pair example at 360 MHz. Left: full 3D arrow plot. Right: 3D contour plot

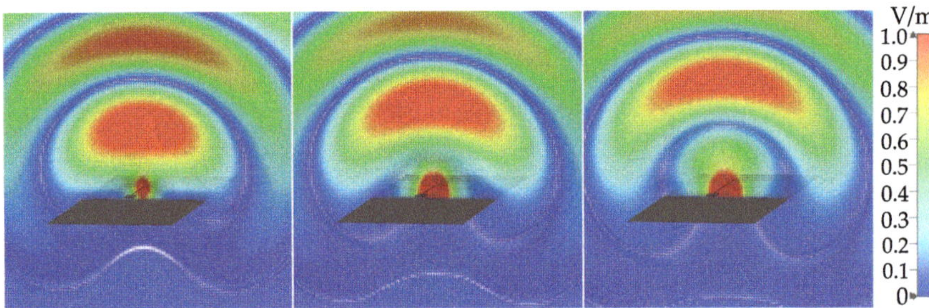

Fig. 6.9 Instantaneous electric field of the PCB pair example at 600 MHz, displayed as a dense arrow plot. Left: 0° phase. Center: 45° phase. Right 90° phase

To visualize fields in the space between CAD shapes, we use field plots on a plane through the computational domain. Compared to full three-dimensional plots, these plots are simpler to navigate. The plane orientation can be chosen freely, but aligning the plane along the Cartesian coordinate system of the simulation model typically simplifies navigation. While multiple planes can be used, employing more than two planes may make the plot difficult to interpret. An example using two planes is shown in Fig. 4.15. On the selected plane, fields can be displayed as contour plots using the absolute field values (Eq. 6.1) or as arrow plots. When plotting the instantaneous field (Eq. 6.3), we can select the phase at which the plot is generated.

In Fig. 6.9, the electric field at 600 MHz is plotted in a plane perpendicular to the direction of the wires, located at the center between both PCBs. The plot uses a dense arrow placement to illustrate the field distribution. The three subplots show the effect of phase rotation on the field:

- In the left plot, the phase is set to 0°.
- In the center plot, the phase is set to 45°.
- In the right plot, the phase is set to 90°.

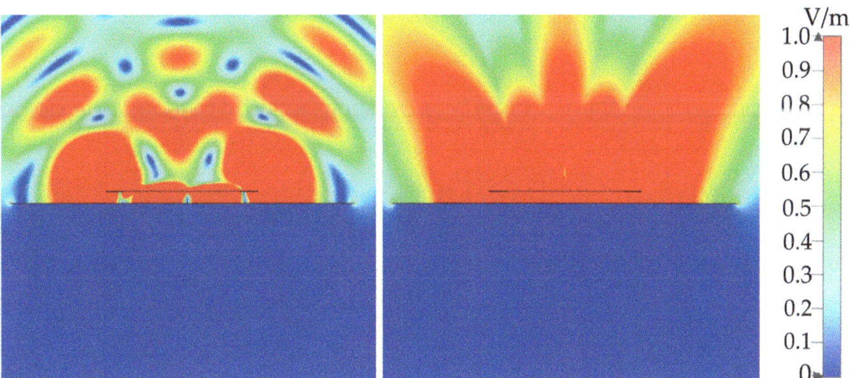

Fig. 6.10 Electric field of the PCB pair at 600 MHz. Left: Instantaneous field with a 0° phase. Right: Maximal electric field

The wave front above the table is observed to propagate upward as the phase increases. Examining the 0° phase plot closely, we note that the field between the wires and the table is relatively low. However, at 90° phase, the red regions indicating higher field intensity are significantly larger. This demonstrates the importance of checking the field distribution at multiple phases when analyzing instantaneous fields. A single phase may show low field values, while another phase may reveal much larger fields.

To remove the effect of phase, the maximal field values can be plotted instead of the instantaneous field. For a maximal field plot, the $e^{j\theta}$ term in Eq. 6.3 is set to one at all positions. However, maximal field plots can only be generated as contour plots. In arrow plots, the absence of phase information eliminates the actual direction at which the field is pointing.

A comparison of the instantaneous and maximal electric field is plotted in Fig. 6.10. The plane is selected perpendicular to the long axis of the table and located in between the two wires. We observe how the absence of the phase indicates the positions at which the field values are high, but it removes the visibility of the wave front.

In Fig. 6.11, we plot the electric and magnetic fields at 150 MHz and 360 MHz. These are instantaneous field plots with a phase of 0°. Analyzing these plots provides insight into the behavior of the probe results depicted in Fig. 6.3.

For the probe results, the probes were positioned near the center axis of the PCB pair. The field plots explain the ratios of the two peaks discussed in Fig. 6.3. At 150 MHz, the electric field exhibits a minimum along the center axis, whereas at 360 MHz, it shows a maximum. For the magnetic field, the behavior is reversed: a maximum is observed at 150 MHz and a minimum at 360 MHz.

Since the probes are offset by 3 cm from the center symmetry axis, the effect of the field minima is less pronounced at close distances. However, at a distance of 2 m and a frequency of 360 MHz, the probe is located in a field minimum. This explains

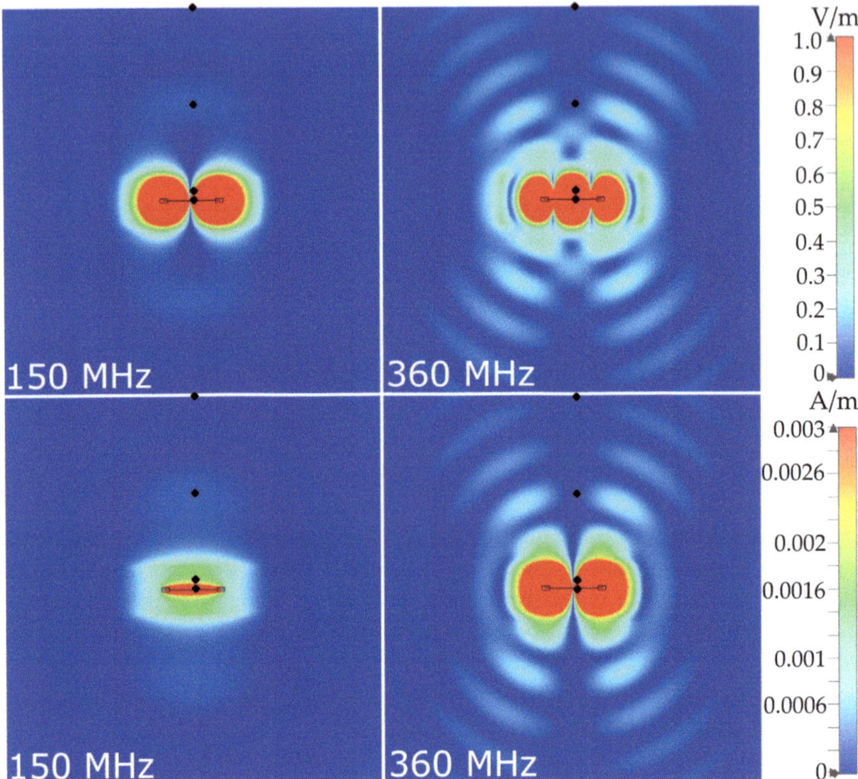

Fig. 6.11 Electric and magnetic field of the PCB pair at different frequencies. Top left: Electric field at 150 MHz. Top right: Electric field at 360 MHz. Bottom left: Magnetic field at 150 MHz. Bottom right: Magnetic field at 360 MHz. The black diamonds indicate the position of the probes we analyze in Sect. 6.2

the significant reduction in the magnitude of the 360 MHz peak in the probe results shown in Fig. 6.3 as the distance increases.

Analyzing fields using probes and monitors simultaneously allows for complementary insights. Probes provide information about the frequencies at which resonances occur, while field monitors visualize the complete spatial distribution of the fields.

The field plots at 360 MHz in the right column of Fig. 6.11 illustrate the transition from the near to the far field. Close to the PCB pair, the electric and magnetic field distributions differ significantly. The electric field is maximal at the center of the wires, while the magnetic field is minimal. As we move further away from the PCB pair—at a distance of 1 m, indicated by the central black diamond—the maxima of the electric and magnetic fields become nearly aligned. At 2 m, at the edges of the plot, the positions of the maxima for both fields coincide.

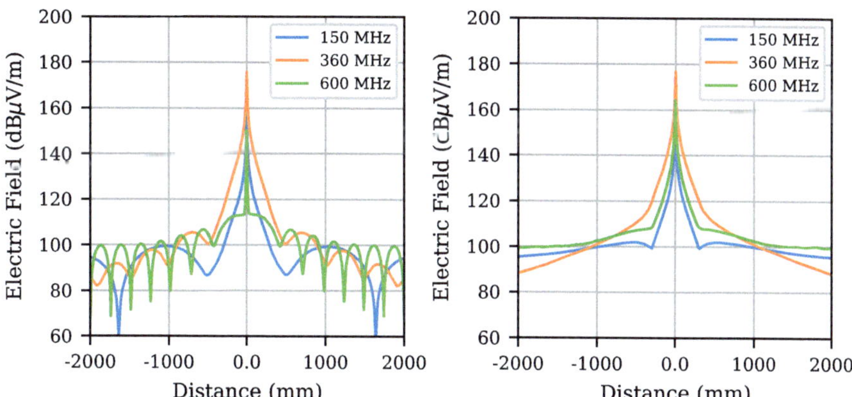

Fig. 6.12 Electric field at the three peak frequencies plotted at a line through the computational domain. Left: Instantaneous field at phase 0°. Right: Maximal field magnitude

The field monitor results can be further postprocessed to obtain local field values at specific positions or along curves through the computational domain. In Fig. 6.12, we plot the electric field at the three peak frequencies of 150 MHz, 360 MHz, and 600 MHz (cf. Fig. 6.7) along a line through the computational domain. The line is aligned with the positions of the field probes shown in Fig. 6.11. At a distance of − 2000 mm, the position corresponds to the bottom of the two-dimensional plot, while a distance of zero corresponds to the field at the center of the model.

When generating this type of plot, we need to define which field component and phase information to display. Here, we select the absolute value and plot the electric field for a phase of 0° on the left-hand side of Fig. 6.12 and the maximum field on the right-hand side. This approach is analogous to the plot type used in Fig. 6.10.

In the phase plot, we observe a wavelength effect as we move away from the center of the model. Since we plot the absolute value, the wavelength can be determined by measuring the distance between every other peak. The peak in between them corresponds to the field pointing in the opposite direction. For the 600 MHz result, we calculate a peak-to-peak distance of 503 mm, which corresponds to a free-space wavelength of 596 MHz, demonstrating good agreement between the field plot and the theoretically expected value.

In the plot of the maximum field amplitude, the wave behavior of the field is no longer visible. The maximum field value plot serves as the envelope of the instantaneous field. This behavior is also observed in Fig. 6.13, where both the instantaneous and maximum field values are plotted together.

With the one-dimensional plots of the electric field, we compare the calculated field decay to the $\frac{1}{r}$ law discussed in Sect. 5.1 for the near field of the loop. In Fig. 6.13, we plot the electric field generated by the PCB pair along the same line described in the previous paragraph at 150 MHz and 600 MHz. This time, we limit the analysis to positions from the center of the computational domain up to the top edge.

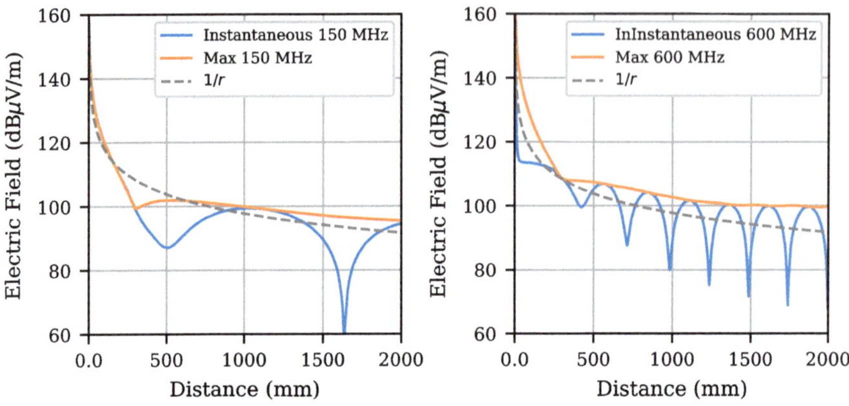

Fig. 6.13 Electric field decay compared to the $\frac{1}{r}$ law. Left: 150 MHz. Right: 600 MHz

The plot reveals that the field decay generally aligns with the $\frac{1}{r}$ law across most positions. Notably, close to the structure, the decay is slightly slower than $\frac{1}{r}$. This behavior highlights the influence of the structure's geometry and near-field effects on the field distribution and decay.

The final remark we would like to make in this section concerns the scaling of field monitors:

All field values in the one-dimensional plots presented in this section are shown in logarithmic scale, whereas all two-dimensional field results are plotted in linear scale. While it is possible to plot two-dimensional field results in logarithmic scale to highlight the behavior of fields at low absolute magnitudes, caution is advised. Visualization in logarithmic scale may emphasize nonrelevant field values or even numerical noise. In such cases, verifying the actual field values is recommended.

The conversion from a voltage to the corresponding electric field is straightforward and can help assess the relevance of the field value. For instance, two conductors spaced 1 mm apart with a voltage of 1 V between them generate an electric field of $1000\,\frac{\text{V}}{\text{m}}$. An electric field of $0.001\,\frac{\text{V}}{\text{m}}$ corresponds to $1\,\mu\text{V}$ (or $0\,\text{dB}\mu\text{V}$) between conductors separated by 1 mm. This value represents the lower end of field strength relevance and is close to the lowest measurable level in an EMC laboratory. Thus, visualizing field distributions with such low field strengths requires careful interpretation.

With the examples presented in this section, we demonstrate how the visualization capabilities of simulation tools can enhance the understanding of the electromagnetic behavior of simulated models. While some may criticize field results as merely "colorful pictures," when used correctly, visualization serves as a powerful tool for analysis.

Up to this point, we have evaluated field values at positions inside the computational domain using probes and monitors. In the next section, we discuss how it is possible to calculate field values outside of the computational domain.

6.5 Field Results Outside of the Computational Domain

Domains: 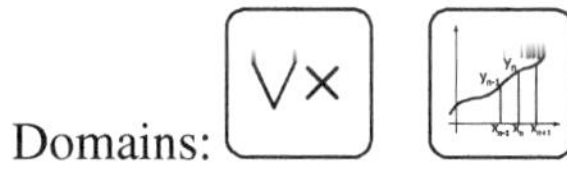

As we have seen in the previous sections, we can calculate field values within the computational domain and evaluate them using probes or field monitors. However, when field values are required at a large distance from the simulated model, using probes or monitors would require a significant increase of the computational domain. This, in turn, would lead to increased simulation time and memory requirements, potentially making the simulation infeasible.

Consider, for example, the FCC radiated emission standard, which specifies limits for the electric field strength at distances of 3 m or 10 m. For the PCB pair example used in this chapter, it is feasible to set up a computational domain with probes at a distance of 3 m. However, at a distance of 10 m, the computational resources required would exceed the capabilities of the available simulation hardware (cf. Sect. 1.5).

6.5.1 *Near to Far Field Transformation*

We require a method that enables the calculation of fields at large distances without the need to resolve these distances with a computational mesh. To achieve this, the surface equivalence theorem (SET), also known as the surface equivalence principle, is utilized (Balanis 2024). The concept of the SET is illustrated in Fig. 6.14. The SET states that the fields $\mathbf{E}_o, \mathbf{H}_o$ in the outer domain V_o, generated by a source in the inner domain, can be computed from electric and magnetic current densities $\mathbf{M}_S, \mathbf{J}_S$ on a surface S that encloses the source. This surface can have an arbitrary shape but must be closed and completely encompass the source. For calculating the fields in the outside domain, the fields on the inside, $\mathbf{E}_i$ and $\mathbf{H}_i$, can be set to zero.

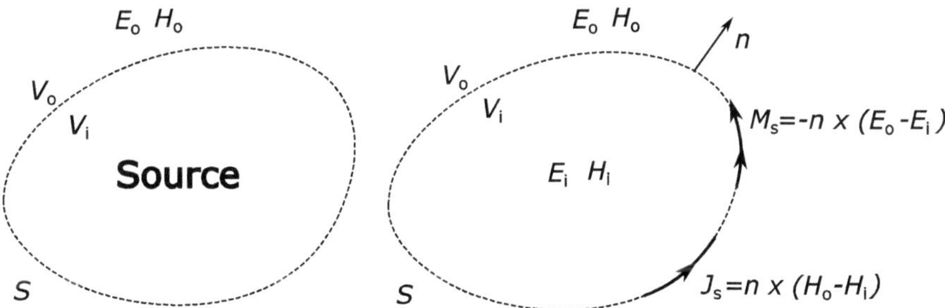

Fig. 6.14 The Surface Equivalence Principle

With this setup, the following relations are obtained:

$$\mathbf{J}_\mathrm{S} = \hat{\mathbf{n}} \times \mathbf{H}_\mathrm{o} \tag{6.5}$$

$$\mathbf{M}_\mathrm{S} = -\hat{\mathbf{n}} \times \mathbf{E}_\mathrm{o} \tag{6.6}$$

These equations enable the calculation of the fields in the outer domain from the surface currents. The SET is a generalization of Huygens' principle. It can be interpreted as a summation of contributions from small current filaments on the surface, where each filament contributes to the field in the outside domain, and the summation of all filaments yields the resulting field.

In simulations, this method can be applied straightforwardly: the fields are recorded on a surface S, and the current densities $\mathbf{M}_\mathrm{S}$ and $\mathbf{J}_\mathrm{S}$ are obtained from these fields. In the next step, the fields in the outside domain are calculated using the surface current densities $\mathbf{M}_\mathrm{S}$ and $\mathbf{J}_\mathrm{S}$.

In the general case, this process requires another full-wave simulation, with the surface current densities used as a source. This concept, known as a field source, is discussed in Sect. 6.9.

However, there is an alternative way to calculate the fields outside the computational domain: the surface S is chosen to be the boundary of the computational domain. Assuming that a far-field approximation is valid in the outer domain, the fields are obtained by inserting the surface currents into an analytical equation. Using this analytical equation reduces computational effort compared to performing a second full-wave simulation. The derivation of the analytical equation relies on a Greens function with suitable approximations valid in the far field. A rigorous derivation can be found in Balanis (2024) .

By applying the far-field approximation, the fields are calculated at an infinite distance from the source, but the approximation still yields finite field values. In the far-field approximation, the fields are purely transverse; no radial components exist. Furthermore, the ratio of the electric field to the magnetic field is constant and equal to the free space impedance of 377 Ω.

For accurate far-field calculations, the distance between the source and the surface where the electric and magnetic current densities are recorded is crucial. If the surface is too close to the source, the resulting far-field calculations may be inaccurate. This requirement aligns with the $\frac{\lambda}{4}$ rule for the open (add space) boundary condition, as discussed in Sect. 4.5. While this is a different issue from the effect of the flat computational domain discussed in Sect. 6.3, both effects emphasize the need for a suitable distance between the boundary and the source.

A special case arises when the open boundary touches a dielectric material. In such scenarios, the analytical equation used to calculate the far-field is modified to account for the dielectric material's presence. However, this type of setup is not utilized in this book.

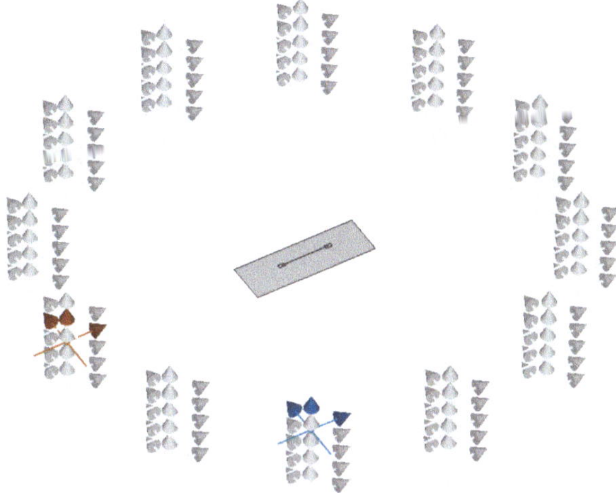

Fig. 6.15 Cylindrical probe arrangement for a comparison of probes inside and outside of the computational domain. Colored probes indicate the position at which the electric field is plotted in Fig. 6.16

6.5.2 *Far Field Probes*

In order to better understand the calculation of the fields outside of the computational domain, let us come back to the PCB pair introduced in Sect. 6.2.1, and compare results of probes located inside the computational domain with results of probes located outside of the domain.

For the comparison, we place a number of probes in a cylinder around the PCB pair. The probes are located at a distance of 3 m from the center of the table, as indicated in Fig. 6.15. The angular resolution in the azimuth is 30°, and at each angle, 5 probes are positioned to sample the height. The highest probe is 50 cm above the table, while the lowest probe is 50 cm below the table.[2]

In the following, we compare the absolute value of the field at the probe positions. One model is set up with the probes inside the computational domain, while the second model uses far-field probes placed outside of the computational domain. Both probe types use a Cartesian coordinate system, allowing for a direct comparison.[3] The boundary of the computational domain is set to a distance of $\frac{\lambda}{4}$ @ 100 MHz, and a signal duration of 100 ns is used, as in Sect. 6.2.

The electric field at the blue and orange probe positions is plotted in Fig. 6.16. The field at the blue probe position is approximately 10 dB higher than at the orange

[2] The probes are placed using a script provided with the installation of CST.

[3] The default setting for probes outside of the computational domain in the CST is a spherical coordinate system.

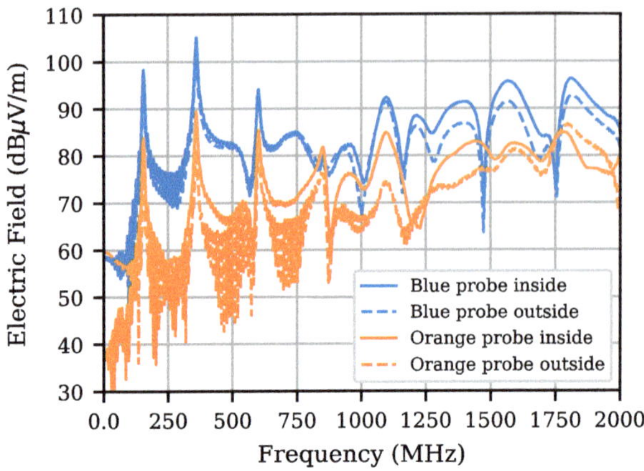

Fig. 6.16 Comparison of the electric field calculated by probes inside and outside of the computational domain. The colors correspond to the placement depicted in Fig. 6.15

position, and at resonances, the difference is even greater. The maximal difference between the blue probe inside and outside the computational domain is around 3 dB. For the orange probe, the difference between the probes inside and outside the computational domain is larger, reaching about 10 dB at 1100 MHz and at the peak at 360 MHz.

This behavior aligns with observations in simulation results discussed so far: for lower field values, the relative accuracy is often worse than for higher field values. On the other hand, as discussed in Sect. 4.3, we know that local values, such as the field strength at a specific location **r**, exhibit higher errors compared to averaged or global values. In Sect. 4.3, this effect was demonstrated on the capacitance between two bodies.

To obtain more relevant results than just the field at individual positions, we can evaluate the highest field values (e.g., the worst-case emission), as this worst case is of primary interest for EMI. The worst-case estimation corresponds to the envelope of all probes, which can also represent a radiated emission measurement performed using a turntable. In a turntable measurement, a max-hold measurement is conducted for various orientations of the DUT and different antenna heights. By calculating the envelope of the probes, we achieve a similar result: it includes all table orientations and antenna heights within a single simulation run because the probes are placed at all required positions in the same simulation model.

The comparison of the probe inside and outside of the computational domain for the worst-case electric field at a distance of 3 m is plotted in Fig. 6.17. In the worst-case estimation, the highest difference between the inside and outside probes is 2 dB, which is significantly less than the differences observed at individual positions. Since the worst case is the primary focus in emission simulations, using the described method allows us to obtain accurate results with improved simulation efficiency:

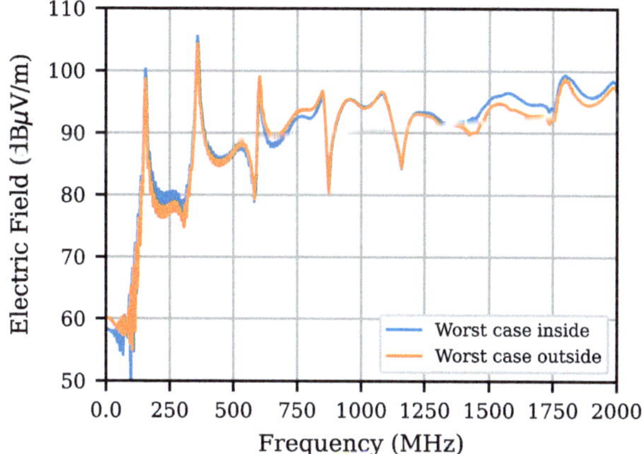

Fig. 6.17 Comparison of the worst case electric field at 3 m distance calculated using probes inside and outside of the computational domain

For the model with the probes inside the computational domain, the simulation time is 56 minutes, requiring 47 million mesh cells. In contrast, the model with the probes outside the computational domain runs in 14 minutes and uses only 9 million cells. By employing probes outside the computational domain, the simulation time in this example is reduced by a factor of four.

The simple geometry of this example permits the above comparison, where both probe types are placed at a distance of 3 m. For more complex models and higher maximum frequencies, probes outside the computational domain are often the only viable option for calculating fields at large distances. In such cases, placing probes within the computational domain would result in models that exceed available hardware resources.

The usage of probes outside of the computational domain also positively affects the performance of open boundary conditions, as it is no longer required to model a large domain and the incidence angle of the field on the boundary is closer to being perpendicular. Recall the effect of the shallow incidence angle in the flat computational domain on the probe results, as discussed in Sect. 6.2 and illustrated in Fig. 6.7.

We test the effect of the bounding box again, this time for probes located outside of the computational domain. The effect of the distance to the boundary for probes outside of the computational domain, positioned at a distance of 3 m, is plotted on the left-hand side of Fig. 6.18. As in the previous section, the distance is characterized by the frequency corresponding to $\frac{\lambda}{4}$. With probes outside the computational domain, the influence of the distance is far less pronounced compared to Fig. 6.7. One reason for this improvement is the reduced effect of the flat incidence angle.

The second reason is the use of worst-case field values in Fig. 6.18, that is, the field value is derived from multiple positions rather than a single local one.

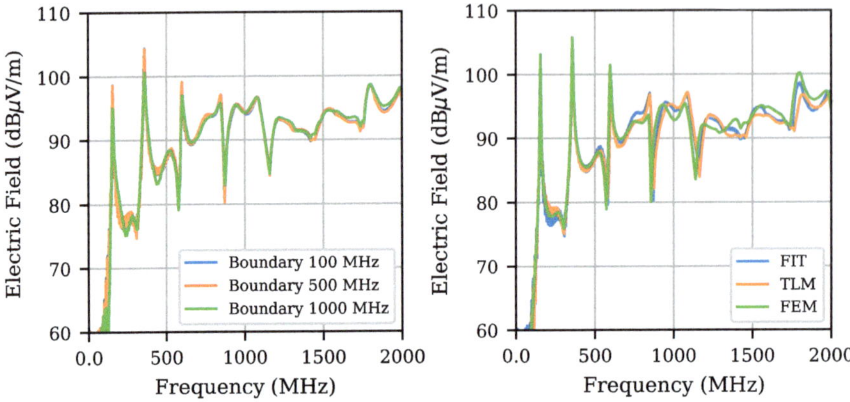

Fig. 6.18 Worst case electric field at 3 m distance. Left: comparison of different distances of the model to the boundary. Right: comparison of results from different solvers

With probes outside the domain, we can reduce the distance to $\frac{\lambda}{4}$@500 MHz = 15 cm and observe no difference compared to results obtained with a distance of $\frac{\lambda}{4}$@100 MHz = 75 cm. Using the smaller bounding box @500 MHz, the simulation time further decreases to just 4 minutes. Even with the smallest distance of $\frac{\lambda}{4}$@1000 MHz = 7.5 cm, the difference remains within a range of 3 dB.

Finally, we compare the worst-case electric field at a distance of 3 m calculated with FIT, TLM, and FEM. The comparison is plotted on the right-hand side of Fig. 6.18. The largest difference among the results from all three solvers is approximately 3 dB. This type of comparison is valuable as it verifies that all simulation settings, such as boundary conditions, signal duration, and mesh density, are correctly configured.

This solver comparison uses a boundary distance set to $\frac{\lambda}{4}$@500 MHz. For this model, the FIT simulation requires 4 minutes, the TLM simulation takes 34 minutes, and the FEM simulation takes 43 minutes. As with all solver comparisons, the differences in computational time are specific to the particular model and solver settings: in the FEM simulation, mesh adaptation is used, while the TLM and FIT meshes are manually defined without refinement. The complete model is driven by only a single port, which is a significant advantage for the two time-domain methods (Sect. 4.9.2).

6.6 Application Example: Coupling from a TWP into an Antenna

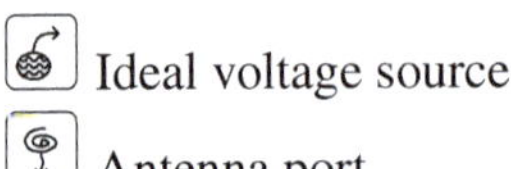

Full-wave, S-parameters, combine results

TLM solver

AC, Transient

0–3 GHz

The application example presented in this section is not a classical EMC measurement scenario but rather addresses the topic of RF interference. With the increasing clock frequencies in digital data transfer in modern electronics, emissions from data buses are increasingly interfering with frequency bands utilized for wireless RF communication. Furthermore, the growing demand for bandwidth in mobile communication has led to a more efficient and tighter utilization of available spectrum resources, with frequencies well below 1 GHz being allocated for mobile phone use. The example presented here serves as an introduction into the topic of platform RF interference. A more detailed discussion of platform RF interference is discussed in detail in Chap. 15.

We simulate the coupling from a wired communication bus connecting two PCBs via the 1000Base-T1 Automotive Ethernet standard into an LTE antenna. This example demonstrates the simulation effort required for a more realistic geometry compared to the simple PCB pair used previously.

In the first step, we require a realistic model of the LTE antenna. As EMC engineers are not typically antenna designers, creating an antenna model from scratch is out of scope. However, using a realistic antenna model increases the accuracy of the sink representation in the simulation. If the antenna design is performed by the same company integrating the antenna into the final product, the design data from the antenna team is the ideal choice for building the simulation model. If the antenna is designed externally, the EMC simulation engineer may not have access to the design. In such cases, an antenna synthesis tool can be used to create a realistic model with minimal effort.

We have used the tool Antenna Magus by Dassault Systèmes to generate a model of a three-band LTE antenna, as shown in Fig. 6.19. The generation of the antenna model using the synthesis tool is straightforward, and the simulation model is used as created by the tool. The antenna type is a Planar Inverted-F Antenna (PIFA), a common choice in consumer electronics products. The antenna is mounted on an FR-4 substrate and connected by a port to a reference plane. The overall size of the antenna, including the reference plane, is 15 cm × 11 cm × 1 cm, resembling the dimensions of a typical mobile phone.

In order to evaluate the performance of the synthesized antenna, the antenna model is simulated using the TLM solver. The resulting S-parameter at the antenna port is plotted in Fig. 6.20. The selected antenna is a three-band design, with the first band located at 750 MHz, the second band around 1600 MHz, and the third band at 2500 MHz, covering several LTE bands. The simulation of the standalone antenna using the TLM solver requires less than one minute. No mesh adaptation is performed, as the synthesis tool not only provides the antenna geometry but also optimized mesh settings.

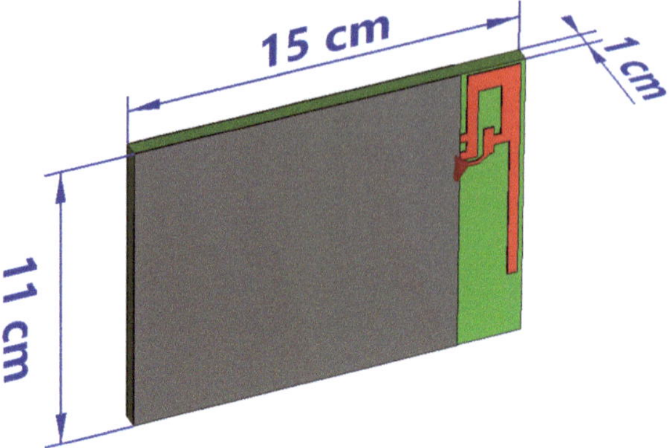

Fig. 6.19 Simulation model of an LTE antenna created by an antenna synthesis tool

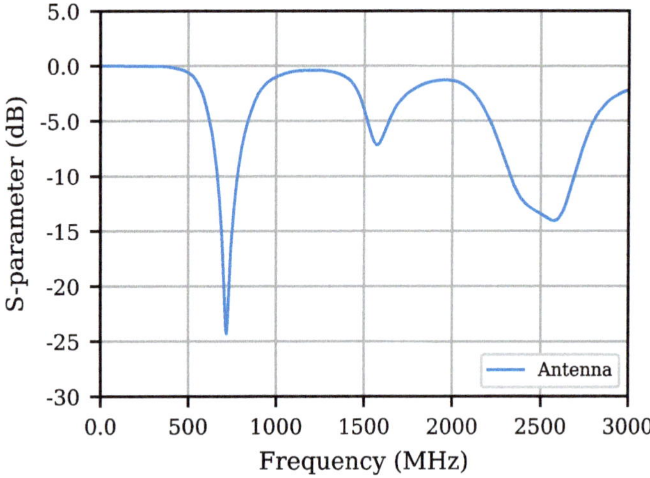

Fig. 6.20 S-parameter of the synthesized antenna

An overview of the simulation model for the coupling into the LTE antenna is shown in the top section of Fig. 6.21. Two identical PCBs are connected by a single unshielded TWP. The TWP is 1.8 m long and smoothly bent along a curve lying in a single plane. The individual wires are made of copper, with a diameter of 1 mm, and are covered by a 0.35-mm-thick polyethylene dielectric, with a permittivity of $\varepsilon_r = 2.3$. This setup results in a characteristic impedance of the TWP of $105\,\Omega$, complying with the Automotive Ethernet standard requirement of $100\,\Omega \pm 10\%$. The twist length of the TWP is selected to be 2 cm, resulting in approximately 90 full twists along the cable's length.

Resolving such a TWP with the mesh in a full-wave simulation presents significant challenges, especially for a hexahedral mesh due to the bending of the cable.

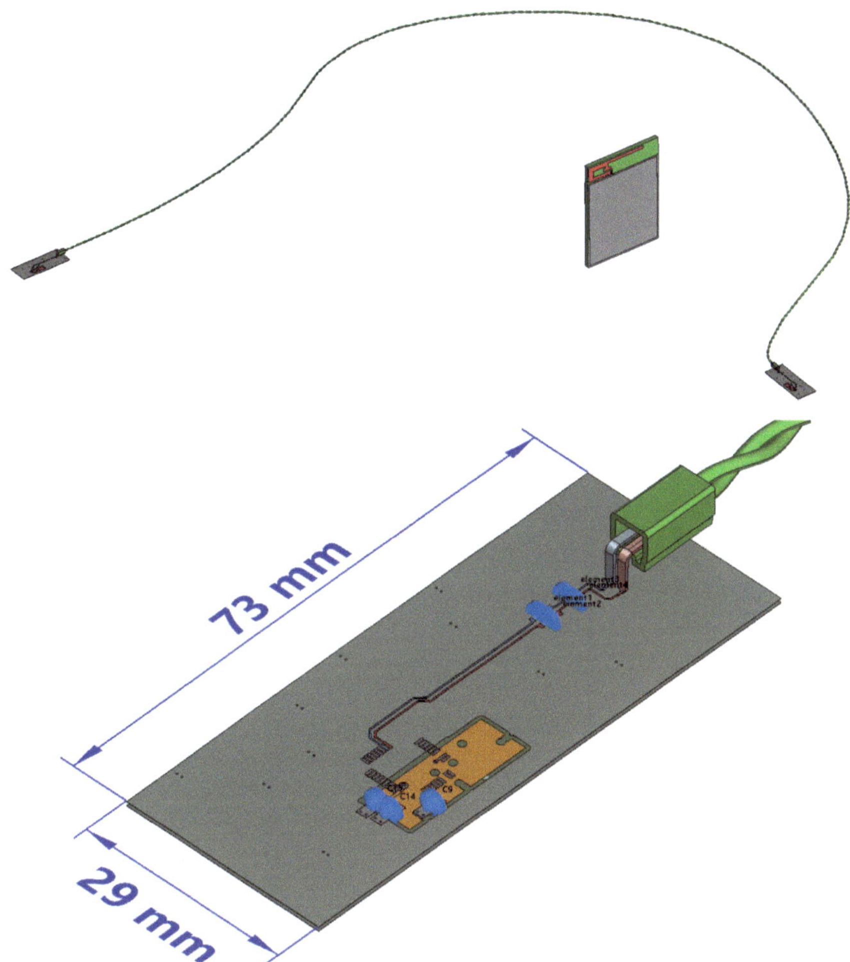

Fig. 6.21 Top: Overview of the simulation model for the coupling of an Automotive Ethernet communication into an LTE antenna. Bottom: Detail view on the Automotive Ethernet PCB

In this simulation model, we use the TLM solver with an octree mesh and benefit from its ability to refine the mesh locally. In Chap. 8, we simulate cables using the multiconductor transmission line method (MTL). The MTL significantly reduces simulation effort compared to full discretization, but it introduces approximations that limit prediction accuracy for certain cases. This example demonstrates that for cable cross-sections with a limited number of conductors, a full-wave simulation is feasible.

On the bottom of Fig. 6.21, a close-up of the PCB is shown. The PCBs in the simulation model on both sides are identical, each measuring 7 cm in length and 3 cm in width, with four metal layers. On the top layer, the differential pair of the

Automotive Ethernet bus is routed, represented by the blue and red traces. The traces are 200 μm wide and separated by 470 μm. The thickness of the substrate between the signal layer and the reference plane is 127 μm. With a substrate permittivity of $\varepsilon_r = 4.3$, this configuration results in a characteristic impedance of 100 Ω for the differential mode.

Just before the connector, two series lumped elements are placed in each signal line. The lumped element closer to the connector is a DC blocking capacitor with a nominal value of 100 nF. The second lumped element serves as a placeholder for a common mode choke that is not used in this simulation and is instead replaced by a shorting resistor with a value of 10 mΩ.

The signal line is driven by a port with an impedance of 100 Ω, placed at the IC pins between the two signal lines. This port configuration ensures purely differential excitation, with no common mode generated by the driving IC. Any common mode generated due to imbalance along the signal path is terminated with an open.

For a detailed study of the behavior of the common and differential modes within a single simulation model, it is more suitable to use single-ended ports along with mode converter blocks. The usage of mode converters is discussed in Sect. 8.3. However, for simplicity and to minimize the port count and simulation time, mode converters are not employed in this example.

The antenna model is positioned between the two PCBs, with the TWP routed around the antenna model, as depicted in Fig. 6.21.

Figure 6.22 depicts the octree mesh used by the TLM solver in this simulation at three different positions in the computational domain. The top left shows a view of the complete computational domain, illustrating the fine mesh resolution at the locations where the TWP is routed. On the top right, a cross section of the TWP is shown, demonstrating that along the entire length of the TWP, at least two mesh cells are placed between the two conductors.

To achieve this resolution, we manually set a mesh step of 300 μm for all axes on the TWP structures. On the PCB, a mesh step of 120 μm is applied, while the signal traces are meshed with a finer step of 80 μm. For the antenna, the mesh generated by the synthesis tool is used without modification. These mesh settings result in an initial mesh size of 2.4 billion cells.

Such a high mesh count makes an FIT simulation unfeasible on the available hardware (cf. Sect. 1.5). In Sect. 4.10, the largest model simulated with the FIT method consists of 400 million mesh cells and, due to a memory requirement of 60 GB, must use the CPU instead of the GPU.

The TLM solver can reduce the initial mesh count by applying mesh lumping. The mesh lumping, which is a step in the octree mesh generation process, reduces the cell count by removing fine cells in regions where they are not required (Sect. 4.10.2). In this example, the mesh count is reduced from the starting number of 2.4 billion to just 4.8 million cells. This allows the simulation to run on a single GPU, requiring only 1.8 GB of RAM. The signal duration for the time domain simulation is set to 50 ns, as discussed in Sect. 6.3. The simulation of the complete S-matrix for the three ports (two ports at the PCBs and one port at the antenna) in the

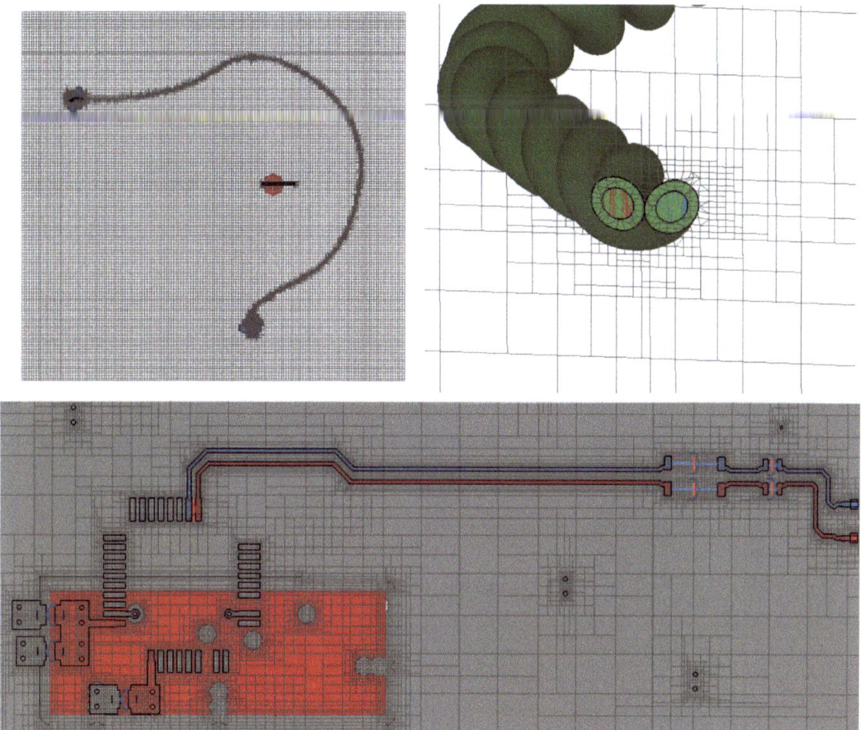

Fig. 6.22 Octree mesh in the simulation of the coupling into the LTE antenna

computational domain requires 5 hours and 40 minutes for a maximum frequency of 3 GHz.

As IBIS models of Automotive Ethernet ICs are not easily available, we construct an idealized signal to drive the transmitter. The 1000Base-T1 version of the Automotive Ethernet standard uses a PAM3 modulation at 333.33 MHz to achieve a data rate of 1 Gb/s.

We define a PRBS PAM3 signal with a peak-to-peak voltage of 1.4 V. The time domain waveforms at the transmitter and receiver are plotted on the left-hand side of Fig. 6.23. The propagation time of 9.1 ns along the 1.8 m long cable is clearly visible in the figure. For the time domain waveforms, a rise/fall time of 1 ns is used, resulting in a waveform with a nearly triangular shape. To perform transient circuit simulations, vector fitting is applied to the S-parameter matrix from the TLM simulation (Sect. 10.3).

A second simulation is performed with a rise/fall time of 333 ps. The time domain waveforms of the faster rise/fall time signal are not plotted. Instead, we compare the voltage spectrum of both transmitter signals on the right-hand side of Fig. 6.23. The magnitude of both spectra is very similar up to approximately 600 MHz, while above 600 MHz, the magnitude of the signal with the faster rise time decays more slowly.

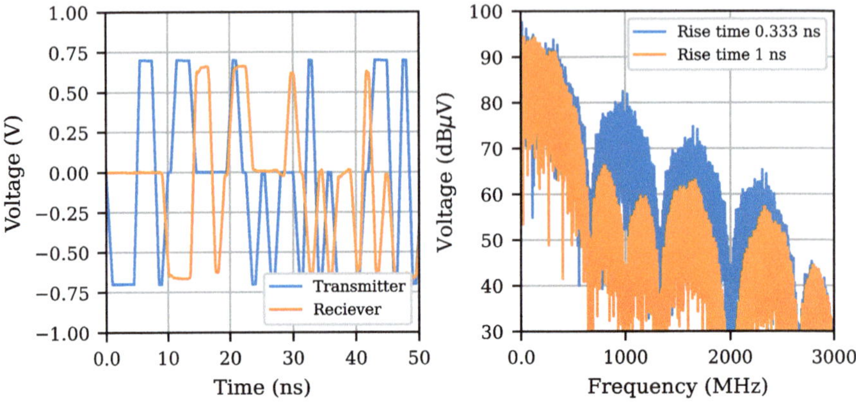

Fig. 6.23 Voltage in the Automotive Ethernet bus. Left: Time domain signals at transmitter and receiver. Right: Voltage spectrum at the transmitter

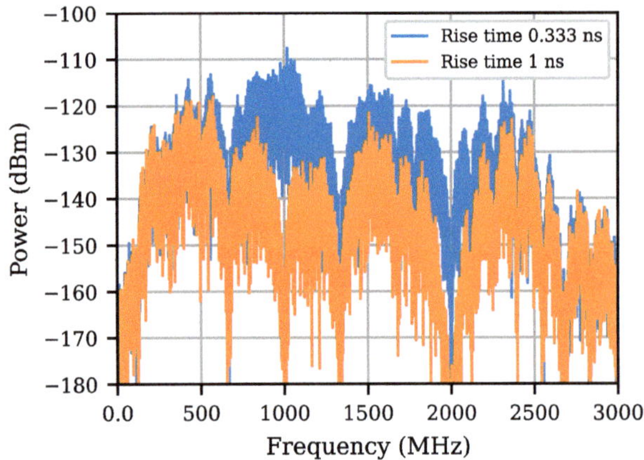

Fig. 6.24 Power received at the LTE antenna for two different rise/fall time setting in the Automotive Ethernet bus

This observation aligns with the plot of Fig. 2.3.2, which shows that the envelope of the spectra are identical up to approximately $\frac{1}{\pi \cdot 1\,\text{ps}} \approx 330\,\text{MHz}$. Beyond this frequency, the slower signal has a lower magnitude.

We use these spectra as sources in an AC simulation and calculate the power received at the antenna. The power received at the LTE antenna for both rise/fall time configurations is plotted in Fig. 6.24. Below 500 MHz and above 2500 MHz, the received power is similar, as expected from the voltage spectra of the transmitter. Around 1000 MHz, the signal with the faster rise time generates a power level at the antenna that is 20 dB higher. The highest received power level is approximately − 110 dBm.

The receiver sensitivity of LTE radios varies across the different channels but typically lies in the range of − 90 dBm to − 105 dBm. With the highest coupling power of − 110 dBm, we are slightly below the sensitivity threshold of the LTE receiver and expect that the signal on the Automotive Ethernet cable does not affect the performance of the LTE receiver.

6.7 Coupling Full-Wave Solvers to Circuit Simulation

Domains: 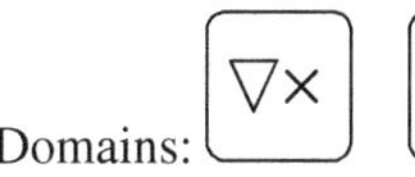 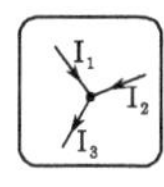 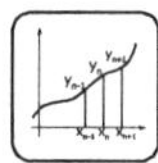

In the previous chapter, we generated circuit models of the coupling path using static EM simulations and incorporated them into the schematic of the decomposition (Fig. 5.4 in Sect. 5.2). In this section, we discuss the generation of circuit models from EM simulations performed with full-wave solvers. A circuit model generated by a full-wave solver allows for the inclusion of a complete description of the coupling path in the schematic simulation: a distinction into resistive, capacitive, and inductive contributions is no longer required, and radiative coupling is inherently included in the generated circuit model.

Additionally, using full-wave solvers to generate the coupling path model provides the opportunity to maintain a link between the EM simulation and the circuit simulation. This link can be leveraged to calculate field-based results, such as probes or monitors, with an excitation defined in the circuit simulation. In CST, this process is referred to as "combine results." The combine results simulation flow is discussed in Sect. 6.7.5.

To include all contributions in the circuit model, the coupling path model obtained from a full-wave simulation is represented using S-parameters instead of RLC elements. Although the S-parameter model is treated as a "black box," the techniques described in Sect. 2.7 and the in following Sect. 6.7.4 allow us to analyze its inner structure when necessary. When generating S-parameter-based circuit models, it is critical to ensure that the EM simulation model remains linear (cf. Sect. 2.4.3). Nonlinear elements, such as diodes or nonlinear materials, must not be included in the EM model if the intention is to use an S-parameter-based coupling path model in a circuit simulation.

The decomposition using S-parameter-based coupling path models is highly versatile in terms of the simulation flow:

For instance, we can exchange the source and the termination and calculate the resulting voltages and currents at the ports. This simulation flow is applied in the LTE antenna example to obtain the waveforms and spectra plotted in Fig. 6.23.

In more advanced setups, we extend this flexibility by not only exchanging sources and terminations but also connecting additional component models to the S-parameter block. The ability to link various sources and component models to the S-parameter coupling path model is a feature utilized throughout all subsequent chapters.

For example in Chap. 9, we use a port to connect a housing and the conductive table. By connecting a resistor to this port on the schematic and varying the resistor value from 1 mΩ to 1 MΩ, we can connect or disconnect the housing and the table.

To fully understand the following sections, readers are expected to be familiar with the concept of ports as discussed in Sect. 2.4.3.1. In the next section, we illustrate the coupled full-wave and circuit simulation based on the same π-filter example introduced in the ports section. This choice simplifies the discussion and highlights that the principles used for coupling full-wave EM and circuit simulations are general circuit simulation concepts for S-parameters. These principles are equally applicable when measured S-parameters are used.

6.7.1 Coupling Path Circuit Model Based on S-Parameters

For the implementation of the π-filter, we use a simple PCB layout, as depicted in Fig. 6.25. The PCB is 6 cm long and 2.3 cm wide. It consists of two layers, with a 0.5 mm thick FR4 substrate in between the layers. On the top layer, 1-mm-wide traces connect the components. All metal shapes are set to copper and use the lossy metal material model.

The input is placed on the left-hand side of Fig. 6.25 and the output on the right-hand side. Input and output are realized by ports, located at positions where connectors are placed in the PCB design. The input and output ports are connected between the signal line and a via connecting to the bottom layer, which acts as a continuous reference plane. The two capacitors and the inductor are also replaced

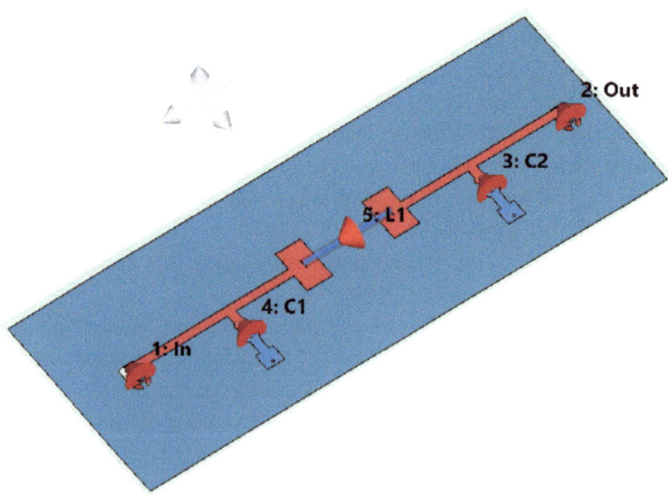

Fig. 6.25 PCB implementation of a simple π-filter

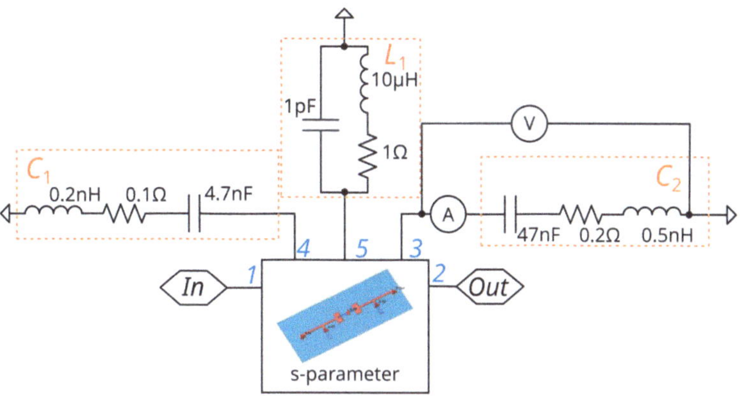

Fig. 6.26 Schematic of the coupled π-filter model using a S-parameter block

by ports. All ports are set to the standard impedance of 50 Ω. Additionally, we place a probe at a distance of 1 cm in the center of the PCB.

The FEM solver calculates the full S-matrix of the model in 43 s, for a frequency band ranging from 1 kHz to 100 MHz. To calculate the filter response, we connect the component models of the inductor and the two capacitors on the schematic. The resulting schematic is shown in Fig. 6.26.

As we discuss in Sect. 2.4.3.1, the individual components are connected to the S-parameter block in a way that differs from the standard SPICE notation. The common interpretation of these schematics is to consider the circuit components located between the block and the ground symbol to be positioned at their respective port locations in the 3D model. This is illustrated by the orange boxes in Fig. 6.26. The placement of the port in the three-dimensional model defines between which traces the component is located and at which specific position. In the π-filter example, Port 5 is connected in series, while all other ports connect from the signal traces to the reference (note the vias close to C_1 and C_2).

A common mistake is interpreting the ground symbols on the schematic as the ground layer of the PCB. This is not the case, as in the full solution of Maxwell's equations, the concept of ground does not exist. All metal shapes are treated equally; referring to some of them as ground is merely an engineering convention.

In this section, we discuss the coupled schematic using simple components consisting of only RLC elements. When integrating circuit models derived from measurements with the S-parameter block generated by the EM simulator, additional considerations must be addressed. These are covered in Sect. 9.4.

In transient circuit simulations, nonsymmetrical components such as diodes or transistors must be connected with the correct orientation. Defining the correct orientation in a coupled schematic is challenging because there are two degrees of freedom: the orientation of the port in the three-dimensional model and the orientation of the component on the schematic.

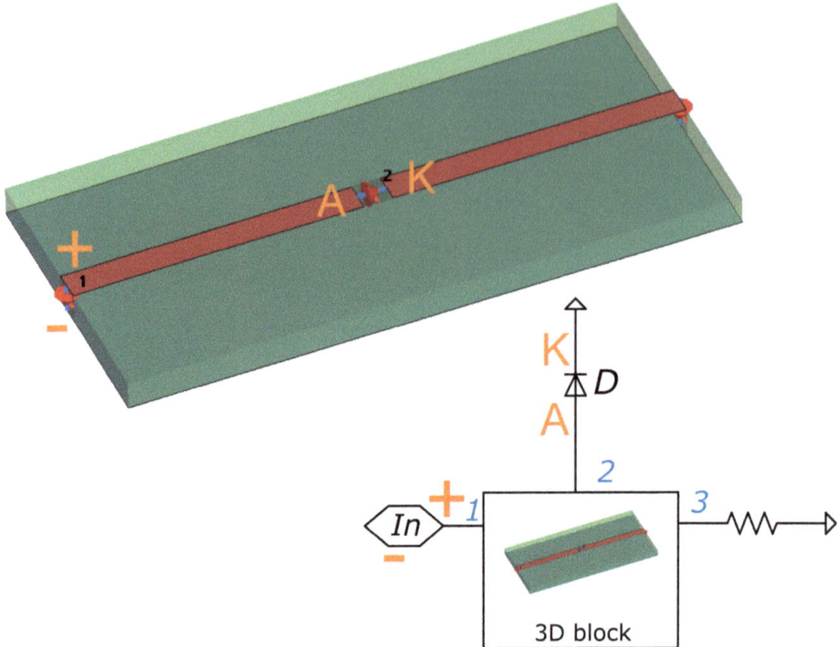

Fig. 6.27 Diagram explaining the orientation of ports and components in a coupled simulation

For discrete ports, there is a convention that is straightforward to understand: In CST, the discrete port is defined in the three-dimensional model by a cone pointing in a specific direction. In other simulation tools, a similar setup applies. The direction of the cone determines the polarity on the circuit (compare to Fig. 2.21).

This convention is illustrated with a simple example in Fig. 6.27. At port 2, a diode is placed. In the combined model, the anode of the diode is positioned where the cone in the three-dimensional port is wide, as the anode side of the diode on the circuit is connected to the 3D block. The cathode is located where the cone is sharp, aligning with the side connected to the ground symbol on the schematic. As mentioned earlier, the correct orientation of the diode in the combined model depends on both the orientation of the diode on the schematic and the orientation of the driving port in the three-dimensional model.

For port 1, the wide side of the cone connects to the signal trace, causing the port's connection to drive a positive polarity on the trace. For components like resistors, the orientation does not play a role.

The schematic also permits connecting two or more ports of the S-parameter block with each other. This is a valid configuration if the distance between the ports in the three-dimensional model is electrically short, and the ports are referenced to the same conductor. In Chap. 8, this approach is used to convert from single-ended to mixed-mode excitation.

However, if the distance between the ports in the three-dimensional model is not electrically short, such a connection on the schematic introduces phase errors because the propagation delay between the ports is neglected.

If the ports in the three-dimensional model are connected to separate conductors but are linked on the schematic, this creates violations of current continuity, as the schematic introduces a connection that does not exist in the physical geometry.

We recommend making connections between ports on the schematic with great caution. In case of uncertainty, it is advisable to revisit the three-dimensional model and consider the physical implications of the connection made on the schematic, for instance, by examining the current flow in the model.

In this example, we use one port for each component. The schematic provides the flexibility to connect multiple components, potentially with complex subschematics, to a single port. However, this approach requires caution, as such models can become difficult to interpret. For beginners in coupled simulations, we recommend adopting the "one port - one component" concept. This simplifies understanding and debugging of the coupled model.

6.7.2 *Lumped Elements vs. Ports*

In the π-filter example, we replace all components mounted on the PCB (two capacitors and one inductor) with ports and define their actual values on the schematic. If the component value changes, only the circuit simulation needs to be rerun, which is much faster than rerunning the EM simulation. This approach works well for simple electronic systems.

However, complex PCBs can include several hundred components, and replacing all components with ports leads to a schematic that is difficult to manage. Instead of replacing a component with a port, we can include it in the EM model as a so-called lumped element. A lumped element is similar to a discrete port but places an impedance value into the EM model. The impedance can be frequency-dependent.

By placing a lumped element into the model, we include its effect in the S-parameter matrix calculated by the full-wave solver. If the impedance value of the component modeled by a lumped element needs to be changed, a new EM simulation is required. A simulation engineer must decide which components should be modeled as lumped elements and which as ports: If the effect of the component's value is required to be studied later, we model this component as a port. If modification of the component's value is not required, we model the component as a lumped element.

The decision on which components should be modeled as ports and which as lumped elements has a significant effect on the simulation time of the EM model. Since time domain solvers require ports to be processed sequentially, keeping the number of ports low results in an efficient simulation model when using FIT or TLM. This is why the DC caps in the LTE example were modeled as lumped elements. On the other hand, when using the frequency domain FEM method with a direct solver, adding ports does not significantly increase the simulation time.

When using time domain solvers, it is also important to understand how a lumped element affects the energy decay used to measure convergence. The values of capacitors and inductors imply a decay time constant relevant to time domain simulation. For example, the time constant of a 1 μF capacitor discharged over a 50 Ω port impedance is $\tau_{RC} = 50\,\mu s$. Thus, a 50 μs signal duration is required to obtain accurate results at low frequencies. Recall the simulations at the beginning of this chapter: The time domain signal duration used there was 100 ns.

When components with time constants much larger than a few nanoseconds are included in a time domain EM simulation, it is more efficient to replace them with ports. The time required to simulate an additional port may be lower than the time required to resolve the component's time constant. A large time constant primarily affects results in low-frequency bands, while results in higher frequency bands remain unaffected, as discussed in Sect. 6.3.

6.7.3 Voltage and Current Probing in Coupled Models

In Fig. 6.26, we add a voltage and a current probe to port 3, which corresponds to the position of C_2. While in the simulation we probe the voltage and current at every component, only one probe is contained in the figure to avoid clutter. The voltages and currents measured in this way are always local quantities, allowing the measurement of the voltage across the port and the current through the port, at the position where the port is placed.

In the π-filter example, the voltage measured at the inductor port represents the voltage drop across the inductor, not the voltage from the signal trace to the reference layer. This result is due to the way the port is connected in the EM model, as discussed in Sect. 2.4.3.1.

If we need to calculate voltages and currents at positions where no components are present, we can place additional ports into the EM model. To measure a voltage, a port is placed between the conductors, and a high-value resistor is used on the schematic. The resistor's value must be much larger than the absolute value of the impedance of any other element to ensure that no current flows through it. A good choice is 1 MΩ. For current measurement, a port is placed into the signal trace, and a low-ohmic resistor is used on the schematic. In this case, the resistor's value must be sufficiently low to avoid altering the current flow in the electronic system. A good choice for current measurement is 1 mΩ.

The above-described technique for probing voltages and currents at arbitrary positions aligns with the circuit model of the decomposition presented in Sect. 5.2.

For the π-filter example schematic in Fig. 6.26, we set up an AC simulation to study the probed voltages and currents. A broadband 1 V source with an inner impedance of 1 mΩ is applied at the "In" port, and a 50 Ω termination is used at the "Out" port.

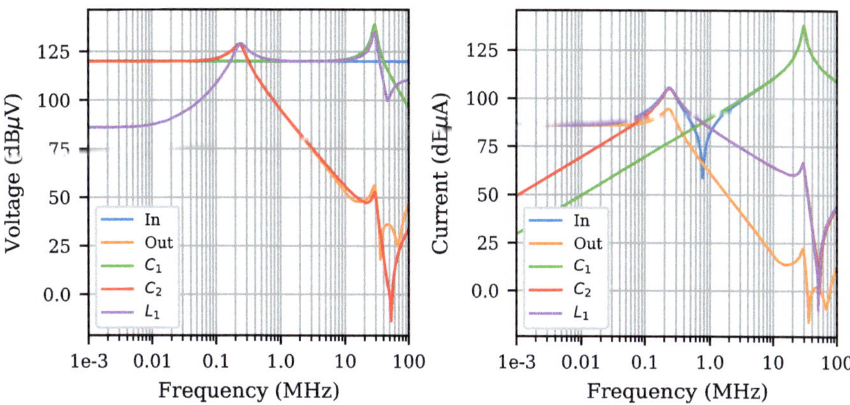

Fig. 6.28 Voltages and currents calculated on the schematic of the π-filter model Left: Voltage. Right: Current

We plot the voltages and currents calculated from the coupled schematic model in Fig. 6.28. As expected from the filter design, the voltage and current at the output port (orange lines) are attenuated starting at around 300 kHz. The current at frequencies below 20 kHz is slightly below 86 dBμA, corresponding to the source voltage divided by the output termination resistance, i.e., $\frac{1\,\text{V}}{50\,\Omega}$.

The effect of the layout on the filter performance in the plotted frequency range is small, and we do not discuss the filter performance in detail, as an in-depth discussion on filter simulations is presented in Chap. 9. At this point, the results are used to explain the possibilities of coupling full-wave and circuit models.

6.7.4 Impedance and RLC Extraction from Full-Wave Solvers

As the S-parameter results from a full-wave solver are black-box circuit models, a direct interpretation of a n-port matrix is cumbersome. Therefore, we discuss the extraction of impedances and RLC models from n-port S-parameter blocks in this section.

This extraction is useful in interpreting the circuit models of the coupling path obtained from a full-wave simulation. While it is not always required when building a simulation model, it can be used to sanity check the circuit model of the coupling path, helping detect possible mistakes in the setup of the EM model.

We use the same methodology as presented in Sect. 2.7, where we discuss the extraction of RLC models from generic network parameter blocks and apply it to the S-parameter block of the π-filter. In this methodology, a desired circuit configuration is selected by applying terminations to the S-parameter block, and RLC values are extracted from the input impedance.

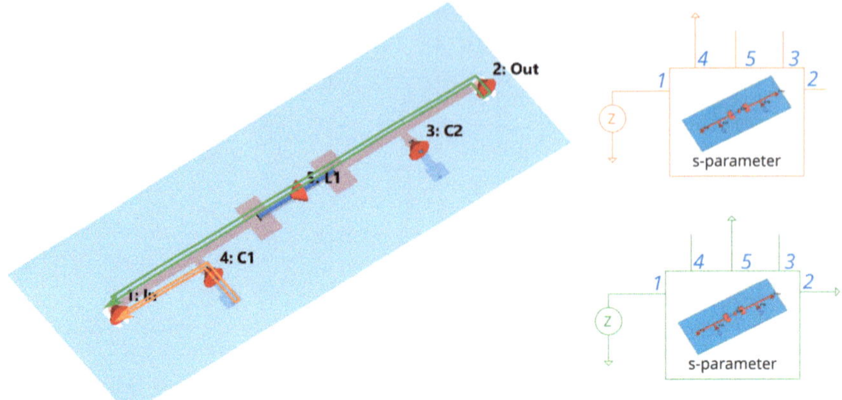

Fig. 6.29 Extraction of the inductance from a S-parameter block obtained from the full-wave solver

In the following examples, we do not use explicit resistor models for the definition of the termination. By connecting a ground symbol to the corresponding S-parameter port, we create a short with the RMIN value in the circuit simulation (cf. Sect. 3.8.1). Leaving the port unconnected creates an open, with the GMIN value.

Recall from Sect. 2.7 that for the extraction of the inductance, a termination with a short is required, creating a current loop. In the π-filter model, we select two loops from which we extract the inductance. The loops are colored in orange and green in Fig. 6.29. The schematics for the extraction are shown on the right-hand side of the same figure.

For the extraction of the inductance of the orange loop, we measure the impedance at port 1 and short port 4. For the extraction of the impedance of the green loop, we measure the impedance at port 1 and short port 5 and port 2. The inductance L is calculated from the impedance using $L = \frac{X}{\omega}$ (Eq. 2.11) and is plotted in Fig. 6.30.

We perform a sanity check on the obtained results: The connection between the ports in the S-parameter block is correct, because by applying the terminations, we obtain inductance values in the 10 nH range. These match the expected inductance for current loops of a few cm length. The inductance of the green loop is slightly more than double the inductance of the orange loop, as we expect from the length of both loops.

For the extraction of the capacitance, we select two areas in the π-filter that are indicated by orange and green colors in Fig. 6.31. Port 1, at which we extract the capacitance, is placed between the signal trace and the reference plane. Thus, the extraction results in the capacitance between the signal trace and the reference plane. For the extraction of the capacitance, we need to use open terminations as discussed in Sect. 2.7. The capacitance of the orange area is obtained by estimating

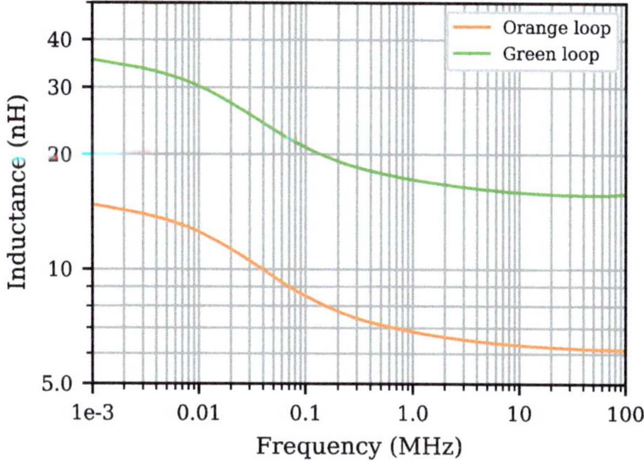

Fig. 6.30 Inductance extracted for two different current loops in the π-filter

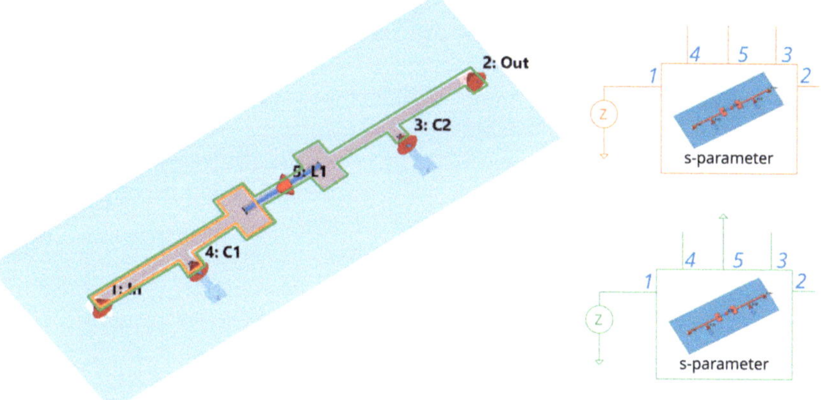

Fig. 6.31 Extraction of the capacitance from a S-parameter block obtained from the full-wave solver

the impedance at port 1 and keeping all other ports open. For the capacitance of the green area, we short port 5, resulting in the connection of both signal traces, but keeping all other ports open.

The extracted capacitance is plotted in Fig. 6.32 and is calculated by the relation $C = -\frac{1}{X\omega}$. As we expect, the capacitance is in the order of pF, and the capacitance of the green area is close to double the value of the capacitance of the orange area.

The extracted inductance and capacitance values are the values of the interconnect geometry that is simulated by the EM solver. They do not include the values of the components, as the component circuit models are not included in the extraction schematic.

The extracted values of the inductance in Fig. 6.30 and capacitance in Fig. 6.32 are plotted only down to 1 kHz and not down to 1 Hz. In the EM model, the extracted

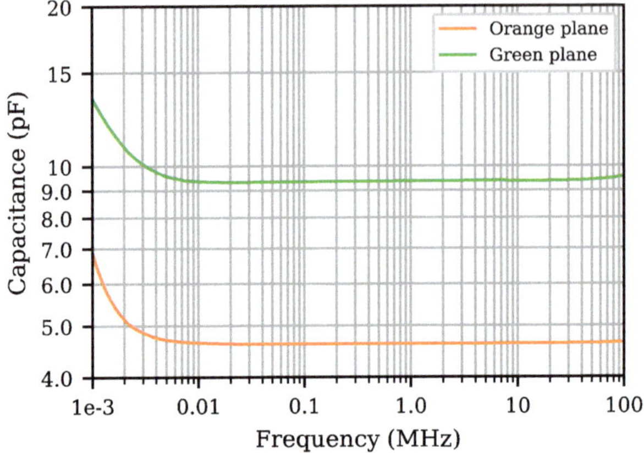

Fig. 6.32 Capacitance extracted for two different areas in the π-filter

values below 1 kHz deviate from the expected values: As we discuss in Sect. 5.6.1, the inductance is composed of an inner and an outer inductance. However, the lossy metal material model we use in the simulation does not model the current flow in the conductor; thus it is not suitable to estimate the inner inductance correctly. This results in the deviation of the inductance from the expected value in the kHz region and below, leading to an increase of the extracted inductance to values above the actual physical value.

The extracted capacitance values plotted in Fig. 6.32 show an increase in capacitance below 10 kHz. This increase is not a physical but a numerical effect (Eller et al. 2017). In the kHz range, the full-wave FEM solver requires additional stabilization that leads to an underestimation of the impedance and the observed increase in the capacitance. This numerical effect is not critical as the capacitance of the interconnects in the kHz range does not influence the characteristics of the filter. The simplest way to circumvent this effect is to read the extracted capacitance in the region in which the capacitance curve is flat.

6.7.5 Combine Results

Combine results is a powerful technique to analyze the details of the coupling path. It allows to plot the field distribution inside the DUT considering the effect of connected circuit elements. It also allows to estimate field values which are not easily accessible by a standalone circuit or standalone EM simulation. Combine results is also efficient from the simulation time point of view. As we are recombining previously calculated field results, a combine results task is much faster than rerunning a full EM simulation.

The term "combine results" is the terminology used in CST. The same functionality can be found under different names in other simulation packages.

We start the discussion of the combine results methodology with frequency domain AC circuit simulations, as they are the foundation of this workflow. At the end of this section, we extend the workflow toward transient circuit simulations.

Combine results is based on the linear superposition principle. For the calculation of the S-parameter matrix, the full-wave solver is required to step through the individual ports as we discuss in Sect. 4.6. This is also how the individual field probe results and field monitors are calculated: For each excited port, the probe and monitor values are calculated and saved. These are the field distributions of a particular port that is excited with a signal normalized to 1 Watt input power. As the port impedance is known, the corresponding voltages and currents at the port can be calculated.

For combine results, we interpret a field probe to be a broadband transfer function from the port to the field value at the probe's position in space. A field monitor is a collection of single-frequency transfer functions from the port to all positions in the computational domain.

We now use a linear superposition of the individual transfer functions to combine (hence the name combine results) them into other excitation patterns. For example, we excite port 1 with a 1 W signal and port 2 with a 2 W signal that is phase-shifted relative to port 1. The superposition of both excitations results in the field value of the simultaneous excitation of both ports. This is the original way the combine results functionality was introduced: it offered a convenient way to calculate the resulting field, for example, in a multiantenna setup.

Instead of manually defining the excitation pattern to, for example, 1 W and 2 W, we use the voltages and currents that are calculated by the coupled schematic. The software monitors the voltages and currents at each port of the S-parameter block that is calculated by the EM solver. They are indicated by the blue-colored numbers in Fig. 6.26. After the AC simulation, power wave amplitudes are calculated from the voltages and currents (Eq. 2.5) and then used to combine the individual field probe or field monitor results.

The electric field generated by the individual ports, as depicted on the left-hand side of Fig. 6.33, is flat over the calculated frequency band. This is as expected, as the structure is electrically short, and for the individual port results, all nonexcited ports are terminated with 50 Ω. Furthermore, the fields generated by a 1 W excitation of the four ports connected between the trace and reference net are very similar. Only the series port at the position of the inductor generates a different electric field level.

For an understanding of the process, recall the plot of all voltages and currents at the five ports of the π-filter in Fig. 6.28. These are used to generate the power wave amplitudes that are combined. The combination of the electric field of the individual ports creates one probe result, plotted on the left-hand side of Fig. 6.33.

In the combined result on the right-hand side, we observe that at frequencies where the voltage and current distributions exhibit peaks, the electric field at the probe also reaches peak values. At 30 MHz, for example, the output voltage at the filter is attenuated by more than 80 dB (cf. Fig. 6.28). However, the electric field at

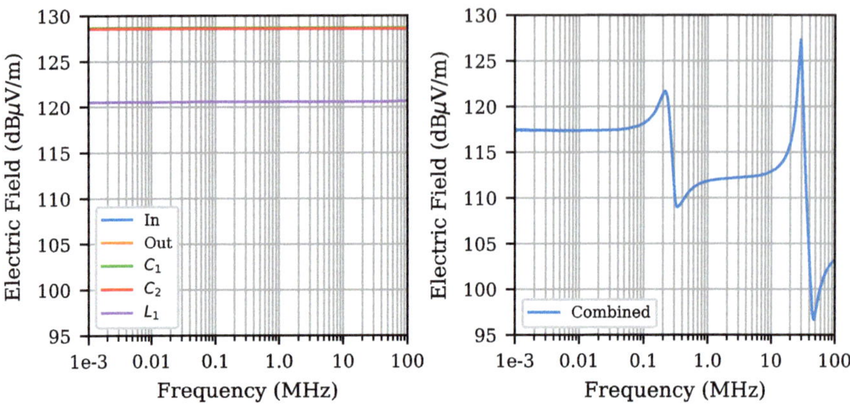

Fig. 6.33 Electric field at a probe close to the π-filter. Left: Field generated by individual ports using a 1 W excitation. Right: Field obtained from the combination with the coupled schematic

the probe shows a peak: At this frequency, the impedance of C_1 is minimal, and the input source is shorted, leading to a high current that results in the high electric and magnetic field close to the filter. We observe this effect in the combined field monitors without any further EM simulation.

We plot the maximum value of the combined surface current at different frequencies in Fig. 6.34. These plots illustrate the frequency-dependent distribution of the current on the PCB and confirm the explanation we have given for the behavior of the electric field at the probe position:

At 1 kHz, the impedance of both capacitors is high; thus the main current path is from the input to the output port. At 1 MHz, the main current flow is through both capacitors. At 10 MHz, the main current flow is through the capacitor on the left-hand side, because the impedance of the inductor has increased. At 30 MHz, which is the resonance frequency observed in Fig. 6.33, the impedance of the capacitor on the left-hand side is very low, leading to a high current in the PCB.

Combine results is based on the linear superposition of results in the frequency domain. Only frequency domain results at the probes and frequency domain monitors can be combined. In the previous paragraphs, the simulation on the schematic that drives the superposition is linear, using an AC simulation. However, in many applications, a transient simulation on the schematic is performed. Typical examples of transient circuit simulation are IBIS models driving a digital signal or PWM signals in power electronics applications.

It is desirable to extend the combine results workflow to transient circuit simulations. This is possible, if we ensure that the transient circuit simulation is in a steady state. When the transient simulation has reached a steady state behavior, it becomes possible to calculate the voltage and current spectra at every port by a Fourier Transform and use these spectra for the linear superposition in frequency domain, as we do for the AC simulation. This workflow was already used to calculate the emission from a digital signal in Fig. 6.5.

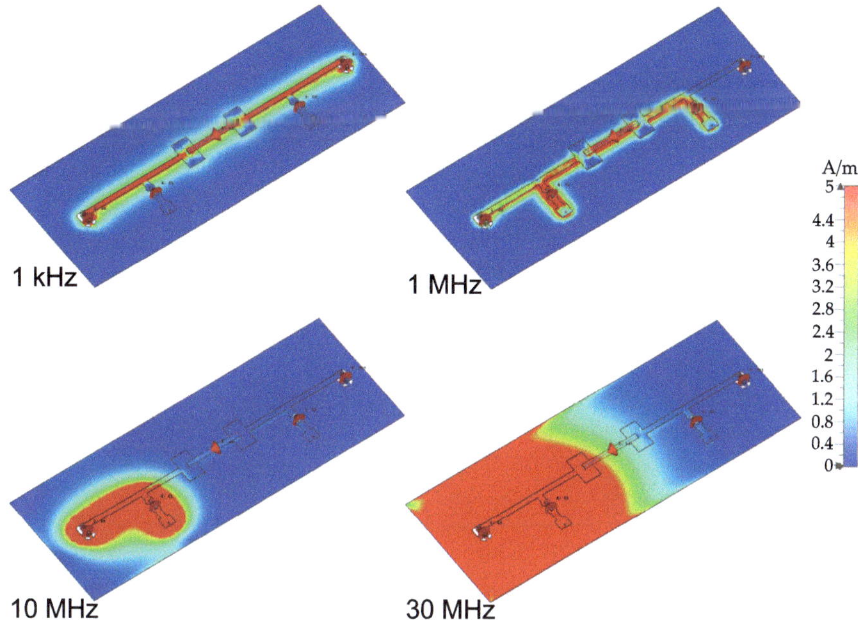

Fig. 6.34 Combined surface current in the π-filter at different frequencies

Digital signals reach a steady-state behavior quickly. In power electronics simulations, in which capacitors and inductors with large time constants are present, the time required to reach steady state might reach orders of milliseconds.

In the simulation of power electronics, we have to check the behavior of the circuit first, before defining the time domain signal window that is Fourier transformed. In the buck converter discussed in Chap. 3, steady state is reached after approximately 1 ms, as can be observed in the right-hand side of Fig. 3.3. For the Fourier transformation that is required for transient combine results, we would only use the time domain signals starting at 1 ms to make sure that the system is in steady state.

The steady state that is required to be reached by the system is a matter of the frequency range of interest:

At higher frequencies, the steady state is reached faster than at lower frequencies. Consider, for example, the lowest frequency of a conducted emission measurement to be 150 kHz: One period at this frequency corresponds to 6.6 μs. Thus, using a signal length of 660 μs in the Fourier transform includes 100 periods, which is a good value for the Fourier transformation.

As transient circuit simulation of a system defined by a coupled circuit and EM model might require significant simulation time, we typically limit it to a few ms. In case of doubt about the accuracy loss caused by the signal truncation, or in cases where the simulated schematic requires a long time to reach steady state, we run the simulation with different signal durations and compare the results to each other.

From this comparison, we select the best compromise between simulation time and accuracy.

If the system behavior is not periodic, such as in an ESD event, the transient combine workflow cannot be applied. In these cases, we can monitor the time domain signals on the schematic, but we cannot combine the frequency domain fields. For these cases, a cosimulation of the EM model with the circuit can be performed. The cosimulation is discussed in Sect. 14.3.

Combine results not only offers the possibility to calculate field values from the combined schematic, it can also be used to speed up simulation tasks. As we are recombining already calculated field results, a combine results task is much faster than rerunning a full EM simulation, for example, in models in which only the excitation signal is varied.

In the case of the π-filter, the combination of one probe and seven field monitors requires 2 seconds. In order to perform combine results, the full-wave solver and the circuit simulation need to exchange data. The full-wave solver passes the S-parameter matrix to the schematic, and the schematic delivers the power wave amplitudes to the full-wave solver. We rely heavily on combine results in the radiated emission Chap. 12.

6.8 Advanced: Far Field Monitors

Domains: 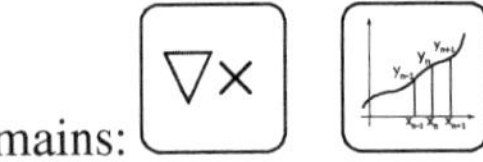

In this chapter, we introduce probes and field monitors inside the computational domain and probes outside of the computational domain, named far-field probes as the far-field approximation is used to calculate the field values outside of the computational domain. A so-called far-field monitor is the equivalent of the field monitor for fields outside of the computational domain. With the calculation method introduced in Sect. 6.5.1, we calculate a spatial distribution of the fields outside of the computational domain.

Far-field monitors are crucial for antenna design as they allow the calculation of the behavior of fields in the far-field region and enable the extraction of antenna characteristics such as gain and directivity. These quantities are important characteristics of an antenna design but are not often required in EMC simulation. Still, we briefly describe the properties of far-field monitors, as they are useful in understanding the spatial distribution of the field emitted from a structure.

A far-field monitor result is plotted in Fig. 6.35. It shows the electric field distribution of the PCB pair from the beginning of this chapter, in the far field at a distance of 3 m. It is common to use a spherical coordinate system for plotting the far field. The far-field distribution is overlaid on the geometry of the structure for

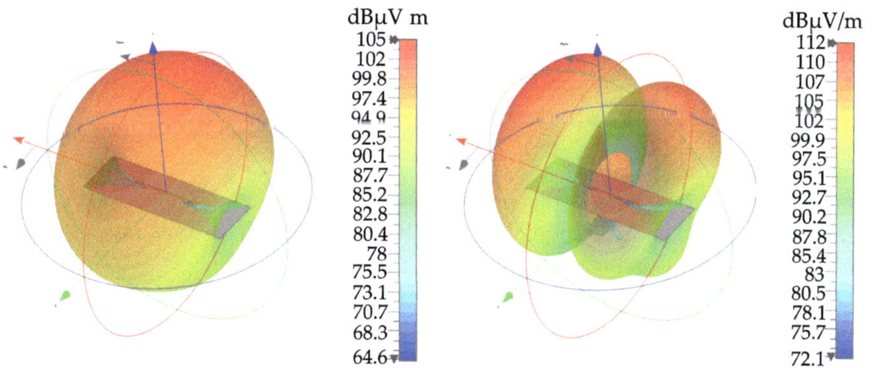

Fig. 6.35 Far field monitor results for the electric field at 3 m distance of the PCB pair. Left: 155 MHz. Right: 360 MHz

better navigation of directions, but we note that the field distribution in the far field corresponds to the distribution at an infinite distance from the structure.

We compare the far-field monitor results to the field monitors in Fig. 6.11 and study their correlation. For the calculation of the field monitors in Fig. 6.11, a large computational domain is required, while for the far-field monitors, a smaller distance to the boundary is used, as discussed in Sect. 6.5.

The far-field distribution in Fig. 6.8 indicates that the highest field values in our PCB example are above the structure, as the table reflects the fields generated by the PCB pair. These field values are not captured by the cylindrical probe arrangement (Fig. 6.15) used to estimate the worst-case field values in Sect. 6.5.

By comparing the maximum field values calculated by the far-field monitor to the worst-case value at the probe, we estimate the difference between both setups. At 155 MHz, the far-field monitor reads 105 $\mathrm{dB}\frac{\mu\mathrm{V}}{\mathrm{m}}$, while the worst-case probe result is 98 $\mathrm{dB}\frac{\mu\mathrm{V}}{\mathrm{m}}$ (Fig. 6.17); at 360 MHz, the far-field monitor reads 112 $\mathrm{dB}\frac{\mu\mathrm{V}}{\mathrm{m}}$ and the field probe 104 $\mathrm{dB}\frac{\mu\mathrm{V}}{\mathrm{m}}$.

To reduce this difference, we can use a spherical distribution of the probes instead of a cylindrical one. The spherical probe arrangement yields results closer to the worst-case estimated by the far-field monitor but might differ from a measurement setup in which the antenna is located only at the side of the table and not above it.

It is up to the simulation engineer to decide which arrangement is more suitable for the task at hand. With the far-field monitor, we can estimate the worst-case emission over the complete frequency band by defining multiple far-field monitors at closely spaced frequency samples and evaluating them for the worst-case field value. However, the authors prefer the method using an envelope of probes as described in Sect. 6.5, because the evaluation of probes is more efficient in terms of the spectral resolution.

6.9 Advanced: Field Sources

Domains: 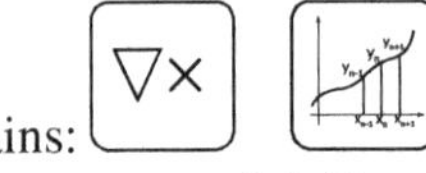

The concept of field sources allows using a field distribution as a source in an EM simulation. The field source can be calculated in a first simulation model or obtained from measurements. It is then used in a second EM model as a source.

The calculation using two subsequent EM simulations can improve simulation time and make a concatenation of solvers possible. The solver that calculates the field source can be different from the solver that consumes the field source in a second simulation. This allows exploiting the different solver properties and using the best-suited solver in each step. For example, a resonant emitter is first simulated using the frequency domain FEM method and then placed into a large computational domain that is effectively simulated using a time-domain or MoM solver.

A hybrid simulation with field sources is used in antenna placement simulations, where a field source around an antenna is generated and then loaded onto a large platform such as an aircraft or a ship. The concatenation reduces simulation time in such scenarios, and the choice of the position of the surface at which the field source is recorded is often straightforward.

In EMC, we often deal with unknown coupling paths, and the choice of the surface at which the source is recorded is less clear. This is the main reason why we mention field sources only in an advanced chapter and do not use them further throughout this book. The authors follow the idea that as long as it is possible to simulate the problem at hand in one simulation model, there is no need to use field sources and deal with the difficulties of finding a suitable surface at which the field source should be defined.

We use the SET in Sect. 6.5 to calculate field values outside of the computational domain by applying a far-field approximation in the outer domain. The SET is also the foundation of the field source concept: We record the fields on an arbitrary closed surface inside the computational domain and use the so-obtained field distribution as a source in a second simulation. These types of sources are often labeled as a “Huygens Box,” as the SET is also known as Huygens’ equivalence principle. The fields on the box must contain both the tangential electric and magnetic field strength, as well as the correct phase relationship of the fields on the box surface. When the field source is calculated by an EM simulation, obtaining the phase information is not a problem. However, when measurements are used to generate the field source, the correct phase information is often not included in the measurement data, particularly when low-cost near-field measurement systems are used.

When using field sources, the box must be defined carefully as it is required to encompass the complete source with the box. If there are sources present in the outer domain, they contribute to the currents $\mathbf{M}_S, \mathbf{J}_S$ in Fig. 6.14, and not including these contributions in the simulation in which the field source is excited leads to incorrect results. In addition, if some structure is present in the outer domain that couples to the source in the inner domain and influences the current distribution on the surface, it needs to be included in the model that generates the source: the SET assumes no interaction between the fields on the box's surface and the outer surroundings.

As an example, let us consider a cable penetrating the box at which we record the field source. The current on the cable outside of the box is a source of fields, and such a setup leads to incorrect results.

A second example is a PCB, coupling capacitively to an enclosure. The current distribution on the surface located between the enclosure and the PCB is influenced by the enclosure. Generating the field source by a model containing only the PCB and then using the field source inside the enclosure does not include this coupling and leads to incorrect results. In order to demonstrate this effect, we come back to the simple PCB pair model and simulate the model with field sources.

Figure 6.36 depicts our setup. A field source monitor is defined in the PCB pair simulation model. The box at which the fields are recorded is shown in the top

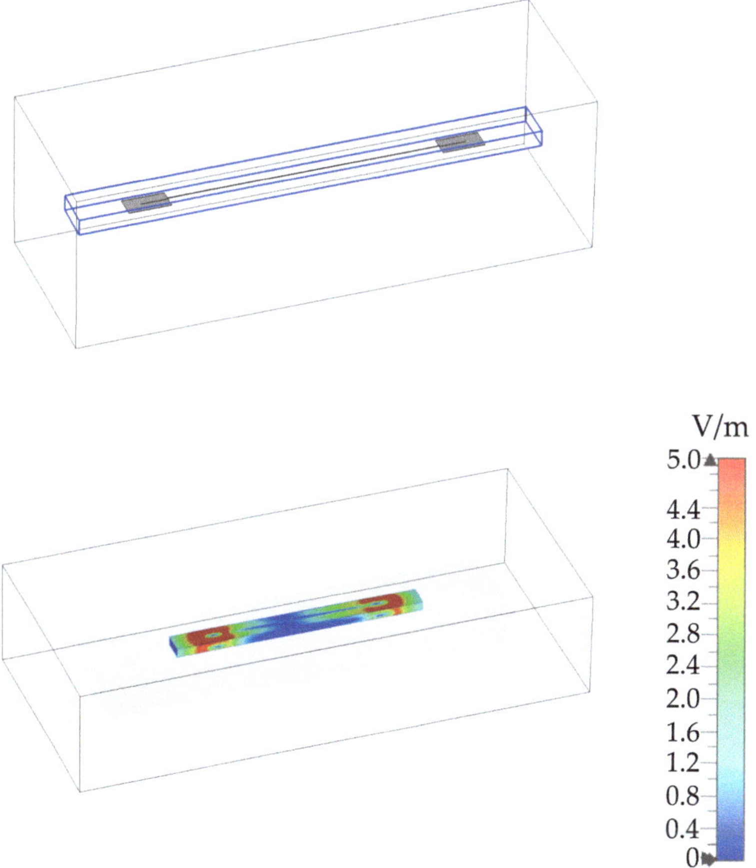

Fig. 6.36 Field source definition. The box depicts the size of the computational domain. Top: Field source monitor for the generation of the field source. Bottom: Field source placed above the table. The picture shows the distribution of the electric field at 150 MHz at the surface of the field source box

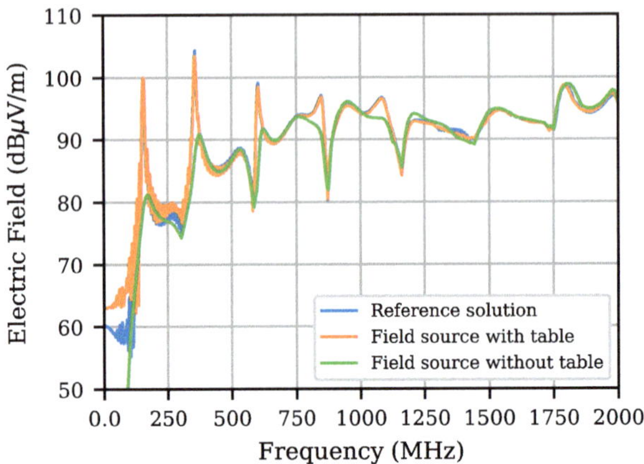

Fig. 6.37 Comparison of the worst case electric field for the simulation using a field source

section. In the first model that generates the field source, the table is not included. The field source is recorded in frequency domain. We define a sampling with a frequency step of 50 MHz, starting at 50 MHz up to 2 GHz.

After the field source is calculated, it is scaled to an AC source with a broadband 1 V excitation and a source impedance of 10 Ω. This setup matches the source definition we used in this model at the beginning of this chapter. The field source is loaded into a model containing the table. We evaluate the worst-case electric field with probes outside of the computational domain in the way we do in Sect. 6.5.

The result of the comparison is plotted in Fig. 6.37. The results of the simulation with the field source agree well for frequencies above 1.2 GHz but do not reproduce the sharp peaks at frequencies below. At the lower frequencies, the coupling of the PCB pair to the table influences the field distribution and cannot be omitted in the model that generates the field source.

To confirm this finding, we use a second field source, which includes the table in the simulation model that generates the field source. The size of the box at which the field source is generated is not changed; it still encompasses only the PCB pair.

The second field source is loaded into exactly the same model that we have used previously. With the table included in the model that generates the field source, we observe that the sharp peaks in the lower frequency band are also correctly calculated. The difference between the reference model and the model with the field source is less than 1 dB. Only below 100 MHz, below the first peak, the difference is higher. This deviation is because of the lacking length of the simulated pulse, that is, the model is not intended to work correctly here (cf. Sect. 6.3).

The above example demonstrates that field sources are an accurate simulation method. However, care has to be taken in the model generating the field source. It requires to include the couplings of the excited system to its surroundings; otherwise inaccuracies can be introduced to the results obtained from the second simulation model.

Further Reading

Near Field to Far Field Coupling

C.A. Balanis, *Advanced Engineering Electromagnetics*, 3rd edn. (Wiley, New York, 2024). ISBN: 978-1-394-18001-1

K.W. Kam, A. Radchenko, D. Pommerenke, On Different Approaches to Combine Cable Information Into Near-Field Data for Radiated-Field Estimation. IEEE Transactions on Electromagnetic Compatibility **56**(2), 276–285 (2014). https://doi.org/10.1109/TEMC.2013.2286201

P. Petre, T.K. Sarkar, Planar near-field to far-field transformation using an equivalent magnetic current approach. IEEE Transactions on Antennas and Propagation **40**(11), 1348–1356 (1992). https://doi.org/10.1109/8.202712

S. Sadeghi et al., An Insight into the Practical Problems of Near-field to Far-field Transformation in the Time Domain for EMI Test. *2024 International Symposium on Electromagnetic Compatibility - EMC Europe* (2024), pp. 104–109. https://doi.org/10.1109/EMCEurope59828.2024.10722100

M. Sørensen et al., Estimate on the uncertainty of predicting radiated emission from near-field scan caused by insufficient or inaccurate near-field data: Evaluation of the needed step size, phase accuracy and the need for all surfaces in the Huygens' box. *International Symposium on Electromagnetic Compatibility - EMC EUROPE* (2012), pp. 1–6. https://doi.org/10.1109/EMCEurope.2012.6396677

A. Taaghol, T.K. Sarkar, Near-field to near/far-field transformation for arbitrary near-field geometry, utilizing an equivalent magnetic current. IEEE Transactions on Electromagnetic Compatibility **38**(3), 536–542 (1996). https://doi.org/10.1109/15.536088

Z. Yu et al., Heat-Sink Modeling and Design With Dipole Moments Representing IC Excitation. IEEE Transactions on Electromagnetic Compatibility **55**(1), 168–174 (2013). https://doi.org/10.1109/TEMC.2012.2210720

Field Circuit Coupling

A. Barchanski, Linking circuit simulation with full-wave solver for board-level EMC design. IEEE Electromagn. Compat. Mag. **4**(3), 52–58 (2015). https://doi.org/10.1109/MEMC.2015.7336756

A. Barchanski, M. Bandinu, A macromodeling approach for emc simulations of power electronics systems, in *2023 International Symposium on Electromagnetic Compatibility – EMC Europe* (2023), pp. 1–5. https://doi.org/10.1109/EMCEurope57790.2023.10274272

A. Barchanski, M. Tröscher, EMC simulation of automotive ethernet systems, in *2018 IEEE International Symposium on Electromagnetic Compatibility and 2018 IEEE Asia-Pacific Symposium on Electromagnetic Compatibility (EMC/APEMC)* (2018), pp. 50–50. https://doi.org/10.1109/ISEMC.2018.8394034

M. Eller, S. Reitzinger, S. Schöps, S. Zaglmayr, A symmetric low-frequency stable broadband Maxwell formulation for industrial applications. SIAM J. Sci. Comput. **39**(4), B703-B731 (2017)

M. Feliziani, F. Maradei, Modeling of electromagnetic fields and electrical circuits with lumped and distributed elements by the WETD method. IEEE Trans. Magn. **35**(3), 1666–1669 (1999). https://doi.org/10.1109/20.767333

M. Feliziani, F. Maradei, FEM solution of time-harmonic electromagnetic fields by an equivalent electrical network. IEEE Trans. Magn. **36**(4), 938–941 (2000). https://doi.org/10.1109/20.877596

M. Feliziani, F. Maradei, Circuit-oriented FEM: solution of circuit-field coupled problems by circuit equations. IEEE Trans. Magn. **38**(2), 965–968 (2002). https://doi.org/10.1109/20.996248

I.C. Garcia, H. De Gersem, S. Schöps, A structural analysis of field/circuit coupled problems based on a generalised circuit element. Numer. Algorithms **83**(1), 373–394 (2020). https://doi.org/10.1007/s11075-019-00686-x

H. de Gersem et al., Solution strategies for transient, field-circuit coupled systems. IEEE Trans. Magn. **36**(4), 1531–1534 (2000). https://doi.org/10.1109/20.877729

H. De Gersem, K. Hameyer, T. Weiland, Field-circuit coupled models in electromagnetic simulation. J. Comput. Appl. Math. **168**(1), 125–133 (2004). https://doi.org/10.1016/j.cam.2003.05.008

H. De Gersem, T. Weiland, Field-circuit coupling for time-harmonic models discretized by the finite integration technique. IEEE Trans. Magn. **40**(2), 1334–1337 (2004). https://doi.org/10.1109/TMAG.2004.824536

I. Munteanu et al., FIT/PVL circuit-parameter extraction for general electromagnetic devices. IEEE Trans. Magn. **36**(4), 1421–1425 (2000). https://doi.org/10.1109/20.877705

D.M. Pozar, *Microwave Engineering*, 4th edn. (Wiley, Hoboken, 2012).

S. Schops, H. De Gersem, A. Bartel, A cosimulation framework for multirate time integration of field/circuit coupled problems. IEEE Trans. Magn. **46**(8), 3233–3236 (2010). https://doi.org/10.1109/TMAG.2010.2045156

S. Schops, H. De Gersem, A. Bartel, Higher-order cosimulation of field/circuit coupled problems. IEEE Trans. Magn. **48**(2), 535–538 (2012). https://doi.org/10.1109/TMAG.2011.2174039

S. Schöps, H. De Gersem, T. Weiland, Winding functions in transient magnetoquasistatic field-circuit coupled simulations. COMPEL: Int. J. Comput. Math. Elect. and Elect. Eng. **32**(6), 2063–2083 (2013). https://doi.org/10.1108/COMPEL-01-2013-0004

P. Zhou et al., A general cosimulation approach for coupled field-circuit problems. IEEE Trans. Magn. **42**(4), 1051–1054 (2006). https://doi.org/10.1109/TMAG.2006.871374

Chapter 7
Shielding

Shielding describes the practice of reducing or redirecting electromagnetic fields through the use of conductive or magnetic materials. When shielding material is placed around a domain, it minimizes field penetration into and out of the shielded area. Shielding electronic systems is a common practice to ensure compliance with emission and susceptibility standards.

At the beginning of this chapter, we introduce shielding effectiveness as a measure of the shielding performance. In the context of EMC problem decomposition (Sect. 1.1), shielding effectiveness represents the coupling path. However, it is also influenced by the characteristics of the source and sink configurations.

For a successful simulation, it is essential to understand how the source and sink configurations impact shielding effectiveness. This chapter is structured accordingly: each section focuses on a specific source/sink configuration. Additionally, we discuss the selection of appropriate material models, as field penetration through the material is crucial for accurately calculating shielding effectiveness, particularly at frequencies in the kHz range and below.

The first configuration we examine is the shielding effectiveness of planar slabs. Due to the simple geometry of a slab, an analytical calculation of shielding effectiveness is possible, allowing a comparison with simulation results. In the second configuration, we calculate the shielding of low-frequency magnetic fields generated by a cable pair in a vehicle, comparing the results to limit values for human safety. In the third configuration, we evaluate the shielding effectiveness of a metallic enclosure with openings, analyzing how these openings affect the shielding performance.

J. Hansen, A. Barchanski, *EMC Simulation of Electronic Systems*,
https://doi.org/10.1007/978-3-032-09237-3_7

7.1 Shielding Effectiveness

In order to describe a shielding effect quantitatively, the shielding effectiveness is introduced as the ratio of the field inside the shielded domain compared to the field at the same position without the shield being present. The shielding effectiveness is defined for the electric and magnetic fields as

$$\mathrm{SE_E} = 20\log_{10}\left|\frac{\mathbf{E}_{\mathrm{wo}}}{\mathbf{E}_{\mathrm{w}}}\right|, \tag{7.1}$$

$$\mathrm{SE_H} = 20\log_{10}\left|\frac{\mathbf{H}_{\mathrm{wo}}}{\mathbf{H}_{\mathrm{w}}}\right|, \tag{7.2}$$

where the subscript wo denotes the incident field when no shielding is present and w denotes the field at the same position transmitted through the shield when it is present. If the shielding effectiveness is positive, the field in the shielded region is lower than the incident field. The shielding effectiveness can also become negative if the field in the shielded domain is larger than the incident field.

The shielding effectiveness is a function of frequency, as the interaction of materials with the electromagnetic field depends on frequency. As already discussed in the introduction of this chapter, it also depends on the properties of the source and sink. The orientation and distance of the source to the shield matter and also the position at which the field is measured inside the shielded domain.

To reduce these dependencies, we can estimate the shielding effectiveness for multiple source and sink configurations in the same model and calculate the worst-case value from all configurations. This methodology is discussed in the section addressing the shielding effectiveness of an enclosure (Sect. 7.4).

7.2 Shielding Effectiveness of Planar Materials

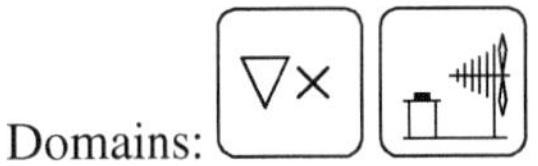

7.2.1 *Theory of Plane Wave Shielding*

The shielding effectiveness of planar slabs can be calculated analytically. Paul (2005) presents a comprehensive derivation of the analytical formulas in his book. In the following, we outline the key points of the derivation.

The shielding effectiveness of a planar slab can be broken into the sum of three terms, as

$$\mathrm{SE}_{\mathrm{dB}} = R_{\mathrm{dB}} + A_{\mathrm{dB}} + M_{\mathrm{dB}}. \tag{7.3}$$

In Eq. (7.3), R_{dB} is the reflection loss. It accounts for the portion of the incident field that is reflected at the illuminated side of the shield and thus does not enter the shielded domain. A_{dB} describes the absorption of the field as it travels through the shielding material. For shields made of good conductors, the absorption loss is related to the skin depth of the material. The term M_{dB} accounts for multiple reflections inside the shield. It is a negative number; thus it reduces the shielding effectiveness. For shields with a thickness significantly greater than the skin depth, it can be disregarded.

For a plane wave incidence, and with the assumption that the shield is a good conductor and significantly thicker than the skin depth, the approximate reflection loss is given by

$$R_{\mathrm{dB}} = 168 + 10\log_{10}\left(\frac{\sigma_{\mathrm{r}}}{\mu_{\mathrm{r}} f}\right), \tag{7.4}$$

where $\sigma_{\mathrm{r}} = \frac{\sigma_{\mathrm{shield}}}{\sigma_{\mathrm{Cu}}}$ is the conductivity of the shield relative to the conductivity of copper ($\sigma_{\mathrm{Cu}} = 5.8 \cdot 10^7\,\frac{\mathrm{S}}{\mathrm{m}}$), f is the frequency, and μ_{r} is the relative permeability of the shielding material. The reflection loss decreases at a rate of $-10\,\mathrm{dB}$ per decade on a decibel scale. It is greatest at low frequencies and for high conductivity materials. Permeable materials reduce the reflection loss.

The approximate absorption loss is given by

$$A_{\mathrm{dB}} = 131.4 \cdot d \cdot \sqrt{f \mu_{\mathrm{r}} \sigma_{\mathrm{r}}}, \tag{7.5}$$

with d being the thickness of the shield. The absorption loss is proportional to $\sqrt{f}$; thus it increases at a rate of 10 dB per decade. Permeable materials increase the absorption loss.

The above equations are valid for the electric field of a plane wave incident on the shield. For the magnetic field of a plane wave, the contribution from the absorption is equivalent, while the reflection obtained at the surface of the shield is different, as we will see in this example.

In the following, we set up a simulation model that aims at reproducing these values and discuss the differences between the simulation and analytical results. This exercise is a good example demonstrating that care is required when simulation and analytical results are compared.

7.2.2 *Shielding of the Electric Field for a Plane Wave Excitation*

Plane wave
Shielding Effectiveness at field probes
Full-wave
FEM solver
None
0–10 MHz

The simulation model of a plane wave incidence on a planar slab is depicted in Fig. 7.1. We define a plate made of copper with a quadratic shape and an edge length of 10 cm. The computational domain is extended to 20 cm in the front and back of the plate. A plane wave excitation is defined on one side of the plate. The plane wave is linearly polarized with the field vectors plotted in the figure. The amplitude of the electric field of the plane wave is set to $1\,\frac{\mathrm{V}}{\mathrm{m}}$. For the shielding effectiveness calculation, the amplitude is not relevant as the material properties are linear. Electric and magnetic field probes are placed in the center axis of the model, at a distance of 1 μm, 1 mm, 10 mm, and 100 mm in front and behind the plate.

In order to model the penetration of the field through the material, we have to select a suitable material model for the plate. Material models are discussed in Sect. 4.4. A lossy metal material completely separates the field on both sides of the plate; thus it is not suitable for this study. We could use a volumetric conductor; however, this requires resolving the skin depth, leading to an increased simulation

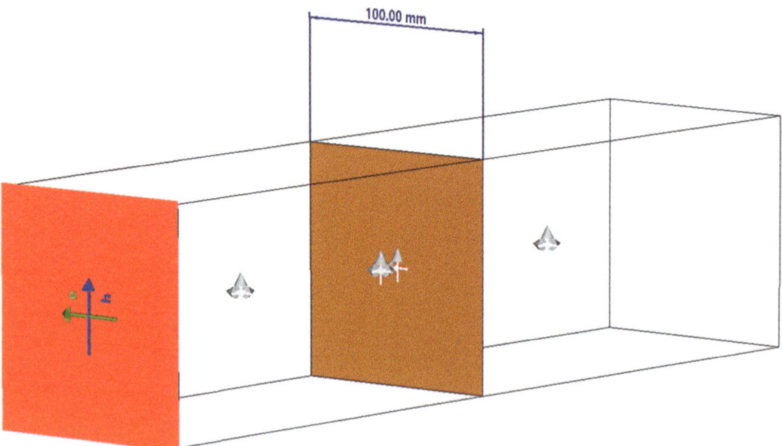

Fig. 7.1 Simulation model to calculate the shielding effectiveness of a planar slab

time. The best choice for this study is the thin panel material. The geometrical thickness of the plate is set to zero; it is modeled as a sheet. The thickness of the plate is defined in the thin panel material model which is described in Sect. 4.4. We set the thickness of the sheet to 500 µm.

The shielding effectiveness at the probe position is calculated for the electric field using Eq. (7.1) and for the magnetic field using Eq. (7.2). In order to obtain the field with and without the shield, as required by the equations, we set up two simulation models. In the first model, the plate is not present. In the second model, the plate is included. The shielding effectiveness is then calculated in a postprocessing step. CST offers a postprocessing script that performs the calculation of the shielding effectiveness at all probes present in both models. In the following, we use the term shielding effectiveness also for results obtained in front of the plate, as the plate affects the field distribution both in front of and behind the plate.

The frequency range in the study is set from 1 Hz up to 10 MHz. At frequencies above 10 MHz, the field behind the plate is practically zero. We use the full-wave FEM solver in this study since the frequency range starts at a very low value. The simulation time for the model without the plate is 3 seconds, while the model with the plate requires 10 seconds. With the usage of the thin panel material, the mesh can be kept coarse, resulting in these very low simulation times.

The shielding effectiveness at the probes located at a distance of 1 mm in front and behind the plate is plotted in Fig. 7.2.

For a plate thickness greater than the skin depth, only the reflection loss affects the field values in front of the plate. Therefore, we plot the simulated shielding effectiveness for the probe in front of the plate against the analytically calculated value of R_{dB} on the left-hand side of Fig. 7.2.

At 1 kHz, the analytical formula (Eq. 7.4) delivers 138 dB, while the simulation yields 135 dB. At frequencies below 1 kHz, the curve obtained from the simulation is flat, while the analytical formula predicts a monotonically increasing shielding

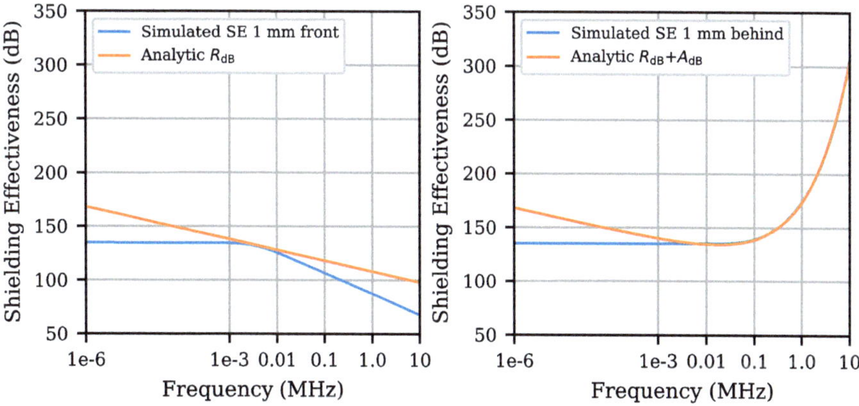

Fig. 7.2 Shielding effectiveness of the copper plate at the probe located 1 mm away from the plate. Left: Field probe in front of the plate. Right: Field probe behind the plate

effectiveness down to DC. The deviation between the simulation model and the analytical result can be attributed to the assumption in the analytical equation that the shield is thicker than the skin depth. Below 1 kHz, this assumption is invalid because the skin depth of copper at 1 kHz is approximately 2 mm, while the plate thickness is only 0.5 mm. For slabs with a thickness less than the skin depth, the multiple reflections term M_{dB} becomes relevant, for which no simple approximate solution is available.

At 10 MHz, the analytical reflection is 98 dB, while the results at the probe 1 mm in front of the plate yield 67 dB. From the analytical equation, the decay of the reflection is −10 dB per decade, whereas the simulation results show a decay of −20 dB per decade. This deviation arises from the different methods used to calculate the analytical R_{dB} and the shielding effectiveness in the simulation. In the analytical equation for R_{dB}, only the reflected field is considered. In the simulation, the shielding effectiveness is evaluated for the combined reflected and incident fields, as they cannot be separated in the full-wave simulation. We analyze this effect in greater detail later in this section, after discussing the simulated shielding effectiveness behind the plate.

The comparison of the simulated shielding effectiveness at the probe behind the plate and the analytical shielding effectiveness obtained from the addition of the reflection and absorption losses is plotted on the right-hand side of Fig. 7.2. Below approximately 10 kHz, the shielding effectiveness behind the plate is dominated by the reflection loss. The difference between the analytical and simulated results of the reflection loss below 1 kHz is explained in the previous paragraph. With increasing frequency, the absorption loss becomes the dominating term, as observed from the increase in the shielding effectiveness at higher frequencies. At 10 MHz, the analytical result is 305.8 dB, while the simulated shielding effectiveness at the probe behind the plate is 306 dB.

For the shielding effectiveness behind the plate, the difference between the analytical and simulated results of R_{dB} observed in front of the plate does not play an important role, as the shielding effectiveness behind the plate, at frequencies above a few kilohertz, is dominated by the absorption loss. Nevertheless, it is instructive to understand the origin of this difference, as it highlights the distinctions between the analytical derivation and the simulation model.

The plot of the simulated shielding effectiveness for different distances of the probes to the plate in Fig. 7.3 shows that the simulated shielding effectiveness depends on the distance of the probe to the plate. The probes are placed at distances of 1 μm, 1 mm, and 10 cm in front of and behind the plate.

Behind the plate, the shielding effectiveness increases with distance (the difference between 1 μm and 1 mm is not noticeable).

In front of the plate, the shielding effectiveness decreases with the distance for most frequencies. The results in front of the plate at a distance of 1 μm are the closest to the analytical estimation: they decay with −10 dB per decade, and the value of the simulated shielding effectiveness at 10 MHz is 103 dB, while the analytical value for R_{dB} at 10 MHz is 98 dB.

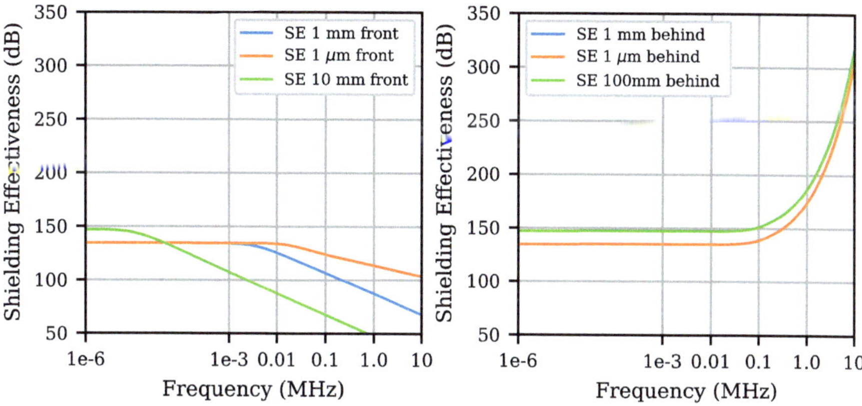

Fig. 7.3 Dependence of the simulated shielding effectiveness on the distance of the probe to the plate. Left: Probes in front of the plate. Right: probes behind the plate

In the analytical Eq. (7.3), there is no distance to the plate; the reflection loss is calculated at the surface of the shield. In the simulation model, it is impossible to place a probe at the surface of the plate: the plate is modeled using a zero-thickness CAD shape. Placing the probe at the position of the plate would give us the field on one side or the other side of the plate, depending on the implementation in the software. Therefore, we move the probes slightly away from the interface.

A second difference between the simulation and the analytical derivation is the separation of the fields into reflected and incident contributions. In the simulation, the field probe in front of the plate records the incident and the resulting fields. To better understand the obtained results, we use field monitors and evaluate the electric field along the center axis perpendicular to the plate. We also extend the frequency range up to 1 GHz, as this helps to understand the behavior of the electric field in front of the plate. The resulting field values are plotted in Fig. 7.4.

In front of the plate, the electric field at -150 mm, as the incident and reflected field form a standing wave in front of the plate. For the lower frequencies, the notch is not visible because the wavelength is larger than the extent of the computational domain. At the position of the plate, a minimum of the standing wave is located because the tangential electric field at the surface of the highly conductive plate is zero. With increasing front distance to the plate, the field increases, and the shielding effectiveness calculated by the simulation is lower than the reflection loss obtained from the analytical equation.

To understand the dependence of the simulated shielding effectiveness at probes located behind the plate, it is necessary to discuss the details of the plane wave excitation in the simulation model without the plate, which is used to calculate the incident field $\mathbf{E}_\mathrm{i}$ and the simulation model with the plate.

The fields of the plane wave are excited on the boundaries of the computational domain. In the model without the plate, the absolute value of the electric field equals $1\,\frac{\mathrm{V}}{\mathrm{m}}$ at every position in space.

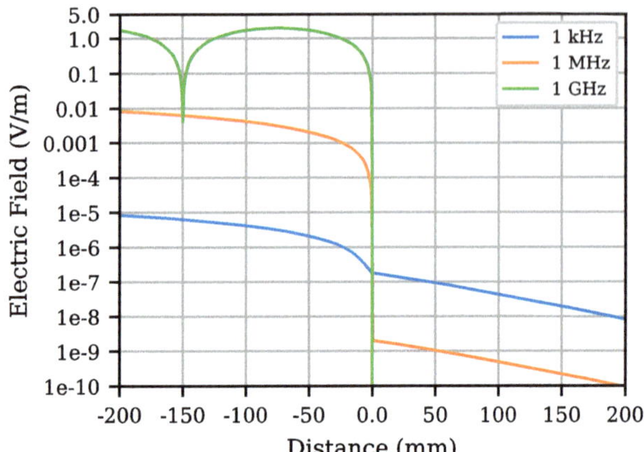

Fig. 7.4 Electric field at the center axis perpendicular to the plate for different frequencies. The copper plate is located at a distance of zero. Negative values of the distance refer to positions in front of the plate, positive value to positions behind the plate

In the model with the plate, the plane wave must not excite the space behind the plate. Because in the EM simulation model the plate touches boundaries perpendicular to the plate, the simulation software places a so-called decoupling plane at the position of the plate. This decoupling plane separates the computational domain into a region in which the plane wave is excited and a region in which no excitation is applied.

This results in the behavior observed on the right-hand side of Fig. 7.4: In the model containing the plate, the field behind the plate decreases with the distance to the plate. If we calculate the shielding effectiveness as the ratio of the constant field value in the free-space model and the decaying field behind the shield, the shielding effectiveness increases with the distance to the shield. This effect is not included in the analytical formula for A_{dB} in Eq. (7.5), as A_{dB} is calculated on the surface of the shield.

This in-depth discussion illustrates that the comparison of analytical results to simulation models requires great care and an understanding of how the simulation setup corresponds to the assumptions required for the derivation of the analytical equations.

Using the simulation model, we explore how material properties influence shielding effectiveness. In Fig. 7.5, the shielding effectiveness is plotted at a distance of 1 mm behind the shield for copper, aluminum, and steel. For aluminum, we use $\sigma_{\mathrm{Al}} = 0.61 \cdot \sigma_{\mathrm{Cu}}$ and $\mu_{\mathrm{r}} = 1$, and for steel, $\sigma_{\mathrm{Steel}} = 0.1 \cdot \sigma_{\mathrm{Cu}}$ and $\mu_{\mathrm{r}} = 1000$ over the entire frequency range. The simulation results align with the analytical predictions: as conductivity decreases, the shielding effectiveness diminishes; with higher permeability, reflection decreases while absorption increases due to the reduced skin depth.

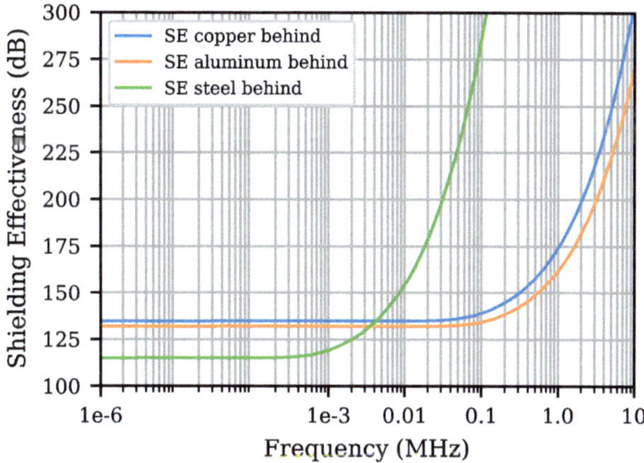

Fig. 7.5 Shielding effectiveness for plates made of different materials

There are alternative approaches to calculating shielding effectiveness using plane waves. For instance, waveguide ports can be placed at the ends of the computational domain, with magnetic and electric boundary conditions applied at opposing boundaries. This method generates a TM or TE mode within the computational domain, enabling shielding effectiveness to be derived from port signals, though it does not support arbitrary incidence angles. Alternatively, periodic boundaries combined with a Floquet port excitation can be used. We used a plane wave and probes because it appears to be most instructive approach to us.

7.2.2.1 Shielding Effectiveness of a Honeycomb Panel

The simulation results for calculating the shielding effectiveness of different materials confirm the theory. The simple plate model is more of a proof of concept than a practical application of simulation. The real strength of simulation lies in handling geometries for which an analytical solution is either not available or is limited by the assumptions needed to derive it.

In the following study, we investigate a honeycomb panel, commonly used in metallic enclosures to allow air circulation while providing good shielding against electromagnetic fields.

A parameterized model of the honeycomb panel is created, as shown in Fig. 7.6. The model allows variation in the side lengths of the individual hexagons and the width of the wires. Such models can be used during the design phase to study and optimize the geometry of honeycomb panels.

The honeycomb panel is modeled as a thin panel material, as in the case of the solid plate. The thickness of the panel is set to zero in the CAD model and defined in the material model, while the geometry of the hexagons is resolved by the mesh. Since the simulation of shielding effectiveness using a plane wave requires a

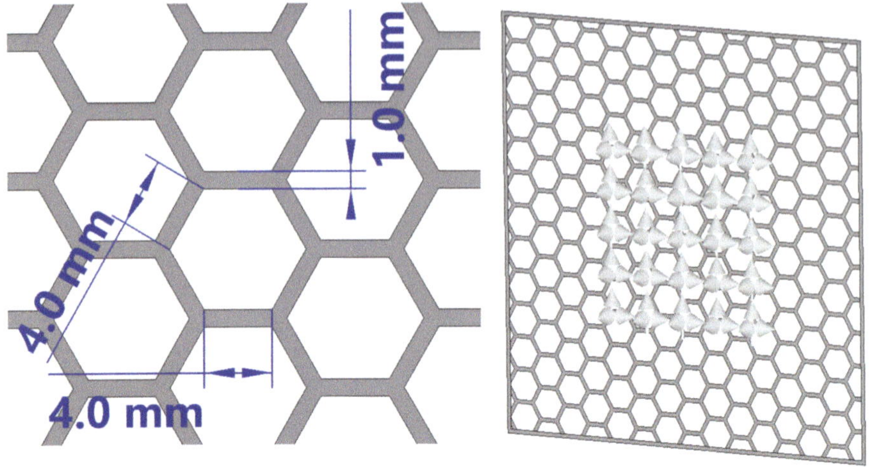

Fig. 7.6 Geometry of the simulated honeycomb panel. Left: Geometry of the panel. Right: Probe placement behind the panel

decoupling plane, we define a 1 mm thick frame around the panel, as shown on the right-hand side of Fig. 7.6.

To investigate the spatial distribution of the shielding effectiveness, an array of electric field probes is placed behind the panel at a distance of 1 mm. For the evaluation of the shielding effectiveness, we calculate the worst-case envelope of all probes. This approach follows the same principle used for estimating the worst-case emission in Chap. 6.

We perform a parameter study on the honeycomb panel. In the first study, the side length of the hexagons is kept at 4 mm while the width of the wire is varied as 0.5 mm, 1 mm, and 2 mm. In the second study, the width of the wires is fixed at 1 mm, and the side length of the hexagons is varied as 3 mm, 4 mm, and 5 mm. The results of both studies are plotted in Fig. 7.7. The simulation time for each parameter combination is approximately 1 minute.

We observe that the shielding effectiveness of the panel decreases with frequency. As frequency increases, the transmission of the field through the holes also increases. Around 100 MHz, the shielding effectiveness drops below 40 dB, a commonly cited requirement for shields. The effect of the wire width is more pronounced at lower frequencies, as seen in the plot on the left-hand side of Fig. 7.7.

In the study of the hexagon side lengths, we observe that reducing the side length from 5 mm to 4 mm results in a small improvement of 2 dB, while further reducing the side length from 4 mm to 3 mm provides a 4 dB improvement. Such studies enable engineers to determine an optimized configuration, balancing material costs with specific shielding requirements.

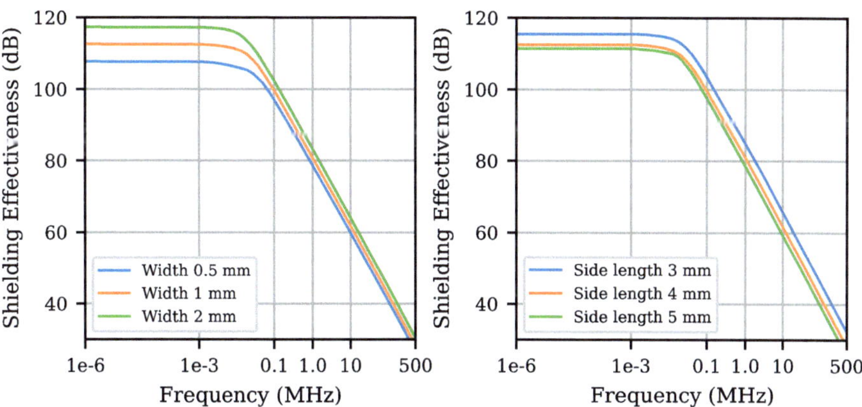

Fig. 7.7 Shielding effectiveness of the honeycomb panel for different geometrical parameters. Left: Variation of the metal width. Right: Variation of the side length

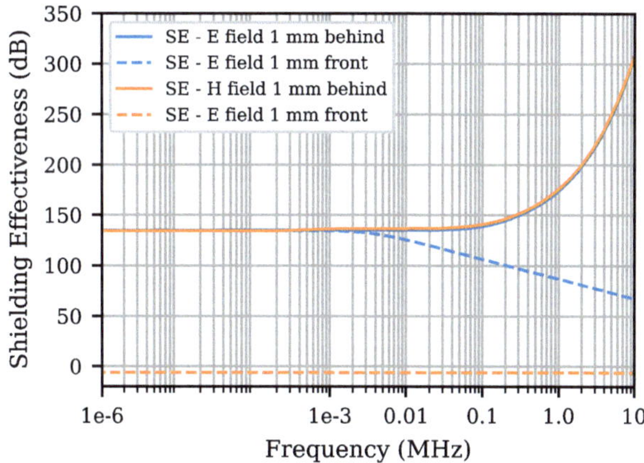

Fig. 7.8 Shielding effectiveness for the electric and magnetic field of a 0.5 mm thick copper plate

7.2.3 Shielding of the Magnetic Field for a Plane Wave Excitation

Up to this point, we have focused exclusively on analyzing the shielding effectiveness for the electric field. Let us now revisit the 0.5 mm-thick copper plate introduced at the beginning of this section.

In Fig. 7.8, we present the simulated shielding effectiveness at a probe positioned 1 mm away from the plate, examining both the electric and magnetic fields. Notably, the electric and magnetic fields in front of the plate exhibit different behaviors. The shielding effectiveness for the magnetic field remains constant at -6 dB, with the magnitude of the magnetic field in front of the plate being twice that of the incident field.

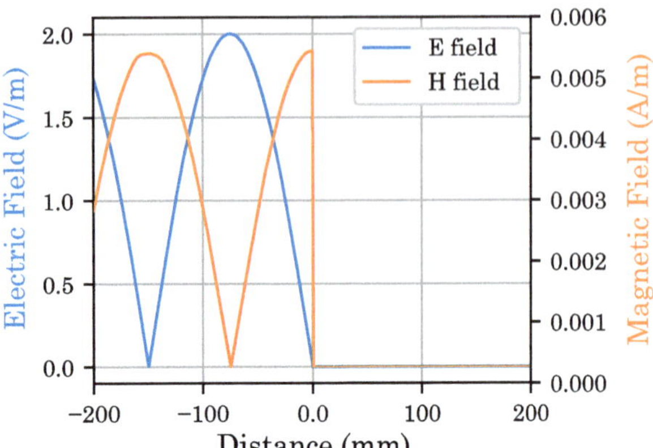

Fig. 7.9 Electric and magnetic fields, at the center axis perpendicular to the plate at 1 GHz. The copper plate is located at a distance of zero. Negative values of the distance refer to positions in front of the plate, positive value to positions behind the plate

The interface condition for the electric field (4.1) requires the tangential electric field to be continuous at the interface, with the magnitude of the electric field being nearly zero since the plate is a good conductor. For the tangential magnetic field, the interface condition does not enforce continuity (4.2) due to the current on the surface of the plate. Similar to the electric field, a standing wave is formed for the magnetic field. However, unlike the electric field, the null of the standing wave is not located at the position of the plate. This behavior is illustrated in Fig. 7.9, which shows the electric and magnetic fields along the center axis at a frequency of 1 GHz.

The shielding effectiveness behind the plate is the same for both the magnetic and electric fields. However, this equivalence is only valid under plane wave excitation, where the incident electric and magnetic fields are coupled by the free space impedance. For a low source impedance, the magnetic field can fully penetrate conductors at very low frequencies, as will be discussed in the next section.

7.2.4 Shielding of the Magnetic Field for a Loop Excitation

To model the magnetic field shielding of a source with low impedance, we keep the plate setup and replace the plane wave with a wire loop that generates a magnetic field close to the surface of the plate. The material of the wire forming the loop is set to PEC. The loop radius is 1 cm, and the loop is positioned 1 cm away from the plate.

The shielding effectiveness of the magnetic field is calculated for copper and steel, using the same material properties as in the previous section, at a probe located 1 mm in front of and behind the plate. The results are presented in Fig. 7.10.

For this simulation, we also used the thin panel material but reduced the plate thickness to 100 μm. This adjustment was necessary as the simulation of the steel plate approaches the limits of numerical precision, evident from the small peaks in

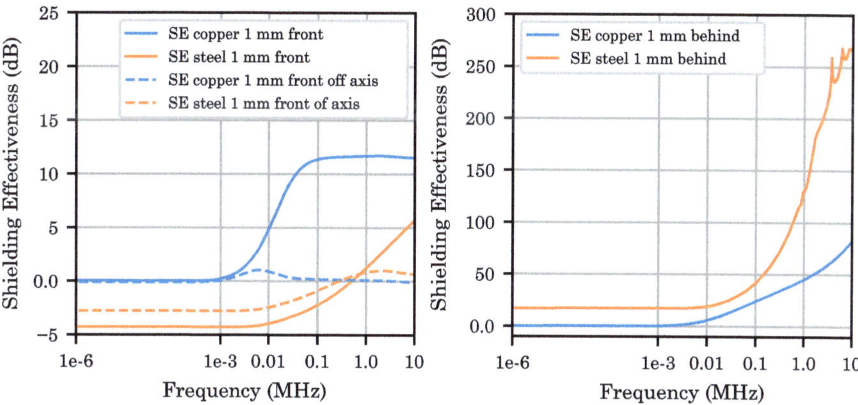

Fig. 7.10 Shielding effectiveness of a plate for the magnetic field when excited by a current loop. Left: Probes located in front of the plate. Right: Probes located behind the plate

the shielding effectiveness of steel behind the plate. The simulation time for each model is approximately 2 minutes.

Behind the plate, from 1 Hz to 1 kHz, the shielding effectiveness of the copper plate is zero. The magnetic field completely penetrates the plate, as shown by the blue curve on the right-hand side of Fig. 7.10. Above 1 kHz, a shielding effect becomes noticeable. Between 5 kHz and approximately 2 MHz, the slope of the curve is smaller compared to frequencies above 2 MHz. This behavior is due to the plate thickness being on the same order of magnitude as the skin depth in this frequency range. Only above 2 MHz we do observe the behavior predicted by the analytical formula, characterized by a steeper increase in shielding effectiveness. For the steel plate, the shielding effectiveness behind the plate remains at 17 dB up to 10 kHz and then increases rapidly, as shown by the orange curve on the right-hand side of Fig. 7.10.

When the plate is made of a conductive material, the shielding effect arises due to induced currents in the plate:
As the frequency increases, currents are induced in the plate. These currents flow in the opposite direction to the current in the source loop, generating a magnetic field that opposes the field produced by the source loop. The result is a reduction in the overall field.

When the material is also highly permeable, such as steel, a second effect contributes to the shielding:
The permeable material affects the distribution of the magnetic field. The magnetic field becomes more concentrated in front of the plate and aligns perpendicular to the plate. This behavior is analogous to the behavior of the electric field in front of a conductor.

Both effects also influence the behavior of the magnetic field in front of the plate, shown for the steel plate as a solid orange line on the left-hand side of Fig. 7.10. When the plate is made of steel, the shielding effectiveness in front of the plate is negative because the permeability causes a higher concentration of the magnetic field in front of the plate. At 500 kHz, the effect of the magnetic field generated by the induced current in the plate becomes comparable to the field concentration effect, resulting in a shielding effectiveness of zero. Above 500 kHz, the magnetic field from the induced current dominates the changes to the field in front of the plate.

For the copper plate, the shielding effectiveness at low frequencies is zero, as the magnetic field is not influenced by the plate. As the frequency increases, the shielding effectiveness rises until it reaches a plateau above 100 kHz.

This shielding effect in front of the plate is highly localized: If the probe is placed 1 mm in front of the plate but positioned off the axis of the loop, at half the radius, the induced currents in the plate have minimal effect on the magnetic field in front of the plate. This can be observed from the dashed lines on the left-hand side of Fig. 7.10.

In the following, we analyze this effect in more detail by examining the magnetic field distribution.

We plot the magnetic field in the presence of the plate for copper and steel at 1 kHz and 1 MHz in Fig. 7.11. At 1 kHz, we observe the complete penetration of the field through the copper plate and the concentration of the field in front of the steel plate.

For the copper plate at 1 MHz, we observe a region in the center of the plate where the magnetic field is lower than in the surrounding areas. This region explains the shielding effectiveness observed at the probe in front of the plate, as shown in Fig. 7.10. The induced currents in the plate are circular, with the highest amplitude occurring at the same radius as the driving loop. At the exact center of the plate, the magnetic field generated by the induced current is maximal, resulting in the region with the reduced field amplitude. When the probe is located in this region, a shielding effectiveness is observed at the center axis in front of the plate, which decreases rapidly as the probe is moved away from the center axis.

In the simulation for the steel material, we assume a fixed permeability across the entire frequency range. If the permeability of the material exhibits a dispersive behavior, the shielding effectiveness at higher frequencies decreases, as an increasing skin depth is associated with decreasing permeability. To demonstrate this effect, we add a Debye dispersion, as described in Sect. 4.4.2, to the steel plate used in this study. Note that this is a synthetic example, as steel typically does not exhibit the dispersion characteristic used in this example. However, it is instructive to observe the difference compared to the constant permeability case.

The assumed properties of the magnetic dispersion are plotted on the right-hand side of Fig. 7.12. A relaxation time of 1 μs was used, resulting in the halving of the initial relative permeability of 1000 at 160 kHz.

The shielding effectiveness for the probe located 1 mm behind the plate is shown on the left-hand side of Fig. 7.12. With increasing frequency, the shielding effectiveness of the dispersive steel is significantly reduced compared to the shielding effectiveness calculated using a fixed permeability.

At 100 kHz, the shielding effectiveness of the dispersive steel is slightly higher than that of the nondispersive steel. This can be attributed to the magnetic losses

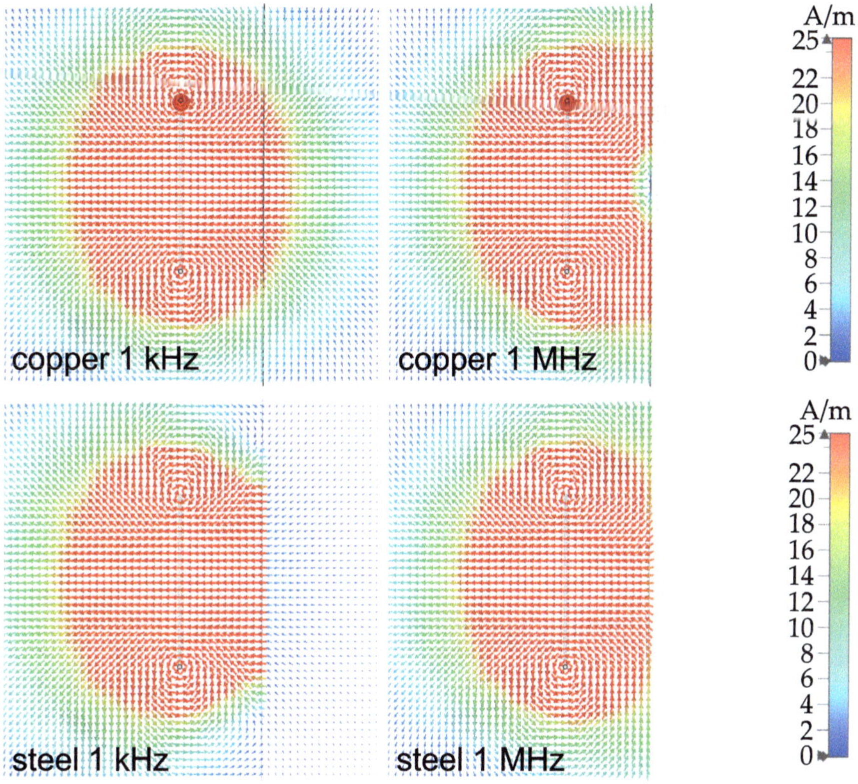

Fig. 7.11 Magnetic field in the presence of the plate, when excited by a current loop. Top row copper, bottom row steel. Left column 1 kHz, right column 1 MHz

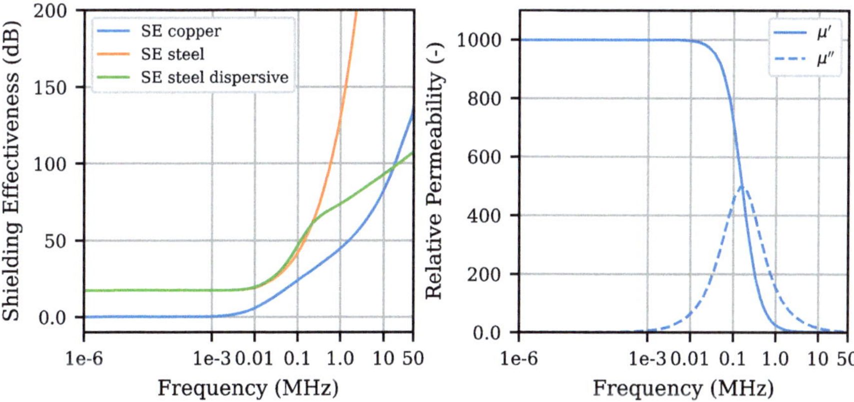

Fig. 7.12 Effect of the magnetic dispersion on the shielding effectiveness. Left: shielding effectiveness at the probe located 1 mm behind the plate. Right: Assumed frequency dependence of the dispersive steel

present in the dispersive model, which are not included in the fixed-permeability model. However, this effect is relatively small.

More importantly, as frequency increases, the copper plate eventually becomes a more effective shield than the steel plate with dispersive material characteristics. To demonstrate this, the frequency in the calculation was increased up to 50 MHz. At 20 MHz, the copper plate surpasses the dispersive steel in shielding effectiveness, as shielding above this frequency is dominated by induced currents, and the conductivity of copper is an order of magnitude higher than that of steel.

In this section, we have simulated canonical shielding examples. By using the thin panel material model, we are able to construct efficient simulation models that produce results comparable to analytical equations. This approach also allows us to include dispersive material characteristics in the simulations, with all simulation times being in the order of a few minutes. We have discussed potential pitfalls when comparing simulation results to analytical equations and analyzed the impact of probe positions on shielding effectiveness. In the following section, we will continue using the thin panel material model in practical examples.

However, it is important to highlight a limitation of the thin panel material model. With this approach, it is not possible to model the saturation of permeability in the shielding material, as described in Sect. 4.4.3. Since the thin panel model requires a zero-thickness CAD shape, it cannot account for local saturation effects within the material.

If modeling the saturation of permeability is necessary—because the incident magnetic field on the shield causes the permeable material to saturate—there are two options:

Advanced option Use a volumetric material and incorporate local saturation effects in a coupled simulation of the magnetostatic solver with the full-wave solver, as described in Sect. 9.7. This approach requires the full-wave solver to resolve the skin depth through the mesh, leading to increased simulation times.

Simplified option Estimate the reduction of permeability due to saturation using the magnetostatic solver, and then apply the reduced permeability uniformly to the entire shield while retaining the thin panel model. This approach provides a worst-case result, preserving the low simulation times offered by the thin panel material model but neglecting local saturation effects.

7.3 Application Example: Magnetic Shielding in a Vehicle according to ICNIRP

Ideal AC current source

Field probes in the computational domain

Full-wave, S-parameters, combine results

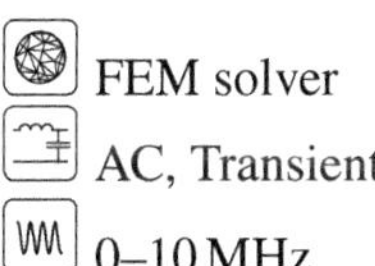

In this example, we simulate the exposure of the human body to low-frequency magnetic fields. The ICNIRP defines reference levels for the maximum field exposure of the general public, which we use to compare with the simulation results. Due to the conductivity of the human body, a time-varying magnetic field induces currents within the body. To safeguard against the effects of these induced currents, the ICNIRP specifies a maximum allowable induced electric field, referred to as the basic restriction. The induced current and the electric field inside the body are related by Ohm's law.

Since it is challenging to measure the field value inside the body, reference levels are derived from the basic restriction. These reference levels are defined in terms of the incident magnetic or electric field at the body's position, unlike the basic restriction, which is defined inside the body. The reference levels are summarized in Table 7.1. As the induced electric field depends on the frequency of the source field, the reference levels are specified for different frequency bands. In the simulation, we compute the magnitude of the magnetic field and compare it with the reference levels.

Since the magnitude of the magnetic field is proportional to the source current, reference levels become particularly important when the current amplitude is high. With the recent increase in the use of electric vehicles, exposure to low-frequency magnetic fields has become a relevant scenario for the general public, as the currents in electric powertrains can reach several hundred amperes.

For instance, an electric motor operating at 100 kW and powered by a 400 V system requires a current of 250 A for several seconds. In combustion engine vehicles, such high currents are not sustained for comparable durations. Manufacturers are required to ensure that the reference levels are not exceeded for all occupants of the vehicle.

Table 7.1 ICNIRP reference levels for the general public exposed to time-varying low-frequency fields

Frequency range	Electric field (kV/m)	Magnetic field (A/m)
1–8 Hz	5	$3.2 \cdot 10^4/f^2$
8–25 Hz	5	$4 \cdot 10^3/f$
25–50 Hz	5	$1.6 \cdot 10^2$
50–400 Hz	$2.5 \cdot 10^2/f$	$1.6 \cdot 10^2$
400 Hz–3 kHz	$2.5 \cdot 10^2/f$	$6.4 \cdot 10^4/f$
3 kHz–10 MHz	$8.3 \cdot 10^{-2}$	21

7.3.1 Magnetic Field of a Pair of Cables

In our example, we use a cable pair driven by a current to investigate the shielding measures required to ensure that the reference levels are not exceeded. We begin with a simple setup, analyze the performance of the cable shield, and then place the cable pair within a vehicle model.

The DC cables connecting the battery to an inverter in an electric vehicle are typically shielded Litz wires. For this simulation, we do not model the individual Litz wires, as the detailed losses are not critical for calculating the magnetic field. Instead, we inject a given current into the cable and represent the inner conductor of the wire as a solid cylinder.

For the cable shield, we assume a thin solid shield made of a good conductor, modeled using the thin panel material. This simplified model of the cable shield does not account for the effects of apertures present in braided shields. The modeling of more realistic shields, including these effects, is discussed in Sect. 8.5.2.

The simulation model is illustrated in Fig. 7.13. Metallic enclosures, defined as PEC, are placed at both ends of the cable pair. These enclosures serve two main purposes:

Field shielding at cable ends The enclosures ensure that the fields at the cable ends are shielded, allowing us to focus on the fields generated by the cable pair rather than those from its terminations.

Connection impedance study The enclosures allow the study of the connection impedance of the shields by incorporating circular ports between the shield ends and the enclosures. These ports provide a 360° connection of the cable shield to

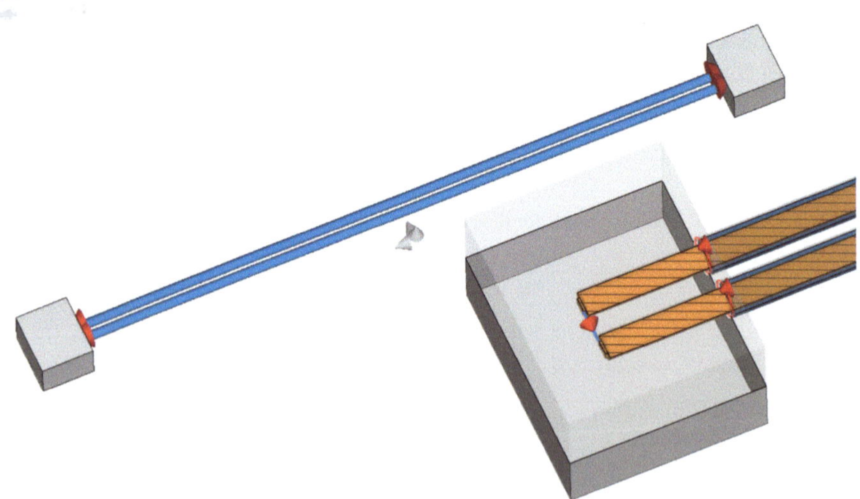

Fig. 7.13 Simulation set up for the calculation of the magnetic field from a shielded wire pair

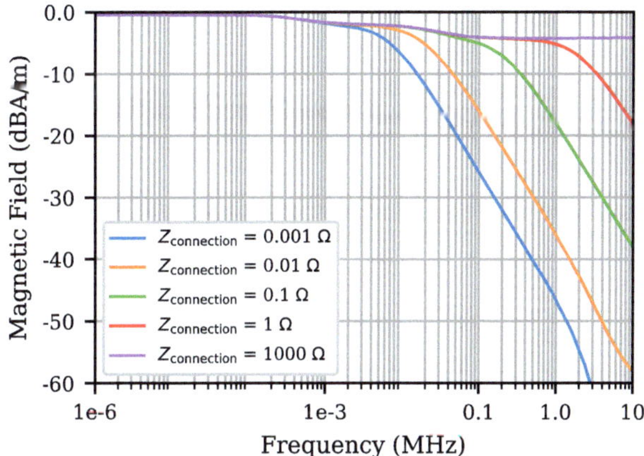

Fig. 7.14 Dependence of the magnetic field at the probe position on the connection impedance of the cable shield to the enclosure

the housing and enable the inclusion of an arbitrary connection impedance in the coupled schematic.

The cable pair is driven by a port positioned between the inner conductors and terminated in the same manner at the other end of the cable. The magnetic field is measured by a probe placed 5 cm away from the center of the cable pair. The diameter of the inner conductor is set to 10.5 mm, while the diameter of the shield is 14.4 mm. The cable cross-section is kept constant throughout this study.

The field values are computed using combined results, with the details of this simulation flow discussed in Sect. 6.7.5. In the first simulation, the cable length is set to 1 m, and the shield is defined as aluminum with a conductivity of $3.5 \cdot 10^7 \, \frac{\mathrm{S}}{\mathrm{m}}$ and a thickness of 0.1 mm. The cable is driven by a current of 1 A. The simulation time for the EM simulation, using the full-wave FEM solver, is 1 minute and 30 seconds.

The magnetic field at the probe position as a function of the shield connection impedance is shown in Fig. 7.14. The connection impedance is defined as the individual value at each connection point. The magnetic field is plotted in $\mathrm{dB}\frac{A}{m}$, allowing a direct comparison of the shielding effect in dB. The curves in Fig. 7.14 exhibit a slight decrease at 1 kHz and 50 kHz, due to the skin effect.

For a connection impedance of 1000 Ω, the shield is effectively floating and provides almost no shielding, as indicated by the purple curve. As the connection impedance decreases, the shielding effectiveness improves. For the lowest connection impedance of 1 mΩ (blue curve), the shielding effect becomes noticeable above 1 kHz and reaches approximately 50 dB at 1 MHz.

As the connection impedance increases, the shielding performance deteriorates. For a connection impedance of 0.1 Ω (green curve), the shielding is approximately 25 dB lower than that achieved with a connection impedance of 1 mΩ.

The effect is illustrated by the sketch in Fig. 7.15. The current in the inner conductor generates an induced current in the shield. When the connection impedance

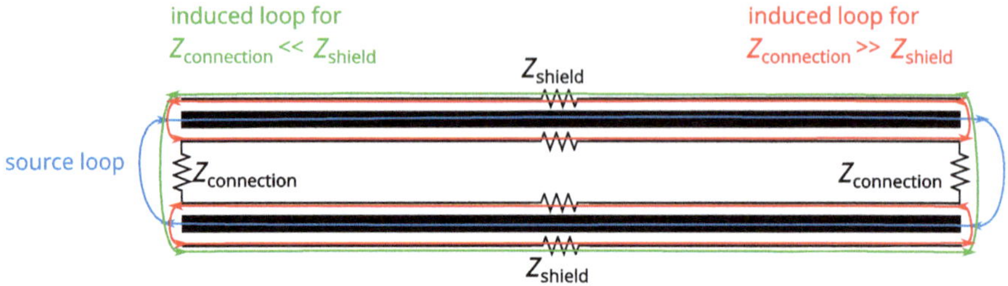

Fig. 7.15 Sketch of the source and induced current loops in a shielded cable pair

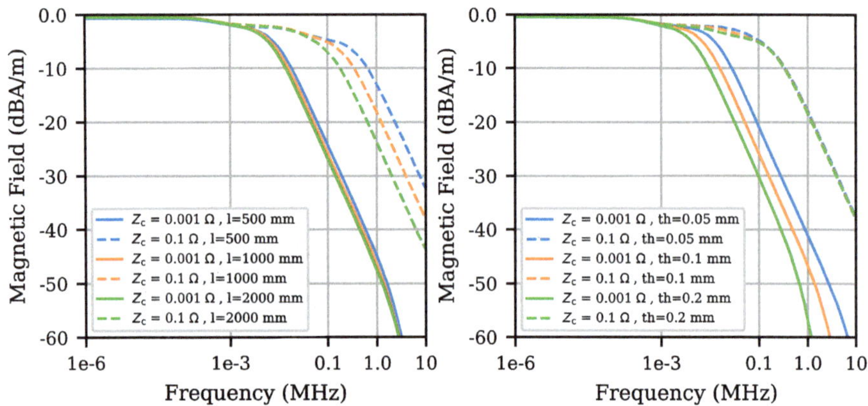

Fig. 7.16 Magnetic field strength at the probe position. Left: Dependence on cable length. Right: Dependence on shield thickness

is much lower than the impedance of the shield, the induced current flows from one shield to the other, forming a current loop that is oriented in the opposite direction (green loop in Fig. 7.15) to the source loop (blue loop in Fig. 7.15). The magnetic field created by this induced current loop opposes the magnetic field generated by the source loop, thereby reducing the resulting field. This phenomenon produces the shielding effect.

When the connection impedance is much higher than the impedance of the shield, the induced current does not flow from one shield to the other but instead forms a closed loop within each individual shield (red loops in Fig. 7.15). In this case, the magnetic field generated by the induced currents within the individual shields no longer opposes the source loop, and no shielding effect is achieved.

When Z_{shield} and $Z_{\text{connection}}$ are of similar magnitude, the current distributes between the two loops. The ratio of these impedances determines the shielding effectiveness.

We demonstrate the effect of the ratio of the two impedances in the following parameter study. In the first study, we increase Z_{shield} by varying the cable length: 0.5 m, 1 m, and 2 m. The connection impedance $Z_{\text{connection}}$ is set to 0.1 Ω and 1 mΩ. The results of this study are plotted on the right-hand side of Fig. 7.16.

For a connection impedance of 1 mΩ, where $Z_{\text{connection}} \ll Z_{\text{shield}}$, the shielding performance is very similar across all cable lengths. However, for a connection impedance of 0.1 Ω, where $Z_{\text{connection}}$ and Z_{shield} are of the same order of magnitude, the increase in Z_{shield} due to the longer cable length affects the shielding effectiveness. This can be observed in the dashed curves: for a cable length of 0.5 m, the shielding at 1 MHz is 5 dB lower compared to the 2 m-long cable, as Z_{shield} is larger for the longer cable.

In a second comparison, we examine the effect of shield thickness, varying it as 0.05 mm, 0.1 mm, and 0.2 mm. The results are plotted on the right-hand side of Fig. 7.16. Increasing the thickness from 0.05 mm to 0.2 mm improves the shielding effectiveness for a connection impedance of 1 mΩ by approximately 10 dB. However, no significant difference in the magnetic field is observed when the connection impedance is 0.1 Ω.

This cable study highlights that the performance of a shield is heavily influenced by the impedance of its connection to the housing, a well-known fact in the EMC community. Increasing the shield thickness does not improve the shielding if the connection impedance is too high. To fully leverage shields with high conductivity, the connection impedance must also be lower than for shields with lesser conductivity.

At high frequencies, this effect is illustrated by a pig-tail connection. In such a connection, the inductance of the pig tail increases the connection impedance between the shield and the housing (or PCB ground). The resulting degradation of shielding effectiveness can also be observed for low-frequency magnetic field shielding, as demonstrated in this study. Ultimately, it is the ratio of the shield impedance to the connection impedance that determines the degree of shielding degradation (Murthy and Rasek 2019).

7.3.2 *Vehicle Model*

Next, we utilize a realistic simulation model of a vehicle's underbody. CAD data for real vehicles is typically large, often in the range of 5 GB. Including the complete CAD data directly in simulations can significantly slow down the process and lead to meshing challenges. To address this, vehicle CAD data is preprocessed before being used in simulation tools.

A common exchange format for preprocessed data is the NASTRAN format. Preprocessed vehicle models are widely used not only for electromagnetic simulations but also for aerodynamic and crash studies. As generating a well-defined vehicle model requires considerable effort, these models are often reused across simulations of various physical effects.

For this study, we use a model originally developed for crash analysis. The model was created at the Center for Collision Safety and Analysis at George

Fig. 7.17 Set up of the vehicle under body with the high voltage cabling

Mason University and is available for download on the university's website.[1] The generation of this model was funded by the Federal Highway Administration.

In our simulations, we utilize the underbody and seat components of the model. The vehicle model represents a Toyota Yaris, which is not an electric vehicle. However, due to the lack of available CAD data for electric vehicles, we adapted the model to resemble a typical setup in an electric vehicle.

The simulation setup is illustrated in Fig. 7.17. Two metallic enclosures are added to the simulation model: a large enclosure below the underbody, representing the battery, and a smaller enclosure in front, representing the traction inverter. The battery and inverter are connected by two shielded cables, with the same cross-section as the cable pair used in the previous section. The excitation of the inner conductors and the connection of the shields to the enclosures are performed in the same manner as before.

Magnetic field probes are positioned in the passenger compartment at various locations. The seats are included solely for proper probe placement and are not part of the simulation. The vehicle underbody is modeled as a thin panel material.

The cable pair is driven with a current of 250 A across all frequencies, and the magnetic field is calculated at the probe positions for three models:

- A model without the underbody,
- A model where the underbody is made of aluminum ($\sigma_{\text{Al}} = 3.5 \cdot 10^7 \, \frac{\text{S}}{\text{m}}$),
- A model where the underbody is made of steel ($\sigma_{\text{Steel}} = 5.8 \cdot 10^6 \, \frac{\text{S}}{\text{m}}, \mu_r = 1000$).

[1] https://www.ccsa.gmu.edu/

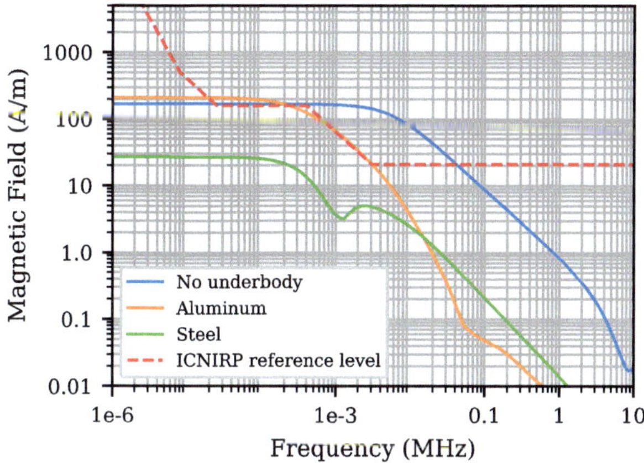

Fig. 7.18 Magnetic field at the position of the leg of the passenger for different under body configurations

The thickness of the underbody is set to 0.7 mm. The simulation time for the model without the underbody is 18 minutes, while for the models including the underbody, the simulation time is 1 hour and 31 minutes using the full-wave FEM solver.

The highest magnetic field is observed at the leg position of the passenger seat, which is the closest probe position to the cable pair. The magnetic field at this position for the three different underbody configurations is plotted in Fig. 7.18.

When the underbody is not present, the magnetic field at the passenger leg position exceeds the ICNIRP reference level by up to a factor of 10 in the frequency range from 500 Hz to 50 kHz. When the underbody is made of aluminum, the shielding effectiveness improves, but the magnetic field remains close to the ICNIRP reference level. This design choice is risky as it leaves little to no safety margin. With the underbody made of steel, the high permeability significantly enhances the shielding, and the magnetic field remains below the reference level.

In this study, we use a constant current of 250 A across all frequencies, representing a worst-case scenario. Employing a more realistic current spectrum, such as one derived from circuit simulation, can improve the accuracy of the prediction. Nonetheless, the presented simulation is valuable for analyzing the magnetic field levels as a function of different materials, cable configurations, and driving currents.

7.4 Application Example: Shielding Effectiveness of an Enclosure

Plane wave, dipole, loop

Shielding Effectiveness at field probes

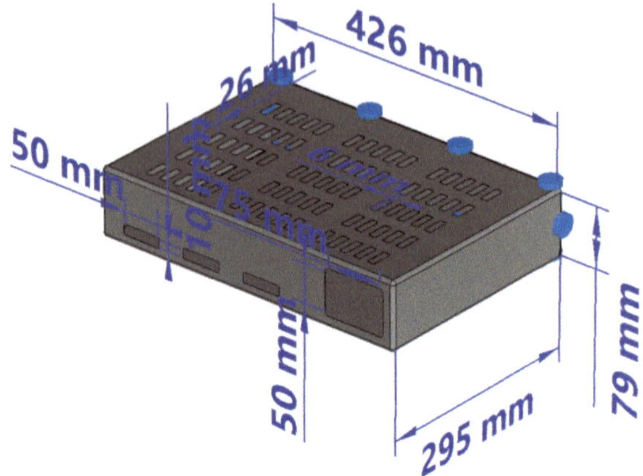

Fig. 7.19 Dimensions of the rack enclosure analyzed in this section

Full-wave
FEM solver
None
0–2 GHz

In this section, we calculate the shielding effectiveness of a rack-mount enclosure. The dimensions of the enclosure are shown in Fig. 7.19. The walls of the enclosure have a thickness of 0.5 mm. On the top of the enclosure, there are numerous ventilation openings. On the back side, three small openings and one large opening are provided for connector placement. This is the main enclosure body. In the front, a panel is positioned 100 μm away from the main enclosure body and connected to the enclosure body via lumped elements. The lumped elements are indicated by blue cones in Fig. 7.19.

The connections of the main body to the front panel consists of four lumped elements on the top side, four on the bottom side, and one on each of the left and right sides of the panel. These lumped elements are configured with a resistance of 1 μΩ, establishing a connection between the panel and the main enclosure body. At the end of this section, we examine how the connection impedance between the front panel and the main enclosure body affects the shielding effectiveness.

In Sect. 7.2.1, we analyze the shielding effectiveness of metallic panels. For example, a 0.5 mm-thick copper panel achieves a shielding effectiveness of approximately 175 dB at 1 MHz (Fig. 7.2). In this section, we study the shielding effectiveness of the enclosure across a frequency range from 1 MHz to 2 GHz. Since field penetration through the material is negligible in this frequency range, we set the enclosure material to lossy metal.

This choice is primarily practical. While the study could be conducted using a thin panel material model, the CAD data for the enclosure is defined with thick solids. Converting the solids in the CAD model to sheets for compatibility with the thin panel material model would require additional effort. This conversion is unnecessary, as it does not yield additional insights compared to using the lossy metal material model.

7.4.1 *Plane Wave Illumination*

One method for measuring the shielding effectiveness of enclosures is to place the enclosure in a reverberation chamber and position a field probe inside the enclosure. This is the setup we model in the following simulation. Three electric field probes are placed in the enclosure, as shown in Fig. 7.20.

The probe ncar the large opening is labeled as "back," the probe near the front panel as "front," and the probe in the middle as "center." The shielding effectiveness is calculated as in the previous sections using Eq. (7.1). In a reverberation chamber, the field is assumed to be homogeneous, impinging on the enclosure from all angles, with the field vector's polarization being random and not fixed.

To model this setup in the simulation without including the entire reverberation chamber, we adopt the following approach:

- A circularly polarized plane wave is used to account for the various polarization directions of the incident field.
- A parameter sweep is conducted over the incidence angle of the plane wave on the enclosure. We apply three elevation angles: $\theta = -45°$, $0°$, and $+45°$, and sample the azimuth angle ϕ with a step width of 60°.

This results in 18 samples for the parameter sweep, requiring simulations both with and without the enclosure.

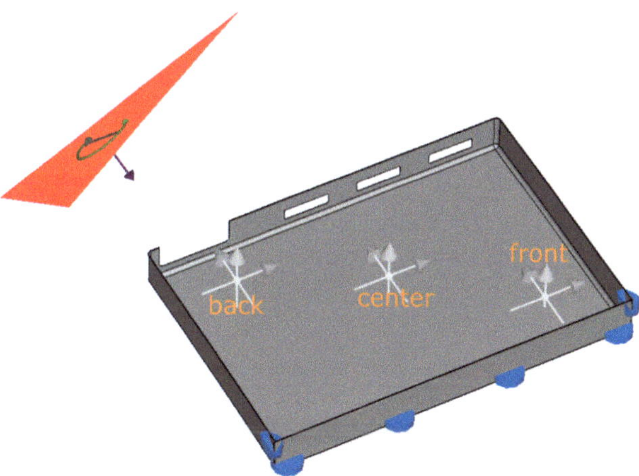

Fig. 7.20 Probes inside the enclosure and incident plane wave

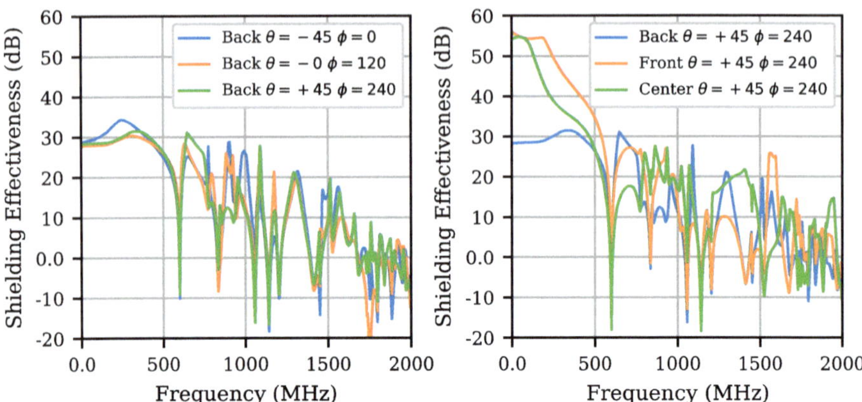

Fig. 7.21 Shielding effectiveness for the plane wave incidence on the enclosure. Left: Results at the back probe for different incidence angles. Right: Results at the three probes for one incidence angle

We use the full-wave FEM solver, as the enclosure dimensions are less than $3\,\lambda$ and resonance effects are expected, as shown later. To improve the spectral resolution, the number of frequency samples is manually increased to 70. The entire parameter sweep takes 1 hour and 13 minutes to complete.

Plotting all 18 graphs in a single plot would result in a cluttered image. To demonstrate the effect of the incidence angle on the shielding effectiveness, we select three different angles. The results for the probe at the back are shown on the left-hand side of Fig. 7.21. For the three incidence angles, the shielding effectiveness at the back probe differs by less than a few dB across most frequencies. However, in certain frequency bands, such as from 500 MHz to 1000 MHz, the shielding effectiveness varies by more than 10 dB.

Sharp dips are observed in the curve at specific frequencies, such as 600 MHz, 844 MHz, and 1049 MHz. At these frequencies, the shielding effectiveness becomes negative, meaning the field within the enclosure is higher at the probe position than the field without the enclosure. These dips correspond to resonant modes within the enclosure, which we will discuss shortly. Above 1600 MHz, the shielding effectiveness is below zero for most frequencies, indicating that the enclosure no longer shields the incident field effectively.

On the right-hand side of Fig. 7.21, the shielding effectiveness is plotted for one incidence angle at the three probe positions. Here, the difference among the probe positions is more pronounced than the difference in incidence angles. The shielding effectiveness varies significantly depending on where the electric field is measured.

At frequencies below 500 MHz, the shielding at the back probe is much lower than at the other two probes. This is because the back probe is located near the large opening on the back side of the enclosure, where the shielding effectiveness is reduced.

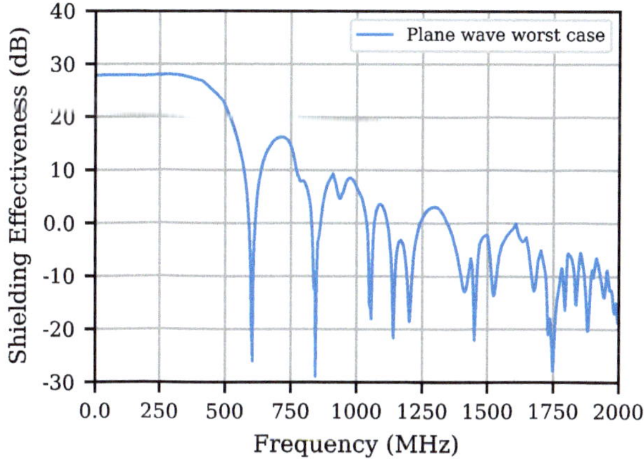

Fig. 7.22 Worst case shielding effectiveness for a plane wave incidence

Another notable observation is at 844 MHz: while the back and front probes exhibit dips in shielding effectiveness at this frequency, the center probe does not (green curve).

We further process the results obtained from the parameter sweep to generate a single shielding effectiveness curve. Naturally, the final result depends on the chosen postprocessing method. One possible approach is to calculate the worst-case shielding effectiveness. For each incidence angle, the lowest envelope of all probes is determined. In a second step, the worst case across all incidence angles is calculated. This produces a single shielding effectiveness curve, as shown in Fig. 7.22.

This method is analogous to the concept of worst-case emission used in Chap. 6. Alternatively, additional probes could be placed inside the enclosure. The optimal setup depends on the simulation engineer's judgment. For instance, including the back probe in the worst-case calculation results in a shielding effectiveness of less than 30 dB, even at low frequencies. This might be an overly pessimistic estimate if the sensitive electronics within the enclosure are not located near the large opening.

We recommend incorporating more information about the actual design of the electronic system before positioning probes inside the enclosure. If the location of the system's sensitive sections is known, the best approach is to place multiple probes in those areas. Alternatively, zones can be introduced within the enclosure, and the shielding effectiveness can be evaluated for each zone individually, rather than calculating the worst-case value for the entire enclosure. After obtaining the probe results, the field distribution can be further analyzed using a field monitor to better understand the internal field behavior within the enclosure.

The electric field at 100 MHz, 600 MHz, 844 MHz, and 1198 MHz is plotted in Fig. 7.23. For this plot, we use an incidence angle of $\theta = 45°$ and $\phi = 120°$. The electric field is displayed in a cut plane close to the bottom of the enclosure, where the three connector openings are located.

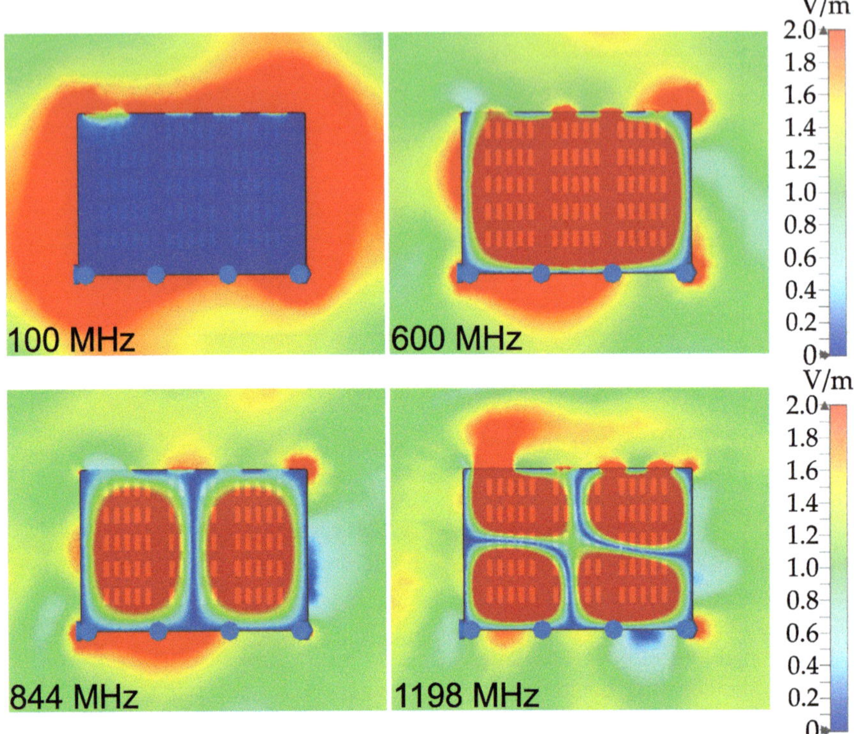

Fig. 7.23 Electric field in the enclosure for a plane wave incidence at different frequencies

At 100 MHz The electric field inside the enclosure is only significant near the openings on the back side.

At 600 MHz This frequency corresponds to the first dip in the shielding effectiveness curve. The enclosure is in resonance, with the maximum field located at the center of the enclosure.

At 844 MHz A higher-order resonance is excited. In this mode, two field maxima are present in the enclosure, with a minimum at the center. This explains why the shielding effectiveness calculation for the probe located at the center does not exhibit a dip, as shown on the right-hand side of Fig. 7.21.

At 1198 MHz The resonance of the enclosure exhibits four field maxima, indicating an even higher mode order.

From these field results, we conclude that all dips observed in the shielding effectiveness curve are due to structural resonances. When the shielding effectiveness falls below 0 dB, it is a typical indication of resonance within the enclosure. To increase the frequencies at which these resonances occur, additional metallic walls can be placed inside the enclosure to reduce the electrical size of the cavity.

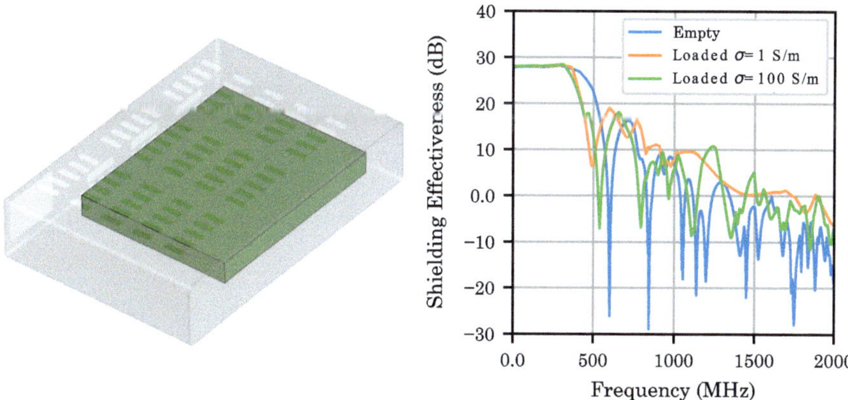

Fig. 7.24 Effect of enclosure loading on shielding effectiveness. Left: Positioning of the loading shape. Right: Shielding effectiveness for different conductivities of the loading shape

In this example, we calculate the shielding effectiveness of an empty enclosure to demonstrate the simulation workflow. When electronic components are present inside the enclosure, they affect the field distribution, reduce the Q factor, and result in less pronounced resonances.

To illustrate the effect of enclosure loading, we introduce a CAD shape inside the enclosure and define it as a lossy volumetric conductor. The shape is hexahedral with dimensions of 29 cm × 23 cm × 3 cm, representing a simplified PCB. The permittivity is set to $\varepsilon_\mathrm{r} = 4.2$, and its positioning is shown on the left-hand side of Fig. 7.24. The worst-case shielding effectiveness is calculated for the loading shape with conductivities of $\sigma = 1\,\frac{\mathrm{S}}{\mathrm{m}}$ and $\sigma = 100\,\frac{\mathrm{S}}{\mathrm{m}}$. The results are plotted on the right-hand side of Fig. 7.24.

With increasing losses, the resonances shift toward lower frequencies. Additionally, the presence of losses in the enclosure results in broader and less pronounced dips. For $\sigma = 1\,\frac{\mathrm{S}}{\mathrm{m}}$, the shielding effectiveness no longer falls below 0 dB at the resonances. If information about the actual placement of the system inside the enclosure is available, we recommend including this information in the simulation.

Placing a probe inside the enclosure and sweeping the incidence angle of a plane wave is a measurement-inspired approach to calculate the shielding effectiveness. However, the main drawback of this method is the requirement for a parameter sweep over the incidence angle, which significantly increases simulation time for geometrically complex structures. If multiple enclosure variants need to be simulated, the computational cost becomes even higher. To address these limitations, we propose a more computationally efficient approach in the next section.

7.4.2 Source Inside the Enclosure

In this section, choose a different source placement to reduce the simulation time in the calculation of the shielding effectiveness. Instead of placing probes inside the enclosure and exciting it from the outside, we place the source inside and measure the field outside of the enclosure.

This approach eliminates the need for a parameter sweep, as required in the previous section, and is computationally more efficient. The choice of probes is straightforward: we define a spherical array of probes outside the computational domain at a distance of 3 m, similar to the cylindrical arrangement used in Sect. 6.5. Since adding more probes has a negligible effect on simulation time, we use fine sampling with probes spaced every 20°, resulting in a total of 172 probes. The setup is shown on the right-hand side of Fig. 7.25. Using this configuration, we obtain the shielding effectiveness for a region far from the enclosure. Alternatively, probes can be placed inside the computational domain, with the choice left to the simulation engineer.

The selection of the source requires additional considerations:
Realistic noise source: If the enclosure is primarily designed to prevent a noise source inside from leaking energy, a realistic source model aligned with the actual noise source is recommended. For example, this could be a signal trace on a PCB within the enclosure.

Broadband shielding effectiveness: To calculate broadband shielding effectiveness for arbitrary sources, the interior of the enclosure must be excited with a rich field distribution covering all orientations and polarization directions. Simple sources, such as loops or dipoles, can be used in this case.

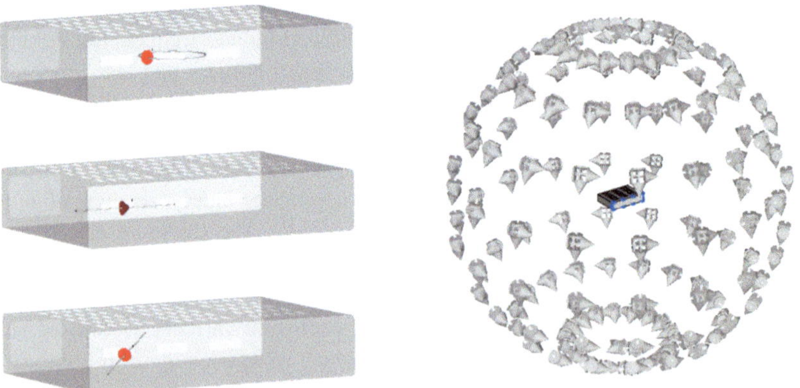

Fig. 7.25 Set up for the shielding effectiveness simulation with the source located inside the enclosure. Left: Loop and dipole orientations. Right: Probes outside of the computational domain located spherically around the enclosure

If the loop or dipole source is electrically small, it generates a low field strength. This is not an issue, as the shielding effectiveness is determined using the ratio of the shielded to unshielded field.

However, the orientation of the source significantly affects the resulting shielding effectiveness. To illustrate this effect, we conduct shielding effectiveness simulations using three different source configurations inside the enclosure:

- A horizontally oriented loop,
- A horizontal dipole,
- A tilted dipole.

These configurations are depicted on the left-hand side of Fig. 7.25. The size and position of the loop and dipole are selected to fit within the enclosure.

For the calculation of the shielding effectiveness, the source is excited in both the model with and without the enclosure. At each frequency, the lowest shielding effectiveness value from all 172 probes is calculated, resulting in a worst-case shielding effectiveness curve for each source configuration.

The results for the different sources located inside the enclosure, along with the worst-case results from the plane wave excitation, are plotted in Fig. 7.26. For the horizontal placement of the sources, both the dipole and the loop fail to resolve the dips in shielding effectiveness at 600 MHz and 844 MHz.

With this source alignment, the resonances are not excited, leading to significant deviations from the results obtained with plane wave excitation.

In contrast, the tilted dipole produces a shielding effectiveness curve that is similar to the response obtained with plane wave excitation. All dips in shielding effectiveness are resolved, and the deviations are comparable to those observed at different probe positions in the plane wave excitation model (right-hand side of Fig. 7.21). With careful source placement, using the source inside the enclosure

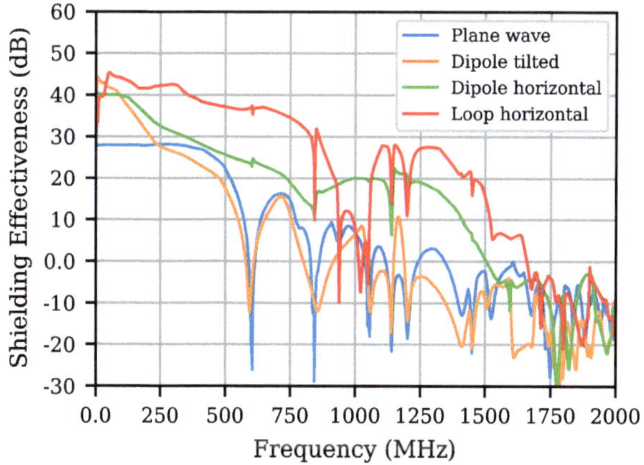

Fig. 7.26 Shielding effectiveness for different source types inside the enclosure compared to the shielding effectiveness calculated by the plane wave excitation

proves to be a valid approach for significantly reducing simulation time. For instance, the calculation of shielding effectiveness using the tilted dipole inside the enclosure required only 7 minutes and 37 seconds, whereas the simulation with plane wave excitation took 1 hour and 13 minutes. Although the difference in simulation times for this example may seem manageable, it can grow significantly when additional parameters of the model need to be analyzed.

Up to this point, we have modeled the screw connection impedance of the front panel with lumped elements set as perfect shorts, having an impedance of 1 μΩ. In this section, we investigate the effect of connection impedance on shielding effectiveness. For this study, connection impedance values of 1 Ω, 10 Ω, and 100 Ω are used. When the source is placed inside the enclosure, the simulations take slightly over 30 minutes. By contrast, using the plane wave approach, the parameter sweep for connection impedances requires nearly 5 hours.

The results of the connection impedance sweep are shown on the left-hand side of Fig. 7.27. These results are obtained using the tilted dipole source. The impact of changing the screw impedance is noticeable below 500 MHz. As the impedance increases, a new resonance appears at 260 MHz.

This resonance is formed by the opening in the front panel. The connection impedance value of 100 Ω, at which the shielding effectiveness begins to significantly deteriorate, seems relatively high. The reason for this high value lies in the construction of the front panel, as illustrated on the right-hand side of Fig. 7.27. The connection is designed as a seam, with the seam positioned 250 μm above the body. The distance between the edge of the body and the front panel is 100 μm, and the overlap of the seam with the enclosure body is 2 mm. This construction significantly enhances the shielding effectiveness of the front panel, making such a high connection impedance necessary to degrade the shielding performance. Investigating the effect of seam dimensions in greater detail is an area well suited for simulation studies.

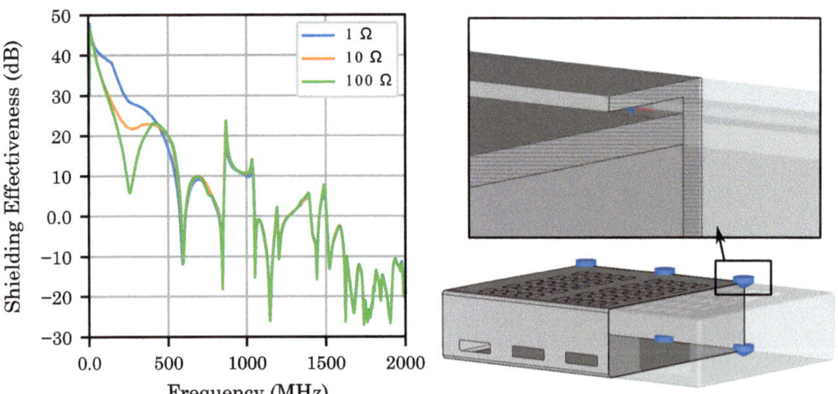

Fig. 7.27 Left: Dependence on the shielding effectiveness on the connection impedance of the screws. Right: close up on the definition of the front panel connection to the enclosure body

Further Reading

Shielding

C. Buccella et al., Magnetic field computation in a physically large domain with thin metallic shields. IEEE Trans. Magn. **41**(5), 1708–1711 (2005). https://doi.org/10.1109/TMAG.2005.846059

P. Dehkhoda, A. Tavakoli, R. Moini, An efficient and reliable shielding effectiveness evaluation of a rectangular enclosure with numerous apertures. IEEE Trans. Electromagn. Compat. **50**(1), 208–212 (2008). https://doi.org/10.1109/TEMC.2007.911922

M. Feliziani, F. Maradei, Fast computation of quasi-static magnetic fields around nonperfectly conductive shields. IEEE Trans. Magn. **34**(5), 2795–2798 (19??1998). https://doi.org/10.1109/20.717650

M. Feliziani, F. Maradei, Finite-difference time-domain modeling of thin shields. IEEE Trans. Magn. **36**(4), 848–851 (2000). https://doi.org/10.1109/20.877577

M. Feliziani, F. Maradei, EMI prediction inside conductive enclosures with attached cables, in *2001 IEEE EMC International Symposium. Symposium Record. International Symposium on Electro-magnetic Compatibility (Cat. No.01CH37161)*, vol. 1 (2001), pp. 167–172. https://doi.org/10.1109/ISEMC.2001.950581

A. Frikha et al., Magnetic shielding effectiveness of enclosures in near field at low frequency for automotive applications. IEEE Trans. Electromagn. Compat. **57**(6), 1481–1490 (2015). https://doi.org/10.1109/TEMC.2015.2463677

L. Klinkenbusch, On the shielding effectiveness of enclosures. IEEE Trans. Electromagn. Compat. **47**(3), 589–601 (2005). https://doi.org/10.1109/TEMC.2005.853162

M. Koledintseva et al., Modeling of shielding composite materials and structures for microwave frequencies. Prog. Electromagn. Res. B **15**, 197–215 (2009). https://doi.org/10.2528/PIERB09050410

M. Li et al., An EMI estimate for shielding-enclosure evaluation. IEEE Trans. Electromagn. Compat. **43**(3), 295–304 (2001). https://doi.org/10.1109/15.942602

Q.-F. Liu et al., Accurate characterization of shielding effectiveness of metallic enclosures with thin wires and thin slots. IEEE Trans. Electromagn. Compat. **51**(2), 293–300 (2009). https://doi.org/10.1109/TEMC.2008.927942

E. Liu et al., Accuracy analysis of shielding effectiveness of enclosures with apertures: a parametric study. IEEE Trans. Electromagn. Compat. **56**(6), 1396–1403 (2014). https://doi.org/10.1109/TEMC.2014.2333520

W. Luo et al., Accurate simulation of shielding effectiveness of metallic cabins using an improved Calderón preconditioner-based time-domain integral equation method. IEEE Trans. Electromagn. Compat. **61**(1), 200–208 (2019). https://doi.org/10.1109/TEMC.2018.2815000

S. Matsuzawa et al., Electromagnetic simulation of low-frequency magnetic shielding of a welded steel plate. IEEE Trans. Electromagn. Compat. **63**(6), 1896–1903 (2021). https://doi.org/10.1109/TEMC.2021.3087187

A.J. McDowell, T.H. Hubing, Analysis and comparison of plane wave shielding effectiveness decompositions. IEEE Trans. Electromagn. Compat. **56**(6), 1711–1714 (2014). https://doi.org/10.1109/TEMC.2014.2332133

H.A. Mendez, Shielding theory of enclosures with apertures. IEEE Trans. Electromagn. Compat. **EMC-20**(2), 296–305 (1978). https://doi.org/10.1109/TEMC.1978.303722

M.S. Murthy, G.A. Rasek, Analytical determination of current distributions in shielded HV cables and ground systems of electric automotive power trains. IEEE Trans. Electromagn. Compat. **61**(3), 911–918 (2019). https://doi.org/10.1109/TEMC.2018.2837121

C.R. Paul, *Introduction to Electromagnetic Compatibility* (Wiley, Hoboken, 2005). https://doi.org/10.1002/0471758159

A. Radchenko et al., Sheet absorbing material modeling and application for enclosures, in *2013 IEEE International Symposium on Electromagnetic Compatibility* (2013), pp. 645–650. https://doi.org/10.1109/ISEMC.2013.6670491

V. Rajamani et al., Validation of modal/MoM in shielding effectiveness studies of rectangular enclosures with apertures. IEEE Trans. Electromagn. Compat. **48**(2), 348–353 (2006). https://doi.org/10.1109/TEMC.2006.873864

M.P. Robinson et al., Analytical formulation for the shielding effectiveness of enclosures with apertures. IEEE Trans. Electromagn. Compat. **40**(3), 240–248 (1998). https://doi.org/10.1109/15.709422

Z.A. Wang et al., Numerical Green's function-based method for modeling radiated emission from PCBs in shielding enclosures. IEEE Trans. Microw. Theory Techn. **69**(12), 5250–5258 (2021). https://doi.org/10.1109/TMTT.2021.3119333

M.-C. Yin, P.-A. Du, An improved circuit model for the prediction of the shielding effectiveness and resonances of an enclosure with apertures. IEEE Trans. Electromagn. Compat. **58**(2), 448–456 (2016). https://doi.org/10.1109/TEMC.2016.2517163

Chapter 8
Coupling Paths: Cables

Cables are typically the longest components in an electronic system. When the DUT is a PCB and the electrical length of the attached cables exceeds that of the DUT, the cables can induce resonances at lower frequencies than the PCB itself does. Due to their greater electrical length, cables radiate more efficiently and are therefore referred to as unintentional antennas.

Simulating cables presents significant challenges because the cable's cross-section requires resolving fine details, while the cable's overall length is orders of magnitude larger. This disparity imposes a substantial computational burden on full-wave solvers.

In Sects. 6.6 and 7.3, we simulate cables using full-wave solvers. These cables are sufficiently simple to be included directly in full-wave three-dimensional models.

For more complex cable bundles, we introduce the multiconductor transmission line method (MTL) in this chapter. The MTL is derived from Maxwell's equations under the assumption of a homogeneous, electrically small cross-section. The resulting equations, known as the Telegrapher's equations, are significantly simpler to solve. This approach is particularly efficient and fast for the calculation of RLCG values per unit length, characteristic impedance, S-parameters, or crosstalk between cables. The details of the MTL are presented in the advanced Sect. 8.7.

Following the introduction, we explore the limitations of the MTL and identify scenarios where full-wave simulations are more appropriate. Additionally, we provide insights into the use of waveguide ports and the emission characteristics of cables.

The first application example in this chapter demonstrates the simulation of crosstalk in long cables located within an aircraft wing using the MTL. This is followed by a second example, which simulates crosstalk into a coaxial cable. In this example, we introduce realistic cable shields into the simulation models. The characterization of cable shields requires the concept of transfer impedance, which is discussed in a dedicated section following the coaxial cable example.

J. Hansen, A. Barchanski, *EMC Simulation of Electronic Systems*,
https://doi.org/10.1007/978-3-032-09237-3_8

Finally, in the last section of this chapter, we discuss the hybridization of cable simulations by coupling the MTL with full-wave time-domain solvers.

Even though the MTL is a long standing theory and standard results of the transmission line method are comparable between implementations, the coupling of the MTL method to the circuit simulation and the hybridization with a full-wave solver include some steps that might be implemented in different ways as a matter of the software package. The following chapters, and especially Sect. 8.6, are written using the methodology implemented in the CST.

8.1 Multiconductor Transmission Line Theory

Domains: 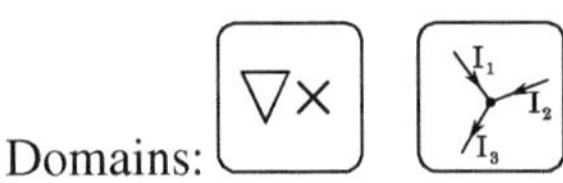

To efficiently simulate EMC problems using the MTL, the simulation engineer must understand how signal propagation is modeled by the MTL. This knowledge enables to judge whether an MTL or full-wave simulation of the cable is the appropriate method for the specific application being analyzed.

In this section, we explore the application of the MTL to common conductor geometries in electronic systems. We introduce the concept of transverse modes and explain how the MTL theory models signal propagation. Building on this foundation, we demonstrate how cascading elementary transmission line segments effectively models resonances in electrically long cables. Additionally, we discuss the characteristic impedance and provide details about the circuit models generated by the CST MTL implementation. Finally, we examine the limitations of MTL models.

8.1.1 Transmission Line Modes

In a two conductor arrangement where the geometry of the cross-section remains constant over a length much greater than the cross-section itself, the electric and magnetic fields between the conductors align according to the cross-sectional geometry. Since the cross-section is assumed to remain unchanged over some distance, the field distribution also remains constant along the conductors.

In electronic systems, many examples of such arrangements exist: a pair of wires, coaxial cables, microstrip lines, or strip lines on a PCB. The electric and magnetic fields for these arrangements are illustrated in Fig. 8.1.

This resulting field distribution is referred to as a (propagation) mode. The directions of the magnetic and electric fields are transverse to the lengthwise

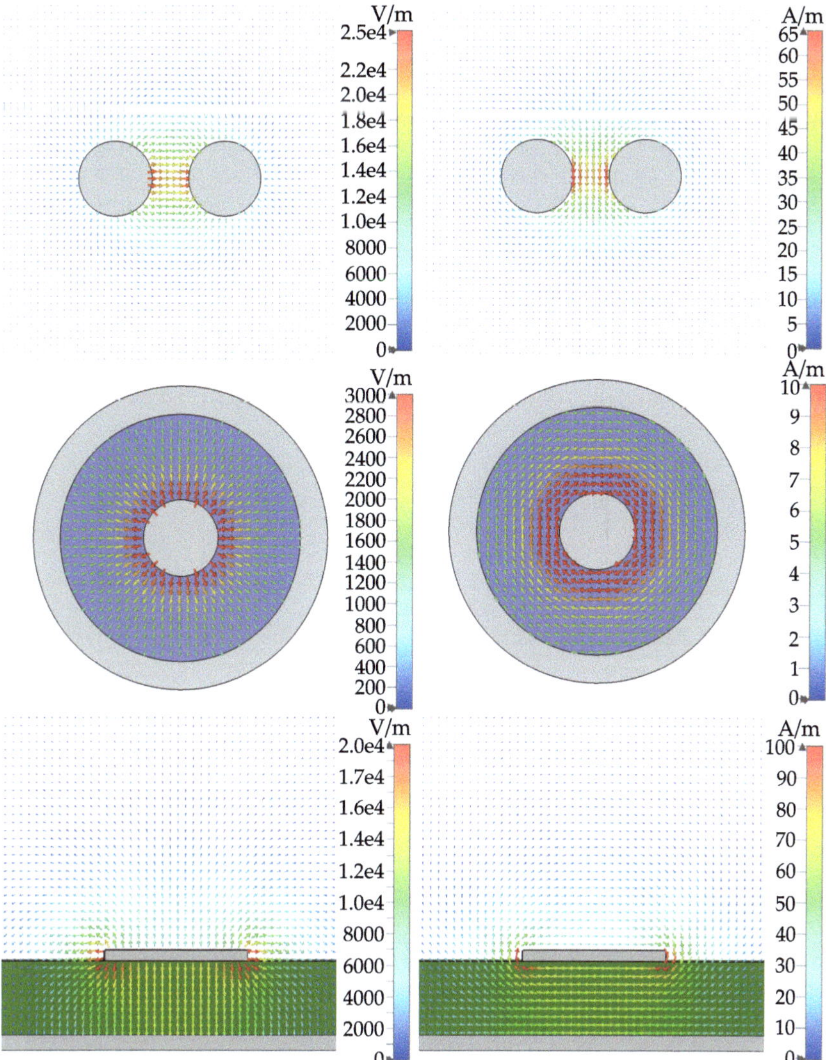

Fig. 8.1 Propagation mode in three different conductor arrangements. Left: Electric field. Right: Magnetic field. Top: Two wires. Center: Coaxial cable. Bottom: Microstrip line on a PCB

direction of the conductors, making the mode a TEM mode.[1] Waveguide modes, such as transverse electric (TE) or transverse magnetic (TM), are not discussed in this book.

[1] The mode on a microstrip line is technically a Quasi-TEM mode, as the substrate does not extend across the entire cross-section. For the purposes of this explanation, this detail is neglected.

An arrangement of two conductors supports only one mode: for instance, in a wire pair, the electric field points from one conductor to the other, while the magnetic field forms circular loops around each conductor, as shown in the top section of Fig. 8.1. Three conductors support two propagating modes. An arrangement of n conductors supports $n - 1$ modes.

To characterize an MTL system, a set of $n - 1$ propagating modes is required. In an MTL system, the individual conductors couple to each other via mutual inductances and capacitances. To obtain the propagating modes and their characteristic impedances, the coupling matrices (cf. Sect. 8.7) must be diagonalized.

In the strict interpretation of MTL theory, only propagating modes are referred to as modes. However, in our application of the MTL, where we calculate only voltages and currents at the ends of the conductors, this strict interpretation is not necessary. Instead, we adopt a relaxed definition: we refer to any given field distribution in the MTL system as a mode, even if it is not a propagating mode.

We define the complete set of $n - 1$ modes (not necessarily propagating) as a mode set. The fields propagating in an MTL system can be described as a superposition of the modes contained within a mode set. This concept is best illustrated using a three-conductor system: two wires above a large metallic reference. Two possible mode sets exist:

Single-Ended Modes Two single-ended modes, where each mode is formed between an individual wire and the reference.

Mixed-Mode Set One mode is formed between the two wires (differential mode), and the other is formed between both wires and the reference (common mode).

A transformation can be applied to convert from one mode set to another. The cases considered in this book are restricted to two modes, namely, common and differential mode. In this case, the transformation is implemented using a simple circuit component, the mode-converter block introduced in Sect. 8.3.

A conductor system supporting the propagation of a mode is referred to as a transmission line. When the transmission line is not electrically short, wave propagation phenomena must be considered in modeling signal propagation. For the simulation of transmission lines, a specialized MTL solver is used, leveraging the computational efficiency of the Telegrapher's Equations.

8.1.2 MTL Circuit Models

The Telegrapher's equations (8.15) provide a means to calculate the voltage and the current along the length of a transmission line. The derivation of these equations is detailed in the advanced Sect. 8.7. In this section, we demonstrate how these equations are applied in simulations and the types of results that can be obtained. The following discussion uses a simple two conductor system, allowing us to omit the complexity of a multiconductor system.

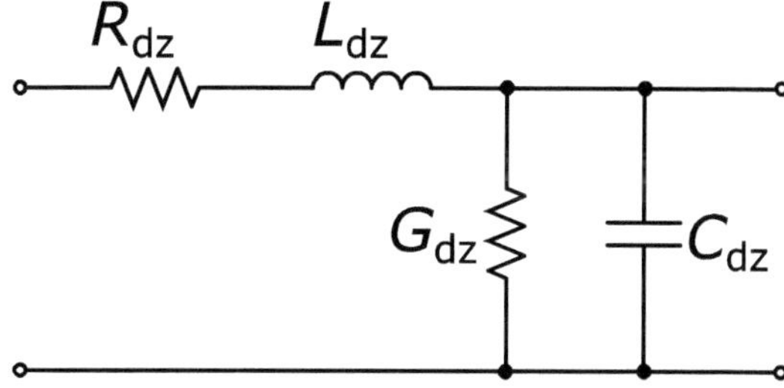

Fig. 8.2 Elementary segment of a two-conductor transmission line model

To calculate the voltages and currents on a transmission line, the per-unit-length parameters $R'L'C'G'$ are required. The terms R, L, and C represent the resistance, inductance, and capacitance of the transmission line, as used throughout this book. The conductance G represents a loss component for the mode propagating on the transmission line, accounting for dielectric losses between the conductors.

Instead of directly solving the Telegrapher's equations, a circuit model of the transmission line can be employed. The circuit model for the complete transmission line is constructed from elementary segments, as shown in Fig. 8.2 for a system with two conductors.

In case of an MTL, the per-unit-length parameters are written in matrices, which are discussed in Sect. 8.7. For n conductors, these matrices are of size $(n-1) \times (n-1)$. In the modal representation, we require $(n-1)$ per-unit-length $R'L'C'G'$ sets to describe the system. The properties of the reference wire are inherently included into this description.

The per-unit-length parameters are used to calculate the $R_{\mathrm{dz}}L_{\mathrm{dz}}C_{\mathrm{dz}}G_{\mathrm{dz}}$ values for a given segment length, as indicated by the subscript $_{\mathrm{dz}}$ in Fig. 8.2. The length of each dz segment must be electrically short, and dz is determined by the geometrical length of the transmission line and the maximum frequency up to which the transmission line is characterized. For a transmission line with an electrical length of $l_\lambda = 1\lambda$, typically at least 10 segments are used. To obtain the $R_{\mathrm{dz}}L_{\mathrm{dz}}C_{\mathrm{dz}}G_{\mathrm{dz}}$ values, the following steps are required:

1. Determine the number of required segments, n, such that $n \geq l_\lambda \cdot 10$.
2. Calculate $RLCG$ by multiplying the per-unit-length parameters $R'L'C'G'$ with the geometrical length of the transmission line.
3. Compute $R_{\mathrm{dz}}L_{\mathrm{dz}}C_{\mathrm{dz}}G_{\mathrm{dz}}$ by dividing $RLCG$ by the number of segments n.
4. Cascade the n segments in the schematic, as shown in Fig. 8.3.

In the following, we demonstrate how cascading elementary transmission line segments can model the resonant behavior of electrically long transmission lines.

In Sect. 2.4.2.5, we discuss the limitations of RLC models by analyzing the input impedance of a wire pair. In this section, we use the same wire pair to analyze how the modeled input impedance changes with the number of transmission line segments. The wire pair consists of two straight, parallel copper wires with a length of $l = 50\,\mathrm{cm}$, a wire diameter of $d = 1\,\mathrm{mm}$, and a center-to-center separation of $s = 1.5\,\mathrm{mm}$. The length of 50 cm corresponds to an electrical length of 1λ at

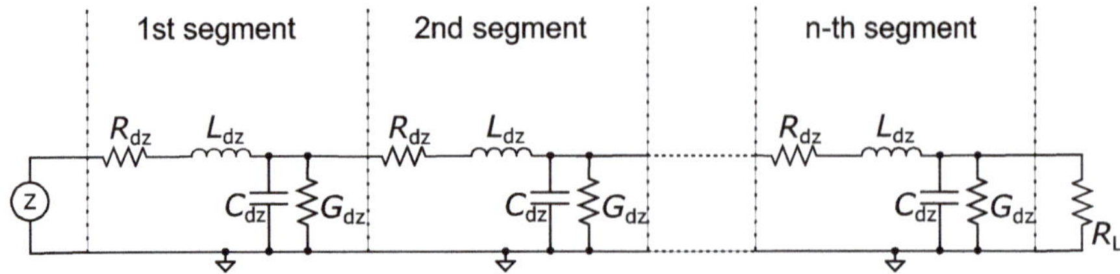

Fig. 8.3 A cascade of elementary transmission line segments used to demonstrate how the input impedance of a wire pair is modeled with an increased number of segments

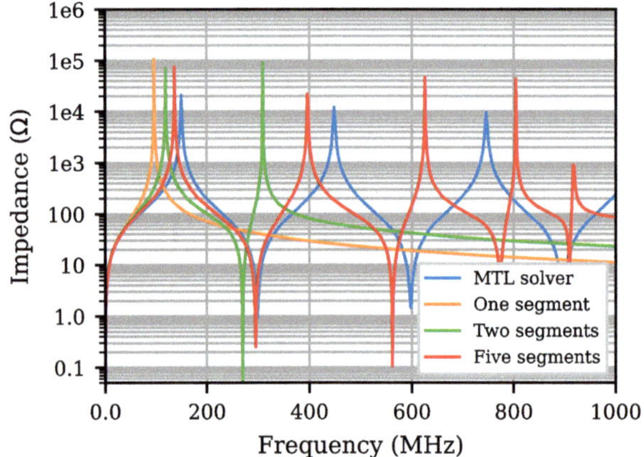

Fig. 8.4 Input impedance of a wire pair modeled with a manually defined, varying number of elementary segments compared to results obtained using a MTL solver

approximately 600 MHz. The RLCG parameters of the wire pair are calculated using the analytical equations provided in Sect. 3.3.

For the wire pair, the calculated values are $R = 21.9\,\text{m}\Omega$, $L = 192\,\text{nH}$, $C = 14.4\,\text{pF}$, and $G = 0\,\frac{\text{S}}{\text{m}}$ (see Fig. 2.19). These values are used to compute the $R_{\text{dz}}L_{\text{dz}}C_{\text{dz}}G_{\text{dz}}$ parameters for the elementary segments.

For the analysis, we calculate the input impedance with a varying number of sections: $n = 1$, $n = 2$, and $n = 5$. These three models are then compared to results obtained using the MTL solver in CST.

The input impedance of the wire pair is plotted in Fig. 8.4. The frequency axis is displayed on a linear scale to facilitate analysis of the frequencies at which resonance peaks occur. The termination of the wire pair is set to $R_{\text{L}} = 1\,\mu\Omega$. The MTL solver automatically selects 33 segments for a maximum frequency of 1 GHz, which is used as the reference solution (blue line).

Using only one segment in the manually defined cascade, only the first antiresonance peak is captured in the solution (orange curve). This result corresponds to the discussion in Sect. 2.4.2.5. When the number of segments is increased to $n = 2$, two antiresonances and one resonance appear in the solution (green curve). Further increasing the number of segments to $n = 5$ results in four resonances and four antiresonances being captured.

As the number of segments increases, both the number of resonances and the accuracy of the resonant frequencies improve. This effect is particularly noticeable at the first resonance at 150 MHz, where the resonance frequency aligns more closely with the reference solution when the number of segments is increased. At higher frequencies, the deviation from the reference solution increases because the chosen number of segments is insufficient to accurately model the transmission line at those frequencies.

Another source of deviation is the method used to calculate the RLCG parameters in the manually defined cascade. In the manual cascade, fixed values for RLCG parameters are used across the entire frequency band, whereas the MTL solver calculates frequency-dependent RLCG parameters. This frequency dependence is discussed in the following section.

8.1.3 MTL Solver

To analyze a transmission line, it is essential to determine the R'L'C'G' parameters. Analytical solutions are only applicable to simple conductor arrangements. For more complex configurations involving multiple conductors, the MTL solver is used to calculate the R'L'C'G' parameters.

The following description focuses on the MTL solver in the CST software; however, other simulation tools that include MTL solvers, for example, EMCoS, operate in a similar manner.

An MTL solver calculates the R'L'C'G' parameters using a quasistatic two-dimensional MoM solution of the cross-section of the conductor arrangement. If the cross-section does not change along the length of the transmission line, the transmission line is uniform. In this case, only one two-dimensional solution is required. In case of a nonuniform transmission line, that is, if the cross-section changes, the solver detects the variations and calculates the R'L'C'G' parameters for each individual cross-section. Once the R'L'C'G' parameters are determined, they are cascaded into the elementary segments, as discussed in the previous section.

For the simulation engineer, understanding how the individual cross-sections are generated is crucial. This process is referred to as (MTL) meshing. In this section, we focus on cable-like geometries. The application of the MTL solver to PCBs is discussed in Sect. 11. When a cable is meshed by the MTL solver in CST, two entities contribute to the cross-section analyzed by the two-dimensional solver:

Cable Bundles User-defined cable bundles specify the cable cross-section, including the conductors present in the cable cross-section.

CAD Shapes Surrounding CAD shapes contribute to the cross-section. The solver detects the materials of these shapes near the cable bundle and includes them in the calculation of the R'L'C'G' parameters. CAD shapes assigned PEC (or lossy metal)

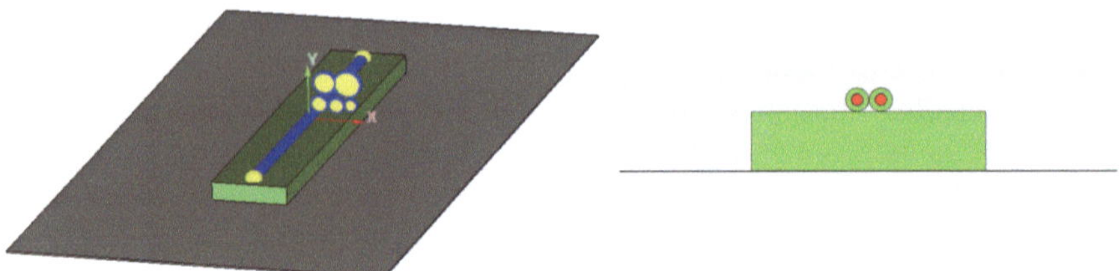

Fig. 8.5 A cable bundle located over a conductive table. Left: CAD model. Right: Cross-section after the meshing step of the MTL solver

material are treated as conductors that can form modes with the conductors of the cable bundle. Conductors defined as CAD shapes are referred to as the reference in the MTL model.

An example of such an arrangement is a cable routed over a conductive table. In this scenario, the cable cross-section is defined as a bundle, while the table is defined as a CAD shape.

For an example model with two wires located above a conductive table, we plot the full CAD model and the cross-section generated by the MTL meshing step in Fig. 8.5. On the left hand side, the CAD model is shown. The cable bundle, indicated in blue, consists of two wires, each with a dielectric coating. The bundle is placed on an insulator modeled as a green box, which rests on a conductive table represented by the gray plane. The cross-section is calculated at the center of the bundle, as indicated by the xy coordinate axis.

On the right-hand side of Fig. 8.5, the cross-section after meshing is displayed. In the cross-section, green shapes represent dielectrics, and red shapes represent metals. Both, the CAD shapes (table and insulator), as well as the individual wires within the cable bundle are included in the cross-section.

When multiple cable bundles are present in the model, the MTL solver determines whether they are included in a single cross-section or meshed into separate cross-sections. This decision is governed by a search distance. If the distance between two cable bundles exceeds the search distance, the bundles are meshed separately, and the coupling between these bundles is neglected in the R'L'C'G' parameters.

Following the discussion on how the MTL solver generates cross-sections, we now examine the R'L'C'G' parameters in more detail. For this discussion, we return to the wire pair analyzed in the previous section. As mentioned earlier, a set of R'L'C'G' parameters corresponds to a specific mode on the transmission line.

The resulting R'L'C'G' parameters for the wire pair are shown in Fig. 8.6. Due to the skin effect, both resistance and inductance exhibit frequency dependence (cf. Sect. 5.6). This frequency dependence causes the resistance to increase, which gradually reduces the amplitude of the resonance peaks at higher frequencies, as illustrated by the blue curve in Fig. 8.4. As the frequency increases, the resonances become more damped due to the rising resistance. These effects are not accounted

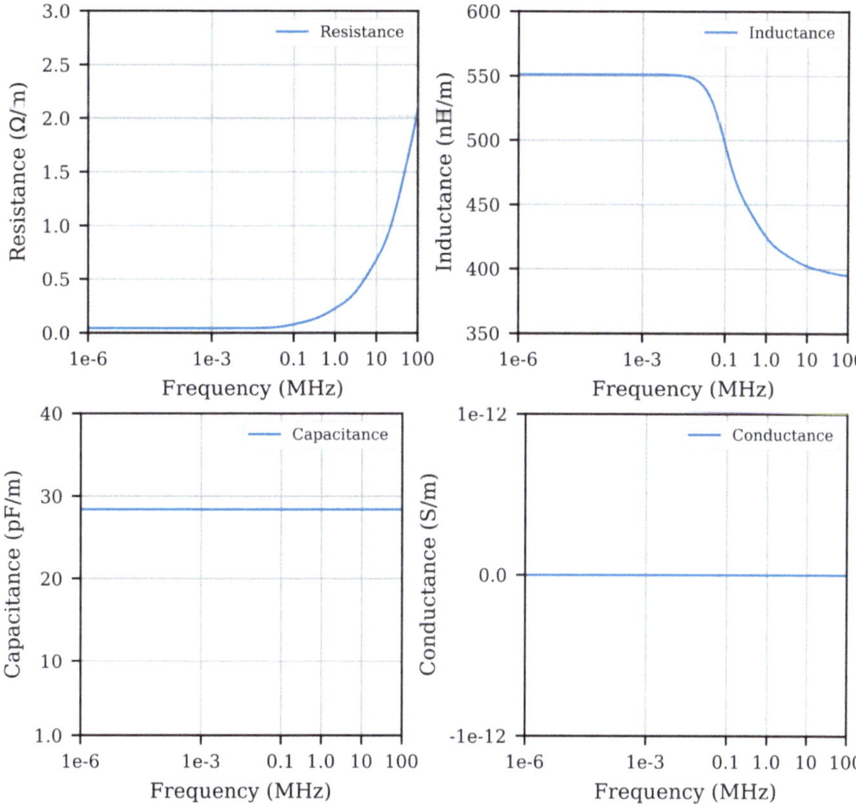

Fig. 8.6 Per-unit-length R'L'C'G' parameters of a wire pair. Top left: Resistance. Top right: Inductance. Bottom left: Capacitance. Bottom right: Conductance

for in the analytical equations used in the cascaded model. The capacitance of the wire pair remains constant over the entire frequency range, and the conductance is zero, as there is no frequency-dependent or lossy dielectric in the model.

As a final note, we point out that a circuit representation of a transmission line model can also be formulated as a modal model, which do not use an RLCG description and model the propagation using the wave impedance and transmission delay. However, we do not cover modal transmission line models in this book, as the RLCG modeling approach is more intuitive. Furthermore, the coupling of the MTL solver to the full-wave solver, discussed in Sect. 8.6, relies on the RLCG description.

8.1.4 Characteristic Impedance

If a transmission line is electrically long, reflections of the propagating mode can occur at its termination. These reflections lead to resonances, as previously discussed in the analysis of the wire pair. At resonance frequencies, a transmission line becomes an efficient radiator. In the EMC community, this phenomenon is referred to as an unintentional antenna.

The reflection coefficient at the termination of the transmission line represents the ratio of the reflected wave to the incident wave at the termination. It is expressed in terms of the voltage at the termination load as

$$\Gamma_L = \frac{V_{\text{reflected}}}{V_{\text{incident}}} = \frac{R_L - Z_0}{R_L + Z_0} \tag{8.1}$$

For the voltage at the driver, the load resistance R_L in the equation is replaced by the driver resistance. When $\Gamma_L = 0$, there is no reflection at the termination. When $\Gamma_L = 1$, the entire incident wave is reflected at the termination.

If the termination resistance R_L equals the characteristic impedance Z_0 of the transmission line, the numerator in Eq. (8.1) is zero, resulting in no reflection. This condition is known as a matching termination. Consequently, the characteristic impedance Z_0 of the transmission line is a critical parameter for understanding and controlling emissions.

For a two-conductor transmission line, the characteristic impedance is given as

$$Z_0 = \sqrt{\frac{R + j\omega L}{G + j\omega C}}. \tag{8.2}$$

For an MTL, there exists a characteristic impedance matrix (Sect. 8.7, Eq. (8.22)).

For the wire pair discussed throughout this section, the characteristic impedance is plotted in Fig. 8.7. The frequency dependence of the characteristic impedance indicates that the characteristic impedance increases as the frequency decreases.

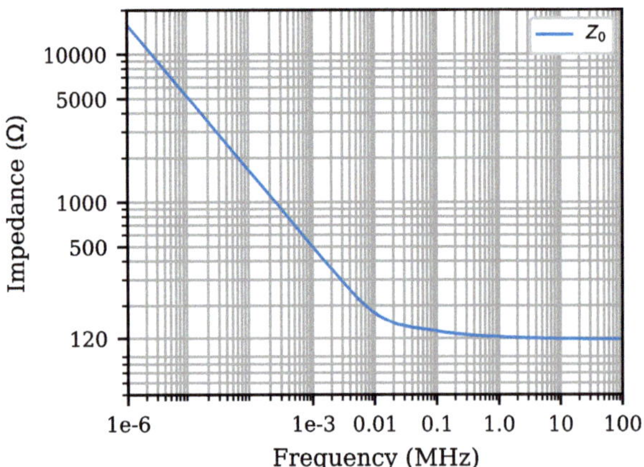

Fig. 8.7 Frequency dependence of the characteristic impedance of a wire pair calculated using the transmission line solver

This behavior is consistent with Eq. (8.2). As ω approaches zero, both the inductive and capacitive contributions diminish. Since the conductance G in our example is zero, the result simplifies to R divided by the vanishing capacitive term in the denominator. Without dielectric losses, the characteristic impedance diverges at DC. However, this effect is not practically relevant, as at very low frequencies, wave propagation modeling is unnecessary, and only the wire resistance is significant.

A common simplification for Eq. (8.2) is the high-frequency limit. When R and G are negligible compared to $j\omega L$ and $j\omega C$, the equation simplifies to:

$$Z_0 = \sqrt{\frac{L}{C}}.$$

At 100 MHz, the characteristic impedance of the wire pair calculated using the MTL solver is 117.95 Ω. Using the analytical values for L and C, the calculated value is 115.47 Ω.

8.1.5 Limits of MTL Models

The MTL solver is efficient for cable simulation as it only requires two-dimensional field solutions to calculate the R'L'C'G' parameters, and the cascading of elementary segments is performed in a circuit simulation. However, it is essential to understand the limitations and assumptions of this method to use it effectively in EMC simulations.

Since the two-dimensional cross-section in the MTL solver is represented by R'L'C'G' models, the distances among the conductors in the transmission line cross-section must be electrically short. This assumption corresponds to a quasistatic formulation. Consequently, radiative coupling between two transmission lines is not included in the MTL model.

Additionally, the MTL solver in CST calculates only the R'L'C'G' parameters for a mode. While a two-dimensional field solution is used to derive these parameters, the field distribution itself is not accessible to the simulation engineer. To obtain the fields generated by the transmission line or to account for radiative coupling between cable bundles, a full-wave method or a hybrid method combining MTL with a full-wave solver is required.

The MTL solver assumes that a mode is always present, which means it cannot model a break in one of the conductors, as no mode exists at the position of the break. For example, consider a wire routed over a metallic plane defined as a CAD shape. If the plane is split into two nonconnected sections, the MTL solver cannot account for this discontinuity and instead assumes a continuous conductor.

This continuity assumption also implies that even when different CAD shapes are used in different two-dimensional cross-sections, they are considered electrically connected. It is not possible to reference the two ends of a cable to electrically

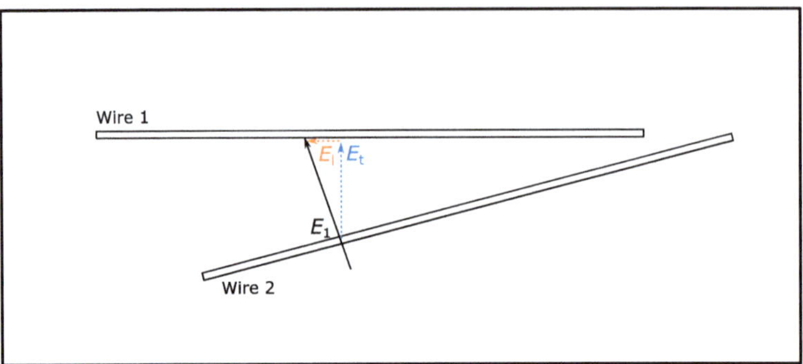

Fig. 8.8 Illustration of modes on transmission lines routed at an angle

disconnected CAD shapes. However, the bidirectional hybrid cosimulation of the MTL solver with a full-wave solver (see Sect. 8.6) eliminates this limitation.

As the coupling between two transmission lines is defined only in a two-dimensional cross-section by transverse field components, inaccuracies can arise in models where neighboring transmission lines are routed at an angle relative to each other. This scenario is illustrated in Fig. 8.8.

In Fig. 8.8, two wires are routed over a reference plane. One mode is formed between Wire 1 and the reference plane and another mode between Wire 2 and the reference plane. The electric field generated by Wire 1, $\mathbf{E}_1$, is used to illustrate the problem. The field $\mathbf{E}_1$ of the mode generated by Wire 1 is perpendicular to Wire 1 but intersects Wire 2 at an angle. At the position of Wire 2, $\mathbf{E}_1$ can be decomposed into a longitudinal and a transverse component.

As the MTL derivation (Sect. 8.7) considers only transverse components, the coupling through longitudinal components is neglected. For small incidence angles, correction terms can be applied to the MTL model to maintain accuracy. However, for larger angles, such corrections are not feasible.

Software implementations handle larger angles in different ways. One approach is to deactivate the coupling above a certain angle. In this case, all cross-sections are calculated, but coupling in cross-sections with high angles is disregarded. Depending on the length of these segments, the loss of accuracy for the overall model may or may not be acceptable. If the solution is significantly affected by coupling in high-angle segments, the MTL method becomes unsuitable, and a full-wave solution is required.

In the MTL solution, the properties of terminations connected to the MTL model are not modeled. This often leads to confusion when comparing inductance values derived from the MTL model to those obtained from a full-wave solution. In full-wave models, if discrete ports are used to excite or terminate the conductors, the length of these ports contributes to the inductance. However, this length is not considered in the MTL results. The MTL model assumes modal excitation, yielding results more directly comparable to full-wave models that use waveguide ports for excitation. This topic is further discussed in Sect. 8.2.1.

8.1.6 MTL Circuit Component Models

For generating a circuit model using the MTL solver in CST, it is necessary to define so-called cable ports. Cable ports simplify the circuit model by allowing the direct specification of the mode to be excited, such as by placing a port between two conductors in a cable bundle. Additionally, the cable port concept enables the use of combined results in hybrid simulations. A cable port functions in the same way as the network parameter ports introduced in Sect. 2.4.3.1.

Cable ports can also be defined relative to the reference of the MTL model, which is always the nearest PEC CAD shape. In the MTL model, the concept of a reference in the cross-section is valid because the cross-section is electrically short. As noted in the previous section, the reference of a transmission line model is assumed to be continuous. Consequently, all cable ports are defined relative to the same reference, even if the CAD shapes are physically disconnected.

Care should be taken when using cable ports defined relative to the MTL reference. For example, in the case of a cable routed over a metallic table, this setup is valid because the table is continuous, and a mode between the conductor in the cable bundle and the table is established along the entire cable length.

However, using cable ports defined relative to the MTL reference in more complex geometries, such as a car chassis, requires careful consideration. The simulation engineer must ensure that the assumption of a continuously connected reference conductor is valid in such cases.

8.2 Cable Simulation: Full-Wave vs. MTL

Domains: 

As discussed at the beginning of this chapter, cables can be simulated using different approaches. In Sects. 6.6 and 7.3, we simulate cables using full-wave solvers. In this chapter, we introduce the MTL method for cable simulation. The choice of the most suitable approach depends on the desired results and the specific effects that need to be captured in the simulation.

When circuit-based results such as voltages, currents, or impedances are required, the MTL method is preferable due to its high computational efficiency. For simulations in which the field distribution is of interest, a full-wave method or a hybrid approach combining the MTL with a full-wave solver is necessary. Similarly, when the limitations of the MTL (discussed in Sect. 8.1.5) are reached, a full-wave method becomes the preferred choice.

In the simulation of electronic systems, the interaction between different parts of the system plays a crucial role. For example, consider a signal routed from a PCB over a connector into a cable. Due to an imbalance on the PCB, a common mode is generated, which is efficiently radiated by the cable. While the emission primarily originates from the cable, understanding the common mode generated on the PCB is essential for modeling the emission levels. The generation of common mode due to imbalance is discussed in Sect. 8.3.2, and the emission from a USB channel is simulated in Sect. 12.5.

8.2.1 Modes in Full-Wave Simulations

At the beginning of this chapter, we plot the field distribution of different transmission line modes in Fig. 8.1. These distributions are calculated using a waveguide port. The concept of a waveguide port is introduced in Sect. 4.6.

A waveguide port is employed in full-wave simulations to determine the modes of a conductor arrangement. For this purpose, a specific area must be defined where a two-dimensional field distribution is calculated. This approach is conceptually similar to the MTL solution, which also relies on a two-dimensional cross-section analysis. However, the waveguide port represents a more general concept, as it can also be used to calculate TE and TM mode patterns in waveguides, which are not supported by the MTL model.

The waveguide port computes the characteristic impedance of a mode directly from the field distribution, rather than relying on the R'L'C'G' parameters. From the field distribution, the voltage and current are determined as described in Sect. 4.7. The characteristic impedance is then calculated using $Z_0 = \frac{V}{I}$.

Waveguide ports are commonly used to evaluate modes and characteristic impedances in the high-frequency regime, often assuming perfect electric conductor (PEC) or lossy metal materials. The mode patterns shown in Fig. 8.1 are calculated using PEC for the conductors, with absolute field values normalized to an input power of 1 W.

At the end of Sect. 8.1.4, we compare the characteristic impedance of a wire pair calculated using the MTL solver with analytically obtained values. The characteristic impedance calculated by the MTL solver is 117.95 Ω, while the analytical calculation yields 115.47 Ω. For the same wire pair, the waveguide port calculates 114.5 Ω. All results agree within 3%.

Nevertheless, it is important to comment on the deviation between the solution obtained from the MTL solver and the waveguide port. For the two-dimensional field solution at the waveguide port, it is necessary to define the area within which the mode fields are calculated. This area is finite; however, for accurate estimation, the area must encompass the complete mode pattern, which is not perfectly achievable. When the distances between the conductors in the cross-section are large, the mode fields can extend over a wide area, and the finite size of the waveguide port introduces some inaccuracies.

The MTL solver employs a Method of Moments (MoM) for analyzing the cross-section. In the MoM, the boundaries of the cross-section are treated as open, meaning the mode area is not constrained, unlike in the waveguide port approach.

As a result, for calculating the characteristic impedance of a mode on a cable, the MTL solver is generally preferable.

In the next section, we discuss a three-conductor setup and analyze it in terms of differential and common modes. Throughout the following section, both approaches are employed: the MTL solver for calculating the characteristic impedance and the FEM full-wave solver for analyzing fields.

8.3 Common and Differential Modes

The three-conductor setup, consisting of two signal lines and a reference conductor, is of particular interest in EMC. Examples of such setups include two power supply wires of an electronic system routed over a metallic table or a differential pair of a communication line on the same table.

A three-conductor setup supports two modes. The system can be described using either two single-ended modes or the so-called mixed-mode description, which consists of a differential mode and a common mode. In the following, we denote one signal conductor as positive (P) and the other as negative (N).

Single-Ended Description In the single-ended description, each signal conductor is referenced to the reference conductor. In our example, this is the metallic table. The voltage is defined between the individual signal conductors and the reference: V_P and V_N. Each conductor carries a current: I_P for the P conductor and I_N for the N conductor.

Mixed-Mode Description In the mixed-mode description, the conductor arrangement is characterized by a differential mode and a common mode:

- **Differential Mode (DM)**: The electric field (and hence the differential voltage, V_{DM}) points from one signal conductor to the other. The differential current, I_{DM}, flows in opposite directions in the two signal lines.
- **Common Mode (CM)**: The electric field (and hence the common-mode voltage, V_{CM}) points from both signal conductors to the ground conductor. The common-mode current, I_{CM}, flows in the same direction in both signal lines and returns through the reference conductor.

In nearly all electronic systems, the differential mode is the desired mode of operation. For example, it is used to supply DC or AC currents or to transmit a differential signal. The common mode propagating on the conductors is not intended for the functionality of the device; it is typically generated by parasitic couplings or imbalances in the system. Only in specialized topologies the common mode might be intentionally used as a design parameter.

We plot a differential and common mode current, I_{DM}, I_{CM} in Fig. 8.9, and the four modes of the electric field in Fig. 8.10. For the plots, we use a wire pair with a wire diameter of 1 mm, a center-to-center distance of 5 mm, and a height above the reference of 5 mm.

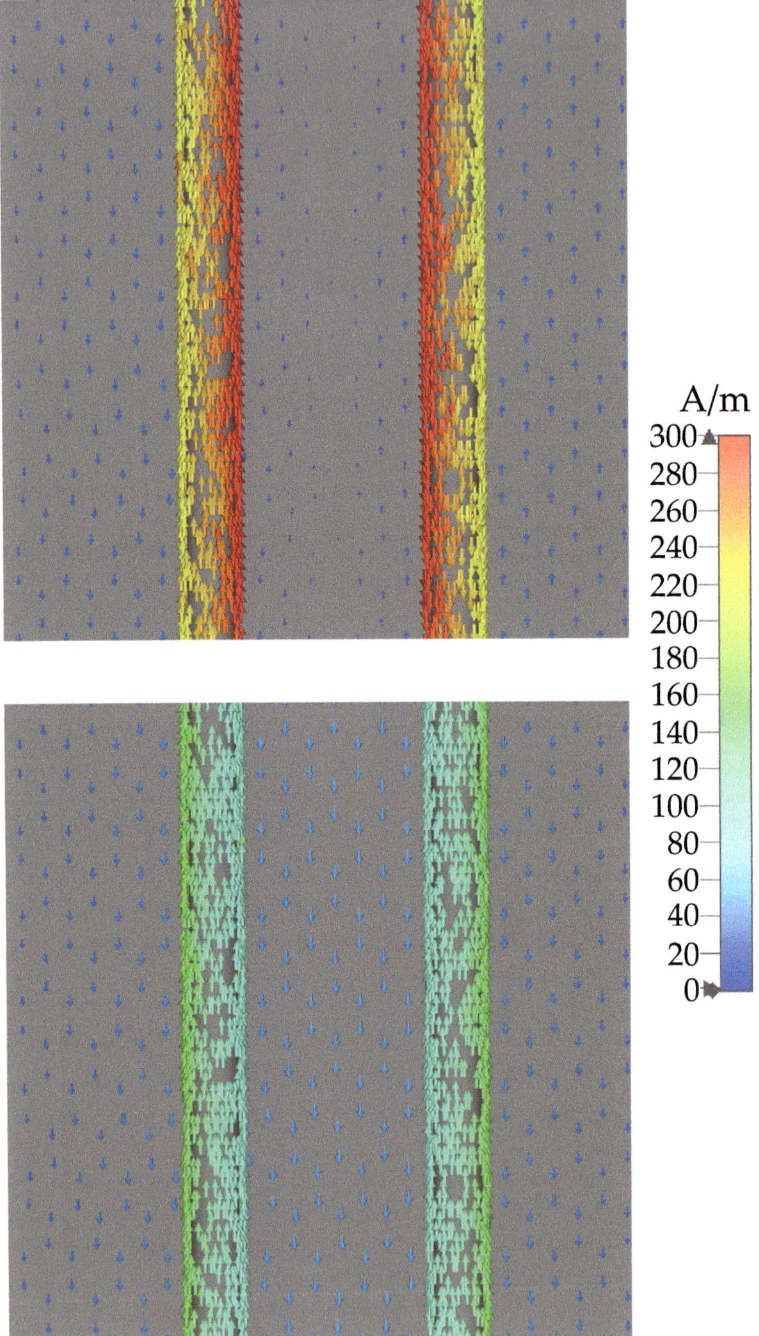

Fig. 8.9 Surface current in a three conductor setup for a CM and DM current of 1 A. Top: differential mode I_{DM}. Bottom: common mode I_{CM}

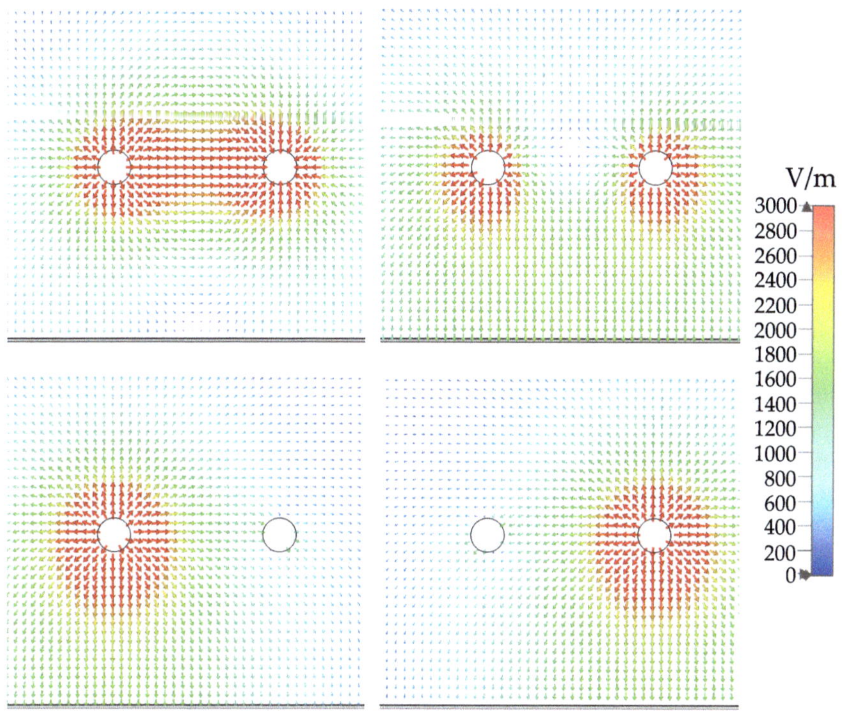

Fig. 8.10 Electric field modes in a three conductor arrangement. Top left: differential mode. Top right: common mode. Bottom left: single-ended mode for conductor 1. Bottom right: single-ended mode for conductor 2. An input power of 1 W is used at the waveguide port

The mixed-mode description is particularly helpful when designing countermeasures against emissions. Different filter topologies are available for differential and common modes, as discussed in Chap. 9.

A single-ended voltage and current can be converted into a mixed-mode voltage and current and vice versa using a circuit element that satisfies the following equations:

$$V_{\mathrm{DM}} = V_{\mathrm{P}} - V_{\mathrm{N}} \tag{8.3a}$$

$$V_{\mathrm{CM}} = \frac{1}{2}(V_{\mathrm{P}} + V_{\mathrm{N}}) \tag{8.3b}$$

$$I_{\mathrm{DM}} = \frac{1}{2}(I_{\mathrm{P}} - I_{\mathrm{N}}) \tag{8.3c}$$

$$I_{\mathrm{CM}} = I_{\mathrm{P}} + I_{\mathrm{N}}. \tag{8.3d}$$

This conversion can, for example, be accomplished by a 4×4 S-parameter matrix as described in Bockelmanand Eisenstadt (1995). In CST, such a circuit element

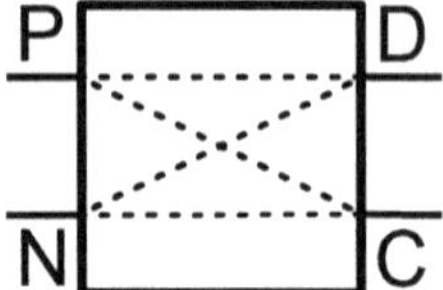

Fig. 8.11 Schematic element for the mode converter block

is referred to as a mode converter block. Mode converter blocks are not standard SPICE components. However, various circuit simulators provide this functionality under different names, such as a balun block (Fig. 8.11).

In this book, we use the circuit element shown in Fig. 8.11 symbol to denote a mode converter block.

The mode converter block does not include any assumptions about the impedances of the individual modes. When both signal conductors are terminated to the reference plane with a single-ended impedance of Z_{SE}, then the differential and common mode terminations correspond to

$$Z_{DM} = 2 \cdot Z_{SE} \tag{8.4a}$$

$$Z_{CM} = \frac{1}{2} Z_{SE}. \tag{8.4b}$$

These equations are the result of a straightforward derivation. For the differential mode, the single-ended terminations are connected in series, while for the common mode, the single-ended terminations are connected in parallel.

When the signal traces are tightly coupled to the reference plane (e.g., by a small spacing) and loosely coupled to each other (e.g., by a large spacing), Eq. (8.4a) and 8.4b also hold true for the characteristic impedances $Z_{0,DM}$ and $Z_{0,CM}$ of the differential and common modes, respectively. Only in this case it is possible to achieve a matched termination for the common and differential mode simultaneously. When the signal lines are tightly coupled, a simultaneous matching of the differential and common mode cannot be achieved, as we demonstrate in the next section.

To illustrate this effect, we calculate the characteristic impedances in the high-frequency limit using the MTL solver. Two separation distances between the signal lines are used: 1.5 mm and 10 mm. The distance of the pair to the reference layer is varied from 1 mm to 50 mm. The results are plotted in Fig. 8.12.

For a wire separation of 10 mm, shown on the left-hand side, Eqs. (8.4a) and (8.4b) accurately predict the characteristic differential and common mode impedances for a reference plane distance smaller than approximately 5 mm. This is indicated by the dashed and dotted lines, which are derived from the single-ended impedance using Eqs. (8.4a) and (8.4b). The solid lines represent results from the two-dimensional impedance calculation. This behavior aligns with coupling levels: when the coupling to the reference conductor is strong, and the coupling between the two wires is weak, Eqs. (8.4a) and (8.4b) accurately compute the characteristic impedances $Z_{0,DM}$ and $Z_{0,CM}$ from $Z_{0,SE}$.

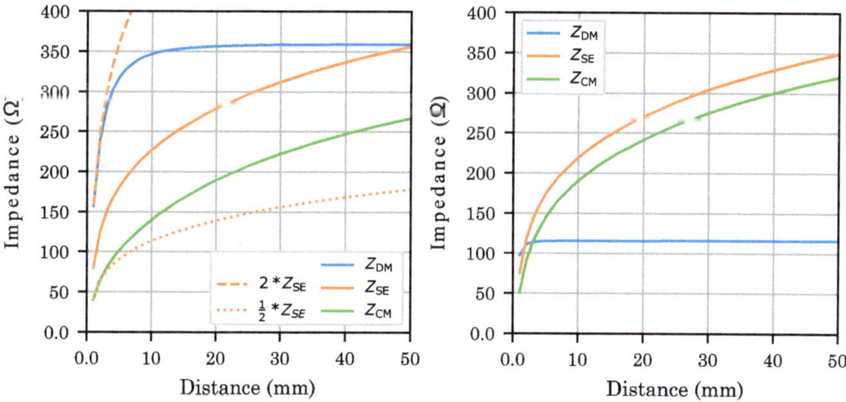

Fig. 8.12 Characteristic mode impedances as a function of the distance of the wire pair to the reference conductor. Left: wire pair distance of 10 mm. Right: wire pair distance of 1.5 mm

In cases where the coupling between the signal wires is strong, for the separation distance of 1.5 mm, as illustrated on the right-hand side of Fig. 8.12, the differential mode impedance is nearly unaffected by the distance to the reference plane. This is a desirable effect. Consider a differential pair transmitting a signal over a cable: the differential impedance of the pair should remain as constant as possible and should not be influenced by metallic objects near the cable. This is achieved by maintaining a short distance between the individual wires of the pair.

8.3.1 Emission from the Common and Differential Mode

After studying the characteristic impedance of the wire pair using the MTL solver, we now demonstrate the effect of the termination impedance on the fields generated by the wire pair. To calculate the fields, we switch from the MTL solver to the full-wave FEM solver. The wire pair, with an individual wire diameter of 1 mm and a wire separation of 1.5 mm, is placed above a reference plane located 50 mm below the wire pair. The reference plane is 1 m wide and 2 m long; the wire pair is located at the center of the plane. One end of the wire pair acts as the transmitter and the other as the receiver. The total length of the wire pair is 1 m.

The wires are referenced in a single-ended configuration to the reference plane at both ends. A mode converter block is used to drive the transmitter with a differential mode voltage of $V_{TxD} = 1\,\text{V}$ and a common mode voltage of $V_{TxC} = 1\,\text{V}$. The excitation is performed sequentially:

Differential Mode Excitation When the differential mode is driven, the common mode at the transmitter side is terminated with Z_{CM}. On the receiver side, the terminations are Z_{CM} for the common mode and Z_{DM} for the differential mode.

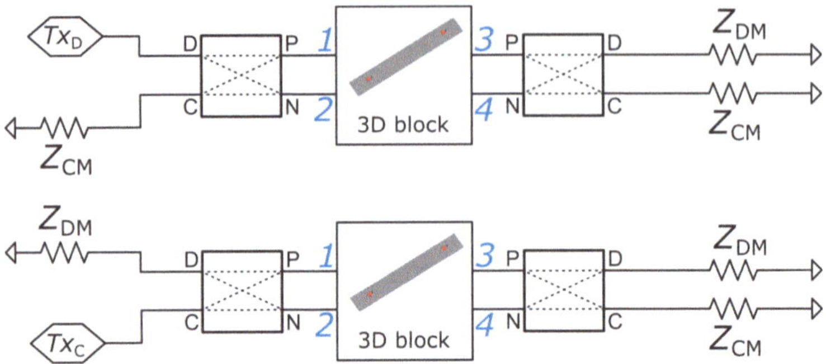

Fig. 8.13 Schematic for emission calculation from common and differential mode. Top: Differential mode excitation. Bottom: Common mode excitation

The schematic for the differential mode excitation is shown in the top section of Fig. 8.13.

Common Mode Excitation When the common mode is driven, the differential mode at the transmitter side is terminated with Z_{DM}, while the receiver-side terminations remain unchanged. The schematic for the common mode excitation is shown in the bottom section of Fig. 8.13.

The characteristic impedance of the differential mode of the wire pair is approximately $Z_{0,DM} = 120\,\Omega$, while the characteristic impedance of the common mode is $Z_{0,CM} = 310\,\Omega$ (right-hand side of Fig. 8.12).

As the signal lines are tightly coupled, it is not possible to achieve a matched termination for both the differential and common mode simultaneously.

Therefore, we simulate two termination scenarios:

- Scenario 1: The differential mode is matched by terminating with $Z_{DM} = 120\,\Omega$ and $Z_{CM} = 30\,\Omega$, corresponding to a single-ended termination of $Z_{SE} = 60\,\Omega$.
- Scenario 2: The common mode is matched by terminating with $Z_{CM} = 300\,\Omega$. This results in $Z_{DM} = 1200\,\Omega$ and $Z_{SE} = 600\,\Omega$.

The EM simulation is performed using the full-wave FEM solver. To measure the emission, we calculate the worst-case electric field at a distance of 3 m using a cylindrical probe arrangement, as described in Sect. 6.5. The frequency sweep employs a convergence criterion based on the field values at the probes. This criterion is satisfied after 51 frequency samples for a maximum frequency of 500 MHz. The total simulation time is approximately 13 minutes.

First, we discuss the effect of the termination impedance. The time-domain signals at the receiver, calculated in the coupled schematic simulation, are plotted on the left-hand side of Fig. 8.14. The wire pair is driven by a voltage source with a source impedance of $R_S = 1\,\mathrm{m}\Omega$, a signal frequency of 100 MHz, and a rise/fall time of 1 ns. To better understand the reflections, only a single rectangular pulse

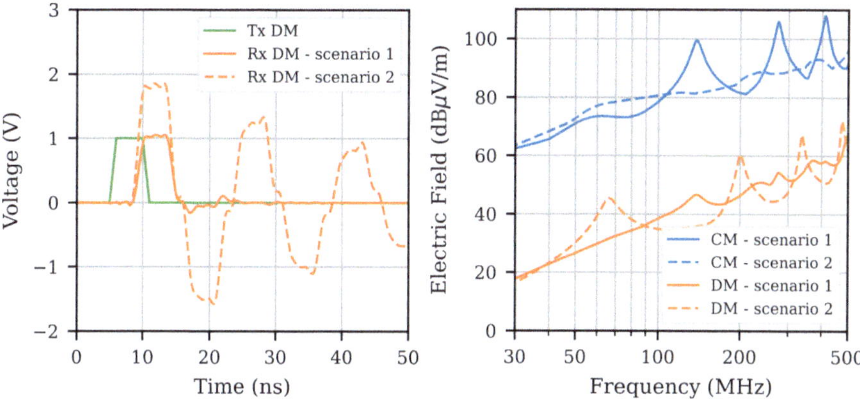

Fig. 8.14 Left: Time domain voltages at the transmitter (Tx) and receiver (Rx). Right: Worst case emission from common and differential mode

is excited. The voltage at the transmitter is measured across the Z_{DM} resistor in Fig. 8.13.

In scenario 1, we observe minor reflections, whereas in scenario 2, the reflected signal travels multiple times along the trace, as shown by the dashed orange line. The termination in scenario 2 leads to severe signal integrity issues and would likely result in functional failure of the system with these signal parameters. Nonetheless, analyzing this scenario provides valuable insights into the radiated emissions from both setups.

The worst-case electric field is plotted on the right-hand side of Fig. 8.14. We observe that when the termination of scenario 1 is applied, the emission from the differential mode does not exhibit any peaks. However, for the same termination, the common mode emission shows three peaks. For the termination used in scenario 2, the situation is reversed: the common mode is almost flat, while the differential mode exhibits four peaks. This demonstrates that for tightly coupled signal lines, it is not possible to achieve a matched termination for both the common and differential modes simultaneously.

For the same driving voltage, the emission from the common mode is approximately 30 dB higher than that from the differential mode. Even when the common mode is matched, the peaks observed when the system is driven in differential mode remain well below the emission levels observed when the common mode is driven. This is a well-known effect, and the intention of this example is to illustrate it clearly.

As the wire pair in this example is balanced, no differential-to-common mode conversion occurs. However, in electronic systems, it is not possible to route perfectly balanced signals. Any imbalance results in differential-to-common mode conversion, increasing the emission compared to the balanced case, even when the system is driven in differential mode.

This topic is explored in the next section.

8.3.2 Generation of Mode Conversion by Imbalance

The significant difference between the emissions of the common mode and the differential mode is a critical aspect of EMI design. Since the differential mode is necessary for functional operation, while the common mode is the primary contributor to radiated emissions, controlling the generation of common mode in electronic systems is essential. The conversion from differential mode to common mode is caused by imbalance in the system. These imbalance can arise from various sources:

Driver Imbalance The driving differential signal is not perfectly symmetric, leading to a common mode at times when the voltages are not exactly opposite. To simulate driver imbalance, a transient circuit simulation is required. In this case, mode converters cannot be used at the driver side. Driver imbalance can be studied by using two single-ended driving signals with a mismatch in polarity or a specific time shift between them.

Termination Imbalance The termination of the transmission line is not symmetric, meaning the single-ended termination impedances are unequal. Termination imbalance can be analyzed by using two single-ended termination resistors with (slightly) different values. Mode converters cannot be used at the termination.

Geometrical Imbalance The geometrical length or the single-ended characteristic impedance of the individual conductors is different. Unlike the previous cases, mode converters can still be used in the circuit simulation to study geometrical imbalance.

The MTL method is an efficient tool for studying geometrical imbalance in cables. Next, we calculate the common mode voltage resulting from the imbalance in the wire pair.

For this calculation, we use the same schematic as in the previous case (Fig. 8.13). The differential mode port denoted by Tx_{D} is excited, and the common mode port is terminated with Z_{CM}. The same termination impedances as in scenario 1 of the previous example are used: $Z_{\mathrm{DM}} = 120\,\Omega$ and $Z_{\mathrm{CM}} = 30\,\Omega$. The common mode voltage is measured at the receiver side across the resistor represented by Z_{CM}.

In the first simulation, the wire pair, with a center-to-center distance of 1.5 mm, is placed 50 mm above the reference plane. The pair is rotated with respect to the reference plane, as shown on the left-hand side of Fig. 8.15. At $\theta = 0°$, the setup is perfectly balanced, as the individual wires are symmetric relative to the reference plane. As θ increases, imbalance is introduced, and the common mode voltage increases, as indicated on the right-hand side of Fig. 8.15.

The results demonstrate the significant impact of imbalance: at an angle of $\theta = 1°$, the common mode voltage is already 100 dB higher than in the perfectly balanced case. While this increase seems large, it is important to note that in the perfectly balanced case, the common mode voltage is zero. From $\theta = 1°$ to $\theta = 10°$, the common mode voltage increases by an additional 20 dB.

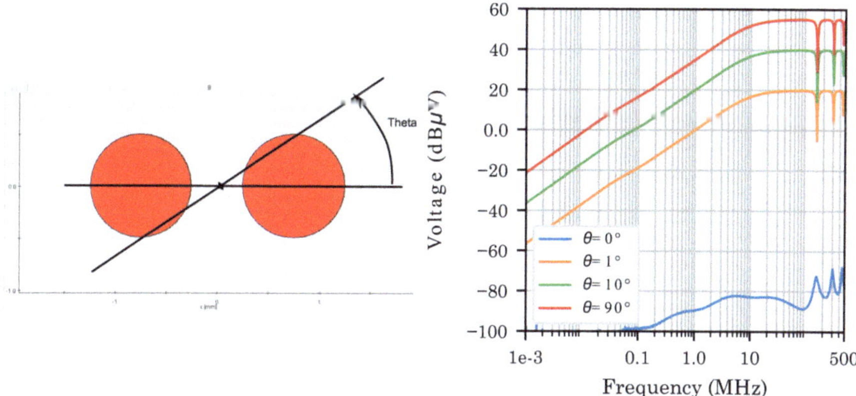

Fig. 8.15 Left: Sketch of the rotation of a wire pair with respect to the reference plane. Right: Common mode voltage due to imbalance in a wire pair rotated with respect to the reference plane

As a physical wire pair is never perfectly symmetric, we recommend introducing some imbalance into simulation models to obtain realistic results.

To obtain the results in Fig. 8.15, we used a very fine two-dimensional mesh for the MTL solver. The solver requires a perfectly symmetric mesh to accurately represent the perfect geometrical symmetry in the solution.

When using a three-dimensional full-wave solvers for the same model, maintaining perfect symmetry is more challenging, especially with the FEM solver. The generation of the tetrahedral three-dimensional mesh often introduces slight asymmetries. Achieving a symmetric mesh is somewhat easier with the FIT method, as it uses a regular hexahedral mesh. However, the cell lumping step in the TLM octree mesh may also result in minor asymmetries.

The effect of imbalance can also be demonstrated using a Twisted Wire Pair (TWP). Rotating the TWP with respect to the reference plane does not introduce imbalance, as the coupling of the TWP to the reference plane remains symmetric over one twist period. To introduce imbalance, we offset the twist center of the pair. With this offset, the individual wires are no longer rotated around their centers of gravity but around a point displaced from this center.

The offset of the twist center is applied to the same 1 mm thick wires, which are coated with a dielectric of $\varepsilon_r = 2.3$. The resulting common mode voltage is plotted on the right-hand side of Fig. 8.16 for various offsets. Even a small offset of 0.01 mm causes the common mode voltage to increase by approximately 30 dB, highlighting the significance of imbalance.

Unlike the straight wire case, the setup without a twist center offset is not perfectly balanced. This is due to how the MTL solver handles twisted pairs. A single twist period is meshed using a fixed number of cross-sections rather than continuously, which introduces truncation in the longitudinal direction. This truncation results in a model that is not perfectly symmetric.

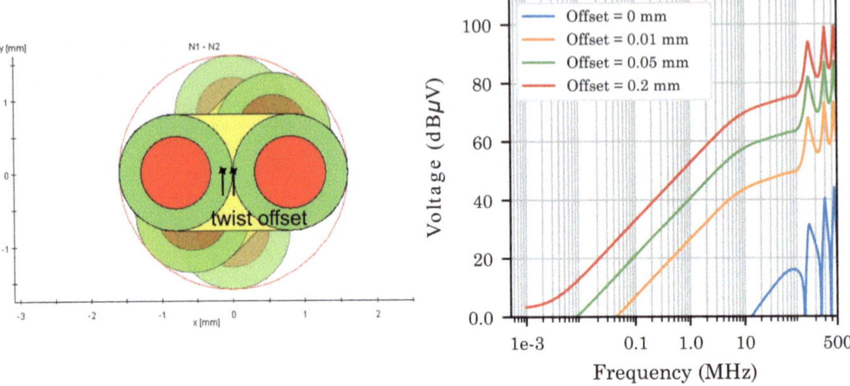

Fig. 8.16 Left: Sketch of the offset of the twist center in a TWP causing imbalance. Right: Common mode voltage due to imbalance in a twisted wire pair with offset of the twist center

Using the MTL method for calculating differential-to-common mode conversion is highly efficient. The simulation of all four parameter sets in the examples in this section requires only 30 seconds. However, the MTL simulation calculates only voltages and currents. A practical approach to simplify the emission calculation for cable bundles with a large number of conductors is to calculate the common mode amplitude in a first simulation and then model the entire bundle as a solid conductor excited with this common mode current. In such a setup, the emission from the differential mode is neglected. However, given the difference in emission levels, as seen in Fig. 8.14, this is a valid approximation. The hybridization of the MTL solver with the FEM solver employs this approximation (Sect. 8.6).

Geometrical imbalance in cables is not the only source of differential-to-common mode conversion in an electronic system. The routing of signals on a PCB and inside connectors are additional sources of imbalance that affect the mode conversion. To achieve realistic simulation results, these factors must be included in the simulation model.

For connectors, obtaining accurate CAD data can be challenging. While most connector manufacturers provide three-dimensional models, these are often placeholders and do not include detailed information about signal routing inside the connector to protect the manufacturer's intellectual property.

For power electronics simulations, the effect of the connector can often be neglected if the electrical length of the connector is short compared to the maximum frequency setting in the simulation. In such cases, simplified connector models are sufficient. However, for fast digital signals, an accurate model of the mode conversion within the connector is typically required. We recommend contacting your connector vendor to discuss the availability of detailed data.

8.4 Application Example: Cable Crosstalk in the Wing of an Aircraft

Ideal voltage source modeling a power transmission

Termination resistors at ends of cables

MTL model, vector fitted MTL model

MTL solver

AC, Transient

0–50 MHz

In this section, we simulate the crosstalk between cables routed within the wing of a large aircraft. This example demonstrates the application of the MTL for complex geometries and explores the limitations of RLC cascades for electrically very long cables.

An overview of the geometry is shown in Fig. 8.17. The CAD geometry of the wing has been created using CAD primitives. The wing is 30 m long and includes one engine connected to it. Inside the wing, three cables are routed. As the wing is not straight and its cross-section changes along its length, manually routing the cables within the simulation tool is challenging. Therefore, the routing of the cables was performed using CATIA Electrical, a specialized cable routing CAD tool. Besides CATIA Electrical, other cable harness routing tools are available on the market, such as Mentor Capital Harness or Zuken E3.

Unfortunately, exchange formats for cable routing are not as advanced as those for PCB layouts. The KBL format is the most widely used format for cable harness data. However, importing cable routing data from the KBL format into CST requires manual definition of certain harness properties, as the KBL format does not include detailed information about individual cable types. For example, the lay of a TWP or the screen properties of shielded cables must be defined manually in the simulation tool.

In the simulation model, we calculate the crosstalk from a power cable routed from the engine to the wing root into two signal cables routed from the wing root to the wing tip. The top part of Fig. 8.17 shows the routing through the entire wing, with cable cross-sections not to scale. The bottom part of Fig. 8.17 provides a zoomed-in view of the wing root, depicting the individual cables to scale.

We designate the TWP routed below the opening to the engine as TWP 1 and the pair routed above the opening as TWP 2. TWP 1 is 34.6 m long, while TWP 2 is 33.0 m long. The cross-section of both pairs is identical, consisting of the same TWP as used in the previous section, but with a lay of 5 cm.

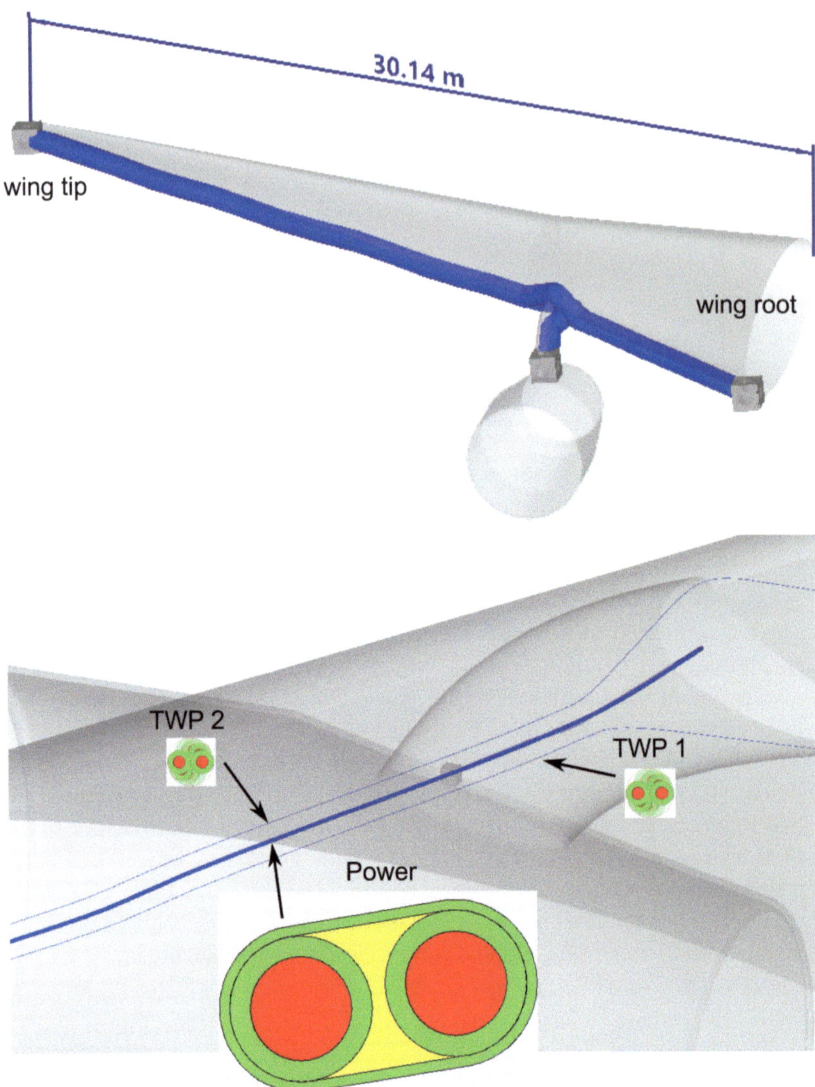

Fig. 8.17 Overview of the simulation model of cable crosstalk in the wing of an aircraft. Top: View of the cable lay in the complete wing. Bottom: Zoom in at the section lose the wing root

The power cable located between the two TWP pairs is 11.8 m long. It consists of two conductors separated by insulation and surrounded by an additional outer insulation layer. Each conductor has a diameter of 10 mm with an insulation layer made of polyethylene ($\varepsilon_r = 2.3$) that is 2 mm thick. The outer insulation layer, also made of polyethylene, has a thickness of 1 mm.

8.4.1 MTL Model and AC Simulations

With a far more complex geometry than in the previous examples, analyzing the details of the MTL model provides valuable insights into how the model operates. The discretization calculated by the MTL method for this example results in 1528 cross-sections along the twisted wire. An overview of the model is shown in Fig. 8.18.

In the top section of Fig. 8.18, the region near the connection between the wing and the engine is depicted. The positions where the cross-sections are calculated are indicated by icons composed of several gray cylinders. For the power cable, the

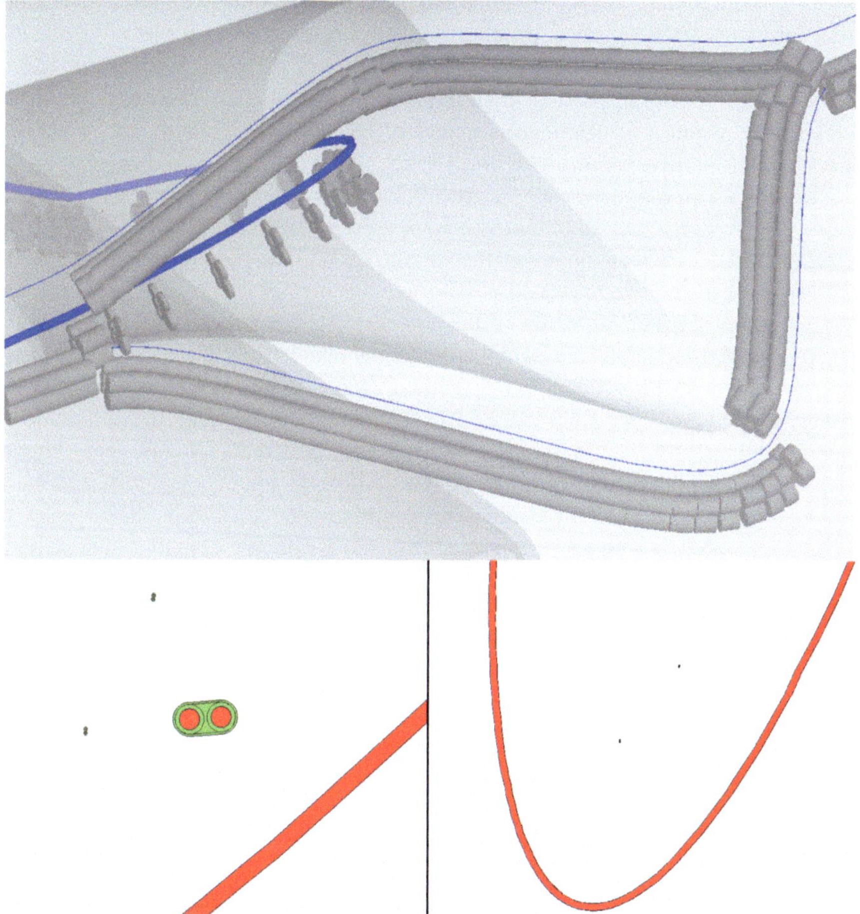

Fig. 8.18 MTL model for the cables in aircraft wing. Top: Overview of the section close to the engine to wing connection. Bottom: Two selected cross-sections, close to the root of the wing shown on the left-hand side and close to the top of the wing on the right-hand side

cross-section spacing is relatively coarse, whereas for the two twisted wires, the spacing is significantly denser.

The close spacing of the cross-sections is required to accurately resolve the lay of a TWP. Using the standard accuracy setting, where each twist of the TWP is resolved by four cross-sections, a large number of cross-sections are generated due to the cable's length. For TWP 1, with a cable length of 34.6 m and a twist length of 5 cm, resolving four cross-sections per twist results in 2768 cross-sections. However, due to the symmetry of the twists, the actual number is reduced to 1528. Even with this reduction, the RLCG model generated by the MTL method poses challenges for circuit simulators, as discussed below.

In the bottom part of Fig. 8.18, two exemplary cross-sections are plotted. On the left-hand side, a cross-section near the root of the wing is shown, zoomed in to the position of the cables. The three individual cables can be seen positioned above the wing (thick red line), which is used as the reference conductor in the MTL model. On the right-hand side, a cross-section near the tip of the wing is depicted. Here, the two twisted wires appear as small dots, while the rounded structure represents the metallic wing. The calculation time for the MTL model with the presented discretization settings is 3 minutes and 49 seconds.

The RLCG model generated by the MTL solver can be exported as a SPICE netlist for use in an external circuit simulator, in case we like to run it in an external circuit simulator instead of using the circuit simulation provided in CST. With 1528 cross-sections, the generated SPICE netlist is 37 MB in size and contains over 400,000 elements. The large size of the netlist is due not only to the large number of cross-sections and conductors in the model but also to the inclusion of dispersive material behavior. When frequency-dependent effects, such as inductance variations and dielectric losses, are included, RLCG ladder models are used. In such models, each mode in a cross-section requires multiple RLCG elements, rather than a single RLCG set. This significantly increases the number of circuit elements, adding a heavy computational load to the circuit simulation.

As a first step, we simulate this model in the frequency domain using an AC analysis. This approach is computationally less demanding compared to a transient analysis.

In the AC simulation, we study the crosstalk from the power cable into the two TWPs using a 1 V source at the engine position. In the MTL model, single-ended ports are defined to the reference at every conductor at all cable ends. This approach is valid because the wing is entirely metallic, and thus the reference is continuous. The single-ended ports are connected to mode converters in the schematic, as shown in Fig. 8.19.

Both TWPs are terminated with $Z_{\mathrm{DM}} = 100\,\Omega$ and $Z_{\mathrm{CM}} = 25\,\Omega$ at their ends. The driving voltage at the engine end of the power cable is differential, while at the root of the wing, the power cable is terminated with $Z_{\mathrm{DM}} = 1\,\Omega$ and $Z_{\mathrm{CM}} = 0.25\,\Omega$. This termination models a high current flow through the power cable. At all terminations, we measure the current and the voltage across the resistors. To avoid excessive clutter in the schematic, the current and voltage probes are only shown as examples for TWP 1 at the tip of thc wing.

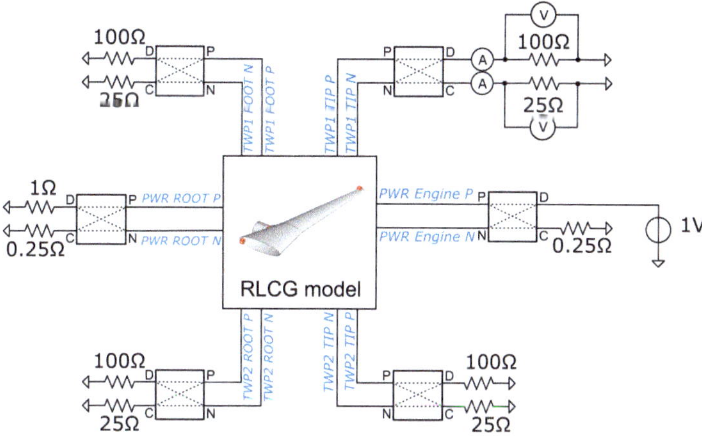

Fig. 8.19 Schematic for the AC analysis of cable crosstalk in the wing of an aircraft

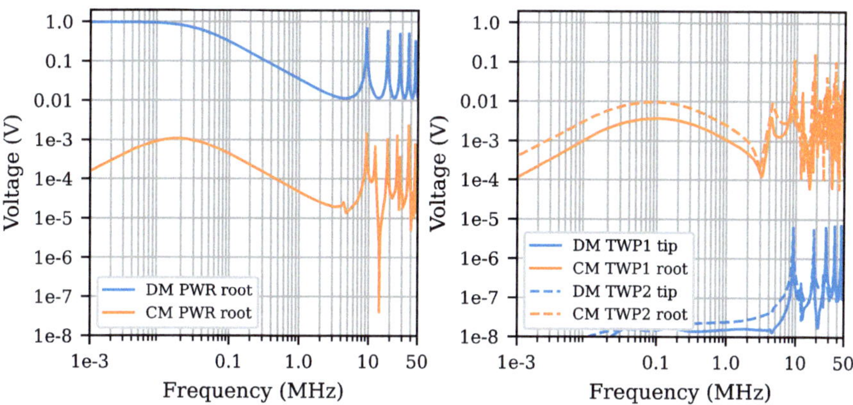

Fig. 8.20 Common and differential mode voltages in cables routed in the wing of an aircraft calculated using an AC analysis. Left: Voltage in the power cable at the root of the wing. Right: Voltage in the TWPs

The AC simulation is performed using the circuit simulator in CST. For a frequency range from 1 kHz to 50 MHz with 5001 frequency samples, the simulation takes 1 minute and 3 seconds.

The results of the AC simulation are plotted in Fig. 8.20. On the left-hand side, the common and differential mode voltages in the power cable at the root of the wing are depicted. Up to 10 kHz, the excitation voltage of 1 V is observed at this position. However, with increasing frequency, resonant behavior due to the cable length begins to appear.

Due to imbalance in the power cable, a common mode voltage is also present. The common mode voltage is approximately 1000 times lower than the differential mode voltage, indicating that the differential-to-common mode conversion in the power cable is −60 dB.

On the right-hand side of Fig. 8.20, the voltages in the TWPs resulting from crosstalk are shown. For both twisted wires, the common mode crosstalk is four orders of magnitude higher than the differential mode crosstalk. Below 3 MHz, the crosstalk into TWP 2 is higher than into TWP 1 because TWP 2 is routed slightly closer to the power wire. Above 5 MHz, resonances appear, making it difficult to evaluate the crosstalk levels based solely on the AC results.

In certain frequency bands, the crosstalk into TWP 1 exceeds that into TWP 2, while in other bands the opposite is true. To better assess the crosstalk and its impact on the signals routed through the TWP, a transient simulation is required.

8.4.2 Transient Simulation

Due to the large number (400k) of RLCG elements generated by the MTL model, performing a transient circuit simulation using the RLCG model is highly challenging. Recall from Sect. 3.8.1 that a nonlinear transient circuit simulation is significantly more computationally demanding than an AC simulation. In this example, the transient circuit simulation in CST becomes unstable after approximately 50 μs of simulation time, while in another circuit simulator, the simulation fails due to floating nodes. This behavior is not surprising given the complexity and size of the RLCG model. To successfully simulate this model in a transient circuit simulation, it is necessary to reduce the complexity while maintaining accuracy.

This complexity reduction can be achieved using vector fitting. As discussed in previous sections, vector fitting is a method that converts S-parameters into a representation suitable for time-domain simulations. In this example, vector fitting is applied to simplify the RLCG model.

In the first step, the RLCG model is simulated in the frequency domain, and S-parameters are calculated for all ports. These S-parameters are then vector fitted, and the transient circuit simulation is performed on the fitted model. This process is fully automated in CST, but it can also be performed manually by exporting the S-parameters, applying vector fitting, and exporting the fitted results into a SPICE netlist.

We perform the manual steps to illustrate the extent of complexity reduction achieved through vector fitting. The resulting netlist after fitting is 900 kB in size and contains 20k lines, reducing the complexity of the circuit model by a factor of 20. The details of the vector fitting process are not discussed here, as they are covered in Sect. 10.3.

Using the vector-fitted model of the cables in the wing, we perform a transient simulation to calculate the crosstalk from the power wires into the differential mode of the TWPs. The power wire is driven by a generator located in the engine. Each phase of the generator consists of an inductor with $L = 200\,\mu\text{H}$, driven by a 120° phase-shifted voltage of 110 V at 400 Hz.

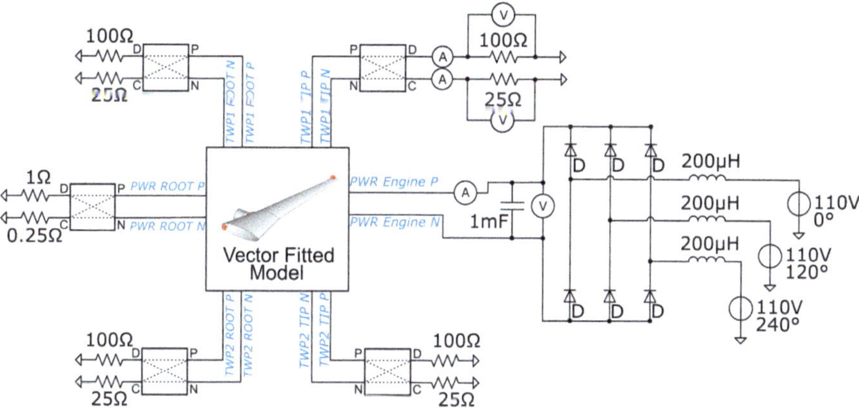

Fig. 8.21 Schematic for the transient analysis of cable crosstalk in the wing of an aircraft

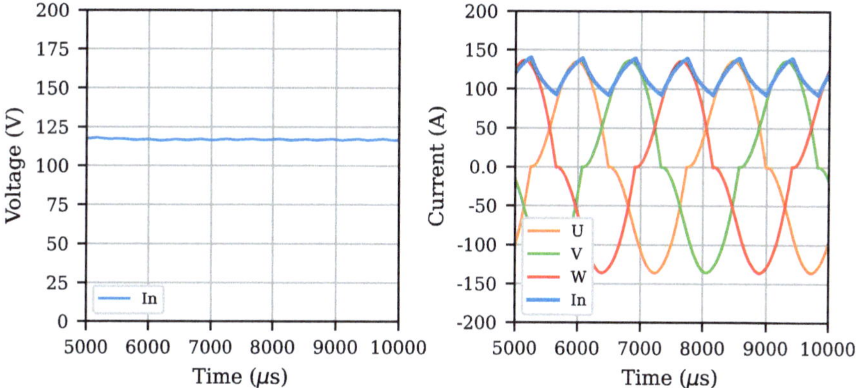

Fig. 8.22 Input voltage and current for the simulation of the crosstalk from a power generator. Left: Input voltage. Right: Input current (blue) and current at the three generator phases U, V, W

The AC voltage is rectified using a diode rectifier. For the diodes, we selected STPS200170TV1 from ST Microelectronics and downloaded the SPICE models from the vendor's website. These diodes are rated for 170 V and 200 A. The output voltage of the diode bridge is smoothed using a 1 mF capacitor and then connected to the engine position in the cable model. The schematic is shown in Fig. 8.21.

The transient circuit simulation, using the vector-fitted model, takes only 30 seconds to simulate a signal duration of 10 ms.

The voltage and current at the engine side of the power cable are plotted in Fig. 8.22. The waveforms are shown after 5 ms of simulation time, once the signals have reached steady-state behavior. The voltage is approximately 120 V with small ripple, while the input current exhibits significant ripple, around 40 A. This ripple highlights the need for a larger smoothing capacitor; however, the current waveform is used here for demonstration purposes.

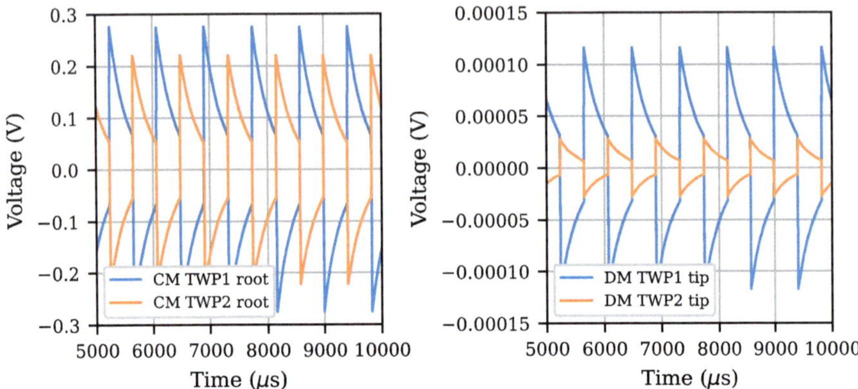

Fig. 8.23 Transient crosstalk into the twisted wire pairs. Left: Common mode voltage. Right: Differential mode voltage

On the right-hand side of Fig. 8.22, the distortion of the currents in the three input phases is shown. The distortion is caused by the inductance values of the generator which are assumed to be 200 μH as indicated in Fig. 8.21.

The transient crosstalk in the twisted wire pairs is plotted in Fig. 8.23. As anticipated from the AC simulation, the common mode crosstalk is significantly larger than the differential mode crosstalk. The peak-to-peak amplitude of the common mode is 0.6 V, while it is only 0.2 mV for the differential mode.

In the transient simulation, the common mode voltage is higher for TWP 1, whereas in the AC results, the crosstalk into TWP 2 is higher. To verify if this discrepancy is an artifact of the vector fitting, we test the fitted model: When exciting the engine side of the power cable with a 1 kHz sine wave, the common mode crosstalk into TWP 2 is higher by a factor of three compared to TWP 1. This behavior matches the AC simulation results, confirming that the fitted model is accurate. The difference in crosstalk levels in the generator case is attributed to the fast rise time of the signals in the power wire.

This example demonstrates the application of the MTL method for complex cable geometries. For such geometrically complex models, the resulting RLCG models are extremely large and cannot be directly simulated in a transient circuit simulation. Vector fitting is therefore essential for enabling transient simulations of such models.

8.5 Shielded Cables

Domains: 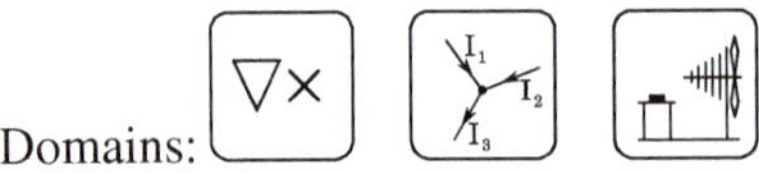

Cable shields are widely used in electronic systems to control both the emission and susceptibility of cables. While shields primarily serve an electrical purpose, they also influence mechanical properties such as bend radius and weight. Thick, solid shields are typically impractical; hence, cable shields are often composed of braided thin wires. This braiding introduces distinct electrical characteristics that must be accurately represented in simulations.

However, resolving the detailed structure of a braided shield over the length of a cable in a full-wave simulation results in an excessively high mesh count, making it computationally highly inefficient. Therefore, a method is needed to incorporate the realistic properties of cable shields into simulation models without excessively increasing computational complexity.

In Sect. 7.3, we simulate the effect of a cable shield and demonstrate how the connection impedance of the shield to the housing influences the shielding effectiveness against magnetic fields. In that section, a thin panel material is used to model a thin, solid shield. Braided shields and combinations of different shields require more advanced modeling than we have presented up to this point. In this section, we discuss the properties of cable shields in more detail. Additionally, we provide a detailed explanation of the modeling of coupling mechanisms through the shield.

8.5.1 *Application Example: Crosstalk into a Shielded Wire*

Ideal broadband current source
Termination resistors from wire to reference
MTL model
MTL solver
AC
0–100 MHz

The performance of a cable shield depends not only on the shield itself but also on how it is connected to other conductors, such as the enclosure discussed in Sect. 7.3. Before discussing the details of cable shield modeling, we study the crosstalk from a single wire into a shielded wire to highlight the dependence on various parameters.

The geometry for this study is illustrated in Fig. 8.24. Two wires are positioned above a conductive table, which serves as the reference plane. A single wire, referred to as the generator wire, has a diameter of 5 mm and is placed 1 cm above the reference plane. A shielded wire is located 5 cm away from the generator wire.

The shielded wire has the same dimensions as those described in Sect. 7.3, with an inner wire diameter of 10 mm and a shield diameter of 14 mm. The shield is

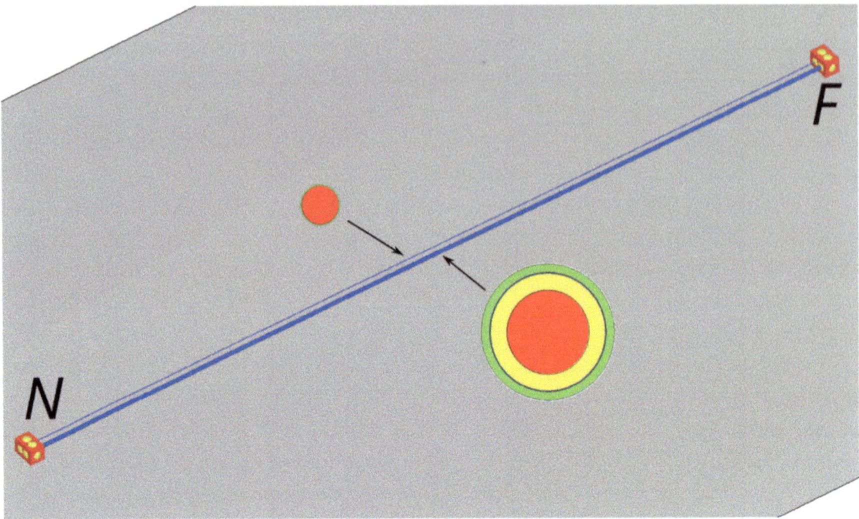

Fig. 8.24 Example of crosstalk into a shielded wire

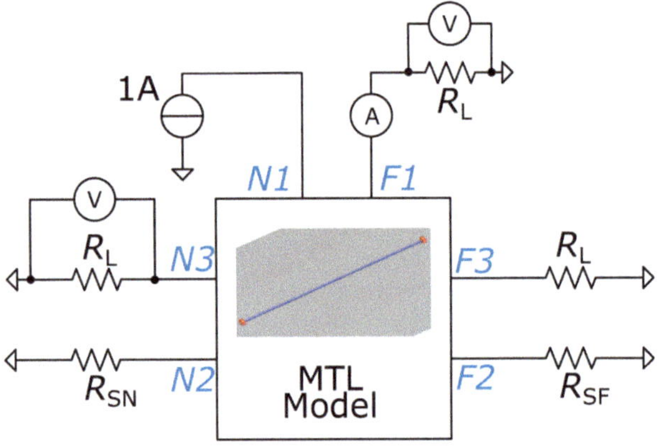

Fig. 8.25 Schematic for the simulation of crosstalk into a shielded wire

modeled as a solid aluminum layer with a thickness of 0.1 mm. Both wires are 1 m in length.

The position marked "N" indicates where the generator wire is excited, and the Near-End Cross Talk (NEXT) is calculated at this location. The position marked "F" indicates where the generator wire is terminated, and the Far-End Cross Talk (FEXT) is calculated there. In the MTL model, cable ports are defined for all three conductors (generator wire, inner conductor of the shielded wire, and the shield) to the reference plane.

The schematic for calculating crosstalk is shown in Fig. 8.25. The generator wire is driven by a 1 A current source, with its other end terminated by a resistor R_L to the reference plane. The shielded wire is similarly terminated to the reference at both ends using resistors of the same value. The shield is connected to the reference plane via R_{SN} at the near end and R_{SF} at the far end.

At the termination of the generator wire, we measure the current and voltage relative to the reference. The Near-End Cross Talk (NEXT) is determined by probing the voltage across the termination resistor of the shielded wire.

For R_L, we select two values to emphasize different coupling mechanisms:

- $R_L = 50\,\Omega$: This setting accentuates inductive coupling.
- $R_L = 1\,k\Omega$: This setting accentuates capacitive coupling.

It is important to note that the MTL model generated by CST inherently accounts for capacitive, inductive, and galvanic coupling simultaneously. This contrasts with analytical solutions, where the different coupling mechanisms can be isolated and studied separately.

The current and voltage at the termination of the generator wire, for both termination resistors, are plotted in Fig. 8.26. Up to a frequency of 1 MHz, the current remains at 1 A, matching the source current. Above 1 MHz, the behavior diverges depending on the termination resistance. For a termination resistance of $50\,\Omega$, the current increases, reaching 2.5 A at 70 MHz. In contrast, for a termination resistance of $1\,k\Omega$, the current decreases to 0.14 A at 70 MHz. This behavior indicates that the generator wire transitions from being electrically short to electrically long, causing changes in the electric and magnetic field coupling along the coupling distance.

The voltage at the termination of the generator wire is shown on the right-hand side of Fig. 8.26. For a termination resistance of $50\,\Omega$, the low-frequency voltage is 50 V, while for $1\,k\Omega$, it is 1000 V. At higher frequencies, the termination voltage begins to vary.

These results demonstrate that the choice of termination resistance for the generator wire has a substantial impact on the fields generated by the wire.

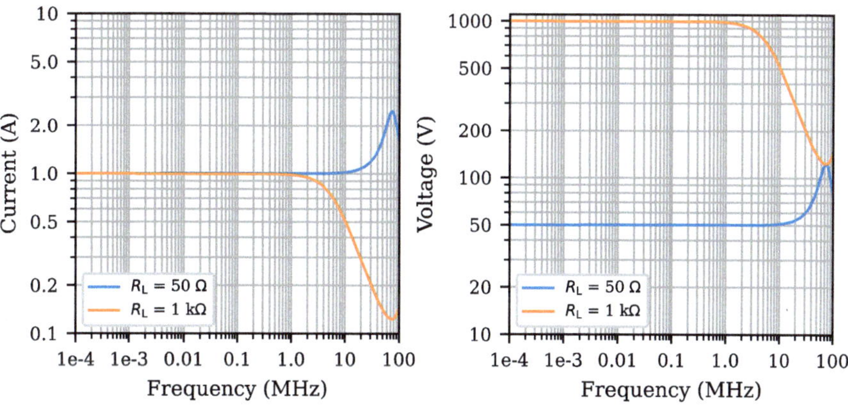

Fig. 8.26 Current and voltage at the termination of the generator wire for the two different loading scenarios. Left: Current. Right: Voltage

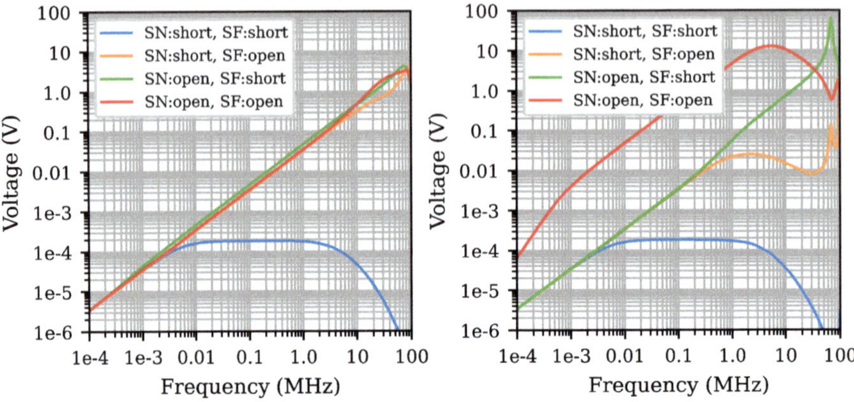

Fig. 8.27 Near-end crosstalk at the shielded wire for different terminations and shield connections. Left: Termination with 50 Ω. Right: Termination with 1k Ω

The near-end crosstalk for the two termination scenarios is plotted in Fig. 8.27. In both scenarios, we vary the connection resistance of the shield to the reference at the near and far ends of the shielded wire. For an open connection, we use $R_S = 1\,\text{M}\Omega$, and for a short connection, $R_S = 1\,\mu\Omega$.

The crosstalk voltage for the 50 Ω termination is shown on the left-hand side of the figure. For this termination resistance, the crosstalk is expected to be dominated by inductive coupling. To effectively shield against inductive coupling, both ends of the shield must be connected to the reference via low-ohmic resistors. The simulation results confirm this: only when both R_{SN} and R_{SF} are set to short, a shielding effect is observed (blue curve). Up to 10 kHz, the crosstalk increases with 20 dB/dec and then remains flat until 2 MHz. Beyond 2 MHz, the crosstalk decreases again, which can be attributed to the solid shield definition. The effect of shield thickness is analyzed later in this section.

For the 1 kΩ termination, the crosstalk behavior differs. When both shield resistors are open, the crosstalk is higher than in the 50 Ω termination scenario. In this case, the crosstalk is predominantly capacitive. Capacitive crosstalk is mitigated by connecting at least one end of the shield to the reference (Paul 2005). This is reflected in the reduced crosstalk for cases where either R_{SN} or R_{SF} is shorted to the reference. Above 1 MHz, the NEXT varies depending on which side of the shield is connected, because the cable is not electrically short. For the short on the near-end, the NEXT is reduced compared to the 50 Ω termination scenario.

When both ends of the shield are shorted to the reference, the crosstalk is identical for both termination scenarios, indicating effective shielding against both inductive and capacitive coupling.

Using the simulation model, we can conveniently vary the cable length and quantify its effect on crosstalk. The variation of the crosstalk for different cable lengths is plotted on the left-hand side of Fig. 8.28. These results are obtained with both ends of the shield shorted to the reference. The length of both cables is swept

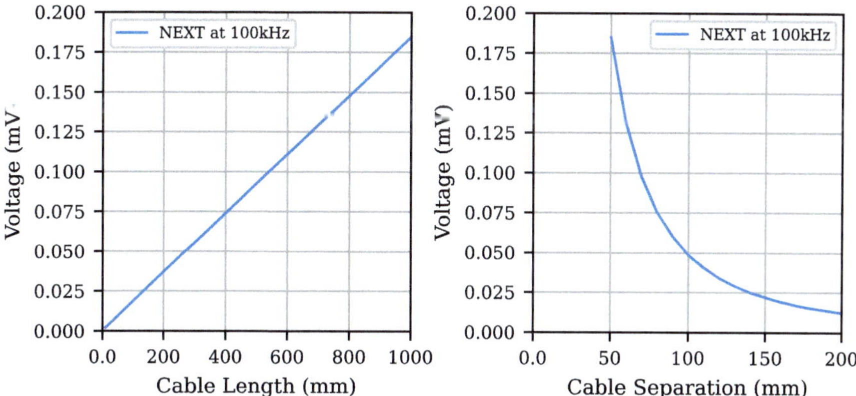

Fig. 8.28 Near-end crosstalk for different wire lengths and separation distances. The shield is shorted to the reference at both end. Left: Dependence on the cable length. Right: Dependence on cable separation

from 10 mm to 1000 mm in 50 mm increments, and the NEXT at 100 kHz is plotted. The results indicate a linear dependence of crosstalk on cable length as long as the cables remain electrically short. For electrically long cables, resonance effects cause shifts in the behavior, and the relationship between cable length and crosstalk becomes more complex.

A similar study is conducted for the separation distance between the generator and the shielded wire. The original cable length of 1 m is fixed and the distance s is varied from 50 mm to 200 mm in steps of 10 mm. The resulting NEXT at 100 kHz is plotted on the right-hand side of Fig. 8.28. The results reveal a $\frac{1}{s}$ dependence of the crosstalk, highlighting the inverse relationship between separation distance and crosstalk.

In another geometrical study, we examine the effect of the distance between the cables and the reference plane. Using a cable length of 1 m and a separation distance of 5 cm, as at the beginning of this section, the NEXT is calculated at 100 kHz for both shield ends shorted to the reference. The results of this study are plotted on the left-hand side of Fig. 8.29. The results show that as the distance between the wires and the reference plane increases, the crosstalk also increases. This behavior can be attributed to the larger loop formed between the cables and the reference plane, which enhances the coupling and thus the crosstalk. The dependence on the distance is logarithmic.

Finally, we investigate the effect of the inductance of the connection between the cable shield and the reference. As discussed at the end of Sect. 8.1.5, the modal description provided by the MTL does not inherently account for the physical properties of the terminations. Up to this point, the connection between the shield and the reference plane is modeled as purely ohmic.

To study the impact of connection inductance within an MTL model, we add inductors in series to the shield connections R_{SN} and R_{SF} in Fig. 8.25. The resulting

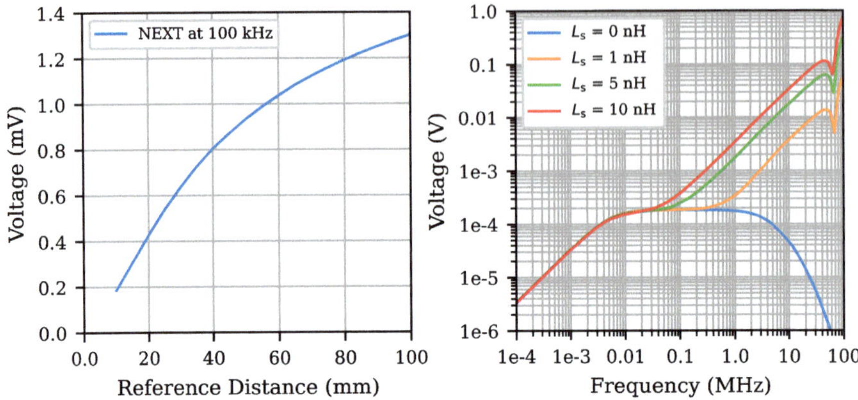

Fig. 8.29 Left: Dependence of the near-end crosstalk on the distance of both wires to the reference plane. Right: Effect of an inductance in series to the shield connections

crosstalk for the cable configuration from the beginning of this section is plotted on the right-hand side of Fig. 8.29. Results are shown for inductances of 1 nH, 5 nH, and 10 nH at each individual cable end, compared to a configuration without inductance. In all results, the resistors are set to a short.

The results reveal that introducing an inductance as small as 1 nH significantly increases the crosstalk above 1 MHz. For higher inductance values, the rise in crosstalk begins at even lower frequencies. This addition of inductance effectively models the impact of a pigtail on the shield connection.

The example in this section demonstrates that even for a relatively simple configuration, such as the crosstalk from a single wire into a shielded wire, the magnitude of the crosstalk is influenced by a complex interplay of factors. These include the source definition, sink termination, and geometrical properties. Simulation proves to be an invaluable tool for studying these effects, and the MTL method provides a highly efficient simulation methodology. The AC simulation for each parameter combination presented here requires approximately 2 seconds. In contrast, measuring such a wide range of parameters in a laboratory would demand a significant amount of time and effort.

8.5.2 Transfer Impedance

Cable shields made of braided wires or combinations of different shield types exhibit a more complex electrical behavior than the simple solid screens treated up to this point. To quantify the coupling through the shield of a cable, we require a formalism that isolates the properties of the shield itself from other aspects of the setup. As demonstrated in the preceding example, crosstalk depends on numerous factors such as termination, cable length, and cable separation. A method is needed to model the coupling through the shield independently of these external parameters.

In this section, we introduce the concept of transfer impedance, which provides a means to characterize the shield's properties without relying on a specific coupling scenario. This formalism allows for a standardized description of cable shielding effectiveness and allows comparisons of different shield designs.

We begin by analyzing how shield properties influence crosstalk and the dependence of shielding performance on shield characteristics. Next, we derive an expression for the shielding effectiveness of a cable shield in terms of its transfer impedance. Finally, we introduce Kley's formula Kley (1993), a widely used method for calculating the transfer impedance of braided shields. Using this formula, we examine the crosstalk through different types of shields.

We continue analyzing the crosstalk into the shielded wire from the previous section. To quantify the effect of the shield type on crosstalk, we vary the thickness of the solid foil shield to 0.01 mm, 0.1 mm, and 1 mm. Two scenarios are selected to study the shield's effect on inductive and capacitive crosstalk.

For inductive crosstalk, we use $R_{\mathrm{L}} = 50\,\Omega$ for the termination of the wires and short both ends of the shield to the reference. The choice of $R_{\mathrm{L}} = 50\,\Omega$ emphasizes inductive coupling, while the connection of the shield ends to the reference ensures the shield effectively mitigates inductive coupling. The near-end crosstalk into the shielded wire for varying shield thicknesses is plotted on the left-hand side of Fig. 8.30. The results show that as the shield thickness increases, the crosstalk is reduced, and the frequency at which the crosstalk flattens (known as the shield-break frequency) shifts to lower values. The shield-break frequency is determined by the shield's inductance and resistance (Paul 2005). For thicker shields, the width of the flat region is reduced, and the crosstalk at higher frequencies is significantly lower for shields with thicknesses of 0.1 mm and 1 mm.

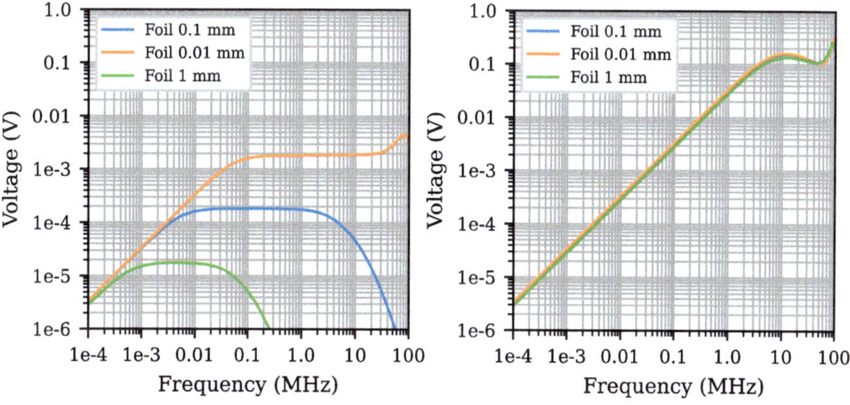

Fig. 8.30 Near-end crosstalk into the shielded wire for different thickness of the solid screen. Left: 50 Ω terminations and both shields end shorted to the reference. Right: 1 kΩ termination with the near-end shorted to the reference and the far-end open

For capacitive crosstalk, we use $R_L = 1\,\mathrm{k}\Omega$ for the terminations, short the near end of the shield to the reference, and leave the far end open. The choice of $R_L = 1\,\mathrm{k}\Omega$ emphasizes capacitive coupling, while shorting only one end of the shield to the reference ensures the shield mitigates capacitive but not inductive coupling. In this configuration, we observe that the shield thickness does not significantly affect the shielding performance against capacitive coupling.

These observations can be explained as follows:

Since the shield is a good conductor, the electric field responsible for capacitive coupling cannot penetrate it. Consequently, the thickness of the shield does not influence its performance against capacitive coupling.

In the case of inductive coupling, an induced current flows on the shield's outer surface. This current penetrates the shield according to the skin depth, and the coupling from one side of the shield to the other depends on the ratio of the shield thickness to the skin depth. For solid shields, the skin effect is the only mechanism that allows current penetration, which is why the crosstalk decreases at higher frequencies, as seen in Fig. 8.30.

However, when the shield is a braided structure or there is a gap in the foil, current can penetrate through the openings, leading to increased crosstalk compared to solid shields.

The concept of transfer impedance in modeling cable shields is illustrated in Fig. 8.31. As an example, we consider an electrically short section of a coaxial cable routed over a reference plane. Since all dimensions are electrically short, voltages and currents can be directly utilized in the following derivation.

In the schematic at the bottom of the figure, the coupling through the shield is modeled as a common impedance coupling, which is equivalent to the galvanic coupling discussed in Sect. 5.3. The primary difference is that in this case, the coupling element Z_T is frequency-dependent. Specifically, Z_T consists of a resistive component, R_T, and an inductive component, L_T, both of which vary with frequency.

An outer loop is formed between the shield and the reference plane (orange color), and an inner loop is formed between the shield and the inner conductor of the coaxial cable (blue color). For simplicity, we assume lossless dielectrics, allowing us to omit the conductance term G in the discussion. Each loop is modeled as an elementary transmission line segment, as shown in Fig. 8.2. Since the section is electrically short, one segment suffices for this analysis.

Both sections are shorted at the far end (right-hand side in the figure). With these shorts in place, the capacitances C_i and C_o do not influence the current and voltage in the loops.

In the outer loop, an ideal current source I_{source} drives a constant current, as the source is ideal. This current in the outer loop induces a voltage in the inner loop through the coupling impedance Z_T. The voltage in the inner loop, V_i, is measured across a resistor R_M. By choosing R_M to be high-ohmic, the impedance of the inner loop ($R_i + j\omega L_i$) can be neglected, as it appears in series with R_M.

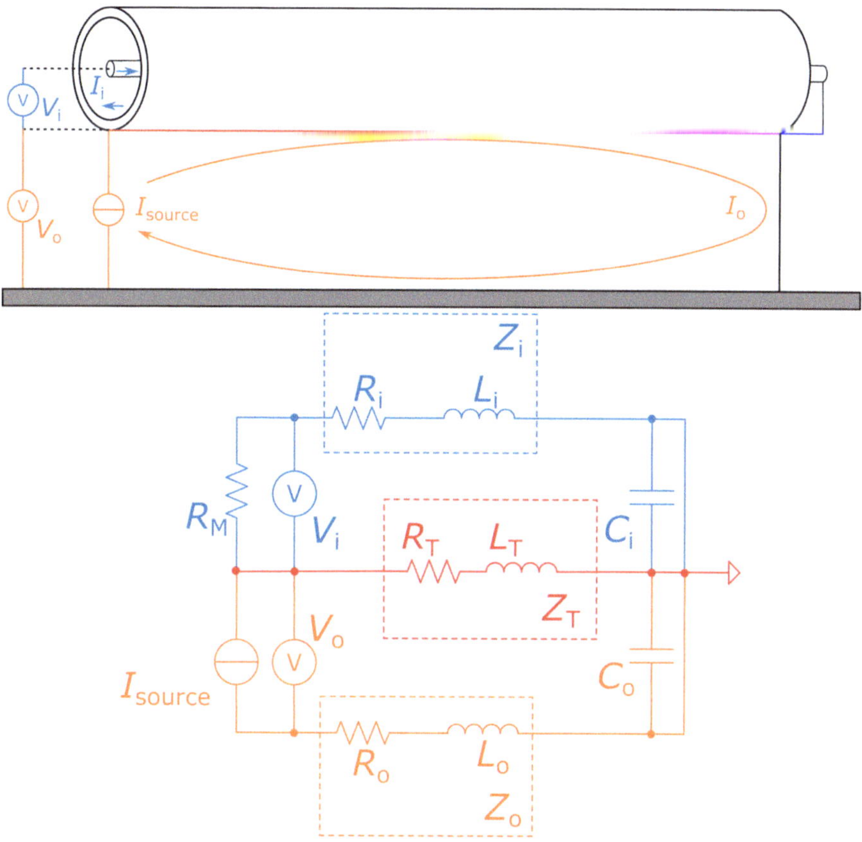

Fig. 8.31 Concept of the transfer impedance Z_T

In this configuration, Z_T couples the outer loop current to the inner loop voltage. It serves as a measure of the shield's performance, depending solely on the shield's properties and not on the setup. The following observations can be made:

- When $Z_T = 0$, the loops are completely isolated, indicating a perfect shield.
- As Z_T increases, the voltage in the inner loop rises, thereby reducing the shield's performance.

The transfer impedance Z_T is a function of length, so it is commonly expressed per unit length (1 m), with the unit $\frac{\Omega}{\text{m}}$.

Although the cable sketch highlights one edge of the shield in red, the current in the shield is distributed across its cross-section. The actual coupling occurs over an area, with the current distribution influenced by the geometry of the outer loop. In this derivation's geometry, the shield's current is concentrated at its bottom, as the return current path lies in the reference plane. In the next section, we introduce the triaxial setup which ensures a more uniform distribution of the current in the cable shield.

The transfer impedance describes the relationship of the current in the outer loop and the voltage in the inner loop. In literature, a shielding factor SF is often used to describe the shield's performance (Paul 2005). The shielding factor relates the voltage in the outer loop to the voltage in the inner loop. We can derive the shielding factor from the schematic in Fig. 8.31. Assuming R_M to be high ohmic, the voltage V_i in the inner loop is:

$$V_\mathrm{i} = (R_\mathrm{T} + j\omega L_\mathrm{T}) \cdot I = Z_\mathrm{T} \cdot I_\mathrm{o} \tag{8.5}$$

The voltage V_o in the outer loop is calculated as:

$$V_\mathrm{o} = (R_\mathrm{T} + j\omega L_\mathrm{T} + R_\mathrm{o} + j\omega L_\mathrm{o}) \cdot I_\mathrm{o} = (Z_\mathrm{T} + Z_\mathrm{o}) \cdot I_\mathrm{o}. \tag{8.6}$$

The shielding factor is then calculated from the ratio of the inner and outer voltage:

$$\mathrm{SF} = \frac{V_\mathrm{inner}}{V_\mathrm{outer}} = \frac{Z_\mathrm{T}}{(Z_\mathrm{T} + Z_\mathrm{o})} \cdot I_\mathrm{o} = \frac{1}{(1 + \frac{Z_\mathrm{o}}{Z_\mathrm{T}})} \cdot I_\mathrm{o} \tag{8.7}$$

When the shielding factor equals 1, the voltage inside and outside are identical, and there is no shielding effect. When $\mathrm{SF} < 1$, the voltage in the inner loop is reduced, indicating that the shield provides a shielding effect. However, this definition of the shielding factor depends on the setup because the calculation of V_o includes the outer inductance L_o, which depends on the mutual inductance between the cable shield and the reference plane, and thus on their separation distance.

8.5.2.1 Modeling of Braided Shields

For solid screens, there exists an analytical formulation for Z_T. For braided screens, Kley's formula (Kley 1993) is a widely used empirical method to calculate the transfer impedance based on several geometrical parameters.

A sketch illustrating how these parameters are defined is shown in Fig. 8.32. A braided screen consists of individual strands of specific thickness. These strands are grouped into a carrier, and the screen is constructed from multiple carriers. The coverage of the braid can be described using one of the following metrics:

- **Optical coverage ratio**: The percentage of the area covered by the braid.
- **Braid angle**: The angle formed between two carriers.
- **Picks per unit length**: The number of braid crossings per unit length of the cable.

Kley's formula is empirical and therefore valid only within specific parameter ranges.

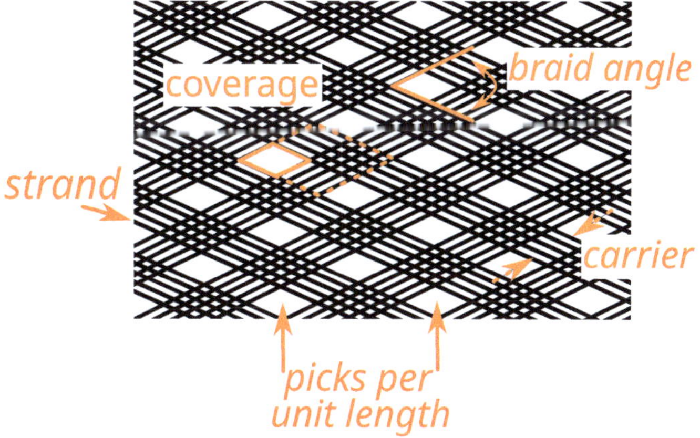

Fig. 8.32 Parameters used to describe a braided screen in Kley's formula

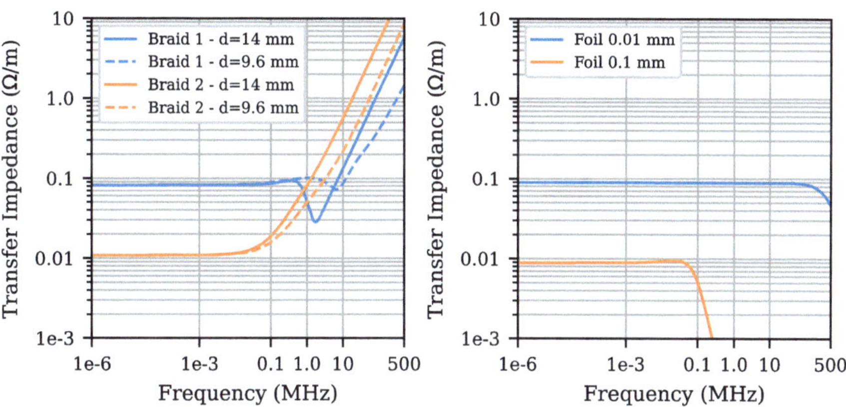

Fig. 8.33 Transfer impedance for different cable shield configurations. Left: Braided shields calculated using the implementation of Kley's formula in the CST. Right: Solid foil calculated using CST

To illustrate the behavior of the transfer impedance calculated using Kley's formula, we plot Z_T on the left-hand side of Fig. 8.33 for two braided screens with parameters specified in Table 8.1. Depending on which parameters are defined in Kley's formula, the resulting transfer impedance may also depend on the shield's diameter.

In our definition in Table 8.1, we use the braid angle to characterize the coverage. When the braid angle is specified, the optical coverage varies with the shield diameter, leading to differences in Z_T for different shield diameters. To demonstrate this dependency, we use two shield diameters:

- $d = 14$ mm, corresponding to the diameter of the cable used in the previous section on crosstalk analysis.

Table 8.1 Parameters for the calculation of the transfer impedance Z_T using Kley's formula for two different braid configurations plotted in Fig. 8.33

	Strand diameter (mm)	Strands in carrier	No. of carriers	(°) Braid angle	Resulting optical coverage	
					d = 14 mm	d = 9.6 mm
Braid 1	0.122	7	16	75	0.82	0.97
Braid 2	0.20	7	16	45	0.57	0.74

- $d = 9.6\,\mathrm{mm}$, corresponding to the diameter of the shielded TWP that is discussed in the next section.

In the following discussion, all cable shields are modeled as aluminum.

We chose the braid parameters such that braid 1 uses thinner strands than braid 2 but has a larger braid angle, resulting in a higher optical coverage. The transfer impedance of a braided screen, as calculated using Kley's formula, consists of three distinct frequency-dependent regions, as illustrated on the left-hand side of Fig. 8.33.

Low-Frequency Region (<100 kHz) In this region, the transfer impedance is flat and dominated by the diameter of the individual strands. Since braid 2 uses thicker strands than braid 1, its transfer impedance is lower. At these frequencies, differences in optical coverage between the two braids do not significantly influence the transfer impedance.

Intermediate-Frequency Region (0.5 MHz to 10 MHz) In this region, a more nuanced behavior is observed. For braid 1 with a diameter of 14 mm, a small dip in the transfer impedance is visible. This dip occurs because the skin depth reduces the penetration distance through the shield, lowering the transfer impedance. At higher frequencies, this effect is overshadowed by the current penetration through the braid apertures, leading to a rise in transfer impedance. For braid 2, which has much lower optical coverage, the transfer impedance increase caused by the braid's apertures begins at a lower frequency, preventing the dip from appearing.

High-Frequency Region (>10 MHz) At high frequencies, the transfer impedance rises due to current penetration through the braid's apertures. The penetration is directly proportional to the optical coverage. For braid 1, with a shield diameter of $d = 9.6\,\mathrm{mm}$ and the highest optical coverage, the transfer impedance is the lowest among all configurations.

On the right-hand side of Fig. 8.33, we plot the transfer impedance of solid foil shields with thicknesses of 0.1 mm and 0.01 mm. The transfer impedance of solid foil shields shows two primary regions:

Flat Region at Low Frequencies The transfer impedance remains constant at low frequencies, dominated by the shield's thickness and material.

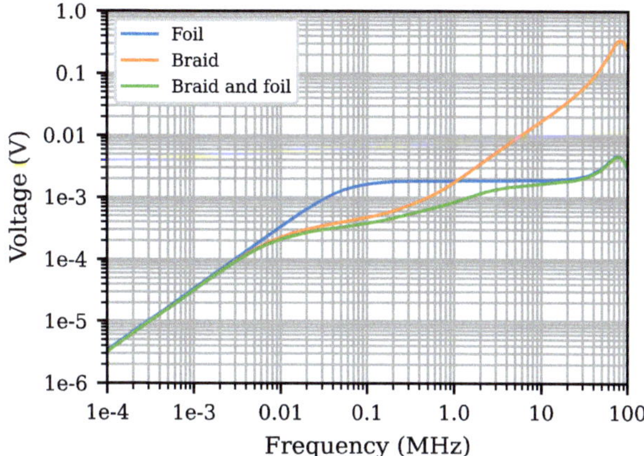

Fig. 8.34 Comparison of near-end crosstalk for different shield configurations: braid only, foil only, and combined braid with foil

Decreasing Region at Higher Frequencies As the skin effect becomes significant, the penetration depth through the shield reduces, leading to a drop in transfer impedance. Even a thin foil provides effective shielding at high frequencies.

In most screens, the braid strands are thicker than the foil, making a braid a more effective shield than a foil at lower frequencies. However, as the frequency increases, the rising Z_{T} of the braid can be mitigated by adding a foil to the screen of the cable. Combining a braid with a foil provides a low transfer impedance across a wide frequency range.

To illustrate the effect of combined shields, we revisit the crosstalk example. Figure 8.34 shows the near-end crosstalk for a 1 m long cable terminated with 50 Ω at both ends, and both shield ends shorted to the reference plane. The braid is configured as braid 2, with parameters defined in Table 8.1. The foil is made of aluminum with a thickness of 0.01 mm. In the configuration where only the braid is present, the crosstalk rises above 1 MHz but remains lower than the foil-only configuration in the range of 10 kHz to 1 MHz. When both the foil and braid are combined and electrically connected, the crosstalk stays low across the entire frequency band. This demonstrates the advantage of using a combined shield to achieve effective shielding over a broad frequency spectrum.

8.5.3 Application Example: Triaxial Method for the Estimation of the Transfer Impedance

Ideal voltage source

Termination at the end of a cable

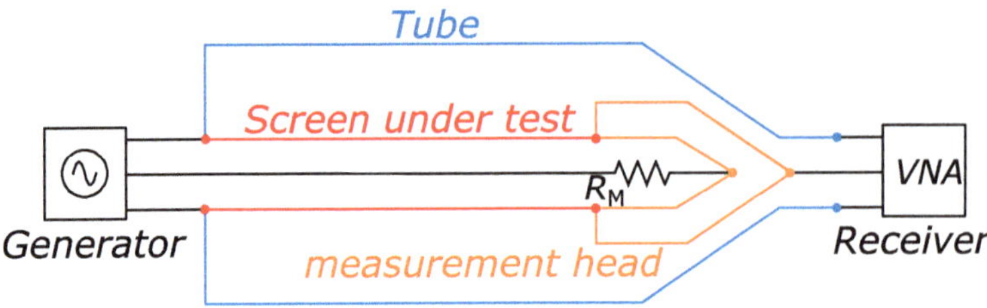

Fig. 8.35 Sketch of the principle behind the triaxial method for the estimation of the transfer impedance Z_T

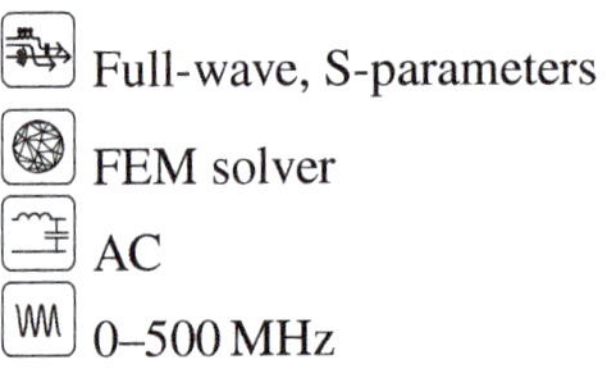

The transfer impedance of a cable screen can be measured using the Triaxial test method, defined by the IEC-62153-4 standard. The principle of this method is illustrated in Fig. 8.35. In this setup, the cable under test is placed inside a cylindrical metallic tube, which acts as the reference for the outer current loop, as discussed in Fig. 8.31. Using a tube instead of a plane[2] ensures that the outer current loop is evenly distributed across the cross-section, rather than being concentrated on one side of the cable. The IEC-62153-4 standard also defines the line injection method for the estimation of the transfer impedance, which is not covered in this book.

In the Triaxial measurement, a signal is injected into the inner loop (between the inner conductor and the screen), while the measurement is performed on the outer loop (between the screen and the tube). This setup is reciprocal to the definition of transfer impedance provided in the previous section. However, since transfer impedance is symmetrical, either configuration yields the same result. On the generator side, the screen is connected to both the tube and the signal generator to drive the signal into the inner loop. On the measurement side, the inner conductor is connected via a matching resistor R_M to a measurement head. The measurement head connects the inner conductor to the inner input of the measurement receiver, while the tube connects to the outer input of the receiver.

Measurements are performed in the frequency domain, typically using a network analyzer. A matching section is inserted between the measurement head and the receiver. This Triaxial test method faces several challenges. For accurate estimation of the transfer impedance Z_T, the cable section under test should be electrically small. However, for very short cable sections, the coupling between the inner and outer loops is minimal, making measurements challenging. With increasing cable

[2] Or a single wire, as done in the line injection method.

length and frequency, undesirable resonances can appear. To address these issues, the IEC-62153-4 standard defines several variations in the measurement setup. An overview of these variations is provided in Halme and Mund (2013). Differences among the variations include calibration of the measurement cables, addition of attenuation components, and usage of a matching resistor. The matching resistor minimizes reflections and results in a smoother transfer impedance curve.

In our simulation setup, we do not include attenuation components or measurement cables but use the matching resistor R_M. With this configuration, the transfer impedance is calculated using the equation:

$$Z_T = \frac{(R_M + Z_0)}{2 \cdot L_C} \cdot S_{21}, \tag{8.8}$$

In Eq. (8.8), R_M is the value of the matching resistor, Z_0 is the reference impedance of the S-parameter measurement, L_C is the length of the cable under test and S_{21} is the S-parameter measured from the generator to the receiver.

For the following discussion, we model a Shielded Twisted Wire Pair (STWP). The individual wires have a diameter of 0.56 mm and are covered by a PE insulator with a thickness of 1.5 mm. This setup results in a characteristic impedance of $Z_{0,DM} = 98\,\Omega$. The twist period is set to 2 cm. The TWP is surrounded by a shield with a diameter of 9.6 mm. The shield is defined as a combination of a braid and a foil. The braid is modeled using Kley's formula with the following parameters: copper material, strand diameter of 0.122 mm, 7 strands per carrier, 16 carriers, and a braid angle of 73°, resulting in an optical coverage of 0.93. The foil is made of copper with a thickness of 20 μm. The braid touches the foil, ensuring a low Z_T across the complete frequency band.

To study the effects of cable length and the matching resistor, we simulate the Triaxial setup using an MTL model. A cylindrical tube is added around the cable, and the matching resistor is connected between the screen and the inner conductors in the schematic. This simulation requires less than 10 seconds.

In a STWP, two inner conductors are present, allowing for two possible definitions of the inner loop. In the differential configuration, the individual wires are terminated with resistors, and a voltage is applied from the screen to the wire pair. In the common mode configuration, both wires are connected to each other and terminated via R_M to the measurement head. The transfer impedance obtained from these two configurations can differ. In the following discussion, we use the common mode configuration.

We plot the resulting Z_T in Fig. 8.36. On the left-hand side, we see the effect of the cable length. For this simulation, $R_M = 1\,\text{k}\,\Omega$ is used. This is a high-ohmic termination, as used in the previous section to measure the voltage in the inner loop during the derivation of the transfer impedance.

Below 1 MHz, the transfer impedance obtained for cable lengths of 1 mm, 25 cm, and 1 m does not differ significantly. However, only for the very short length of 1 mm is the transfer impedance obtained from this setup as expected. For the other two cable lengths, above 2 MHz, the curves start to deviate, and resonances are observed

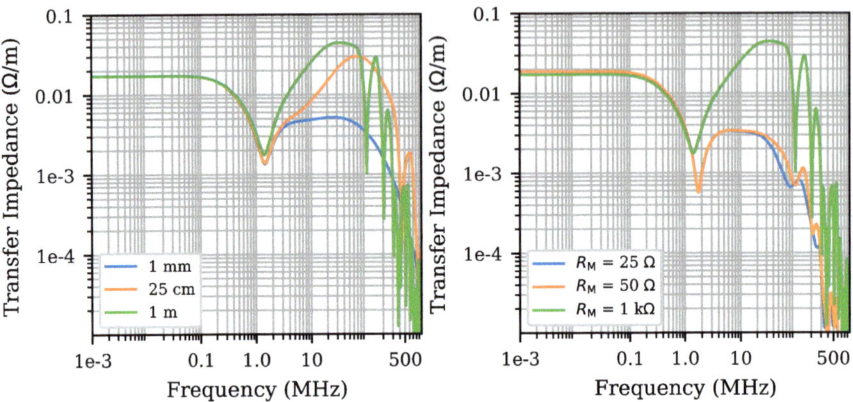

Fig. 8.36 Transfer impedance obtained from a MTL model of the Triaxial method. Left: dependence on cable length. Right: Dependence on matching resistor R_M

in the transfer impedance. This is the effect of poor matching of the transmission line.

In the derivation of the transfer impedance in the previous section, we assume an electrically short section. In an electrically short section, only one elementary segment is required to describe the inner and outer loop. For a cable that is not electrically short, multiple segments are required, and the capacitance between the screen and the inner conductors can no longer be simply omitted, as it is in the previous section.

The matching resistor R_M is used to suppress these resonances. To study the effect of the matching resistor, the 1 m long cable is simulated with R_M set to 50 Ω, 25 Ω, and 1 kΩ. The results are plotted on the right-hand side of Fig. 8.36. By using an appropriate matching resistor, resonances in the cable are suppressed, and the transfer impedance obtained from the triaxial setup exhibits the expected behavior. The difference between the matching with 50 Ω and 25 Ω is not significant.

Next, we investigate the effect of an inline connector on Z_T. Connectors introduce discontinuities in both the individual wires and the screen of a cable. This analysis focuses on the connector's impact on the screen by examining Z_T. Since the field distribution within a connector is not necessarily a transverse mode, we cannot use the MTL method and instead employ a full-wave solver. To create a complete three-dimensional model of the Triaxial test setup, we first construct the cable model. This task is complex, as modeling a twisted wire pair requires advanced CAD modeling skills. CST provides the "cable to three-dimensional" functionality, which converts a cable defined by its cross-section and path (as required for the MTL solver) into a three-dimensional CAD representation. Additionally, this functionality converts the screen definition from a transfer impedance into a thin panel material, optimizing it for full-wave simulations.

Once the cable is created, we define the connections to the connector and to the tube at the cable ends using CAD modeling tools. The resulting simulation model,

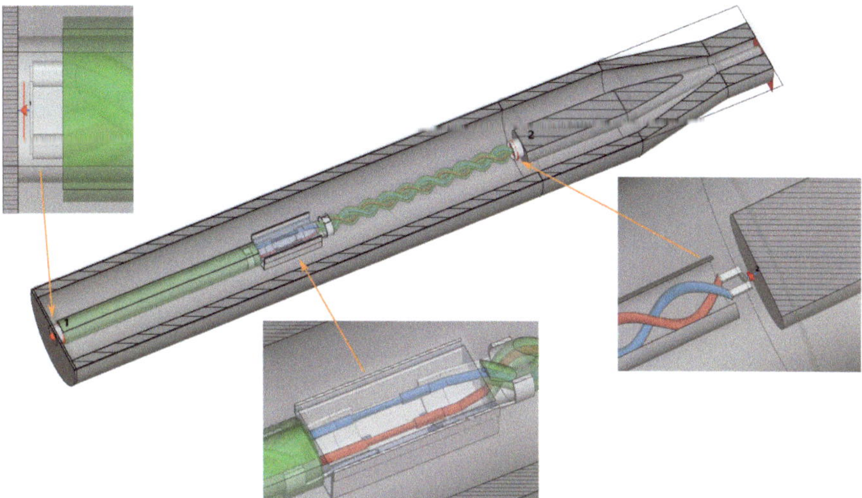

Fig. 8.37 Three-dimensional simulation model of the triaxial tube with a STWP and connector

including the connector, is shown in Fig. 8.37. For clarity, the screen on the right-hand side of the cable is hidden to reveal the twisted wire pair. The total cable length, including the connector, is 25 cm. The cable is enclosed within a PEC tube with an inner diameter of 4 cm. At the tube's right end, a straight section represents the measurement head and connects via a matching region to the measurement port, as illustrated in the inset on the right. On the left, the two wires are joined and connected to the tube via a port, as depicted in the top-left inset of Fig. 8.37. The bottom inset shows the details of the connector.

The connector design is intentionally simplified. In real-world applications, connectors often include additional features for mechanical stability and wire crimping. This basic design serves to demonstrate the connector's influence on Z_{T}. The cable shields are attached to the connector shell using small clamps. Inside the connector, the clamps are semicircular and connected via small protrusions to the lower section of the connector shell. A block of dielectric material with $\varepsilon_{\mathrm{r}} = 2.3$ is included within the connector to maintain the characteristic impedance of the differential mode near 100 Ω.

We perform Triaxial measurement simulations using a full-wave solver, both with and without the connector, to validate the EM simulation results against those obtained from the MTL-based Triaxial model. For all comparisons, R_{M} is set to 25 Ω. Since the analysis includes results in the kHz range and with the selected maximal frequency of 200 MHz, the model is electrically short, the FEM solver is used. The simulation time for the model containing only the cable is 28 minutes, while the model including both the cable and the connector takes 34 minutes.

The transfer impedance calculated using the full-wave model and the MTL model is plotted in Fig. 8.38. When comparing the results of the MTL model to the three-

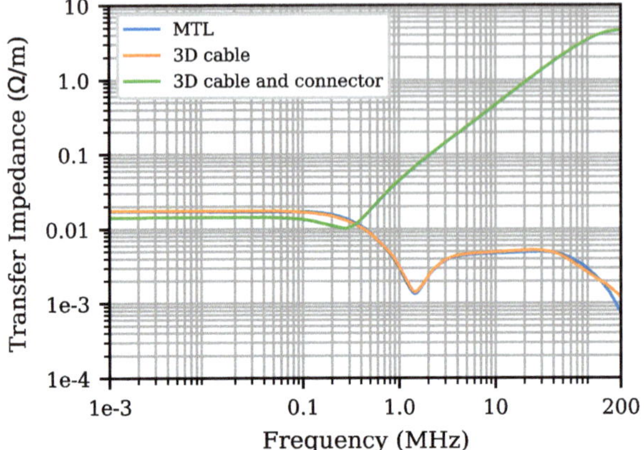

Fig. 8.38 Transfer impedance for an STWP calculated using the Triaxial method. Comparison of results obtained using an MTL based model and a full-wave model using the thin panel material for the screen and a full three-dimensional model including an inline connector

dimensional model containing just the STWP, we observe excellent agreement. Up to approximately 150 MHz, the curves overlap, demonstrating that the thin panel material effectively models complex transfer impedance behavior in a full-wave simulation. This confirms its capability for efficiently simulating cable shields.

When comparing the results of the model containing the connector to the model with only the cable, we observe that below 100 kHz, the transfer impedance of the connector model is lower. This occurs because the connector section has been modeled using PEC, which reduces the overall resistance of the cable connector assembly along the connector's length.

Above 400 kHz, the transfer impedance of the standalone cable decreases, while that of the model with the connector increases. The holes in the connector act similarly to those in a braided shield, diminishing the shielding effectiveness at higher frequencies. At 10 MHz, the transfer impedance of the connector and cable assembly is a factor of 100 higher than that of the cable alone. As highlighted in previous sections, the connector design significantly impacts the transfer impedance of the cable connector assembly. The shielding performance of a well-designed cable shield can be easily compromised by a poorly designed connector.

8.6 Hybrid Transmission Line to Full-Wave Solver Coupling

Domains: 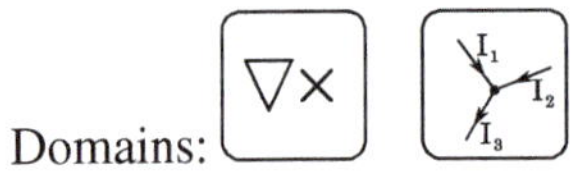

The MTL method is fast and accurate as long as the fields on the cables are predominantly transverse modes. In contrast, a full-wave simulation of cables enables the analysis of transitions in cable sections where the fields are no longer transverse, albeit at the cost of increased simulation time. By combining a full-wave method with the MTL method, a hybrid simulation workflow is achieved that promises to benefit from the strengths of both approaches.

The implementation of such a hybrid method, which integrates the MTL method with a full-wave solver, can be approached in various ways. The specific details of the hybrid method depend on the type of full-wave solver to which the MTL method is coupled.

Several commercial tools on the market (FEKO, EMCoS) couple the MTL method with a MoM-based full-wave solver. Because the two-dimensional cross-sectional calculations in the MTL method are performed using a MoM-based solver, the three-dimensional MoM and MTL methods share certain fundamental properties.

However, the coupling of the MoM and MTL methods is not available in CST, and the authors are not familiar with the specific aspects of this hybridization. Therefore, this section focuses exclusively on the hybridization of the MTL method with the TLM solver.

8.6.1 *Sequential Coupling: Unidirectional*

The simplest coupling of the MTL method to a full-wave solver is a sequential coupling. Sequential coupling is inherently unidirectional, requiring a decision before the simulation begins as to whether the cable acts as a source of fields or a sink of fields.

Cable as a Source If the cable acts as a source, the first step of the hybrid method involves performing a standalone MTL simulation. From this simulation, the common mode current I_{CM} is calculated for every segment of the cable and passed on to the full-wave solver. Although the MTL solver computes the current in both the cable path and the reference, it assumes an ideal reference and cannot provide a localized current on the reference to the full-wave solver. Consequently, the full-wave solver uses only the common mode current at the cable's centerline as an excitation source. To ensure the current loop is closed in the full-wave simulation, connections to the nearest PEC/lossy metal shapes are added at the cable ends. With this setup, the fields generated by I_{CM} flowing on the cable are calculated.

Cable as a Sink If the cable acts as a sink, the first step of the hybrid method involves a full-wave simulation. During this simulation, the common mode voltage along the cable path is calculated from the tangential electric field in each mesh cell. After the full-wave simulation is completed, the common mode voltage is passed to the MTL model, where it serves as the excitation source for the transmission line calculation.

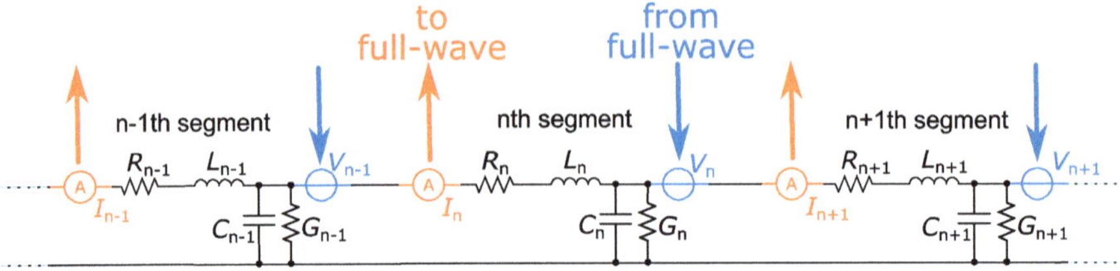

Fig. 8.39 Sketch of the transfer of voltages and currents between the full-wave method and the MTL in a hybrid simulation

Both coupling scenarios are illustrated in Fig. 8.39. When the cable is treated as a source, only the orange-colored sections are used. When the cable is treated as a sink, only the blue-colored sections are employed.

In the sequential coupling approach, the cable is not physically present in the full-wave simulation. The common mode current for the source configuration is calculated using the MTL method, while the common mode voltage in the sink configuration is based on the CAD geometry surrounding the cable. In this setup, the cable does not influence the field distribution in the full-wave simulation. This method is effective when the cable is only loosely coupled to the environment, such as when a cable is routed over a metallic table.

However, there are scenarios where this assumption does not hold. For instance, consider a metallic enclosure where a cable is connected to the enclosure walls. This connection alters the resonance frequencies within the enclosure (as discussed in Sect. 7.4), an effect that is not accounted for in sequential coupling. Another example involves cable entry susceptibility. In this case, a shielded cable located outside a metallic enclosure is illuminated by an external source, picking up the external field. The cable is then routed into the enclosure, with its shield connected to the enclosure wall. The current on the inner conductor can reradiate the energy picked up outside into the interior of the enclosure.

In such setups, the cable must act as both a sink and a source simultaneously, which cannot be accommodated by the sequential coupling method.

8.6.2 Simultaneous Cosimulation: Bidirectional

To overcome the limitations of unidirectional coupling, the hybrid scheme can be extended to a simultaneous coupling that enables bidirectional interaction between the MTL method and the full-wave solver. Achieving this requires a simultaneous exchange of both voltages and currents between the full-wave solver and the MTL method, as illustrated in Fig. 8.39.

At each mesh cell in the three-dimensional model through which a cable is routed, the electric field is converted into a voltage. This voltage is passed to the circuit simulation, where a time step is computed. The current calculated by the circuit simulation is then transferred back to the full-wave solver, which proceeds with the next time step, and the process repeats. We describe the cosimulation methodology without the coupling to the MTL in Sect. 14.3, where it is used for ESD simulations.

This scheme establishes a separation, with different parts of the model being handled by their respective methods: The MTL solver models the coupling within the cable bundle, while the current on the reference is modeled by the full-wave solver. This approach allows cable bundles to be referenced to multiple, disconnected conductors—something not possible with the MTL method alone, as it treats all reference conductors as interconnected.

The described approach is performed using a time-domain solver and introduces inaccuracies due to the mapping of the thickness of the entire cable bundle onto a single mesh line. These must be corrected when cables are thicker than a mesh cell. Similar to the unidirectional simulation, the ends of a cable bundle must be connected to the nearest PEC/lossy metal shape to close the current loop.

Traditionally, bidirectional coupling is performed using only common mode coupling. This makes the method well-suited for shielded cables, where the common mode current on the shield is the dominant contributor to emissions.

However, for nonshielded cables carrying differential signals, such as a TWP, the absence of differential mode coupling in the scheme results in an inaccurate estimation of differential-to-common mode conversion along the cable path. The imbalance between the individual conductors in a TWP is not accounted for when only the common mode current is used to couple the full-wave solver and the MTL solver. For such cables, this scheme is not recommended.

Recent developments have extended the coupling scheme to support the coupling of individual conductors within the cable cross-section. These advancements were released around the time this book was written, and the authors were unable to investigate the performance and limitations of this new approach.

For cable simulations, several approaches are available, and the most efficient one depends on the specific model setup and the desired results. In Chap. 1, we emphasized the importance of clearly defining the goals of a simulation before building the actual model. This consideration is especially crucial for cable simulations.

The MTL method is both fast and accurate for simulating crosstalk and cable parameters. While it can be coupled to the TLM solver for emission analysis, hybrid methods have their limitations, as discussed earlier.

For cables with a relatively small number of conductors (though there is no strict threshold) and an electrical length that can be managed by a full-wave solver (see Sect. 4.9), a complete full-wave simulation is also feasible. Despite common claims in the literature that this approach is computationally prohibitive, it is particularly useful in cases where field values are required, such as for radiated emission analysis.

The use of thin panel material provides an efficient way to simulate cable shields, and resolving the cable cross-section with a three-dimensional mesh is achievable, as demonstrated in several examples throughout this book. At low frequencies, ranging from kHz to MHz, full-wave simulations using the FEM solver are a viable option. This is especially relevant since the hybrid MTL-to-full-wave coupling method in CST is currently only available for transient solvers, which perform poorly in this low frequency range.

8.7 Advanced: MTL Theory

Domains: 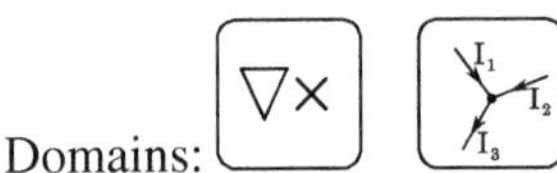

A transmission line is structure that conducts electromagnetic waves, and which has a cross-section which is uniform along its length. The structure can be electrically long, that is, the propagation delay of the traveling wave affects the properties of the voltage and currents along the line. The theory of MTL is extensively described in the book Paul (2007), and we give a summary in the first part of this section. In the second part, we extend these results to nonuniform transmission lines, which vary the cross-section stochastically along their length.

8.7.1 Solution of the Telegrapher's Equations for MTLs

The fundamental assumption of a transmission line is that the magnetic and electric field strengths satisfy the field TEM structure. The electric and the magnetic field lines are perpendicular to each other, and there is no component of **E** or **H** parallel to the direction of propagation. We denote these transverse components as $\mathbf{E}_{\mathrm{tr}}$, $\mathbf{H}_{\mathrm{tr}}$. Further, the cross-section is electrically small.

Structures like a bundle of coated wires are not homogeneous, because there are both the coating and air in the cross-section. There is a small field component which is not transverse to the direction of propagation, and the corresponding field pattern is called quasi-TEM. One can solve this issue by approximating the two different materials by a homogeneous one with suitable effective material properties. We do not discuss this case further.

We assume perfect conductors which are surrounded by a homogeneous medium. We derive the properties of a single TEM mode by writing the first two of Maxwell's Equations as

$$\nabla \times \mathbf{E}_{\mathrm{tr}} = -\mu \frac{\partial \mathbf{H}_{\mathrm{tr}}}{\partial t}, \tag{8.9a}$$

$$\nabla \times \mathbf{H}_{\mathrm{tr}} = \sigma_{\mathrm{c}} \mathbf{E}_{\mathrm{tr}} - \varepsilon \frac{\partial \mathbf{E}_{\mathrm{tr}}}{\partial t} \tag{8.9b}$$

where all terms parallel to the direction of propagation vanish. The transversal fields $\mathbf{E}_{\mathrm{tr}}$, $\mathbf{H}_{\mathrm{tr}}$ decompose into a scalar time-variant component $e(z,t)$, $h(z,t)$ with $\hat{\mathbf{n}}_z$ being the direction of propagation and a static component $\psi_e(x,y)$, $\psi_h(x,y)$ according to

$$\mathbf{E}_{\mathrm{tr}} = e(z,t)\nabla_{\mathrm{tr}}\phi_e(x,y) \tag{8.10a}$$

$$\mathbf{H}_{\mathrm{tr}} = h(z,t)\nabla_{\mathrm{tr}}\psi_h(x,y) \tag{8.10b}$$

with

$$\nabla_{\mathrm{tr}} = \mathbf{e}_{\mathrm{x}}\frac{\partial}{\partial x} + \mathbf{e}_{\mathrm{y}}\frac{\partial}{\partial y}. \tag{8.11}$$

For any transmission line, the property of the field in the direction of propagation is decoupled from its properties in the transversal plane.

The scalar components $\phi(x,y)$, $\psi(x,y)$ obey the Laplace Equation

$$\nabla_{\mathrm{tr}}^2\psi_e(x,y) = \frac{\partial^2\psi_e(x,y)}{\partial x^2} + \frac{\partial^2\psi_e(x,y)}{\partial y^2} = 0 \tag{8.12a}$$

$$\nabla_{\mathrm{tr}}^2\psi_h(x,y) = \frac{\partial^2\psi_h(x,y)}{\partial x^2} + \frac{\partial^2\psi_h(x,y)}{\partial y^2} = 0. \tag{8.12b}$$

The electric and magnetic fields in the transverse plane are static. To solve for it, time-efficient static solvers are applied.

The longitudinal part obeys for the scalar fields $e(z,t)$, $h(z,t)$

$$\frac{\partial^2 e(z,t)}{\partial z^2} = \mu\sigma_{\mathrm{c}}\frac{\partial e(z,t)}{\partial t} + \mu\varepsilon\frac{\partial^2 e(z,t)}{\partial t^2} \tag{8.13a}$$

$$\frac{\partial^2 h(z,t)}{\partial z^2} = \mu\sigma_{\mathrm{c}}\frac{\partial h(z,t)}{\partial t} + \mu\varepsilon\frac{\partial^2 h(z,t)}{\partial t^2}. \tag{8.13b}$$

In longitudinal direction, the field distribution of the TEM mode is that of a damped one-dimensional wave, the equation of which is the superposition of a forward and a backward traveling damped sinusoidal wave. The mode has a propagation constant of

$$\gamma = \sqrt{j\omega\mu\left(\sigma_{\mathrm{c}} + j\omega\varepsilon\right)} \tag{8.14a}$$

$$= \alpha + j\beta, \tag{8.14b}$$

with α the attenuation constant and β the phase constant, which determines the TEM mode's velocity.

The fact that the fields are static in the transversal plane has an important consequence. We can uniquely define voltages and currents by the integration over the electric fields between and along the magnetic fields around the conductors. Because of Ohm's law, the integral over the electric field along the conductors is proportional to their resistance. The magnetic flux through a surface between the conductors is proportional to their (mutual) inductance. If the medium surrounding the conductors has itself a (low) conductance σ_c, for example, as a lossy insulator, there is a current along the transversal electric field between the conductors which is proportional to σ_c, and the equation of continuity relates this transversal current with the change of current along the conductors. This change of current is also proportional to the capacitances and the time-variant voltages in between the conductors.

In the shown derivation, we have assumed that the transverse fields are static. However, the derivation can also be applied to quasistatic fields, where the per-unit-length parameters become frequency dependent, as demonstrated in Sect. 8.1.3.

For multiple conductors, the mathematically rigorous combination of these effects yields the two Telegrapher's equations for MTLs,

$$\frac{\partial}{\partial z}\mathbf{V}(z) = -\mathbf{R}'\mathbf{I}(z,t) - \mathbf{L}'\frac{\partial}{\partial z}\mathbf{I}(z,t), \tag{8.15a}$$

$$\frac{\partial}{\partial z}\mathbf{I}(z) = -\mathbf{G}'\mathbf{V}(z,t) - \mathbf{C}'\frac{\partial}{\partial z}\mathbf{V}(z,t). \tag{8.15b}$$

The equations relate the rate of change of the voltage and the current along each conductor with their resistance along the conductors, with their self-inductance, and with the conductance, capacitance, and mutual inductance in between them. The presence of a reference conductor which carries the return current is required. If the reference conductor is a metallic plane below the wire bundle, the vectors $\mathbf{V}$, $\mathbf{I}$ have $N = n_\mathrm{W}$ entries with n_W the number of wires. If there is no metallic reference plane, they have $N = n_\mathrm{W} - 1$ entries. The vector $\mathbf{I}$ contains all currents but that of the reference conductor. Because of Kirchhoff's current law, the reference conductor carries the sum of the currents of all other conductors. The influence of the reference conductor is contained in the matrices $\mathbf{R}'$, $\mathbf{L}'$, $\mathbf{C}'$, and $\mathbf{G}'$. For example, the matrix $\mathbf{R}'$ contains the resistances of all wires, including the return wire.

For a linear, time invariant system, we can rewrite Eq. (8.15) using phasors as

$$\frac{\mathrm{d}}{\mathrm{d}z}\underline{\mathbf{V}}(z) = -\underline{\mathbf{Z}}'\underline{\mathbf{I}}(z) \tag{8.16a}$$

$$\frac{\mathrm{d}}{\mathrm{d}z}\underline{\mathbf{I}}(z) = -\underline{\mathbf{Y}}'\underline{\mathbf{V}}(z) \tag{8.16b}$$

with the impedance and admittance matrices $\underline{\mathbf{Z}}'$ and $\underline{\mathbf{Y}}'$ as

$$\underline{\mathbf{Z}}' = \mathbf{R}' + j\omega\mathbf{L}' \tag{8.17a}$$

$$\underline{\mathbf{Y}}' = \mathbf{G}' + j\omega\mathbf{C}'. \tag{8.17b}$$

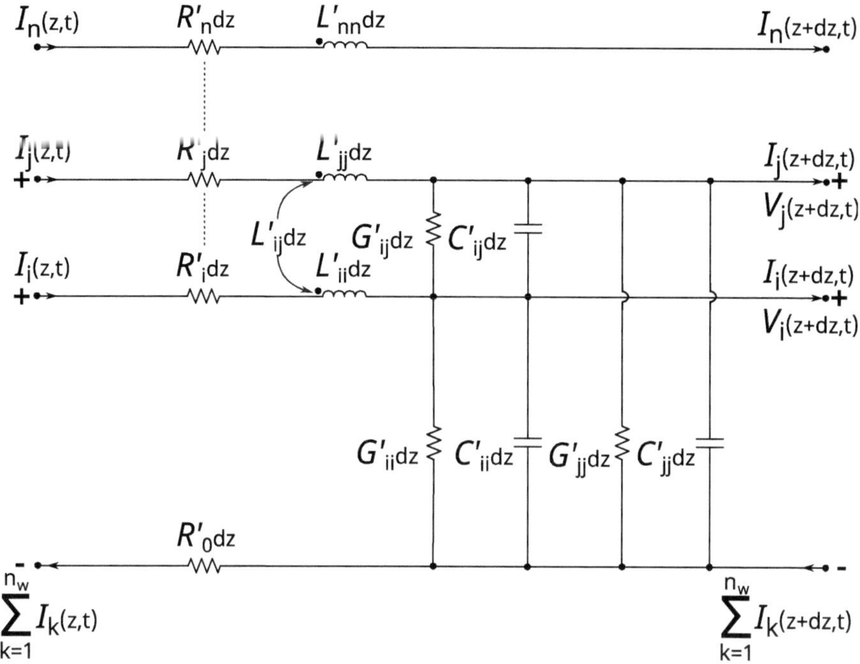

Fig. 8.40 Circuit representation of the Telegrapher's Eqs. (8.16) and (8.17)

This system of equations is identical to a circuit of the form displayed in Fig. 8.40. It represents a system of n_w wires and a reference wire or ground plane, at the position z. In each wire i, there flows a current I_i, and it has a voltage V_i with respect to the reference. Each wire segment of length dz has the per-unit-length resistance R'_i and a per-unit-length self-inductance L'_{jj}. To any other conductor j, it has the per-unit-length mutual inductance of L'_{ij}, the capacitance C'_{ij}, and the per-unit-length conductance G'_{ij}. The coupling to the reference conductor is characterized by C'_{ii} and G'_{ii}. The reference conductor itself has the per-unit-length resistance R'_0. Due to the coupling, the currents and voltages $I_i(z + \mathrm{d}z)$, $V_i(z + \mathrm{d}z)$ differ from those at position z. Using a SPICE simulator, such a circuit block can be connected to other SPICE elements. An MTL system with sources and loads can be simulated in a circuit simulator. For a single conductor with a reference, the circuit shown in Fig. 8.40 reduces to the circuit shown in Fig. 8.2.

Equations 8.16 are two coupled partial differential equations of first order which we can decouple into two partial differential equations of second order,

$$\frac{\mathrm{d}^2}{\mathrm{d}z^2}\underline{\mathbf{V}}(z) = \underline{\mathbf{Z}}'\underline{\mathbf{Y}}'\underline{\mathbf{V}}(z) \tag{8.18a}$$

$$\frac{\mathrm{d}^2}{\mathrm{d}z^2}\underline{\mathbf{I}}(z) = \underline{\mathbf{Y}}'\underline{\mathbf{Z}}'\underline{\mathbf{I}}(z). \tag{8.18b}$$

Since the matrices $\underline{\mathbf{Z}}'\,\underline{\mathbf{Y}}'$ are symmetric, there exist, for both the voltage $\underline{\mathbf{V}}$ and the currents $\underline{\mathbf{I}}$, similarity transforms $\mathbf{T}_V$ and $\mathbf{T}_I$ which diagonalize $\underline{\mathbf{Z}}'\underline{\mathbf{Y}}'$ and $\underline{\mathbf{Y}}'\underline{\mathbf{Z}}'$, and the diagonal elements γ_i, $i = 1, \ldots, N$,

$$\mathbf{T}_V^{-1}\underline{\mathbf{Z}}'\underline{\mathbf{Y}}'\mathbf{T}_V = \text{diag}\left(\gamma_1^2, \ldots, \gamma_N^2\right) \tag{8.19a}$$

$$\mathbf{T}_I^{-1}\underline{\mathbf{Y}}'\underline{\mathbf{Z}}'\mathbf{T}_I = \text{diag}\left(\gamma_1^2, \ldots, \gamma_N^2\right). \tag{8.19b}$$

Applying the matrices $\mathbf{T}_V$, $\mathbf{T}_I$ we finally obtain the modal voltages and currents $\underline{\mathbf{V}}_m(z)$, $\underline{\mathbf{I}}_m(z)$

$$\underline{\mathbf{V}}(z) = \mathbf{T}_V \underline{\mathbf{V}}_m(z) \tag{8.20a}$$

$$\underline{\mathbf{I}}(z) = \mathbf{T}_I \underline{\mathbf{I}}_m(z). \tag{8.20b}$$

There are as many modes as there are conductors without the reference conductor.

In terms of the modal voltages and currents $\underline{\mathbf{V}}_m(z)$, $\underline{\mathbf{I}}_m(z)$, the Eq. (8.18) decouple into

$$\frac{\mathrm{d}^2}{\mathrm{d}z^2}\underline{\mathbf{V}}_m(z) = \text{diag}\left(\gamma_1, \ldots, \gamma_N\right)\underline{\mathbf{V}}_m(z) \tag{8.21a}$$

$$\frac{\mathrm{d}^2}{\mathrm{d}z^2}\underline{\mathbf{I}}_m(z) = \text{diag}\left(\gamma_1, \ldots, \gamma_N\right)\underline{\mathbf{I}}_m(z). \tag{8.21b}$$

Equation (8.21) explain the term of a propagating mode introduced in Sect. 8.1.1. Each row i in Eq. (8.21) is solved by the superposition of a forward and a backward traveling damped wave, which is the mode i with its propagation constant γ_i.

The characteristic impedance matrix of the MTL relates the modal voltages with the modal currents, $\underline{\mathbf{V}}(z) = \underline{\mathbf{Z}}_C'\underline{\mathbf{I}}(z)$ and can be expressed in terms of the propagation constants as

$$\underline{\mathbf{Z}}_C' = \underline{\mathbf{Z}}'\mathbf{T}_I \text{diag}\left(\gamma_1, \ldots, \gamma_N\right)^{-1}\mathbf{T}_I^{-1}. \tag{8.22}$$

For perfect conductors ($\mathbf{R} = \mathbf{0}$) in lossy, homogeneous media all eigenvalues γ_i, $i = 1, \ldots, N$ degenerate to the same value

$$\gamma = \sqrt{j\omega\mu\left(\sigma_\mathrm{c} + j\omega\varepsilon\right)} \tag{8.23}$$

which is the propagation constant of a propagating wave in a lossy medium assuming perfect conductors and identical to Eq. (8.14).

We conclude that under the assumption made for the MTL, the Maxwell equations reduce to a concatenation of identical circuit elements which consist of the resistance, inductance, conductance, and capacitance matrix. If these assumptions are valid, the computational effort to solve for currents and voltages on the conductors is reduced significantly compared to a full-wave solution.

8.7.2 Stochastic Wire Harnesses

In large cable bundles, the position of the individual wires is random, which contradicts the assumption of the MTL to have a uniform cross-section. The wire positions may even change if the cable bundle is removed from the setup and measured again. A measurement based on such a random configuration is also random. The consideration of mean values and standard deviations of several measurements ensures that the investigation of the electronic system under test does not refer to a particularly harmful or gentle realization of the wire bundle. Stochastic methods are required.

Several groups worked on the characterization of wire harnesses with random position of the wires, which results in the definition of the randomness of the wire positions in the individual cross-section. In this section, we present exemplarily one of these approaches. Tool suites such as CST and EMCoS offer to model stochastic wire harnesses.

In case of nonhomogeneous transmission lines, we can still apply the Telegraphers equations. We choose the length of the individual segments not according to $\frac{\lambda}{10}$ but according to the nonhomogeneity. Since the wires in the bundle change their positions, we compute the matrices $\mathbf{R}'$, $\mathbf{L}'$, $\mathbf{G}'$, $\mathbf{C}'$ individually for each wire configuration and stack the resulting circuit segments.

Next, we need to account for the randomness of the positions. We refer to the model described in Gonser et al. (2014), which suggest a wire-hole model. A number of n_{w} wires is distributed in a hexagon. The wires are not densely packed; there is a number of n_{h} holes. The wire bundle is partitioned in a number of N_{S} segments. If next to a wire there is a hole, there is a probability of $p_{\mathrm{h}} \in [0, 1]$ that, from one segment to the next, the wire changes its position and moves to the adjacent hole. The probability p_{h} is an intrinsic property of the wire bundle. A very random bundle has a large p_{h}, in case of an orderly one, p_{h} is low. The number of holes also is an intrinsic property of the bundle. The density ρ_{w} of the wire bundle is the ratio of the number of wires to the sum of the wires and the holes, $\rho_{\mathrm{w}} = \frac{n_{\mathrm{w}}}{n_{\mathrm{w}}+n_{\mathrm{h}}}$. A plot of three adjacent cross-sections is shown in Fig. 8.41. From the first to the second cross-section, the yellow wire changes its position, from the second to the third, the blue, green and red wires. The wire bundle is generated using a stochastic algorithm which decides, for each wire and based on the probability p_{h}, whether a hole is filled or not. If the hole is filled, an arbitrary wire adjacent to the hole is chosen to fill it. The bundle is generated segment after segment. After its generation, the per-unit-length matrices $\mathbf{R}'$, $\mathbf{L}'$, $\mathbf{C}'$, and $\mathbf{G}'$ are computed for each segment, and the circuit model of the wire bundle is created by concatenation of all segments.

There are many types of wire bundles. They have different numbers of wires n_{w}; they can be densely packed to have a high density ρ_{w} or loosely spread. They

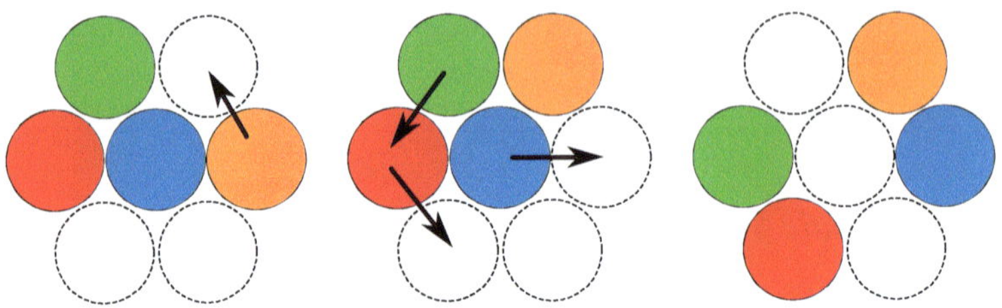

Fig. 8.41 Wire-hole model on a hexagonal cross-section (Gonser et al. 2014)

may have a large cross-section per wire A_W or a low one. The number of wires n_W is known by construction and also the cross-section of each wire. The number of holes n_h can be estimated from the cross-section of the entire bundle A_B and the cross-section of all wires, that is, $\rho_W \approx \frac{n_W \cdot A_W}{A_B}$. The only unknown parameter is the probability p_h that a hole is filled by an adjacent wire. The authors in Gonser et al. (2014) perform Monte Carlo simulations of generated MTL models over a field of four different n_W, two different cross-sections A_W, and a grid of ρ_W and p_L. Examining the S-parameters of 1 million samples, they find that above $n_W \geq 9$, the probability p_h is a constant which does not depend on the number of wires n_W or the cross-section of the wires A_W. It is also constant for a variety of different bundle types. This finding is also supported by measurement data. The results indicate that the proposed model of wires and holes on a hexagon and a transition probability p_h can be parameterized to result in realistic S-parameters. Commercial solvers provide algorithms to parameterize the bundles and their randomness, but they do not suggest suitable parameters. These are left to be defined by the user.

We investigate exemplarily a bundle of five wires using the stochastic harness implementation in CST. The wires are of equal diameter of 1 mm and a coating of 0.2 mm thickness and calculate the near-end crosstalk into one of the wires. The bundle has a length of 1 m and is placed 50 mm above a PEC ground plane. One of the wires is excited by an AC voltage source with a flat 1 V amplitude and a impedance of 0.1 Ω and terminated at the other end of the wire by 1 Ω. All other wires are terminated with 50 Ω at both ends. This represents a situation in which a wire with a comparably large current couples into adjacent wires with higher impedance.

In the simulation model, we set a change length of 100 mm. Thus, every 10 cm, the cross-section of the bundle changes, which results in ten different cross-sections for one realization of the bundle.

In Fig. 8.42, we show the near-end crosstalk voltage V_{NEXT} for $N_S = 50$ realizations of the random bundle (gray lines). At low frequencies, below 1 MHz, the coupling within the bundle does not depend on the positions of the individual wires. Above 1 MHz, the voltages vary because the individual realizations of the wires influence the crosstalk. Above 30 MHz, the variation is so large that the behavior of the entire set requires a stochastic description. Based on the data, we calculate for each frequency sample f an estimation of the mean value $\hat{\mu}(f)$ and the standard

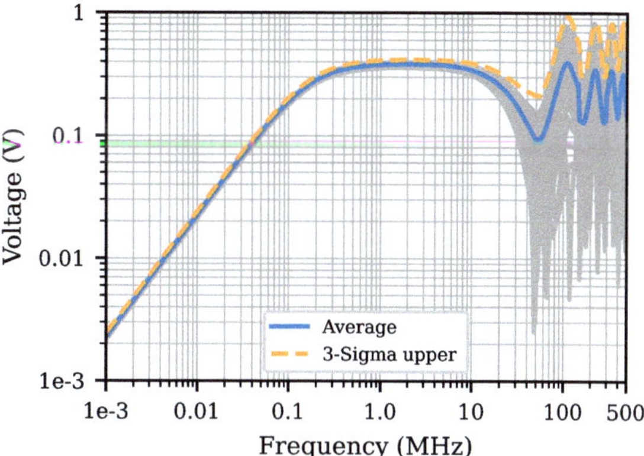

Fig. 8.42 Near-end crosstalk of $N_S = 50$ realizations of a stochastic wire harness (gray lines). Estimated average value (blue line) and average plus three times the estimated standard deviation (dashed orange line)

deviation $\hat{\sigma}(f)$ as

$$\hat{\mu}(f) = \frac{1}{N_S} \sum_{i=1}^{N_S} V_{\mathrm{NEXT,i}}(f), \tag{8.24a}$$

$$\hat{\sigma}(f) = \sqrt{\frac{1}{N_S - 1} \sum_{i=1}^{N_S} \left(V_{\mathrm{NEXT},i}(f) - \hat{\mu}(f)\right)^2}. \tag{8.24b}$$

The figure shows the estimated mean value $\hat{\mu}(f)$ in blue. Since for the estimation of crosstalk rather the worst-case than the mean is of interest, we also require the standard deviation. The orange line in Fig. 8.42 displays the value $\hat{\mu}(f) + 3\sigma(f)$, which estimates the upper bound of, statistically, about 99% of all realizations of $V_{\mathrm{NEXT},i}(f)$.

A statistical investigation is not restricted to a wire bundle only. Any quantity of an electronic system, part of which is a stochastic wire bundle, can be computed multiple times in order to derive statistically meaningful results about its electric or electromagnetic properties.

Further Reading

Transmission Line Theory

A.K. Agrawal, H.J. Price, S.H. Gurbaxani, Transient response of multiconductor transmission lines excited by a nonuniform electromagnetic field. IEEE Trans. Electromagn. Compat. **EMC-22**(2), 119–129 (1980). https://doi.org/10.1109/TEMC.1980.303824

G. Antonini, A. Orlandi, S.A. Pignari, Review of Clayton R. Paul studies on multiconductor transmission lines. IEEE Trans. Electromagn. Compat. **55**(4), 639–647 (2013). https://doi.org/10.1109/TEMC.2013.2265038

P. Besnier, S. Chabane, M. Klingler, Some limiting aspects of transmission line theory and possible improvements. IEEE Electromagn. Compat. Mag. **3**(2), 66–75 (2014). https://doi.org/10.1109/MEMC.2014.6849549

D. Bockelman, W. Eisenstadt, Combined differential and common-mode scattering parameters: theory and simulation. IEEE Trans. Microw. Theory Techn. **43**(7), 1530–1539 (1995). https://doi.org/10.1109/22.392911

S. Caniggia, F. Maradei, Equivalent circuit models for the analysis of coaxial cables immunity, in *2003 IEEE Symposium on Electromagnetic Compatibility. Symposium Record (Cat. No.03CH37446)*, vol. 2 (2003), pp. 881–886. https://doi.org/10.1109/ISEMC.2003.1236725

X. Dong et al., Approximation of worst case crosstalk at high frequencies. IEEE Trans. Electromagn. Compat. **53**(1), 202–208 (2011). https://doi.org/10.1109/TEMC.2010.2081676

M. Feliziani, F. Maradei, Capacitance matrix calculation of a wire conductor line: a new FEM approach. IEEE Trans. Electromagn. Compat. **40**(3), 262–270 (1998). https://doi.org/10.1109/15.709425

J. Guo et al., High-frequency electromagnetic field coupling to transmission line with discontinuity above a lossy ground. IEEE Trans. Electromagn. Compat. **66**(2), 615–625 (2024). https://doi.org/10.1109/TEMC.2024.3352037

M.S. Halligan, D.G. Beetner, Maximum crosstalk estimation in weakly coupled transmission lines. IEEE Trans. Electromagn. Compat. **56**(3), 736–744 (2014). https://doi.org/10.1109/TEMC.2014.2304735

M. Magdowski, S.V. Tkachenko, R. Vick, Coupling of stochastic electromagnetic fields to a transmission line in a reverberation chamber. IEEE Trans. Electromagn. Compat. **53**(2), 308–317 (2011). https://doi.org/10.1109/TEMC.2010.2097267

I. Oganezova et al., Extension of a hybrid MTL/MoM approach for automotive applications. IEEE Trans. Electromagn. Compat. **66**(3), 821–828 (2024). https://doi.org/10.1109/TEMC.2024.3366274

C.R. Paul, *Analysis of Multiconductor Transmission Lines*, 2nd edn. (Wiley-IEEE Press, Hoboken, 2007).

A. Vukicevic et al., On the evaluation of antenna-mode currents along transmission lines. IEEE Trans. Electromagn. Compat. **48**(4), 693–700 (2006). https://doi.org/10.1109/TEMC.2006.884511

Shielded Wires

G. Antonini, A. Orlandi, Spice equivalent circuit of a two-parallel-wires shielded cable for evaluation of the RF induced voltages at the terminations. IEEE Trans. Electromagn. Compat. **46**(2), 189–198 (2004). https://doi.org/10.1109/TEMC.2004.826887

S. Caniggia, F. Maradei, SPICE-like models for the analysis of the conducted and radiated immunity of shielded cables. IEEE Trans. Electromagn. Compat. **46**(4), 606–616 (2004). https://doi.org/10.1109/TEMC.2004.837841

S. Caniggia, F. Maradei, Two-wire shielded cable modeling for the analysis of conducted transient immunity, in *International Symposium on Electromagnetic Compatibility – EMC EUROPE* (2012), pp. 1–6. https://doi.org/10.1109/EMCEurope.2012.6396760

K. Feldhues, M. Diebig, S. Frei, Analysis of the low frequency shielding behavior of high voltage cables in electric vehicles, in *2014 International Symposium on Electromagnetic Compatibility* (2014), pp. 408–413. https://doi.org/10.1109/EMCEurope.2014.6930941

O. Gassab et al., Efficient analytical model for the transfer impedance and admittance of noncoaxial/twinax braided-shielded cables. IEEE Trans. Electromagn. Compat. **62**(6), 2725–2736 (2020). https://doi.org/10.1109/TEMC.2020.2996204

L. Halme, B. Mund, EMC of cables, connectors and components with triaxial test set-up. Transfer **62153**, 83 90 (2013).

T. Kley, Electromagnetic coupling to and from shielded cables-optimizing factors. IEEE Trans. Electromagn. Compat. **34**(1), 39–46 (1992). https://doi.org/10.1109/15.121665

T. Kley, Optimized single-braided cable shields. IEEE Trans. Electromagn. Compat. **35**(1), 1–9 (1993). https://doi.org/10.1109/15.249390

K.S.H. Lee, C.E. Baum, Application of modal analysis to braided-shield cables. IEEE Trans. Electromagn. Compat. **EMC-17**(3), 159–169 (1975). https://doi.org/10.1109/TEMC.975.303403

N. Mora et al., Numerical simulation of the overall transfer impedance of shielded spacecraft harness cable assemblies. IEEE Trans. Electromagn. Compat. **57**(4), 894–902 (2015). https://doi.org/10.1109/TEMC.2015.2404928

S. Mortazavi et al., EMC analysis of shielded twisted pair and shielded parallel pair transmission lines for automotive multi-Gig Ethernet, in *2019 IEEE International Symposium on Electromagnetic Compatibility, Signal & Power Integrity (EMC+SIPI)* (2019), pp. 193–198. https://doi.org/10.1109/ISEMC.2019.8825261

M.S. Murthy, G.A. Rasek, Analytical determination of current distributions in shielded HV cables and ground systems of electric automotive power trains. IEEE Trans. Electromagn. Compat. **61**(3), 911–918 (2019). https://doi.org/10.1109/TEMC.2018.2837121

A. Mushtaq et al., Analysis of shielding effectiveness of HV cable and connector systems used for electric vehicles, in *2013 International Symposium on Electromagnetic Compatibility* (2013), pp. 241–246.

R. Otin, J. Verpoorte, H. Schippers, Finite element model for the computation of the transfer impedance of cable shields. IEEE Trans. Electromagn. Compat. **53**(4), 950–958 (2011). https://doi.org/10.1109/TEMC.2011.2146257

C.R. Paul, *Introduction to Electromagnetic Compatibility* (Wiley, 2005). ISBN: 9780471758150. https://doi.org/10.1002/0471758159

M. Raya, Circuit models of shielded single and multiconductor cables for EMC analyses. Res. Electr. Magdeburg. **88** (2022).

S.A. Schelkunoff, The electromagnetic theory of coaxial transmission lines and cylindrical shields. Bell Syst. Tech. J. **13**(4), 532–579 (1934). https://doi.org/10.1002/j.1538-7305.1934.tb00679.x

E.F. Vance, Shielding effectiveness of braided-wire shields. IEEE Trans. Electromagn. Compat. **EMC-17**(2), 71–77 (1975). https://doi.org/10.1109/TEMC.1975.303389

Stochastic Harnesses

B. Cannas, A. Fanni, F. Maradei, Crosstalk prediction in twisted bundles by a neural approach, in *2002 3rd International Symposium on Electromagnetic Compatibility* (2002), pp. 638–641. https://doi.org/10.1109/ELMAGC.2002.1177512

B. Cannas, A. Fanni, F. Maradei, Neural characterization of wire bundles multiconductor transmission lines. IEEE Trans. Magn. **38**(2), 785–788 (2002). https://doi.org/10.1109/20.996203

M. Gonser et al., Simulation of automotive EMC emission test procedures based on cable bundle measurements, in *IEEE/MTT-S International Symposium* (2012).

M. Gonser et al., Advanced simulations of automotive EMC measurement setups using stochastic cable bundle models, in *Asia-Pacific Symposium on EMC* (2014), pp. 90–93.

T. Sekine, S. Usuki, K.T. Miura, Stochastic modeling and analysis of automotive wire harness based on machine learning and polynomial chaos method, in *2022 International Symposium on Electromagnetic Compatibility – EMC Europe* (2022), pp. 879–884. https://doi.org/10.1109/EMCEurope51680.2022.9901033

S. Sun et al., Hand-assembled cable bundle modeling for crosstalk and common-mode radiation prediction. IEEE Trans. Electromagn. Compat. **49**(3), 708–718 (2007). https://doi.org/10.1109/TEMC.2007.897142

M. Wu et al., Statistical prediction of "reasonable worst-case" crosstalk in cable bundles. IEEE Trans. Electromagn. Compat. **51**(3), 842–851 (2009). https://doi.org/10.1109/TEMC.2009.2026740

E. Yavolovskaya et al., A study of stochasticity effects of cable bundles in automotive EMC problems, in *9th International Symposium on EMC Joint with 20th International Wroclaw Symposium on EMC, EMC Europe 2010* (2010), pp. 197–202. https://doi.org/10.23919/EMC.2010.10826241

Chapter 9
EMI Filters

EMI filters are essential for suppressing conducted emissions to ensure a DUT successfully passes conducted emission tests. Additionally, these filters prevent noise from propagating through the electronic system, which could otherwise cause radiated emissions, for example, via wires. A filter must be compact and, in the case of high-volume production, cost-effective. The role of the EMC engineer involves designing the filter, which includes determining the filter topology, selecting components, designing the mechanical structure, and integrating it into the electronic system.

In this section, we examine the filter of a 300 W motor drive. The layout of this drive is discussed in Sect. 5.3, which covers galvanic coupling and shunt design for monitoring the machine current. The motor drive, in combination with the filter, is further explored in Chap. 10.

We examine the possibilities and limitations of the simulation of the filter only. We use circuit simulation to investigate the choice of the filter components and their parasitics. We select source models which are suitable for filter examination even if there is no information about the noise sources of the DUT that is to be filtered. We describe how to include EM models of the layout and the mechanical design efficiently to perform filter design at higher frequencies. The parasitic properties of common mode chokes can be obtained either by measurement or by simulation. Both approaches and their inclusion into an EM model are discussed. Eventually, we design the EMI filter for the 300 W drive unit using both circuit and EM models. We conclude with a section on EM modeling of common-mode chokes and an advanced section on saturation in magnetic materials.

J. Hansen, A. Barchanski, *EMC Simulation of Electronic Systems*,
https://doi.org/10.1007/978-3-032-09237-3_9

9.1 Filter Design

Domains: 

Filter design is a fundamental task of EMC engineering. It begins early in the design process and accompanies the development of an electronic system throughout all design stages. Many modeling approaches exist, each tailored to the information available about the system at a given point in time. Filter design starts with circuit models to predict filter attenuation. These models include the filter components and any known parasitic properties.

As soon as information about the layout and the mechanical design is available, EM simulation is employed, extending the accuracy of the filter attenuation prediction to higher frequencies. Once details about the noise source to be filtered are known, the process advances from simulating filter attenuation to simulating emissions (Chap. 10). This step significantly improves the accuracy of the simulation, particularly because the source impedance is now known. However, it also increases the complexity of the simulation and the computation time.

The filter discussed in this chapter is shown in Fig. 9.1. The figure depicts a section of the layout of the 300 W motor drive covered in Chap. 10. The layout

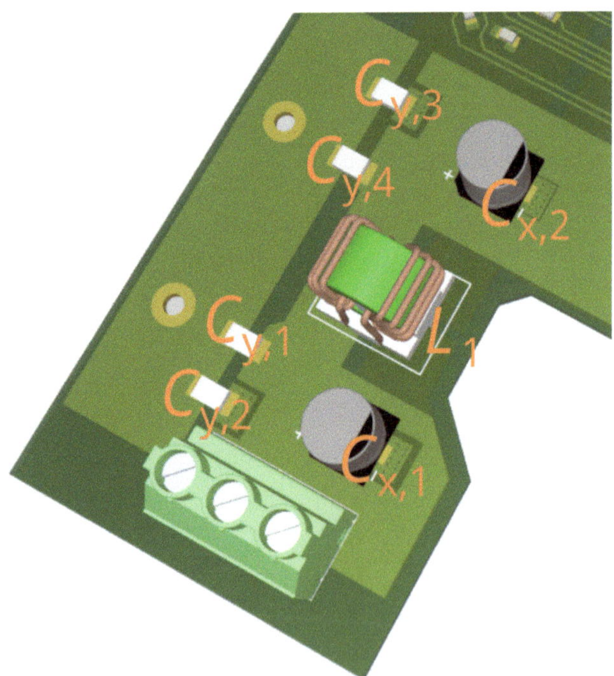

Fig. 9.1 Filter section of a 300 W motor drive

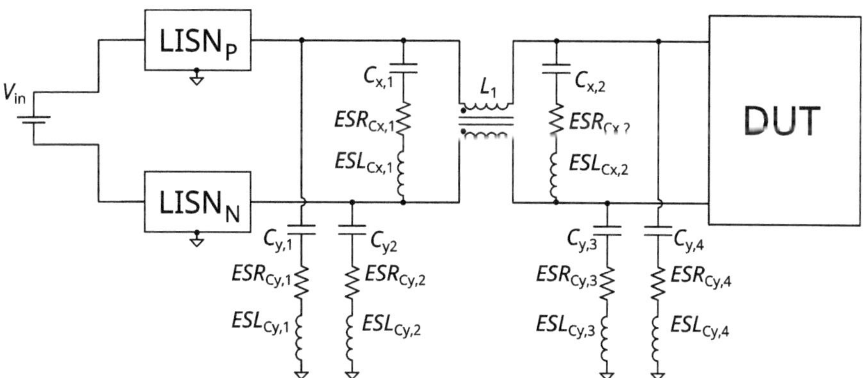

Fig. 9.2 Filter topology and its application between the DUT and the LISNs

includes the connector to the power supply, a toroidal common mode choke, two aluminum electrolytic C_x capacitors (Panasonic), and four SMD C_y capacitors (Murata). The PCB consists of two layers: the positive branch of the power supply is on the top layer, and the negative branch is on the bottom layer. The four C_y capacitors are connected to a common ground. We assume the system is tested in a CISPR 25 setup, that is, the DUT is placed 50 mm above a metallic table plate. Its housing is connected to the plate via a ground strap. This is a crucial assumption, as it influences the grounding of the filter and the propagation of the common mode. The DC link capacitor of the drive is considered not to be part of the filter.

The topology of the filter and its application between the DUT and LISNs is illustrated in Fig. 9.2. There are two LISNs: LISN_P in the positive power supply path and LISN_N in the negative power supply path. The DUT is supplied with a voltage V_{in}.

The filter includes two C_y capacitors, $C_{y,1}$ and $C_{y,2}$, located on the LISN side, and two C_y capacitors, $C_{y,3}$ and $C_{y,4}$, on the DUT side. To each C_y capacitor $C_{y,i}$, a parasitic inductance $\text{ESL}_{\text{Cy},i}$ and a parasitic resistance $\text{ESR}_{\text{Cy},i}$ is added in series. Similarly, the C_x capacitor $C_{x,1}$ with its parasitic inductance $\text{ESL}_{\text{Cx},i}$ and parasitic resistance $\text{ESR}_{\text{Cx},i}$ is on the LISN side, while the capacitor $C_{x,2}$ with corresponding parasitic elements is on the DUT side. The common mode choke in the center of the filter is represented as L_1.

The C_y capacitors and the common mode choke attenuate the common mode voltage $V_{\text{LISN,CM}}$ at the LISNs, while the C_x capacitors and the stray inductance of the common mode choke attenuate the differential mode voltage at the LISNs, $V_{\text{LISN,DM}}$. Therefore, we analyze the common-mode voltage and the differential mode voltage at the LISNs using Eqs. 8.3a and 8.3b. Specifically, the differential mode voltage $V_{\text{LISN,DM}}$ and the common-mode voltage $V_{\text{LISN,CM}}$ at the LISNs are given as

$$V_{\text{LISN,DM}} = V_{\text{LISN,P}} - V_{\text{LISN,N}}, \tag{9.1a}$$

Table 9.1 Reference parameter values for the common mode choke L_1, the C_x and C_y capacitors, and their parasitic elements

L_1 (μF)	k_coup	$C_{\mathrm{x},i}$ (μF)	$C_{\mathrm{y},j}$ (nF)	$\mathrm{ESL}_{\mathrm{Cx},i}$ (nH)	$\mathrm{ESR}_{\mathrm{Cx},i}$ (Ω)	$\mathrm{ESL}_{\mathrm{Cy},j}$ (nH)	$\mathrm{ESR}_{\mathrm{Cy},j}$ (Ω)
16	0.95	10	4.7	1	0.1	0.5	0.1

$$V_\mathrm{LISN,CM} = \frac{1}{2}\left(V_\mathrm{LISN,P} + V_\mathrm{LISN,N}\right). \tag{9.1b}$$

We begin the filter investigation with the following elements: the choke has an inductance of $L_1 = 16\,\mu\mathrm{H}$ and a coupling coefficient of $k_\mathrm{coup} = 0.95$, corresponding to 5% stray inductance. The two C_x capacitors have a capacitance of $C_{\mathrm{x},i} = 10\,\mu\mathrm{F}$. All C_y capacitors are selected to have $C_{\mathrm{y},j} = 4.7\,\mathrm{nF}$. We also include the parasitic elements of the assembly components (see Sect. 2.4.2). The equivalent series inductance of the selected aluminum electrolytic capacitors is estimated as $\mathrm{ESL}_{\mathrm{Cx},i} = 1\,\mathrm{nH}$ and their resistance as $\mathrm{ESR}_{\mathrm{Cx},i} = 0.1\,\Omega$. The SMD C_y capacitors are estimated to have a parasitic inductance of $\mathrm{ESL}_{\mathrm{Cy},j} = 0.5\,\mathrm{nH}$ and a resistance of $\mathrm{ESR}_{\mathrm{Cy},j} = 0.1\,\Omega$. As explained in Sect. 2.4.2.3, the ESL and ESR are attached in series to the nominal capacitance values.

At this stage, we do not consider any parasitic elements of the choke model. Neglecting these elements overestimates the performance of the choke at frequencies where resistive and capacitive effects dominate (cf. Sect. 2.4). The values of all nominal and parasitic elements of the filter are summarized in Table 9.1.

Filter design is carried out to ensure compliance with emission limits in the frequency domain. All simulations in this chapter are performed using frequency domain simulation, as it is both fast and stable.

9.2 Filter Analysis by Circuit Simulation

Domains: 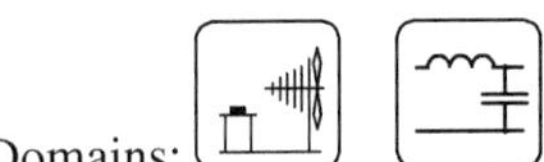

Filter analysis can begin as early as the quotation phase of a new project, when only the customer request and system requirements are available, with no detailed information about the electronic system to be developed. At this stage, the size and cost of the EMI filter must be estimated to support the quotation process. During the requirements analysis, the necessary filter attenuation can be determined. To achieve this, the required EMI limits and the anticipated conducted emissions of the system to be designed must be considered. The expected emissions can be estimated based on a similar existing electronic system or through a circuit simulation of the emission sources, as discussed in Chap. 3. The difference (in decibels) between the expected emissions and the EMI limits provides an estimate of the required filter attenuation.

For this estimate, we construct a circuit model of the filter that includes the circuit components and their parasitic elements while neglecting explicit mechanical details. This approach is permissible because cost and size of the filter are primarily determined by the low-frequency behavior of the circuit elements, such as the choke and the C_x and C_y capacitors. The impact of the mechanical design becomes significant at higher frequencies, which is addressed in Sect. 9.3. Using a simplified circuit model enables us to estimate the size and quantity of the required filter elements, forming the basis of any cost estimation. It also aids in deciding for or against a specific filter topology, such as a second- or third-order filter or the necessity of including one or more bulky common mode chokes.

To compute $V_{LISN,CM}$ and $V_{LISN,DM}$, we require two circuit models, as depicted in Fig. 9.3. The distinction between the two lies in the implementation of the noise

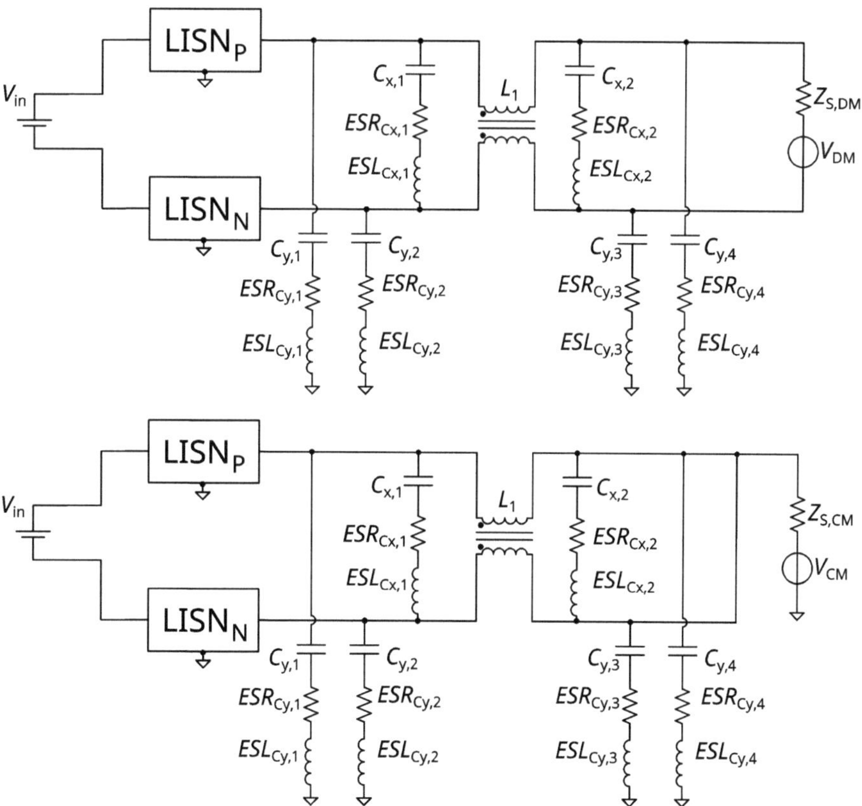

Fig. 9.3 Circuit model for the calculation of LISN voltages in the filter model. Top: Calculation of the LISN voltage from a differential mode excitation $V_{LISN,DM}$. Bottom: Calculation of the LISN voltage from a common-mode excitation $V_{LISN,CM}$

source. In the upper circuit, the EMI noise is generated by a voltage source V_{DM} with a source impedance $Z_{S,DM}$. This source is connected between the positive and negative power supply planes through the filter, mimicking the differential mode noise emitted by the noise sources of the electronic system yet to be developed.

In the circuit on the bottom, a common-mode noise source V_{CM} with a source impedance $Z_{S,CM}$ is implemented. Here, the positive and negative supply planes of the filter are shorted, allowing a common mode to be injected, which returns either via the LISNs or the C_y capacitors. For both the common mode and differential mode sources, an AC source with an amplitude of $V_{CM} = V_{DM} = 1\,\text{V}$ is used.

The choice of the source impedances $Z_{S,DM}$ and $Z_{S,CM}$ has a significant impact on the filter attenuation and requires careful consideration. This filter is designed for a 300 W motor drive. The differential mode corresponds to the path of the load current, which has a low resistance of several hundred mΩ. Therefore, we select $Z_{SDM} = 0.1\,\Omega$, which mimics the impedance of the DC link capacitor.

The common mode is generated by the displacement current that propagates via capacitive coupling from the noise sources, primarily the power transistors, to the housing and the metallic table on which the electronic system is tested. The dominating capacitance in a power drive is formed within the electrical machine between the rotor and the stator or the capacitance between the transistor and the PCB ground. This capacitance is a suitable choice for $Z_{S,CM}$.

Figure 9.4 shows $V_{LISN,DM}$ and $V_{LISN,CM}$ across a frequency range from 1 kHz to 200 MHz. The differential mode voltage $V_{LISN,DM}$ (blue line) is attenuated by 35 dB at 1 kHz. A resonance occurs at 43 kHz, after which $V_{LISN,DM}$ decreases, with the filter attenuation being more effective at higher frequencies.

The common-mode voltage $V_{LISN,CM}$ is plotted for two different source impedances $Z_{S,CM}$: a resistive impedance of $0.1\,\Omega$ (dashed orange line) and a capacitive impedance of 100 pF (continuous orange line). The impact of $Z_{S,CM}$ is significant. At low frequencies, the capacitive source impedance reduces $V_{LISN,CM}$

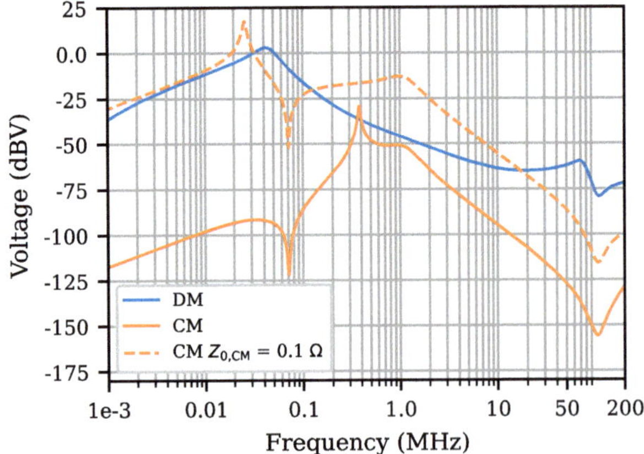

Fig. 9.4 Voltages $V_{LISN,DM}$ and $V_{LISN,CM}$ at the LISNs, with $Z_{SDM} = 0.1\,\Omega$. For $V_{LISN,CM}$, we use $Z_{SCM} = 100\,\text{pF}$. A value of $Z_{SCM} = 0.1\,\Omega$ yields the dashed line

by 100 dB. This is expected, as the effect of capacitive coupling to the housing approaches zero at lower frequencies. As the frequency increases, the coupling also increases.

Above 2 MHz, the $V_{LISN,CM}$ curves for the resistive and capacitive impedances exhibit the same shape, with an offset of 40 dB at 2 MHz and 30 dB at 200 MHz.

Using the capacitance to describe the capacitive coupling of the common-mode results in a very large $Z_{S,CM}$ at low frequencies. Below 5 MHz, $Z_{S,CM}$ exceeds 1 kΩ. Since it is in series with the voltage source V_{CM}, it significantly reduces the current compared to an impedance of $Z_{SCM} = 0.1\ \Omega$, leading to a correspondingly low $V_{LISN,CM}$. Estimating $Z_{S,CM}$ accurately is challenging if the capacitive coupling of the DUT is not known.

When $Z_{S,CM}$ dominates the circuit impedance, the common-mode voltage $V_{LISN,CM}$ scales inversely with $Z_{S,CM}$. For the circuit shown in Fig. 9.3, increasing the capacitance from 100 pF to 400 pF results in a fourfold increase in the current through the source and an increase in $V_{LISN,CM}$ by 12 dB. While estimating $Z_{S,CM}$ accurately may be difficult, its impact on $V_{LISN,CM}$ can be quantified if $Z_{S,CM}$ is much larger than any other impedance in the circuit.[1]

A detailed analysis of the filter components is provided in Sect. 9.5.1.

9.3 Coupled EM-Circuit Simulations

Domains:  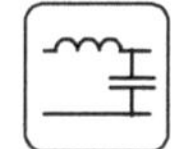

A circuit simulation cannot efficiently account for the electromagnetic properties of the mechanical design. Once CAD data for the filter is available, an EM model must be created to include all parasitic effects induced by the geometry. In Sect. 6.7, we develop such an EM model by equipping the geometry model with multiple ports. After performing the EM computation, we obtain a circuit description of the mechanical geometry, which can then be integrated with circuit elements in a circuit simulation. In this section, we apply this method to simulate the attenuation of the filter, incorporating both its geometry and electrical components.

Given a particular geometry, we can still vary the filter components in a circuit simulation without needing to rerun the EM simulation, which would require significantly more computational effort. This approach combines the accuracy of the EM computation of the mechanical design with the efficiency of circuit simulation for the filter components.

The choice of sources in this circuit simulation follows the same principles as those described for the circuit model in Sect. 9.2. Two common procedures are

[1] A discussion about the choice of source models and their impact on simulation results is also found in Sect. 3.1.

outlined in that section. The first uses S-parameter ports and is well suited for comparison with network analyzer measurements. The second provides a more detailed description of the sources and sinks using circuit models. Both procedures utilize the same underlying EM model.

9.3.1 EM Model Generation

The geometry used to compute the EM model is shown in Fig. 9.5. The layout can be imported using standard layout data formats (cf. Sect. 4.2). It is reduced to the section containing the filter elements, similar to what is depicted in Fig. 9.1. The positive power supply plane of the filter, connected to LISN_P in the system, is displayed in red and the negative plane in blue, and the common ground, to which the C_y capacitors are connected, is shown in orange.

We assume that we are in an early design stage and that the CAD data for the housing is not yet available. In EMC design, the housing serves as the return path for currents, and the inductance of this return path is critical. This inductance is influenced by the distance between the housing and the PCB, as well as the number and positions of the connection points between them.

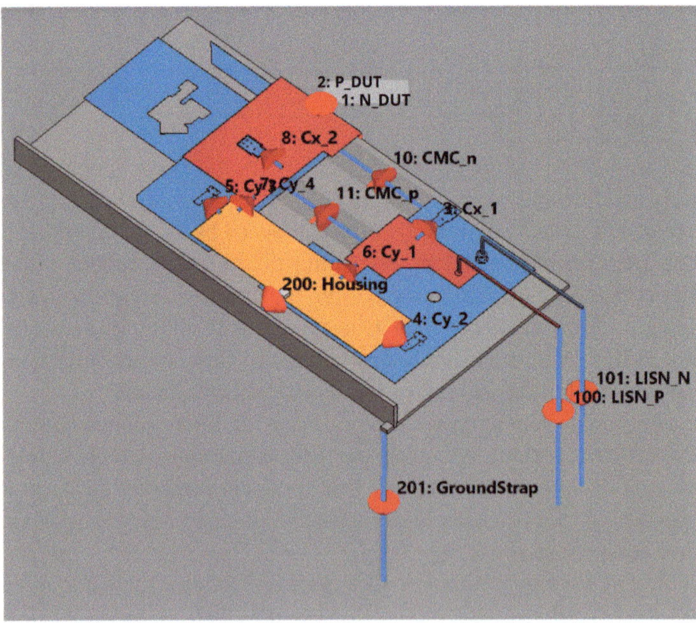

Fig. 9.5 CAD model of the mechanical design of the EMI filter, including a self-made sketch of the housing, connector pins, and ports

The EMC engineer must sketch the relevant parts of the housing, approximating its electromagnetically significant features. In the figure, it is represented as a gray plane located 1 mm from the PCB. One side of the housing is connected to the common ground of the PCB, representing just one of many possible designs. The impact of this design on the filter attenuation is analyzed later.

The connector pins are also self-created, as this approach is typically faster than obtaining accurate CAD data for the connector. The exact design of the connector is not relevant in the frequency range of interest, which is the CISPR 25 conducted emission test that is performed below 200 MHz. Below the PCB, the gray structure represents the metallic table on which the electronic system is tested for conducted emissions according to the CISPR 25 standard. This table is positioned 50 mm below the housing of the PCB, as specified by the standard.

The table is modeled as an infinitely large plane using an electric boundary condition, assuming that its finite size has no significant impact on the return path or the inductance of the common-mode currents.

In this model, no specific circuit elements are defined. Instead, the model is equipped with a total of n S-parameter ports (cf. Sect. 2.4.3.2). The subsequent EM simulation operates as follows: Each port is sequentially excited while all other ports are terminated with 50 Ω. The EM simulation computes the voltages V_i across each port and the currents I_j through each port for $i, j = 1, \ldots, n$. Using Eq. 2.5, it calculates the incoming and outgoing power waves for all excitations and terminals.

From this data, the complete S-parameter matrix is assembled, providing a multiport network description of the electromagnetic properties of the CAD structure in terms of voltages and currents. This network description is subsequently used in a circuit simulation.

In Fig. 9.5, the model includes a total of $n = 14$ ports. Two ports are assigned to the C_X capacitors, $C_{\mathrm{X},1}$ and $C_{\mathrm{X},2}$, which connect the positive and negative sections of the layout (red and blue). Two additional ports, $\mathrm{CMC_P}$ and $\mathrm{CMC_N}$, connect the positive and negative branches of the CMC to the layout. There are four ports for the C_Y capacitors. One port, named 200:Housing, connects the common ground (orange) to the housing. Two ports, 100:$\mathrm{LISN_P}$ and 101:$\mathrm{LISN_N}$, are located between the connector pins and the table, while the port 201: Groundstrap connects the housing to the table. Finally, the ports 1:$\mathrm{N_{DUT}}$ and 2:$\mathrm{P_{DUT}}$ terminate the model on the side of the filter connected to the motor drive's half-bridges. The first port, 1:$\mathrm{N_{DUT}}$, represents an excitation or termination between the negative layout section (blue) and the housing, while the second port, 2:$\mathrm{P_{DUT}}$, represents an excitation or termination between the positive layout section (red) and the housing.

We run the respective EM simulation to compute the voltages and currents at the ports and subsequently the S-parameter matrix. The EM model employs lossy metal for all layers of the PCB and PEC material for the housing and the table. Below the table, an electric boundary condition is applied, while open boundary conditions are used elsewhere. For the simulation, we select the direct solver in the frequency domain, as the model contains many ports (here: 14) (cf. Sect. 4.9). The element order is variable, with a maximum order of 2. The simulation time for the EM model is 30 seconds.

9.3.2 S-Parameter Simulation

The impact of termination impedances on voltages and currents in a circuit is significant, as demonstrated in Sect. 9.2 and Fig. 9.4. When computing S-parameters, the termination impedances are set to 50 Ω. This choice is motivated by RF measurement practices, where 50 Ω is the standard termination used, for instance, in network analyzers. While these terminations may not accurately represent the EMI characteristics of the electronic system under test, S-parameter simulations are nevertheless valuable. They provide a means to compare filter simulation results with measurements obtained using a network analyzer.

When applied to a filter, an assessment based on S-parameters yields the filter insertion loss, denoted by S_{21} for a single-line filter. In the present case, there are two input lines and two output lines, resulting in a 4-port configuration. The corresponding circuit model is shown in Fig. 9.6, the ways to monitor the insertion loss are discussed below.

The EM model contains $n = 14$ ports, with all filter components attached. These include the two C_x capacitors and four C_y capacitors, along with their parasitic elements as specified in Table 9.1. Additionally, the ideal model of the CMC L_1, consisting of two coupled inductors, is connected to two ports.

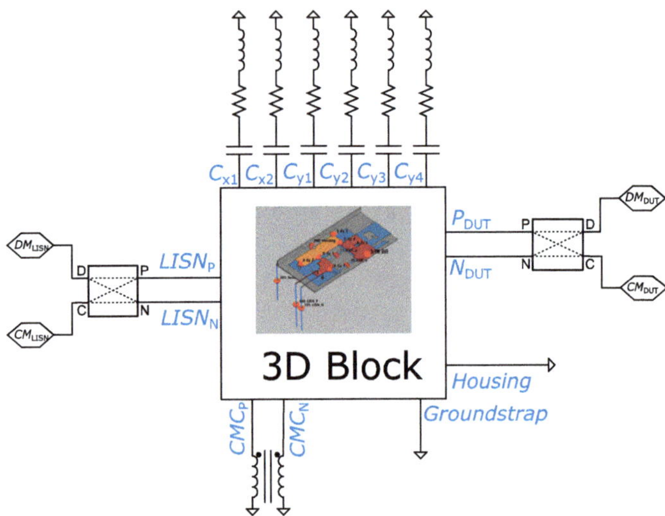

Fig. 9.6 EM model to compute the S-parameters for common and differential mode voltage $V_{LISN,CM}$, $V_{LISN,DM}$

In the model, we assume the DUT to be connected to the metallic table using a groundstrap. We model the strap to be ideal by shorting its pin to ground. There is no voltage drop along the groundstrap. Similarly, the port 200: Housing is connected to the common ground ideally, with zero impedance, to the housing. We make use of these ports later in the circuit simulation and also in the system simulations in Chap. 10.

Four ports remain: the positive and negative power supply planes toward the LISNs and the DUT. In hardware, this setup could be measured using a four-port network analyzer, allowing us to extract S-parameters such as $S_{\mathrm{LISN_P,P(DUT)}}$ or $S_{\mathrm{LISN_N,N(DUT)}}$, which describe the insertion loss along the power supply planes.

However, since our focus is on analyzing the common and differential mode, we use the mode converters introduced in Sect. 8.3, as shown in Fig. 9.6. With these converters, we obtain the insertion losses for the common and differential modes, $S_{\mathrm{CM(LISN),CM(DUT)}}$ and $S_{\mathrm{DM(LISN),DM(DUT)}}$, as well as the mode conversion parameters, such as $S_{\mathrm{CM(LISN),DM(DUT)}}$ and $S_{\mathrm{DM(LISN),CM(DUT)}}$. These are four elements out of the complete 4×4 S-parameter matrix, which has 16 elements in total.

The impedances of the four sources shown in Fig. 9.6 are not 50 Ω. To measure the S-parameters, all lines are terminated with 50 Ω. This configuration means that the differential mode propagates through two 50 Ω resistors in series, while the common-mode propagates through two 50 Ω resistors in parallel. When using mode converters, in this book we use a source impedance for differential mode sources of 100 Ω and of 25 Ω for common-mode sources.

The S-parameters are shown in Fig. 9.7. In the common mode (orange curve), the insertion loss is 0 dB up to a frequency of 300 kHz, increasing to a minimum of −40 dB at 4 MHz. The insertion loss for the differential mode (blue) is significantly better, reaching values as low as −100 dB.

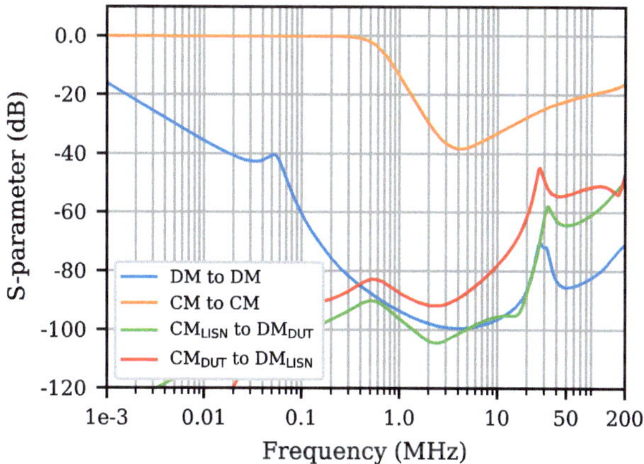

Fig. 9.7 S-parameters of the filter with regard to common and differential modes and mode conversion

Below approximately 30 kHz, the mode conversion remains below −100 dB. However, at higher frequencies, the mode conversion is stronger than the differential mode insertion loss. Notably, for $S_{DM(LISN),CM(DUT)}$, this indicates that common mode noise generated at the DUT contributes more to the differential mode noise at the LISNs than differential mode noise generated by two single-ended 50 Ω sources at the DUT (the red curve exceeds the blue curve).

The results computed here could be compared to four-port network analyzer measurements obtained from the same filter in hardware. However, there are situations where a four-port network analyzer is not available, and only a two-port analyzer is accessible. In such cases, a common procedure to measure the common mode insertion loss involves shorting the power supply layers on both the input and output sides and measuring the resulting two-port response. This approach does not yield the same results as those obtained here. In the two-port case, the differential mode is shorted, whereas in the four-port setup used in the simulation, the differential mode port is terminated with 100 Ω.

In order to compare a two-port with a four-port setup, the termination impedances of the four-port setup must be chosen appropriately.

9.3.3 System Description

In a CISPR 25 setup, the power supply wires are terminated by two LISNs. We investigate filter components and their effect on the common and differential mode voltages, $V_{LISN,CM}$ and $V_{LISN,DM}$. Instead of terminating the power supply wires with 50 Ω ports, as done in Sect. 9.3.2, we use LISN models for termination in this section. Similarly, instead of source impedances of 50 Ω at the DUT side, we employ the differential and common mode source impedances defined in Sect. 9.2, that is, $Z_{SDM} = 0.1\,\Omega$ for the differential mode and a series capacitance of $C_{CM} = 100\,\text{pF}$ in series to the common mode source. This value is a typical capacitance between the stator windings and the housing of a small electric motor.

The resulting circuit is depicted in Fig. 9.8. The EM model is identical to the one used in the previous section for the computation of the S-parameters. The assembly with the filter components, including the capacitances and the CMC, as well as the shorts for the connections to the housing and the ground strap, also remains unchanged. No additional EM computation is required to transition from the S-parameter simulation to this setup, which represents the geometry and the components of the filter as a part of the entire electronic system.

The resulting $V_{LISN,CM}$ and $V_{LISN,DM}$ are plotted in Fig. 9.9. The blue line represents the differential mode voltage $V_{LISN,DM}$, as excited by the differential mode voltage source V_{DM}. The orange line indicates the common mode voltage $V_{LISN,CM}$, as generated by the source V_{CM}. For comparison, the results from the circuit simulation in Sect. 9.2 are also included.

Below 2 MHz, the parasitic inductances and capacitances have no impact on the voltages at the LISNs. However, above this frequency, the filter attenuation begins

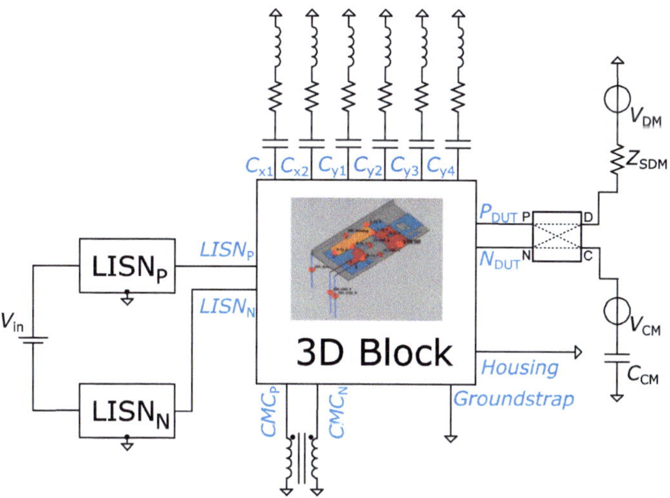

Fig. 9.8 EM model with connected filter elements, sources and sink models

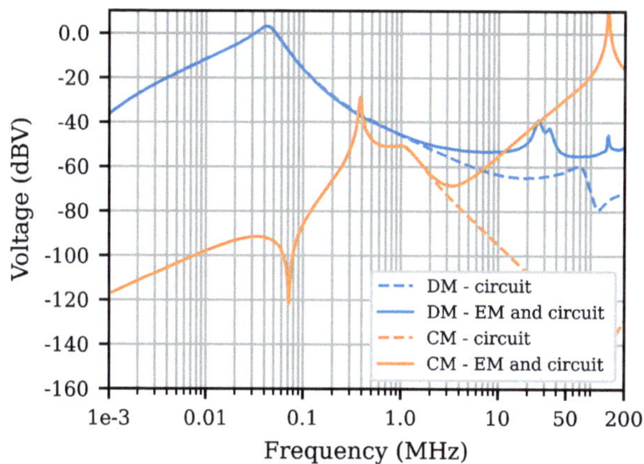

Fig. 9.9 $V_{LISN,CM}$, $V_{LISN,DM}$ of the system. Comparison of the results with the EM model and the circuit simulation of Sect. 9.2

to degrade. For $V_{LISN,DM}$, the degradation reaches up to 20 dB, with two resonances appearing around 30 MHz. The common mode voltage $V_{LISN,CM}$ deteriorates even stronger. While the circuit model predicts an improvement in filter attenuation up to 100 MHz, the EM model shows a decline in performance above 2 MHz.

The impact of the layout and mechanical design on the common mode is examined in detail in Sect. 9.5.2.

Comparing the results from Fig. 9.9 with the S-parameters in Fig. 9.7, we observe much more structure in the curves of the voltages. The S-parameters monitor the transmission through the filter, which behaves as a through at low frequencies. In the system simulation, we see resonances due to interaction with the circuit elements of the LISNs and C_{CM}.

We plot the voltages starting from 1 kHz, even though the CISPR 25 standard specifies measurements only above 150 kHz. Examining frequencies as low as 1 kHz or even 1 Hz is valuable for understanding the model topology and verifying connections between elements. At these lower frequencies, capacitors behave as open circuits, while inductors act as short circuits, making the model behavior intuitive. As shown in Fig. 9.9, $V_{\mathrm{LISN,CM}}$ decreases to 0 V at low frequencies because the series capacitance with the source V_{CM} causes the impedance to increase toward infinity. Conversely, the differential mode voltage, $V_{\mathrm{LISN,DM}}$, increases at approximately 20 dB per decade due to the inductors in the LISNs behaving as short circuits at low frequencies. Above 150 kHz, this behavior is obscured by resonances.

Extending the frequency range above 108 MHz to 200 MHz or 300 MHz provides insight into resonances that occur beyond the upper frequency range defined by the CISPR 25 conducted emission standard. These resonances are influenced by parasitic elements in the mechanical design or layout. If the design is altered later, these resonances could shift into the frequency range below 108 MHz. It is advisable to account for them early in the modeling process to anticipate and address potential issues that may arise later.

9.4 Using Measured S-Parameters in EM Simulations

Domains: 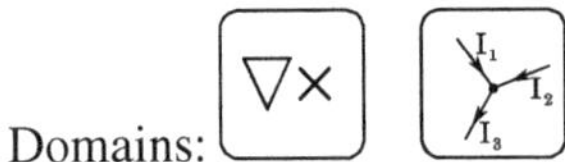

In the previous section, we utilize a simplified model of the CMC, consisting solely of two coupled inductors and neglecting the parasitics of the CMC. This simplified model is overly optimistic and tends to overestimate the filter's attenuation above the frequencies where parasitics degrade the choke's inductive properties.

S-parameters provide a convenient and more precise characterization of frequency-dependent behavior compared to ideal circuit elements. These parameters can be obtained through measurements or simulations and are often provided by component vendors for download. A common mode choke is typically represented by a four-port S-parameter model, characterized using the choke's four ports with respect to a common reference.

In the EM model developed in Sect. 9.3.1, we only consider two ports, labeled as port 11 and port 10 in Fig. 9.20. During circuit simulation, an inductor representing one branch of the choke is connected to each of these ports (e.g., Fig. 9.6). This inductor induces a voltage drop at the respective port, as described in Sect. 6.7. However, the behavior of a component extends beyond the interaction between the two ports where it is connected. When modeled or measured as a 4-port device, there

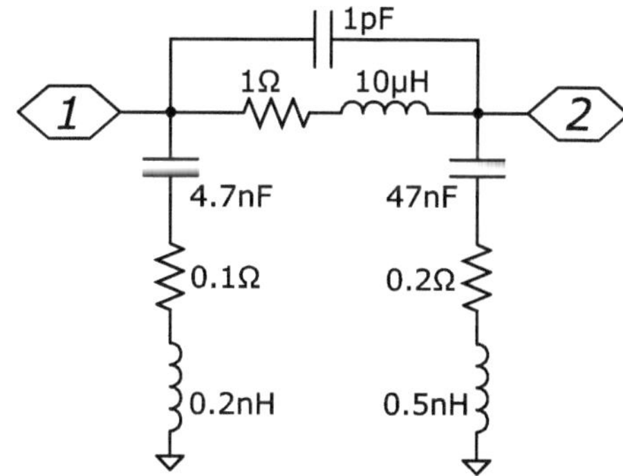

Fig. 9.10 π-filter schematic for the export of S-parameters used to connect to an EM model in this section

is a ground reference shared by all four ports—a feature absent in the EM model to which the choke is attached (refer to Sect. 9.6).

In this section, we discuss the limitations of circuit component models generated in isolated setups and used in coupled EM-circuit simulations. To simplify the discussion, we begin with the π-filter introduced in Sects. 2.4.3.1 and 6.7. After analyzing the π-filter configuration, we extend the discussion to the common mode choke.

9.4.1 π-Filter

For this study, we calculate the S-parameters of the π-filter from Sect. 2.4.3.1 and export them into a Touchstone file. The circuit of the π-filter and its port definition are illustrated in Fig. 9.10. In the next step, we connect this Touchstone file to an EM model.

The EM model is based on the filter PCB that we also use in Sect. 6.7. Here, instead of connecting the two capacitors and the inductor individually, we connect the complete filter to the EM model. For this purpose, the traces used to connect the capacitors are removed from the geometry model, and ports are placed only at the position of the inductor. Since the distances between the capacitor connections and the inductor are electrically short within the frequency range of this study, we can lump the entire filter circuit onto a single position without expecting significant deviations.

We study two different setups. The first setup consists of a single port placed in series, connecting the two sides of the signal trace. In the second setup, two ports are placed at the ends of the signal trace, between the signal trace and the reference layer. Both setups are illustrated at the top of Fig. 9.11. The resulting schematics, which connect the Touchstone file of the filter to the EM model, are shown below the EM models.

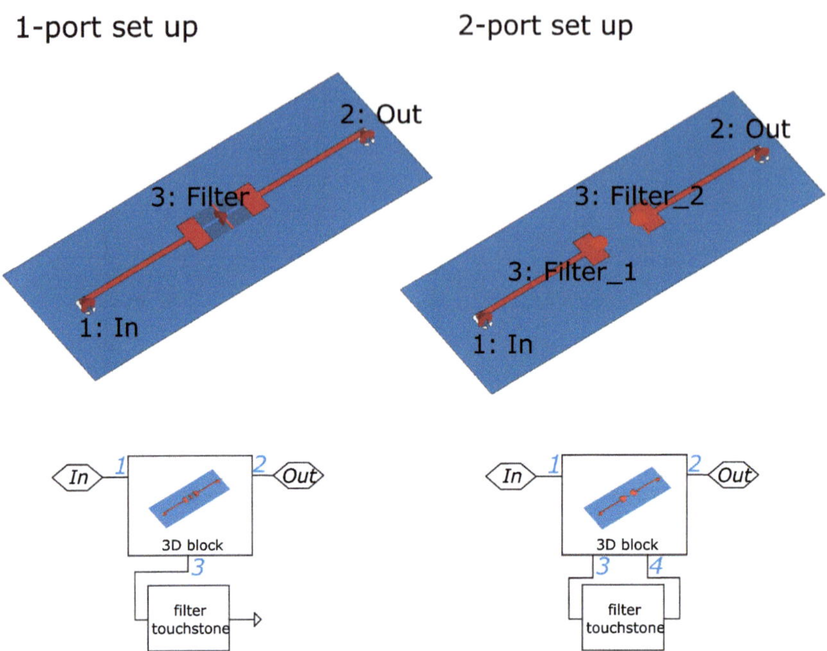

Fig. 9.11 Connecting an S-parameter block to an EM block. Left: 1-port setup. Right: 2-port setup

In the one-port setup, one pin of the Touchstone file is connected to the EM block, while the other pin is connected to the circuit reference. In contrast, the two-port setup connects the Touchstone file to the two filter ports of the EM model.

At this point, we can already deduce that the one-port setup cannot deliver correct results. In the circuit model, the capacitors are connected to ground. However, the one-port setup in the EM model does not allow any current flow to the reference plane because the port is placed in series. This intuition is confirmed by comparing the resulting S-parameters of the one-port setup, the two-port setup, and the standalone Touchstone file of the filter. The results are plotted in Fig. 9.12, where a reference impedance of 50 Ω is used in the calculation of the S-parameters.

The return loss at the "In" port is plotted on the left-hand side of Fig. 9.12. We observe no difference between the standalone Touchstone file of the π-filter and the two-port setup, while the results of the one-port setup deviate from the other two above 1 kHz. Beyond this frequency, the current path through the capacitors to the reference becomes significant in the S-parameter results, rendering the one-port setup invalid. A similar observation can be made for the insertion loss from the "In" port to the "Out" port, as shown on the right-hand side of Fig. 9.12. Only above 20 MHz do the results of the standalone Touchstone file and the two-port setup begin to differ. This divergence is due to the three-dimensional geometry effects, which are minor, as the difference is only noticeable below -80 dB.

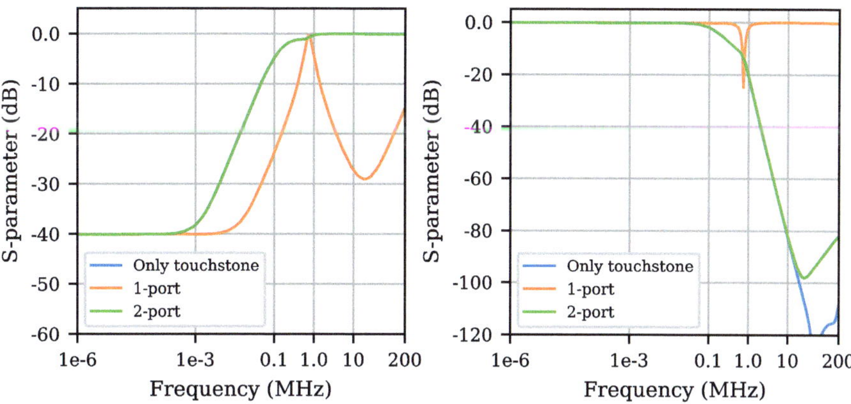

Fig. 9.12 Comparison of S-parameters of the π-filter for the one-port setup, two-port setup, and touchstone standalone. Left: Return loss at "In" port. The blue curve is completely covered by the green curve. Right: Insertion loss of the filter from port "In" to port "Out"

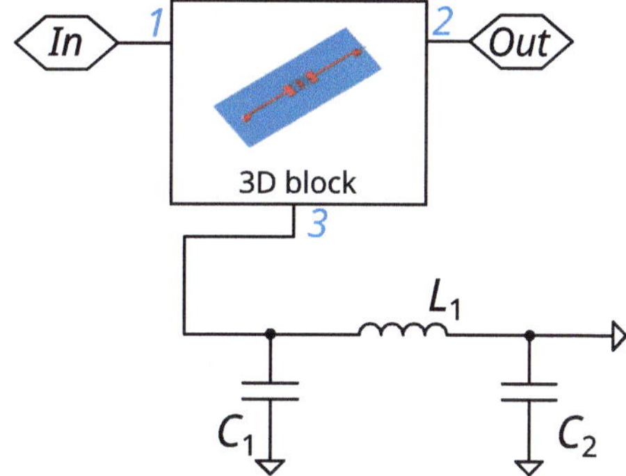

Fig. 9.13 Schematic of the one-port setup with circuit elements instead of the touchstone file

The results in Fig. 9.12 confirm that connecting a two-port Touchstone file to a one-port EM model leads to incorrect results when the ratio of the shunt impedance to the through impedance of the network contained in the Touchstone file decreases. As this ratio becomes smaller, the current to the reference increases. It is worthwhile to analyze the one-port setup in greater detail to understand the effect of the one-port configuration on the attached network represented by the Touchstone file.

Because the Touchstone file of the π-filter is obtained from another circuit simulation and not from a measurement, we can replace the Touchstone file in the one-port setup with the actual circuit. The resulting circuit is shown in Fig. 9.13. For clarity, we omit the parasitics of the filter in Fig. 9.13, though they can be included if necessary. When analyzing the one-port setup with the actual circuit, we observe the following: the capacitor C_2 is shorted because both terminals of C_2 are connected to the circuit reference. Capacitor C_1 is connected in parallel with the inductor L_1, as one side of C_1 and L_1 is connected to the EM block, while the other side is connected to the circuit reference. This circuit configuration results in the S-parameters observed for the one-port setup in Fig. 9.12.

While connecting a two-port Touchstone file to a serial one-port in the EM model can lead to incorrect results, this approach is still commonly used because it simplifies the setup and reduces the number of ports. Since the deviation of the one-port setup from the two-port setup arises from the current flowing to the reference, the one-port setup can still be used if this current is negligible in the network represented by the Touchstone file. Many component models provided by vendors are measured as two-port models, and it is important to understand the conditions under which these models can be used with a one-port setup. For capacitors, when characterized by a series measurement, the one-port setup is often suitable: the through impedance of the actual capacitor is much smaller than the shunt impedance to the reference of the measurement fixture. As a result, the error introduced by the one-port setup is typically small for capacitors. For inductors, the suitability depends on the specific measurement setup. If uncertain, the Touchstone model should be tested in the same manner as the filter in this section. For measured filters with significant current flowing to the reference, the one-port setup is not suitable.

In this discussion, we only consider S-parameter models loaded as Touchstone files, but the same effects apply to SPICE netlists. Occasionally, component vendors provide both a measurement-based Touchstone file and a SPICE netlist for the same component, which may produce different results. The network contained in the SPICE netlist can be generated by the vendor using various methods. A common example is the modeling of common mode chokes: the measurement-based Touchstone file includes intrinsic common-to-differential mode conversion in the common mode choke, while this conversion is omitted in the SPICE model. Additionally, the parasitic capacitance of the common mode choke to the measurement reference is included in the measurement but omitted in the SPICE netlist. When in doubt, it is recommended to test the component model provided by vendors in a simple EM model, as done in this section, before incorporating it into the simulation model of the DUT. Methods to extract RLC values and impedances from component models are presented in Sect. 2.7.

9.4.2 Common Mode Choke

For the π-filter in the previous subsection, we use both a one-port setup and a two-port setup. The common mode choke that we use in Sect. 9.3 was measured as a four-port, but in the EM model of the filter, we use it as a two-port. In this section, we discuss the differences between a two-port setup and a 4-port setup for a common mode choke.

The primary reason for using a two-port setup is the absence of a nearby reference plane in the filter layout (see Figs. 9.5 and 9.20). In the filter model, the reference is the common ground, represented by the orange plane. Connecting four ports from the choke's footprint to the orange conductor in the geometry model would result in long ports across the filter, creating a current path that does not physically exist. The issue arises because the choke was measured with a reference plane present, whereas the layout lacks this reference.

When using the simplified circuit model of the choke, which consists of two coupled inductors, this problem does not occur because the inductors only represent a through impedance. However, for the measured Touchstone file, we need to evaluate the error caused by the two-port setup in the same way we tested the π-filter.

The Touchstone file containing the S-parameters of the common mode choke was obtained from a four-port measurement using a reference plane of a small PCB specifically manufactured for impedance and S-parameter measurements of the choke.[2] The S-parameters obtained from our measurement align with the S-parameter measurements provided by the vendor.

We use mode converter blocks to calculate the common and differential mode transmission of the choke directly from the Touchstone file. To compare these results with a coupled simulation using an EM model and the measured Touchstone data, we use the same simple PCB layout as in the π-filter case but add a second signal trace required to connect the common mode choke. The layout and the resulting schematics are shown in Fig. 9.14.

The resulting transmission S-parameters for the common and differential modes are plotted in Fig. 9.15. In S-parameter notation, these are $S_{\mathrm{InCM,OutCM}}$ and $S_{\mathrm{InDM,OutDM}}$.

We compare three models: the standalone Touchstone block (without coupling to an EM model) and the models using a two-port and a four-port connection to the layout depicted in Fig. 9.15.

Comparing the results of the simulation using the standalone Touchstone file with the four-port setup, we observe only minor differences, with a maximum deviation of 2 dB above 200 MHz. These differences can be attributed to the signal traces. However, when comparing the two-port setup to the other two models, a downshift in the resonance frequency is evident. This deviation becomes noticeable above a few MHz and is caused by the capacitance to the reference plane included in the measurement. In the two-port setup, this capacitance appears in parallel with the inductor, as described in the previous section (see Fig. 9.13).

Overall, the deviation between the two-port and four-port setup for the common mode choke is less significant than in the case of the π-filter, but it still impacts the accuracy of the results. It is ultimately up to the simulation engineer to determine whether this deviation is acceptable for the specific application.

For the simulation of the complete filter of the motor drive discussed in this chapter, the deviation between the two-port model and the four-port model is actually lower than in the standalone test. In the frequency band where the port setup affects the choke impedance, the common mode current flows mainly through

[2] The measurements were performed at the Institute of Electronics, Graz University of Technology.

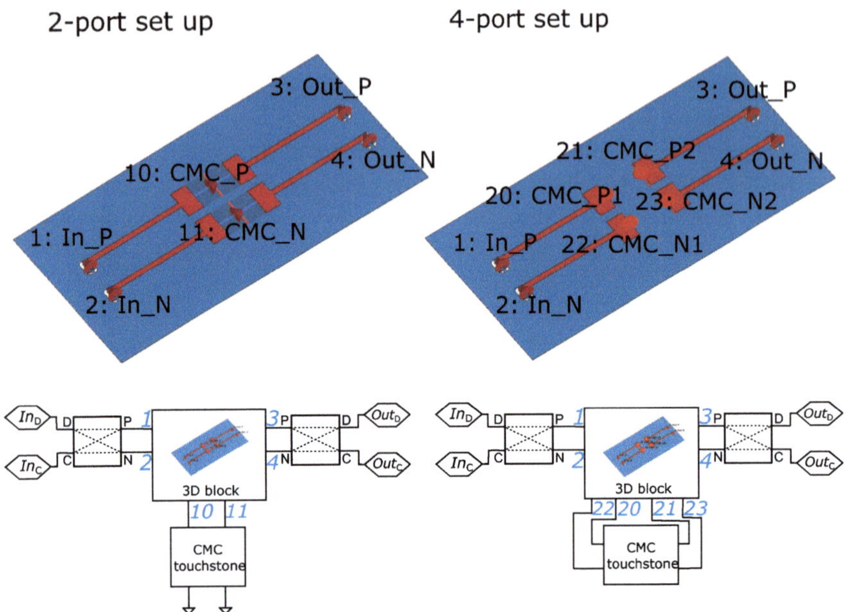

Fig. 9.14 Connecting the common mode choke S-parameter block to an EM block. Left: two-port setup. Right: four-port setup

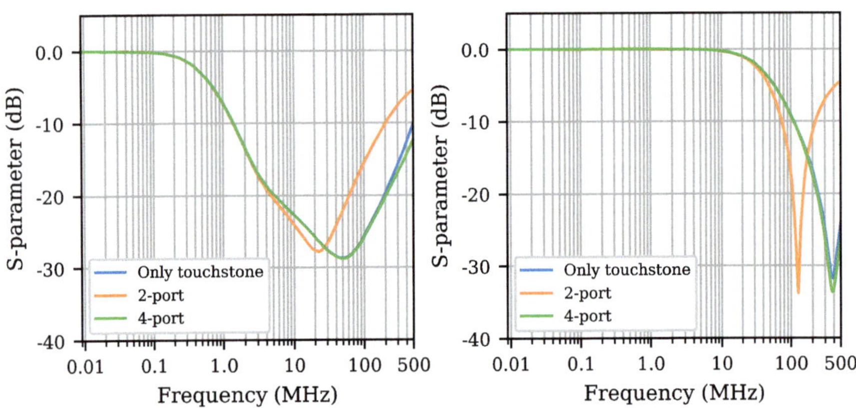

Fig. 9.15 Comparison of S-parameters of the common mode choke for the one-port setup, two-port setup, and measured touchstone standalone. Left: $S_{\mathrm{InCM,OutCM}}$. Right: $S_{\mathrm{InDM,OutDM}}$

the C_y capacitors, and thus the filter response is not significantly influenced by the incorrect port setup of the choke.

For models where a reference close to the mounting position is present, we recommend using the four-port setup when a four-port Touchstone file is available. For instance, in the LTE example in Sect. 6.6, we use a four-port setup. Similarly, the USB channel simulations in Sects. 11.1.2 and 12.5 also employ a four-port setup.

One possible way to mitigate the impact of the measurement fixture is to create an RLC-based model of the common mode choke. The RLC elements are selected to match the measured data for the through impedance, while omitting the capacitance to ground. However, modeling components above their resonance frequency using an RLC-based model can be challenging.

Another option is to include the common mode choke as a CAD model, making it part of the EM model of the DUT, thereby avoiding the need for a circuit description entirely. While the creation of such an EM model introduces additional overhead, it provides the most accurate solution to address the discrepancies between the measurement fixture and the DUT model. This approach integrates the coupling of the inductor with other parts of the DUT into a single EM model. The construction of EM models for chokes is discussed in Sect. 9.6. Due to the effort required for model creation, we recommend this approach only if simpler models fail to deliver satisfactory results or if the accuracy of these simple models is in question.

A streamlined approach is to start with coupled inductors and, during the further development of the electronic system, assess whether more detailed models are necessary. This approach is followed throughout this chapter.

9.5 Application Example: EMI Filter Design for a 300 W Motor Drive

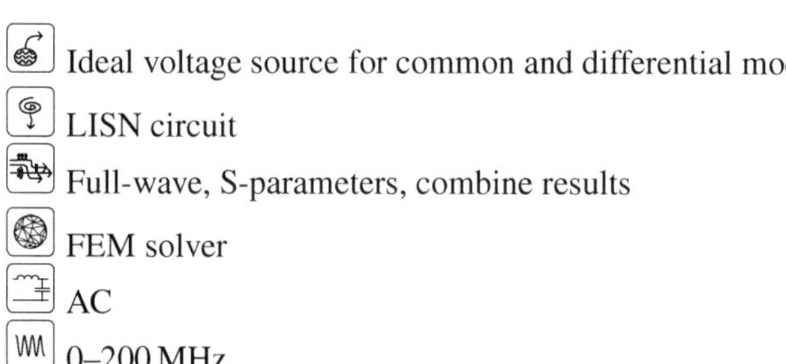

In the previous sections of this chapter, we outline how to design EMI filters through simulation. In this section, we apply this approach to analyze the EMI filter of the 300 W electric drive described in Sect. 9.1. We develop an improved design and use this design in the next Chap. 10 to study the conducted emissions of the electric drive.

We begin by investigating the filter components and their parasitics. Using circuit simulation, we analyze the common and differential mode voltages, $V_{\mathrm{LISN,CM}}$ and

$V_{\mathrm{LISN,DM}}$. Once the impact of the components on the filter attenuation is understood, we proceed with EM simulation to evaluate the layout and refine its design. Finally, we conclude with a simulation that incorporates the CAD model of the CMC in the EM model, validating that the simplified CMC model used so far does not compromise the filter design.

9.5.1 Component Investigation

In Sect. 9.2, we simulate the EMI filter using circuit simulation. Based on the parameter values of the filter components and their parasitics, as listed in Table 9.1, we obtain the voltages shown in Fig. 9.4. From these results, we observe that the common mode voltage is significantly lower than the differential mode voltage. The differential mode attenuation is 0 dB around 50 kHz and improves to better than − 60 dB above 5 MHz.

At an early stage in the system design, the layout and filter components are not yet finalized. The EMC engineer can propose using larger filter elements if this is expected to significantly improve filter performance or smaller ones if the reduction in filter attenuation does not risk exceeding EMI limits. Since frequency-domain circuit simulations complete within seconds, such studies can be conducted efficiently.

In the following, we conduct such an investigation in the form of a rudimentary sensitivity analysis. For each circuit element, we study two cases: using its defined element value as the reference and using ten times that value. This approach provides a quick way to determine if, and in which frequency range, a circuit element has an impact. At this stage of the analysis, we do not consider whether ten times the reference value is realistic.

The purpose of this investigation is not to make precise predictions but to identify whether and in which frequency range a circuit element is effective. We do not allow arbitrary parameter combinations. Instead, we distinguish between components that affect $V_{\mathrm{LISN,CM}}$ (e.g., the common mode choke and the C_{y} capacitors) and those that affect $V_{\mathrm{LISN,DM}}$ (e.g., the C_{x} capacitors and the stray inductance of the CMC).

Additionally, we separate the parameters into nominal values of the components and their parasitic values for more structured analysis.

For reference, we repeat the nominal component values of the filter and those of its parasitic elements in Table 9.2. These values are identical to those listed in Table 9.1.

Table 9.2 Reference parameter values of the common mode choke L_1, the C_{x} and C_{y} capacitors and their parasitic elements

L_1 (μH)	k_{coup}	$C_{\mathrm{x,i}}$ (μF)	$C_{\mathrm{y,j}}$ (nF)	$\mathrm{ESL}_{\mathrm{Cx},i}$ (nH)	$\mathrm{ESR}_{\mathrm{Cx},i}$ (Ω)	$\mathrm{ESL}_{\mathrm{Cy},j}$ (nH)	$\mathrm{ESR}_{\mathrm{Cy},j}$ (Ω)
16	0.95	10	4.7	1	0.1	0.5	0.1

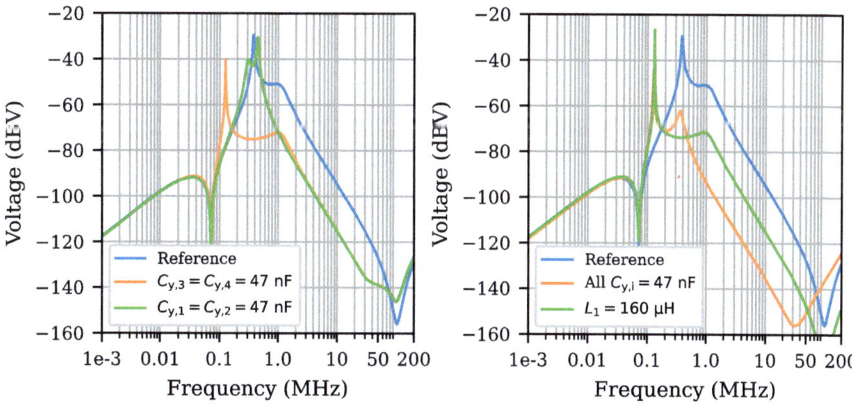

Fig. 9.16 Parameter sweep of the common mode voltage $V_{\mathrm{LISN,CM}}$, changing nominal component values. Left: change of individual capacitor values. Right: change of all capacitors and of the common mode choke

Changing Nominal Component Values Which Affect $V_{\mathrm{LISN,CM}}$ We investigate the common mode voltage $V_{\mathrm{LISN,CM}}$ in Fig. 9.16 across the frequency range from 1 kHz to 200 MHz. On the left side of the figure, we individually increase the values of the C_{y} capacitors on the DUT side, $C_{\mathrm{y},3}$ and $C_{\mathrm{y},4}$, or those on the LISN side, $C_{\mathrm{y},1}$ and $C_{\mathrm{y},2}$. On the right side of the figure, we increase all C_{y} capacitors collectively by a factor of 10, from 4.7 nF to 47 nF, and the value of L_1 by a factor of 10, to 160 µH. On both sides of the plot, the reference scenario described in Table 9.2 is shown in blue. Changing the nominal values of the C_{y} capacitors or the common mode choke has no effect below 100 kHz (left-hand side plot). In the reference scenario, $V_{\mathrm{LISN,CM}}$ reaches its maximum at 400 kHz and decays at a rate of 40 dB per decade above 1 MHz, reaching a minimum at 100 MHz. When the values of $C_{\mathrm{y},3}$ and $C_{\mathrm{y},4}$ are increased to 47 nF (left-hand side, orange curve), the maximum shifts to approximately 130 kHz, which is about $\frac{400\,\mathrm{kHz}}{\sqrt{10}}$. Since the resonance frequency f_{r} of an LC circuit obeys the relation $f_{\mathrm{r}} = \frac{1}{2\pi\sqrt{LC}}$, we conclude that the maximum of $V_{\mathrm{LISN,CM}}$ is caused by an LC resonance involving the two parallel capacitors $C_{\mathrm{y},3}$ and $C_{\mathrm{y},4}$.

Above 1 MHz, the orange curve for $V_{\mathrm{LISN,CM}}$ also decays at 40 dB per decade and is consistently 20 dB ($20 \log_{10} 10$) lower than the reference. This indicates that the decay is determined by the C_{y} capacitors.

For the green curve, where $C_{\mathrm{y},1}$ and $C_{\mathrm{y},2}$ are set to 47 nF, the behavior above 1 MHz matches the orange curve, indicating that capacitors on both sides of the choke are equally effective. However, changing the values of $C_{\mathrm{y},1}$ and $C_{\mathrm{y},2}$ does not shift the resonance at f_{r}, suggesting that these capacitors are not responsible for generating the resonance.

On the right-hand side of Fig. 9.16, all C_{y} capacitors are increased by a factor of 10 (orange curve). Above 1 MHz, the filter attenuation improves by an additional 20 dB compared to the reference, resulting in a total improvement of $20 \log_{10} 100 =$

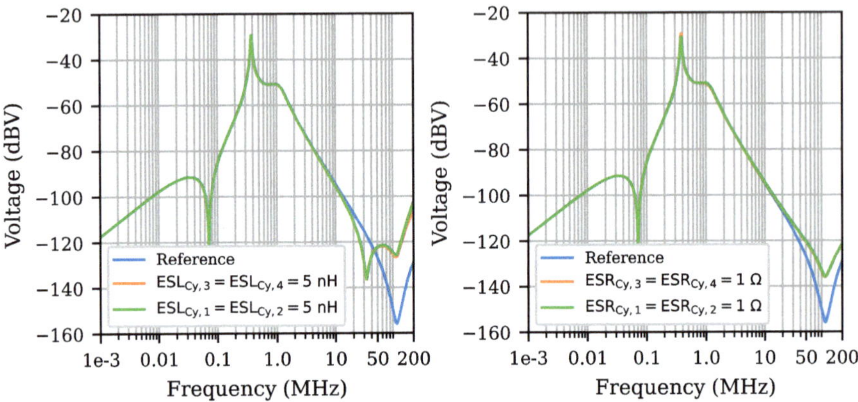

Fig. 9.17 Parameter sweep of the common mode voltage $V_{\text{LISN,CM}}$ changing parasitic component values. Left: variation of ESL. Right: variation of ESR

40 dB. The resonance shifts to 130 kHz, consistent with the behavior observed when increasing $C_{\text{y},3}$ and $C_{\text{y},4}$. Additionally, the minimum at 100 MHz shifts by a factor of $\frac{1}{\sqrt{10}}$ to 33 MHz.

Finally, when the value of L_1 is increased by a factor of 10 (green curve, right-hand side), the filter attenuation above 1 MHz also improves by 20 dB compared to the reference. Since the resonance frequency f_r shifts by a factor of $\frac{1}{\sqrt{10}}$, we conclude that L_1 constitutes the inductive part of the resonance.

As a summary, these four simulations indicate the following:

- The maximum at 400 kHz is a resonance caused by L_1 and $C_{\text{y},3}$, $C_{\text{y},4}$.
- The voltages $V_{\text{LISN,CM}}$ above 1 MHz depend on the nominal values of all C_y capacitors and L_1.
- The minimum of $V_{\text{LISN,CM}}$ depends on L_1.
- There is no effect of the filter elements below 100 kHz.

Changing the Parasitic Values Which Affect $V_{\text{LISN,CM}}$ In Fig. 9.17, we investigate the impact of the parasitic elements of the C_y capacitors. On the left-hand side, the ESL values of the C_y capacitors are increased by a factor of 10, while on the right-hand side, their ESR values are increased by the same factor. In both plots, the blue curve represents the reference scenario as defined in Table 9.2.

Below 10 MHz, none of the parameters visibly impacts $V_{\text{LISN,CM}}$. The curves representing the increase in the ESL values of either $C_{\text{y},3}$, $C_{\text{y},4}$ or $C_{\text{y},1}$, $C_{\text{y},2}$ (green and orange curves on the left-hand side) overlap. Increasing these values results in a new minimum of $V_{\text{LISN,CM}}$, which is a factor of $\frac{1}{\sqrt{10}}$ lower than the reference minimum (blue curve) at 100 MHz. However, the filter attenuation deteriorates above 50 MHz.

On the right-hand side of the plot, the parasitic resistance (ESR) of either $C_{\text{y},3}$, $C_{\text{y},4}$ or $C_{\text{y},1}$, $C_{\text{y},2}$ is increased by a factor of 10, resulting in a value of

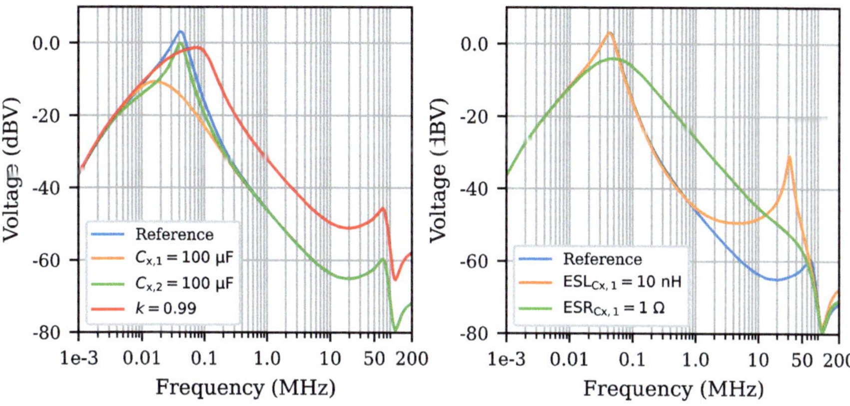

Fig. 9.18 Differential mode voltage $V_{\mathrm{LISN,DM}}$. Left: sweep of nominal component values and stray inductance. Right: sweep of parasitic values

$1\,\Omega$. The corresponding curves (orange and green) again overlap. Qualitatively, their shape matches the blue reference curve, but the minimum is a factor of $20 \log_{10} 10 = 20\,\mathrm{dB}$ higher at $-135\,\mathrm{dB}$.

From this analysis, we conclude:

- The parasitic elements of the components have no impact below 10 MHz.
- Changing the parasitic elements on either the DUT side or the LISN side has the same effect.
- Increasing the parasitic inductances by a factor of 10 deteriorates $V_{\mathrm{LISN,CM}}$ above 50 MHz.
- Increasing the parasitic resistances by a factor of 10 raises the minimum of $V_{\mathrm{LISN,CM}}$ by a factor of 10.

Changing the Nominal Values Which Affect $V_{\mathrm{LISN,DM}}$ On the left-hand side of Fig. 9.18, we investigate the differential mode voltage $V_{\mathrm{LISN,DM}}$ for the nominal values of the C_{x} capacitors and the coupling coefficient k of the common-mode choke.

According to Eq. 5.14, k quantifies the stray inductance of L_1. If $k = 1$, the magnetic field generated by one of the coil pairs in the CMC couples entirely into the other coil. In this case, the coil functions as a perfect common mode choke but does not filter any differential mode. For $k < 1$, part of the magnetic field penetrates the area outside the core, creating a stray inductance that attenuates the differential mode.

The blue curve represents the reference scenario. In the orange curve, the value of the capacitor $C_{\mathrm{x},1}$ on the LISN side is increased tenfold to $100\,\mu\mathrm{F}$. As a result, the resonance of the blue curve at approximately 42 kHz shifts by a factor of $\frac{1}{\sqrt{10}}$ to about 15 kHz. Additionally, the quality factor of the resonance decreases.

When the value of $C_{\mathrm{x},2}$ on the DUT side is increased (green curve), the impact is minimal, with a maximum deviation of around 4 dB at approximately 20 kHz.

This limited effect is due to the source impedance of the differential mode voltage source V_{DM}. With zero source impedance, the voltage across $C_{\mathrm{x},2}$ would equal V_{DM} because the source is an ideal voltage source, rendering $C_{\mathrm{x},2}$ ineffective. The chosen source impedance of $Z_{\mathrm{SDM}} = 0.1\,\Omega$ is still low, resulting in $C_{\mathrm{x},2}$ having little impact.

Increasing the coupling coefficient k of the CMC from 0.95 to 0.99 (red curve) reduces the stray inductance from $0.05 \cdot 16\,\mu\mathrm{H}$ to $0.01 \cdot 16\,\mu\mathrm{H}$, a reduction by a factor of 5. Consequently, the resonance frequency shifts by a factor of $\sqrt{5} = 2.24$, from 42 kHz to 94 kHz. Additionally, the voltage $V_{\mathrm{LISN,DM}}$ above 300 kHz increases by $20 \log_{10} 5 = 14\,\mathrm{dB}$.

Changing the Parasitic Values Which Affect $V_{\mathrm{LISN,DM}}$ On the right-hand side of Fig. 9.18, we plot $V_{\mathrm{LISN,DM}}$ while varying the parasitic components. The reference curve is shown in blue. For the orange curve, the parasitic inductance of $C_{\mathrm{x},1}$ on the LISN side is increased by a factor of 10. Below 1 MHz, there is no difference between the blue and orange curves. However, a strong resonance appears at 30 MHz in the orange curve, causing the filter attenuation at this frequency to decrease by more than 30 dB. The green curve represents the result of increasing the ESR of $C_{\mathrm{x},1}$ by a factor of 10. In this case, the voltage $V_{\mathrm{LISN,DM}}$ increases by 20 dB in the frequency range from 200 kHz to 10 MHz.

We do not plot any variation of the parasitic elements of $C_{\mathrm{x},2}$ on the DUT side. Due to the low source impedance of $Z_{\mathrm{SDM}} = 0.1\,\Omega$, the difference compared to the blue reference curve remains below 5 dB.

We summarize the findings of the five circuit simulations for $V_{\mathrm{LISN,DM}}$ as follows:

- The circuit components have no impact below 100 kHz.
- The capacitor $C_{\mathrm{x},2}$ has little impact due to the low source impedance $Z_{\mathrm{SDM}} = 0.1\,\Omega$.
- The resonance at approximately 40 kHz scales with $C_{\mathrm{x},1}$ and the stray inductance of L_1.
- Above 100 kHz, $V_{\mathrm{LISN,DM}}$ scales with the stray inductance of L_1.
- Above 1 MHz, the parasitic inductance $\mathrm{ESL}_{C_{\mathrm{x},1}}$ causes a resonance.
- The parasitic resistance $\mathrm{ESR}_{C_{\mathrm{x},1}}$ increases $V_{\mathrm{LISN,DM}}$ above 200 kHz.

The effects observed in this study are exaggerated because a component variation by a factor of 10 is significant, especially for parasitic values. These simulations aim to provide a clear fingerprint of the cause-effect relationships within the circuit. The large variations amplify the consequences, making any impact distinctly visible.

From this study, we conclude that, for the parameter ranges used

- There is no impact of the filter elements below 100 kHz.
- The nominal values of the C_{y} capacitors and the choke L_1 have a significant impact on the filter resonance in the 100 kHz range, as well as on $V_{\mathrm{LISN,CM}}$ above 1 MHz.
- The parasitic elements of the C_{y} capacitors have an impact above 10 MHz.

- The effect of the C_X capacitor and its parasitics on the DUT side is minimal due to the low source impedance chosen for the differential mode source.
- The stray inductance of L_1 has a significant impact on $V_{LISN,DM}$ above 100 kHz.
- The C_X capacitor on the LISN side is effective between 10 kHz and 100 kHz.
- The ESR of the C_X capacitor on the LISN side must be small.
- The ESL of the C_X capacitor on the LISN side can generate a resonance above 10 MHz.

A comprehensive filter analysis requires an extensive set of simulations. In this section, we describe only a portion of what is necessary to thoroughly understand the interactions between the filter elements. The goal is to demonstrate that, with an initial set of approximately a dozen simulations, many essential relationships between the filter attenuation and the filter components can be identified.

9.5.2 Layout and Component Model Refinement

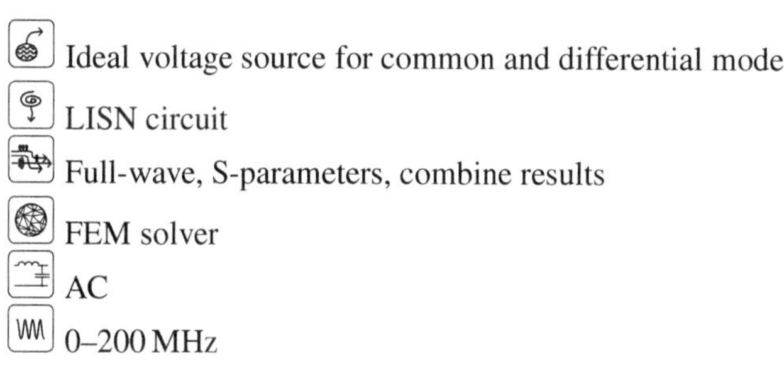

Once the impact of the components on the filter attenuation is understood, the layout and other geometric features must be investigated. We model the filter as part of a CISPR 25 conducted emission test, as described in Sect. 9.3.3. It is assumed that an initial layout has been produced, but the mechanical data is still incomplete. At this stage of the design process, EMI-relevant changes to the layout and mechanical design, such as the placement of screws, still have a good chance of being implemented.

As discussed in Sect. 3.3, layout and mechanical elements introduce additional parasitic elements into the system. The difference between a simulation with and without the layout is addressed in Sect. 9.3, where the EM model is introduced.

Figure 9.9 demonstrates that for $V_{LISN,CM}$, with the layout and housing shown in Fig. 9.5, this difference is significant. The circuit model with component parasitics and the EM model begin to deviate at 2 MHz, which is lower than the 10 MHz observed as the minimum frequency at which the parasitic elements of the C_Y capacitors begin to have an impact (Fig. 9.17).

This suggests that the current return path for the CM current via the housing is not well designed. Examining the layout shown in Fig. 9.5, the CM current flows

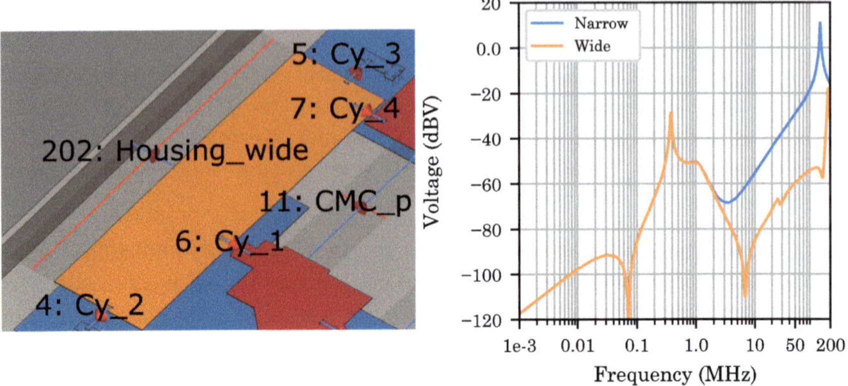

Fig. 9.19 A low-inductive connection between the common ground (orange) and the housing improves the filter by 20 dB above 5 MHz

through the C_y capacitors to the common ground (orange) and then through the connection between the common ground and the housing (port 200 : Housing). The current returns via the housing to the ports 1 : N_{DUT}, 2 : P_{DUT}, which connect the PCB to the housing.

The parasitic inductance of this path determines its impedance. A higher inductance results in a lower frequency threshold above which the C_y capacitors lose effectiveness. To address this, we implement a simple modification to improve the effectiveness of the C_y capacitors: replacing the narrow port 200 : Housing with the wider port 202 : Housing_wide. This new port extends along the entire edge of the common ground, connecting it to the housing, and significantly reduces the partial self-inductance of this part of the return path.

The geometry of port 202 : Housing_wide and the resulting $V_{LISN,CM}$ are shown in Fig. 9.19 for the frequency range from 1 kHz to 200 MHz. The impact of the port's reduced partial self-inductance becomes evident above 2 MHz. Above 5 MHz, it improves the filter attenuation by 20 dB.

Such a connection between a PCB and a housing is not mechanically feasible. However, in simulations, mechanical feasibility is no necessity in the initial stages. Replacing a narrow port with a wide one is implemented quickly, whereas engineering and implementing a realistic geometry require significantly more time and coordination with layout and mechanical design engineers. This model change is made primarily to investigate whether reducing the return path inductance has an impact on lowering $V_{LISN,CM}$.

Encouraged by this finding, we design and investigate a mechanically feasible modification, shown in Fig. 9.20. The common ground is attached to the housing using two conductive screws placed near the C_y capacitors. The four C_y capacitors are symmetrically arranged. In the EM model, the screws are represented as S-parameter ports (50 : Screw1, 51 : Screw2).

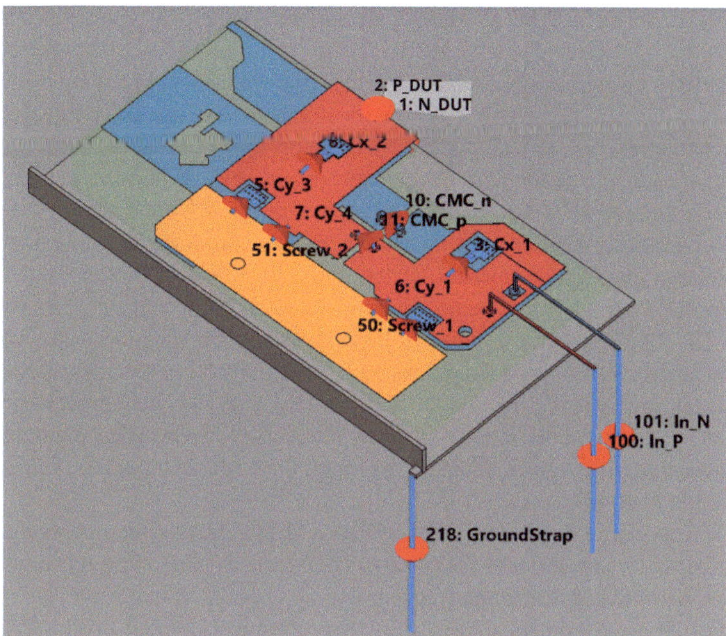

Fig. 9.20 Improved mechanical and layout design: two screws, 50 : Screw1, 51 : Screw2, connecting the PCB common ground (orange) and the housing

In the subsequent circuit simulation using the EM model, the terminals representing these ports are shorted, meaning they are considered ideal and do not produce any voltage drop. The refined model is solved using a frequency-domain FEM simulation, with a simulation time of 43 seconds.

Using the more realistic mechanical design, the next issue to address is the model of the CMC. The current assumption is a common mode inductance of 16 μH with a coupling coefficient of $k = 0.95$. It is modeled as two coupled inductors without any parasitic elements. To improve this model, detailed information about the choke's design is required. A suitable choice depends on additional specifications beyond the inductance, such as the stray inductance, saturation characteristics, and physical size.

For the improved model, we use the CMC described in (Würth Electronics). This is a wound choke with four windings on each side and a toroidal NiZn core. At 100 kHz, it has an inductance of 16 μH, a rated current of 10 A, and a DC resistance of 3 mΩ.

There are several options to model the choke. In Sect. 2.4.2.4, we describe a generic SPICE model of a coil. A more complex schematic can be constructed to represent a CMC. In Sect. 9.4.2, we discuss how to incorporate touchstone files with measured S-parameters into an EM model. The third option is to create a geometry model using CAD along with the material properties, as explained in detail in Sect. 9.6.

A SPICE circuit model represents the low-frequency behavior well, as it primarily reduces to resistive behavior with a resistance in the mΩ range. However, at higher frequencies, typically in the MHz range, the coupling between geometric elements becomes significant, resonances occur, and SPICE modeling is less accurate compared to measurements represented in a touchstone file.

An S-parameter dataset from measurements, encapsulated in a touchstone file, describes the electrical properties of the choke in terms of voltage and current. However, network analyzers do not operate below a certain frequency, for example, 9 kHz. The DC resistance is not available from such a measurement. Additionally, the instrument has an impedance of 50 Ω, which makes measurements of impedances in the mΩ range—typical for a choke at low frequencies—prone to inaccuracies. Further inaccuracies can arise due to S-parameter measurements taken in a fixture that differs from the PCB on which the CMC is mounted. This effect is discussed in Sect. 9.4.2.

An EM model derived from the CAD data of the choke circumvents the issues associated with measurement-based models. However, generating these models is complex, particularly due to the frequent lack of data about magnetic core material properties. If measurement data is available, effective material properties can be determined by matching the EM model to the values provided in the data sheet. This represents the most intricate method of choke modeling, and the details are described in Sect. 9.6.

In this section, we compare the modeling approaches of the filter with three variants of the CMC:

- The circuit model of two coupled inductors, which is a functional model without parasitics. This model is described in Fig. 9.8.
- A touchstone model based on measurements of the choke using a four-port network analyzer. This model is shown in Fig. 9.21.
- The geometric model of the choke developed in Sect. 9.6. The geometry model of the choke on the layout is shown in Fig. 9.22.

The first two approaches require only modifications to the circuit simulation that uses the EM model of the filter. For the third approach, a new EM model is created, incorporating the geometry and material properties of the choke. This model is computed in the frequency domain, requiring 2 minutes and 53 seconds. Once the EM model is solved, we attach circuit elements to it as in Figs. 9.8 or 9.21, except that the ports for the CMC are missing, because the choke is included in the geometrical design.

The results of the model variants are plotted in Fig. 9.23 for the common mode voltage at the LISNs, $V_{\mathrm{LISN,CM}}$ (left side of the plot), and for the differential mode voltage, $V_{\mathrm{LISN,DM}}$ (right side). Each plot contains four curves. The blue curves represent the ideal, coupled inductor model of the choke, attached to the EM model of the layout with the wide ports, as shown in Fig. 9.19. The orange curves represent the ideal, coupled inductor model attached to the refined layout shown in Fig. 9.20

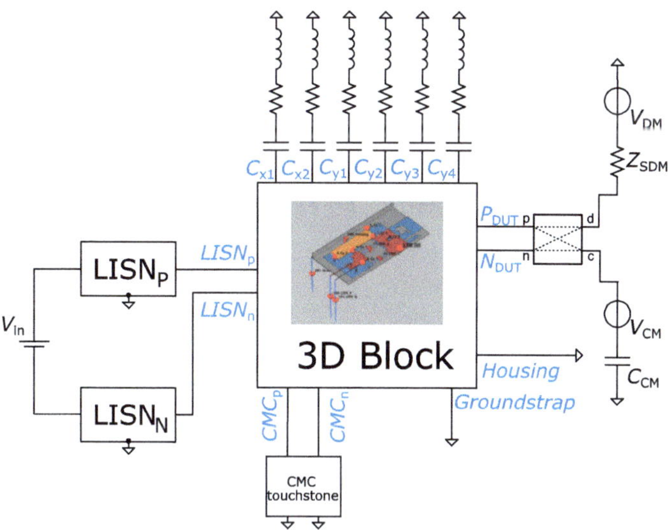

Fig. 9.21 EM model with connected filter elements, sources, and sink models. The CMC is attached as a touchstone model

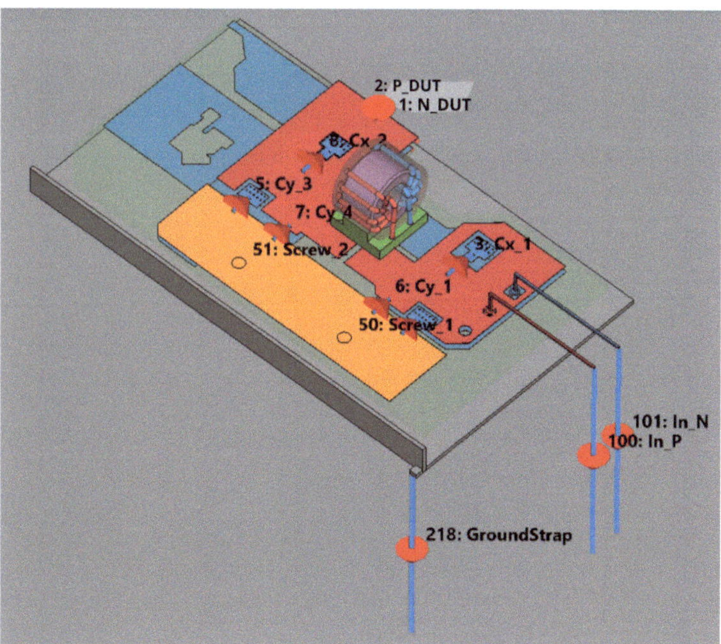

Fig. 9.22 Layout and mechanical model using the EM model of the choke instead of two ports

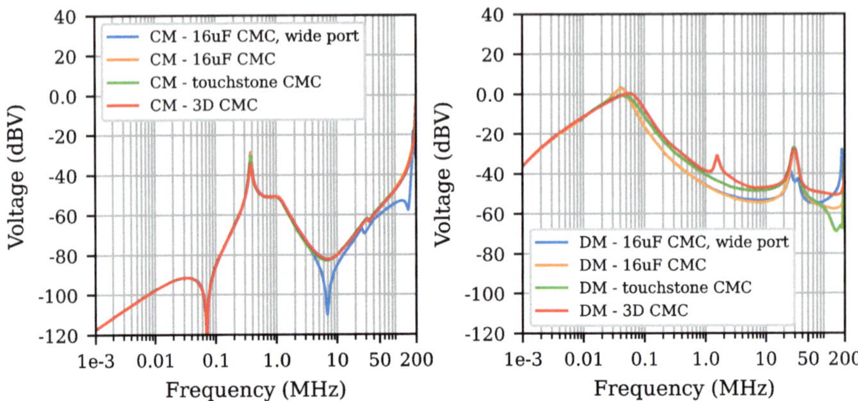

Fig. 9.23 Voltage for the choke model of 16 μH for two layout variants, the touchstone model, and the CAD model of the choke. Left: $V_{\text{LISN,CM}}$. Right: $V_{\text{LISN,DM}}$

with the two screws. The green curves correspond to the touchstone model of the choke attached to the refined layout. The red curves display the results of the simulation with the CAD model of the choke. The green curves start at 9 kHz, as this is the minimum frequency of the network analyzer used to measure the choke.

For $V_{\text{LISN,CM}}$, all four curves coincide below 5 MHz. Above this frequency, the curve with the largest deviations is the one corresponding to the layout with the wide port, which exhibits a notch at 7 MHz. All models based on the refined layout with the two screws differ by only a few dB. For the given layout, the model of two coupled inductors with an inductance of 16 μH each is sufficient. There is no visible impact of parasitic components, and neither the touchstone model nor the CAD model affects $V_{\text{LISN,CM}}$. This outcome is not a priori obvious. Subtle cases and their causes are discussed in Sect. 9.4.2.

Whether differences in the voltage are observed depends not only on the layout and the choke model but also on other components in the circuit. At high frequencies, the choke has a high impedance, and the common mode current flows mainly through the C_y capacitors. In this frequency range, the details of the choke are no longer relevant; instead, the properties of the capacitors dominate.

For the differential mode voltage $V_{\text{LISN,DM}}$, the blue and orange curves deviate above 25 MHz. Since both models use the ideal choke model of coupled inductors, this deviation reflects the impact of differences in the two layouts. When the touchstone model and the CAD model of the choke are applied, deviations from the coupled inductor model appear much earlier, starting at 30 kHz.

In the model using coupled inductors, the coupling coefficient of $k = 0.95$ is an assumption. From these plots, it becomes evident that this assumption does not align with the characteristics of the choke ultimately selected. The curves of the touchstone model and the CAD model coincide up to 40 MHz, except for a resonance at 1.5 MHz, which is exhibited only by the CAD model. The differential mode is more subtle to model, as it depends on the stray fields in between the core windings and around the choke.

9.6 EM Models of Inductors

Domains: 

Inductor models based on S-parameters or SPICE networks are available on the websites of many magnetic component suppliers. These models enable the computation of voltages and currents at the ports of EM models. However, inductors also generate magnetic fields that can couple to adjacent components. These fields are influenced not only by the properties of the inductor itself but also by the nearby geometry. For example, a metal housing can alter the magnetic field around an inductor and thereby affect its self-inductance.

An EM model of an inductor is essential if the coupling between the inductor and its surroundings impacts the filter response. Geometric models of inductors are available on supplier websites or can be manually created. The challenging part of this process is estimating the parameters that describe the material properties for these models.

In this section, we investigate the common mode choke used in the EMI filter described in Sect. 9.5.2. We demonstrate how to determine the magnetic core material parameters of the choke by reverse-engineering effective magnetic and dispersive properties from available impedance curves. Among the various methods to model magnetic elements, this approach offers a good compromise between accuracy and efficiency.

The prerequisite for this approach is the availability of S-parameter data for the choke. If this data is not available, it can be measured using a four-port network analyzer. The data used in this study was obtained from measurements conducted at the Institute of Electronics of Graz University of Technology. The measurements were conducted with a four-port network analyzer in the frequency range from 9 kHz to 1 GHz.

Core saturation is not considered here and is addressed separately in Sect. 9.7.

9.6.1 Modeling Geometry and Materials

Figure 9.24 shows the imported CAD geometry of the 16 μH common mode choke (Würth Electronics). After importing, the geometry is slightly adapted to define distinct geometric components: the pair of wound wires (red and blue wires: plus and minus), the cylindrical plastic cover of the core (olive), and the plastic body onto

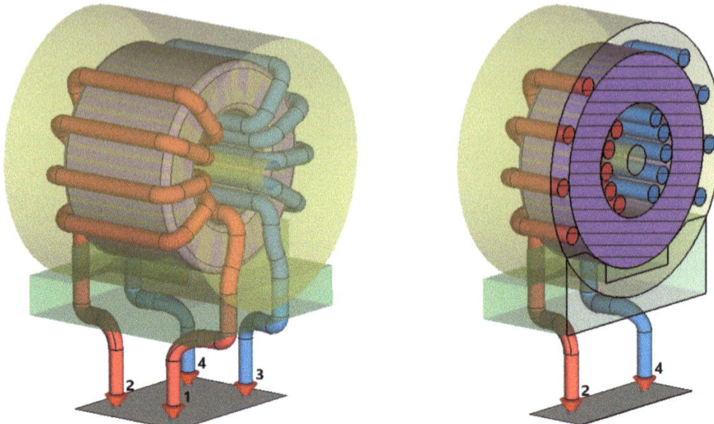

Fig. 9.24 CAD model of the choke (left) and its cross section (right). The magnetic core (purple) is inside the plastic cover (olive color)

which the choke is mounted (green). The cross-sectional view on the right-hand side of the figure reveals that inside the plastic cover is the magnetic core (purple).

An auxiliary metal sheet (gray) is added after the import as a reference for the four ports. These ports are used to compute the electrical characteristics of the choke as a function of its geometry and material parameters.

In the low-frequency region, the impedance is primarily resistive, determined by the conductivity of the choke's wires. To model this resistance, the wires are represented as normal material with the conductivity of copper, $\sigma_{\mathrm{Cu}} = 5.8 \cdot 10^7\,\frac{\mathrm{S}}{\mathrm{m}}$. A fine mesh is used to ensure that skin and proximity effects are correctly accounted for (Sect. 4.4).

Above the frequency range where the choke behaves resistively, its impedance is dominated by its common mode and differential mode inductance. The values of these inductances are primarily influenced by the winding scheme, the core dimensions and the magnetic core material for the common mode, and the stray inductance for the differential mode. According to the Debye model of magnetic dispersion (Eq. 4.11), the static permeability decreases when the relaxation frequency $\frac{1}{2\pi\tau_{\mathrm{r}}}$ is approached. In this frequency range, the impedance exhibits a less-than-linear increase or may even flatten out.

At even higher frequencies, the choke becomes capacitive because the impedance of the interwinding capacitances becomes lower than the impedance of the inductor wound around the core. To accurately compute the inductive part, the core geometry, as well as its complex magnetic permeability and permittivity as functions of frequency, must be provided in the choke's data sheet.

Frequently, this data is not available, as the dimensions of the core are concealed beneath the plastic cover. The capacitive behavior of the choke is influenced by the dielectric permittivity of the core and the plastic cover, as well as the distance between the windings. For densely wound chokes, particularly those with enameled wire, this distance is in the order of micrometers. Such small distances are difficult to measure accurately and a challenge for mesh generation.

We address this issue by using effective material parameters. We apply reverse-engineering techniques to the available S-parameter data of the choke, rather than directly measuring the material parameters and the dimensions of the core, plastic cover, and wires. While material measurements are possible, they require specialized equipment that is often unavailable in EMC engineering groups. Furthermore, measuring the core and cover would involve destructively disassembling the choke, which is feasible but does not necessarily yield a more accurate model than one derived from reverse engineering.

We rely on the measured curves to produce the common mode and differential mode impedances Z_{CM}, Z_{DM}, which can be obtained from measurements. Alternatively, they may be part of the data sheet of the choke. The complex permeability of the magnetic core is used as a tuning parameter to match the inductive behavior of the EM model of the choke, while the complex permittivity serves as a tuning parameter to match the model's capacitive behavior. This process yields effective permeability and permittivity values that may deviate from the actual frequency-dependent material properties. Since our primary interest lies in reproducing the currents and impedances of the choke, rather than in accurately characterizing the material properties themselves, this approach is both convenient and permissible. The magnetic field inside and around the choke is also influenced by the resulting permeability. As the tuned effective permeability ensures the correct current through the windings, this current, in turn, generates an appropriate magnetic field.

As discussed in Sect. 4.4.2, we use a first-order Debye model to describe the frequency-dependent relationship between the real and imaginary parts of the permeability. The magnetic core of the choke is a NiZn ferrite. A review of data sheets available online shows that these ferrites exhibit a wide range of relative permeabilities μ_r, from a few hundred to more than 1000. The magnetic relaxation time τ_r is less commonly provided in data sheets. When no such information is available, it must be estimated from the impedance data.

The complex relative permeability is plotted in Fig. 9.25 over the frequency range from 1 kHz to 200 MHz, using $\mu_s = 500$, $\mu_\infty = 1$, and $\tau_r = 10\,\text{ns}$. At the frequency $\frac{1}{2\pi\tau_r} \approx 16\,\text{MHz}$ (cf. Eq. 4.11), the imaginary part of the permeability, μ_r'', reaches its maximum, and the real part, μ_r', decreases to 50% of the static value μ_s. The permittivity of the ferrite is set to $\varepsilon_r = 1$. At this stage, these parameters are used as an educated engineering guess.

We apply these material data to the CAD model of the choke, as shown in Fig. 9.24. For the plastic housing, we assume a constant relative permittivity of $\varepsilon_r = 5$ over the entire frequency range. We run an EM simulation using a full-wave FEM direct solver. The total simulation time, including mesh adaptation, is 5 minutes and 15 seconds.

We use four ports at the reference plane, resulting in an EM simulation producing a four-port network. This model is then used to compute the common mode and differential mode impedances, Z_{CM} and Z_{DM}, which are functions of the chosen

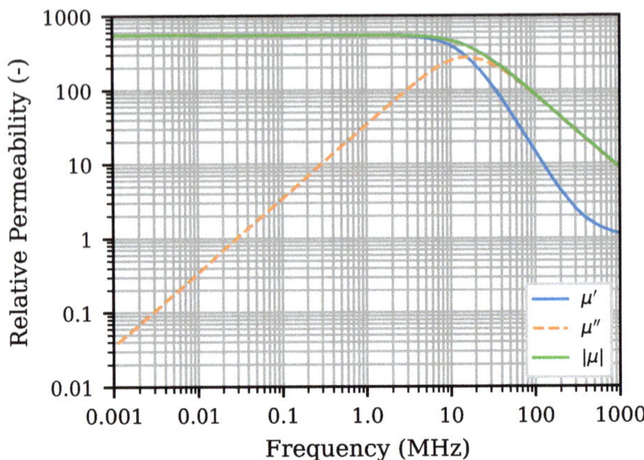

Fig. 9.25 Debye model of the relative permeability (Eq. 4.11), with $\mu_s = 500$, $\mu_\infty = 1$, and $\tau_r = 10\,\text{ns}$

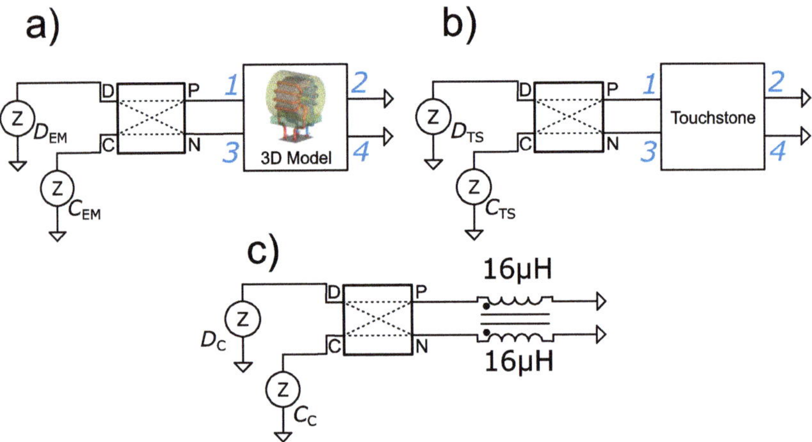

Fig. 9.26 Three circuit models to the CMC: (**a**) EM model, (**b**) measured touchstone, and (**c**) coupled inductors

material parameters. The circuit model is shown in Fig. 9.26a. In parts b) and c) of this figure, we see the same circuit topology, but with different descriptions for the choke: the EM model is replaced by a touchstone file containing the measurement data b) and by a circuit model of two coupled inductors, each with an inductance of 16 μH and a coupling coefficient of $k = 0.95$, as used in Sect. 9.1 c).

In each model, ports 1 and 2 represent the input and output of the positive branch of the choke, while ports 3 and 4 represent the input and output of the negative branch. This numbering is consistent with the setup of the ports in the EM model shown in Fig. 9.24. To ports 1 and 3, we attach the mode converter block (Sect. 8.3). To the left of the mode converter block (ports 1 and 3), we connect both

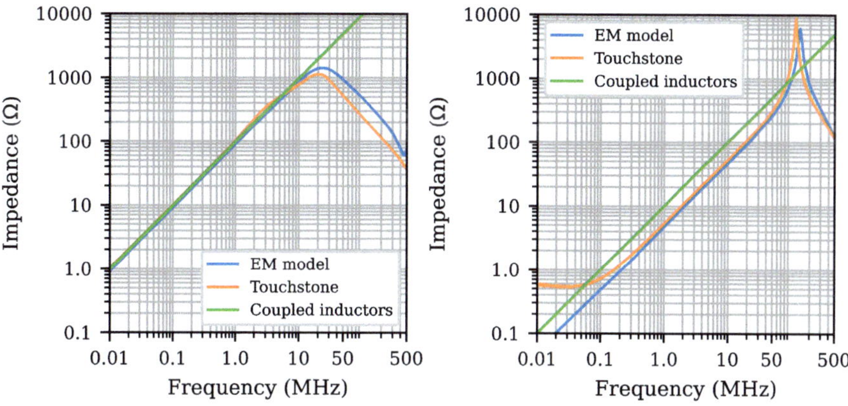

Fig. 9.27 Impedance of the three choke models of Fig. 9.26. Left: Common mode impedance Z_{CM} Right: Differential mode impedance Z_{DM}

a differential mode and a common mode source. Ports 2 and 4 are shorted, that is, ideally connected to the reference plane defined in the CAD model.

The differential mode excites a current between ports 1 and 3, with the current loop closing on the reference plane between ports 2 and 4. The common mode excites two equal currents flowing through ports 1 and 3 toward ports 2 and 4, with the current returning via the reference plane. Using models a), b), and c), we evaluate the common mode and differential mode impedances, Z_{CM} and Z_{DM}, for the EM model, the measured touchstone data, and the coupled inductor model.

Figure 9.27 shows, on the left-hand side, the common mode impedance Z_{CM} for the three different models and, on the right-hand side, the differential mode impedance Z_{DM} over the frequency range from 1 kHz to 500 MHz. In both plots, the blue curve is the impedance calculated by the EM model, the orange curve that of the S-parameter model, and the green curve that of the coupled inductors.

The common mode impedance Z_{CM} coincides for all three models with no visible deviation up to 10 MHz. Both the EM and the touchstone models exhibit a maximum at 20 MHz. Above this frequency, the impedance decreases, showing capacitive behavior. At 100 MHz, the impedance of the EM model is 475 Ω, while that of the touchstone model is 235 Ω. Using the relation $|Z| = \frac{1}{\omega C}$, the effective capacitance of the touchstone model is at this frequency calculated as 6.8 pF and that of the EM model is 3.4 pF. This difference indicates that the initial guess of the permittivity in the EM model, $\varepsilon_{\mathrm{r}} = 5$, results in a parasitic capacitance approximately a factor of 2 too low. The coupled inductor model, using the choke's nominal inductance of 16 μH, describes Z_{CM} very well up to the resonance. However, since this model does not account for any capacitive behavior, its curve continues to rise beyond the resonance point.

For Z_{DM}, the choke model of coupled inductors again produces a straight line. In the frequency range from 200 kHz to 50 MHz, it overestimates the impedance by approximately a factor of 2. The differential mode impedance Z_{DM} is determined by the stray inductance, which depends on the coupling coefficient $k = 0.95$. The stray inductance in the coupled inductor model is too high. Since increasing k reduces the stray inductance, the result shows that a coupling coefficient matching the touchstone model is higher than 0.95. The value of k chosen in Sect. 9.1 served as a working hypothesis. The comparison with the measurements reveals that this hypothesis requires correction.

Both the Z_{DM} of the touchstone file and the EM model exhibit a resonance at approximately 130 MHz. At higher frequencies, the impedance decays due to capacitive behavior, which cannot be captured by the model of coupled inductors. Below 100 kHz, the impedance of the touchstone model flattens, showing a resistance of 0.6 Ω. This value is too high to be attributed solely to the conductance of the wire. At frequencies close to their minimum frequency (here 9 kHz) and small resistances (mΩ), network analyzer measurements often show a flattening of the measured impedance at a too large value. For the filter design performed in this Chap. 9, this deviation is not an issue because we use an AC simulation in the frequency domain and target a CISPR 25 setup, where the lowest frequency considered is 150 kHz.

9.6.2 *Tuning Effective Material Parameters*

To produce Fig. 9.27, we use an educated guess for the parameters of the Debye model of the core and the plastic cover of the choke. In this section, we investigate the frequency ranges in which these parameters influence Z_{CM} and Z_{DM}. This analysis allows for parameter tuning in case the accuracy of the EM model needs improvement. We use three tuning parameters for the EM choke model:

Static permeability μ_{s} Affects the inductive part of Z_{CM}.
Permittivity ε_{r} Influences the capacitive parts of Z_{CM} and Z_{DM}.
Relaxation time τ_{r} Impacts the maximum of Z_{CM}.

The inductive part of Z_{DM} behaves approximately like that of an air coil. Only the geometry of the windings is significant. This geometry should be defined by the CAD data of the choke model. The permeability μ_∞ of the Debye model has an impact only at very high frequencies, where the choke model exhibits capacitive behavior.

In the following, we vary the parameters μ_{s}, τ_{r}, and ε_{r} one at a time. As in Sect. 9.5.1, we define a reference parameter set. We use the parameters from the educated guess, which are repeated in Table 9.3.

Table 9.3 Reference parameter set for tuning the EM model of the choke

	μ_{s}	τ_{r} (ns)	ε_{r}
Reference	500	10	5

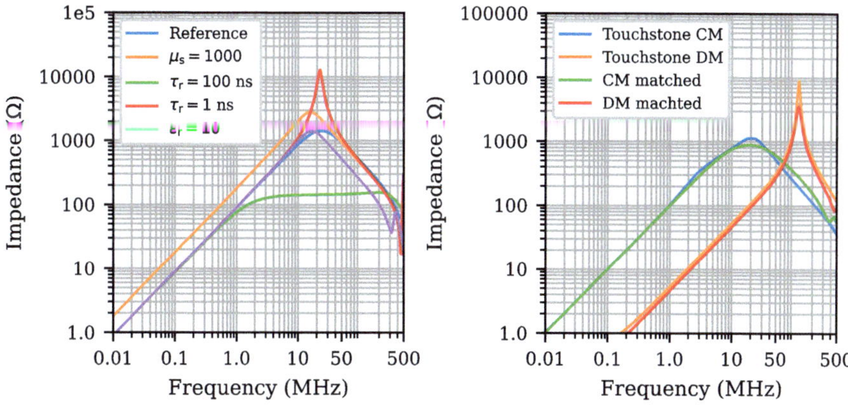

Fig. 9.28 Impact of μ_s, τ_r and ε_r on Z_{CM}. Left: comparison to the reference scenario. Right: Z_{CM} and Z_{DM} for $\mu_s = 560$, $\tau_r = 18\,\text{ns}$, $\varepsilon_r = 7.5$, compared to measured touchstone data

We perform the following four additional simulations and investigate their impact on Z_{CM}:

- $\mu_s = 1000$, twice the reference value.
- $\tau_r = 100\,\text{ns}$, 10 times the reference value.
- $\tau_r = 1$ ns, one-tenth of the reference value.
- $\varepsilon_r = 10$, twice the reference value.

The parameter ranges are chosen to illustrate their impact and do not correspond to a specific magnetic material. Each simulation requires a new EM simulation. Since only qualitative results are investigated, the FEM solver is used without mesh refinement.

The results are shown on the left-hand side of Fig. 9.28. The reference curve, based on the material data provided in Table 9.3, is shown in blue. Using twice the static magnetic permeability, μ_s, doubles the impedance of the inductive part of Z_{CM} (orange curve) and slightly shifts the resonance to the left. Increasing τ_r to 100 ns (green curve) flattens the maximum; μ_r'' reaches its peak at 1.6 MHz, resulting in much higher attenuation above this frequency. Conversely, with $\tau_r =$ 1 ns, the magnetic resonance shifts to 160 MHz, leaving the resonance peak of Z_{CM} essentially unattenuated (red curve). For $\varepsilon_r = 10$, the capacitance between the windings increases, causing the capacitive decay to set in at a lower frequency.

On the right-hand side of Fig. 9.28, we plot values of μ_s, τ_r, and ε_r that yield a good match between Z_{CM} and Z_{DM} as obtained from the EM model and the touchstone data. This match is achieved for $\mu_s = 560$, $\tau_r = 18\,\text{ns}$, and $\varepsilon_r = 7.5$.

Using the magnetic and dielectric materials of the choke as tuning parameters enables the reengineering of a suitable EM model based on the CM and DM impedance of the choke. This procedure works beyond the first resonance up to the next minimum, above which the parameters that determine the second resonance play a role. However, for chokes that exhibit multiple resonances within the frequency range of interest, reconstructing suitable material parameters for an EM model becomes significantly more complex.

We use the impedances Z_{CM}, Z_{DM} as tuning measures. This choice reflects that modeling the inductive part of the NiZn core investigated in this section is relatively straightforward, as there is a clear 20 dB per decade increase in Z_{CM}. Using the common and differential mode S-parameters of the choke as tuning measures also is a feasible approach that yields similar but not identical material parameters.

Other magnetic materials, such as nanocrystalline cores, exhibit more complex behavior. For such materials, a first-order Debye model may not adequately describe their permeability. More tuning parameters, such as the conductivity of the magnetic core, may be required, further complicating the simulations.

However, for EMC system simulation, it is often not necessary to have a choke model with impedance accuracy over a wide frequency range. In an EMI filter, the impedance of nearby capacitors can dominate that of the choke, as demonstrated in Sect. 9.23. Before investing significant engineering time into creating an overly accurate choke model, parameter studies of the system under test should be conducted to determine the frequency range in which such accuracy is actually required.

9.7 Advanced: Core Saturation

Domains: 

Magnetic materials are utilized in EMI filters to suppress conducted emissions. Since the impedance of an inductor is $|Z| = \omega L$, it serves, with sufficient L, at high frequencies as an effective high-impedance element. Magnetic materials are a critical component of inductive elements, with their inductance being proportional to their relative permeability μ_{r}. The relative permeability μ_{r} describes the relationship between the applied magnetic field and the flux density within the magnetic material (Eq. 4.4) and is a macroscopic quantity. Despite decades of research, there does not appear to be a universally applicable relation between the microscopic properties of a magnetic material and its macroscopic description that can be used to characterize the core materials for EMC engineering. The permeability is typically derived from measurements, and the availability of such measurement data remains a bottleneck for modeling the performance of magnetic materials in EMC simulations. There are numerous types of core materials, and their magnetic properties vary significantly based on the manufacturing process, temperature, and the specific current profiles applied to the inductor.

Two properties of magnetic materials[3] are general: (1) their relative permeability decreases in large magnetic fields and (2) at high frequencies. The relationships governing these behaviors have been previously explained in Sect. 4.4.2 (frequency dependency) and Sect. 4.4.3 (saturation).

In this section, we apply these relations to study the saturation and the common mode impedance Z_{CM} over frequency under DC-bias for the common mode choke discussed in the previous Sect. 9.6. After reviewing the properties of magnetic materials, we first examine the magnetic saturation of the choke under the application of a DC current. Subsequently, we extend the approach to arbitrary frequencies.

The lack of an applicable theory limits our discussion to simplified models. The approach described here depends on the availability of suitable data, which is often lacking. We encourage the reader to test this approach with various available datasets to determine which design decisions can be reliably made based on the given data.

9.7.1 Modeling Saturation of Magnetic Materials

We repeat the constitutive relation between the flux density **B** and the magnetic field **H** (Eq. 4.4),

$$\mathbf{B}(\mathbf{r}, \omega) = \mu_0 \mu_\mathrm{r}(\mathbf{r}, \omega) \cdot \mathbf{H}(\mathbf{r}, \omega). \tag{9.2}$$

The product $\mu_0\mu_\mathrm{r}$ equals the ratio of the absolute values of the flux density **B** to the absolute value of the magnetic field **H**. A high flux density **B** is desirable because it results in a large self-inductance (cf. Eq. 5.12) of the magnetic component. Since both $\mathbf{H}(\mathbf{r})$ and $\mathbf{B}(\mathbf{r})$ are functions of the spatial coordinate $\mathbf{r}$, the relative permeability $\mu_\mathrm{r}(\mathbf{r}, \omega)$ also depends on this coordinate.

The magnetization curve, or BH-curve, describes the relationship between the magnetic field strength $H = |\mathbf{H}|$ and the magnetic flux density $B = |\mathbf{B}|$ in the static case ($\omega = 0$). Figure 9.29 schematically illustrates a BH-curve for a soft magnetic material. The initial magnetization curve (blue) starts at the origin and describes the relationship between B and H in the absence of premagnetization. In the presence of magnetic hysteresis, a magnetic remanence B_r remains when an initial magnetization is reduced to $H = 0$. The coercivity H_c is the magnetic field strength required to reduce the flux density to zero.

For EMC engineering of electronic systems, the following properties are important:

Decrease of $\mu_\mathrm{r}(H)$ with increasing H With increasing H, the slope of the BH-curve decreases. The relative permeability $\mu_\mathrm{r}(H)$ decreases as a function of

[3] The large permeability desired for EMI suppression restrict the discussion in this section to ferromagnetic materials. Other magnetic materials exhibit different properties.

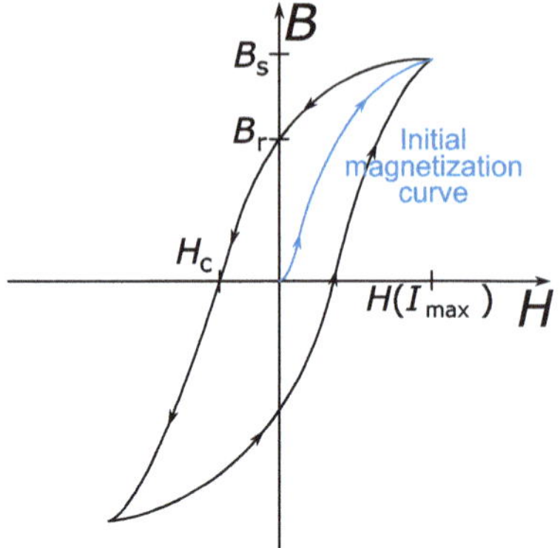

Fig. 9.29 Sketch of the BH-curve of a soft magnetic material. The slope of the blue curve corresponds to the starting permeability $\mu_r(H = 0)$ (Eq. 9.3)

H. The field $H(I_{DC})$ is generated by a current I_{DC}. As I_{DC} increases, $\mu_r(H)$ decreases, and the effectiveness of the core material deteriorates for high currents.

Validity range of the BH-curve The BH-curve is valid for a chosen $H(I_{max})$, which results in a B_{max}. Increasing H beyond H_{max} generates a different BH-curve that circumvents the given one. Additionally, BH-curves exist inside the shown one, representing $B(H)$ for smaller ranges of H.

Saturation and B_s threshold For large $H(I_{DC})$, $\mu_r(H)$ approaches values so low that the magnetic material becomes ineffective, and the magnetic flux density B increases at a very low rate. Core manufacturers specify this value as the saturation magnetic flux density B_s. For a magnetic core to be effective in an EMI filter, the flux density B inside the core must remain much smaller than B_s. The current I_{DC} must satisfy $B(H_{DC}(I_{DC})) < B_s$.

The choke model designed in Sect. 9.6 assumes no saturation and no initial magnetization. With $I_{DC} \approx 0$, we have $H_{DC} \approx 0$. In this case, the magnetic permeability μ_r is the slope of the BH initial magnetization curve through the origin,

$$\mu_r(H = 0) = \frac{1}{\mu_0} \frac{\mathrm{d}B}{\mathrm{d}H}\bigg|_{B=0, H=0}, \tag{9.3}$$

which is also referred to as the magnetic starting permeability (Kaltenbacher 2015). In Sect. 9.6, we reverse-engineer an effective value of $\mu_r(H = 0) = 560$ as a suitable parameter to describe the choke's common and differential mode impedances, Z_{CM} and Z_{DM}.

In the presence of a DC magnetic field intensity H_{DC}, the BH-curve yields a magnetic flux density $B(H_{DC})$. Now, consider that H_{DC} is superimposed with a small-scale HF magnetic field $H_{AC}(f)$ of frequency f, where $H_{AC}(f) \ll H_{DC}$. For $H_{AC}(f)$ sufficiently small, the magnetic permeability corresponds to the tangent to the BH-curve at H_{DC} (Kaltenbacher 2015) , given by

$$\mu_r(H_{DC}) = \frac{1}{\mu_0} \lim_{\Delta H \to 0} \frac{\Delta B}{\Delta H}(H_{DC}). \tag{9.4}$$

Both Eqs. 9.3 and 9.4 are an incremental permeability, that is, measured as a tangent on the BH-curve at a particular H. The apparent permeability starts at the origin, and its slope coincides with the incremental permeability only for a linear BH-curve (it is the secant through a given H). Whenever a BH-curve is linearized, the incremental and apparent permeabilities are equal.

9.7.2 *Core Saturation Due to a DC Current*

In this section, we investigate the saturation of a magnetic material due to a magnetic field H_{DC} generated by a DC current I_{DC} passing through a CMC. Core saturation can be caused by both common mode and differential mode currents through the CMC. For example, in a motor drive application, the differential mode current I_{DM}, which powers the motor, is typically much larger than the common mode current I_{CM}, which is generated by mode conversion and parasitic elements in the mechanical design. Even if $I_{DM} \gg I_{CM}$, investigating core saturation for both cases is valuable. The winding configuration of a CMC leads to a constructive superposition of the magnetic fields from the two wires in the case of I_{CM} and a destructive superposition in the case of I_{DM}. Both I_{DM} and I_{CM} have the potential to saturate the core. Once the core is in saturation, be it by a differential mode or a common mode current, it loses its effectiveness in attenuating the common mode and Z_{CM} decreases with increasing current through the choke. The differential mode impedance Z_{DM} is much less affected because Z_{DM} depends on the stray inductance outside the magnetic core.

We focus, exemplarily, on I_{DM}, using the CMC model designed in Sect. 9.6. Additionally, we assume that the CMC is applied in an EMI filter for a motor drive, as described in Sect. 9.5 and applied in Chap. 10. The I_{DM} of such a motor drive exhibits a saw-tooth waveform around a mean value $\hat{I}$. For simplicity, we further assume $I_{DM} \approx \hat{I}$. Under these premises, we solve the magnetostatic equation to compute the magnetic field $H(I_{DM})$ and the flux density $B(H(I_{DM}))$ inside and around the CMC. This is done by solving the magnetostatic Eq. 5.11 with the current I_{DM}. The details of the magnetostatic solver are discussed in Sect. 5.5. From the results, we determine the threshold I_{DM} above which core saturation begins to degrade the performance of the magnetic material.

The model setup is shown in Fig. 9.30 (left-hand side). The positive wire of the choke is depicted in red and the negative in blue. A differential mode current I_{DM} is applied between the two ports on one side of the choke, defining a coil segment (red line). The other side is shorted using a PEC sheet (gray connection). The differential current flows through the red winding packet and the short and then through the blue

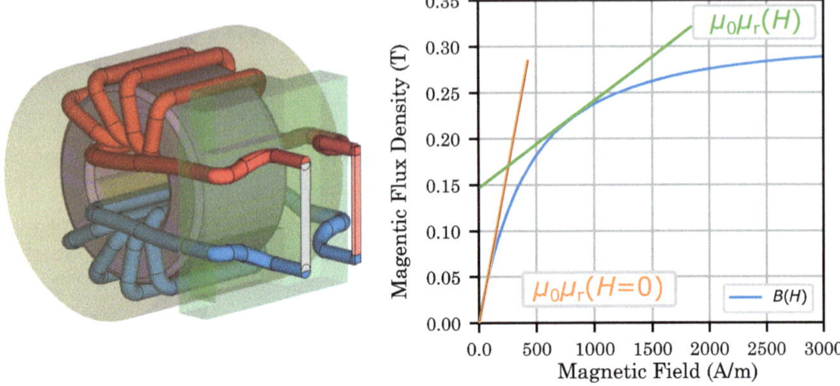

Fig. 9.30 Left: Choke model based on the geometry from (Würth Electronics), with an applied differential current. Right: BH-curve with $\mu_r(H = 0) = 560$, $B_s = 0.3\,\text{T}$, and an arbitrary derivative $\mu_0\mu_r(H)$ for intermediate values

winding packet and back to the source. The wire material is defined as normal, with a conductivity of $\sigma_{\text{Cu}} = 5.8 \cdot 10^7\,\frac{\text{S}}{\text{m}}$, and open boundaries are used.

To compute the magnetic flux density, the BH-curve of the core material is required. In rare cases, the BH-curve of a core material is provided. If appropriate equipment is available, a BH-curve can be measured, but in most EMC departments, such equipment is not accessible. Alternatively, measured BH-data can be requested from the supplier of the magnetic material. However, if this data is unavailable or there is insufficient time to wait for it, a BH-curve must be assumed, and the implications of these assumptions must be understood. This is the approach we take in this section.

The choke used in Sect. 9.6 has a NiZn ferrite core. Comparison with measured data for $H = 0$ is reengineered as $\mu_r(H = 0) = 560$. The saturation flux density of NiZn ferrite can be obtained from data sheets (TDK Corporation) or scientific literature, e.g. Suarez et al. (2018), and we estimate it as $B_s = 0.3\,\text{T}$. These two values provide the corner points for $H = 0$ and large H.

A corresponding curve is shown in Fig. 9.30 on the right (blue curve). It is generated using a macro for creating BH-curves for soft magnetic materials available in CST. At $H = 0$, the curve has a derivative of $\mu_0\mu_r$ with $\mu_r(H = 0) = 560$, as described by Eq. 9.3 (orange line). For large H, the curve asymptotically approaches the flux saturation of $B_s = 0.3\,\text{T}$. In this range of H, the slope of the BH-curve becomes low, and the magnetic material has minimal influence on the performance of the CMC. Even at very large H, the curve retains a minimum slope of at least μ_0. In this saturation range, further details about μ_r are unnecessary for assessing the performance of the CMC.

According to Eq. 9.4, the derivative of the BH-curve serves as a measure of core saturation for any H (green curve). At the selected point $H = 750\,\frac{\text{A}}{\text{m}}$, the derivative of the green curve is approximately $\frac{1}{6}$ of that of the orange curve. This implies that the relative permeability $\mu_r(750\,\frac{\mathrm{A}}{\mathrm{m}})$ of the core material has decreased to about $\frac{560}{6} \approx 90$. At this point, the designer must decide whether a reduction in the impedance of the CMC by a factor of six is still acceptable.

The shown curve is an example. It is possible that the slope of the BH-curve could be steeper between 0 and $750\,\frac{\mathrm{A}}{\mathrm{m}}$, meaning that $\mu_r(H)$ would remain closer to $\mu_r = 560$ in this range. Beyond $750\,\frac{\mathrm{A}}{\mathrm{m}}$, the curve could flatten more abruptly, as the saturation flux density remains 0.3 T. In such a case, core saturation would occur more abruptly. The core would behave more like an unsaturated one below $H = 750\,\frac{\mathrm{A}}{\mathrm{m}}$, and it would reach saturation more rapidly above this value.

Since I_{DM} is a design requirement for the electric motor drive and $H(I_{\mathrm{DM}})$ is initially unknown, the system designer needs to determine the current I_{DM} at which core saturation occurs. We compute the magnetic field intensity H and the flux density B for differential mode currents $I_{\mathrm{DM}} = 1\,\mathrm{A}$, 10 A, 50 A, and 100 A. These calculations model various load cases to which the choke is exposed when attached to the electric motor drive. The model described in Fig. 9.30 (left-hand side) is used. The computation times are 3 minutes 11 seconds for $I_{\mathrm{DM}} = 1\,\mathrm{A}$, 6 minutes 8 seconds for $I_{\mathrm{DM}} = 10\,\mathrm{A}$, 8 minutes 21 seconds for $I_{\mathrm{DM}} = 50\,\mathrm{A}$, 11 minutes 30 seconds for $I_{\mathrm{DM}} = 100\,\mathrm{A}$. With higher saturation, the computation time increases because more Newton iterations are needed (cf. Sect. 4.4.3).

The resulting vector fields for the magnetic field and the flux density are shown in Fig. 9.31 for the case of $I_{\mathrm{DC}} = 50\,\mathrm{A}$. The plot is restricted to the plane parallel to and centered within the core. The choke is excited in differential mode. The flux density (left-hand side) is strongest in the sections where the choke is wrapped by the wires. The direction of the flux density on one side is opposite to that on the other side, canceling in the sections where there are no wires (top and bottom of the choke). The normal component of **B** across any interface between two media is continuous (Eq. 4.2). At the top and bottom, the flux crosses the interface between the core material and vacuum perpendicularly to the surface and remains continuous. To the left and right, the flux density is parallel to the surface, resulting in a small stray field. Outside the core, in vacuum, we have $\mu_r = 1$, and the flux density **B** is equal

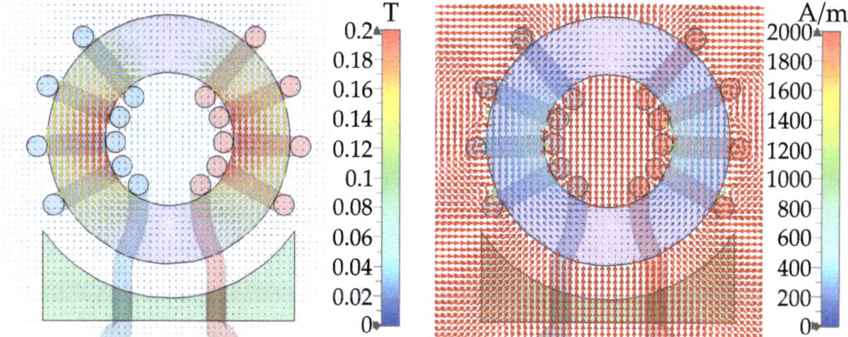

Fig. 9.31 Flux density **B** (left) and magnetic field **H** (right) within and around the magnetic core

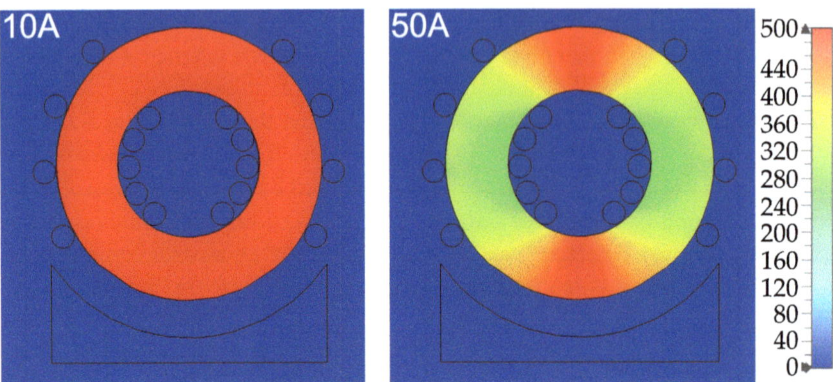

Fig. 9.32 Relative magnetic permeability $\mu_r(\mathbf{r})$ for $I_{DM} = 10\,\text{A}$ (left) and $I_{DM} = 50\,\text{A}$ (right). At 50 A, saturation is evident in the regions where the wires of the coil induce a high field

to $\mu_0\mathbf{H}$. However, the transition of the normal component of the $\mathbf{H}$ across the surface of discontinuity is discontinuous.

Since the orthogonal component of the flux density $\mathbf{B}$ is continuous, the orthogonal component of the magnetic field $\mathbf{H}$ is not. Inside the core, this component is lower by a factor of $\frac{1}{\mu_r(\mathbf{r})}$. The absolute value of the magnetic field is largest in the area where the wire is wrapped around the core, reaching approximately $500\,\frac{\text{A}}{\text{m}}$. Due to the winding scheme of the CMC, the direction of the field lines on one side is opposite to that on the other side. At the top and bottom of the core, the magnetic fields cancel each other, and $H(\mathbf{r})$ is close to zero.

Since both $\mathbf{B}(\mathbf{r})$ and $\mathbf{H}(\mathbf{r})$ are functions of the spatial location $\mathbf{r}$, this also applies to $\mu_r(\mathbf{r})$, as indicated in Eq. 9.2. We plot $\mu_r(\mathbf{r})$ in the same plane of the choke in Fig. 9.32, for $I_{DM} = 10\,\text{A}$ (left) and $I_{DM} = 50\,\text{A}$ (right). At $I_{DM} = 10\,\text{A}$, the permeability is homogeneous throughout the core and close to its maximum value of $\mu_r(H = 0) = 560$. Only near the equator of the ring core, where the field intensity is largest, are the values reduced by about 10%. At $I_{DM} = 50\,\text{A}$, the permeability values near the equator and closer to the inner boundary of the core are reduced to 50% of the unsaturated value of $\mu_r(H = 0) = 560$. At the north and south poles of the core, where the field contributions from the currents on each side of the coil cancel, the core does not saturate. From the simulation, we conclude that there is no core saturation at $I_{DM} = 10\,\text{A}$, but the core performance degrades significantly at $I_{DM} = 50\,\text{A}$.

A conservative design of a choke avoids core saturation. To determine whether saturation occurs, information about the BH-curve of the core material is essential. If only the starting permeability $\mu_r(H = 0)$ and the saturation flux density B_s are known (e.g., from a data sheet), a magnetostatic simulation using an assumed BH-curve can indicate the current above which saturation occurs, enabling a conservative choke design. If the magnetostatic simulation shows a magnetic field intensity sufficient to produce a flux density close to B_s, saturation occurs, and the inductance of the magnetic component is reduced. Countermeasures include increasing the size of the choke or selecting a magnetic material with a higher B_s. To design a choke that guarantees a certain impedance for any given current, measured BH-data must be available.

9.7.3 Core Saturation at High Frequencies

A magnetic component driven into saturation exhibits reduced impedance. In the previous Sect. 9.7.2, we investigate the decrease of $\mu_r(H, \mathbf{r})$ at each location $\mathbf{r}$ for H generated by a DC current. The current is applied to simulate a differential mode, flowing from the positive to the negative pin on one side of the choke, while the other side of the choke is bridged by a short.

For the design or analysis of a CMC, we need to extend this procedure to predict the degradation of its common mode impedance $Z_{CM}(\omega)$. This requires combining two effects. Firstly, referring to Eq. 9.4, we assume a dominant DC current with a relatively small AC current of frequency ω superimposed. In this case, we use the tangent to the BH-curve, that is, $\mu_r(H_{DC})$, as a local approximation for the deviation of $B(H_{AC})$. Secondly, the relative permeability decreases with frequency, as described in detail in Sect. 4.4.2 and in Eq. 4.11. This equation applies for zero bias current ($H_{DC} = 0$).

In this subsection, we modify the Debye model (Eq. 4.11) to describe the behavior of the magnetic material over frequency under a DC bias current $H_{DC} > 0$. The initial permeability μ_s is replaced by the spatially varying $\mu_r(\mathbf{r}, H_{DC})$ from Eq. 9.4. We extend the Debye model as follows,

$$\mu_r(\mathbf{r}, \omega, H_{DC}) = \mu_\infty + \frac{\mu_r(\mathbf{r}, H_{DC}) - \mu_\infty}{1 + j\omega\tau_r}. \tag{9.5}$$

This is an empirical approach that assumes the DC current reduces the static permeability but has no impact on τ_r.

We apply Eq. 9.5 to calculate $Z_{CM}(\omega)$ for the choke subjected to the differential mode current I_{DM}. As in Sect. 9.7.2, this current is either static or has a very low frequency. With increasing current, the magnetic core saturates, and the inductance is expected to decrease. As the frequency increases, the impedance is expected to exhibit a maximum and then decrease due to the frequency-dependent reduction in permeability and the capacitive coupling between the windings of the choke (Fig. 9.28).

We apply the relative magnetic permeability $\mu_r(\mathbf{r}, H_{DC})$, computed in the previous subsection for $I_{DM} = 1\,\text{A}, 10\,\text{A}, 50\,\text{A}, 100\,\text{A}$, in Eq. 9.5 to compute the core impedance over frequency with the core saturation included for these values of I_{DM}. The parameters $\tau_r = 18\,\text{ns}$ and $\mu_\infty = 1$ are chosen as in Sect. 9.6. We use the four-port EM model explained in Sect. 9.6 and solve it using the frequency domain

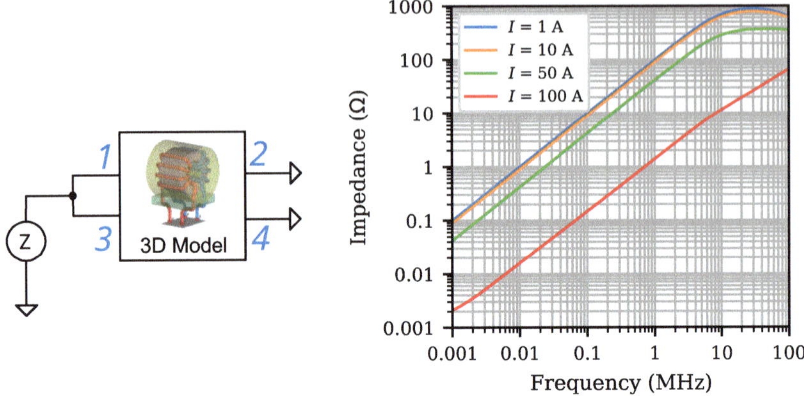

Fig. 9.33 Left: Network representation of the EM model of the choke with saturated core. Right: Z_{CM} as a function of frequency for various values of I_{DM}

solver. The computation time is 7 minutes, 10 seconds, regardless of the value of I_{DM}.

The resulting EM model has four ports. To compute Z_{CM}, the connections are set up as shown in Fig. 9.33 on the left. The input ports 1 and 3 of the CMC are connected to each other, while the output ports 2 and 4 are both grounded. The impedance Z_{CM} is calculated between these two pairs. The resulting impedance curves are shown on the right-hand side of the figure in the frequency range from 1 kHz to 100 MHz. At $I_{DC} = 1$ A, the choke exhibits unsaturated behavior, as seen in Fig. 9.28 on the right. At 1 MHz, the choke has $Z_{CM} = 100\,\Omega$, corresponding to an inductance of $L = 16\,\mu$H, as specified in (Würth Electronics) and Sect. 9.6. Above 30 MHz, Z_{CM} begins to decrease.

At $I_{DC} = 10$ A, there is negligible change in Z_{CM}, consistent with Fig. 9.32, which shows that $\mu_r(\mathbf{r})$ remains close to the starting permeability $\mu_r(H = 0)$ across the core. At $I_{DC} = 50$ A, the impedance drops by about 50%, caused by the lower $\mu_r(\mathbf{r})$ in the regions near the windings. At $I_{DC} = 100$ A, the choke impedance is reduced to nearly 1% of its original value, effectively behaving like an air-coil. The maximum in the impedance at approximately 30 MHz, visible for $I_{DC} = 1$ A, 10 A, 50 A due to magnetic dispersion over frequency, is no longer present at $I_{DC} = 100$ A.

Based on the common mode impedance, we compute the common mode inductance of the choke at $f = 1$ MHz using the relation $|Z| = \omega L$. The results are tabulated in Table 9.4. These results illustrate the decrease in $L(I_{DM})$ with increasing I_{DM}. At 1 A, the inductance is at 99% of the nominal value of 16 μH. At 10 A, the inductance it is at 88%. At 50 A, the inductance is reduced to 40% of the original value, and at 100 A, it drops to nearly 1% of the unsaturated case.

Table 9.4 Computed inductance at 1 MHz for various I_{DM} and remaining inductance compared to unsaturated value of $L = 16\,\mu$H

I_{DM} (A)	1	10	50	100
L(1 MHz)(μH)	15.8	14.1	6.5	0.2
Reduction of L to	99%	88%	40%	1%

Further Reading

Filter Design

A. Asmanis, et al., High-frequency modelling of EMI filters considering parasitic mutual couplings, in *2016 ESA Workshop on Aerospace EMC (Aerospace EMC)* (2016), pp. 1–6. https://doi.org/10.1109/AeroEMC.2016.7504554

M. Hartmann, H. Ertl, J.W. Kolar, EMI filter design for a 1 MHz, 10 kW Three-Phase/Level PWM Rectifier. IEEE Trans. Power Electron. **26**(4), 1192–1204 (2011). https://doi.org/10.1109/TPEL.2010.2070520

R. He, et al., Modeling strategy for EMI filters, IEEE Trans. Electromagn. Compat. **62**(4), 1572–1581 (2020). https://doi.org/10.1109/TEMC.2020.3006127

H. Huang, T. Lu, A cancelation method of mutual inductance between capacitors in EMI filter. IEEE Trans. Power Electron. **37**(10), 11974–11984 (2022). https://doi.org/10.1109/TPEL.2022.3173082

I.F. Kovačević, et al., 3-D electromagnetic modeling of EMI input filters. IEEE Trans. Ind. Electron. **61**(1), 231–242 (2014). https://doi.org/10.1109/TIE.2013.2242422

I.F. Kovačević, et al., 3-D electromagnetic modeling of parasitics and mutual coupling in EMI filters. IEEE Trans. Power Electron. **29**(1), 135–149 (2014). https://doi.org/10.1109/TPEL.2013.2254130

Murata, *Ceramic Capacitor GR443DR73D472KW01L*. https://www.murata.com/products/productdetail?partno=GR443DR73D472KW01# (visited on 2025)

B. Narayanasamy, F. Luo, A survey of active EMI filters for conducted EMI noise reduction in power electronic converters. IEEE Trans. Electromagn. Compat. **61**(6), 2040–2049 (2019). https://doi.org/10.1109/TEMC.2019.2953055

S. Negri, et al., Black-box modeling of EMI filters for frequency and time-domain simulations. IEEE Trans. Electromagn. Compat. **64**(1), 119–128 (2022). https://doi.org/10.1109/TEMC.2021.3105735

Panasonic, *Aluminium Electrolytic Capacitors*. https://industrial.panasonic.com/ww/products/pt/aluminum-cap-smd/models/EEE2AA100UP (visited on 2025)

B. Pintér, A. Bingler, M. Csörnyei, High frequency measurement and simulation of electromagnetic interference filters, in *2023 International Symposium on Electromagnetic Compatibility—EMC Europe* (2023), pp. 1–6. https://doi.org/10.1109/EMCEurope57790.2023.10274188

Common Mode Chokes, Transformers, and Magnetic Materials

F. Baig, et al., Electrical model validation of magnetic cores for current sensing and loop filter applications, in *2024 International Symposium on Electromagnetic Compatibility—EMC Europe* (2024), pp. 122–127. https://doi.org/10.1109/EMCEurope59828.2024.10722481

A. Bartel, et al., Quantification of uncertainty in the field quality of magnets originating from material measurements. IEEE Trans. Magn. **49**(5), 2367–2370 (2013). https://doi.org/10.1109/TMAG.2013.2241041

L. Dalessandro, W.G.H. Odendaal, J.W. Kolar, HF characterization and nonlinear modeling of a gapped toroidal magnetic structure. IEEE Trans. Power Electron. **21**(5), 1167–1175 (2006). https://doi.org/10.1109/TPEL.2006.880357

D. Eremyan, et al., PEEC-based wideband micro-model of inductive components for power electronics applications, in *2023 International Symposium on Electromagnetic Compatibility—EMC Europe* (2023), pp. 1–6. https://doi.org/10.1109/EMCEurope57790.2023.10274336

R. Fischbacher, et al., On the difficulties to determine the intrinsic material parameters for MnZn ferrites, in *2023 International Symposium on Electromagnetic Compatibility—EMC Europe* (2023), pp. 1–6. https://doi.org/10.1109/EMCEurope57790.2023.10274159

R. Huang, D. Zhang, K.-J. Tseng, Determination of dimension-independent magnetic and dielectric properties for Mn-Zn ferrite cores and its EMI applications. IEEE Trans. Electromagn. Compat. **50**(3), 597–602 (2008). https://doi.org/10.1109/TEMC.2008.926914

H. Jie, et al., VNA-based fixture adapters for wideband accurate impedance extraction of single-phase EMI filtering chokes. IEEE Trans. Ind. Electron. **70**(8), 7821–7831 (2023). https://doi.org/10.1109/TIE.2023.3239906

S. Jing, et al., Extraction of permittivity and permeability for ferrites and flexible magnetodielectric materials using a genetic algorithm. IEEE Trans. Electromagn. Compat. **57**(3), 349–356 (2015). https://doi.org/10.1109/TEMC.2015.2389332

W. Konrad, H. Hackl, M. Stoiber, Extraction of ferrite material properties and application to simulation-based common-mode filter design, in *2024 IEEE International Conference on Electrical Systems for Aircraft, Railway, Ship Propulsion and Road Vehicles & International Transportation Electrification Conference (ESARS-ITEC)* (2024), pp. 1–6. https://doi.org/10.1109/ESARS-ITEC60450.2024.10819890

Z. Kutchadze, A. Gheonjian, R. Jobava, Investigation of magnetic permeability of toroidal cores, in *2019 International Symposium on Electromagnetic Compatibility—EMC EUROPE* (2019), pp. 498–503. https://doi.org/10.1109/EMCEurope.2019.8872112

Q. Li, et al., A general analytical model of single-layer common-mode chokes. IEEE Trans. Power Electron. **39**(6), 6591–6596 (2024). https://doi.org/10.1109/TPEL.2024.3372297

X. Margueron, J.-P. Keradec, Design of equivalent circuits and characterization strategy for n-input coupled inductors. IEEE Trans. Ind. Appl. **43**(1), 14–22 (2007). https://doi.org/10.1109/TIA.2006.887001

X. Margueron, J.P. Keradec, Identifying the magnetic part of the equivalent circuit of n-Winding Transformers. IEEE Trans. Instrum. Meas. **56**(1), 146–152 (2007). https://doi.org/10.1109/TIM.2006.887402

I. Mayergoyz, Mathematical models of hysteresis. IEEE Trans. Magn. **22**(5), 603–608 (1986). https://doi.org/10.1109/TMAG.1986.1064347

N. Moonen, et al., 3-D full-wave high frequency common mode choke modeling. IEEE Trans. Electromagn. Compat. **62**(3), 707–714 (2020). https://doi.org/10.1109/TEMC.2019.2914371

K. Nomura, T. Kojima, Y. Hattori, Straightforward modeling of complex permeability for common mode chokes. IEEJ J. Ind. Appl. **7**(6), 462–472 (2018). https://doi.org/10.1541/ieejjia.7.462

Y. Pei, et al., Study of high frequency characteristics modeling and EMI suppression of common mode chokes. Appl. Comput. Electromagn. Soc. J. **38**(2), 91 (2023)

S. Sadeghi, et al., A simple method to build a linear circuit model for a multiwinding transformer used in EMI/EMC analysis. IEEE Lett. Electromagn. Compat. Pract. Appl. **6**(4), 138–143 (2024). https://doi.org/10.1109/LEMCPA.2024.3459795

G.R. Skutt, F.C. Lee, Characterization of dimensional effects in ferrite-core magnetic devices, in *PESC Record. 27th Annual IEEE Power Electronics Specialists Conference*, vol. 2 (1996), pp. 1435–1440. https://doi.org/10.1109/PESC.1996.548770

I. Stevanović, S. Skibin, Behavioral circuit modeling of single-and three-phase chokes with multi-resonances, in *8th International Conference on Power Electronics—ECCE Asia* (2011), pp. 435–439. https://doi.org/10.1109/ICPE.2011.5944573

I. Stevanovic, et al., Behavioral modeling of chokes for EMI simulations in power electronics. IEEE Trans. Power Electron. **28**(2), 695–705 (2013). https://doi.org/10.1109/TPEL.2012.2203319

R. Suárez, et al., Influence of complex magnetic permeability on 3-D simulation of MnZn common-mode chokes, in *2023 International Symposium on Electromagnetic Compatibility – EMC Europe* (2023), pp. 1–6. https://doi.org/10.1109/EMCEurope57790.2023.10274356

A. Suarez, et al., Characterization of different cable ferrite materials to reduce the electromagnetic noise in the 2-150 kHz frequency range. MDPI J. Mater. **11**, 174 (2018). https://doi.org/10.3390/ma11020174

A. Takacs, et al. *Physical Modeling of Saturated Common Mode Choke* (2025). arXiv: 2504.01900 [physics.app-ph]. https://arxiv.org/abs/2504.01900

TDK Corporation, *Basics of Ferrite and Noise Countermeasures*. Internet. https://product.tdk.com/system/files/dam/doc/content/emc-guidebook/en/eemc_basic_06.pdf (visited on 2025)

M. Tousignant, et al. Incorporation of a vector Preisach-Mayergoyz Hysteresis model in 3-D finite element analysis. IEEE Trans. Magn. **55**(6), 1–4 (2019). https://doi.org/10.1109/TMAG.2019.2900690

D. Vanoost, et al., Magnetic hysteresis at the domain scale of a multiscale material model for magneto-elastic behaviour. J. Magn. Magn. Mater. **414**, 168–179 (2016). ISSN: 0304-8853. https://doi.org/10.1016/j.jmmm.2016.04.028. https://www.sciencedirect.com/science/article/pii/S0304885316303420

Würth Electronics. *WE-CMB NiZn Common Mode Power Line Choke*. Internet. https://www.we-online.com/components/products/datasheet/7448421016.pdf (visited on 2025)

X. Yao, G. LuoA simple method for extracting complex permeability of magnetic core in common-mode choke, in *2022 Asia-Pacific International Symposium on Electromagnetic Compatibility (APEMC)* (2022), pp. 289–292. https://doi.org/10.1109/APEMC53576.2022.9888460

Chapter 10
Conducted Emission

Testing compliance for conducted emissions is essential to protect the power supply from electrical noise. Simulating the conducted emission test enables to estimate the conducted emission without the need for hardware.

These simulations should be initiated early in the design process. At this stage, there is limited information about the electronic system to be built, and both the layout and mechanical assembly are subject to fewer constraints than in later stages. This flexibility allows for easier modifications. In this section, we demonstrate how to construct simulation models that represent the conducted emission test. Specifically, we examine the conducted emission voltage test as defined by CISPR 25.

As the system under test, we use a 300 W electric motor drive, whose filter is discussed in Sect. 9. To perform the conducted emission test, we must model the entire drive, treating the switching transistors as noise sources. Modeling the electric drive, including its PWM sequence at the gates of the switching transistors, requires a transient simulation. We perform this simulation at the circuit level and integrate it with the EM model of the system's geometry (Sect. 10.2). Additionally, we must describe the drive's load, which is the electric motor. Various approaches to motor modeling for EMC simulation are explored in Sect. 10.4.

We introduce the basics of the technique of vector fitting in Sect. 10.3, which is essential to include S-parameter data in a transient circuit simulation. Further details are explored in the advanced Sect. 10.9 at the end of the chapter. Vector fitting enables us to eventually generate and run the conducted emissions model of the motor drive (Sect. 10.5).

With this model, we continue the EMI filter design of the motor drive that is started in Chap. 9. With the model of the drive from Sect. 10.5 and the limit lines of the CISPR 25 test (Sect. 10.1), we deduce the filter requirements. We investigate several configurations of the drive system with respect to common mode propagation. We test several filter variants with respect to the CISPR 25 class 3 limit

J. Hansen, A. Barchanski, *EMC Simulation of Electronic Systems*,
https://doi.org/10.1007/978-3-032-09237-3_10

for the average detector and compare the EMI spectra generated by different model variants (Sect. 10.6).

In Sect. 10.7, we investigate the conducted emission of the DC-DC converter that is investigated in Sect. 5.6.1. In this chapter, we incorporate an S-parameter-based EM model and compare the results to those of the circuit model that contains less precise information about the geometry.

10.1 The Conducted Emission Test

Domains: 

Various conducted emission tests exist, each tailored to different product sectors. For multimedia equipment, CISPR 32 applies, while CISPR 11 covers industrial, scientific, and medical electrical equipment, and CISPR 14-1 applies to household appliances. In this chapter, we specifically focus on CISPR 25, which is designed for vehicles, boats, and internal combustion engines.

A schematic representation of the CISPR 25 conducted emission voltage test is shown in Fig. 10.1. The DUT is placed on a metal table, positioned on a 50 mm thick support with a relative permittivity ε_r close to one. The positive and negative supply lines of the DUT are each connected to a LISN using wires with a minimum length of 200 mm and a maximum of 400 mm. The LISNs are grounded to the metal table, and the load is placed directly on the table. The DUT may or may not be connected to the table via a ground strap, as a matter of how the DUT will eventually be grounded in its target application.

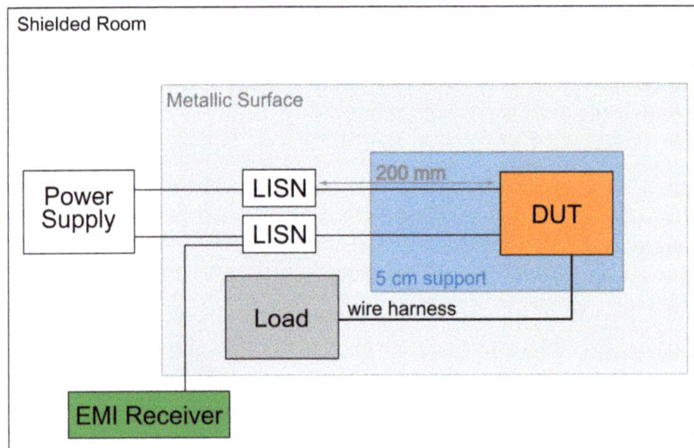

Fig. 10.1 Sketch of the CISPR 25 setup

Table 10.1 CISPR 25 emission limits in dBμV for classes 1–5 for the average detector in the frequency range from 150 kHz to 108 MHz (Bogdan Adamczyk 2018)

Service / Band	Frequency (MHz)	Class 1	Class 2	Class 3	Class 4	Class 5
LW	0.15 – 0.30	90	80	70	60	50
MW	0.53 – 1.8	66	58	50	42	34
SW	5.9 – 6.2	57	51	45	39	33
FM	76 – 108	42	36	30	24	18
TV Band I	41 – 88	48	42	36	30	24
CB	26 – 28	48	42	36	30	24
VHF I	30 – 54	48	42	36	30	24
VHF II	68 – 87	42	36	30	24	18

The system's power supply is connected to the LISNs and may also power the load if necessary. The test is conducted inside a shielded enclosure, and an EMI receiver is used to measure the emission spectrum at the LISN.

The CISPR 25 standard defines emission limits for the peak, average, and quasipeak detectors (cf. Sect. 2.5.4) across various frequency bands within the range of 150 kHz to 108 MHz. At the lower end of the spectrum, the standard includes the LW band (150 kHz–300 kHz), followed by MW (0.53 MHz–1.8 MHz), SW (5.9 MHz–6.2 MHz), and CB (26 MHz–28 MHz). Additionally, there are two VHF bands: VHF I (30 MHz–54 MHz) and VHF II (68 MHz–87 MHz). The standard also includes the TV I band (41 MHz–88 MHz) and the FM band (76 MHz–108 MHz).

The standard defines five different limit classes, with class 5 imposing the strictest requirements. An overview of the limits for the average detector is provided in Table 10.1, while the limits for the peak and quasi-peak detectors can be found in Bogdan Adamczyk (2018).

A simulation of the CISPR 25 test tracks the time-domain voltage at the LISN (Sect. 2.5.2) and applies postprocessing using an EMI receiver script (Sect. 2.5.4) to compute the voltages $V_{\mathrm{LISN,P}}$ and $V_{\mathrm{LISN,N}}$ over frequency. Comparing these results with the limit lines determines whether the DUT meets or fails a given limit class.

10.2 EMI Model for Conducted Emission of a Motor Drive

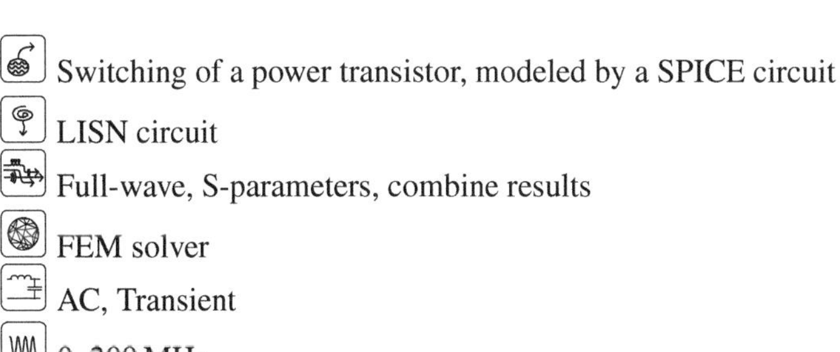

To calculate the conducted emission voltage at the LISNs, we must create a model of the DUT that includes the attached load, the connecting wires to the LISNs and power supply, and the metallic table.

Exemplarily, we use the 300 W motor drive, whose EMI filter is analyzed in Chap. 9. The motor drive contains switching transistors. We investigate both transient and frequency-domain simulations.

We begin by reviewing the functional circuit of the electric drive. Next, we develop the EM model to represent its geometry and the surrounding measurement setup. Finally, we integrate these elements to construct a circuit model for transient simulation, incorporating all relevant electromagnetic properties of the conducted emission test setup.

The electrical drive operates at a supply voltage of 48 V and utilizes a B6 bridge topology . The drive unit is standalone, with an external brushless motor attached. A schematic representation of the DUT, including the power supply, LISNs, and the electrical motor as the load, is shown in Fig. 10.2. At the center of the metallic table is the PCB of the motor drive. On the upper left side of the PCB, the EMI filter is located, consisting of the CMC, two C_x capacitors, and four C_y capacitors. The components of the filter and their values are analyzed in Chap. 9.

To the left of the PCB, near the connector, are the two supply wires leading to the LISNs and the power supply. On the bottom side of the PCB, the commutation cell is located. This cell includes three 100 μF aluminum electrolytic DC link capacitors, a B6-bridge composed of six BUK9Y72-80E N-channel MOSFETs (Nexperia), and a connector for the three-phase wires to the electric motor on the right side of the figure. Additionally, the gate driver IC is situated in the upper part of the PCB.

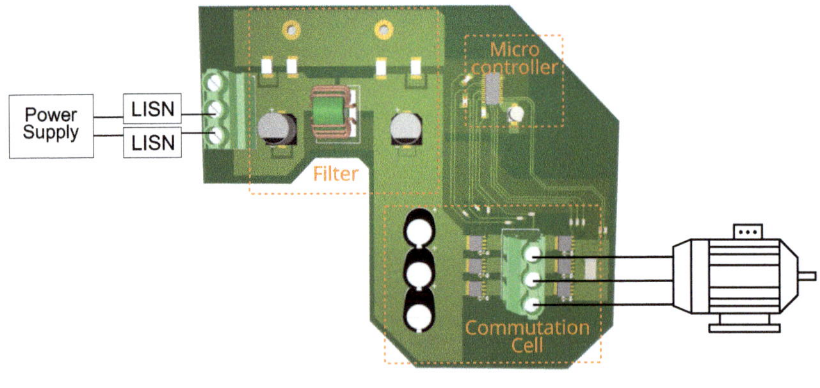

Fig. 10.2 Layout of the 300 W motor drive with EMI filter, the commutation cell, and the microcontroller with the gate drives, as situated in the CISPR 25 conducted emission voltage test

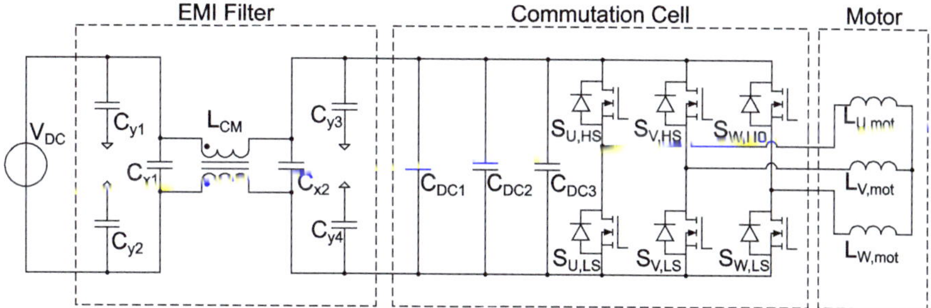

Fig. 10.3 Functional circuit of the motor drive with EMI filter

10.2.1 Functional Circuit

The circuit diagram of the power unit is shown in Fig. 10.3. At the center, the B6 bridge consists of three phases (U, V, W), each comprising a high-side switch ($S_{i,\mathrm{HS}}$) and a low-side switch ($S_{i,\mathrm{LS}}$), with a parallel diode for each switch. The three DC link capacitors are denoted by C_{DCj}. The phase wires U, V, W connect to the electric motor, which is represented by its inductances $L_{\mathrm{i,mot}}$. The motor follows a star topology. To the left of the B6 bridge and the DC link capacitors, the EMI filter is located. As discussed in Chap. 9, it includes a common mode choke, along with, on each side, a C_{x} capacitor and two symmetrically placed C_{y} capacitors. The model is functional; it does not contain any parasitic elements.

The motor drive is powered by a DC voltage of 48 V, which is isolated from the ground. The C_{y} capacitors are referenced to this ground. The drive employs sine-triangle PWM with a switching frequency of 10 kHz. The gate voltage varies between 0 V and 10 V, and a modulation index of 0.8 is applied. We set the fundamental frequency of the PWM to 500 Hz, that is, the motor current has a period of 2 ms. We use ideal switches which step from low to high in 0.5 μs and have a blanking time of 1 μs.

We limit the model to the transistors, the parallel diodes, and the connecting lines. The lines connect on one end to the motor and on the other to the power supply. The transistors and their parallel diodes are the primary noise sources in the motor drive. Other potential noise sources, such as the gate driver or additional ICs on the PCB, may also contribute to emissions. While incorporating these elements into the model can improve accuracy, it also increases complexity. Our focus here is to determine which EMI filter components enable the motor drive to meet which emission limits, regardless of additional noise sources that may contribute to further interference.

We conduct a transient circuit simulation of the network shown in Fig. 10.3 up to 15 ms. To verify proper circuit operation, we analyze the gate voltage pattern and the resulting phase currents, as shown in Fig. 10.4. On the left we see the gate voltages V_{UH}, V_{VH}, V_{WH} for the three high-side switches over a time interval from 10 ms to 10.4 ms. The pattern exhibits a PWM sequence with center-aligned pulses. The low-side gate voltages, which are not displayed here, follow the inverse pattern:

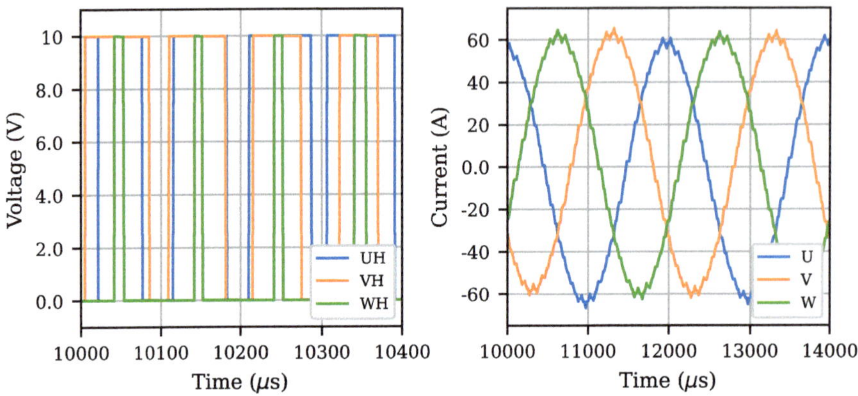

Fig. 10.4 High-side gate voltages and phase currents of the model shown in Fig. 10.3

whenever a high-side switch is on, the corresponding low-side switch is off, and vice versa.

On the right-hand side of Fig. 10.4, we plot the three phase currents I_{U}, I_{V}, I_{W} of the motor phases. These currents are phase-shifted by 120° relative to each other. With the modulation index of 0.8, the peak current reaches 60 A. The plot covers a time range from 10 ms to 14 ms. A settling time of approximately 10 ms is required for all phase currents to reach a steady-state condition, becoming symmetric around 0 A and achieving equal amplitude. Each phase current completes one period in 2 ms.

This simulation does not include current control. Instead, the modulation index is fixed, determining the resulting peak current. The PWM period of 2 ms leads to a spectral content significantly lower than the minimum frequency of 150 kHz, above which the CISPR 25 standard imposes EMI limits. Consequently, the PWM does not influence the frequency spectrum relevant for CISPR 25. The same applies to current control, which operates at a much lower rate than the 10 kHz switching frequency and does also not affect the frequency range of interest for CISPR 25 compliance.

10.2.2 EM Model

The EM model of the conducted emission setup incorporates the electromagnetic effects of the CISPR 25 configuration shown in Fig. 10.1. In addition to the PCB and the electric motor, the setup includes the LISNs, the measurement receiver, the wiring and the table plate.

However, not all of these components require geometric modeling. The geometric model must include all components that exhibit significant electromagnetic coupling. These include the PCB, the wires connecting to the LISNs and the electric

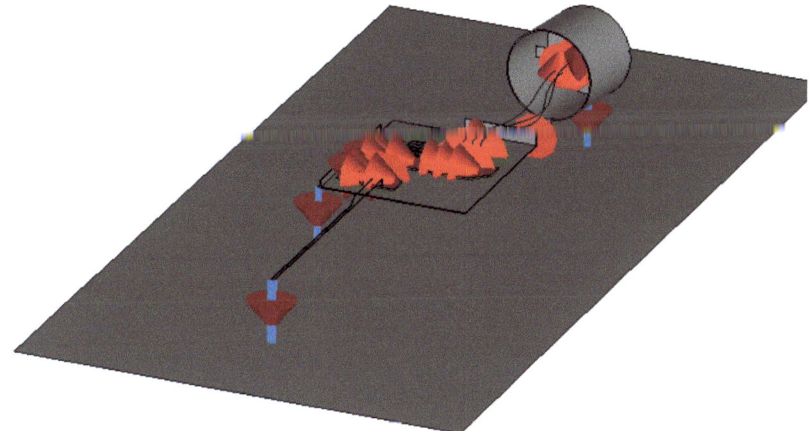

Fig. 10.5 EM model of the conducted emission setup for the motor drive

motor, the housing of the electric motor, and the table plate. When the interaction between two components can be represented purely by voltages and currents, we introduce ports and connect these elements at the circuit level.

Figure 10.5 illustrates the geometry used to construct the EM model. At the center, the PCB is located with the simplified housing model introduced in Sect. 9.3. In the front, two 20 cm-long wires connect the PCB to the LISNs. At the back, three phase wires lead to the electric motor, for which only the empty housing and some internal connection elements are modeled. The table plate is positioned 50 mm below the housings of both the PCB and the motor. The numerous red ports in the figure indicate additional elements incorporated into the EM model at the circuit level. The LISNs are not modeled and neither their housing.

In the following, we describe the individual components of the EM model and outline the function of the ports.

PCB Model The PCB model includes the B6 bridge, the EMI filter, and multiple connections to ground and other surrounding elements. The B6 bridge features ten ports: six for the transistors with parallel diodes, three for the DC link capacitors, and one for the shunt used to measure the current (cf. Sect. 5.3.1). The ports on the EMI filter are defined as described in Sect. 9.5.2. Specifically, there are four ports for the C_y capacitors and two for the C_x capacitors. There are two additional ports for integrating a circuit-level model of the common mode choke. (Section 9.4.2 discusses limits of and alternatives to this choice.) Furthermore, two ports are assigned to model the screws that connect the common ground of the PCB to the housing. In total, the PCB model comprises 20 ports.

PCB Housing and Wires In addition to the screws connecting to the PCB, the housing and wires feature four additional ports. One port models the ground strap between the housing and the table, while another represents the galvanic connection between the PCB housing and the electric motor housing.

At the ends of the 20 cm wires leading to the LISNs, two ports connect the wire ends to the metallic table. Each of these ports is linked to a circuit model of the LISNs (Sect. 2.5.2) and the 48 V voltage source representing the battery. This configuration assumes there is no noticeable electromagnetic interaction between the metallic housing of the LISNs and the DUT. Instead, the interaction between the LISNs and the DUT is modeled solely through the LISNs' impedance, ensuring accurate calculation of voltages and currents at the wire ends.

Electric Motor Apart from its connection to the PCB housing, the electric motor includes seven ports: three for the motor inductances, three for the rotor-stator capacitances, and one for the ground strap between the motor and the table. The details of this modeling approach are discussed in Sect. 10.4.

Computation and Boundary Conditions The complete EM model consists of 31 ports. The FEM frequency-domain solver is used to compute the model, applying a perfect electric boundary at the side of the table while leaving all other boundaries open. Including mesh adaptation, the computation takes 12 minutes and 22 seconds. The frequency range considered spans from 1 Hz to 200 MHz.

10.2.3 Coupled EM-Circuit Model

We attach suitable circuit elements to the EM model generated in Sect. 10.2.2. Each element in this setup enforces a specific current through its pin and a corresponding voltage across it, determined by the element's impedance. A detailed explanation of the interaction between this network block and the attached elements is provided in Sect. 6.7. The resulting combined circuit-EM model is illustrated in Fig. 10.6.

We describe the individual circuit elements attached to the ports in a clockwise manner, starting from the upper side of the EM block.

Upper Side: DC Link and Filter Capacitors The three pins on the left correspond to the three DC link capacitors, each with a capacitance of 100 μF. These capacitors have an ESR of 0.3 Ω and an ESL of 1 nH (Panasonic). The next six pins to the right are assigned to the four C_{Y} capacitors (SMD-type) and the two C_{X} capacitors (electrolytic) of the EMI filter (cf. Sect. 9.5). In the figure, these pins are left open, indicating that the filter components have not yet been populated in this configuration.

Right Side: AC-Side Elements At the uppermost position, there is the shunt resistor, configured with a resistance of 1 mΩ. Below this, the three motor inductances and resistances are connected, one for each motor phase. Each has a value of 100 μH and 100 mΩ, representative of an electric motor in a 300 W motor drive. Each motor phase also exhibits a parasitic capacitance (see Sect. 10.4). The total capacitance is approximately 100 pF, meaning that each phase wire has a parasitic capacitance of about 33 pF. This is a typical value for the capacitance between the stator and the housing of a small electric motor. The value can be derived from the mechanical

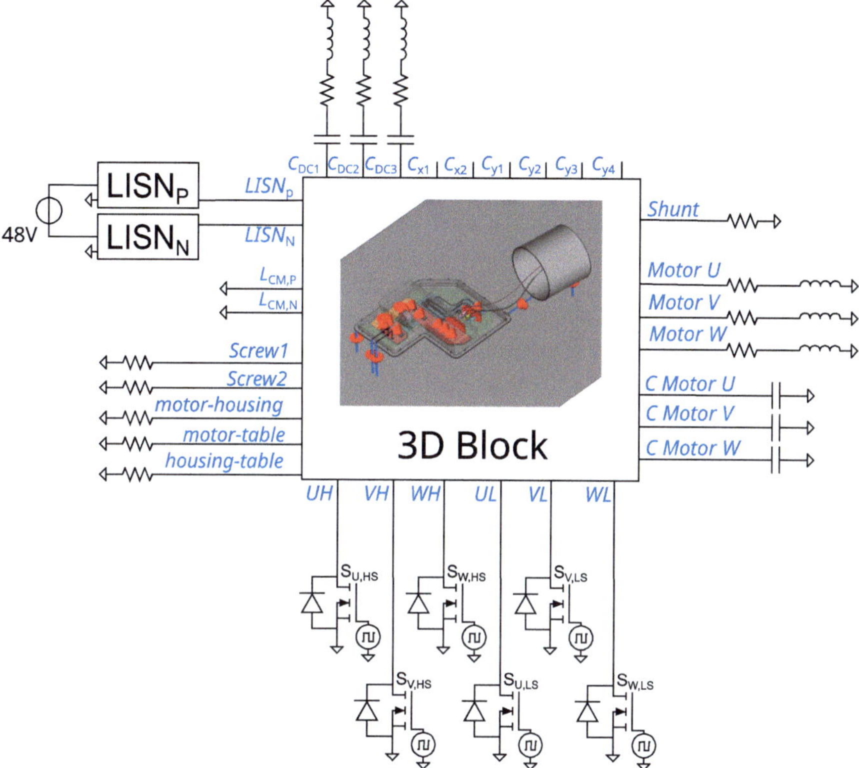

Fig. 10.6 Combined circuit-EM model with connected source and sink models, without the EMI filter

properties of the motor through direct measurement if a prototype is available or via electrostatic simulation using CAD data. Alternatively, capacitance values from similar motors can be used as a reference.

If no prior capacitance data is available, the impact of this parameter can be analyzed using the current simulation model. Conducting this sensitivity analysis is recommended regardless, as a new motor design with a different capacitance may be introduced later in the development process. Such a simulation study indicates the impact of a new motor design.

Lower Side: Transistors and Diodes On the lower side of the network block, the six MOSFETs (BUK9Y72-80E) (Nexperia) are located. Each MOSFET model consists of three pins: one for the gate voltage, one for the drain, and one for the source voltage. To apply the drain-source voltage to the network model, the drain pins are connected to the network, while the source pins are grounded (cf. Sect. 2.4.3.1).

At the six gate pins, appropriate gate voltage signals with suitable rise and fall times must be applied. These voltage patterns can be obtained from predefined libraries. The CST provides a macro for their generation. Additionally, each transistor is connected in parallel to an ideal diode.

Left Side: Connecting Straps, CMC, and Power Supply On the lower left side, there are pins for the ground strap at the PCB housing, the ground strap of the motor housing, and a connection between the PCB housing and the motor housing. If a pin is left open, no connection is established. If it is shorted or assigned a low-ohmic resistance (e.g., 10 mΩ), a galvanic connection is formed, modeling, for example, a metal strap. It is important to note that any metal strap has an inherent self-inductance. In the model, this self-inductance is represented as a port, where the port's geometry determines its electromagnetic properties as if it were a metal strap. If the geometry of a ground strap is known, it should be explicitly modeled by an appropriately shaped port to ensure accurate electromagnetic behavior.

The next two pins correspond to ports positioned at the locations of the screws that connect the common ground of the filter to the housing (cf. Sect. 9.5.2). The pins labeled $L_{\mathrm{CM,P}}$ and $L_{\mathrm{CM,N}}$ represent the ports designated for connecting the CMC of the filter (cf. Sect. 9.3 for the discussion of this choice). Since the filter is not installed in Fig. 10.6, these two pins must be shorted to establish a galvanic connection between the power supply and the B6 bridge. Finally, on the upper left, the LISNs are connected to the pins representing the ports at the ends of the 20 cm supply wires. As illustrated in Fig. 10.1, the 48 V voltage source supplies power through the LISNs to the motor drive.

The EM block represents the electromagnetic properties of the conducted emission test geometry in the frequency domain. It can be directly utilized for an AC simulation, as demonstrated in the design of the EMI filter in Sect. 9.5. However, to efficiently integrate this block into a transient simulation, vector fitting must be applied. The fundamental concepts of vector fitting are introduced in Sect. 10.3, with a more in-depth discussion provided in Sect. 10.9. The application of vector fitting to compute the conducted emissions of the model shown in Fig. 10.6 is detailed in Sect. 10.6.

10.3 Vector Fitting Basics

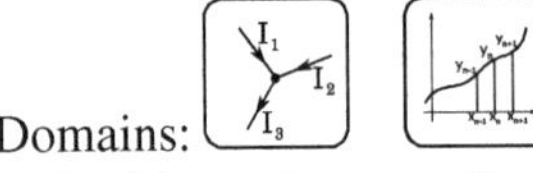

In this section, we discuss the fundamentals of vector fitting, an essential step in simulation workflows that utilize S-parameter-based circuit models in transient circuit simulations. In CST, vector fitting is performed automatically when a transient circuit simulation involving S-parameter models is initiated. Consequently, this step may be overlooked by simulation engineers. However, the accuracy of the vector-fitted model has a significant impact on the results obtained from transient circuit simulations.

For complex electronic system models, it is crucial for simulation engineers to understand the vector fitting process, assess the quality of a fitted model, and improve its accuracy if it is found to be inadequate.

Throughout this book, we employ vector fitting in several examples; the calculation of emissions from the simple board pair in Sect. 6.2, the LTE example in Sect. 6.6, the cables-in-wing example in Sect. 8.4, and the motor drive in this chapter.

10.3.1 Passive Macromodeling

Vector fitting is a method to convert frequency domain data into a circuit representation that can be used in transient circuit simulations. It can be applied to general frequency domain data; however, in this book, we limit the discussion to S-parameters that characterize linear passive interconnects.

Even if we use the term "vector fitting" throughout this book, a more accurate wording is "passive macromodel generation" as vector fitting is only one step of the entire process. The passive macromodel generation process consists of four steps as depicted in Fig. 10.7.

To use the macromodel generated by this process in transient circuit simulations, two critical properties must be satisfied: causality and passivity. Causality ensures that the output of the macromodel does not precede its input, a common issue when using the Inverse Fourier Transform instead of vector fitting. Passivity requires the macromodel not to amplify the input signal, as any amplification would result in instabilities during the transient circuit simulation (see also Sect. 2.6).

The first step in the macromodeling process involves importing the S-parameter data. This can be done manually by importing a Touchstone file or automated in a coupled EM and circuit simulation. In CST, the full-wave solver transfers the computed S-parameters to the circuit simulator, which automatically triggers the macromodel generation if it is needed for a transient simulation. If the circuit schematic includes S-parameter models from external sources, such as imported measurements, the macromodel generation is also performed on these models.

In the second step, the input data is vector fitted. Each S-parameter is interpreted as a transfer function, $\mathcal{H}_{ij}(\omega_m)$, defined at a set of discrete frequency samples ω_m. During vector fitting, these discrete transfer functions are approximated using a sum

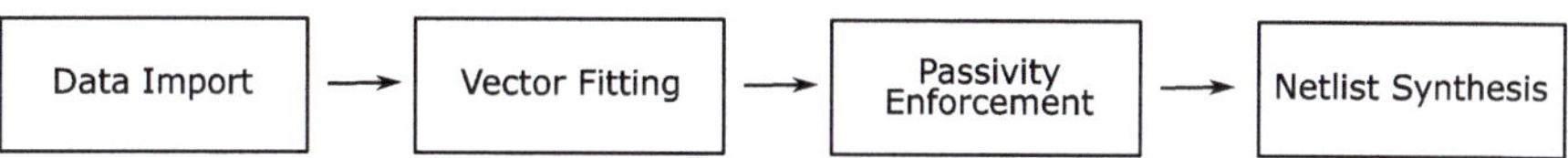

Fig. 10.7 Passive macromodel generation flow

of rational functions of the form

$$\hat{\mathcal{H}}_{ij}(s) = \sum_{k=1}^{K} \frac{c_{k,ij}}{s - p_{k,ij}} + \mathcal{H}_{\infty,ij} \tag{10.1}$$

where the $p_{k,ij}$ are the poles, the $c_{k,ij}$ complex coefficients, and $\mathcal{H}_{\infty,ij}$ an offset term. In the fitting procedure, we use Eq. 10.1 to approximate each S-parameter at each frequency by

$$\hat{\mathcal{H}}_{ij}(\omega_m) \approx S_{ij}(\omega_m), \quad \forall m. \tag{10.2}$$

Essentially, this process involves applying a complex-valued rational functions fit. The fitting procedure determines the values of $c_{k,ij}$ and $p_{k,ij}$. The number K of partial fractions required to achieve a specified accuracy is determined through an iterative scheme. Initially, a starting number of partial fractions is selected based on the input data, the fitting is applied, and the fitting error is calculated. If the resulting model does not meet the desired accuracy, the number of partial fractions is increased, and the fitting process is repeated. These steps continue until the required accuracy is achieved, or further increasing the number of partial fractions no longer improves the fitting accuracy.

The accuracy threshold is typically set to 10^{-3}, a value we use consistently in all examples presented in this book. In the context of vector fitting, the number of partial fractions is referred to as the number of poles, which serves as a measure of the complexity of the generated macromodel.

The required number of poles depends significantly on the electrical size of the structure described by the S-parameter set. As the electrical size increases, so does the number of resonances in the structure. A general rule of thumb is that two poles are needed to model a single resonance in the S-parameters. For electrically short structures, the number of poles typically ranges from 10 to 20. However, for electrically large structures, this number can exceed 100.

With the appropriate choice of poles during the vector fitting process, the causality condition is ensured. However, this does not guarantee passivity. The passivity of the model is enforced in the third step of the macromodel generation process. This step employs a passivity enforcement scheme that perturbs the system parameters (specifically, the state-space matrices of the model) to ensure that the resulting modified system is passive.

This process is also iterative and is performed under the condition that the deviation of the passivity-enforced model from the vector-fitted model—or from the input data—is minimized. However, if the input data itself contains passivity violations, some deviations from the input data are inevitable, as it is not possible to create a passive macromodel with an identical response to nonpassive input data.

Passivity violations are commonly observed in measurement data, often arising from improper calibration, de-embedding errors, or measurement noise. Also S-parameters derived from simulations can exhibit passivity violations due to issues such as poor meshing, low solver accuracy, nonphysical material models, or improper interpolation.

As long as the passivity violations in the input data are minor, the passivity enforcement step corrects them without a significant loss in accuracy. We discuss the measure of passivity in the advanced Sect. 10.9.

The final step in macromodel generation is the synthesis of the circuit model. The passivity enforced pole-residue representation is converted into a netlist that can be processed by a circuit simulator. Several circuit simulators, such as the one in CST, Synopsys H-SPICE®, or VHDL simulators, directly use the pole-residue representation in transient circuit simulations by applying convolution. This approach is an efficient method for including macromodels in simulations.

Standard SPICE simulators, however, do not support the pole-residue representation. Instead, it is necessary to convert the pole-residue model into a netlist constructed from dependent sources. These sources have values that correspond—up to arbitrary normalization factors—to the poles and residues of the model. Although these values do not represent physical properties of real components, they accurately reproduce the behavior of interconnects in system-level simulations.

This description highlights the inherent complexity of macromodel generation. The macromodel generation methods are relatively recent compared to EM field computation, with the first works published in the late 1990s (see References at the end of this chapter). Today, macromodel generation has become a de facto standard in the SI community, particularly for fitting electrically large interconnects.

In the EMC and power electronics simulation, RLC models are still often preferred, and macromodel generation using vector fitting is sometimes met with skepticism. By providing a detailed explanation of the steps involved and the challenges encountered, the authors aim to address this skepticism. We offer simulation engineers practical guidance for their generation, because macromodels allow the characterization of wideband S-parameters with many ports with an accuracy that by far exceeds the possibilities of RLC models.

10.3.2 *Accuracy of Macromodel Generation*

With automatic settings, the macromodel generation process described here produces accurate models for scenarios of simple to medium complexity. For more complex scenarios, however, we recommend verifying the accuracy of the fitted model before using it in a transient simulation.

A straightforward method for testing the accuracy of the macromodel is to compare its frequency-domain response to the original data. This is achieved by performing an AC simulation for both the macromodel and the original data and comparing the results. Examples of such testing scenarios are presented in the advanced Sect. 10.9.

Before discussing specific examples, we examine some implications of the macromodel generation process and their effects on model accuracy.

The iterative scheme applied during vector fitting controls the fitting process by calculating an error value that quantifies the deviation between the fitted model and the input data. Understanding the role of this error control is crucial, as it has impact on the accuracy of the generated model. The fitting error is a single value that evaluates the quality of the fitting across the frequency band covered by the input data and for all entries of the S-matrix.

A key point to note is that the fitting error is calculated based on the reference impedance specified in the input data. While the fitting process may achieve high accuracy for the given reference impedance, this accuracy is preserved in circuit simulations only if the same impedance is used for the loads in the simulation. If the component models connected to the macromodel have impedances that deviate significantly from the reference impedance used during fitting, the fitting error can be amplified, potentially leading to inaccuracies in the circuit simulation. For this reason, renormalizing the S-parameter dataset to a more suitable reference impedance before vector fitting enhances the accuracy of the resulting circuit simulation.

In digital systems, a reference impedance of 50 Ω is typically a reliable choice. However, for power electronics simulation models, such as the motor driver discussed in this chapter, using a lower reference impedance improves accuracy. Unfortunately, no single reference impedance is universally optimal, as the impedance of components connected to the fitted model is frequency-dependent, while the S-parameter dataset can only use a fixed reference impedance in the macromodel generation process.

Based on the analysis of numerous simulation models, we recommend reference impedance values between 1 and 10 Ω for the power components of the circuit when performing power electronics emission simulations up to approximately 200 MHz. We demonstrate the effect of the reference impedance on the fitting results in Sect. 10.9.

The error amplification caused by a mismatch between the reference impedance and the connected component model increases with the number of ports in the S-parameter model for which the macromodel is generated. When vector fitting is applied to an S-parameter circuit model with many ports, the small errors from individually fitted entries may accumulate in the circuit simulation. Therefore, models with a high port count should be evaluated in terms of their system response before use in transient simulations.

Below, we provide some heuristic guidelines for managing error amplification:

Up to 20 ports The effect of error amplification is typically negligible.

20–60 ports A system response check is recommended.

60–100 ports Manual adjustments can become necessary, and a thorough check of the system response is recommended.

More than 100 ports Vector fitting accuracy requires accuracy control. A possibility to reduce the port count is to model only critical components as ports while representing others as lumped elements in the EM model (refer to Sect. 6.7.2).

When fitting very small couplings with high precision, the calculation of the fitting error can be switched from an absolute to a relative scheme.

Using these techniques, the accuracy of macromodel generation via vector fitting is well-controlled, enabling the creation of accurate models for transient circuit simulations. For readers seeking further details, we explore examples that demonstrate these guidelines in Sect. 10.9.

10.4 Modeling Electric Motors

Domains: 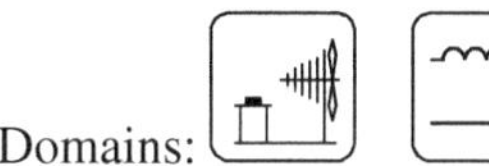

The electric motor serves as the load in an electrically commutated motor drive. Its impedance influences both the differential mode and common mode currents that propagate within the motor and toward its housing. Brushed motors, which are known sources of electromagnetic emissions, are not considered in this book.

There exists a wide variety of electric motor types, which we do not describe here. Instead, we treat the motor as a passive termination for the three-phase wires of the B6 bridge described in Sect. 10.2. Our primary focus is on the impedance characteristics of the motor over frequency, independent of its specific type.

Several approaches can be used to determine these impedances, which we briefly outline in the following subsections. A common method is to model the electric motor as a network of discrete RLC elements, as discussed in Sect. 10.4.1. Alternatively, the motor's impedance characteristics can be derived from measurements (Sect. 10.4.2). If CAD data and material properties are available, an EM model can be constructed (Sect. 10.4.3). Electric motors are inherently complex, and each modeling approach has its own limitations. Open challenges in motor modeling are discussed in Sect. 18.1.

For all motor types, there are three frequency ranges: at low frequencies, the resistance of the motor wires dominates, resulting in a constant impedance. At higher frequencies, the inductance becomes dominant, causing the impedance to increase. Once the impedance of the inductive part of the windings exceeds the impedance due to the capacitances between them, the impedance decreases. Beyond this frequency, the motor impedance exhibits resonant behavior. Additionally, there are capacitances between the rotor and the stator of the motor, as well as between both components and the motor housing. These capacitances contribute to common mode currents.

A view inside an electrical motor is sketched in Fig. 10.8. This motor is a permanent magnet synchronous machine (PSM). For this motor, the outer ring is the stator, which is usually made of laminated steel plates (stator iron). The winding packets of the three-phase wires are shown in blue (phase U), orange (phase V), and green (phase W). Each winding packet is wrapped around a tooth of the stator.

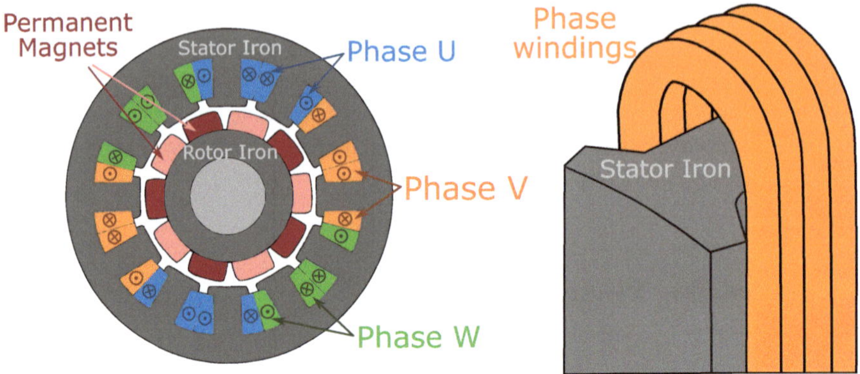

Fig. 10.8 Sketch of an electric motor. Left: Cross-sectional view of an electric PSM motor. Right: example of a winding scheme inside the stator

Inside the stator, there is the rotor, which is equipped with permanent magnets. On the right, a more detailed view of the stator windings is provided. A coated wire is wound around the stator.

There are many winding possibilities. Hairpin windings are very regular. There are also winding schemes which look very random, even though, due to the manufacturing process, the windings of all electrical motors produced on the same manufacturing line are identical. Motor modeling methods must cope with all these different cases.

We are not concerned with the functional parameters of the electric motor, such as power, torque, or losses. Instead, our focus is on its impedance, the modeling of which presents two key challenges. The first challenge lies in the fine details of the windings. Their small spacing relative to each other and to the stator influences the network of differential and common mode capacitances.

The second challenge involves material properties. In the motor, thin laminated steel plates are used to reduce eddy currents. As a result, the magnetic permeability of this structure is anisotropic, and its values are typically not available from datasheets. Detailed modeling of these plates is required to account for the skin effect unless an equivalent material model is applied. If permanent magnets are present in the motor, their material properties must also be determined.

In the following, we discuss three approaches to motor modeling. Both RLC models and EM models are effective at low frequencies, where capacitive effects are negligible. However, at higher frequencies, where the impedance exhibits resonant behavior, motor modeling using RLC or EM models remains a challenge. At these frequencies, an S-parameter measurement provides a reliable alternative. This measurement method, however, does not capture the DC resistance of the motor.

The ultimate goal of all these approaches is to represent the motor as a passive network, which can then serve as a passive termination in a system-level simulation of the motor drive.

10.4.1 RLC Motor Models

Both the DC resistance and the inductance of an electric motor are functional parameters, determined by design. The DC resistance affects the conduction losses of the motor, impacting its overall efficiency. The inductance is a design parameter for the magnetic flux. Both parameters are typically known in the early stages of motor design. An RL network effectively models the differential mode impedance of the motor up to the first resonance frequency.

An RL model can be extended to higher frequencies by introducing additional discrete capacitances. The higher the desired frequency range, the more capacitances need to be determined and incorporated. The values of these capacitances can be computed using static electromagnetic solvers or, if the length-dependent properties of the wires are considered, MTL solvers. The capacitances between the wires and the stator, as well as between the stator and the rotor, can be calculated without modifying the model. However, to compute interwinding capacitances, the motor windings must be partitioned into electrically isolated segments.

RLC models remain effective up to the second resonance of the impedance. Beyond this point, model generation becomes increasingly complex. Moreover, the placement of discrete capacitor networks is not unique, as discussed in Sect. 10.4.4.

There are multiple ways to parameterize an RLC model. The RL values can be provided by the motor designer. If hardware is available, RL values, along with the most dominant capacitances, can be measured using an impedance analyzer (Sect. 10.4.2). If CAD data and material properties are available, magnetoquasistatic simulations (Sect. 10.4.3) can be used to determine resistances and inductances. Electrostatic simulations compute the dominant capacitances between electrically isolated metal components. Alternatively, RLC parameters can be extracted from a full-wave solution, which requires meshing the motor geometry or parts of it.

10.4.2 Measured Motor Models

If a motor is available as hardware, its electrical properties can be determined through measurement.

The intricate resonant behavior of an electric motor can be analyzed using a network analyzer. Measuring a three-phase motor requires a three-port network analyzer. Each phase wire is connected to a separate port. The ground of each port is connected to the motor housing to capture the common mode behavior. This procedure yields an accurate S-parameter model of the motor. In a circuit simulation, this model can be used with appropriately defined ports to compute the differential mode impedance between pairs of phase wires and the common mode impedance, as illustrated in Fig. 10.9.

Network analyzers do not operate effectively at frequencies close to DC. Near their minimum measurement frequency, the measured impedance may be in the

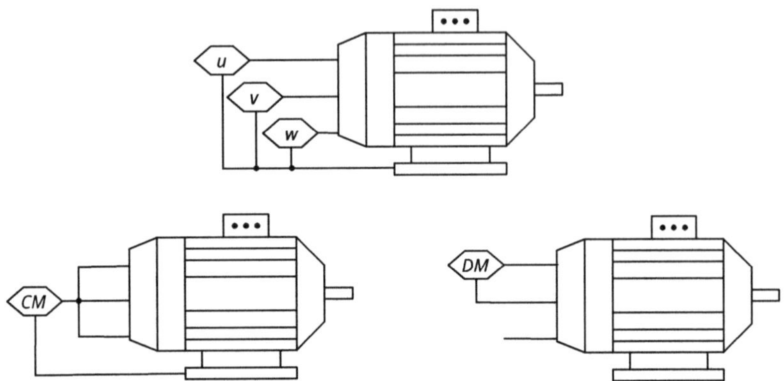

Fig. 10.9 Circuit simulation to extract, from a measured three-port network (top), the common mode impedance Z_{CM} (bottom left), and the differential mode impedance Z_{DM} between two phase wires (bottom right)

range of Ω rather than mΩ (cf. the measurement presented in Fig. 9.27). If the measured impedance is too high, it artificially limits the current through the motor to unrealistic values. Consequently, the operating point of the electric drive is incorrect, potentially leading to inaccurate emission calculations.

To measure impedance down to DC, an impedance analyzer can be used to characterize the individual impedances between phase wires or between the phase wires and the motor housing. However, a two-port impedance analyzer cannot determine the complete impedance matrix for all phase wires and the housing. Additionally, the maximum measurable frequency of an impedance analyzer is significantly lower than that of a network analyzer.

A possible approach is to manually merge the low-frequency impedance data obtained from an impedance analyzer with high-frequency data from a network analyzer. However, this may introduce discontinuities at the transition frequency, which can cause vector fitting (cf. Sect. 10.3) to fail.

S-parameter data obtained from network analyzer measurements is the preferred choice for frequency-domain simulation of electric drives. Since these simulations have limitations (cf. Sect. 10.5.2), an alternative approach is mandatory which is applicable for transient simulation.

The common approach is to use measurements from an impedance analyzer. Differential measurements, as shown on the bottom right of Fig. 10.9, yield the RL values and the most significant capacitance for the differential mode. A common mode measurement, as depicted on the bottom left, determines the common mode capacitance. The resulting model remains valid up to the first resonance. A more broadband model can be generated if the data is vector-fitted and exported as SPICE model. Such a vector fit can also be applied to S-parameter models. Here, it must be cross-checked if the extrapolated DC resistance is realistic. Details on the vector-fitting are given in Sect. 10.9.

Most often, measurements are conducted on a nonrotating motor. In some motor types, the measured S-parameters and impedances are a function of the rotor-stator position. Recalling the left side of Fig. 10.8, we see that the magnetic flux between the tooth of the stator and the rotor can vary whether on the stator side; there is a gap or a permanent magnet or parts of both. Together with the current, this flux determines the measured inductance (Eq. 5.12).

10.4.3 EM Motor Models

With given material properties and CAD data of the electric motor, an EM model can be constructed. An EM model for the computation of EMI should not be confused with existing EM modeling strategies used for functional motor design. Many commercial vendors offer tool suites for functional motor design, which apply the magnetoquasistatic Maxwell equations (cf. Table 5.1) to compute the currents and magnetic fields inside the motor. These models are used to derive functional properties such as torque. However, they neglect the coupling between the magnetic and electric fields, meaning that capacitances are not computed. Results from a magnetoquasistatic solver can be used to parameterize the R-L matrix of an RLC motor model, as described in Sect. 10.4.1.

High-frequency EM models of electric motors are computationally expensive due to the fine geometric details that must be considered, such as the capacitances between coated wires and the material properties. A time-domain simulation is infeasible because the magnetic properties of the steel sheets and other high-permeability materials are in general unknown, requiring detailed BH-curves. The modeling of magnetic materials using a BH-curve is discussed in Sect. 9.7, particularly in the context of common mode choke modeling.

In Sect. 9.6, a frequency-domain model of a common mode choke is constructed based on S-parameter measurements, allowing the derivation of effective material properties. While electric motors are significantly more complex, a similar approach can be employed to generate an EM model. Due to the intricate geometry of electric motors, solving these models can require substantial computational resources. For certain motor designs, particularly those with a large number of tightly packed windings, generating an accurate EM model is extremely challenging.

At present, EM models of electric motors are not widely used in the EMI simulation of motor drives. However, advancements in computing power, high-performance mesh generation, and improved modeling techniques for high-permeability materials may eventually enable EM modeling of electric motors to become a standard approach in industrial EMC simulations. Whether this will become a practical reality remains an open question.

10.4.4 Connection of an RLC Motor Model in an EM Simulation

For the conducted emission simulation in this chapter, we use an RLC-based motor model. In an RLC model, the resistance, inductance, and capacitance of the motor are lumped into discrete elements, whereas in physical reality, these properties are distributed over the motor conductors. Nevertheless, an RLC model serves as a good starting point for incorporating the motor into the EMC simulation of the electronic system. There are several options to implement RLC elements in an EM model, and we discuss two approaches in this section.

In the first option, we model the common mode capacitance of the three phases individually and place it between the bridge and the motor inductance. This configuration results in a worst-case common mode impedance because the capacitances are in parallel with the motor inductance.

In the second option, we model the common mode capacitance of all phases as a single capacitor and place it between the motor inductance and the housing. This configuration results in a best-case common mode impedance because the capacitance is in series with the motor inductance.

The terms “worst-case” and “best-case” refer to the common mode impedance. A low common mode impedance (worst-case) leads to increased emissions, while a high common mode impedance (best-case) reduces emissions. The corresponding schematics of both options are shown in the top section of Fig. 10.10.

For the connection of the motor inductances (and resistances), we define an auxiliary shape in the CAD model and use this shape as the reference for three ports, each defined from the individual phase wires to the auxiliary shape. The auxiliary shape is best visible on the right-hand side of Fig. 10.10 and is indicated by a dashed blue line. In our example, the definition models a motor with a star connection, where the auxiliary shape represents the neutral point.

The auxiliary shape in the EM model introduces an additional capacitance between the phases and the windings, but it is necessary for the connection of the lumped inductances. In our model, this auxiliary shape is relatively large due to the significant separation of the wires. The simulation engineer must choose an appropriate size for the auxiliary shape. However, since we also include a lumped common mode capacitance of the motor in the simulation model, the capacitance of the auxiliary shape to the housing is smaller than the lumped capacitance and thus does not affect the simulation results.

For the connection of the lumped common mode capacitance, we first discuss the best-case scenario, where the capacitance is in series with the motor inductances. The corresponding setup is illustrated on the right-hand side of Fig. 10.10. A port is placed between the auxiliary plate and the motor housing, and the motor's common mode capacitance C_{CM} is connected in the coupled schematic.

To realize the worst-case common mode impedance, where the individual motor phase capacitances are in parallel with the inductors, we use a second auxiliary shape that connects to the housing, as shown on the left-hand side of Fig. 10.10.

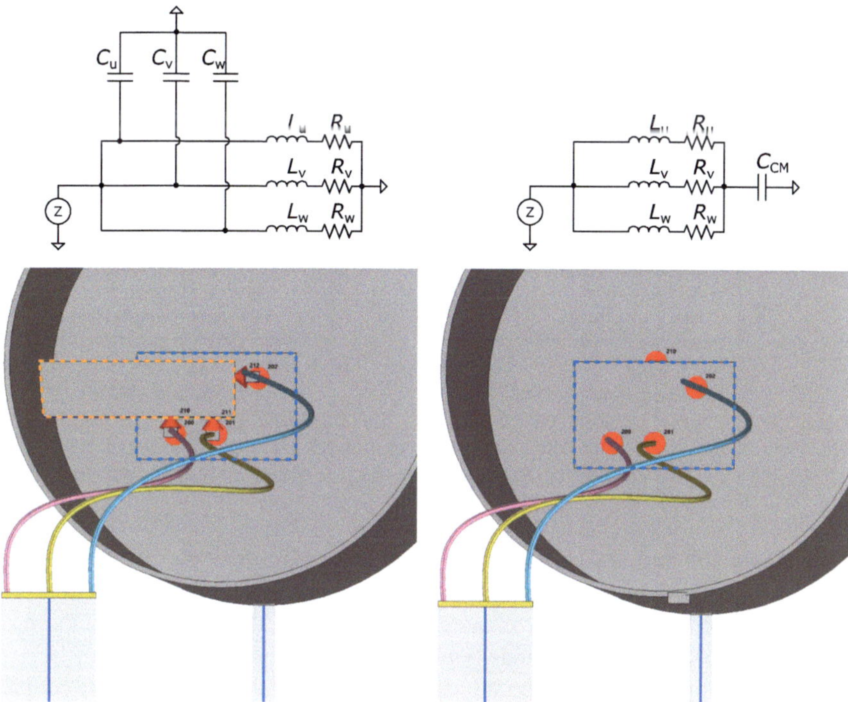

Fig. 10.10 Implementation of an RLC motor model in an EM simulation. Left: Individual phase capacitance. Right: A single common mode capacitance. The top of the figure depicts the resulting schematics and the bottom of the figure the port setup in the EM simulation

This second auxiliary shape is indicated by a dashed orange line. We connect a port from each wire to the second auxiliary shape. There is no port between the housing and the first auxiliary shape.

We calculate the input impedance of the common mode at the position where the wires connect to the bridge. At this position, the three wires are connected to each other, and a port is placed from the connection to the metallic table. This connection is indicated by the yellow shape in the foreground of Fig. 10.10. The metallic table, used as the reference, is located 5 cm below the wires and the housing, though it is not shown in the figure. A second port is placed from the motor housing to the metallic table, allowing the housing to be either connected to the table or left floating.

For the individual phase inductances L_X, we use $100\,\mu\text{H}$, and for the resistances R_X, we set $0.1\,\Omega$. The common mode capacitance of the motor is set to $C_{CM} = 100\,\text{pF}$. In the worst-case model, the individual phase capacitances are set to $\frac{1}{3}C_{CM}$.

On the left-hand side of Fig. 10.11, we compare the best and worst cases with the motor housing connected to the metallic table. Additionally, we compare the standalone circuits shown in the top section of Fig. 10.10 to the coupled EM-circuit simulation. The dashed lines represent the results of the standalone circuit

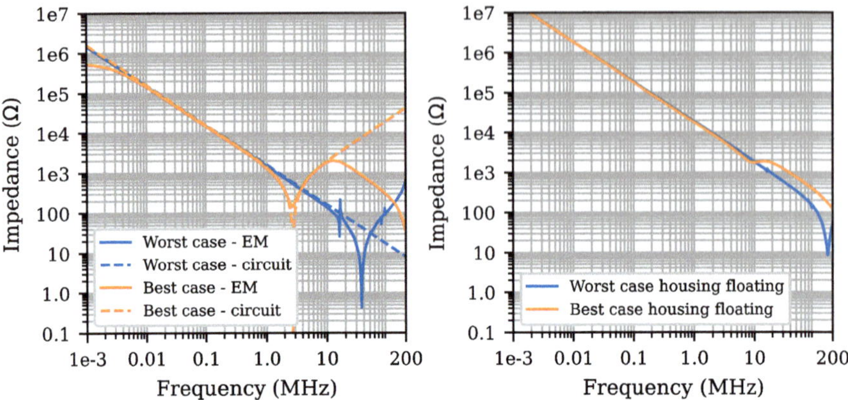

Fig. 10.11 Comparison of the common mode input impedance at the bridge output. Left: Motor housing connected to the reference. Right: Motor housing disconnected from the reference

simulations for both cases, while the solid lines represent the results obtained from the coupled EM-circuit simulation. In the standalone circuit simulation, both cases begin to diverge at approximately 5 MHz, with the best case exhibiting a significantly higher common mode impedance, as expected.

Up to approximately 10 MHz, the standalone circuit simulation and the coupled EM-circuit simulation coincide; however, they deviate from each other above this frequency. This deviation is caused by the parasitics of the wires and the realistic geometry of the housing. Nevertheless, the common mode impedance in the best case remains higher than in the worst case up to a frequency of approximately 140 MHz. Above 140 MHz, the impedance of the best case becomes lower than the impedance of the worst case, due to the resonant behavior at higher frequencies.

The common mode impedance for the motor housing disconnected from the metallic table is plotted on the right-hand side of Fig. 10.11. These results are obtained from the coupled simulation. When the motor housing is disconnected from the table, the common mode impedance increases, as the common mode loop can only be closed through the capacitance formed between the motor housing and the table. The best- and worst-case setups begin to diverge at 10 MHz, with the worst case exhibiting a lower common mode impedance across the entire frequency band up to 200 MHz. While the difference in impedance between the two setups is less pronounced than in the case where the motor housing is connected to the table, it noticeably impacts the conducted emission results, as discussed in Sect. 10.6.1.

The two models studied in this section represent corner cases that help us understand the effect of the motor's common mode capacitance when defined as a lumped RLC network. For a more accurate description of the motor, an EM model is required to account for the distributed capacitance in detail. Such models are briefly introduced in Sect. 10.4.3, but due to their complexity, they are not discussed extensively.

10.5 Simulation of Conducted Emission of a Motor Drive

Domains: 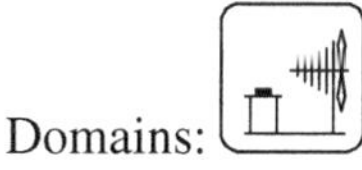

We summarize the findings of the previous sections to introduce two simulation models suitable for simulating the conducted emission voltage test according to CISPR 25. The first model operates in the time domain, while the second is a simplified variant that functions in the frequency domain.

In Sect. 10.6, both models are applied to assess the effectiveness of the EMI filter discussed in Sect. 9.5. The simulations use the EM model depicted in Fig. 10.5, with circuit elements as shown in Fig. 10.6. Specifically, the filter elements, that is, the C_X and C_Y capacitors, are not yet included in this section. Instead of the common mode choke, both the plus and minus supply lines are shorted. For the subsequent simulations, the port of screw1 is shorted. The second screw is not used, as the filter investigation in Sect. 9.5.2 indicates that it does not enhance the filter's insertion loss. As a result, the port of screw2 remains open. The housing of the PCB is connected to the metallic table using a ground strap, meaning that the port housing–table is shorted. Conversely, the motor housing is neither connected to the PCB housing nor to the table, leaving both ports, motor–table and housing–table, open.

The motor has a common mode capacitance of 100 pF, which is divided into three equal capacitances of 33 pF each. These capacitances are positioned at the motor inductance, on the side of the phase wires facing the motor drive (cf. Sect. 10.4.4). The parameters of the DC link capacitors and the electric motor are as described in Sect. 10.2.3. Each DC link capacitor has a capacitance of 100 µF, with an ESR of 300 mΩ and an ESL of 1 nH (Panasonic). The motor windings have an inductance of 100 µH and a resistance of 100 mΩ each.

10.5.1 Transient Simulations

For the transient simulation, we run the model at the operating point described in Sect. 10.2.1, using sine-triangle modulation PWM with a modulation index of 0.8 and a switching frequency of $f_{sw} = 10\,\text{kHz}$. The PWM period is 2 ms. Automated vector fitting is applied to the EM model prior to the circuit simulation. In the EM model, all ports have an impedance of 1 Ω, as explained in Sect. 10.3. For the transistors, we use the BUK9Y15-100E (Nexperia), downloaded from the vendors homepage. The time-domain simulation runs for 3000 µs.

We monitor the voltage at the positive LISN, $V_{LISN,P}$, in both the time and frequency domains, as shown in Fig. 10.12. The voltage $V_{LISN,P}$ has a mean value of 0 V because the DC voltage is blocked by the 100 nF capacitor of the LISN (cf.

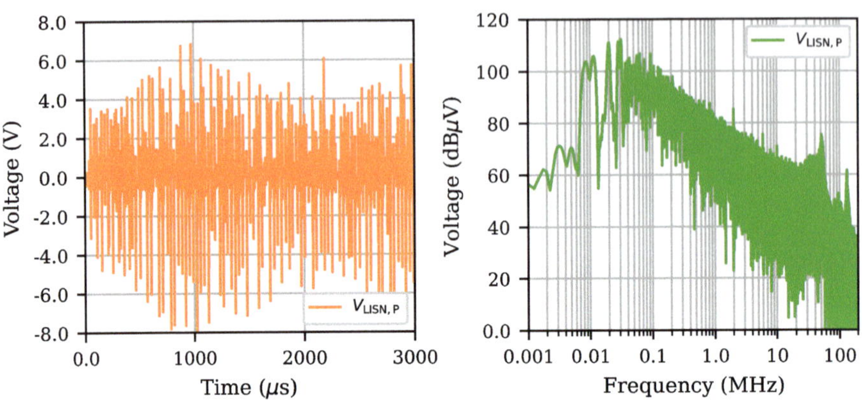

Fig. 10.12 Voltage at $V_{\mathrm{LISN,P}}$. Left: Time domain signal. Right: spectrum

Sect. 2.5.2). The amplitude of $V_{\mathrm{LISN,P}}$ varies. The peak amplitude reaches 8 V, and there is a minimum at 2 ms.

The reason for the variation is that the simulation has not yet reached the steady state. This is revealed by an inspection of the motor currents (not shown here). Within the simulation time of 3 ms, the motor currents remain asymmetric around 0 A. We recall that the functional simulation shown in Fig. 10.4 is monitored for times beyond 10 ms to ensure steady-state conditions.

However, to assess the EMI of the motor drive, reaching the steady-state condition of the motor currents is less essential. For the EMI, we analyze the spectrum of $V_{\mathrm{LISN,P}}$. Relevant spectral components include all frequencies above multiples of the PWM switching frequency. While the slow, 500 Hz fundamental frequency of the phase current does not directly influence the spectrum at the switching frequency, its amplitude affects the differential mode current, which in turn impacts the EMI spectrum. Comparing EMI spectra computed using short (millisecond-scale) and long (above 10 ms) transient times will reveal differences between results from various simulation times. To optimize simulation efficiency, it is advisable to start with short simulation times of a few milliseconds. Once the model and its analysis are finalized, a validation simulation should be performed with a simulation time long enough to reach steady-state conditions.

The DFT of $V_{\mathrm{LISN,P}}$ in Fig. 10.12 is plotted over the frequency range from 1 kHz to 200 MHz. The fundamental harmonic appears at 10 kHz, which corresponds to the switching frequency of the PWM. Sidelobes are present at the 10 kHz harmonic and its multiples due to the modulation of the PWM sequence with the PWM period of 2 ms, which corresponds to 500 Hz. The spectrum decreases at a rate of approximately 20 dB per decade, similar to the frequency response of a rectangular waveform. A strong resonance occurs at 50 MHz, beyond which the spectral decay rate increases, exceeding 20 dB per decade. Another pronounced resonance is observed at 130 MHz.

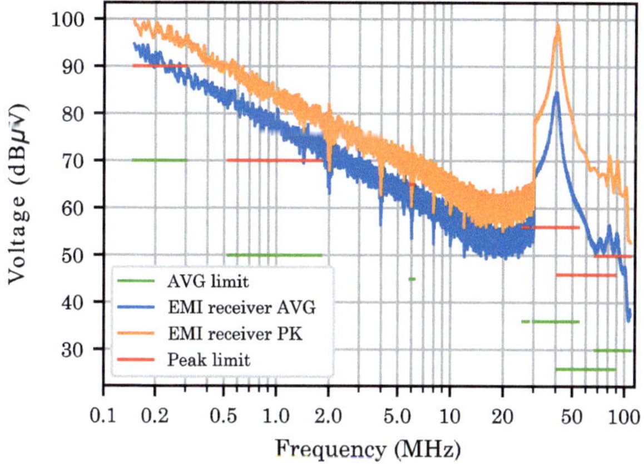

Fig. 10.13 Spectrum of $V_{\mathrm{LISN,P}}$ for average and peak detector and limit lines for CISPR 25 class 3

To accurately resolve the lowest harmonic of the switching frequency of 10 kHz, the transient simulation must extend significantly beyond 0.1 ms. Additionally, the lowest frequency computed by the frequency-domain EM solver should be well below 10 kHz. The EM model of Sect. 10.2.2 is computed with the lowest frequency of 1 Hz.

After performing a sanity check on the model results, we proceed to analyze $V_{\mathrm{LISN,P}}$ as monitored by the EMI receiver in Fig. 10.13. The plot displays measurements taken with both the average detector (blue) and the peak detector (orange).

As per the CISPR 25 standard, the frequency range for this plot extends from 150 kHz to 108 MHz. Consequently, the fundamental harmonic at 10 kHz and the prominent resonance at 130 MHz are not visible. However, the spectral decay of approximately 20 dB per decade is evident in both spectra. The change in receiver bandwidth from 9 kHz to 120 kHz at 30 MHz is also noticeable as a kink in the spectrum (cf. Sect. 2.5.4). The resonance at 50 MHz is visible, along with smaller resonances at 80 MHz and near 100 MHz.

Additionally, we include the limit lines for the CISPR 25 conducted emission voltage test, class 3,[1] for the eight frequency bands listed in Table 10.1. Alongside the average detector limit lines (green) from the table, we also show the peak detector limit lines (red). Since the limits for the FM band and the VHF II band are identical, they appear as a single limit line.

[1] Class 3 is suitable to demonstrate filter design by simulation without studying too intricate coupling effects and layout modifications, which is beyond the scope of this book.

Since the motor drive does not include a filter, the spectra obtained for both the average and peak detectors exceed all specified limits. To meet the class 3 peak detector limits, the filter must provide attenuation of approximately

- 10 dB at 150 kHz
- 20 dB at 500 kHz
- 12 dB in the SW and CB bands
- More than 30 dB for frequencies above 30 MHz

For the average detector, the required attenuation is approximately:

- 25 dB at 150 kHz
- 35 dB at 500 kHz
- 25 dB in the SW and CB bands
- More than 30 dB above 30 MHz

These attenuation values are critical for EMI filter design, as they define the necessary filter performance requirements. Moreover, they serve as reference points for further EMI filter simulations, such as those conducted in Sect. 9.5.

10.5.2 Simulation in Frequency Domain

Compared to the time-domain approach, frequency-domain simulations are significantly less computationally expensive. This makes them useful for quickly assessing the EMI of the motor drive through parameter sweeps, even if the results from a frequency-domain simulation neglect nonlinearities and use simple source definitions.

In this section, we demonstrate how to convert the transient model into a model that operates in the frequency domain. While this simplification is substantial, it provides the advantage of a simulation model that runs within seconds.

The modified model is shown in Fig. 10.14. The transient sources, which consist of six transistors, are removed. The transistor between the high side and phase U is replaced by an ideal AC voltage source of 1V. The low-side transistors of phase V and phase W are shorted, while all other transistors remain open. This configuration mimics one of the possible transistor states in a PWM cycle.

The open transistors are replaced by their drain-source capacitance, C_{DS}, in the open state. This capacitance value is provided in the transistor's datasheet (Nexperia), but it may differ from the value in the SPICE model provided by the vendor. Since we aim to compare transient and frequency-domain simulations, we extract C_{DS} directly from the SPICE model.

To determine C_{DS}, we follow the methodology described in Sect. 2.7. We calculate the impedance of the transistor's drain-source channel, while maintaining 0 V at the gate. The impedance follows the relation $|Z_{DS}| = \frac{1}{\omega C_{DS}}$. The resulting capacitance is $C_{DS} = 2\,\text{nF}$, which is incorporated into the model shown in Fig. 10.14.

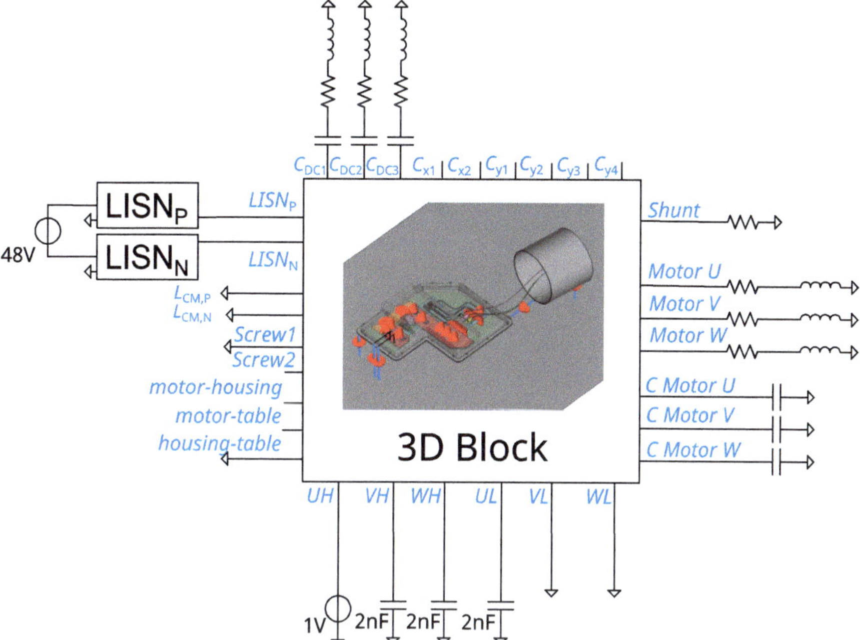

Fig. 10.14 Simplified simulation in frequency domain

For the following simulations, we use the setup described at the beginning of this section. No filter elements are attached, the choke and the port of screw1 are shorted, and the port connecting the PCB housing to the table is also shorted. The motor housing remains unconnected to both the table and the PCB housing. In this simulation, the DC source acts as a short.

We run the simulation over a frequency range from 10 kHz to 200 MHz. The voltage source drives a current through the motor windings. The model represents a switching state in which three transistors are conducting while all others remain nonconductive. Any other possible transistor states are not considered; however, we expect that the coupling path from a particular state is similar to the coupling path of other states of the inverter. Additionally, the voltage source does not exhibit the time-variant impedance profile characteristic of a transistor. As a result, the spectral decay of a trapezoidal waveform, as discussed in Fig. 2.5, is not taken into account. The focus is to excite and analyze the resonances within the system.

On the left side of Fig. 10.15, we plot $V_{\mathrm{LISN,P}}$ and $V_{\mathrm{LISN,N}}$ as obtained from the frequency-domain simulation. Up to a minimum at 280 kHz, these voltages have the same magnitude, as well as at the system's resonant frequencies. The identified resonances occur at 17 MHz, 28 MHz, 34 MHz, and 140 MHz. Between these resonances, the magnitude of $V_{\mathrm{LISN,P}}$ is larger than that of $V_{\mathrm{LISN,N}}$. On the right side of the figure, we present the differential and common-mode voltages at the LISNs, $V_{\mathrm{LISN,DM}}$, and $V_{\mathrm{LISN,CM}}$, as defined in Eqs. 9.1a, 9.1b. Up to the first

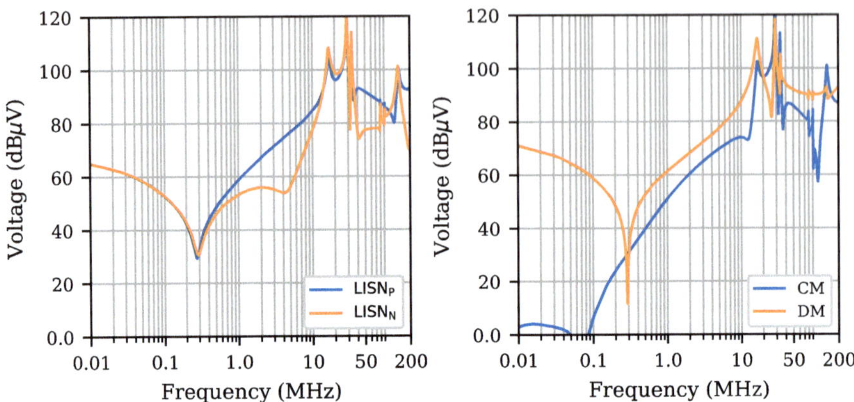

Fig. 10.15 Evaluation of the frequency domain model. Left: $V_{\mathrm{LISN,P}}$, $V_{\mathrm{LISN,N}}$, Right: $V_{\mathrm{LISN,CM}}$, $V_{\mathrm{LISN,DM}}$

resonance at 17 MHz, the differential mode voltage dominates the contributions to $V_{\mathrm{LISN,P}}$ and $V_{\mathrm{LISN,N}}$. Beyond this frequency, $V_{\mathrm{LISN,CM}}$ becomes comparable in magnitude to $V_{\mathrm{LISN,DM}}$.

The voltage source in the model is differential, meaning that any $V_{\mathrm{LISN,CM}}$ observed is generated through mode conversion in the EM model. A variation in C_{DS} reveals that this parameter influences the minimum at 280 kHz. Increasing C_{DS} by a factor of four shifts the minimum frequency to half its original value, that is, 140 kHz. This minimum is not observed in the transient simulation because, in the transient simulation results, multiple transistor states—both open and conducting—contribute to the overall response.

Resonances observed in the frequency-domain simulation may or may not match to those in the transient simulation. Any results obtained from frequency-domain simulations should be validated through corresponding transient simulations. The key advantage of frequency-domain simulations is their speed, allowing for efficient investigations into the root causes of resonances. Also, because only one source excites the model, the coupling path is easier to analyze. In Sect. 10.6, we use this model to study different grounding scenarios for the motor housing.

10.6 Application Example: 300 W Motor Drive

Switching of a B6 bridge, modeled by a SPICE circuit

LISN circuit

Full-wave, S-parameters

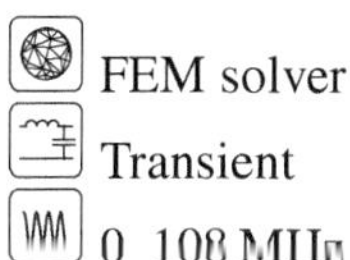

The conducted emission model describes the emissions generated by the motor drive. By incorporating the filter elements described in Sect. 9.5 into the model shown in Fig. 10.6, we can analyze the voltage levels at the LISNs. We investigate different system configurations in terms of grounding and assess the impact of various filter components and load models on the voltage detected by the EMI receiver (cf. Sect. 2.5.4). Unlike the filter attenuation analysis in Sect. 9.5, where relative attenuation is examined, this study provides absolute voltage values, which we compare against the CISPR 25 class 3 limits. For this analysis, we again use the motor drive model, which delivers approximately 300 W of power (Sect. 10.2) when connected to the motor model described in this chapter.

10.6.1 Impact of Load and Grounding Concepts

As a first step, we investigate various grounding scenarios for the housings of the PCB and the motor. The objective is to determine which scenarios generate which level of emissions and to understand the underlying reasons. At this initial step of the analysis, we use frequency-domain simulations as described in Sect. 10.5.2, because for the better/worse evaluation of the grounding concepts, the absolute voltage values at the LISNs are not yet required.

The system includes three grounding straps,

- the ground strap connecting the PCB housing to the table,
- the ground strap between the motor housing and the table,
- the conducting strap between the PCB housing and the motor housing.

We assume that the ground strap between the PCB housing and the table is fixed (see schematic Fig. 10.14), as this is typically dictated by system requirements. This leaves four possible grounding configurations, determined by whether the motor grounding and the strap between the PCB housing and motor housing are connected or not.

For each scenario, we analyze the common mode voltage $V_{\mathrm{LISN,CM}}$ at the LISNs. The model does not include a common mode choke, meaning the ports where the choke is mounted are shorted. However, we include the four C_{y} capacitors, as omitting them would eliminate the connection between the PCB and its housing, leading to an underestimation of common mode propagation.

We use the C_{y} capacitor values from Sect. 9.5, setting each capacitor to $C_{\mathrm{y},i} = 4.7\,\mathrm{nF}$ with an ESL of 0.5 nH and an ESR of 0.1 Ω. The presence or absence of C_{x} capacitors does not affect this analysis since we focus solely on $V_{\mathrm{LISN,CM}}$.

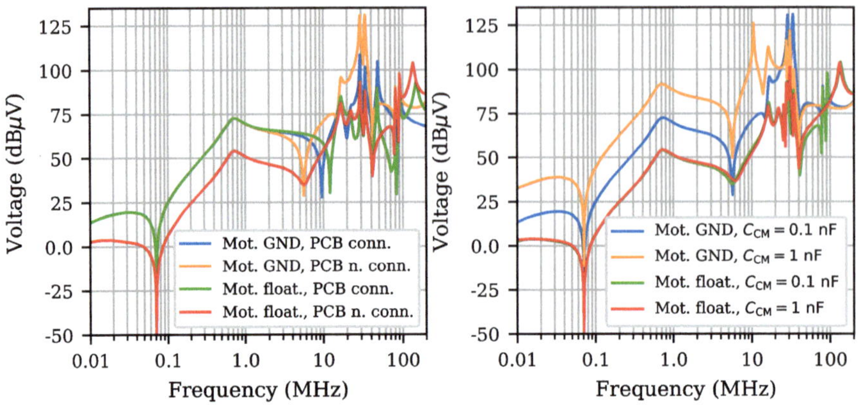

Fig. 10.16 Common mode voltage at LISN for different configurations. Left: Impact of grounding concepts. Right: Impact of the load capacitance

The four resulting graphs, each representing a different grounding configuration, are shown on the left-hand side of Fig. 10.16, covering the frequency range from 10 kHz to 200 MHz.

- The blue curve represents the case where the motor housing is grounded and connected to the PCB housing.
- The orange curve corresponds to the scenario where the motor housing is grounded but not connected to the PCB housing.
- The green curve depicts the case where the motor housing is not grounded but is connected to the PCB housing.
- The red curve represents the situation where the motor housing is isolated from both the table and the PCB housing.

Up to a frequency of 2 MHz, the first three curves in the left plot coincide, whereas the fourth curve (corresponding to the isolated motor housing) is 20 dB lower.

The excitation in this model is a differential mode source, meaning that any common mode voltage is generated through mode conversion. At low frequencies, the attached motor housing significantly contributes to the common mode. When the motor housing is connected to the metal table—either indirectly through the PCB housing and its ground strap or directly via its own strap—the common mode voltage is higher than in the isolated case. Above 10 MHz, each configuration produces resonances of varying amplitude, and no single grounding configuration appears to be definitively preferable.

The capacitance between the motor and the motor housing serves as a critical coupling path for common mode emissions. To analyze this coupling path, we examine the results shown on the right-hand side of Fig. 10.16. We consider four cases, all of which assume that the motor housing and the PCB housing are not connected. Instead, the capacitance between the motor and the motor housing is varied by a factor of 10, ranging from 100 pF to 1 nF. Additionally, the motor housing is either attached to the table or isolated. In the frequency range from 10 kHz to 20 MHz, the common mode voltage $V_{LISN,CM}$ is highest when the motor housing is grounded and the capacitance between the motor and its housing is 1 nF. Reducing this capacitance by a factor of 10 to 0.1 nF lowers emissions by 20 dB. If the motor housing is not connected to the table, emissions decrease by another 20 dB (blue curve). This reduction is observed for both capacitance values of 0.1 nF (green curve) and 1 nF (red curve).

When the motor housing is not connected to the table, the coupling between the housing and the table becomes capacitive, with a value much smaller than 100 pF. Since this capacitance is in series with the capacitance between the motor and its housing, the smaller capacitance dominates, rendering the larger capacitance ineffective in influencing $V_{LISN,CM}$.

From this study, we conclude that for minimizing $V_{LISN,CM}$ in the given setup, an isolated motor housing is the most effective configuration in the frequency range up to 20 MHz. However, the various resonances occurring above this range must be analyzed individually.

Whether isolating the motor housing from the PCB housing is feasible depends on the mechanical constraints of the electronic system. If isolation is not practical, reducing the capacitance between the motor and the motor housing lowers $V_{LISN,CM}$.

10.6.2 EMI Filter Design

Finally, we apply the filter investigated in Sect. 9.5 to reduce the conducted emissions of the 300 W motor drive. We compare the conducted emission at $V_{LISN,P}$ against the CISPR 25 class 3 limits for the average detector (Table 10.1). To better understand and analyze the results, we also examine the common mode and differential mode voltages, $V_{LISN,CM}$ and $V_{LISN,DM}$, and compare emissions with and without the filter.

For this study, we use the simulation model shown in Sect. 10.2.2, which connects a lumped model of the common mode choke via ports and incorporates a simplified housing model introduced in Sect. 9.3. To assess the impact of these model simplifications, we compare the resulting emissions against those obtained using a more detailed model that includes the CAD representation of the choke and a mechanically finalized housing model.

We begin with the same configuration used in Sect. 10.5.2, based on the model shown in Sect. 10.2.2. One screw connects the PCB to the housing, while the second screw remains unused. The common mode choke is not mounted; instead, its ports are shorted. In contrast to Sect. 10.2.2, all filter capacitors are included, using the values specified in Table 9.1. The two C_x capacitors have a capacitance of 10 μF, an ESL of 1 nH and an ESR of 0.1 Ω. The four C_y capacitors have a capacitance of 4.7 nF, an ESL of 0.5 nH and an ESR of 0.1 Ω.

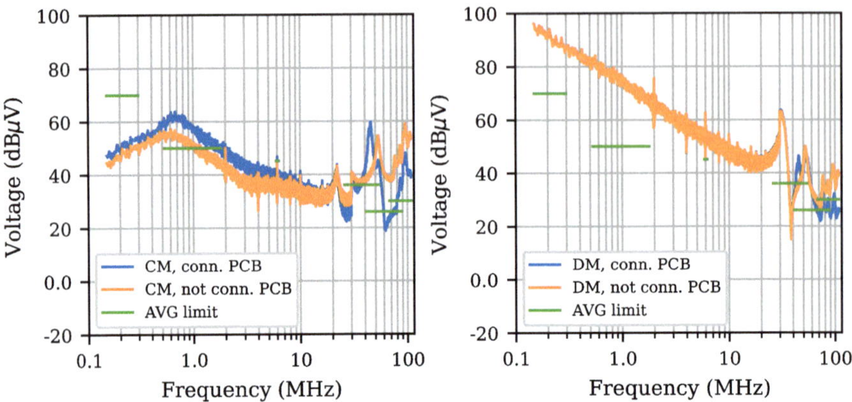

Fig. 10.17 LISN voltage monitored by the average detector, compared to class 3 limits. Two different grounding scenarios. Left: Common mode. Right: Differential mode

First, we analyze the common mode and differential mode voltages, $V_{\mathrm{LISN,CM}}$ and $V_{\mathrm{LISN,DM}}$. We consider two grounding configurations,

- the PCB housing connected to the motor housing,
- the PCB housing disconnected from the motor housing.

These scenarios are also illustrated in Fig. 10.16, using a frequency-domain simulation with a single excitation source. We plot $V_{\mathrm{LISN,CM}}$ and $V_{\mathrm{LISN,DM}}$ as detected by the average detector of the EMI receiver. To compute these values, we first monitor the voltages in the time domain (following Eqs. 9.1a, and 9.1b) and then feed the data into the EMI receiver. The resulting plots are shown in Fig. 10.17. On the left-hand side, we present $V_{\mathrm{LISN,CM}}$, while on the right-hand side, we show $V_{\mathrm{LISN,DM}}$. The green lines indicate the class 3 limit lines. The computation time of the transient model is 4 minutes and 33 seconds including vector fitting, for a simulation time range from 0 ms to 3 ms.

Similar to the results in Fig. 10.16, the configuration where the motor housing is connected to the PCB housing produces higher emissions up to 10 MHz. The difference reaches up to 10 dB; it is smaller than in the frequency-domain simulation. For $V_{\mathrm{LISN,CM}}$, the LW, SW, and CB limits are met even with this filter configuration, which includes only C_{y} capacitors and no common mode choke. However, all other limits are exceeded. Up to 40 MHz, the differential mode voltage $V_{\mathrm{LISN,DM}}$ is higher than $V_{\mathrm{LISN,CM}}$. For $V_{\mathrm{LISN,DM}}$, all EMI limits are exceeded. We conclude that the filter design must focus on improving differential mode attenuation to ensure compliance with emission limits.

Next, we insert the common mode choke into the model. For simplicity, we use the model of the two coupled inductors of 16 μH each (Würth Electronics) with $k = 0.95$, which is also used in Sect. 9.5. Additionally, we investigate the effect of snubbers connected in parallel to each transistor. The snubbers consist of a series

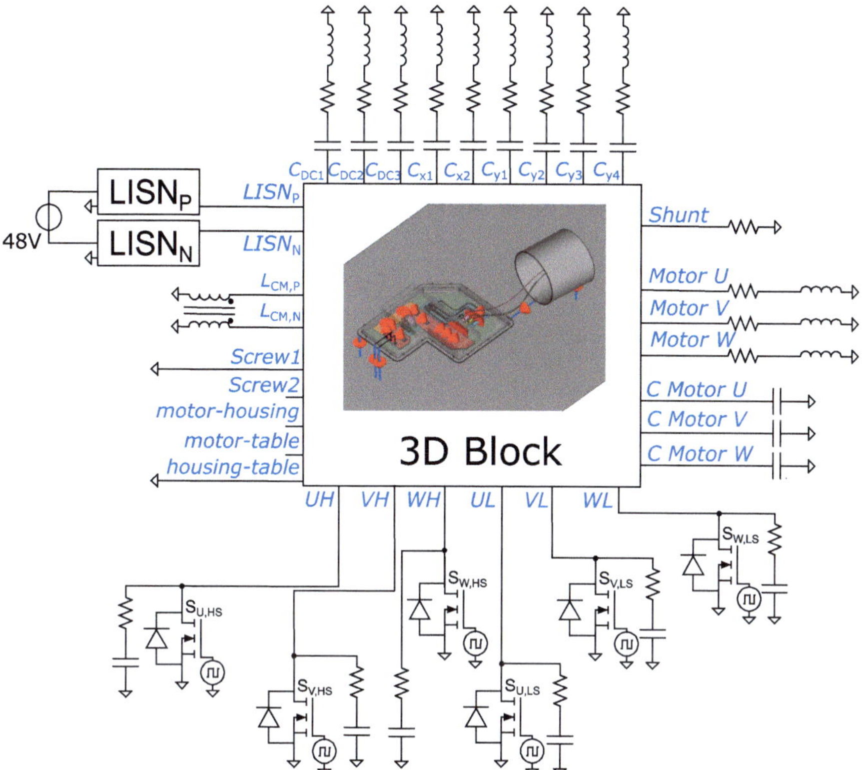

Fig. 10.18 Circuit schematic with filter capacitors, circuit model of common mode choke, snubbers, loads, and ground connections

circuit with a capacitor of 1 nF and a resistor of 1 Ω. The resulting circuit is depicted in Fig. 10.18, showing the mounting of all components and interconnects.

We again plot $V_{\mathrm{LISN,CM}}$ and $V_{\mathrm{LISN,DM}}$, as detected by the average detector, comparing them against the class 3 limits. Although the pass/fail criteria in the standard are based on the LISN voltages $V_{\mathrm{LISN,P}}$ and $V_{\mathrm{LISN,N}}$, analyzing common mode and differential mode voltages provides a clearer assessment of the filter components' effectiveness. As long as one mode dominates the other, $V_{\mathrm{LISN,CM}}$ and $V_{\mathrm{LISN,DM}}$ serve as useful approximations of $V_{\mathrm{LISN,P}}$ and $V_{\mathrm{LISN,N}}$ (cf. Fig. 10.16 and Eqs. 9.1a, 9.1b).

The common mode voltage $V_{\mathrm{LISN,CM}}$ is displayed on the left of Fig. 10.19, $V_{\mathrm{LISN,DM}}$ on the right. For $V_{\mathrm{LISN,CM}}$, the blue curve shows the result if only the filter capacitors are mounted. The LW, the SW, and the CB limits are fulfilled. With the insertion of the common mode choke (orange curve), the emission is reduced above 500 kHz, and also the MW limit is fulfilled. The choke shifts and increases a resonance at 360 kHz which is also found and investigated in the common mode

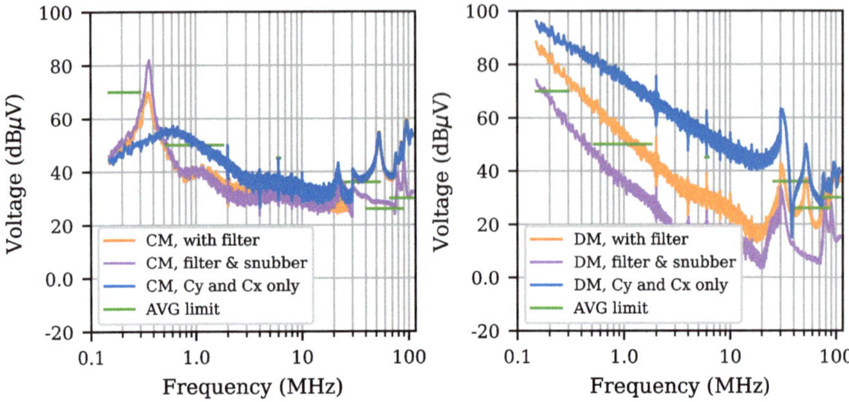

Fig. 10.19 LISN voltage for three assemblies of filter components. Left: Common mode. Right: Differential mode

voltage of the EMI filter, Fig. 9.16. The addition of the snubbers (purple curve) does not change $V_{\mathrm{LISN,CM}}$, except that the quality factor of the resonance is increased.

In the differential mode $V_{\mathrm{LISN,DM}}$, the application of the common mode choke reduces the emission from 10 dB at 150 kHz to almost 30 dB at 20 MHz (blue curve and orange curve). This is due to the stray field of the choke, which exists because $k < 1$ (see Fig. 9.18). The application of the snubbers result in another reduction of more than 10 dB (purple curve). In this configuration, the LW limit is exceeded by less than 5 dB; the MW and the SW and CB limits are passed. In the VHF and FM range, the limits are still exceeded.

The EM model used so far contains two approximations, namely, two coupled inductors to describe the common mode choke, and a rudimentary housing geometry, introduced in Sect. 9.3. We improve the EM model by inserting the CAD model of the choke, which is developed in Sect. 9.6. The choke is now characterized by its material parameters, which are $\mu_r = 560$, $\tau_r = 18\,\mathrm{ns}$, $\varepsilon_r = 2.5$, using a first-order Debye model to describe the frequency-dependence of the magnetic core material (Eq. 4.11). Further, the geometrical model of the housing is updated to a mechanically reasonable design. The resulting EM model is shown in Fig. 10.20.

The results are shown in Fig. 10.21. We plot the voltage $V_{\mathrm{LISN,P}}$ against the class 3 limit of the average receiver. The blue curve corresponds to the simulation with the simplified housing geometry used so far, and the coupled inductors. The orange curve corresponds to the updated model shown in Fig. 10.20.

Additionally, we investigate the impact of different motor model connections. In Sect. 10.4.4, we introduce two different methods for connecting the discrete-component motor model to the EM model. The orange and the blue curve correspond to the model shown on the left of Fig. 10.10 (using three motor capacitances), the purple curve to that on the right (using a single motor capacitance with three times the capacitance value).

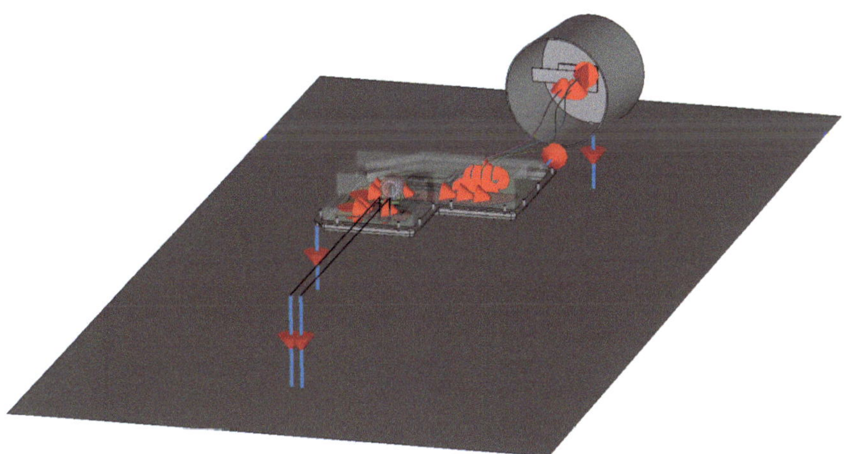

Fig. 10.20 EM model with mechanically sound housing and CAD model of the common mode choke included

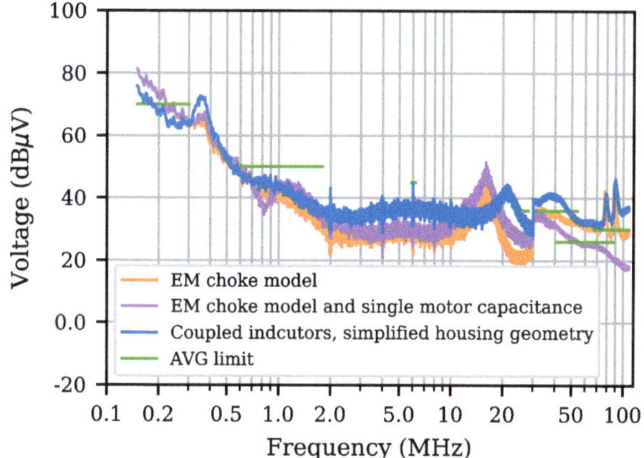

Fig. 10.21 LISN voltage $V_{LISN,P}$ and limits of the average detector class 3 for three different simulation model configurations

Between 150 kHz and 300 kHz, the model using the discrete representation of the common mode choke yields 5 dB lower emissions compared to the model that includes the EM representation. At these frequencies, differential mode emissions dominate (see Fig. 10.19). From Fig. 9.27, we recall that the coupling coefficient $k = 0.95$ is underestimated, causing the discrete model to overestimate Z_{DM} compared to the EM model of the choke. As a result, the discrete model predicts lower a $V_{LISN,P}$ voltage. At 360 kHz, the common mode resonance is less pronounced when using the EM model of the choke along with the updated housing geometry.

At frequencies above 30 MHz (in the VHF and FM ranges), the motor model influences emissions. The orange and blue curves (representing the motor model with three capacitances) exhibit resonances at 70 MHz and 90 MHz. These resonances do not appear in the model with a single capacitor, as this capacitor is connected on the grounded side of the motor model, where high-frequency noise is blocked by the motor inductors. Consequently, emissions at these frequencies are lower when using the single-capacitor motor model. Between 2 MHz and 30 MHz, both the common mode choke and the housing geometry influence emissions.

Conducted emissions are influenced by multiple components within the system model. Understanding the contributions of these components and their interactions is crucial for analyzing the voltage at the EMI receiver. This study highlights the impact of the common mode choke model, the motor model, and housing geometry to the simulated conducted emission voltage.

10.7 Application Example: DC-DC Converter

- Switching of a power transistor, modeled by a SPICE circuit
- LISN circuit
- Full-wave, S-parameters
- FEM solver
- Transient
- 0–300 MHz

In Chap. 3 we study the conducted emission of a buck converter using a circuit simulation. We rely on analytical equations and best guesses to estimate the parasitic values on the PCB. The converter is inspired by an actual design of a DC-DC converter in the digital PCB. With the methodology introduced in the current chapter, we simulate the conducted emission of this DC-DC converter including its electromagnetic properties due to layout and interconnects.

The EM model of the buck converter is depicted is Fig. 10.22. As prescribed by CISPR 25, the PCB is placed at a distance of 5 cm to a metallic reference plane. The wires on the left-hand side are the supply wires referenced to the metallic plane. The wires on the right side are terminated with the load resistor of the DC-DC converter. Port 2 is the location of the input capacitor, the switch is located at port 5. At port 4, the diode is connected. Port 6 is the output inductor and port 1 and 3 the output capacitors. The model is simulated using the FEM solver up to a frequency of 300 MHz. The simulation time for the EM part, including 6 mesh adaptation steps, requires 9 minutes and 29 seconds. The circuit simulation, up to 2 ms, requires 6 minutes.

A comparison of the conducted emission spectrum of the pure circuit model and the model with the EM simulation is plotted in Fig. 10.23. At the fundamental switching frequency of 500 kHz, both model agree up to 0.5 dB. With increasing

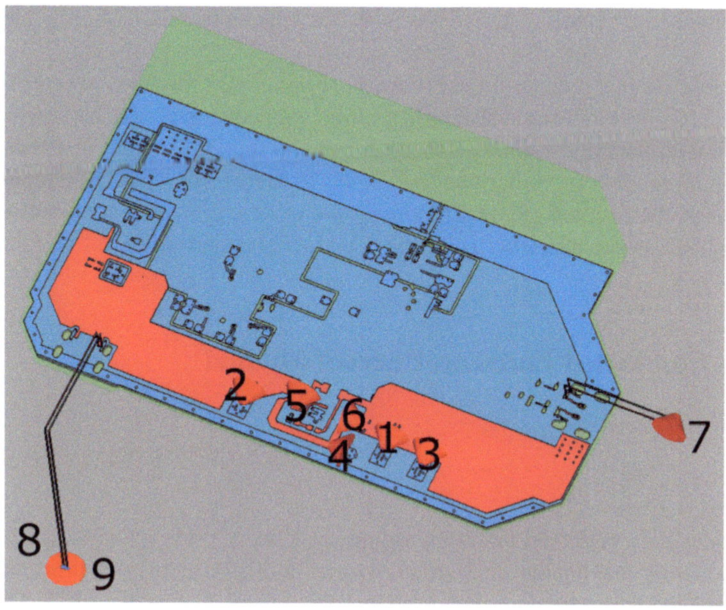

Fig. 10.22 Simulation model for the calculation of conducted emission from a of a buck converter

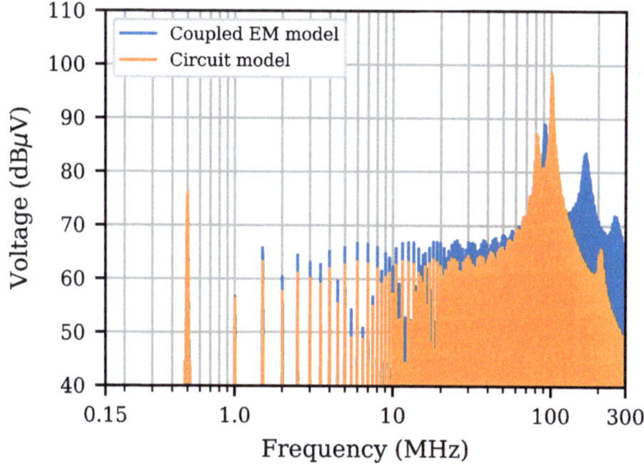

Fig. 10.23 Comparison of the conducted emission voltage $V_{\mathrm{LISN,P}}$ for the pure circuit model from Chap. 3 and a coupled EM-circuit model

frequency, the deviation is in the range of 3–5 dB. Starting with 80 MHz, where a first resonance is located, the deviation becomes larger and is around 10 dB at the peak. The circuit model calculates two peaks, the first is at 80 MHz and the second at 100 MHz. The second peak has a very high-quality factor, because there is no lossy material modeled in the circuit simulation. The coupled circuit-EM model produces one resonance at 90 MHz. Above this peak, the spectra begin to differ by more than 10 dB.

For this simple geometry, the accuracy of the circuit model is acceptable, as it can calculate the emission sufficiently accurate up to the first peak. In more complex layouts, with parasitic couplings to other parts in the PCB, building a reasonable, pure circuit model for prediction above some 10 MHz turns into a difficult task. For such cases, the coupled circuit-EM simulation in time domain that we introduce in this chapter is more efficient to calculate conducted emission.

10.8 Conducted Emission Current Method

Domains: 

Besides the conducted emission voltage test, the CISPR 25 standard defines a current measurement method (Bogdan Adamczyk 2017). In this test, the disturbance current on the wire harness is measured instead of the voltage at the LISN. The setup includes a long wire harness of up to 2 m in length, through which both supply and load wires are routed toward the LISNs. The current is measured using current clamps, which are positioned at distances of 50 mm and 750 mm from the DUT.

As in the voltage test, the DUT is isolated from the metallic table plate by a 50 mm insulating layer with $\varepsilon_r \approx 1$. As in Table 10.1 for the voltages, there are five classes of limits defined for the three detector types average, peak, and quasipeak. The limits are given in dBμA over the frequency bands LW, MW, SW, FM, TV I, CB, VHF I, and VHF II.

The conducted emission current method is simulated in the same way as described in this chapter for the conducted emission voltage test, with two key differences:

- The EM model must include the longer wire harness. This can be done within the EM model. Alternatively, for a more complex wire harness, an MTL model can be used to represent the harness and connected to the EM model at the circuit level. This procedure is explained in the immunity chapter (Sect. 13.2.3).
- The current on the wire bundle is monitored. At the locations of the current clamps, the wire is cut, a port is inserted into each wire, and the corresponding EM model is computed. At the circuit level, the respective pin is connected to a low-ohmic resistor, and the current through this element is measured. If the simulation needs to account for the transfer function of the current clamp, an EM model of the current clamp can be used. Both approaches are discussed in Sects. 13.1 and 13.2.

If these differences are considered, the procedures shown in the chapter for the conducted emission voltage test equally apply to the current method.

10.9 Advanced: Vector Fitting

Domains: 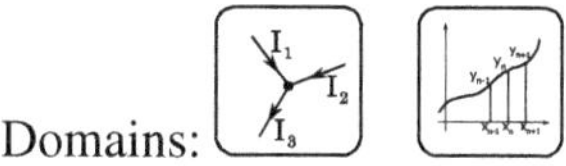

In this section, we explore the properties of vector fitting introduced in Sect. 10.3. S-parameters, as computed from EM simulations over discrete frequency samples, cannot be used in a transient circuit simulation without the application of further mathematical operations. Vector-fitting has evolved as the most commonly used procedure to convert frequency domain data into accurate causal and passive circuit models that can be used in transient simulations. We discuss a number of examples and also present test cases that allow to estimate the accuracy of vector fitting for a system level response.

10.9.1 Reference Impedance

We begin the discussion with an example that illustrates the dependence of fitting accuracy on the reference impedance of the S-parameter data used in the vector fitting procedure. For this demonstration, we use a simple circuit, depicted in the top section of Fig. 10.24. The test circuit is a ladder-type network consisting of four RL sections, each with 1 mΩ and 3 nH, and three shunt capacitors with a nominal value of 47 nF placed between these sections. This represents a simplified model of a low-pass filter, where the RL sections model trace parasitics.

We evaluate the fitting accuracy as follows: on the input side of the network, we connect an ideal 1 V voltage source. At the output, we measure the current through a short. With increasing frequency, we expect the current to bypass the output through the capacitors, resulting in a decrease in the measured output current.

The circuit that is to be vector-fitted is shown in the center of Fig. 10.24. Ports are placed at the input and output, as well as across the capacitors, effectively replacing the capacitors with ports. This configuration generates a five-port S-parameter matrix that includes only the RL sections.

Two models are created for comparison: In the first model, the S-parameter matrix is vector fitted with a reference impedance of 50 Ω. In the second model, the S-parameters are renormalized to a reference impedance of 1 Ω, and vector fitting is applied after renormalization.

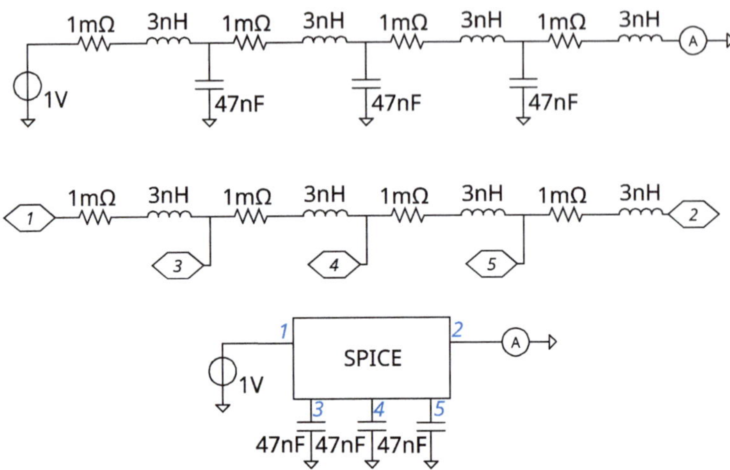

Fig. 10.24 Effect of the reference impedance on vector fitting. Top: Test schematic with source and current measurement. Center: Schematic used for the calculation of S-parameters used in the vector fitting. Bottom: Schematic with the vector fitted model exported as a SPICE netlist

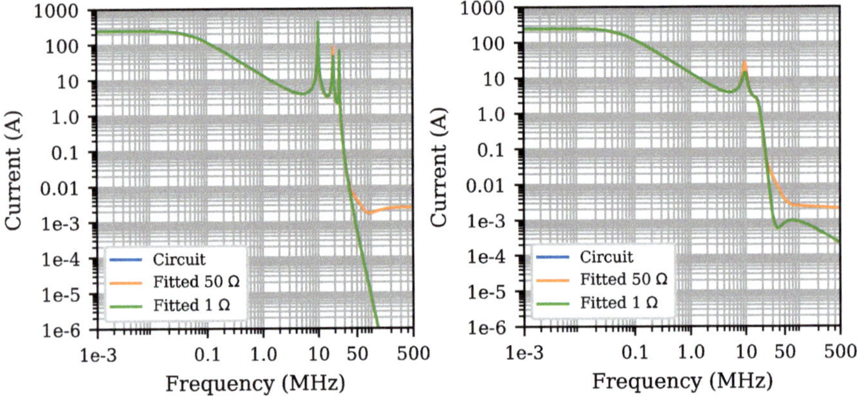

Fig. 10.25 Current measured for different reference impedances of S-parameters in the vector fitting process. Left: Ideal capacitors. Right: Capacitors with ESL and ESR. In both plots, the blue curve is completely covered by the green curve

Both models are fitted using the standard requested accuracy value of 10^{-3}. The synthesized netlist is then connected to the input and output ports, as well as to the capacitors, as depicted in the lower section of Fig. 10.24.

The results comparing the fitted model to the reference network are shown in Fig. 10.25. On the left-hand side of the figure, the output current is plotted for the case where all three 47 nF capacitors are ideal. When vector fitting is performed using S-parameters with a reference impedance of 50 Ω, deviations between the fitted results and the reference solution are observed at the second resonance peak and above 30 MHz. These deviations are caused by error amplification due to the mismatch between the reference impedance of the S-parameters and the impedance of the connected capacitors.

The iterative vector fitting scheme evaluates the fitting error based on the 50 Ω system, while the impedance of the connected capacitors decreases with frequency, reaching 0.11 Ω at 30 MHz. This mismatch amplifies fitting errors, as we demonstrate in the following example.[2] Still, the fitted model is accurate over five orders of magnitude in the measured current.

This error amplification can be mitigated by renormalizing the S-parameters prior to vector fitting. When the fitting is performed on S-parameters renormalized to a reference impedance of 1 Ω, the fitted model matches the reference results so closely that no visible difference can be seen between the green and blue curves on the left-hand side of Fig. 10.25. With the reduced impedance mismatch, the vector-fitted model achieves accuracy over the nine orders of magnitude shown in Fig. 10.25.

To evaluate the impact of fitting error on more realistic scenarios, we perform the same test using nonideal capacitor models. For this second test, an ESL of 0.5 nH and an ESR of 0.05 Ω are added to the capacitors. The resulting current at the output is shown on the right-hand side of Fig. 10.25.

With the inclusion of parasitic values, the impedance of the connected capacitors no longer decreases monotonically with increasing frequency. As a result, the output current does not approach zero at higher frequencies. The deviation between the reference model and the model fitted with a reference impedance of 50 Ω is significantly smaller than in the previous case with ideal capacitors. This indicates that, in scenarios where the capacitors are nonideal, the system-level accuracy is actually improved because the minimum impedance of the capacitors is bounded.

This example highlights two key points:

- The reference impedance of the S-parameters used in the vector fitting procedure can influence the fitting accuracy.
- The system-level accuracy of the fitted model depends on the termination impedances applied to the fitted block, which are dictated by the system characteristics.

Therefore, testing the fitting accuracy must be conducted with a meaningful test case that reflects the intended termination scheme dictated by the application. In this section, we explore this for a simple system. In the next section, we discuss how to construct test cases for more complex models.

[2] This is similar to the measurement errors of a network analyzer when measuring low impedances in a 50 Ω system.

10.9.2 Test Scenarios

A test case for evaluating the accuracy of a vector-fitted model must satisfy two requirements. It must represent the impedances of the components connected to the vector-fitted model and define a meaningful quantity for comparison between the original data and the fitted model. Since the S-parameter models operate in the frequency domain, AC simulations are performed.

For the inverter model discussed in this chapter, a test case is constructed as follows: The primary noise source in the inverter is the switching of the transistors, and the emission is measured at the LISN. The schematic should allow for the comparison of coupling from the transistors to the LISN. This is the same schematic used in Sect. 10.6.1 for system-level EMI estimation in the frequency domain. In these simulations, one phase is excited with a flat 1 V spectrum, and the voltage is measured at the LISN. We use this scenario as a reference to examine the accuracy of the vector-fitted model. We compute the same LISN voltage by exciting the vector-fitted model with a flat 1 V spectrum, and we compare the two results. Any difference in the LISN voltage is caused by an error in the vector-fitting procedure and indicates an inaccuracy of the macromodel.

We use the inverter model that includes the realistic housing and an EM model of the common mode choke (Fig. 10.20). As the reference impedance of the ports in the full-wave model is already set to 1 Ω, no manual renormalization is necessary. The vector-fitting is performed with default settings using the automatically triggered fitting procedure, which includes the passivity enforcement step. The vector fitting accuracy of the automatic setting is 10^{-3}.

A comparison of the common and differential mode voltages $V_{\mathrm{LISN,CM}}$, $V_{\mathrm{LISN,DM}}$ at the LISN for the reference and fitted models is shown in Fig. 10.26. The results demonstrate that the fitted model matches the reference data. Deviations

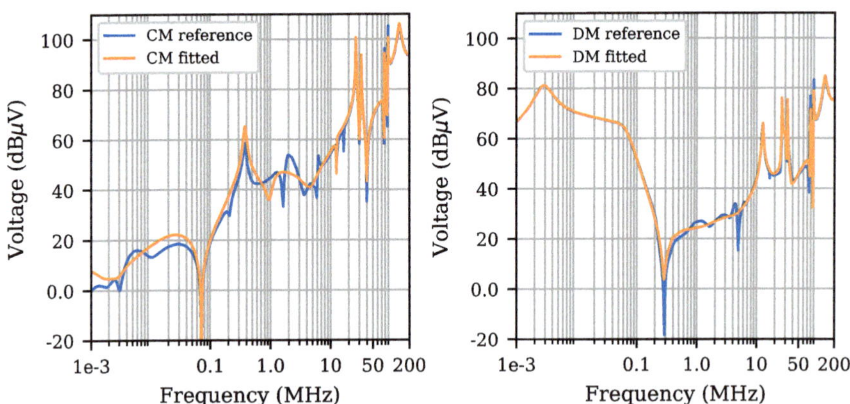

Fig. 10.26 Comparison of the voltage at the LISN calculated using the S-parameters obtained from the full-wave solver and a fitted model. Left: Common mode voltage $V_{\mathrm{LISN,CM}}$. Right: Differential mode, $V_{\mathrm{LISN,DM}}$

are observed in the common mode within the frequency range from 100 kHz to 20 MHz, primarily around 2 MHz.

These results highlight the accuracy achievable when vector fitting larger S-parameter matrices. The model in this example consists of 29 ports, requiring the fitting of the full S-matrix, which includes a total of $29 \times 29 = 841$ S-parameter curves across a 200 MHz bandwidth. The vector fitting process delivers a model with a dynamic range of 100 dB with automatic settings.

This test is conducted on one inverter phase. It compares the voltage at the LISN when this phase is excited. Additional tests could be performed for other phases, but in our design, all phases share identical components, such as filters, motors, and housing connections.

Based on this reasoning, we conclude that the fitted model also performs accurately in transient circuit simulations. However, it is essential to ensure that the frequency range of the frequency-domain test case aligns with the frequency range of the model used in the transient simulation. In cases in which the comparison of the reference and fitted model shown significant deviations, manual adjustments to the reference impedance are necessary. Additionally, the used error measure may need modification, which is discussed in the next section.

10.9.3 Error Measure

When very small couplings need to be fitted, the calculation of the fitting error can be modified from an absolute scale to a relative scale.[3] By default, the fitting error in vector fitting is measured on the S-parameters in an absolute scale, which may lead to inaccuracies in small couplings within the S-parameter matrix. In such cases, applying a frequency-dependent weighting scheme to enable relative error control is beneficial. This approach is effective for systems with responses that exhibit significant dynamic variation across the frequency range of interest. Prior to fitting, each S-parameter S_{ij} is scaled by the factor

$$\frac{1}{\hat{\mathcal{H}}_{ij}(\omega_k)^{\alpha}}. \tag{10.3}$$

The parameter α determines the balance between an absolute fitting error, $\alpha = 0$, and a fully relative fitting error $\alpha = 1$. Using the relative error measure $\alpha = 1$ is not recommended, as it can amplify noise in the input data. A recommended starting range for α is between 0.3 and 0.5.

[3] This adjustment requires manual control of the fitting process using the IDEM package.

10.9.4 Passivity Enforcement

Passivity enforcement is essential if the generated macromodel is intended for use in transient circuit simulations. This makes the passivity enforcement a critical step in macromodel generation.

It is possible to generate vector-fitted models without performing the passivity enforcement step. Such models can be used to reduce the complexity of the circuit, as demonstrated in the example of the wires in the wing of an aircraft in Sect. 8.4. However, nonpassive models are restricted to frequency-domain circuit simulations.

In this section, we discuss the interplay between vector fitting and passivity enforcement, using the S-parameters of the inverter discussed in the previous section.

The passivity condition for an S-parameter matrix requires that the largest singular value does not exceed one. To check the passivity of an S-parameter matrix, we calculate the singular values at all frequency samples and compare them to the unit threshold. For an n-port matrix, there are n singular values, and the S-parameter matrix is nonpassive if any singular value exceeds one at any frequency. This property can also be tested on the input data to identify existing passivity violations.

We perform a passivity check on the S-parameter matrix of the inverter, with results plotted on the left-hand side of Fig. 10.27. Although the S-parameter matrix obtained from the full-wave FEM simulation is not fully passive, the all passivity violations are smaller than 10^{-3}. These minor violations are not critical and allow the data to be fitted and passivity enforced without introducing significant deviations from the original data.

Next, we perform the vector fitting step with an accuracy of 10^{-3}. The process requires 10 poles to achieve this accuracy. The passivity of the vector-fitted model is

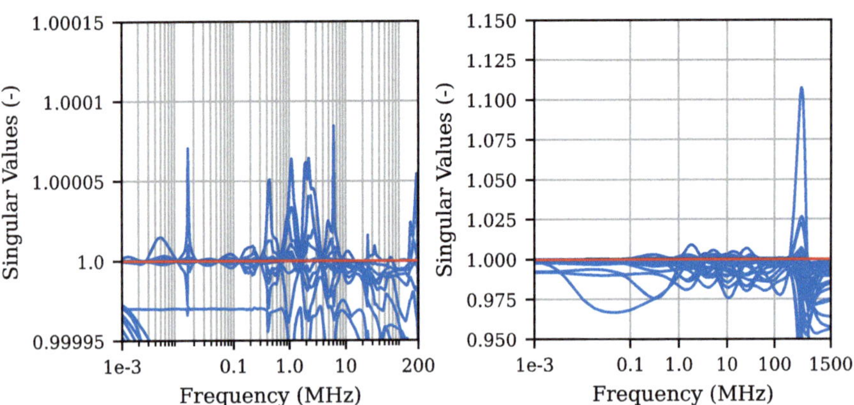

Fig. 10.27 Singular values of the S-parameter matrix of the inverter. Left: Data generated by the EM solver. Right: After vector fitting

displayed on the right-hand side of Fig. 10.27. The passivity violations of the fitted model are significantly greater than those of the original data, with a large peak observed around 350 MHz. This peak lies outside the frequency band provided in the input data, where the upper limit is 200 MHz.

This behavior is typical: vector fitting operates only within the frequency band of the input data and does not account for frequencies beyond this range. While the fitting scheme ensures asymptotic passivity as $f \to \infty$, passivity violations can still occur outside the provided band, as well as within it. These violations must be corrected during the passivity enforcement step.

Tools for macromodel generation can choose whether to enforce passivity across all frequencies or only within the input data's frequency band. The passivation scheme in CST enforces passivity over the entire frequency range, ensuring that the resulting macromodels remain stable for all transient circuit simulations.

As described in Sect. 10.3, the passivity enforcement process is an iterative scheme that perturbs the parameters of the system (specifically, the state-space matrices of the model) to ensure that the resulting system is passive. This iterative process is conducted such that the deviation between the perturbed model and the original data is minimized.

For the S-parameters of the inverter model, the passivity enforcement process requires seven iteration steps. On the left-hand side of Fig. 10.27, we plot the singular values after the fourth iteration. At this stage, some passivity violations persist, but others have already been removed. This progress can be observed by comparing the left-hand side (4 iterations) with the right-hand plot of Fig. 10.28 which shows the singular values after the seventh iteration. At this final step, all passivity violations are eliminated. The passive model has all singular values below the unit threshold.

The passivity enforcement process introduces deviations from the vector-fitted model. These deviations are small for small passivity violations.

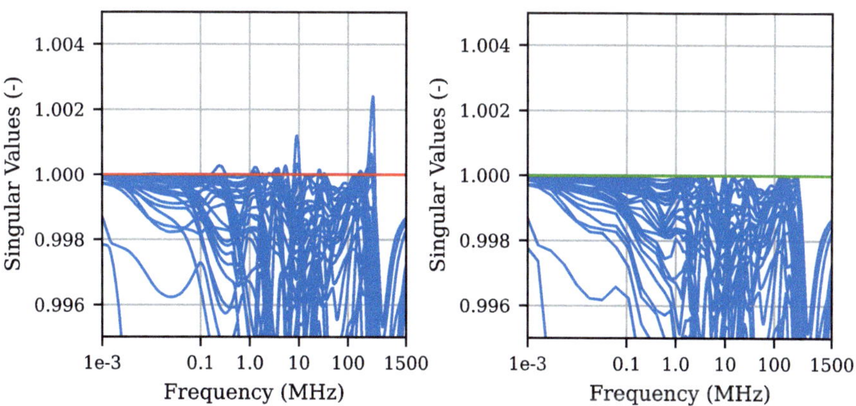

Fig. 10.28 Singular values during passivity enforcement. Left: Singular values at step four of the passivity enforcement. Right: Singular values at final step 7 of the passivity enforcement

This interplay between the vector fitting and the passivity enforcement step is critical. The S-parameters of the inverter are fitted with an accuracy of 10^{-3}, resulting in a model with 10 poles and requiring seven passivity enforcement iterations. If a higher accuracy is requested, the resulting model has more poles, which require additional iterations in the passivity enforcement step and potentially lead to a lower overall accuracy.

For instance, requesting an accuracy of $0.5 \cdot 10^{-4}$ produces for the inverter a model with 20 poles and requires 26 iterations for passivity enforcement. The final macromodel generated under these conditions may be less accurate than the model created with the original accuracy of 10^{-3}. It is not recommended to increase the accuracy of the vector fitting step beyond the default setting of 10^{-3}.

A heuristic measure of the quality of passivity enforcement is the number of iterations required to enforce passivity. If more than 50 iterations are needed, it may be beneficial to reduce the value for the accuracy. This adjustment can reduce the number of iterations required for passivity enforcement and improve the overall efficiency of the process.

In the case of the inverter model, which consists of 29 ports, these adjustments are unnecessary. However, as the number of ports increases, such tuning of the macromodel may become essential. General guidelines for the number of ports and their impact on fitting accuracy are outlined in Sect. 10.3.

10.9.5 Vector Fitting of Measured Data

In this section, we discuss macromodel generation from measured data. Component measurements are performed using a vector network analyzer, resulting in S-parameter data. As an example, we use the measurement data of the common mode choke introduced in Sect. 9.4.2.

A common method for representing the electrical behavior of a component is to generate RLC-based models fitted to the measurement results as shown by Hillenbrand et al. (2015).

Macromodel generation using vector-fitting enables the automated creation of models from measurement data, resulting in passive and causal black-box models. Additionally, the calculation of the singular values of the measured S-parameter matrix serves as a quality check for the measurement. Since the measurement results of a passive component are also expected to be passive, any significant passivity violation in the data indicates measurement errors.

In Fig. 10.29, we plot the singular values obtained from the measurement of the common mode choke. At lower frequencies, we observe a passivity violation in the data. When comparing the magnitude of these singular values with those obtained from the simulation model in Fig. 10.27, the passivity violation in the measured data is greater.

However, the singular values from the measurement data are sufficiently close to one to not introduce significant deviations between the macromodel and the original data.

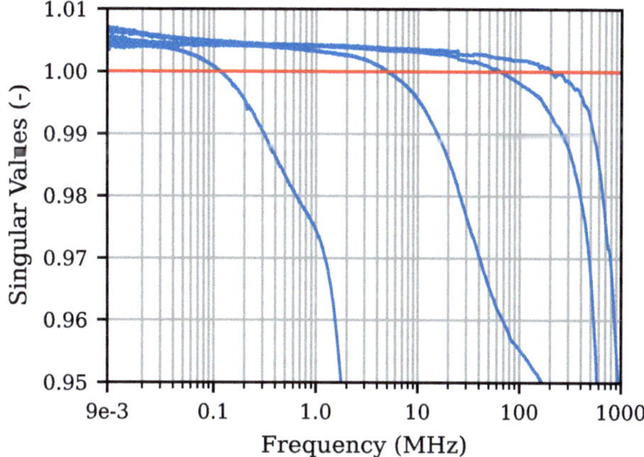

Fig. 10.29 Singular values of the S-parameter data from a measurement of a common mode choke

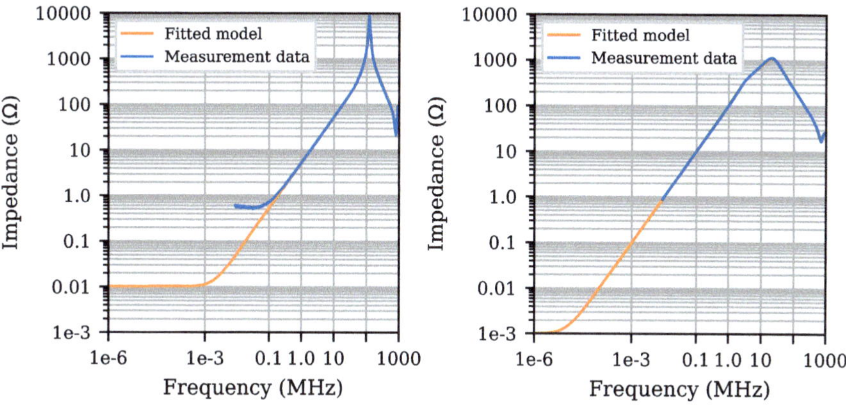

Fig. 10.30 Comparison of the measured and fitted differential and common mode impedances. Left: Z_{DM}. Right: Z_{CM}

In Fig. 10.30, we plot the measured and vector-fitted differential mode impedance Z_{DM} and the common mode impedance Z_{CM} of the common mode choke. The results are obtained using the same schematic described in Sect. 9.6, Fig. 9.26. The vector fitting process is performed with an accuracy of 10^{-3}.

The fitted model aligns with the measured data for Z_{CM}. For Z_{DM}, however, a deviation is observed below 100 kHz. In this case, the fitted model more closely represents the actual behavior of the common mode choke than the measured data. Below 100 kHz, Z_{DM} drops below 1 Ω, where measurements using a 50 Ω reference impedance show an impedance which is too large to be a realistic DC resistance. By applying the macromodel generation methodology, these measurement inaccuracies are removed, resulting in a more accurate macromodel: In order for the fitted model to be passive, the imaginary part of the impedance is required to be zero at DC.

Because this condition is enforced during the fitting, the fitted model is able to remove the measurement error.

The macromodel generation process yields a differential mode resistance of 10 mΩ close to DC (1 Hz). The data sheet of the common mode choke specifies a DC resistance of 3 mΩ. While it is possible to fix the DC resistance value in the vector fitting process to match the data sheet specification—since this value is not captured by the vector network analyzer measurement—we do not use this functionality, as the differences of the macromodel and the data sheet values do not affect the quality of the emission simulations.

The methodology described in this section enables the automated generation of circuit component models from measured data for EMC simulations. This approach saves simulation engineers time compared to the manual creation of RLC models, with controlled accuracy of the generated models.

Further Reading

Conducted Emission Modeling

B. Adamczyk, *In Compliance Magazine*, Internet (2017) https://incompliancemag.com/conducted-emissions-measurements-voltage-method/ (visited on 2025)

B. Adamczyk, *In Compliance Magazine*. Internet (2018). https://incompliancemag.com/bulk-current-injection-bci-substitution-method-and-closed-loop-method-with-power-limitation/

M. Hampe, A. Stieler, K.-D. Tieste, Damping of high-frequency oscillations in power devices using optimized snubber circuits, in *2018 International Symposium on Electromagnetic Compatibility (EMC EUROPE)* (2018), pp. 1–6. https://doi.org/10.1109/EMCEurope.2018.8485056

P. Hillenbrand et al., Understanding conducted emissions from an automotive inverter using a common-mode model, in *2015 IEEE International Symposium on Electromagnetic Compatibility (EMC)* (2015), pp. 685–690. https://doi.org/10.1109/ISEMC.2015.7256246

P. Hillenbrand, et al., Frequency domain EMI-simulation and resonance analysis of a DCDC-converter, in *International Symposium on Electromagnetic Compatibility (EMC Europe)* (Wroclaw, Poland, 2016), pp. 176–181

P. Hillenbrand, et al., Transient co-simulation of electromagnetic emissions caused by a SiC traction inverter, in *International Symposium on Electromagnetic Compatibility (EMC Europe)* (Angers, France, 2017)

IEC, *Vehicles, Boats, and Internal Combustion Engines—Radio Disturbance Characteristics—Limits and Methods of Measurement for the Protection of On-Board Receivers Apparatus* (2021)

R. Keller, S. Landolt, I. de Mendizábal, *Academy of EMC*. https://www.academyofemc.com/ (visited on 2025)

Nexperia, *BUK9Y72-80E N-channel 80 V, 78 mOhm logic level MOSFET Product Data Sheet*. Internet. https://www.nexperia.com/products/mosfets/automotive-mosfets/BUK9Y72-80E.html (visited on 2025)

Panasonic Industry, *Aluminum Electrolytic Capacitors*. https://industrial.panasonic.com/cdbs/www-data/pdf/RDF0000/ABA0000C1259.pdf (visited on 2025)

C. Riener, et al., Broadband modeling and simulation strategy for conducted emissions of power electronic systems up to 400 MHz. Electronics **11**(24), 4217 (2022). ISSN: 2079-9292. https://www.mdpi.com/2079-9292/11/24/4217

Panasonic, *Conductive Polymer Hybrid Aluminum Electrolytic Capacitors*. Internet. https://industrial.panasonic.com/ww/products/pt/hybrid-aluminum/documents (visited on 2025)

Rohm Semiconductors, *Snubber Circuit Design Methods*. Internet. https://fscdn.rohm.com/en/products/databook/applinote/discrete/sic/mosfet/sic-mos_snubber_circuit_design_an-e.pdf (visited on 2025)

Modeling Electric Motors

H. De Gersem, et al., Transmission-line modelling of wave propagation effects in machine windings, in *2008 13th International Power Electronics and Motion Control Conference* (2008), pp. 2385–2392. https://doi.org/10.1109/EPEPEMC.2008.4635621

H. De Gersem, et al., Simulation of wave propagation effects in machine windings. COMPEL—The Int. J. Comput. Math. Electr. Electron. Eng. **29**(1), 23–38 (2010). ISSN: 0332-1649. https://doi.org/10.1108/03321641011007948

H. De Gersem, A. Muetze, Finite-element supported transmission-line models for calculating high-frequency effects in machine windings. IEEE Trans. Magn. **48**(2), 787–790 (2012). https://doi.org/10.1109/TMAG.2011.2172197

H. De Gersem, S. Vanaverbeke, G. Samaey, Three-dimensional-two-dimensional coupled model for eddy currents in laminated iron cores. IEEE Trans. Magn. **48**(2), 815–818 (2012). https://doi.org/10.1109/TMAG.2011.2172924

V. Karakaşlı, et al., High-frequency modeling of delta-and star-connected induction motors. IEEE Trans. Electromagn. Compat. **64**(5), 1533–1544 (2022). https://doi.org/10.1109/TEMC.2022.3181337

M. Schinkel, et al., Efficient HF modeling and model parameterization of induction machines for time and frequency domain simulations, in *Twenty-First Annual IEEE Applied Power Electronics Conference and Exposition, 2006. APEC '06* (2006). https://doi.org/10.1109/APEC.2006.1620689

S.-P. Weber, et al., Modeling induction machines for EMC-Analysis, in *2004 IEEE 35th Annual Power Electronics Specialists Conference (IEEE Cat. No.04CH37551)*, vol. 1 (2004), pp. 94–98. https://doi.org/10.1109/PESC.2004.1355720

B. Wex, et al., Efficient frequency-domain evaluation of transient voltage effects in electric machines, in *2022 International Power Electronics Conference (IPEC-Himeji 2022-ECCE Asia)* (2022), pp. 2733–2740. https://doi.org/10.23919/IPEC-Himeji2022-ECCE53331.2022.9807048

Vector Fitting

A. Carlucci, T. Bradde, S. Grivet-Talocia, Improving robustness to termination conditions in passivity enforcement of rational macromodels. IEEE Trans. Compon. Packag. Manuf. Technol. **15**(1), 64–74 (2025). https://doi.org/10.1109/TCPMT.2024.3407526

D. Deschrijver, T. Dhaene, Stability and passivity enforcement of parametric macromodels in time and frequency domain. IEEE Trans. Microwave Theory Tech. **56**(11), 2435–2441 (2008). https://doi.org/10.1109/TMTT.2008.2005868

D. Deschrijver, T. Dhaene, DC-preserving passivity enforcement for S-parameter based macromodels. IEEE Trans. Microwave Theory Tech. **58**(4), 923–928 (2010). https://doi.org/10.1109/TMTT.2010.2042556

B. Gustavsen, A. Semlyen, Rational approximation of frequency domain responses by vector fitting. IEEE Trans. Power Delivery **14**(3), 1052–1061 (1999). https://doi.org/10.1109/61.772353

Chapter 11
Signal and Power Integrity

In this and the next chapter, we shift our focus from power electronics applications to high-speed digital design. In high-speed electronics, ensuring signal integrity (SI) and power integrity (PI) is crucial for the functionality of the product. We briefly introduce SI and PI in Sect. 2.3:

Power Integrity (PI) The analysis of the section of the electronic system from the power supply to the transmitter

Signal Integrity (SI) The analysis of the section of the electronic system from the transmitter to the receiver

SI and PI are closely related to EMC, and in many hardware design teams, engineers handle SI, PI, and EMC in parallel. Ensuring SI and PI is a prerequisite for achieving EMC compliance, as systems with poor SI/PI performance often fail to meet EMC requirements. However, the reverse is not always true: a system that passes SI/PI testing is not necessarily EMC compliant, as some effects critical for EMC are less significant for SI/PI.

In this discussion, we limit SI/PI analysis to simulations at the PCB level, excluding on-chip topics. This focus allows us to employ specialized simulation methods that are more efficient than full-wave methods for PCB-level analysis; however, they come with certain limitations that are briefly discussed in the next section.

The upcoming sections address key aspects of SI/PI simulation: In the next section, we discuss how the MTL method introduced in Chap. 8 can be augmented with via models to simulate SI. In the section on PI, we introduce a 2.5D simulation method that efficiently analyzes copper planes typically used as power and ground nets in a PCB. Finally, we explore how SI and PI simulations can be combined into a single simulation model, a technique commonly known as simultaneous switching noise (SSN) analysis.

Several dedicated tools are available for SI/PI simulations, including Ansys SIwave, Cadence Sigrity, Mentor Graphics (Siemens) HyperLynx, Keysight ADS, and Simbeor. These tools often use specialized simulation methods, many of which

J. Hansen, A. Barchanski, *EMC Simulation of Electronic Systems*,
https://doi.org/10.1007/978-3-032-09237-3_11

are proprietary to their respective vendors. In the next two sections, we discuss SI/PI simulations performed in CST, noting that the methodologies may differ in detail from those used in other tools.

11.1 Signal Integrity

Domains: 

SI refers to a set of measures used to evaluate the quality of an electrical signal, providing insights into the performance of data transfer within a data link. We can separate an SI analysis into a characterization of the physical transmission channel in frequency domain and the signal quality and timing characterization in time domain. An overview of the common outputs and metrics used to characterize the signal qulaity of a data link is summarized in Table 11.1.

Modern high-speed data interfaces, which operate at high data rates, employ various signal processing techniques at both the Tx and Rx. For a throughout analysis of signal quality and timing, it is essential to include these signal processing steps in the simulation.

This differs from EMC simulations, where only the voltage, current, and electromagnetic fields of the data link are relevant. Beyond the properties of the termination on the Rx side, signal processing steps like equalization or error correction at Rx side have no influence on emissions. In this book, we focus on the transmission channel characterization and introduce only briefly the basics of signal quality and timing considerations. For further information on these topics, we recommend, e.g., Bogatin (2004).

A way to plot signal quality metrics in SI simulations is the eye diagram, an example of which is shown in Fig. 11.1. An eye diagram is generated from transient simulations using a PRBS, where the time-domain signals of individual bit periods are overlaid. The quality of the data link is deduced from the eye opening, measured in terms of eye width and eye height.

Table 11.1 Frequency- and time-domain characterization in SI analysis. With permission from Radu Voina

	Transmission channel characterization	Signal quality and timing characterization
Outputs	S-parameters Voltages/currents in frequency domain	Voltages/currents in time domain
Metrics	Return loss, insertion loss NEXT, FEXT	Signal quality: max, min, mean, RMS, jitter, undershoot/overshoot Timing: rise/fall/setup/hold time, pulse width, duty cycle, period

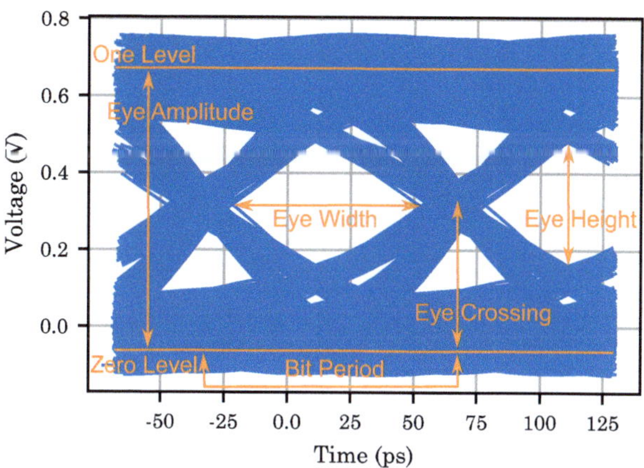

Fig. 11.1 Eye diagram

By incorporating the electrical properties of passive interconnects and components (the passive section of the digital channel), the impact of the passive channel on the eye opening can be estimated. The passive channel model is extracted from EM simulations, similar to how EMC problems are decomposed into source-coupling-path-sink segments. For realistic eye diagram calculations, the signal processing at Tx and Rx must also be included. The example eye diagram in Fig. 11.1 does not account for signal processing and is therefore used solely as a demonstration.

Another distinction between SI and EMC simulations is the number of simulations required and the degree of automation involved. In the EMC simulations discussed in this book, we focus on specific source-coupling-path-sink combinations, preselected for analysis. In contrast, SI simulations often involve a large number of physical traces within the channel. For example, a DDR4 memory module includes 288 pins, necessitating simulations of the entire channel. Simulating the complete interface requires a high level of automatization in both model setup and result evaluation.

11.1.1 MTL for Signal Integrity Simulation of PCBs

In this section, we discuss how the MTL method can be extended for SI simulations of PCBs. In Chap. 8, we demonstrate the efficiency of the MTL method in cable simulations. By segmenting the interconnects on a PCB into MTL sections and incorporating via models, we can create a highly efficient simulation method for analyzing the SI of PCBs.

To illustrate the details of this approach, we use a USB 3.0 channel as an example. An overview of the simulation model is shown in Fig. 11.2. The USB transmitter

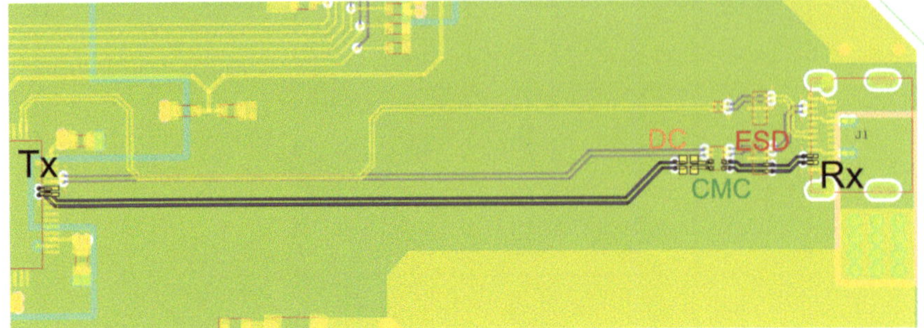

Fig. 11.2 Overview of the routing of the USB 3.0 channel

(Tx) is located on the left-hand side of the figure. At the Tx, the differential pair is routed through vias to inner layer 3 of the PCB, where it continues as a differential stripline for approximately 5 cm.

After the stripline section, the differential pair transitions to the top layer, where it passes through two DC blocking capacitors, followed by a common mode choke and a diode array for ESD protection. The receiver (Rx) is located at the two differential pins of the connector. As this analysis focuses solely on the PCB, the connector is not included in the simulation. Simulations including connector and cable models of the USB channel are discussed in Sect. 12.6.

The electromagnetic field at signal traces on a PCB forms a TEM mode, with the electric and magnetic field vectors perpendicular to the trace's longitudinal axis, is shown on the left-hand side of Fig. 11.3. In this figure, the electric and magnetic fields at 1 GHz are plotted for a four-layer board. The signal trace is colored red, and the ground layers are colored blue. The trace transitions from layer one to layer three via a signal via, while a grounding via, connecting the ground layers, is positioned nearby, as visible in the bottom-right part of the figure. The frequency of 1 GHz is selected to lie in the bandwidth of high-speed digital interfaces. In the following discussion, the exact field distribution is not relevant, only the difference between the field at the signal trace and the field in the vicinity of the via.

In Chap. 8, we introduce the MTL method, which is used to calculate the electrical properties of a transmission line. This method can also be applied to simulate signal lines of a PCB.

However, at via transitions, the fields are no longer transverse, as illustrated on the right-hand side of Fig. 11.3. In these regions, the MTL method is not applicable, and specialized via models are required. Numerous via modeling approaches can be found in the literature, ranging from simple RLC models to extrapolated datasets.

Different SI simulation tools employ different via models, and many details of these models are proprietary and not disclosed to the public. While simple via models are sufficient for most standard layouts, their accuracy diminishes with increasing layer count, via length, and frequency.

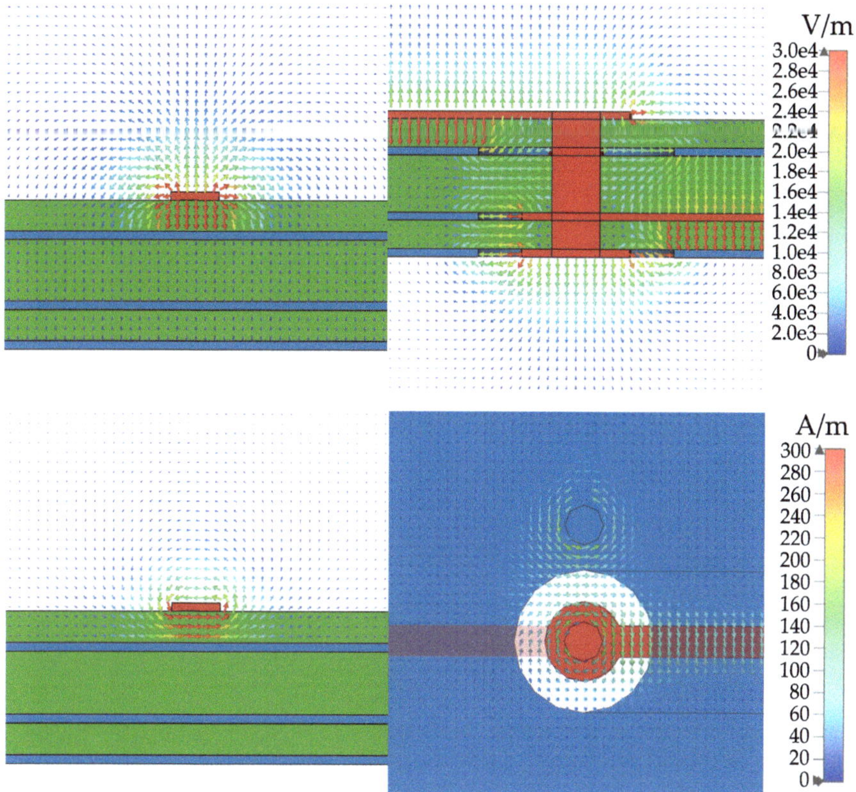

Fig. 11.3 Electric and magnetic field at 1 GHz on a microstrip line and at a via transition. Top: electric field. Bottom: magnetic field

For the simulation of high-speed signals, the dispersive material properties of the substrate and the loss in the metal layers are required to be modeled accurately. By incorporating dispersive material properties into the MTL model, it is possible to extend the applicability of the model toward higher frequencies.

As we see, not will see in the following sections, an MTL-based solver for SI simulations is highly efficient. However, it has several inherent limitations that depend on the following factors:

- Electrical size of the cross section
- Applicability of via models for the given stack-up
- Models used for the dispersive characteristics of materials

Due to the interplay of these factors with the geometry of the PCB and the bandwidth to be simulated, it is not straightforward to estimate the accuracy of the MTL-based solver in advance. For PCBs with a low layer count or routing without layer transitions, the MTL-based solver using simple via models can provide

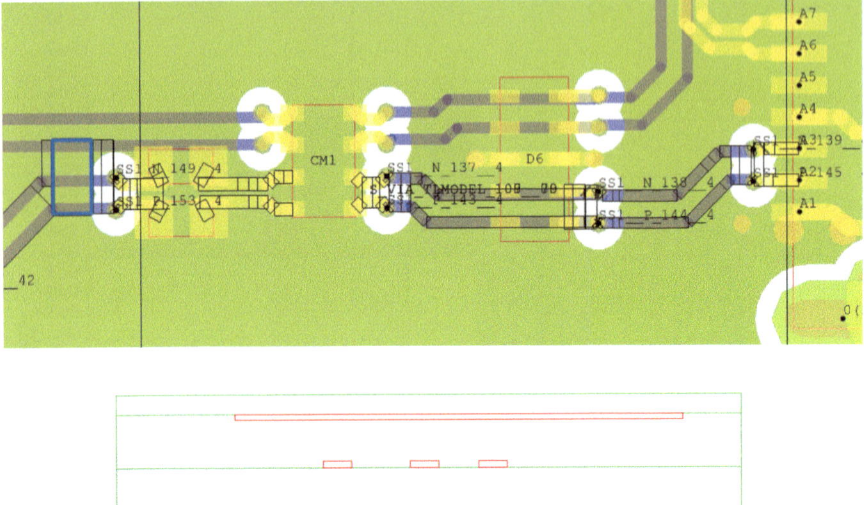

Fig. 11.4 Transmission line mesh on a PCB. Top: overview of segments close to the position of the common mode choke in the USB channel. Bottom: cross section at the position indicated by the blue box

accurate results even at double-digit GHz frequencies. However, for complex layer transitions, simple via models may become inaccurate, requiring a full-wave simulation of the via structure.

In the USB 3.0 example, we apply transmission line meshing to the PCB. The resulting mesh, zoomed into the region near the common mode choke, is shown in Fig. 11.4. The small black boxes along the USB signal traces represent the MTL models generated for each segment of the signal trace. The boxes with a small black dot indicate the positions where via models connect adjacent MTL sections.

On the bottom part of Fig. 11.4, a cross section generated by the MTL solver is depicted. This corresponds to the cross section indicated by the blue box on the left-hand side of the figure. In the cross-sectional view, metal is colored red, while the boundaries of the substrates are outlined in green. The two large red boxes represent the ground layers located above and below the stripline. Within this cross section, three signal traces are visible because the second USB pair lies within the search distance (as introduced in Sect. 8.1.3) to neighboring nets. For the simulation, we can define how non-excited nets are treated. They can be set to grounded or floating. In our simulation, we set them to floating.

The entire channel is meshed into 187 MTL sections and 6 differential via models. As discussed in Chap. 8, the cross section of an MTL model must be electrically short, limiting the model's validity to a certain maximum frequency. The MTL meshing algorithm estimates this maximum frequency by analyzing the distances and materials in the PCB. For the simulation model in this example, the estimated maximum valid frequency is 5.2 GHz. The entire meshing process is completed in just a few seconds.

11.1.2 Application Example: Signal Integrity of a USB 3.0 Channel

Digital signal modeled by an IBIS model
Pin voltage terminated by IBIS model
Vector-fitted MTL model, vector-fitted S-parameters
MTL with via models, FEM, TLM
Transient
0–5 GHz

In this application example, we utilize the MTL solver for PCBs to simulate the USB channel. Specifically, we calculate the differential-to-common mode conversion on the PCB. As discussed in Sect. 8.3, common mode currents on a cable are a common source of radiated emissions. By analyzing mode conversion on the PCB, we can estimate potential emission issues caused by imbalances in the PCB layout.

At the end of this section, we compare the results from the MTL solver with those from a full-wave simulation. This comparison evaluates the limitations of the MTL approach and provides insights into the simulation times required by the different solvers.

The MTL block and its connected schematic are depicted in Fig. 11.5. The USB driver and termination are modeled using an IBIS model obtained from the vendor's website. This model represents the Cypress Semiconductor (now Infineon) CYUSB3014-BZXI IC. On the Tx side, the IBIS driver is controlled by a PRBS sequence, with pulse lengths of 400 ps, corresponding to a base frequency of 2.5 GHz. At the connector location, the differential pair is terminated by the IBIS Rx model from the same IC.

The actual component footprint at the Tx side is designed for a different IC that does not support USB 3.0. However, this does not pose an issue for the simulation, as we can use a different driver model to evaluate the performance of the routing.

The DC blocking capacitors have a nominal value of 10 nF and are modeled using a SPICE model provided by Murata for the part number GCM155R71H103KA55 (Murata n.d.a). The common mode choke is also modeled using a SPICE model from Murata, corresponding to part number DLM0NSN900HY2 (Murata n.d.b). We use the SPICE model of the common mode choke instead of the measured S-parameters.

The ESD protection diode array is an NXP component with part number PUSB3FR4 (Nexperia). Although the SPICE model provided by NXP only includes

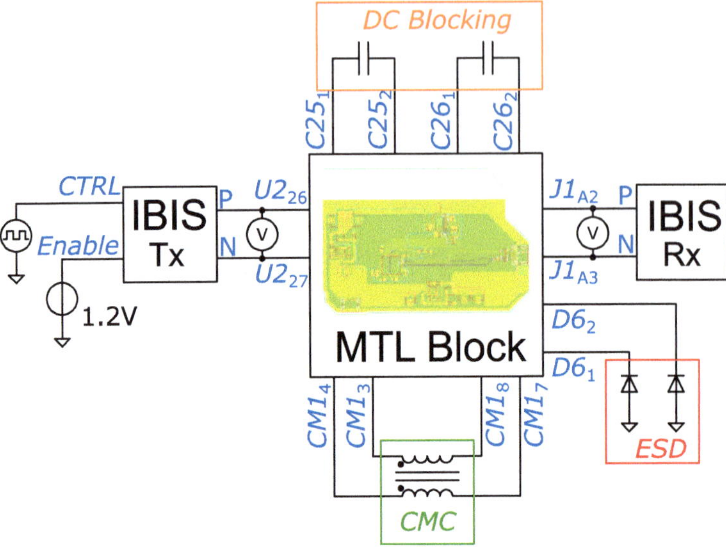

Fig. 11.5 Schematic for the simulation of the USB 3.0 channel

individual diodes rather than the full array, this is not an issue. Since the schematic is defined manually, we connect two diode SPICE models to the MTL block to represent the diode array.

When using the MTL method, the model generated by the EM solver includes a global reference. Consequently, the schematic differs from those used in coupled simulations with full-wave solvers, where S-parameters are employed. This difference is evident at the capacitors, where two pins are used in the MTL-based schematic, whereas in S-parameter-based schematics, we recommend using a single pin for a capacitor.

Time-domain voltages from the MTL simulation are plotted in Fig. 11.6. On the left-hand side of the figure, we show the single-ended voltages at the Tx and Rx. In this simulation, only the DC blocking capacitors are included; the common mode choke is omitted, with the pins at the position of the CMC shorted, and the ESD diode pins left unconnected.

The single-ended voltages highlight the effect of the DC blocking capacitors: the Tx voltages are centered around 0.5 V, while the Rx voltages are centered around 0 V, as the DC component is filtered out by the capacitors. To observe this centering, the transient simulation must run for several microseconds to allow the capacitors to fully charge. The simulation is performed using a brute-force transient method, simulating bit by bit.

For statistically significant results, such as calculating the bit error rate, an even larger number of bits would be required. This task is typically addressed using statistical methods where signal waveforms are derived from the unit response of the channel. However, this technique is beyond the scope of this book.

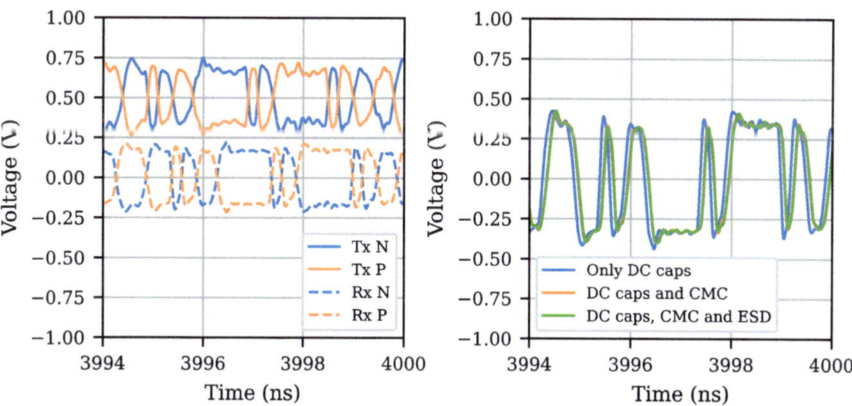

Fig. 11.6 Voltage levels in the USB 3.0 simulation. Left: single-ended voltages at the Tx and the Rx for a simulation including the DC capacitors, no common mode choke, and no ESD diodes. Right: differential voltage at the Rx for three different component setups

On the right-hand side of Fig. 11.6, we plot the differential voltage at the Rx side for various component configurations. The blue curve is obtained from a simulation with only the DC blocking capacitors, the orange curve from a simulation with added common mode choke, and the green curve from a simulation with the common mode choke and additional ESD protection diodes. The addition of the common mode choke introduces a small delay to the received signal due to the additional inductance. The inclusion of the ESD diodes does not produce a visible effect on the time-domain differential voltages, as the orange curve is entirely overlapped by the green curve.

For the simulation, vector fitting is applied to the MTL model. Generating the MTL model with vector fitting takes approximately 2 minutes. Simulating the full charging of the capacitors requires a time signal of 4 µs, resulting in a circuit simulation runtime of 16 minutes and 40 seconds.

For SI, the differential time-domain signals are the primary metric of interest. In EMC, however, we also need to consider the common mode signals generated in the channel. As discussed in Sect. 8.3, emissions due to common mode currents are typically much higher than those caused by differential mode currents. Using the MTL simulation described above, we can estimate the level of common mode voltage V_{CM} generated on the PCB.

The spectrum of V_{CM} at the Rx side is plotted in Fig. 11.7. Adding the common mode choke significantly reduces V_{CM} above 2 GHz. However, introducing the ESD diodes adds additional capacitance from the traces to the PCB ground. Since the ESD diodes are mounted between the common mode choke and the connector, this capacitance results in an increase of the common mode at the connector.

With the very low simulation times offered by the MTL method, we can quickly predict the V_{CM} generated by the imbalance introduced by the Tx and the PCB layout. However, for a comprehensive emission simulation of the entire system, this

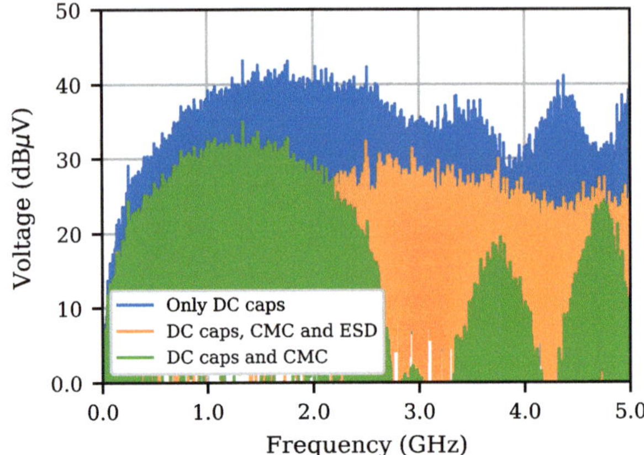

Fig. 11.7 Common mode voltage V_{CM} at the Rx for different component configurations

estimation is not sufficient, as it does not account for the imbalances at the connector or along the cable. Nevertheless, this type of simulation is valuable even in EMC analysis, as it allows for quick comparisons of different PCB layouts.

As the MTL method focuses only on the surroundings of the signal traces of interest and employs specialized models for the vias, it is insightful to compare the results of MTL simulations with those obtained from full-wave solvers. Specifically, we compare the spectrum of V_{CM} at the Rx obtained from full-wave simulations with the results from the MTL model.

For this purpose, we set up a full-wave simulation model of the USB 3.0 channel. The full-wave model includes the USB signal traces of interest as well as the complete power and ground structure of the PCB. The model consists of 12 ports: 2 ports at the Tx, 2 ports for the DC blocking capacitors, 4 ports for the common mode choke, 2 ports for the ESD diodes, and 2 ports at the Rx.

Since the MTL solver reports a maximum frequency limit of 5.2 GHz, we test this frequency limit by running the circuit simulation up to 10 GHz, exceeding the MTL solver's frequency range. Comparing the results of the MTL model with full-wave simulation allows us to estimate the quality of the MTL results above the maximum frequency limit. Full-wave simulations are conducted up to both 10 GHz and 20 GHz, with the 20 GHz models used solely to benchmark EM simulation times.

We simulate the full-wave model using both the TLM and FEM solvers, allowing us to compare results from these two different full-wave methods. In the TLM simulation, the mesh is defined manually with a mesh step of 50 μm for the USB traces and 200 μm for the rest of the board. For the FEM simulation, the mesh is automatically adapted through five adaptation steps. The energy criterion in the TLM solver is set to -50 dB, and GPU acceleration is used. The FEM simulation employs a direct solver. In both models, PML open boundaries are used with a distance of $\frac{\lambda}{4}$ at 2 GHz.

Table 11.2 Simulation time and mesh count for the full-wave simulations of the USB 3.0 channel

	10 GHz		20 GHz	
	TLM	FEM	TLM	FEM
Mesh count	1.6M	637k	3.2M	1.1M
Simulation time	2h03m	1h00m	3h03m	12h24m

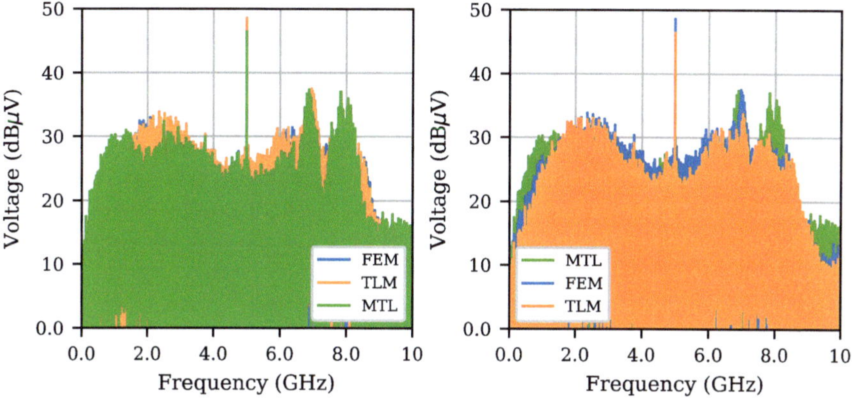

Fig. 11.8 Common mode voltage V_{CM} at the connector for different simulation methods. The results on both sides are identical; however, the order of the curves has been modified for a better insight into the dense spectrum

The simulation times and resulting mesh counts are summarized in Table 11.2. For the model with a maximum frequency of 10 GHz, the FEM solver is faster than the TLM solver. However, at a maximum frequency of 20 GHz, the simulation time for the FEM solver increases more than tenfold compared to the 10 GHz model. In contrast, the TLM solver for the 20 GHz model requires only 50% more simulation time than the TLM model at 10 GHz.

The resulting common mode voltage V_{CM} from the simulation using different simulation methods is plotted in Fig. 11.8. All components are included in the model. We observe that below 5 GHz, the results from the three methods agree with a maximal deviation of around 6 dB. Above 5 GHz the agreement is still satisfactory; however, the results from the MTL method exhibit a peak around 8 GHz, which is not present in the results obtained from the full-wave solvers. This is expected as the MTL model is valid only up to 5.2 GHz. Even with these differences, the MTL model can be used as a very fast option for the simulation of SI on PCBs.

11.2 Power Integrity

Domains: 

In the previous section, we performed an SI analysis by placing an IBIS model at the transmitter and providing an ideal voltage source to the Tx (Fig. 11.5). However, in a real PCB layout, the power source is not located at zero distance to the transmitter IC. The power required to drive the digital signal onto the channel must be delivered to the IC via the power delivery network (PDN) on the PCB.

The current consumption of modern ICs is highly dynamic. During idle periods, the power consumption is very low, but when the IC is highly active, its current draw can increase significantly. In these periods of high current demand, the PDN must respond rapidly to the fast-changing current consumption. Achieving this requires a PDN with low impedance across a broad frequency range.

Typically, the impedance of the complete PDN can be divided into three frequency-dependent regions:

Low frequency (kHz to a few hundred kHz) The impedance is dominated by the response of the voltage regulator module.

Mid frequency (hundreds of kHz to approximately 100 MHz) The impedance is governed by the power-ground system on the PCB.

High frequency (above 100 MHz) The impedance is primarily determined by the power-ground system within the IC.

In the following discussion, we focus on the physical properties of the PCB section between the output of the voltage regulator module and the input of the IC, as this section can be efficiently simulated using EM models. More comprehensive studies, including the voltage regulator and the IC, are available in the references at the end of this chapter.

The typical metric for characterizing the PDN is the impedance between the power and ground pins of the consumer IC. This impedance curve is compared with a target impedance. The target impedance is a predefined maximum allowable impedance for the PDN specific to a particular IC. In its simplest version, the target impedance is estimated using the formula:

$$Z_{\text{target}} = \frac{\Delta V}{I_{\max}}, \tag{11.1}$$

where ΔV is the maximum allowed deviation from the nominal supply voltage (as specified in the IC datasheet) and $I_{\max}$ is the maximum current drawn by the IC. For example, an IC with a maximum allowed voltage deviation of 50 mV and a current draw of 2 A would have a target impedance of 12.5 mΩ (Bogatin 2004).

For PI simulation, a specialized solver is available in CST. This solver is based on the frequency-domain finite element method, combined with a domain decomposition approach. It uses special basis functions tailored for the planar structures of PCBs. With this solver, the example presented in the next section can be solved in just 15 seconds.

11.2.1 Application Example: PI of a Small PDN

- Power supply noise modeled by an ideal voltage source
- Pin voltage
- S-parameters
- Specialized FEM method
- AC, Transient, S-parameter
- 0–2 GHz

In our PCB example, a 12 V power supply is delivered to the PCB through a connector. The 12 V input is down-converted by two SMPSs. One SMPS converts the 12 V to 5 V, which powers the USB output. This converter is analyzed in Chap. 3 (see Fig. 3.10). The second SMPS converts the 12 V input to 3.3 V, providing power to the digital components on the board.

In the following analysis, we focus on the PDN between the second buck converter and a digital IC.

The PDN is shown in Fig. 11.9. The VDD power net, located on the top layer of the PCB, is depicted in red. The GND net, shown in blue, is routed partially on the top layer and partially on the second inner layer. Other nets are represented in orange.

The output of the buck converter supplying 3.3 V is located at L1, on the left-hand side of the figure. Directly following the inductor are two bulk capacitors, C2 and C3, which are 22 μF capacitors manufactured by Panasonic (part number: EEEFK1H220P (Panasonic Industry)). Panasonic does not provide simulation models for this part, but the datasheet specifies an ESR value of 0.88 Ω. For the ESL, we assume a value of 5 nH due to the capacitors' large size. When analyzing the complete PDN with all capacitors, the ESL of these bulk capacitors has minimal influence, as the impedance is primarily governed by the ESL of smaller capacitors positioned closer to the IC.

At IC U4 (top right), a WiFi IC from H&D Wireless (model:[1] HDG104-DN-2), six capacitors are placed at the power pins. The capacitors at U4 are a combination of 10 nF and 100 nF components manufactured by Murata (part numbers: GCM188R71H103KA37D (Murata n.d.c) and GRM188R71H104KA93D

[1] We choose this outdated model because it serves our example analyzed in Sect. 5.4.1, as it is based on a single-ended line.

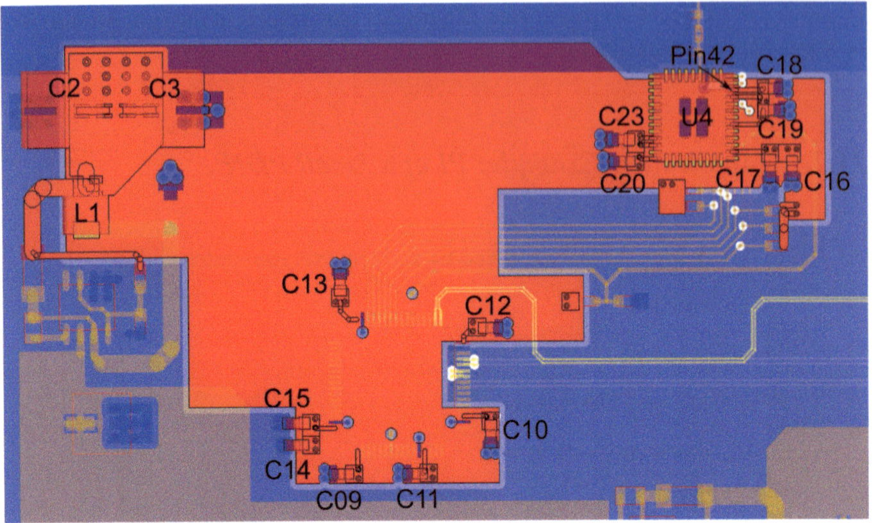

Fig. 11.9 Overview of the PDN used in the simulation example

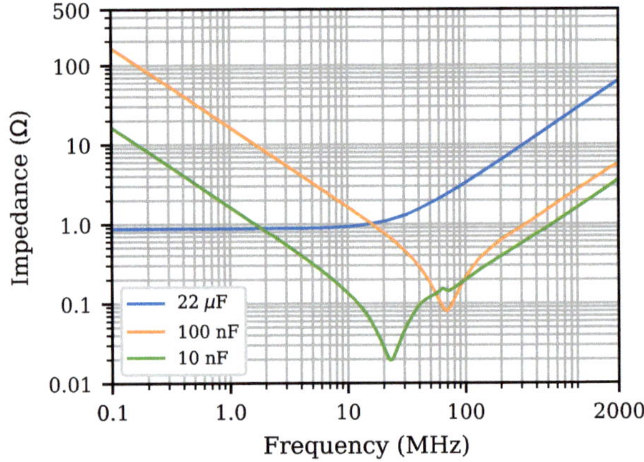

Fig. 11.10 Impedance of the three capacitor types used in the example PDN

(Murata n.d.d)). For the simulation, we use the (extended) SPICE models provided by Murata.

In the lower section of the VDD net, a second IC is placed along with the same 10 nF and 100 nF capacitors. Although we do not specifically analyze the impedance of this second IC, the nearby capacitors influence the overall impedance of the PDN.

Before running the PI simulation, we calculate the impedance of the three capacitor types used in the PDN. The impedance results are plotted in Fig. 11.10.

At frequencies below 2 kHz, the two 22 μF capacitors exhibit the lowest impedance. Between 2 MHz and 60 MHz, the 100 nF capacitors provide the lowest

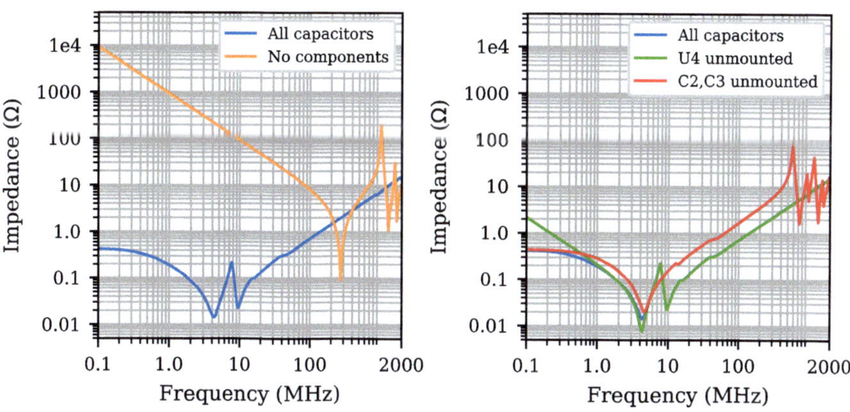

Fig. 11.11 Impedance between GND and VDD at pin 42 of U4

impedance, while above 60 MHz, the 10 nF capacitors demonstrate the lowest impedance. This behavior is explained by the typically smaller ESL of smaller capacitors, which results in lower impedance at higher frequencies.

The comparison of the impedances of these three capacitor types highlights the principle behind the often-cited "three-capacitor rule": by employing capacitors of varying sizes, it is possible to achieve a consistently low impedance across a wide frequency range.

The impedance between pin 42 (VDD) and pin 3 (GND) at U4, calculated using the specialized PI solver, is plotted in Fig. 11.11.

The orange curve on the left-hand side of the figure shows the impedance when no capacitors are mounted. In this case, the impedance is high at low frequencies due to an open circuit at the pin. As the frequency increases, the impedance decreases due to the capacitance between the VDD and GND planes, reaching a minimum at 270 MHz. Above this frequency, the inductance of the pins becomes relevant for the impedance.

The blue curve represents the impedance when all capacitors are mounted, resulting in the lowest overall impedance.

The green curve in the right-hand side of the figure shows the impedance when the bulk 22 μF capacitors are unmounted. In this configuration, the impedance increases in the frequency band below 1 MHz, as the bulk capacitors are no longer present.

The red curve depicts the impedance when all capacitors near U4 are unmounted. At higher frequencies, the impedance increases and resonances appear, with the first impedance maximum occurring at 590 MHz.

The overall increase in impedance after removing all capacitors near the IC is less than might be expected. This is because the simulation still includes the capacitors near the second IC, which contribute to reducing the overall impedance.

The simulation also enables visualization of the spatial distribution of the impedance, as shown in Fig. 11.12. Such plots are valuable for identifying areas

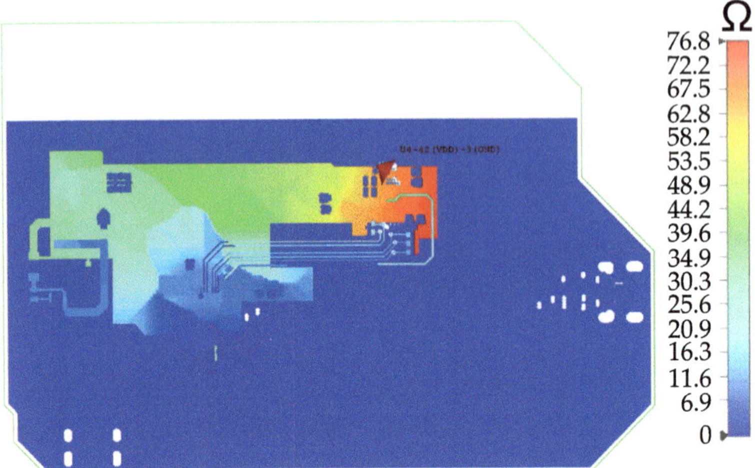

Fig. 11.12 Spatial impedance plot on the VDD at 579 MHz in the case where the capacitors at U4 are not mounted

of high impedance within the PDN. Adding capacitors at these locations can significantly improve the overall impedance of the PDN.

The spatial impedance distribution in Fig. 11.12, plotted at 579 MHz (the frequency of the first impedance peak), aligns with the analysis above. At the position of U4 in the upper-right corner, the impedance is high due to the absence of mounted capacitors in that region. Conversely, in the lower section of the VDD plane near U2, the impedance remains low, as capacitors close to U2 are mounted.

While impedance analysis is the standard approach for evaluating a PDN, the results from the PI simulation can also be used in transient circuit simulations to demonstrate the interplay between frequency- and time-domain results. The PI solver generates S-parameters for the PDN, which can be converted into a time-domain model using vector fitting (see Sect. 10.3).

The results of such a time-domain simulation are shown in Fig. 11.13. To model the current draw of the IC, a current source is placed at pin 42. The current waveform is a simple rectangular shape, with the following pattern:

- From 0 to 500 ns, no current is drawn.
- At 500 ns, a 250 mA current is drawn for 500 ns, with a rise and fall time of 5 ns.
- From 1000 ns to 2000 ns, the current draw is zero.
- At 2000 ns, the current draw remains constant at 250 mA for the remainder of the simulation.

The voltage at pin 42 is calculated for three configurations:

- With all capacitors mounted
- With the capacitors at U4 removed
- With both bulk 22 μF capacitors removed

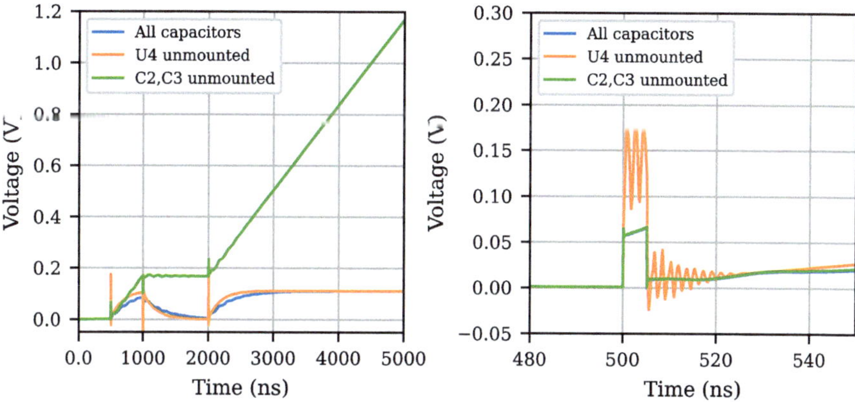

Fig. 11.13 Voltage between GND and VDD at pin 42 for a current draw of 250 mA. Left: overview of the voltage. Right: zoom in on the voltage during the current ramp

These configurations correspond to those analyzed in Fig. 11.11. When the 22 μF capacitors are removed, the voltage at pin 42 rises above 1 V during the long current draw at the end of the simulation, as seen in the green curve. Removing the bulk capacitors leads to a significant voltage ripple at low frequencies, particularly during extended periods of current draw.

The effect of removing the capacitors at U4 is evident on the right-hand side of Fig. 11.13, where we zoom in on the voltage during the ramp of the current draw. This ramp excites higher frequencies, causing voltage ringing due to resonances in the PDN impedance. Such ringing can be triggered not only by the current ramp but also by signal trace vias passing through the PDN, such as the SDIO traces in our example.

To minimize such issues, a well-designed PDN should avoid resonances. By performing PI simulations, potential resonances can be identified and mitigated by placing capacitors in optimal positions, as indicated by the spatial impedance plot.

11.3 Simultaneous Switching Noise

Domains: 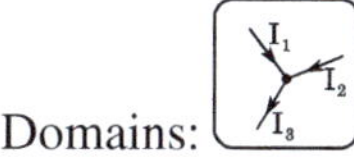

In the previous sections, we simulate signal integrity and power integrity independently. This separation allows us to focus on optimizing the respective layout sections. However, in real products, SI and PI are closely intertwined. A transmitter draws current to drive signals onto the signal traces, and if the PDN cannot provide a stable power supply, the signal quality on the traces is degraded. This phenomenon is referred to as simultaneous switching noise (SSN).

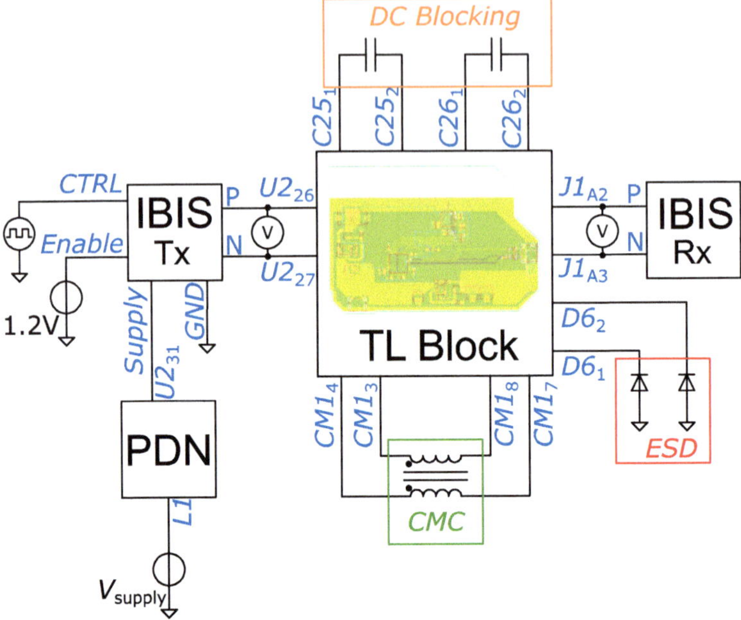

Fig. 11.14 Schematic for the simulation of SSN

To simulate the influence of the PDN on SI, a power-aware IBIS model is required as introduced in Sect. 2.3.5. We modify the SI schematic used earlier in this chapter as follows: In the initial SI simulation, the power supply to the IBIS model is assumed to be ideal, and the power supply pins are not included in the schematic (Fig. 11.5). For the SSN simulation, we enhance the schematic by incorporating the power pins, as shown in Fig. 11.14.

The layout used for the simulation is shown in Fig. 11.15. Using the PI simulation, we calculate a 2-port S-parameter block to model the PDN. The power supply is represented by a port between VDD and GND, located at the position of L1, the output inductor of the SMPS. The supply to U2, assumed to be the USB 3.0 driver, is located at pin 31 of U2. All capacitors between VDD and GND are included in the simulation. The SI section of the USB 3.0 channel is modeled with all key components, including the DC blocking capacitors, the common mode choke, and the ESD protection diodes.

As noted previously, the footprint of U2 does not correspond to the actual footprint of the IBIS model. Additionally, while the board is designed for $V_{\text{VDD}} = 3.3\,\text{V}$, the USB 3.0 IBIS model is intended for $V_{\text{VDD}} = 1.2\,\text{V}$. This discrepancy does not affect the simulation, as the supply voltage is set to the required value for the IBIS model.

Before running the full SSN simulation with the PDN model, we analyze the effect of V_{VDD} on the Tx signals using a simplified voltage waveform for V_{VDD}. This approach provides a clearer understanding of the interplay between SI and

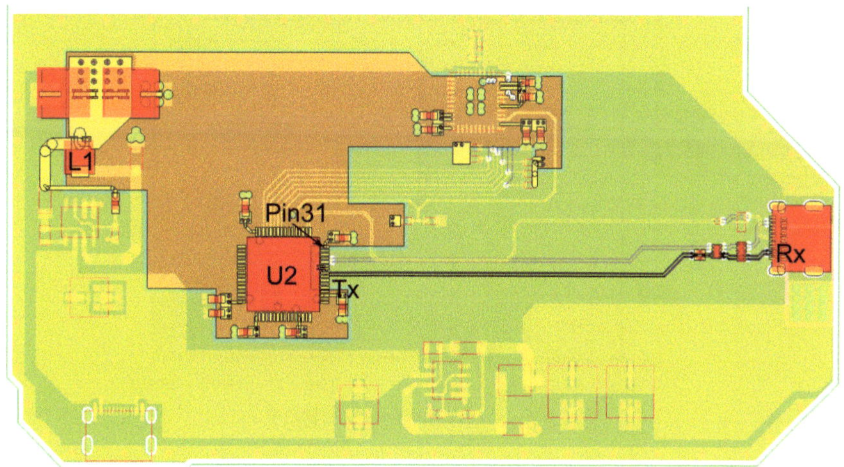

Fig. 11.15 Overview of the simulation model for SSN

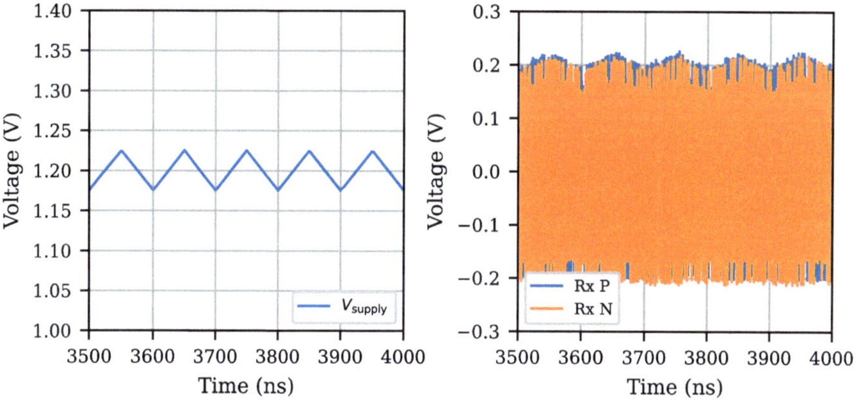

Fig. 11.16 Voltages in the SSN simulation using a voltage source with a triangular ripple. Left: supply voltage. Right: single-ended voltages at the Rx

PI before incorporating the complete PDN model. For this simplified analysis, we connect a voltage source V_{VDD} directly to the IBIS Tx block, omitting the PDN block. The supply voltage is modeled as a triangular waveform centered at 1.2 V, with a peak-to-peak ripple of 50 mV and a frequency of 10 MHz.

The results of the simulation are shown in Fig. 11.16. On the left-hand side, V_{VDD} is plotted, while on the right-hand side, the single-ended voltages at the Rx are depicted. The ripple introduced to the supply voltage is also reflected in the PRBS signal sequence.

Although the introduced ripple might still be too small to significantly affect the SI, it has a noticeable impact on the differential-to-common mode conversion. The variation in V_{VDD} introduces additional imbalance to the single-ended signals, as

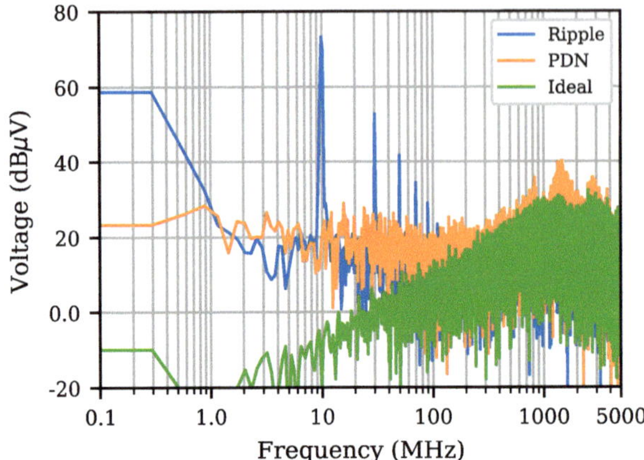

Fig. 11.17 Spectrum of V_{CM} at the Rx calculated in the SSN simulation

their voltage levels are no longer symmetric. This is in contrast to the case with a constant V_{VDD}, where the supply voltage is ideal. The imbalance caused by the ripple is further illustrated by analyzing the spectrum of the common mode voltage at the Rx.

The spectrum of the common mode voltage V_{CM} at the Rx is plotted in Fig. 11.17. This plot is similar to Fig. 11.7, but the x-axis is logarithmic to better observe effects at lower frequencies. When the supply voltage is constant, as shown by the green curve, V_{CM} decreases with decreasing frequency, and its overall amplitude remains below 30 dBμV.

By introducing the 10 MHz ripple to the supply voltage (Fig. 11.16), the common mode voltage at the Rx exhibits an almost 80 dBμV peak at 10 MHz. The higher harmonics decay at 40 dB/dec, consistent with the spectral decay of a triangular waveform. These peaks could contribute to increased emissions.

The orange curve represents the results of including the PDN model in the SSN simulation, using an ideal 1.2 V source. In this case, no supply voltage ripple is present, so no peaks are observed. However, the common mode voltage increases because the PDN introduces an impedance to the supply voltage that is absent in the ideal case. The effect of the PDN is noticeable at low frequencies and at higher frequencies, particularly around 1.5 GHz. At this frequency, a small peak in the PDN impedance becomes visible in the common mode spectrum.

In this chapter, we have demonstrated how specialized solvers for SI and PI enable fast simulations on a PCB. These methods provide voltages and currents as outputs but do not calculate the fields generated by the PCB. Although tools exist to extract a near-field source from an SI simulation and use it in a full-wave simulation, this approach does not account for the effects of connectors, cables, or interactions between the PCB and its surroundings.

For reliable radiated emission simulations, both the connectors and cables, as well as their interactions with the PCB, must be included. This is the topic of the next chapter.

Further Reading

Signal Integrity

E. Bogatin, *Signal Integrity—Simplified* (Prentice Hall Professional, New York, 2004)

E. Bogatin, What's new in signal integrity and high-speed serial links: approaching the fundamental limits of copper interconnects. IEEE Microw. Mag. **23**(5), 84–95 (2022). https://doi.org/10.1109/MMM.2022.3148810

E. Bogatin, Essential principles of signal integrity. IEEE Microw. Mag. **12**(5), 34–41 (2011). https://doi.org/10.1109/MMM.2011.941411

J. Fan, et al., Signal integrity: efficient, physics-based via modeling: Return path, impedance, and stub effect control. IEEE Electromagn. Compat. Mag. **3**(1), 76–84 (2014). https://doi.org/10.1109/MEMC.2014.6798802

Y. Hassab, et al., Machine learning based data validation for signal integrity and power integrity using supervised time series classification. IEEE Trans. Electromagn. Compat. **66**(6), 2150–2158 (2024). https://doi.org/10.1109/TEMC.2024.3474917

S. Müller, et al., Accuracy of physics-based via models for simulation of dense via arrays. IEEE Trans. Electromagn. Compat. **54**(5), 1125–1136 (2012). https://doi.org/10.1109/TEMC.2012.2192123

S. Müller, et al., Complete modeling of large via constellations in multilayer printed circuit boards. IEEE Trans. Compon. Packag. Manuf. Technol. **3**(3), 489–499 (2013). https://doi.org/10.1109/TCPMT.2012.2234211

S. Müller, et al., Energy-aware signal integrity analysis for high-speed PCB links. IEEE Trans. Electromagn. Compat. **57**(5), 1226–1234 (2015). https://doi.org/10.1109/TEMC.2015.2427362

Murata, *Chip Multilayer Ceramic Capacitors for Automotive GCM155R71H103KA55* (n.d.a). https://www.murata.com/products/productdetail?partno=GCM155R71H103KA55#

Murata, *Common Mode Choke Coil DLM0NSN900HY2* (n.d.b). https://www.murata.com/products/productdetail?partno=DLM0NSN900HY2#

Nexperia, *PUSB3AB4. ESD Protection for Ultra High-Speed Interfaces*. https://assets.nexperia.com/documents/data-sheet/PUSB3AB4.pdf

T. Reuschel, S. Müller, C. Schuster, Segmented physics-based modeling of multilayer printed circuit boards using stripline ports. IEEE Trans. Electromagn. Compat. **58**(1), 197–206 (2016). https://doi.org/10.1109/TEMC.2015.2481001

K. Scharff, H.-D. Brüns, C. Schuster, Efficient crosstalk analysis of differential links on printed circuit boards up to 100 GHz. IEEE Trans. Electromagn. Compat. **61**(6), 1849–1859 (2019). https://doi.org/10.1109/TEMC.2019.2894536

Power Integrity

H. Barnes, S. Sandler, Decoupling capacitor optimization for flat Z PCB power distribution networks, in *2018 IEEE Symposium on Electromagnetic Compatibility, Signal Integrity and Power Integrity (EMC, SI & PI)* (2018). https://doi.org/10.1109/EMCSI.2018.8495286

Murata, *Chip Multilayer Ceramic Capacitors for Automotive GCM188R71H103KA37* (n.d.c). https://www.murata.com/products/productdetail?partno=GCM188R71H103KA37#.

Murata, *Chip Multilayer Ceramic Capacitors for General Purpose GRM188R71H104KA93D* (n.d.d). https://www.murata.com/products/productdetail?partno=GRM188R71H104KA93

Panasonic Industry, *EEEFK1H220P: Aluminum Electrolytic Capacitors (Surface Mount Type)*. https://industrial.panasonic.com/ww/products/pt/aluminum-cap-smd/models/EEEFK1H220P

J.B. Preibisch, X. Duan, C. Schuster, An efficient analysis of power/ground planes with inhomogeneous substrates using the contour integral method. IEEE Trans. Electromagn. Compat. **56**(4), 980–989 (2014). https://doi.org/10.1109/TEMC.2013.2292095

L. Ren, et al., Prediction of power supply noise from switching activity in an FPGA. IEEE Trans. Electromagn. Compat. **56**(3), 699–706 (2014). https://doi.org/10.1109/TEMC.2013.2293872. https://api.semanticscholar.org/CorpusID:36578807

S.M. Sandler. *Power Integrity: Measuring, Optimizing, and Troubleshooting Power Related Parameters in Electronics Systems*. 1st. McGraw-Hill Education, 2014. ISBN: 9780071830997

M. Schierholz, Y. Hassab, C. Schuster, Engineering-informed design space reduction for PCB-based power delivery networks. IEEE Trans. Compon. Packag. Manuf. Technol. **13**(10), 1613–1623 (2023). https://doi.org/10.1109/TCPMT.2023.3292577

Chapter 12
Radiated Emission

In this chapter, we discuss the simulation setup for radiated emission tests. Radiated emission (EMI) limits are defined in terms of a maximum allowed field strength at a particular distance from the source. In Chap. 6, we explain how field values can be estimated in simulations using field probes or field monitors. In actual electromagnetic compatibility (EMC) measurements, the field cannot be accessed directly, so antennas are used to capture the field emitted by the device under test (DUT). The signal measured at the antenna terminal is scaled using the antenna factor to estimate the field strength and compare it to the specified limits.

Simulating a receiving antenna alongside the DUT is possible, and software vendors often demonstrate their tools by showcasing large models that include the entire EMC chamber with absorbers, antennas, tables, and the DUT. However, adding such details significantly increases computational effort. This chapter explores the differences between comprehensive simulation models and those with simplified setups, offering insights into selecting the most efficient modeling approach.

It is a common belief that including an antenna in the simulation is essential to replicate measurement results accurately. However, this step requires a precise model of the measurement antenna. While some antenna vendors provide CAD models upon request, the examples in this chapter use generic antenna models based on specifications. Section 12.2 discusses the simulation of receiving antennas in EMC applications and compares results obtained using these antenna models with those from field probes. This comparison enhances understanding of the differences in the resulting field values.

In all simulation models, we do not detail the absorbing walls of the chamber. Publicly available data on materials used as absorbers is limited, and including such details further increases computational demands. Instead, we rely on open boundary conditions to represent the chamber's side walls and ceiling.

When the test standard requires a metallic table, as specified in the CISPR 25 standard, it is included in the simulation. For the chamber floor, we apply electric

J. Hansen, A. Barchanski, *EMC Simulation of Electronic Systems*,
https://doi.org/10.1007/978-3-032-09237-3_12

boundary conditions to account for reflections. In Sect. 12.5.2 we discuss the effect of the chamber floor in more details.

All simulations presented in this chapter are performed using full-wave solvers.

12.1 Monopole Antenna

Domains: 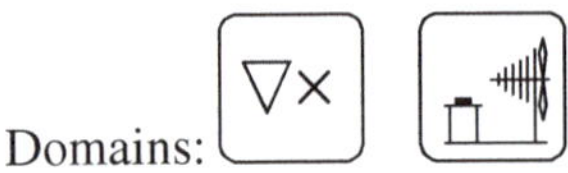

Monopole antenna measurements are used in various test standards, such as CISPR 25 and MIL-STD-461, for measurements in the kHz and low MHz bands. In the CISPR 25 standard, the monopole measurement is performed in the frequency band from 150 kHz to 30 MHz. We have sketched the CISPR 25 measurement setup in Fig. 12.1. According to this standard, the monopole is placed at a distance of 1 meter from the DUT, and the counterpoise of the monopole is connected to the measurement table via a wide metallic connection.

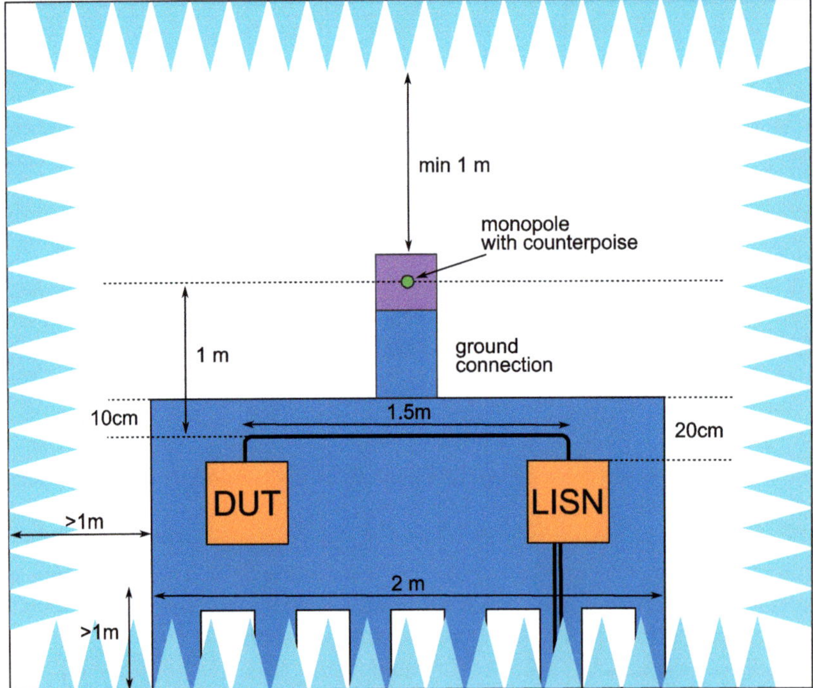

Fig. 12.1 Radiated emission measurement using a monopole antenna in the frequency band from 150 kHz to 30 MHz according to CISPR 25

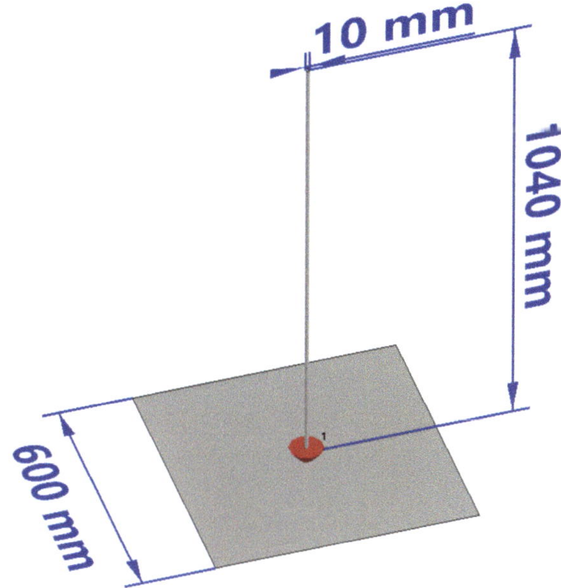

Fig. 12.2 Geometry of the monopole antenna used in the antenna calibration study

At 30 MHz, the antenna operates in the near field of the DUT. Although referred to as a radiated emission measurement, it primarily measures differences in the capacitive coupling of the DUT to the table and to the antenna rod. Nevertheless, we use a full-wave solver for these simulations, as it is more convenient to employ antenna ports or lumped elements in the simulation setup. Due to the low-frequency range, the FEM method is selected. As a side note, the MoM method is also well-suited for the simulation of monopole antennas.

In this section, we describe the calibration procedure for antennas in a simulation. To illustrate this, we have selected a simple simulation model of a monopole antenna, as shown in Fig. 12.2. The model consists of a 1.04 m long cylindrical rod with a diameter of 1 cm, positioned 1 cm above a metallic plane measuring 60 cm × 60 cm. A port with a reference impedance of 1 MΩ is placed between the rod and the ground plane. Although measurements with a monopole are typically conducted in a 50 Ω system, this impedance reflects the output of the amplifier connected to the monopole. The impedance between the rod and the plane is high-ohmic at low frequencies.

During calibration, a well-defined field $\mathbf{E}_{\mathrm{inc}}$ illuminates the antenna, and the voltage at the antenna is measured. In physical measurements, the calibration source is usually another antenna. In simulations, however, the process is simplified as we can use a plane wave with a specified field strength to illuminate the antenna. For this simulation, we define a plane wave with vertical linear polarization and a field strength of $|\mathbf{E}_{\mathrm{inc}}| = 1\,\frac{\mathrm{V}}{\mathrm{m}}$, then calculate the voltage V_{ant} at the port located at the base of the antenna. Using the FEM solver, the simulation takes 1 minute and 40 seconds to compute 50 frequency samples in the range from 150 kHz to 30 MHz. In a post-

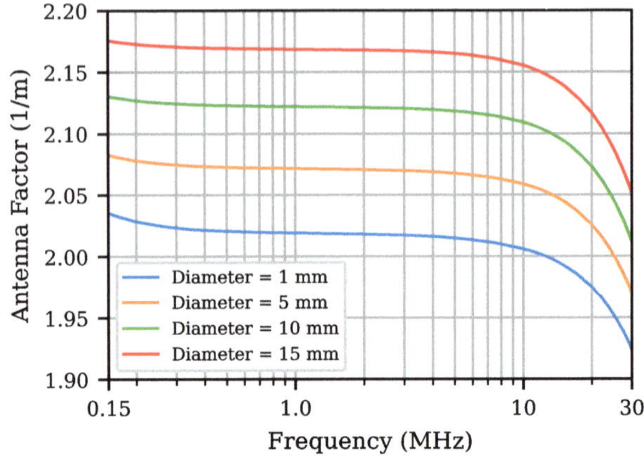

Fig. 12.3 Antenna factor of the monopole antenna for different rod diameters

processing step, the antenna factor AF is calculated as:

$$\mathrm{AF} = \frac{|\mathbf{E}_{\mathrm{inc}}|}{V_{\mathrm{ant}}} = \frac{1\,\frac{\mathrm{V}}{\mathrm{m}}}{V_{\mathrm{ant}}} \tag{12.1}$$

at each frequency.

The antenna factor is then used to estimate the field emitted by the DUT. This is done by calculating the voltage V_{ant} at the antenna in the full-wave model and multiplying it by AF. This methodology is applied to both the monopole in this section and the log-periodic antenna described in the next section. Notably, the antenna factor of a circular monopole rod depends on the rod's diameter, as the diameter affects the rod's capacitance to the plane. We calculated the antenna factor for four different rod diameters and plotted the resulting AF values over frequency in Fig. 12.3.

In this simulation, all boundaries are set to open. If a metallic chamber floor is present beneath the plate, the antenna factor is also affected. However, this influence is relatively small and is not discussed in detail.

It is important to note that the effect of the chamber floor becomes more pronounced when a calibration antenna, rather than a plane wave, is used. The field emitted by the calibration antenna can reflect off the chamber floor, leading to variations in the voltage measured at the monopole which are frequency dependent due to interference between the direct and reflected signal.

When using a plane wave for calibration, the situation is simpler. Due to the image theory, incorporating an electric boundary at the bottom of the computational domain and setting the plane wave's electric field vector perpendicular to the boundary with an amplitude of $1\,\frac{\mathrm{V}}{\mathrm{m}}$ result in a field strength of $2\,\frac{\mathrm{V}}{\mathrm{m}}$ in the computational domain. To achieve a field strength of $1\,\frac{\mathrm{V}}{\mathrm{m}}$ in the computational domain, the plane wave amplitude can simply be set to $0.5\,\frac{\mathrm{V}}{\mathrm{m}}$.

Given the frequency band in the kHz and low MHz range, including the monopole antenna at a distance of 1 m does not significantly increase the electrical size of the simulation model containing the DUT. The overall model remains electrically small and can be solved efficiently using the FEM method. Additionally, the monopole's geometry is simple, resulting in only a moderate increase in the number of mesh elements due to the antenna geometry. Consequently, we recommend including the monopole antenna in such simulations. If the antenna factor is not estimated in advance, a value of $2\,\frac{1}{\mathrm{m}}$ can be used across the entire frequency band without a significant loss of accuracy.

12.2 Biconical Log-Periodic Antenna

Domains: 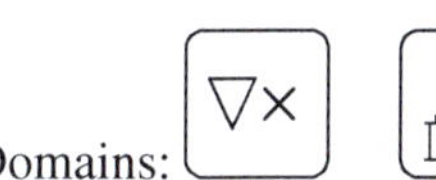

In this section, we simulate the CISPR 25 radiated emission setup in the frequency range from 30 MHz to 1 GHz. The CISPR 25 defines the absorber-lined shielded enclosure (ALSE) method for the measurement of the radiated emissions from a DUT. The measurement setup in a semi-anechoic chamber is illustrated in Fig. 12.4. For the measurement, the DUT is connected to the LISN via a 1.5 m long cable. The power supply to the LISN is located outside the measurement chamber. The DUT and the LISN are placed on a conductive table positioned 1 m above the chamber floor. The LISN is electrically connected to the table, while the electrical connection between the DUT and the table depends on the specific configuration of the DUT. The measurement is performed for the horizontal and vertical field component by rotating the receiving antenna relative to the table.

In the following, we discuss the properties of the measurement antenna and compare the received signal to the field obtained using field probes. By using field probes, the simulation time can be significantly reduced. Understanding the differences between simulation models that include the antenna and those that use field probes allows simulation engineers to select the most suitable approach for the current stage of the design process.

In the filter design chapter (Chap. 9), we observe that during the design process, a large number of simulations are required to evaluate how design modifications affect emissions. At this stage, simulations are compared relative to one another, and absolute comparisons are not necessary. Therefore, it is advantageous to use fast simulation models, which can be achieved by employing field probes instead of the receiving antenna. Only in the final stage of the design process the antenna model can be included in the simulation, ensuring results that are closer to actual measurements.

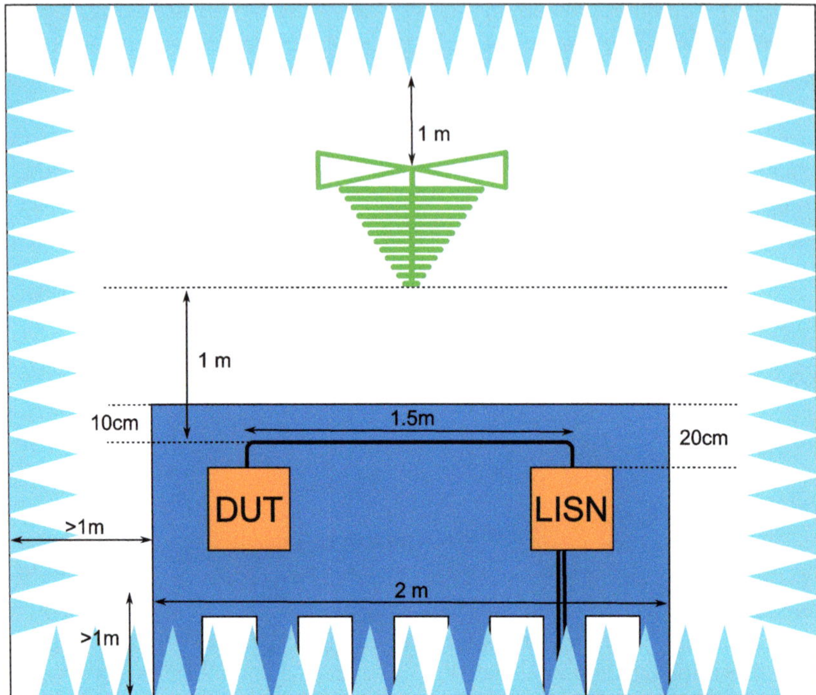

Fig. 12.4 Radiated emission measurement using a biconical log-periodic antenna in the frequency band from 30 MHz to 1 GHz according to CISPR 25

12.2.1 Simulation Model of the Radiated Emission Measurement Setup

The setup simulated in this section is illustrated in Fig. 12.5. The DUT consists of the same pair of simple PCBs used in Chap. 6 (cf. Fig. 6.1). However, in this instance, the cable connecting the PCBs is 1.5 m long. Different from the CISPR 25 standard, the 20 cm long sections of the wire that are perpendicular to the long axis of the table are not included. The dimensions of the metallic table are as follows: length, 2 m; width, 1 m; and height, 1 m.

Five metallic connections to the floor are positioned at the back side of the table. The floor is modeled as an electric boundary, making it perfectly conductive. The DUT and the cable are placed at a distance of 10 cm from the edge of the table. The DUT is excited on one side with a 1 V broadband source operating in the frequency range of 30 MHz to 1 GHz. The signal trace on the PCB at the opposite side is terminated by a 50 Ω lumped element to minimize the port count.

At a distance of 1 m from the DUT, a receiving antenna is positioned. The details of the antenna model are discussed in the next section.

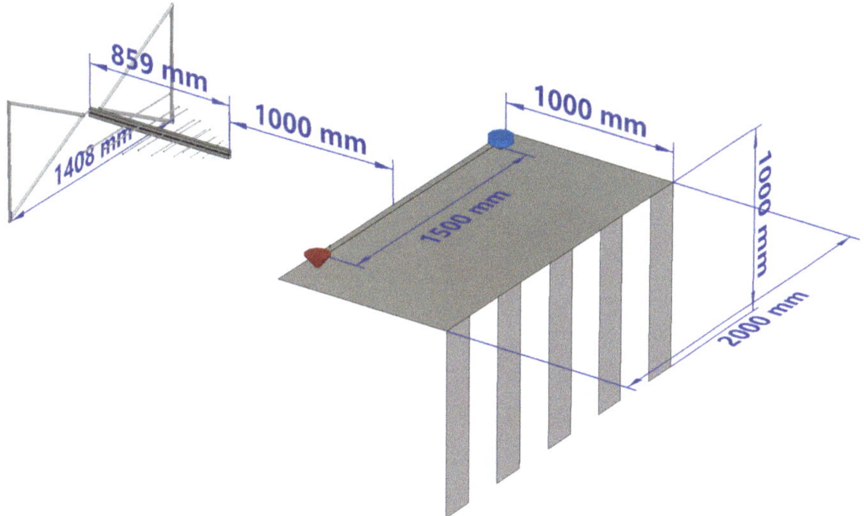

Fig. 12.5 CISPR 25-inspired radiated emission simulation setup for the frequency band from 30 MHz to 1 GHz

12.2.2 Antenna Model

As already mentioned, we do not use a vendor antenna model and perform the comparison using a generic antenna. Similar to Sect. 6.6, we have designed an antenna using the Antenna Magus software of CST. The design is a hybrid log-periodic antenna with a biconical section at the far end of the antenna. The antenna is designed to function from 30 MHz to 1 GHz. It is 85.8 cm long, and the biconical section is 140.8 cm wide. The log-periodic section consists of 16 elements on each side. The antenna port is located at the front of the antenna and is set to 50 Ω.

In the first simulation, the antenna is simulated separately from the DUT, and the S_{11} of the antenna is calculated and compared among the three different full-wave solvers: FIT, FEM, and TLM. In this simulation, the "open (add space)" boundary condition is used with $\frac{\lambda}{4}$ set to 30 MHz. The S-parameter of the antenna is plotted in Fig. 12.6. The solver comparison serves as a benchmark for the different full-wave solvers.

The S-parameter of the antenna calculated using the different full-wave methods agrees within approximately 3 dB. For the time-domain solvers, the default mesh with a lines-per-wavelength setting of 15 is used, with a manual mesh step of 0.5 mm applied only in the section between the antenna booms. The energy criterion is set to −50 dB. Both the TLM and the FIT solvers require less than 16 minutes for the simulation (using a GPU). In the FEM model, an adaptive mesh refinement is used, which converges after three iteration steps to an accuracy of 0.5%. Due to the wide bandwidth of the antenna, 53 frequency samples are required, resulting

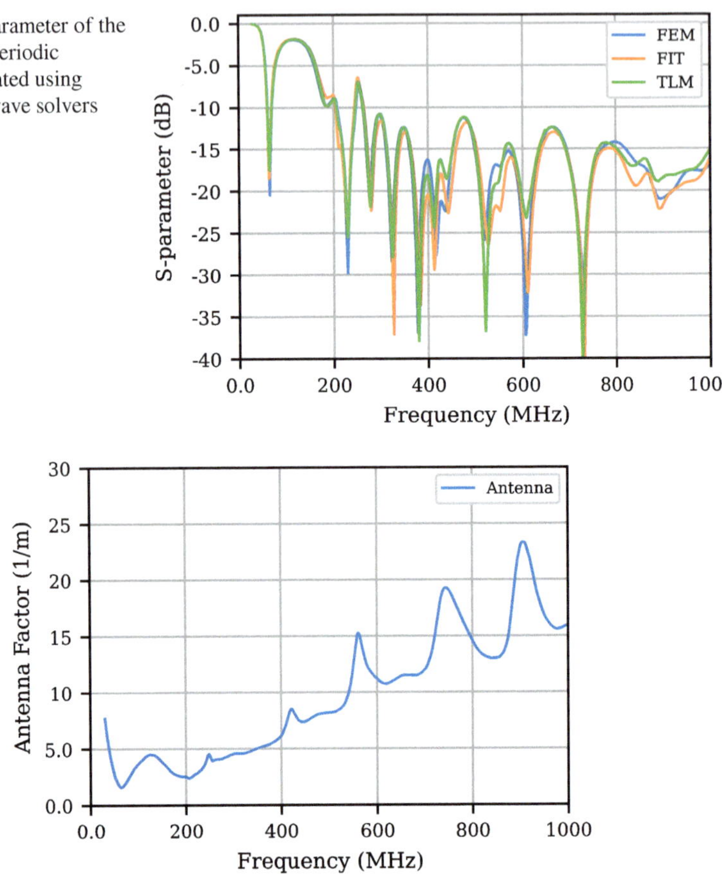

Fig. 12.6 S-parameter of the biconical log-periodic antenna calculated using different full-wave solvers

Fig. 12.7 Antenna factor of the biconical log-periodic antenna calculation using the FIT

in a simulation time of 9 hours and 51 minutes. Thus, the time-domain solvers are approximately 35 times faster than the FEM solver for this antenna.

For the calculation of the antenna factor required in the radiated emission setup, we illuminate the antenna with a linearly polarized plane wave with an amplitude of $1\,\frac{\mathrm{V}}{\mathrm{m}}$. This approach is identical to the one used for the monopole. The simulation with the plane wave requires approximately the same simulation time as the calculation of the S-parameters. The magnitude of the antenna factor calculated using the FIT solver is plotted in Fig. 12.7. This antenna factor is used to calculate the field values in the following sections.

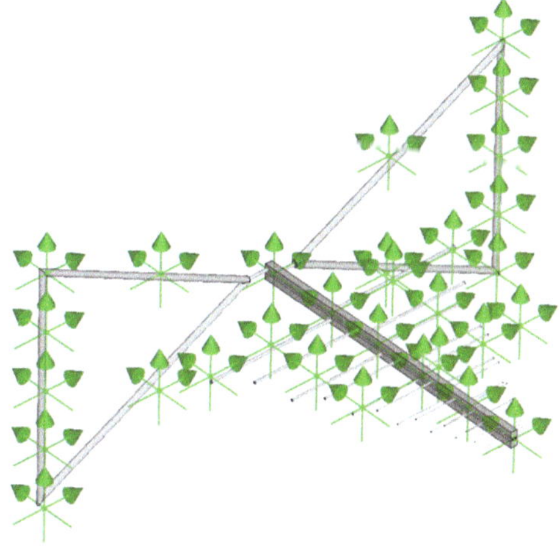

Fig. 12.8 Probe placement over the volume of the antenna

12.2.3 Comparison of Antenna and Probes

In order to compare the electric field calculated using probes to results obtained from the simulation model with the antenna, probes are placed over the volume of the antenna, as indicated in Fig. 12.8. This placement is performed in the full simulation model, as depicted in Fig. 12.5. In Fig. 12.8, we zoom in on the antenna to better illustrate the probe placement.

In the model containing the probes, the antenna is not included, as the antenna would disturb the emitted field from the DUT. However, the dimensions of the computational domain are kept constant in both the model with the antenna and the model with the probes. The distance from the table to the walls and ceiling is set to $\frac{\lambda}{4}$ at 100 MHz, which equals 75 cm. PML open boundaries are used. This results in a computational domain of 3.5 m × 4.3 m × 2.84 m, or, in terms of wavelength at 1 GHz, $11.6\,\lambda \times 14.3\,\lambda \times 9.47\,\lambda$. This corresponds to a volume of $1570\,\lambda^3$, which is electrically large.

We perform a study on the electrical size of the computational domain in Sect. 4.10 and compare the performance of the FIT and FEM solvers (Table 4.6). From this study, we conclude that the FEM solver, even in the model without the antenna, leads to higher simulation times compared to the time-domain solvers.

The simulation times of the FIT solver compared to the TLM solver are difficult to judge a priori. Therefore, we run the simulation model in both solvers and compare the run times. For this comparison, a total signal duration of 150 ns is selected. This is longer than the 100 ns used in Sect. 6.2 for a similar example (cf. Fig. 6.6), due to the increased length of the cable connecting the PCBs in the example in this section.

Even though the TLM solver reduces the mesh count by 80%, the FIT solver is still faster: it requires 1 hour and 34 minutes, while the TLM solver requires 3 hours and 50 minutes. These values are valid for the simple PCB geometry in this example.

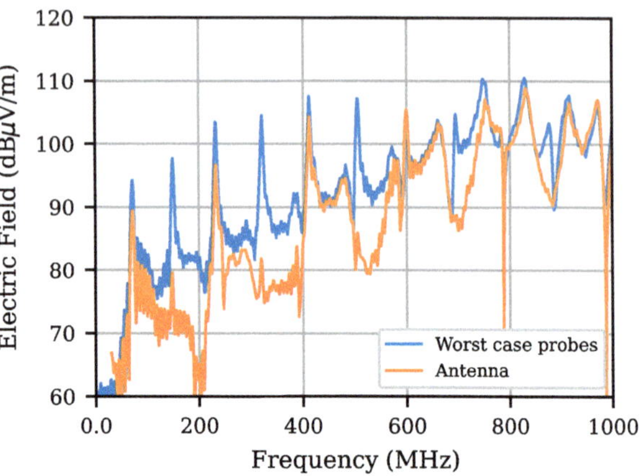

Fig. 12.9 Comparison of the horizontal electric field at the antenna and the worst-case probe results

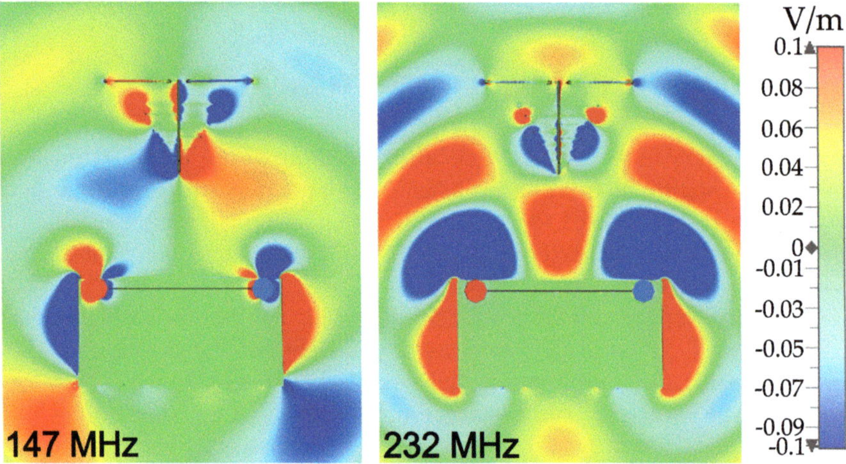

Fig. 12.10 Instantaneous horizontal component of the electric field in the computational domain for two different frequencies. Left: 147 MHz, no field detected by the antenna. Right: 232 MHz, field detected by the antenna

For a realistic PCB layout, the simulation time shifts in favor of the TLM solver, as a complex PCB requires a finer local mesh. In the following discussion we use the FIT solver.

In Sect. 6.5, we introduce the idea of calculating an envelope of a number of probes to obtain the worst-case emission results. If we apply this method to the probes placed over the antenna volume, we obtain the results plotted in Fig. 12.9. While the worst-case probe results exhibit peaks in the electric field every 75 MHz, the antenna only captures every second peak. We can understand this effect by analyzing field monitors at the peak frequencies.

We plot the horizontal component of the electric field at 147 MHz and 232 MHz in Fig. 12.10. The plane in the plot is aligned with the conductive table. We observe

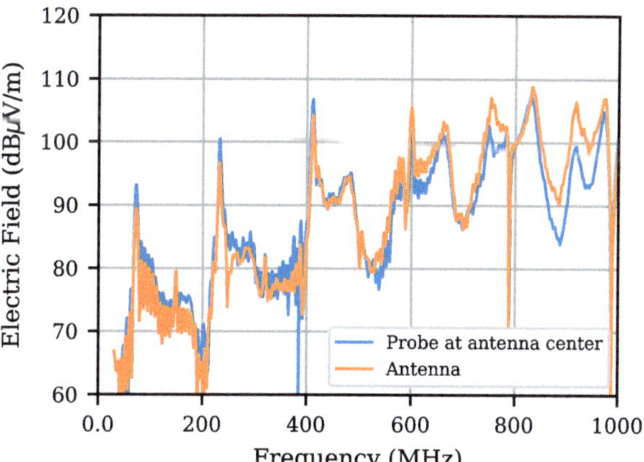

Fig. 12.11 Comparison of the horizontal electric field at the antenna and at a probe located at the center of the antenna

different field patterns emitted from the PCB pair for the different frequencies. At 147 MHz, the antenna does not detect the electric field because the electric field distribution is symmetric to the axis of the antenna, with a field zero in the center of the antenna. At 232 MHz, the field is also symmetric; however, the maximum of the field is located at the center of the antenna, allowing the antenna to detect the field. When we use the worst-case envelope post-processing of the probe results, we also measure the field away from the antenna axis, which leads to the detection of all peaks.

We confirm this by comparing the antenna results to a probe located at the center of the antenna, as depicted in Fig. 12.11. For this comparison, we select a probe located at the center of the antenna, at a distance of 40 cm from the front of the antenna. This is the part of the antenna closest to the DUT. The probe located at the symmetry axis can only record the same peaks as the antenna, because at every second peak the field at the probe position cancels out. By using only one probe in the center of the antenna, the same peaks are detected by the antenna and the probe.

In Fig. 12.11, we can also observe the influence of the antenna length: in the lower frequency band, the field at the probe is higher than the field recorded by the antenna, whereas in the higher frequency band, the opposite is true. In the log-periodic antenna, the higher frequency range is detected closer to the DUT, as this is where the short dipoles are located. Conversely, the field in the lower frequency band is detected further from the DUT, as the long dipoles are located at the back of the antenna. As we are comparing to a probe located at a fixed distance from the DUT, this frequency-dependent discrepancy between the field probe and the antenna results is observed.

The canceling out of the fields at the antenna in this example is very pronounced due to the perfect symmetry of the setup. The DUT is completely symmetric, as the same PCB is used on both sides. The center of the antenna is perfectly aligned with the center of the table.

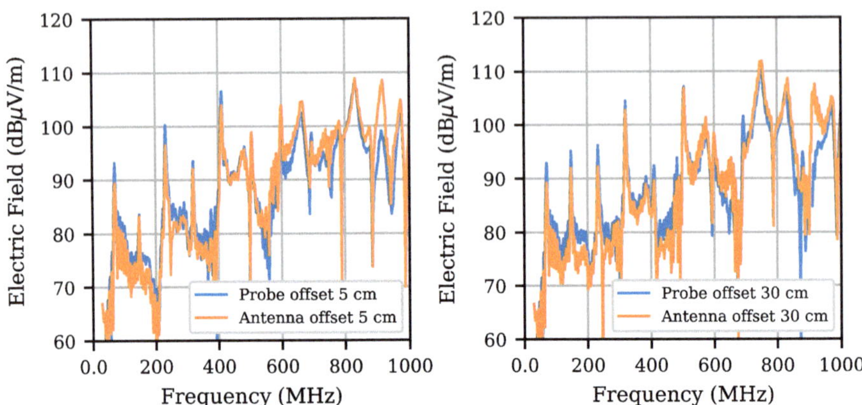

Fig. 12.12 Comparison of the horizontal electric field at the antenna and a probe located at the center of the antenna. Left: offset of 5 cm. Right: offset of 30 cm

In a real measurement, some misalignment is expected. For this reason, we have simulated the setup with the antenna and the center probe moved horizontally by 5 cm and 30 cm off the symmetry axis.

The results of the simulation with the antenna and the probe with a horizontal offset from the symmetry axis are plotted in Fig. 12.12. We observe that moving the antenna further away from the symmetry axis leads to an increase in the peaks that are not detected when the antenna is placed on the symmetry axis. The comparison of the probe and the antenna still matches within 3 dB; however, around 900 MHz, the deviation is larger. This larger deviation is due to the distance of the probe from the front of the antenna. For a probe located at the very front of the antenna, the difference reduces to 4 dB. This probe result is not plotted to reduce clutter in the plots.

Up to this point, we have only discussed the results for the horizontal polarization. For the simulation of the vertical polarization, we rotate the antenna around its long center axis by 90° and re-run the simulation. The results comparing the vertical antenna alignment with the probe results are plotted in Fig. 12.13.

For the plot of the vertical electric field component in Fig. 12.13, we maintain the same y-axis scaling as in the plot of the horizontal field. We observe that the vertical component is approximately 15 dB lower than the horizontal component of the electric field. Nevertheless, it is still worth discussing why the match between the probe and the antenna results for the vertical component is not as good as it is for the horizontal component.

We observe that the small peaks at 232 MHz and 410 MHz are recorded by the antenna but are not visible in the probe results. Similar to the horizontal polarization, we plot the vertical component of the electric field using field monitors in Fig. 12.14.

At 147 MHz, the antenna and the field probe record a peak in the vertical component of the field. This peak is generated by the lobe produced by the DUT, as observed in the top section of Fig. 12.14. In the bottom section of Fig. 12.14, we plot

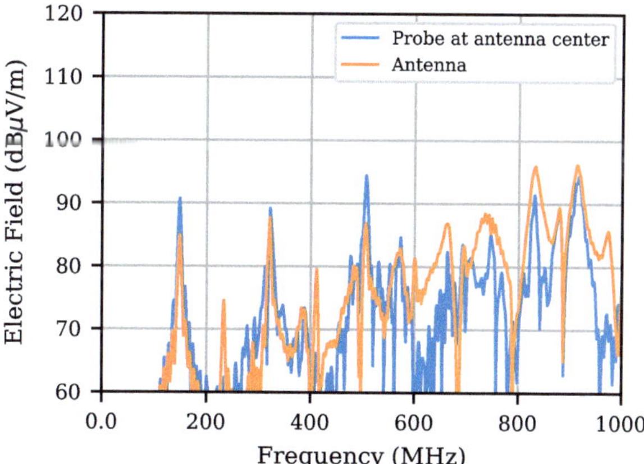

Fig. 12.13 Comparison of the vertical electric field at the antenna and a probe located at the center of the antenna

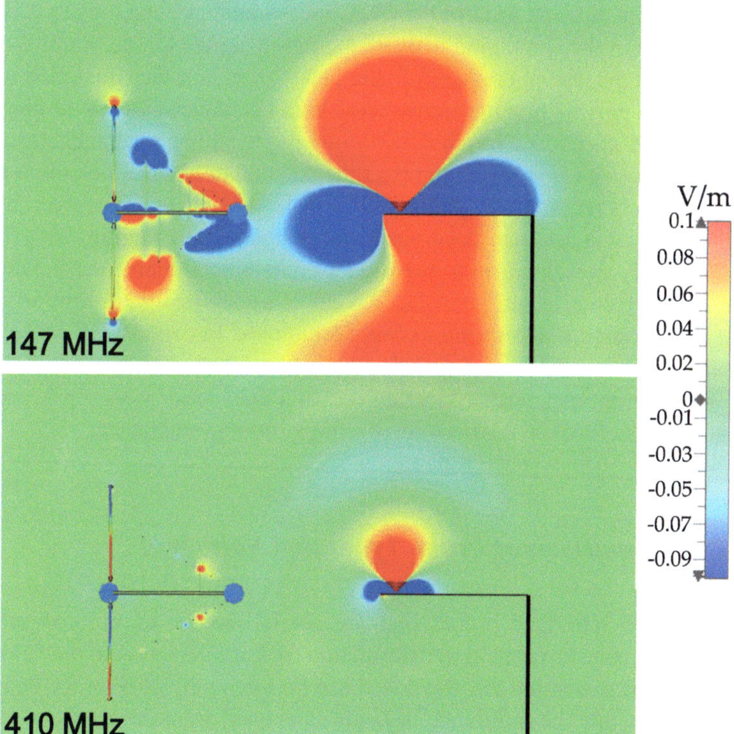

Fig. 12.14 Vertical component of the electric field in the computational domain for two different frequencies. Left: 147 MHz, antenna and probe both record a peak. Right: 410 MHz, only the antenna records a small peak

the field at 410 MHz, which is different from the frequency used in the horizontal field monitor plots. The peak at 232 MHz is so small that we would need to adapt the color ramp for visibility; otherwise, the field distribution would be almost invisible. This is why we chose 410 MHz instead.

For the vertical polarization, the antenna records a peak at 410 MHz, while the field probe does not. Although it is difficult to discern in the figure, this difference is due to the non-symmetry in the vertical polarization of the field caused by the chamber floor. The chamber floor reflects the field as it is set to PEC. The large size of the antenna means that the antenna dipoles are closer to the floor and detect a difference in the field that is not captured by the probe located exactly in the center of the antenna. Even though the difference is visible in Fig. 12.13, the horizontal component of the field at 410 MHz is 22 dB higher than the vertical field component. From the plots of the vertical field component, we also conclude that the highest field values are actually located above the DUT, where they are not measured at all.

We conclude that most of the differences observed are due to physical effects and limitations of the antenna used in the EMC measurements. The biggest advantage of using probes in the simulation is the reduction in simulation duration. While the simulation with the antenna requires 1 hour and 34 minutes, the simulation with the probes requires only 37 minutes, as the mesh count is reduced due to the absence of the antenna. This difference in simulation times might still be acceptable for a single simulation. However, if we wish to offset the antenna or change the polarization, a new simulation is required. By using probes instead of the antenna, we can obtain both polarizations at arbitrary positions from a single simulation run.

For the study presented here, the model containing the antenna is calculated four times (horizontal center, horizontal 5 cm offset, horizontal 30 cm offset, and vertical), requiring a total of 6 hours and 16 minutes. In contrast, all probe results are obtained from a single simulation that takes only 37 minutes. For a realistic PCB layout, simulation times further increase due to a finer mesh at the DUT and more ports, making the time-saving benefits of using probes even greater.

12.2.4 Comparison of Probe Types and Solvers

In Sect. 6.5, we introduce probes that are placed outside of the computational domain; thus, a near-to-far-field transformation for the calculation of the field values is applied. We demonstrate that the difference between individual probe results can be significant, while the difference in fields calculated from the envelope is small. These conclusions also hold for the model in this section, where we use a single probe to detect the field, in order to mimic the measurement of an antenna. A comparison of the individual probe results using probes inside and outside of the computational domain is plotted in Fig. 12.15.

Particularly below 100 MHz, the probe outside of the computational domain deviates from the probe inside of the computational domain. The criterion for

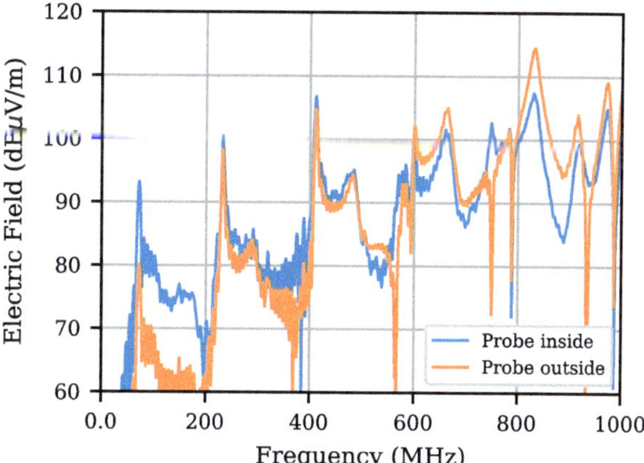

Fig. 12.15 Comparison of the horizontal electric field calculated by a probe inside a larger and a probe outside of a smaller computational domain

the distance used at open (add space) boundaries of $\frac{\lambda}{4}$@100 MHz in the model with probes outside of the computational domain is not sufficient for these low frequencies, and we observe the deviation.

For the radiated emission setup with a log-periodic antenna at a distance of 1 m from the measurement table, we recommend using probes inside the computational domain, as the usage of probes outside of the computational domain requires a boundary distance that is similar to the actual distance of the antenna. With these distances, there is no benefit from using probes outside of the computational domain.

Up to this point, we have only used the FIT solver in a model that requires excitation of only a single port. For the simulation of power electronics systems, the FEM solver is a better choice as it can solve many ports simultaneously. Therefore, it is desirable to perform the simulation of this setup using the FEM solver. At the beginning of this section, we note that the simulation domain is electrically large, which poses a challenge for the FEM solver.

In order to run the simulation model using the FEM solver, we slightly reduce the size of the computational domain. The open boundary behind the probes is moved by 75 cm and is now at a distance of 2 m from the table. The other boundaries remain unchanged. With this setup, we simulate the model using the FEM solver and compare the results to the FIT solver in Fig. 12.16. We observe that the overall agreement in the location of the peaks is good, but some deviations are present in the higher frequency range close to 800 MHz. This deviation can be attributed to the mesh density, which is kept as coarse as possible to ensure the FEM model runs within a reasonable simulation time.

In the low-frequency band, for the peak around 75 MHz, which is zoomed in on the right-hand side of the figure, we observe the effect of the signal duration in the FIT simulation. For a signal duration of 150 ns, which we have used throughout this section, the low-frequency signal has not yet fully decayed. Increasing the signal duration to 450 ns yields much better agreement but increases the simulation time of the FIT model by a factor of 3, from 37 minutes to 1 hour and 51 minutes. This finding aligns with the results presented in Fig. 6.6.

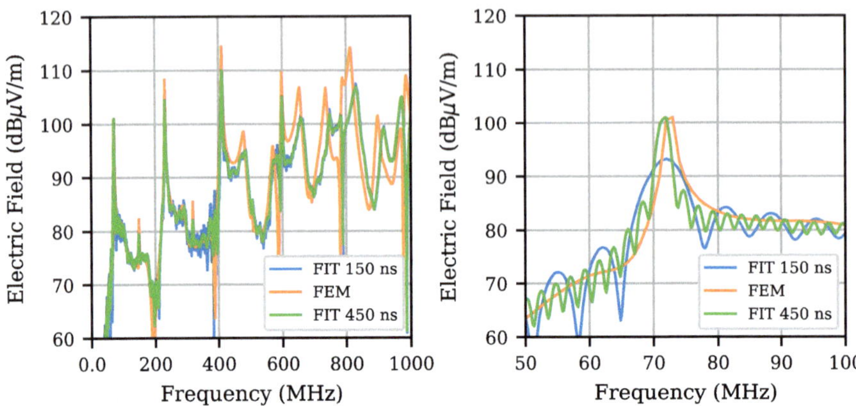

Fig. 12.16 Horizontal electric field at a probe calculated using FIT and FEM. Two different signal durations are used in the FIT simulation. Left: complete frequency band. Right: zoom in on the first peak

Table 12.1 Simulation times of the FEM solver with different maximal frequencies

Maximal frequency	250 MHz	500 MHz	750 MHz	1000 MHz
Simulation time	5 m 0 s	17 m 17 s	3 h 32 m 09 s	18 h 01 m 06 s
Mesh elements	236 k	307 k	525 k	850 k
Memory requirement (GB)	4.6	9.9	33.6	77.7
Frequency samples	26	38	65	88

When working on the model and iterating the design, a shorter signal duration can be used. Once the design is finalized and higher accuracy is required, the signal duration can be increased.

For the FEM simulation, it is necessary to reduce the computational domain due to the memory requirements of the direct solver. In the following, we discuss the simulation times of the FEM solver for the (slightly) reduced computational domain with different maximal frequencies. The simulation times are summarized in Table 12.1.

For the comparison to the FIT solver shown in Fig. 12.16, the simulation using the FEM solver requires 18 hours, whereas the FIT simulation requires only 27 minutes. This significant difference arises from the varying performance of the FIT and FEM solvers for an electrically large computational domain. For the FEM, at 1 GHz, the electrical size of the computational domain substantially increases the required number of mesh elements and frequency samples needed to resolve the electric field at the probes.

Reducing the maximal frequency has a significant impact on the simulation time of the FEM solver. Using a maximal frequency of 500 MHz reduces the simulation time to just 17 minutes. It might seem surprising that halving the maximal frequency results in a reduction in simulation time by a factor of 60, but this is explained by the model properties. The mesh must resolve the volume of the computational domain, and reducing the maximal frequency by a factor of 2 decreases the electrical volume by a factor of 8, as this dependence is cubic. This reduction, combined with fewer required frequency samples and shorter solving times per sample, leads to the substantial difference in simulation times.

For the simulation of power electronics systems, a high maximal frequency of 1 GHz is often unnecessary. The frequencies of PWM signals used in power electronics are typically on the order of a few MHz, and their spectral content near 1 GHz is minimal (Fig. 2.5). For such simulations, reducing the maximal frequency in FEM models significantly decreases simulation time. We demonstrate this in the next section, where we calculate the radiated emission from the inverter analyzed for conducted emissions in the previous chapter. For simulations of digital electronics with higher signal frequencies and a small number of ports, time-domain solvers generally perform better.

In the following two sections, we present two application examples: radiated emission from the inverter studied in Chap. 10 and radiated emission from the USB channel discussed in Chap. 11. We use probes to calculate the emitted electric field. The field at the probes is calculated using combined results, and we do not apply the EMI receiver post-processing to these results.

When using combined results, the spectrum at the probe is calculated through a linear superposition in the frequency domain. Thus, no time-domain signals are calculated at the probes, which are required by the EMI receiver. As an approximation of the detector's influence, we could use the time-domain results at the LISN, calculate a scaling factor from the peak detector to the average detector, and apply this scaling factor to the field probe result, as offered by the software. However, this process introduces additional steps to the simulation workflow without providing further insight into the results.

Therefore, all radiated emission field values are presented as peak values with the RBW of the Fourier transform performed after the transient circuit simulation. We do not consider this a significant limitation: all the simulations we perform are excited by deterministic signals, making the average values directly proportional to the peak values.

For the conducted emission results in the next section, the EMI receiver is used in the same way as in Chap. 10.

12.3 Application Example: Radiated Emission from an Inverter

- Inverter switching modeled by SPICE
- Field probe inside the computational domain
- Full-wave, S-parameters, combine results
- FEM method

Transient

0–500 MHz

In this section, we discuss the radiated emission simulation of an inverter. It is the same device that we study in terms of conducted emission in Chap. 10. An overview of the radiated emission measurement setup is provided at the beginning of the previous Sect. 12.2. The measured horizontal and vertical polarization of the electric field is compared to the limits imposed by the CISPR 25 standard. The standard defines five classes, with class 5 being the most stringent. The limits are summarized in Table 12.2.

In this example, we simulate only the frequency band covered by a measurement with a biconical log-periodic antenna. A simulation above 1 GHz, which would require the use of a different antenna, is not performed. As we see in the following, the emission from the inverter is not significant at frequencies above 1 GHz.

In the first study, we modify the connection between the housing of the inverter and the housing of the motor to analyze the interplay between radiated and conducted emissions. In the second study, we simulate the effect of a snap-on ferrite on the emission levels.

12.3.1 Simulation Model

The simulation model is depicted in Fig. 12.17. For the inverter, we use the final model introduced at the end of Chap. 10. It includes the 3D model of the common mode choke and the inverter housing. The modifications required to run a radiated emission simulation based on the conducted emission model are minor: we place the inverter model on the table used in the previous radiated emission studies and

Table 12.2 Limits in dBμV/m for different service/band classes and frequency ranges for the CISPR 25 ALSE radiated emission test (Adamczyk 2017)

Service/band	Frequency (MHz)	Class 1	Class 2	Class 3	Class 4	Class 5
LW	0.15–0.30	86	76	66	56	46
MW	0.53–1.8	72	64	58	52	46
SW	5.9–6.2	62	56	50	44	38
FM	76–108	62	56	50	44	38
TV band I	41–88	52	46	40	34	28
TV band III	174–230	56	50	44	38	32
DAB III	171–245	50	44	38	34	28
TV band IV	468–944	63	57	51	47	41
DTTV	470–770	69	63	57	51	45
DAB L band	1447–1494	52	46	40	34	28
SDARS	2320–2345	58	52	46	40	34

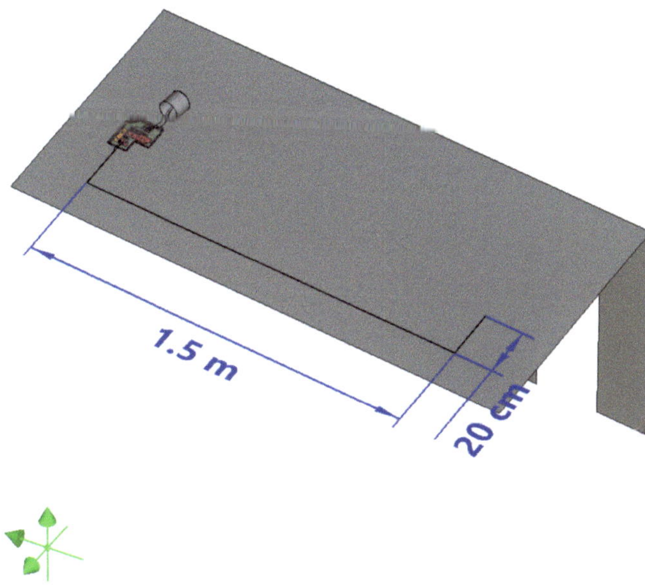

Fig. 12.17 Overview of the radiated emission simulation model of the inverter

connect it to a 1.7 m long power cable. The ports at the end of the power cable, to which the DC source and the LISN are connected, are moved to the end of the long cable.

A probe is placed at a distance of 1.4 m from the table, which is the same distance used in the comparison with the biconical log-periodic antenna in the previous section. We calculate the emitted electric field value at this probe using transient combined results.

The transition from the conducted emission simulation model to the radiated emission model is straightforward due to the use of ports and S-parameters in the conducted emission simulation. If RLC models had been used for the conducted emission simulation, creating a radiated emission simulation model would have required significantly more effort. This would involve redefining all sources in the 3D model (from RLC nodes to ports) and redefining the schematic (from an RLC node schematic to a port schematic).

Before proceeding to the RE simulation, we examine the impact of cable length on the CE results. Figure 12.18 shows the voltage at the LISN for two different cable lengths: 20 cm, as used in the CE simulation, and 1.9 m, required for the RE simulation. The plot illustrates the emission detected using EMI receiver post-processing at the LISN connected to the positive supply line, similar to what is shown in Fig. 10.13.

Changing the cable length significantly affects the conducted emission. In the frequency range from 5 MHz to approximately 50 MHz, the longer cable results in reduced emissions. However, above 50 MHz, the CE for the longer cable increases

Fig. 12.18 Conducted emission from the inverter for different cable lengths. The limit lines indicate the CISPR 25 class 3 conducted emission limits for the AVG detector

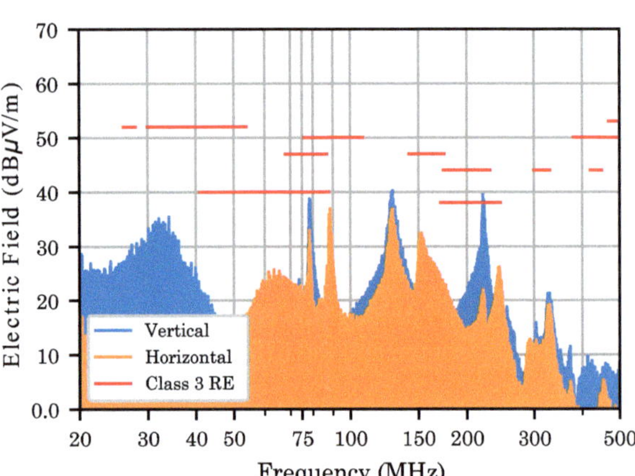

Fig. 12.19 Radiated emission from the inverter with 1.7 m cable length. The limit lines indicate the CISPR 25 ALSE class 3 emission limits. DFT results without using a detector

by about 10 dB, exceeding the CISPR 25 class 3 limit used in the conducted emission chapter.

While the CE simulation results for the 1.9 m cable are not directly relevant for comparison to CE emission limits, the analysis underscores that resonance frequencies differ between the CE and RE spectra. The longer cable introduces additional resonances, which amplify the emission relative to a shorter cable. When these cable resonances are excited by the inverter's emission spectrum, an increased radiated emission level can be expected.

The results of the RE simulation are plotted in Fig. 12.19. The radiated electric field exhibits resonant behavior between 20 MHz and 300 MHz. It decreases above 300 MHz and falls below 10 dB$\frac{\mu V}{m}$ at 500 MHz, due to the low PWM frequency and slow rise/fall time. Consequently, the simulation is run only up to 500 MHz instead of 1 GHz, as defined for this antenna. Reducing the frequency band significantly decreases the required simulation time, as discussed at the end of Sect. 12.2.3. The EM simulation of the RE model for the inverter requires 21 minutes. The circuit simulation, which includes inverter switching, is run for a signal duration of 3 ms, as in the chapter on CE. It requires 50 minutes, which is longer than in the conducted emission chapter, because the higher maximum frequency of 500 MHz requires a smaller time step in the circuit simulation. Additionally, the model generated by vector fitting requires more poles to fit the broader frequency range, resulting in a circuit with an increased number of unknowns.

The radiated electric field at the probe is compared to the CISPR 25 ALSE class 3 limit for the horizontal and vertical field components separately. In the DAB band, around 230 MHz, the vertical polarization exceeds the class 3 limit. In the TV I band, at 78 MHz and 88 MHz, the electric field is close to the limit. For most frequencies, the emission in the vertical polarization is larger than in the horizontal polarization. This behavior differs significantly from the simple PCB example discussed earlier in this chapter, where the horizontal polarization is approximately 15 dB larger than the vertical polarization (Figs. 12.11 and 12.13).

In the simple PCB case, the primary source of emission in the horizontal polarization is the cable. However, in the inverter example, where the vertical polarization is dominant, this suggests that the source of the emission is not the power cable but other parts of the system.

12.3.2 Impact of the Connection Between the Housings

To further understand the RE from the inverter and the ratio of horizontal to vertical polarizations, we plot the electric field at the peak frequency of 88 MHz in Fig. 12.20. The plot aids in identifying the source of the emission and supports the hypothesis that the RE above 50 MHz is not primarily governed by the emission from the supply wires.

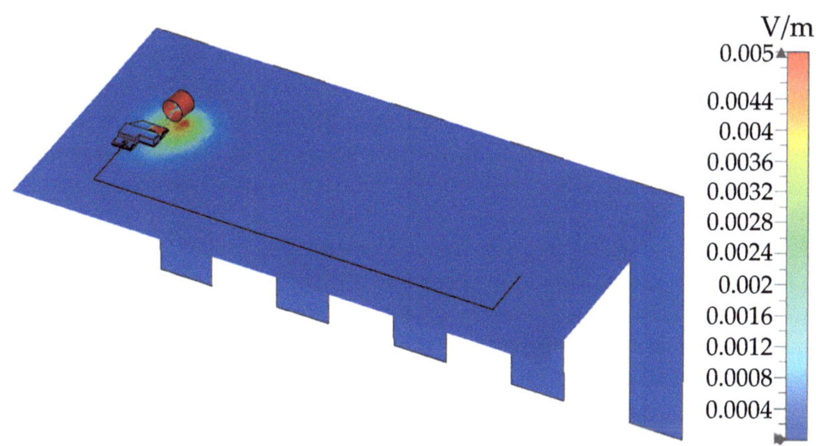

Fig. 12.20 Electric field in the simulation of radiated emission from the inverter at 88 MHz

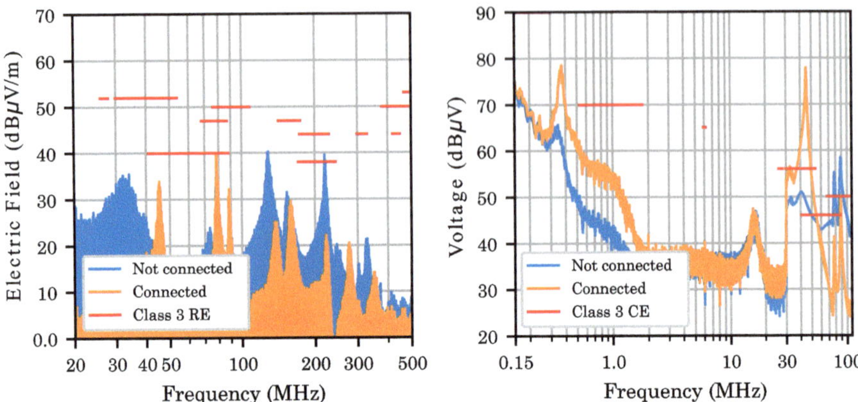

Fig. 12.21 Dependence of the conducted and radiated emission from the inverter on the connection between the PCB housing and the motor housing. Left: vertical polarization of the RE for a cable length of 1.7 m. Right: CE for a cable length of 20 cm

We observe that the field at 88 MHz is concentrated around the motor housing and near the inverter's PCB housing. In Chap. 10, where the CE of the inverter is simulated, the PCB housing and the motor housing are not electrically connected. Although a port exists to connect them (motor housing in Fig. 10.6), a 1 MΩ resistor is used in the schematic of the CE model, effectively isolating the two housings.

Since the motor housing is suspected to be a source of the RE, we test how connecting the PCB housing to the motor housing affects radiation levels. As the port is already included in the simulation model, it is not necessary to re-run the EM model. Instead, we simply replace the 1 MΩ resistor in the schematic with a 1 mΩ resistor and re-run the circuit simulation along with the combined results.

The effect of connecting the two housings is shown on the left-hand side of Fig. 12.21. Only the vertical polarization of the electric field is displayed. By connecting the two housings, the vertical polarization in the DAB band is reduced by almost 20 dB, and the emission no longer exceeds the limits. In most frequency bands, the emission is also reduced; however, at 88 MHz, the radiated field remains close to the limit. Thus, connecting the two housings appears to be a promising approach to reduce RE levels, but we also need to evaluate how this modification impacts the CE.

For this evaluation, we return to the conducted emission model and connect the two housings. A comparison of the peak CE voltage, $V_{\mathrm{LISN,P}}$, with and without the connection between the two housings is plotted on the right-hand side of Fig. 12.21. Connecting the housings improves the CE above 60 MHz, consistent with the improvement observed for RE. However, in the frequency range from 160 kHz to 2 MHz, the CE increases, though it remains below the class 3 limit.

At 35 MHz, a significant peak in the CE voltage emerges, approximately 20 dB higher than the class 3 CE limit. This peak is also visible in the RE results but remains below the RE limits. Consequently, the simple connection of the two housings is not a viable solution to achieve compliance with class 3 limits for both CE and RE.

12.3.3 Effect of a Snap-On Ferrite

The field plot in Fig. 12.20 suggests that the source of the RE is located between the PCB housing and the motor housing. We proceed to test the emission from the phase wires leading to the electric motor. A straightforward method to reduce common mode emissions from a cable bundle is to place a snap-on ferrite onto the bundle. In laboratory measurements, this is an effective and noninvasive approach to determine whether common mode currents on a cable bundle contribute to radiated emissions. We replicate this approach in the simulation model.

For this test, we require a model of the snap-on ferrite, which can be constructed using the principles outlined in Sect. 9.6.

The snap-on ferrite for which we build the simulation model is inspired by the Würth Elektronik WE 78271111 component (Wurth Electronic). According to the data sheet of this component, the impedance of the ferrite is measured for a 10 cm long wire. We construct a simple hollow cylinder to model the ferrite material and place a 10 cm long wire through the cylinder. The ends of the wire are connected to electric boundaries, which create the return current path in the simulation. A port is inserted into the wire, as shown on the left-hand side of Fig. 12.22.

A Debye model (cf. Sect. 9.6) is used for the ferrite material, and μ_s and τ_r are swept to match the impedance curve to the data sheet. Using $\mu_s = 600$ and $\tau_r = 2.8 \cdot 10^{-8}$ s, we obtain the impedance plotted on the right-hand side of Fig. 12.22. The simulated curve aligns well with the minimal and typical values provided in the data sheet.

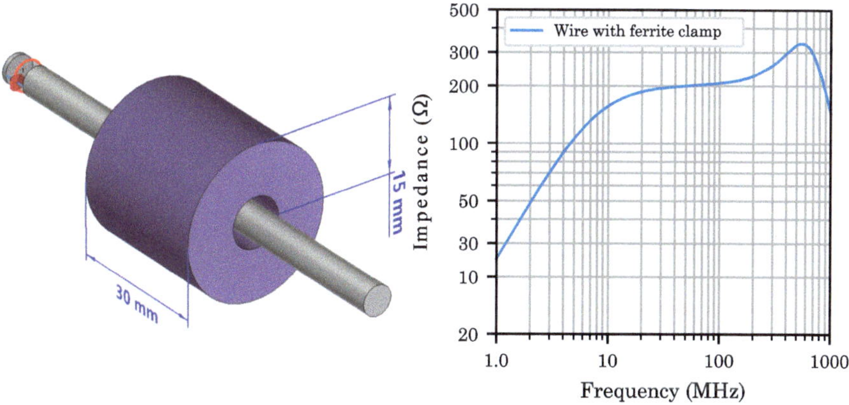

Fig. 12.22 Left: model of a snap-on ferrite used for the calibration of the ferrite material. Right: impedance of a 10 cm long wire with the snap-on ferrite

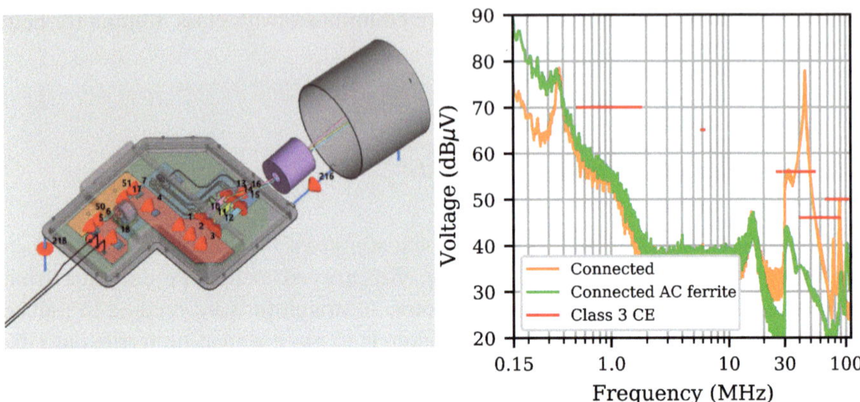

Fig. 12.23 Left: close-up on the inverter with a snap-on ferrite on the phase wires in the RE simulation model. Right: CE levels with and without the snap-on ferrite

In the frequency band from 30 MHz to 300 MHz, the impedance of the snap-on ferrite is approximately 200 Ω. Therefore, if the common mode on the phase wires is the primary source of the emission, we expect a noticeable reduction in emission levels.

We place the snap-on ferrite into the simulation models for CE and RE to evaluate its effect on both types of emissions. To do this, the phase wires must be remodeled. In the original model, the three wires connecting the inverter to the motor are loosely arranged, as shown in Fig. 10.5. In this configuration, it is not possible to place the snap-on ferrite around all three wires. We redefine the wires to form a bundle, allowing the ferrite to be placed around them. No additional modifications are made to the model.

The connection between the two housings is retained, and the CE is calculated for the model with a cable length of 20 cm. The peak CE voltage, $V_{\mathrm{LISN,P}}$, is plotted on the right-hand side of Fig. 12.23. With the ferrite, the CE level decreases by more than 40 dB above 20 MHz. Below 400 kHz, the emission level increases slightly but remains well below the limit of 90 dBμV. By adding the snap-on ferrite, the CE complies with the class 3 limits.

The effect of the snap-on ferrite can also be observed in the RE levels plotted in Fig. 12.24. By adding the ferrite, the emission is significantly reduced, with both polarizations being approximately 20 dB below the class 3 limit, even meeting the most stringent class 5 requirements. Adding a snap-on ferrite is a well-known and effective method for suppressing common mode emissions, though it comes with increased costs. Through simulation, we efficiently quantify whether the improvement justifies the additional expense.

In this section, we demonstrate how to simulate the radiated emission from a power electronics system, specifically an inverter. The simulation times for the model are well below 2 hours for the combined EM and circuit simulation.

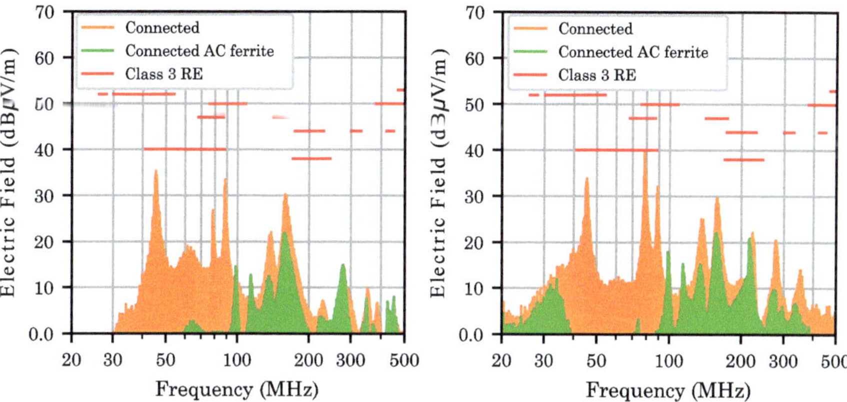

Fig. 12.24 RE from the inverter at a distance of 1 m compared to CISPR 25 class 3 limits. Left: horizontal polarization. Right: vertical polarization

Using field plots, we identify the source of the emission and test various model modifications to reduce emission levels.

In this book, we only present the effects of connecting the enclosure and adding a snap-on ferrite. However, during the development of this content, several other countermeasures, such as adding additional C_{y} capacitors or placing the snap-on ferrite on the supply wire, were tested without achieving the desired effect. We encourage simulation engineers to explore different scenarios when running EMC simulations. Modifications to an existing model are simple to implement, even those that are not feasible in hardware. By systematically modifying the model and analyzing the effect of these changes on emission levels, engineers can gain a deeper understanding of the physical effects in the DUT.

12.4 Turntable Measurements

Domains: 

Certain EMC limits, such as those defined in FCC part 15, are specified in terms of radiation limits for antenna distances of 3 m or even 10 m. Measurements are conducted in a semi-anechoic chamber. In the turntable-based measurement described by the FCC part 15 standard, the DUT is placed on a wooden table 80 cm above the chamber floor. For one turntable orientation, the antenna height is varied from 1 m to 4 m above the floor in 1 m steps. The set up is indicated in Fig. 12.25. The turntable is then rotated by 90°, and the measurement with the four antenna heights is repeated. At each frequency, the maximum value from every measurement is retained to obtain the final emission curve. The variation in antenna heights

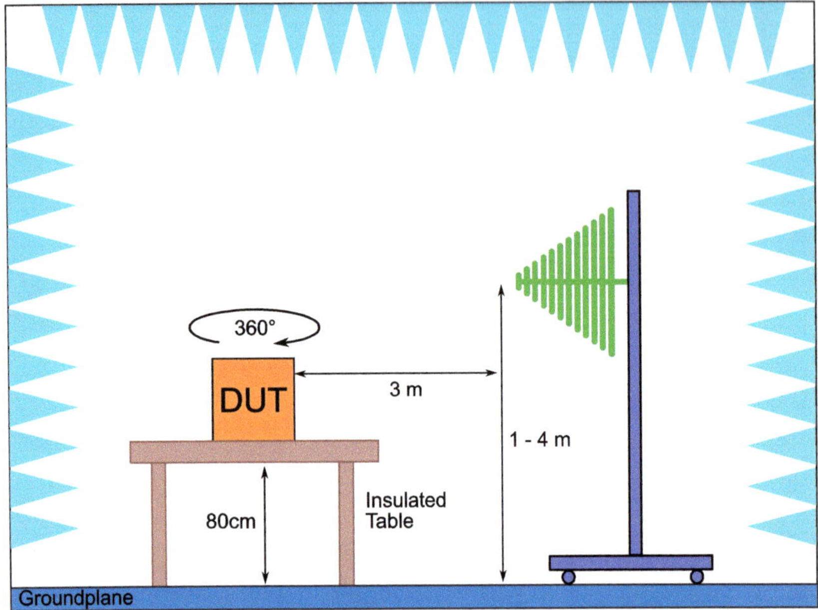

Fig. 12.25 Radiated emission measurement with the DUT placed on a nonconductive turntable

ensures that any destructive interference of the emission from the DUT caused by reflections on the chamber floor does not affect the results, as the interference does not occur at different antenna heights.

With volumetric full-wave methods used throughout this book, modeling a complete antenna at such distances is highly inefficient. Therefore, in the following, we do not compare antenna simulations to probes as we did in Sect. 12.2. Instead, we rely on probes placed outside of the computational domain to examine the RE. As discussed in Sect. 6.5, probes outside the computational domain yield results comparable to those inside the computational domain when the envelope over a set of probes is calculated.

The calculation of the envelope of the probes aligns with the methodology of turntable measurements. Both approaches aim to estimate the worst-case emission from the DUT. This is the method we employ in the following example.

12.5 Application Example: Radiated Emission from a USB channel

Digital signal from an IBIS model

Field probe outside of the computational domain

Full-wave, S-parameters, combine results

FEM method

Transient

0–5 GHz

For the demonstration of the simulation of a turntable measurement, we use the USB 3.0 channel from the SI study in Chap. 11 and augment it with a model of the connector and cable. The radiated emission is simulated up to a maximum frequency of 5 GHz.

In a turntable measurement, the chamber floor is located at a distance of 80 cm from the DUT. At a maximum frequency of 5 GHz, the electrical size of the model increases significantly compared to the simulations performed in the previous section. Consequently, the initial simulations for this example estimate the effect of the chamber floor on the emission levels. By quantifying the impact of the chamber floor, we derive simplifications to the model that substantially reduce simulation time.

Using these simplifications, we compare the emission from the board alone to the emission from the assembly of the board, connector, and cable. Additionally, we study the effect of components on the PCB and various cable designs on the radiated emission.

12.5.1 Simulation Model

In this section, we describe the simulation model of the USB 3.0 channel. The model setup of the PCB and the schematic are identical to those used in the SI chapter. A close-up of the PCB traces is shown in Fig. 11.2, and the schematic is depicted in Fig. 11.5. In the following simulations, we use the fully populated channel, including the DC blocking capacitors, the common mode choke, and the ESD protection diodes. The excitation signal is the same PRBS sequence driven by the IBIS model as used in the SI chapter.

A connector model is obtained as a CAD file from the PCB layout performed in the Altium CircuitMaker online software (Molex). The metallic parts of the connector are assigned PEC material, and the dielectric filling inside the connector is set to $\varepsilon_r = 3$. This value is determined by performing a TDR simulation (Schmidt 2002; Smail et al. 2010) of the connector and sweeping the permittivity value until a differential impedance of 90 Ω is achieved.

A 50 cm long cable is constructed using the “cable to 3D” functionality, as introduced in Sect. 8.5.3. The cable cross section is simplified compared to a realistic cable, including only a ground wire and one TWP.

The diameter of the ground wire is 0.6 mm, and the individual TWP wires are modeled with a diameter of 0.37 mm and a center-to-center distance of 0.76 mm.

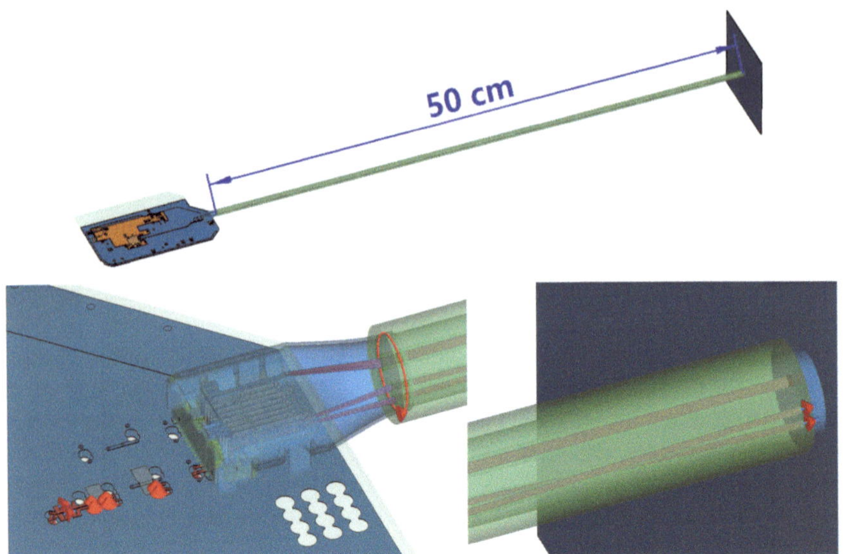

Fig. 12.26 Simulation model of radiated emission from a USB 3.0 channel

The TWP is not individually shielded; instead, a shield is placed around the entire cable. The shield is modeled using a thin panel material model with the conductivity of aluminum, $\sigma_{\text{Al}} = 3.3 \cdot 10^6 \frac{\text{S}}{\text{m}}$, and a thickness of 20 μm. An insulating material surrounds the cable shield. The diameter of the shield is 4.5 mm, and the thickness of the insulating material is 1.2 mm. An overview of the model is shown in Fig. 12.26.

The transition from the connector to the cable is modeled using the CAD functionality of CST. The connection between the connector pins and the connector shell is created using the CAD loft functionality, which allows for a smooth transition between two CAD faces. On the cable side, a short section of the cable screen is converted into a solid ring, enabling the connector shell to connect to the shield. A circular port is placed within the ring to modify the connection impedance between the shell and the screen. While modern USB cables often include a small PCB inside the connector, we omit this part as no model of such a board is available. Additionally, since a USB device must function with various USB cable assemblies, our connector-to-cable transition remains simple and generic.

Our focus is on the driver PCB and the PCB-connector-cable transition. The other end of the cable is modeled using a simple metallic plate as a reference. A second PCB is not used to terminate the cable. Instead, the reference consists of a square metallic plate with a side length of 8 cm. At the center of the plate, the shield is connected via a PEC ring to the plate. The ground wire is also connected to the plate using a short PEC section, while the two wires of the TWP terminate through ports to the plate.

This setup represents a trade-off between accuracy, available data, and computational effort. Below, we summarize the simplifications compared to a USB system in hardware:

- The transition from the inner pins of the connector to the wires in the cable is geometrically simplified and does not include a PCB inside the connector.
- The transition from the connector shell to the shield is simplified, omitting crimp contacts.
- The cable includes only one TWP, along with the VCC and GND wires. The inner shield of the TWP is introduced later during the study.
- The far-end cable termination uses a simplified configuration instead of a second PCB.

12.5.2 Effect of the Chamber Floor

In the first simulation, we study the effect of the chamber floor on the emission results. Given the maximum frequency of 5 GHz, the distance of 80 cm from the DUT to the chamber floor significantly increases the electrical size of the computational domain. By comparing a simulation model with and without the chamber floor, we estimate the effect of the chamber floor on the emission results. This estimation serves as an error measure in the subsequent simulations, enabling us to save significant simulation time.

The setup of the electric field probes in the simulation model for the turntable measurement is depicted on the left-hand side of Fig. 12.27. Probes are placed outside of the computational domain at a distance of 3 m from the center of the PCB model. In the azimuth direction, the probes are positioned every 30°. In the height direction, ten probes are located at heights ranging from 1 m to 4 m. The chamber floor is explicitly included in the EM model, as indicated by the small black

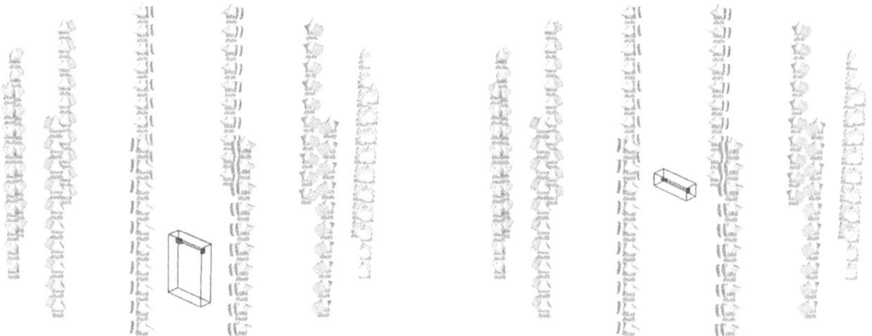

Fig. 12.27 Placement of electric field probes in the RE simulation model. Left: probes in a model with the chamber floor. Right: probes in a model without the chamber floor

box representing the bounding box of the simulation model. The lower boundary condition is set to electric in order to model the conductive chamber floor.

Explicitly including the chamber floor in the simulation leads to an increased computational effort. While techniques exist to incorporate a reference plane (such as the chamber floor) into the near-to-far field transformation without explicitly modeling it in the computational domain, the FEM solver in CST does not support this functionality. Therefore, the floor must be explicitly included in the EM model.

To analyze the effect of the chamber floor on emission results and computational time, we set up a second model where the floor is not included. In this setup, probes are placed from -4 m to $+4$ m in the height dimension. This configuration is shown on the right-hand side of Fig. 12.27. The probes are defined to calculate the horizontal and vertical components of the electric field, similar to a receiving antenna in a physical measurement. The placement and orientation of the probes are defined by a script provided with the software installation.

The simulation is run up to a maximum frequency of 5 GHz. The FEM solver is selected because it provides the fastest simulation time, even though the electrical size of the model is approximately $10\,\lambda$. The primary reason why the FEM solver outperforms the time-domain solver is the port count: there are 13 ports in the model, which are used to represent the driver, receiver, DC blocking capacitors, common mode choke, ESD protection diodes, and the connection of the cable screen to the connector.

At the boundaries of the computational domain, we apply PML open boundaries at a distance of $\frac{\lambda}{4}$ at 1 GHz. The adaptive mesh refinement is performed for five steps, resulting in 1.7 M tetrahedrons for the model with the chamber floor and 1.1 M tetrahedrons for the model without the floor. To ensure convergence across the frequency band at all probes, the model with the floor requires 217 frequency samples, while the model without the floor requires 184 samples.

The simulation times for these models are significantly different: the model with the floor takes 35 hours, whereas the model without the floor requires only 7 hours. Although the sample count and mesh size differ by less than a factor of 2, the simulation time for the model with the floor is four times larger. This discrepancy arises due to the adaptive meshing process with variable element order. Higher-order elements are better suited for modeling the open space between the DUT and the chamber floor. Consequently, the mesh adaptation algorithm selects more higher-order elements, which increases the solver time for each frequency sample. Given the large number of required samples, the cumulative solving time becomes significantly longer.

The circuit simulation of the PRBS sequence, including the combine results of all probes, requires 7 minutes and 17 seconds.

To compare the RE in both models, we calculate the envelope of all electric field probes for both models and both polarizations. The results are plotted in Fig. 12.28. The emitted field values are higher in the model with the chamber floor by approximately 3–5 dB. This outcome is expected because reflections at the chamber floor increase the field values recorded by the probes above the floor. In contrast, in the model without the floor, the emission is spread over a larger elevation

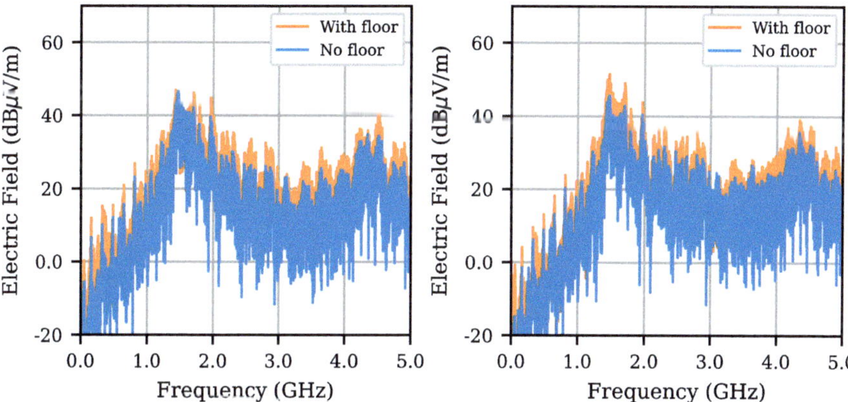

Fig. 12.28 Comparison of the RE from a model containing the chamber floor and a model without the floor. Left: horizontal polarization. Right: vertical polarization

angle. However, the spectral distribution of the emission remains unchanged, and the positions of the peaks are unaffected by the presence of the floor.

If many model variants need to be simulated, the extended simulation time of the model with the floor makes such simulations impractical for electronic system design. By quantifying the difference in emission between the model with and without the chamber floor, we determine the trade-off: a model that is closer to the measurement setup versus a model that simulates significantly faster but deviates from the measurement setup.

For all subsequent results presented in this example, we use the model without the chamber floor, keeping in mind a difference of approximately 3–5 dB compared to the model with a setup closer to the measurement. Additionally, we focus only on the horizontal polarization in the following sections, as no significant differences are observed between the two polarization directions.

12.5.3 Complete Model vs. Standalone Board

Once we have established the model and the results we aim to analyze, we can begin comparing different variations of the DUT to determine which sections of the system influence the RE.

In the first simulation, we compare the RE from the complete model, which includes the connector and the cable, to the emission from the standalone PCB board. In the standalone model, the USB signal traces are terminated to the reference at the position of the connector, as in the models simulated in Chap. 11.

A comparison of the horizontal component of the emitted electric field is plotted on the left-hand side of Fig. 12.29. As expected, the emission from the complete model is higher than the emission from the standalone board. For most frequencies,

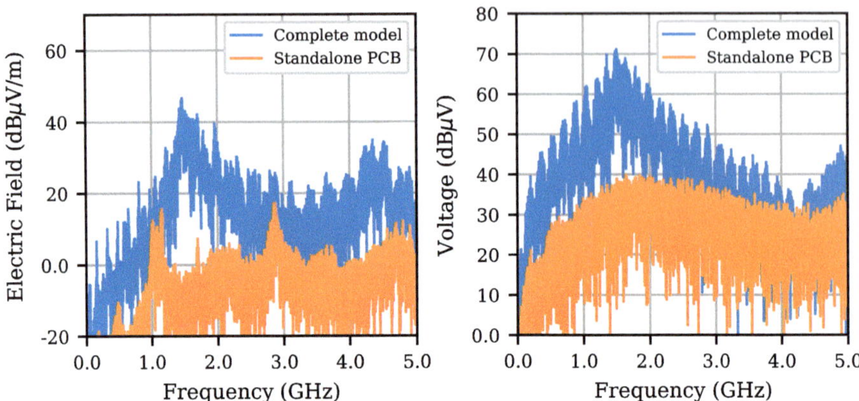

Fig. 12.29 Left: horizontal component of the RE from the complete model including PCB, connector, and cable, compared to a model containing only the PCB. Right: comparison of the common mode voltage at the receiver for the complete model vs. PCB only

the emission from the complete model is at least 20 dB higher than that from the standalone PCB board. However, the emission from the standalone board exhibits two peaks: one slightly above 1 GHz and another slightly below 3 GHz. These peaks indicate resonances on the board that cannot be identified in the complete model, as they are masked by the higher emission levels of the complete model.

On the right-hand side of Fig. 12.29, we plot the common mode voltage at the receiver. The common mode voltage at the receiver correlates well with the emitted electric field for the complete model: both exhibit a peak around 1.5 GHz. This suggests that, in the complete model, the common mode is the primary source of the emission.

For the standalone board model, the resonances observed in the emitted field are not visible in the common mode voltage, indicating that these peaks are unrelated to the common mode. As already mentioned in Chap. 11, predicting the emission from the complete model cannot rely solely on the common mode voltage calculated using the standalone board model. The inclusion of the connector and the cable in the simulation is essential to accurately capture the emission from the entire system.

To further confirm our findings, we repeat the study of the effect of different component mountings performed in Chap. 11. We calculate the emission from the complete model for three different component combinations:

- Only the DC capacitors
- DC capacitors and the common mode choke
- DC capacitors, common mode choke, and ESD protection diodes

In Chap. 11, we calculate the common mode on the board for these three combinations and observe differences among them (Fig. 11.7). However, as the common mode in the system is governed by the connector and the cable, changes to

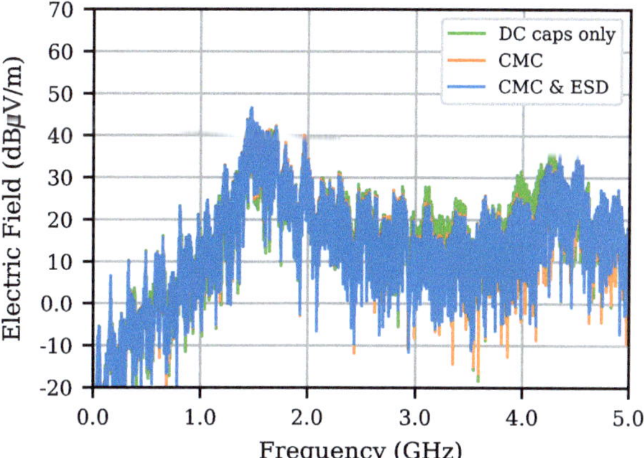

Fig. 12.30 Horizontal component of the RE from the complete model with different mounted components

the components on the board have only a minor effect on the emission, as illustrated in the results plotted in Fig. 12.30.

Modifying the components only requires re-running the circuit simulation with combine results, as all components are modeled as ports in the EM model. Therefore, the EM simulation does not need to be repeated in this case.

12.5.4 *Effect of the Cable Screen and the Cable Layout*

In this section, we analyze the effect of the cable shield and the cable layout on the emission. First, we demonstrate the impact of the shield material. Since we use a solid shield without any holes, the thickness of the shield does not influence the results: at 1 GHz, the skin depth in aluminum is approximately 2.5 μm, which is much smaller than the shield thickness of 20 μm. To confirm this, we simulate the complete model with a shield thickness of 5 μm and with a solid PEC shield. The emission results are plotted on the left-hand side of Fig. 12.31.

As expected, changing the shield material and thickness has no effect on the RE results.

We also simulate the complete model with the shield removed. Although shielding is mandatory for USB 3.0 cables, comparing the emission with and without the shield provides useful insights. As shown on the right-hand side of Fig. 12.31, completely removing the shield increases the emission by approximately 20 dB.

As the screen thickness does not affect the RE, we proceed to test other modifications to the cable design. From the study in the previous section, where we compared the complete model to the standalone PCB, we suspect that differential-to-common mode conversion in the cable significantly influences the calculated emission. To test this hypothesis, we modify the layout of the cable. The modifications are depicted in Fig. 12.32.

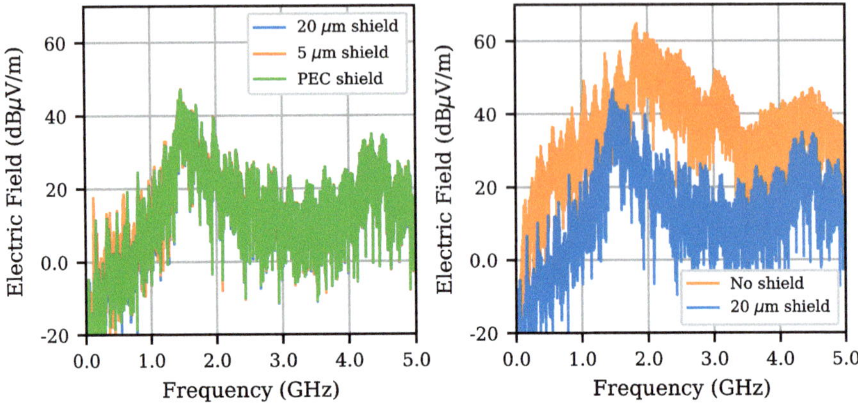

Fig. 12.31 RE in the horizontal polarization for different screen configurations. Left: different screen thicknesses and a PEC screen. Right: simulation with and without the screen

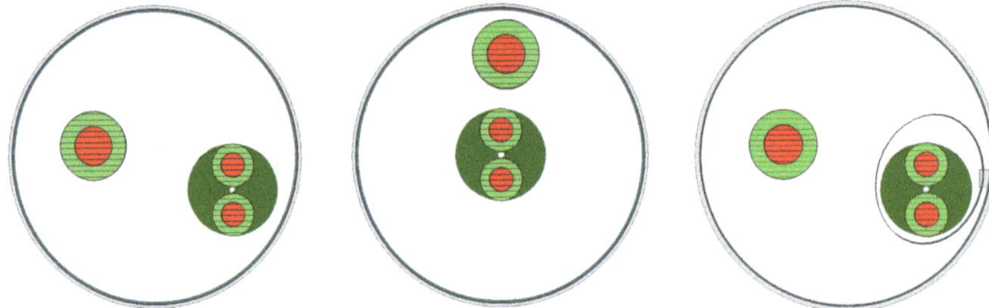

Fig. 12.32 Different cable cross sections used in the study of the cable design. Left: reference cable layout used in this section. Center: symmetric cable layout. Right: cable with added inner screen

The cross section on the left-hand side of Fig. 12.32 represents the design we have used so far. Since common mode emission is influenced by imbalance, we create a more symmetric setup by relocating the TWP to the center of the cable and positioning the ground wire above the TWP, as shown in the center of Fig. 12.32. For the third cable layout variation, we introduce an elliptically shaped inner screen around the TWP, which is a common feature in real USB 3.0 cables. The inner screen is connected to the outer screen at both ends of the cable.

The results of the study with different cable layouts are plotted in Fig. 12.33. Both cable modifications decrease the emission. In particular, a noticeable reduction is observed in the frequency band from 2 GHz to 3 GHz. Adding the inner screen reduces the emission across a broad frequency range by approximately 5 dB. The symmetrization of the cable layout proves more effective at the peak frequency: the emission at the 1.5 GHz peak drops by 10 dB.

From the analysis in this chapter, we conclude that for the simulation model we have built, the cable design has the strongest influence on the RE. In most

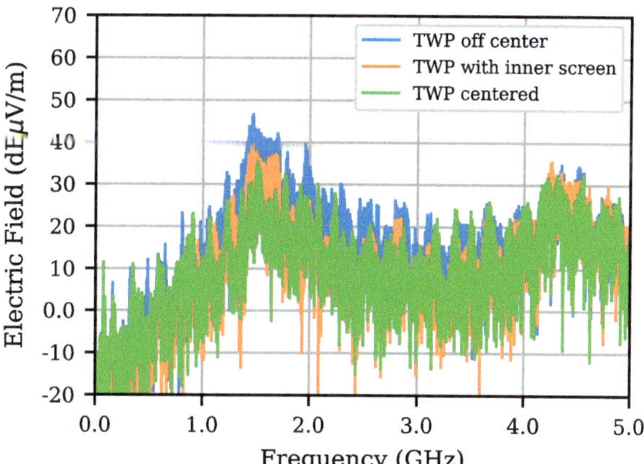

Fig. 12.33 RE results for three different cable layouts depicted in Fig. 12.32

products incorporating a USB interface, designers have little control over cable specifications, as the product must function with third-party cables. Thus, there is limited room for optimization in this model. However, this is largely due to the design used in this study: the PCB layout is highly symmetric because of the ample space on the example board. In a realistic PCB with densely packed components, the differential-to-common mode conversion on the board can be much more significant. Additionally, the connector-to-cable transition in our model is ideal, featuring a 360° connection. A more realistic connector might introduce higher mode conversion.

The simulations performed in this section are brute-force simulations, where all parts of the system are modeled using a full-wave solver, and the electric field at a 3 m distance is obtained from the envelope of the probes. The advantage of a full-wave simulation is that it inherently includes all coupling effects and directly delivers the radiated electric field. However, it is computationally expensive. Each modification to the geometry requires an EM simulation, which takes approximately 7 hours on the machine used for this study.

When the simulation engineer has access to large simulation resources, this approach is valid. However, if results are needed with reduced simulation time, a segmented approach can be applied, where individual segments are cascaded to obtain the system results. The cascaded approach is discussed in the next section.

12.6 Advanced: Cascading of Model Sections

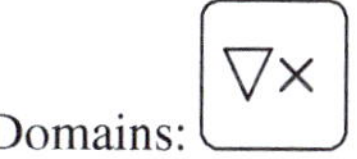 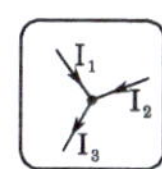 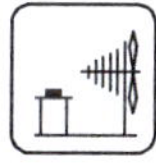

 Digital signal from an IBIS model

Voltages at circuit pin
Full-wave, S-parameters, transmission line
Cascade of FEM method and transmission line method
Transient
0–5 GHz (limited accuracy above 3 GHz)

The analysis in the previous section models the complete signal channel using a full-wave method. While this approach allows for the inclusion of all couplings between parts of the system, it is computationally expensive. Segmenting the system into sections and cascading these sections results in a significantly faster simulation.

In this section, we divide the complete model into segments that are analyzed separately using appropriate EM simulation methods, and then combine the segments in a circuit model. While the circuit model can only provide voltages and currents, not the emitted field, we correlate the common mode voltage to the emitted electric field obtained from the full-wave model and discuss the limitations of this cascading approach.

When performing segmentation, it is critical to identify suitable positions where the model can be divided. For the segmented models to combine correctly in the system-level circuit simulation, the interface at the individual segments must be identical. Suitable segmentation positions require the structure to be length-homogeneous and the field to exhibit a TEM mode. Segmenting the structure at positions where the field is not transversal would lead to inaccuracies in the cascaded model.

In the USB example, we segment the model into three regions:

- A PCB segment, which includes the signals on the PCB from the driver up to the connector
- A connector segment, which includes the transition from the PCB to the cable
- A cable segment, which contains the cable itself

The PCB segment and the cable segment are efficiently solved using the MTL method. The connector segment, however, requires a full-wave simulation. Since we calculate S-parameters for the individual segments and cascade them in the schematic, it is essential to ensure that the reference for the S-parameters matches across the models. This can be challenging, as the reference for segments modeled using a transmission line solver is determined by the solver settings, whereas the reference in the full-wave simulation is explicitly defined by the port (Fig. 12.34).

In the MTL model of the PCB segment, all ports are referenced to the PCB ground. For the connector segment, we define a waveguide port close to the transition from the PCB to the connector. At this waveguide port, we calculate the single-ended modes of the two signal traces relative to the reference layer below them. On the cable side, we also use a single-ended waveguide port; in this case, the cable screen serves as the reference, and we calculate three S-parameters of the inner wires relative to the screen.

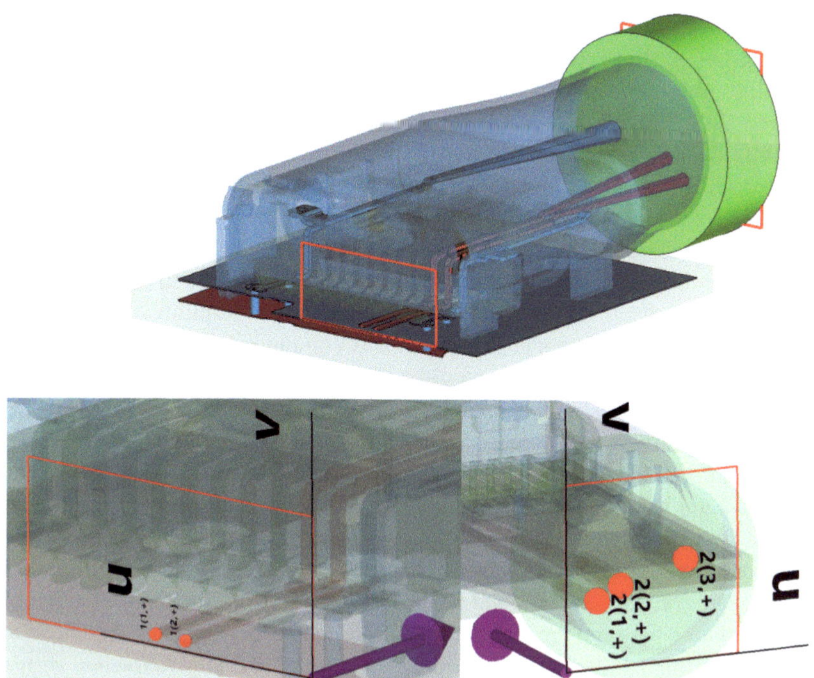

Fig. 12.34 Overview of the segmented connector model with a close-up on the waveguide ports at both ends

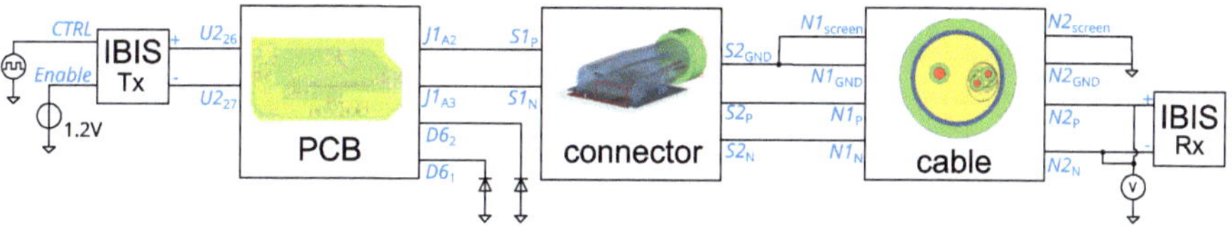

Fig. 12.35 Cascading of the three segments on the circuit schematic

The MTL model of the cable is calculated as an 8-port network. At each end of the cable, the three inner conductors and the screen are referenced to the MTL model's reference. This setup requires special treatment in the MTL solver, as there is no explicit reference conductor present in the model. In CST, the MTL solver references all ports of the MTL model to a virtual reference placed outside of the computational domain.

The three segments are then cascaded on a schematic, as depicted in Fig. 12.35. In the PCB model, we include the DC blocking capacitors and the common mode choke into the S-parameter block. The diodes, however, are retained in the cascaded schematic because they are non-linear elements and cannot be straightforwardly included in the S-parameters of the PCB segment.

At the interface between the connector and cable models, we connect the port corresponding to the ground wire of the connector model to the port of the ground wire of the cable model and to the cable screen. This connection mimics the physical transition. At the second end of the cable, the screen and ground wire are connected to the reference. This setup replicates the termination in the 3D model, where we used a metallic plate to terminate the cable. However, it is not identical: in the 3D model, the reference is explicitly defined close to the cable end, while in the MTL model, the transmission line solver defines the reference, as described earlier.

The assembly of the individual segments in the cascaded model must be performed carefully to ensure precise matching, which requires additional setup time. However, once the model is created, it is computationally very efficient. The EM simulation for each segment requires less than 2 minutes, while the cascaded circuit simulation takes approximately 1 minute. Vector fitting (cf. Sect. 10.3) is applied to the individual segments to enable transient circuit simulations. If changes to the geometry are required, they can be implemented within the individual segments, significantly reducing the simulation time for geometrical variations.

In the first step, we compare the common mode voltage V_{CM} at the receiver of the cascaded model to the complete 3D model calculated at the beginning of this example. The comparison is plotted on the left-hand side of Fig. 12.36. We observe a good correlation between the results of the cascaded model and the complete model. The cascaded model slightly overestimates V_{CM}, and a minor frequency shift can be observed. Nevertheless, the cascaded model is highly efficient for comparing different designs, as its simulation time is on the order of minutes.

The cascaded model also enables the study of individual segments by selectively removing them from the schematic. For the comparison on the right-hand side of Fig. 12.36, we simulate the complete cascade, a schematic including only the PCB and connector, and a schematic including only the connector and cable. As

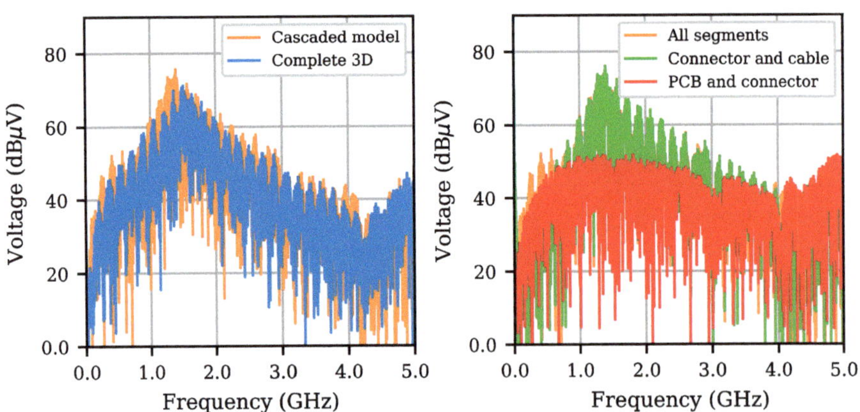

Fig. 12.36 Left: common mode voltage V_{CM} calculated at the receiver for the cascaded model and the model completely in 3D. Right: common mode voltage V_{CM} at the receiver in the cascaded model for different combinations of the segments

expected, when only the PCB and connector are included, the peak in V_{CM} at 1.5 GHz disappears. However, above 3 GHz, the results of the complete model and the model with only the PCB and connector begin to match. This indicates that, starting from this frequency, the cable's contribution to common mode generation diminishes, and the connector becomes the dominant source of mode conversion.

In the final comparison, we use the cascaded model to calculate the common mode voltage and test how geometrical changes influence the radiation in the complete 3D model. In Sect. 12.5.4, we analyze several cable layout variants that reduce the common mode at the receiver. Here, we modify the pitch of the twists in the TWP.

The original twist pitch of the TWP is set to 6 cm. We simulate the effect of reducing the twist pitch to a denser value of 2 cm. A more tightly twisted pair is expected to reduce the differential-to-common mode conversion in the cable, thereby lowering the radiated emission.

The common mode voltage V_{CM} at the receiver, calculated by the cascaded model for different pitch lengths, is plotted on the left-hand side of Fig. 12.37. By reducing the pitch length to 2 cm, the peak at 1.5 GHz is reduced by approximately 30 dB. However, above 3 GHz, the common mode voltage increases by 10 dB. To validate the cascaded model's prediction, the TWP in the complete 3D model is modified to the new pitch length, and the electric field at the probes is evaluated. The results are plotted on the right-hand side of Fig. 12.37. Reducing the pitch length of the TWP lowers the emitted field at the peak around 1.5 GHz by approximately 20 dB. While this is less than predicted by the cascaded model, it still represents a significant improvement. The increase in emission above 3 GHz is also observed in the emitted electric field; however, it is much more pronounced than the corresponding increase of the common mode voltage in the cascaded model.

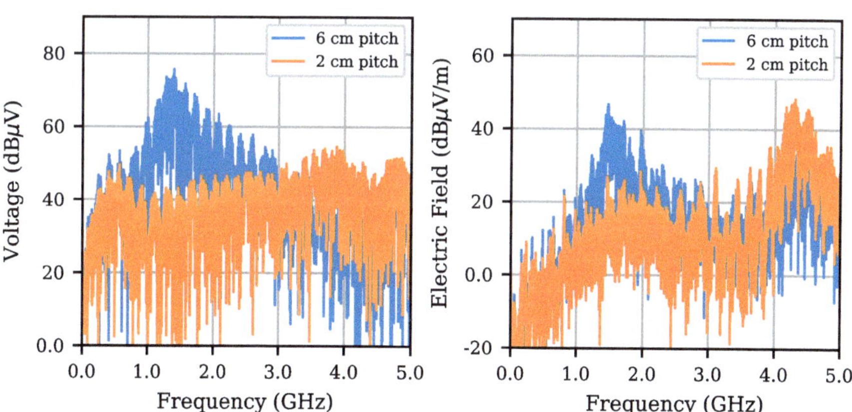

Fig. 12.37 Left: common mode voltage at the receiver calculated by the cascaded model for different pitch lengths of the TWP. Right: horizontal polarization of the RE calculated by the complete 3D model for different pitch lengths

This comparison highlights the accuracy that can be expected from cascaded models. Up to approximately 2–3 GHz, the cascaded model predicts the emission reasonably well. Above 3 GHz, the predictions become less reliable. One possible explanation for this discrepancy is the absence of a well-defined reference in the MTL cable model: since no metallic reference exists along the cable path, the MTL model assumes the reference is outside the computational domain. This limitation could be addressed by using a full-wave simulation for the cable section (not shown in the results). However, performing a full-wave simulation for the cable would lead to simulation times comparable to those of the complete model, which defeats the primary motivation for using the cascaded approach—reducing simulation time.

The cascaded model provides a fast and efficient method for predicting trends. However, once promising modifications to the system are identified, they should be validated using the complete 3D model. In Sect. 13.2.3, a cascaded model of a BCI test is presented. The frequency range of the BCI test lies well below 1 GHz, and a metallic reference plane is present. In that case, the results of the cascaded model are comparable to those of the complete 3D model for the complete frequency band of interest.

Further Reading

Radiated Emission Modeling

B. Adamczyk, In Compliance Magazine (2017). https://incompliancemag.com/radiated-emissions-measurements-oats-and-alse-methods/

S. Deng, T. Hubing, D. Beetner, Estimating maximum radiated emissions from printed circuit boards with an t-tached cable. IEEE Trans. Electromagn. Compat. **50**(1), 215–218 (2008). https://doi.org/10.1109/TEMC.2007.915288

M. Dhara, G.A. Rasek, Analytical method to estimate radiated magnetic field emissions in automotive electric drives, in *2022 International Symposium on Electromagnetic Compatibility—EMC Europe* (2022), pp. 445–450. https://doi.org/10.1109/EMCEurope51680.2022.9901062

H. Hackl, G. Winkler, B. Deutschmann, Simulation of radiated emission during the design phase based on scattering parameter measurement, in *2015 10th International Workshop on the Electromagnetic Compatibility of Integrated Circuits (EMC Compo)* (2015), pp. 228–231. https://doi.org/10.1109/EMCCompo.2015.7358362

D.M. Hockanson et al., Investigation of fundamental EMI source mechanisms driving common-mode radiation from printed circuit boards with attached cables. IEEE Trans. Electromagn. Compat. **38**(4), 557–566 (1996). https://doi.org/10.1109/15.544310

H. Inoue, Y. Kayano, Prediction of EM radiation from a PCB driven by a connected feed cable. IEICE Trans. Commun. **E92-B**(6), 1920–1928 (2009). ISSN: 1745-1345, https://doi.org/10.1587/transcom.E92.B.1920

G.-H. Li et al., Estimating the radiated emissions from cables attached to a switching power supply in a MIL-STD 461 test, in *2013 IEEE International Symposium on Electromagnetic Compatibility* (2013), pp. 626–631. https://doi.org/10.1109/ISEMC.2013.6670487

G. Li et al., Prediction of radiated emissions from cables over a metal plane using a SPICE model. IEEE Trans. Electromagn. Compat. **57**(1), 61–68 (2015). https://doi.org/10.1109/TEMC.2014.2364405

Z. Ma et al., Modeling, analysis and mitigation of radiated EMI due to PCB ground impedance in a 65 W high-density active-clamp flyback converter. IEEE Trans. Ind. Electron. **70**(12), 12267–12277 (2023). https://doi.org/10.1109/TIE.2023.3239904

Z. Ma et al., A review of radiated EMI research in power electronics systems. IEEE J. Emerging Sel. Top. Power Electron. **12**(1), 675–694 (2024). https://doi.org/10.1109/JESTPE.2023.3335972

S. Mee, B. Adamczyk, *In Compliance Magazine*. Internet (2018). https://incompliancemag.com/bulk-current-injection-bci-substitution-method-and-closed-loop-method-with-power-limitation/

Molex, *Universal Serial Bus (USB) Shielded I/O Receptacle, Type C* (visited in 2025). https://www.molex.com/en-us/products/part-detail/2171830001

H. Muniganti, B. Nayak, D. Gope, Diagnosis of radiating elements for CISPR 25 RE test setup using huygens box method. IEEE Lett. Electromagn. Compat. Pract. Appl. **2**(2), 40–45 (2020). https://doi.org/10.1109/LEMCPA.2020.2985342

L. Niu, T.H. Hubing, Rigorous derivation of imbalance difference theory for modeling radiated emission problems. IEEE Trans. Electromagn. Compat. **57**(5), 1021–1026 (2015). https://doi.org/10.1109/TEMC.2015.2422678

I. Oganezova et al., Simulation of conductive and radiated emissions from a wiper motor according to CISPR 25 standard, in *2015 IEEE International Symposium on Electromagnetic Compatibility (EMC)* (2015), pp. 963–968. https://doi.org/10.1109/ISEMC.2015.7256296

Open Alliance, *Automotive Ethernet Specifications*. Internet (2024). https://opensig.org/automotive-ethernet-specifications/

H.W. Ott, *Noise Reduction Techniques in Electronic Systems*, 2nd edn. (Wiley-Interscience, New York, 1988). ISBN: 978-0471850687

H.H. Park, H.-B. Park, H.S. Lee, A simple method of estimating the radiated emission from a cable attached to a mobile device. IEEE Trans. Electromagn. Compat. **55**(2), 257–264 (2013). https://doi.org/10.1109/TEMC.2012.2219587

A. Radchenko et al., Transfer function method for predicting the emissions in a CISPR-25 test setup. IEEE Trans. Electromagn. Compat. **56**(4), 894–902 (2014)

H. Rezaei et al., Analyzing the influence of imbalanced two-or three-wire VHF LISN on radiated emissions from AC cables. *IEEE Trans. Electromagn. Compat.* **64**(2), 327–337 (2022). https://doi.org/10.1109/TEMC.2021.3111136

H. Rezaei et al., Influence of variations in imbalanced LISN termination impedances on radiated emissions. IEEE Trans. Instrum. Meas. **74**, 1–11 (2025). https://doi.org/10.1109/TIM.2024.3500049

C. Rostamzadeh et al., Investigation of electromagnetic field coupling from DC–DC buck converters to automobile AM/FM antennas, in *2016 IEEE International Symposium on Electromagnetic Compatibility (EMC)* (2016), pp. 364–369. https://doi.org/10.1109/ISEMC.2016.7571674

M. Schmidt, *Use of TDR for Cable Testing*, Logan, UT (2002)

S. Shinde et al., Modeling EMI due to display signals in a TV. IEEE Trans. Electromagn. Compat. **58**(1), 85–94 (2016). https://doi.org/10.1109/TEMC.2015.2502589

S. Shinde et al., Radiated EMI estimation from DC–DC converters with attached cables based on terminal equivalent circuit modeling. IEEE Trans. Electromagn. Compat. **60**(6), 1769–1776 (2018). https://doi.org/10.1109/TEMC.2017.2782659

M.K. Smail et al., Detection of defects in wiring networks using time domain reflectometry. IEEE Trans. Magn. **46**(8), 2998–3001 (2010). https://doi.org/10.1109/TMAG.2010.2043720

E. Song, H.-B. Park, H.H. Park, An evaluation method for radiated emissions of components and modules in mobile devices. IEEE Trans. Electromagn. Compat. **56**(5), 1020–1026 (2014). https://doi.org/10.1109/TEMC.2014.2307915

V. Sreenivasulu Reddy, P. Kralicek, J. Hansen, A novel segmentation approach for modeling of radiated emission and immunity test setups. IEEE Trans. Electromagn. Compat. **59**(6), 1781–1790 (2017)

G. Tsintsadze et al., Predicting radiated emissions from an electrical drive system, in *2023 Joint Asia-Pacific International Symposium on Electromagnetic Compatibility and International Conference on ElectroMagnetic Interference & Compatibility (APEMC/INCEMIC)* (2023), pp. 1–4. https://doi.org/10.1109/APEMC57782.2023.10217615

Wurth Electronic, *WE-CAR-TEC Snap Ferrite*. Internet (2025). https://www.we-online.com/components/products/datasheet/78271111.pdf

J. Yao et al., Modeling and reduction of radiated EMI in a GaN IC-based active clamp flyback adapter. IEEE Trans. Power Electron. **36**(5), 5440–5449 (2021). https://doi.org/10.1109/TPEL.2020.3032644

J. Yao, S. Wang, Z. Luo, Modeling, analysis, and reduction of radiated EMI due to the voltage across input and output cables in an automotive non-isolated power converter. IEEE Trans. Power Electron. **37**(5), 5455–5465 (2022). https://doi.org/10.1109/TPEL.2021.3128628

Y. Zhang, S. Wang, Y. Chu, Investigation of radiated electromagnetic interference for an isolated high-frequency DC–DC power converter with power cables. IEEE Trans. Power Electron. **34**(10), 9632–9643 (2019). https://doi.org/10.1109/TPEL.2019.2892706

Chapter 13
Immunity

The simulation-based prediction of immunity failures enables the optimization of electronic systems to prevent malfunctions. We introduce a range of immunity tests inspired by established standards for conducted and radiated immunity.

For these tests, we describe the process of modeling and calibrating the laboratory equipment used as an interference source. We do not model a specific hardware component from a vendor. Instead, we present generic models and explain their capabilities and limitations. Whenever measurement data is available, the models presented here can be adjusted or extended to match existing hardware.

Exemplarily, we apply the source models to various DUTs. We analyze the properties of the coupling path and its impact on the immunity of the electronic system under test. The presented DUTs are intentionally simple. We do not incorporate functional models of ICs that could simulate operational failures as observed in measurements. Instead, the DUTs in this chapter illustrate the types of analyses that can be performed even when only limited information about the system under test is available. All procedures can be extended to incorporate more advanced models of DUTs, enabling a quantitative prediction of immunity.

The models discussed in this chapter are based on various immunity standards, namely, bulk current injection (BCI, ISO 11452-4); burst, surge, and pulse disturbances (IEC 61000-4-4, IEC 61000-4-5, ISO 7637-2); radiated immunity (RI, ISO 11452-2); as well as TEM and stripline methods (ISO 11452-3, 11452-5).

We do not consider electromagnetic pulses (EMP), high-intensity radiated fields (HIRF), or lightning. Protection against electromagnetic environmental effects (E3) is not treated in this book.

J. Hansen, A. Barchanski, *EMC Simulation of Electronic Systems*,
https://doi.org/10.1007/978-3-032-09237-3_13

13.1 Bulk Current Injection

Domains: 

In the BCI test (ISO 11452-4), a current clamp is used to inject an AC or modulated HF current into the cable harness connected to a DUT. The injected current is a common mode current.[1] The harness and DUT are placed on a metal table, which serves as the return path for the injected current. The DUT must continue operating correctly despite the presence of this current. If the DUT has a differential input, conversion from common mode to differential mode produces a differential voltage at this input. If the electronic system under test were perfectly balanced, there would be no mode conversion and no disturbance of differential mode signals would occur.

In BCI testing, there are two methods: open-loop BCI, also known as the substitution method, and closed-loop BCI. Both methods monitor the forward power P_{fwd}, which is the power injected into the harness, as a measure of the injected current.

The forward power P_{fwd} is related to the current in the wire harness via the termination impedances at the wire ends. In open-loop BCI, these impedances are not determined by the electronic system under test but by a calibration jig. This jig uses a single wire terminated at both ends by 50 Ω. The injection clamp is mounted onto this jig. During calibration, a defined current profile over frequency is injected into the wire, and the forward power P_{fwd} required to produce this current profile is recorded. After calibration, this stored P_{fwd} value is used to drive the injection clamp during testing of the system under test.

Since the system under test has an impedance different from 50 Ω, the actual current injected during testing differs from the calibrated current profile. For instance, if the DUT attached to the wire harness is floating above the metal table, only displacement currents close the loop for the injected common mode current. The termination impedance is capacitive and, at low frequencies, significantly larger than 50 Ω. The forward power P_{fwd} determined during calibration with the 50 Ω termination results in a much lower actual injected current compared to the calibration jig scenario.

In closed-loop BCI, the current injected into the system under test is monitored using a second current clamp. The current profile over frequency, which is utilized during the calibration procedure of the open-loop test, represents the target current that must be generated by the injection clamp. Thus, in closed-loop BCI, the current is directly monitored within the actual system under test and is not determined through calibration.

Furthermore, the injected power is constrained. Considering the example of a floating DUT mentioned in the previous paragraph, capacitive coupling at low

[1] As with other EMC tests, variants exist but are not discussed here.

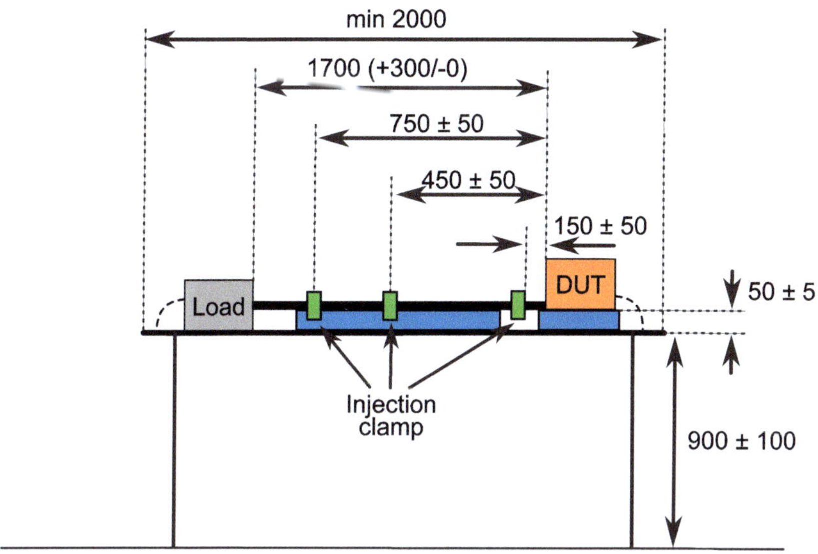

Fig. 13.1 Sketch of the open-loop BCI test. All distances in mm

frequencies would necessitate excessive power to achieve the desired current within the wire harness. For this reason, the maximum forward power is limited to, typically, $4\,P_{\mathrm{fwd}}$, that is, four times the forward power derived from the clamp calibration conducted in the jig. If this limit of forward power is reached, it is not to be increased further, even if the specified current limit has not been attained.

In the following, we describe the setup of both open-loop and closed-loop BCI tests. Figure 13.1 illustrates the open-loop test arrangement. The harness used in this test has a length of up to 2 m. The DUT is placed on a support of 50 mm thickness, with a relative permittivity ε_{r} close to unity. The harness is terminated by an artificial load, and both the DUT and the load may or may not have galvanic connections to the metal table. The injection clamp must be positioned at three distinct points along the harness, specifically at distances of 150 mm, 450 mm, and 750 mm from the DUT.

Figure 13.2 shows the closed-loop BCI test setup. Similar to the open-loop configuration, the DUT is positioned 50 mm above the metal table. The DUT can either float above the metal table or be connected to it. The wire harness measures approximately 1 m in length and is terminated by a load simulator, which can, as the DUT, be either grounded or ungrounded. The current injection clamp is placed 900 mm from the DUT, while the monitoring clamp is positioned at a distance of 50 mm.

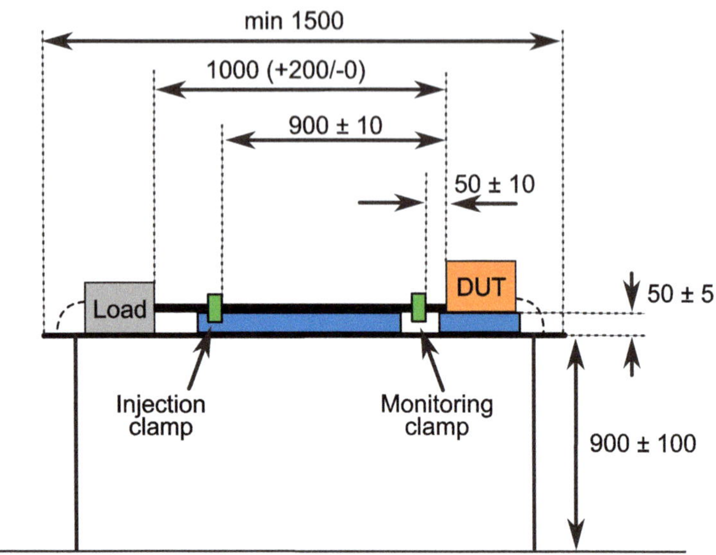

Fig. 13.2 Sketch for closed-loop BCI test. All distances in mm

13.1.1 BCI Clamp Models and Calibration Jig

A current clamp acts like a transformer, with the wire harness being its secondary winding. Although a simple transformer circuit model can represent the clamp, this model deviates from the actual clamp behavior at high frequencies. A more accurate EM model needs to include the magnetic core, the primary winding that excites the field in the magnetic core, and, as secondary winding, a section of the wire harness that is sufficiently long to correctly describe the coupling between the clamp and the harness.

Figure 13.3 illustrates both models. The circuit model represents an ideal coupler, consisting of two coupled inductors with a coupling coefficient $k = 1$. In this circuit, the wire ends are represented by port 2 and port 3, and the voltage at the clamp is injected at port 1. The primary and secondary sides are isolated from each other.

The EM model includes a single primary winding that excites a magnetic field in four adjacent ferrite rings. The current in the primary winding can be excited by the clamp voltage V_C. The ferrite rings are characterized by magnetic permittivity parameters of $\mu_\mathrm{s} = 450$, $\mu_\infty = 10$ and relaxation time $\tau_\mathrm{r} = 4.5$ ns. Each ring has an inner diameter of 52.5 mm, an outer diameter of 74.5 mm, and a height of 10.75 mm. The clamp housing, made of PEC material, contains a narrow slit on the inner side designed to couple the magnetic field to the wire. The outer diameter of the housing measures 95 mm, and the feed-through diameter is 32 mm. This model is taken from the CST model library. If excited by V_C, a current I_W flows in the wire. The amplitude of this current along the wire is a function of the wire's terminations.

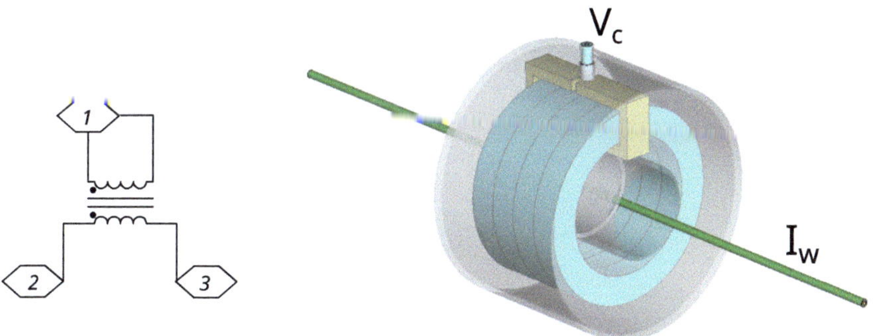

Fig. 13.3 Models of a current clamp with a single wire through it. Left: circuit model. Right: EM model

Current clamps are characterized either by their insertion loss, defined as $-20\log_{10}|S_{21}|$, or by their transfer impedance, given by $Z_T = \frac{V_C}{I_W}$. The relation between Z_T and S_{21} is as follows. We can compute the transfer impedance by monitoring the voltage V_C at the clamp connector (primary winding) while exciting a current I_W in the wire. Introducing the incident and reflected power waves at port i and j as a_i and b_j (cf. Sect. 2.4.3.2 and Eq. 2.5), we obtain

$$Z_T = \frac{V_C}{I_W} = Z_0 \frac{a_1 + b_1}{a_2 - b_2} = Z_0 \frac{\frac{b_1}{a_2}}{1 - \frac{b_2}{a_2}} = Z_0 \frac{S_{12}}{1 - S_{22}} \tag{13.1}$$

where we set $a_1 = 0$ because the current is injected into the wire, and the primary side is not excited. The transfer impedance Z_T is proportional to the insertion loss scaled by reflection loss inside the clamp. Since we only consider magnitudes, we use $|S_{21}| = |S_{12}|$, i.e., the insertion loss can be computed from either of the two S-parameters.

Next, we analyze S_{21} and Z_T of the circuit and EM clamp models. The geometric model of the clamp is placed into a jig consisting of a single wire passing through the clamp. Both ends of the wire are connected via 50 Ω terminations to a metal plate serving as a reference (see Fig. 13.4). The plates on both sides are connected to each other by a plate below the clamp. This jig is also used in Sect. 13.1.2 to determine the forward power P_{fwd} required for both open-loop and closed-loop tests. The EM model of the jig computes in 3 minutes 25 seconds using the FEM solver.

The S_{21} parameter and the transfer impedance Z_T of both models are presented in Fig. 13.5, with the left side showing S_{21} and the right side displaying Z_T. The frequency range spans from 10 kHz to 400 MHz.

The transformer model uses $k = 1$ and assumes identical inductances on both the primary and secondary sides. The inductance value is $L = 1.35\,\mu$H, chosen to match the low-frequency behavior of the EM model. The transition between the low-frequency and high-frequency behavior of the clamp occurs at approximately

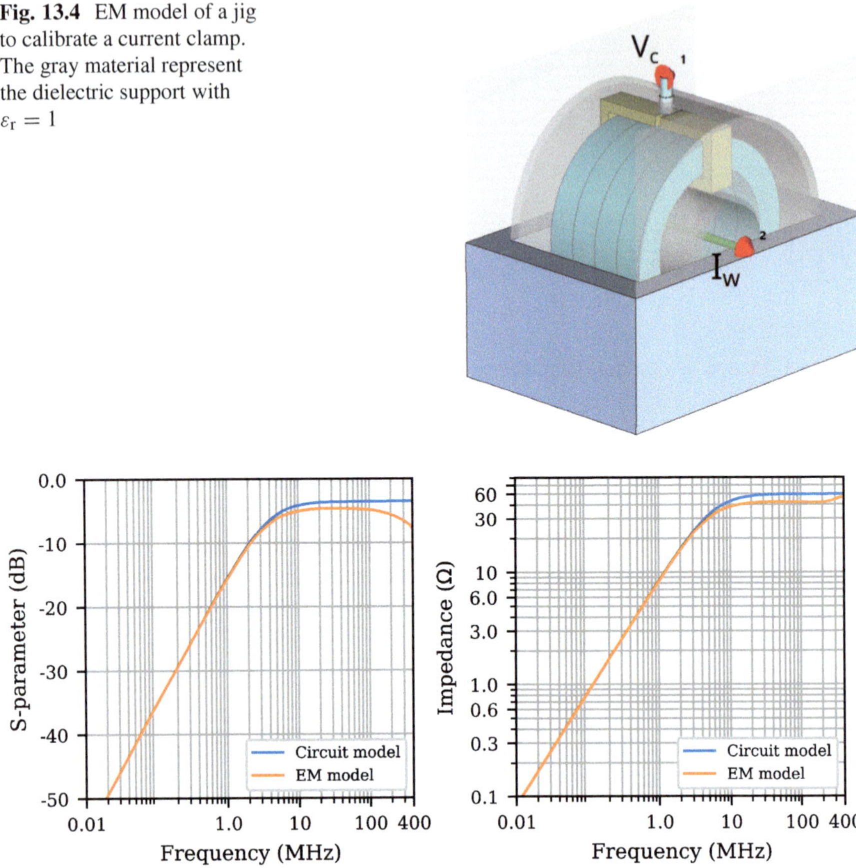

Fig. 13.4 EM model of a jig to calibrate a current clamp. The gray material represent the dielectric support with $\varepsilon_r = 1$

Fig. 13.5 Comparison of results of the BCI clamp modeled as a transformer and as an EM model. Left: S-parameter S_{21}. Right: transfer impedance

10 MHz. Above 10 MHz, the circuit model exhibits $S_{21} = -3.6$ dB, while the EM model is 1 dB lower. The transfer impedance Z_T on the right-hand side follows a qualitatively similar trend. At 10 kHz, both models show approximately 0.1 Ω. Above 10 MHz, the ideal transformer model reaches 50 Ω, whereas the EM model is 10 Ω lower.

At high frequencies, parasitic effects such as capacitive coupling arise. The investigation reveals that above 10 MHz, a current clamp cannot be accurately modeled as a simple transformer. We see in Fig. 13.6 in Sect. 13.1.2 that Z_T has an impact on the forward power P_{fwd} in the jig. To account for these effects, either an EM model or a more sophisticated circuit model (see the references at the end of this chapter) must be developed and calibrated against measurement data. Such a model then accurately represents the S_{21} or Z_T of the specific clamp under study.

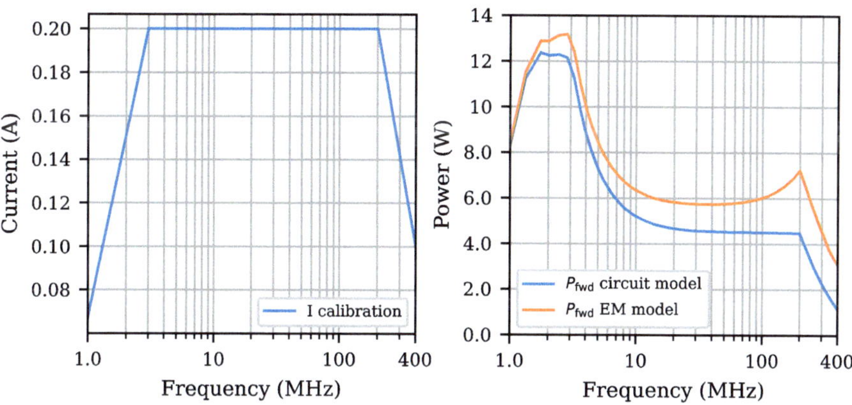

Fig. 13.6 Results obtained from Eq. 13.4. Left: calibration current. Right: required forward power P_{fwd}

However, if the primary interest is solely in the physical effect of magnetic coupling, the ideal coupler model is sufficient.

13.1.2 Open and Closed Loop BCI Calibration

In the calibration procedure for open-loop BCI, the calibration jig is used to adjust the forward power P_{fwd} to achieve a specified target current level, I_{target}, in the wire passing through the clamp. This calibrated power is then applied to the clamp during the BCI test with the DUT.

We obtain P_{fwd} from the simulation of the clamp in the jig (Fig. 13.4). The straightforward procedure consists of two steps: 1. Excite the clamp with a voltage of 1 V and measure the current I_{W} in the wire. 2. Scale the voltage source at the clamp by a factor of $\frac{I_{\text{target}}}{I_{\text{W}}}$ and determine the forward power injected into the clamp at the target current. We present here a more efficient approach that requires only the S-parameter S_{21} to directly compute P_{fwd} from I_{target} in a single step (Grassi et al. 2013).

We calculate the amplitude a_1 of the incident power wave at the clamp connector (primary winding) assuming a matched system, i.e., no reflection. In this case, the relationship between a_1 and the voltage V_{C} at the port of the clamp is given by

$$a_1 = \frac{V_{\text{C}}}{2\sqrt{Z_0}} \tag{13.2}$$

with $Z_0 = 50\,\Omega$. At port 2, located at the end of the wire (cf. Fig. 13.4), the outgoing power wave has an amplitude b_2 given by

$$b_2 = S_{21}a_1 = \sqrt{Z_0}\, I_{\mathrm{W}}. \tag{13.3}$$

From Eq. 13.3, we deduce the relation between P_{fwd} and the prescribed target current I_{target} as

$$P_{\mathrm{fwd}} = |a_1|^2 = \frac{Z_0 \left|I_{\mathrm{target}}\right|^2}{|S_{21}|^2}. \tag{13.4}$$

Equation 13.4 enables the computation of P_{fwd} from I_{target} and the insertion loss of the clamp. This calculation assumes a linear system, meaning that the magnetic material does not reach saturation even when a large current flows through the clamp.

A typical current profile I_{target} is shown on the left-hand side of Fig. 13.6. It begins at 1 MHz with a ramp-up phase, reaching a constant value of 200 mA at 3 MHz. Beyond 200 MHz, the current decreases linearly over the log of the frequency. BCI tests are typically conducted up to 400 MHz. This specific current profile is taken from Bogdan Adamczyk (2018), though other limits also exist.

The simulated forward power P_{fwd} is shown on the right-hand side of Fig. 13.6 for both the circuit model of the clamp (blue) and the EM model (orange). In the case of the circuit model, we observe a relatively flat region, where P_{fwd} remains between approximately 4 and 4.5 W in the frequency range of 10–200 MHz. Above 200 MHz, the power decreases due to the decay in I_{target}. Additionally, there is a peak of 12 W between 1.5 MHz and 3 MHz. The EM model shows a similar curve and takes up to 2 W more power with a peak at 200 MHz.

More than twice the forward power, 12 W instead of 6 W, is required for both models to achieve the target current I_{target} below 3 MHz, even though the target current remains below 200 mA in this range. This is due to the insertion loss of the clamp model. Figure 13.5 shows that for the given clamp models, S_{21} is relatively low. Specifically, the magnitude of $S_{21}(2\,\mathrm{MHz})$ is only one-third of that of $S_{21}(3\,\mathrm{MHz})$, while $I_{\mathrm{target}}(2\,\mathrm{MHz})$ is $\frac{3}{4}$ of $I_{\mathrm{target}}(3\,\mathrm{MHz})$. According to Eq. 13.4, this results in a higher forward power at 2 MHz than at 3 MHz, highlighting the sensitivity of P_{fwd} to variations in the clamp model's S_{21}.

For comparison, in the datasheet of a commercially available clamp (Fischer Custom Communications) of similar size, the insertion loss remains fairly constant above 1 MHz. As a result, a measurement using this clamp would not exhibit such a peak in P_{fwd}. Additionally, the forward power P_{fwd} of the EM model of the clamp is slightly higher than that of the circuit model (Fig. 13.6, right-hand side, orange curve) because the S_{21} of the EM model is lower.

Once the forward power P_{fwd} is estimated, it is used to excite the BCI injection clamp during testing with the DUT. The source excitation for open-loop BCI is straightforward: the clamp is driven by a 50 Ω source with the forward power P_{fwd}.

The closed-loop BCI test utilizes a monitoring clamp to measure the common mode current in the wire harness. The power at the injection clamp is adjusted to ensure that the current measured by the monitoring clamp follows the target profile

I_{target}, which equals the calibration current shown in Fig. 13.6. Additionally, a power limit of typically 4 P_{fwd} is set, where P_{fwd} is the forward power obtained from the calibration measurement in the jig during the open-loop calibration.

The simulation of closed-loop BCI follows a two-step procedure. First, the injection clamp is excited with a forward power of 4 P_{fwd}. In this simulation we monitor the common mode current I_{CM} in the wire harness at the location of the monitoring clamp. This can be done either by using a model of the monitoring clamp or by directly probing the current in the wire. We compare the target current I_{target} with the absolute value of I_{CM}. If $|I_{\text{CM}}| < I_{\text{target}}$, the target current is not reached, even though the system is already excited with $4 \cdot P_{\text{fwd}}$. In this case, the excitation with $4 \cdot P_{\text{fwd}}$ correctly represents the BCI measurement. However, if the common mode current I_{CM} meets or exceeds the target current limit I_{target}, the forward power of the source must be reduced.

This adjustment is achieved using the scaling factor γ_{CL}, defined as

$$\gamma_{\text{CL}} = \min\left(1, \frac{I_{\text{target}}}{|I_{\text{CM}}|}\right). \tag{13.5}$$

The factor γ_{CL} remains unity at all frequencies where the BCI simulation is constrained by $4 \cdot P_{\text{fwd}}$. At frequencies where the common mode current I_{CM} in the monitoring clamp exceeds the target current I_{target}, γ_{CL} linearly scales down the excitation to ensure the correct current level is achieved.

After computing γ_{CL}, the closed-loop simulation is finalized with a second simulation, where the injection clamp is excited with an incoming power wave of amplitude $\gamma_{\text{CL}} \cdot a_1$. Equation 13.5 assumes a linear system behavior, meaning that the applied currents remain sufficiently low to prevent saturation of the clamp.

Figure 13.7 illustrates the scaling factor γ_{CL} (red) in conjunction with the target current I_{target} (blue), the simulated common mode current I_{CM} at the

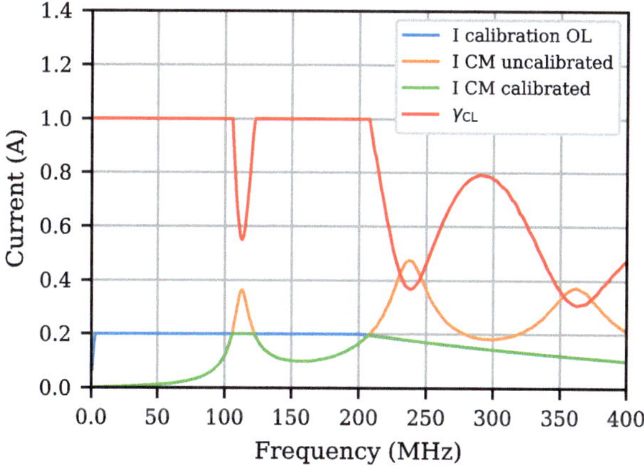

Fig. 13.7 Current and power limitation in closed-loop BCI

monitoring clamp location (orange), and the corrected common mode current in the final simulation (green). The curves are produced using the application example discussed in Sect. 13.2. Whenever $|I_{CM}| > I_{target}$, the factor γ_{CL} scales I_{CM} down to ensure it does not exceed I_{target}. At lower frequencies, the power limitation of $4 \cdot P_{fwd}$ is the dominant constraint. However, above 200 MHz, the available power is sufficient to generate the target common mode current, achieving $I_{CM} = I_{target}$.

13.2 Application Example: Closed-Loop BCI on 100BASE-T1 Automotive Ethernet

- EM model of BCI clamp
- Voltages at PCB pins
- Full-wave, S-parameters
- FEM method
- AC, transient
- 0–400 MHz

We illustrate the simulation of closed-loop BCI using the example of a 100BASE-T1 transceiver pair. The PCB introduced in the radiated coupling section (Sect. 6.6) is used as the basis for this setup. In the closed-loop BCI configuration, two of these boards are interconnected with an unshielded TWP cable.

To model this setup, we first construct the EM model, incorporating both the injection and monitoring clamps. This EM model is then combined with a circuit simulation. We analyze the differential voltage at the receiver IC in the time domain for an injected BCI disturbance at 90 MHz. Next, we transform the time-domain model into the frequency domain, allowing for efficient evaluation of the BCI disturbance's impact across a broad frequency range. Finally, we demonstrate how the EM model can be partitioned into multiple segments, optimizing computational efficiency when investigating different DUT variants.

The schematic of the transceiver is shown in Fig. 13.8. In the line towards the transceiver, there is a common mode choke along with two 100 nF DC blocking SMD capacitors. The common mode choke used in this setup is the ACT1210L-201-2P-TL00 from TDK (TDK Product Center). In the model, the differential termination of the transceiver is 100 Ω, while the common mode termination is 25 Ω, as proposed by Austermann and Frei (2022).

Both transceivers are identical and positioned floating above a metal table. The TWP wire connecting them has a length of 1 m. An injection clamp is placed 900 mm from one of the transceiver boards to introduce a common mode current I_{CM} into the system. Additionally, two ports are positioned 50 mm from the same board in line with the two cable wires to measure the current in the TWP wires between the boards (cf. Fig. 13.2). These ports function as idealized monitoring

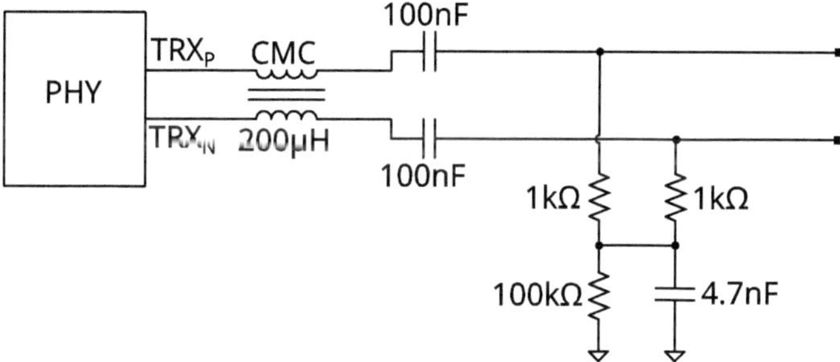

Fig. 13.8 The 100BASE-T1 transceiver is protected by a common mode choke and 100 nF series capacitors

clamp. One end of the system transmits a PAM3-modulated digital signal with a clock frequency of 33 MHz (Open Alliance 2024). After propagating through the TWP wire, the signal is received by the receiving IC. This signal is disturbed by the current injected via the injection clamp. The common mode current in the TWP wire is monitored using the monitoring clamp and adjusted according to the BCI open-loop criteria, taking into account current and power limitations, as shown in Fig. 13.7. At the receiving IC, we observe the differential voltage across the 100 Ω termination to assess the impact of the disturbance.

13.2.1 Coupled EM-Circuit Model

First, we describe the geometry of the EM model, which is subsequently used in a circuit simulation. The transceiver is mounted on a four-layer board, measuring approximately 7 cm in length and 3 cm in width (Fig. 13.9). A differential pair, about 40 mm in length, runs from the connector to the IC. Near the connector, there are designated footprints for the DC blocking capacitors and the common mode choke.

In the EM model, the differential pair at the IC is terminated with one port on each trace, with each port connecting to the ground plane of the PCB. Two ports are assigned to the CMC choke, and two additional ports are designated for the DC blocking capacitors. Since the board contains no other circuits, its layout is simple. The connector is also modeled by simple CAD elements, as it does not include any holes or additional mechanical contact elements to interface with its counterpart. To analyze the impact of layout and mechanical asymmetries, two additional ports, $8 : Cy_p$ and $9 : Cy_n$, are introduced at the positions of the SMD capacitors. These ports connect each trace to the ground plane and serve as placeholders for future studies, allowing for the artificial introduction of capacitive asymmetry in the layout. The board is positioned 50 mm above the metal table. At the connector, the outgoing pins extend into a pair of parallel cylindrical wires, each 9 mm in length. A coated TWP wire of 1 m length is then attached to these stubs.

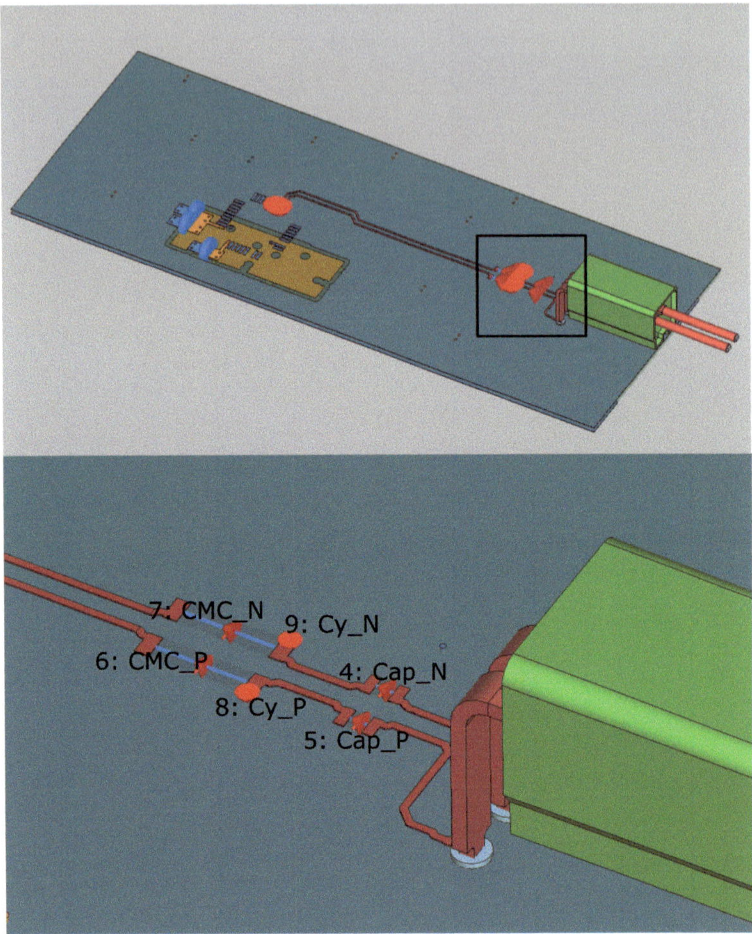

Fig. 13.9 PCB layout. Top: complete board. Bottom: details of ports to mount the CMC, the DC block capacitors, and two capacitors to artificially introduce layout imbalance

The complete EM model is shown in Fig. 13.10. The PCB on the left side, referred to as PCB A, is duplicated and rotated to create an identical counterpart, PCB B. This duplicate is positioned at the opposite end of the model, ensuring that a 1 m TWP wire connects both boards. At a distance of 90 cm from PCB B, the EM model of the injection clamp, as described in Sect. 13.1.1, is placed. There is no explicit EM model for the monitoring clamp. Since the role of the monitoring clamp is to measure the common mode current in the TWP wire, the wire is cut at the clamp's position, which is 50 mm from PCB B. At this location, two in-line ports are introduced along the direction of the individual wires. These ports are later used in the circuit simulation to measure the current. The model is defined with open

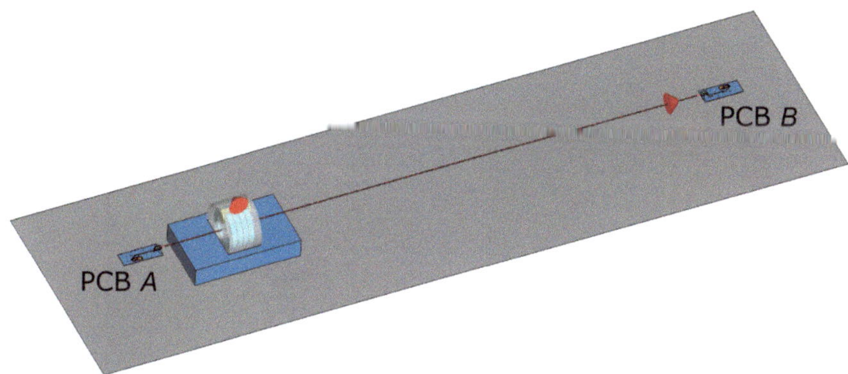

Fig. 13.10 Complete EM model with geometric model of the injection clamp and two in-line ports to monitor the common mode current. Ports on the PCBs are hidden

boundary conditions in all directions except for the ground plane, represented by the metal table, which is modeled as an electrical boundary.

In total, the EM model consists of 19 ports. The computation is performed in the frequency domain, covering a frequency range from 1 Hz to 1 GHz, which allows to apply a maximum rise time of $t_r = 0.333$ ns (cf. Sect. 6.6). Using the FEM solver, the model computes in 3 hours 17 minutes including mesh adaption.

After completing the EM simulation, the EM model is connected to the circuit elements, as illustrated in Fig. 13.11. On the left side of the schematic, the component models of PCB A are shown, while those of PCB B are on the right. The pin for connecting the injection probe is located at the top, while at the bottom, the pins correspond to the two ports in the TWP wire, which are used to monitor the CM current. Mode converters (Sect. 8.3) are used where it is more convenient to analyze or utilize common mode and differential mode signals instead of referencing them to a ground plane. In particular, mode conversion is applied to monitor the common mode current at the position of the monitoring clamp, using the ports InLine_P and InLine_N.

The transmitter is located on PCB A. To model the PAM3 modulation, we use two time-domain voltage sources in series, each generating a digital signal with a 50% duty cycle, a clock frequency of 33 MHz, and an amplitude of 0.65 V. These digital signals are random and phase-shifted by π, effectively producing a random, PAM3-like digital signal. A mode converter is used to differentially feed this signal into the IC pins at PCB A. The common mode at the IC pins is terminated with 25 Ω.

On both PCB A and PCB B, the SPICE simulation model of the common mode choke has a common mode impedance of 12 kΩ at 10 MHz, which corresponds to an inductance of 190 μH. The two 100 nF DC blocking SMD capacitors are modeled as ideal elements. The ESL and ESR are neglected since they are significantly lower than those of the common mode choke, which is in series with them. The ports A_CyP

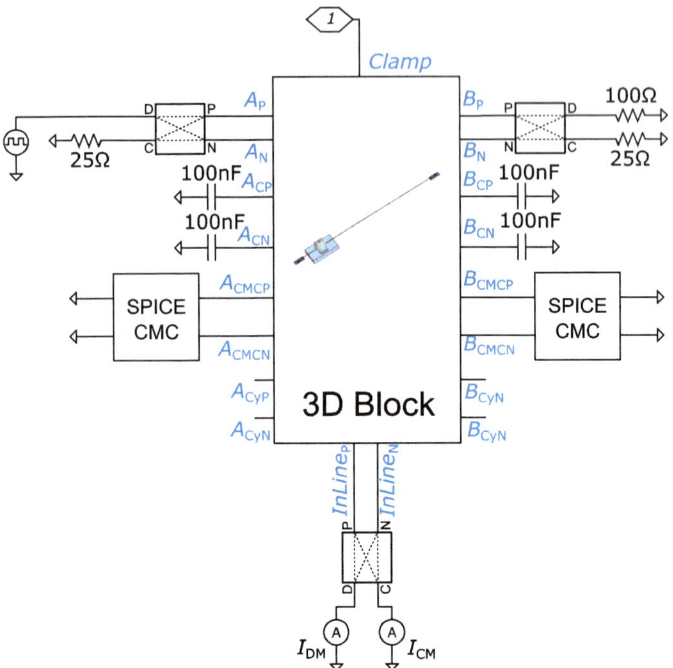

Fig. 13.11 In the circuit model, there is a transient source on PCB *A*. The clamp is equipped with a source injecting a sinusoid, and the received voltage is monitored at the 100 Ω differential port of B_P and B_N

and A_{CyN} are left open and reserved for later use. On PCB *B*, the same component models are used. However, instead of an excitation source, the differential port at the receiver IC pins is terminated with 100 Ω, while the common mode port is terminated with 25 Ω. The high-ohmic resistors to ground which are shown in Fig. 13.8 are neglected in this circuit model.

The pin that connects the BCI injection clamp is excited with a sinusoidal voltage. A single-tone injection is used, with a frequency of $f_{inj} = 90\,\text{MHz}$. The voltage amplitude at the clamp is manually adjusted to ensure that the current limit in the monitoring clamp is met. In the BCI test at 90 MHz, the current limit is 200 mA (RMS value). This corresponds to a peak current of $\sqrt{2} \cdot 200\,\text{mA}$. The power limit is not yet considered.

Finally, we compute the received differential voltage V_{DM} at the IC pin of PCB *B*. The result is presented in Fig. 13.12. On the left side, the common mode current at the in-line ports (which simulate the monitoring clamp) is shown. The current waveform is sinusoidal, with an amplitude slightly (10%) larger than the required $\sqrt{2} \cdot 200\,\text{mA}$. On the right side, the ideal 33 MHz, 50% duty cycle PAM3 signal generated on PCB *A* is displayed in orange. The differential voltage received at the IC pins of PCB *B* is plotted in green. The received signal experiences a delay of a few nanoseconds due to the signal propagation along the 1 m TWP wire.

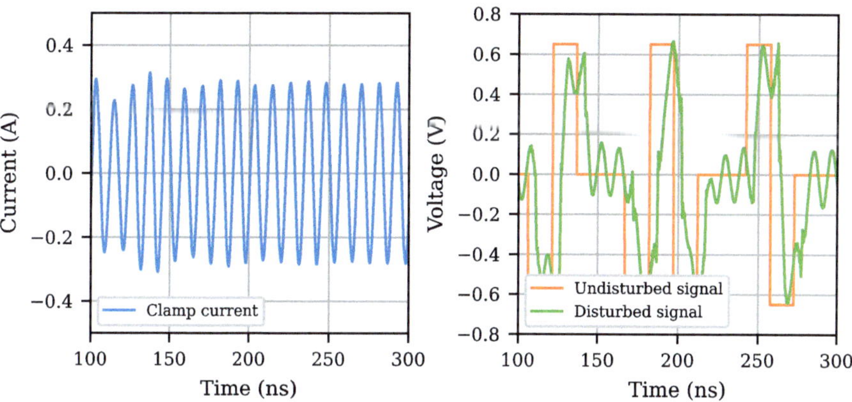

Fig. 13.12 For $f_{\text{inj}} = 90\,\text{MHz}$ (left side, CM current in the monitoring clamp), we see the signal generated on PCB A and the disturbed signal at the receiver on PCB B

Additionally, the superimposed 90 MHz sinusoidal disturbance introduces distortion into the received signal.

There is no strict pass-or-fail criterion for evaluating the impact of the disturbance. A visual inspection suggests that at 90 MHz, the receiver is able to correctly decode the transmitted signal. A receiver typically processes the average of the received signal over the duration of one bit. In this case, the distorting amplitude is approximately $1/3$ of the amplitude of the PAM3 signal. Since the oscillating component averages out over time, the overall distortion is reduced. For a more detailed assessment, the received signal could be further analyzed within the signal-processing chain of the transceiver system. This chain incorporates not only the modulation scheme but also coding and decoding processes to determine the bit error rate of the communication channel under BCI-induced interference. However, implementing such a system is beyond the scope of this book. An alternative procedure is the assessment of a potential failure by recording the voltage at the IC pins and comparing them to specifications in the IC's data sheet. This approach is shown, e.g., in Nayak et al. (2020) and Benz et al. (2019).

13.2.2 Frequency-Domain Model and Impact of Imbalance

The study in Sect. 13.2.1 is conducted in the time domain and provides illustrative results. Since the disturbance in the BCI test is a single-tone AC signal,[2] spanning a broadband range from the kHz domain up to 400 MHz, numerous time-domain

[2] Extending the analysis to amplitude-modulated signals is possible for linear components but is not considered here.

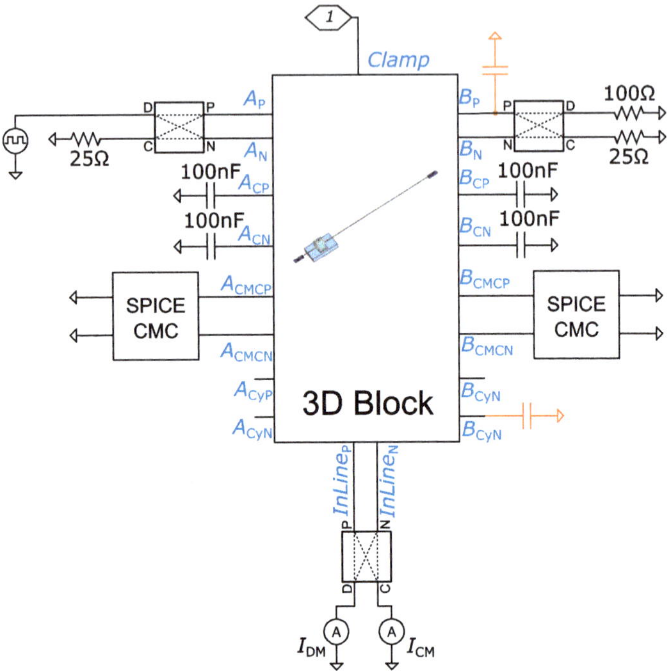

Fig. 13.13 Frequency-domain combined circuit-EM model of the closed-loop BCI test

simulations would be required—one for each injected frequency—to complete the BCI test. Therefore, modeling BCI in the frequency domain is a significantly more efficient approach.

In this section we use this approach to investigate the impact of the injected BCI current onto the differential voltage received at a pair of IC pins.

The corresponding model is shown in Fig. 13.13. Unlike the time-domain circuit, we use an AC source at the pin where the injection clamp is attached. The amplitude of this voltage is set such that the CM current detected by the monitoring probe aligns with the BCI closed-loop specifications described in Sect. 13.1.2. To compute the voltages and currents of the schematic in Fig. 13.13, we excite the clamp with four times the computed forward power P_{fwd} from Eq. 13.4. We then determine the scaling factor from Eq. 13.5 and perform a second simulation that accounts for both the current and power limitations of the closed-loop BCI test. No signal is generated at the transmitting PCB A. Any signal at the receiving IC of PCB B is solely due to the injected disturbance. The simulation is conducted over a frequency range from 10 kHz to 400 MHz.

First, we examine the CM current in the wire harness, monitored at the position of the monitoring clamp. This is represented by the solid blue curve in Fig. 13.14, plotted over the frequency range from 1 MHz to 400 MHz.

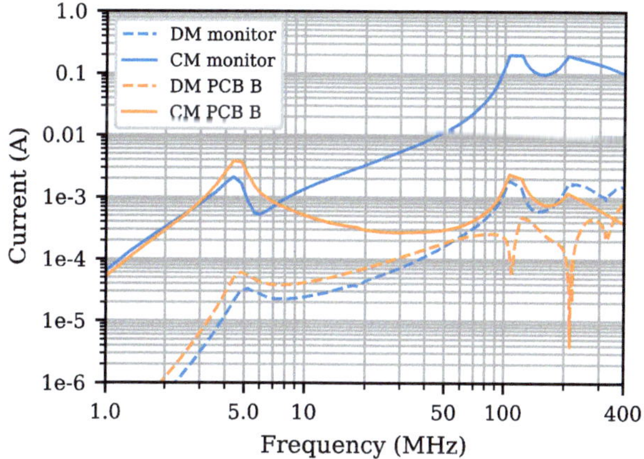

Fig. 13.14 Common and differential mode currents in the current monitors at the monitoring clamp and through the IC pins of PCB *B*

The CM current I_{CM} injected by the clamp remains below 0.2 A up to 100 MHz, where the forward power $4P_{fwd}$ acts as the limiting factor. At 100 MHz, I_{CM} reaches the threshold of 200 mA, forming a visible plateau where the current limit becomes dominant. Beyond 200 MHz, I_{CM} decreases due to a progressively lower current limit between 200 MHz and 400 MHz.

Additionally, the mode converter attached to the in-line ports at the monitoring clamp position enables the investigation of the differential mode current I_{DM} in the TWP (dashed blue line). The magnitude of I_{DM} is approximately two orders of magnitude smaller than that of I_{CM}. Since the BCI clamp injects only common mode current and no differential mode source is present, the differential mode observed originates from mode conversion. Due to the symmetric layout of the PCBs and the TWP, mode conversion remains low.

Figure 13.14 plots the differential mode current (dashed orange curve) and the common mode current (solid orange curve) through the two IC pins of the receiver on PCB *B*. Up to 4 MHz, the common mode current at the IC pins matches the current measured at the monitoring clamp. Beyond this frequency, it remains relatively flat, while I_{CM} in the wire continues to increase. This behavior is due to the ACT1210 common mode choke, whose increasing common mode impedance suppresses common mode currents toward the IC pins.

The dashed orange curve represents the differential mode current through the IC pins. Up to 60 MHz, its shape closely follows that of the differential mode current in the wire, albeit at a slightly higher magnitude. The additional path length between the monitoring clamp and the IC pins introduces some mode conversion, even though the PCB layout is nearly ideal: apart from the section at the connector and at the IC, the pair of traces is symmetric.

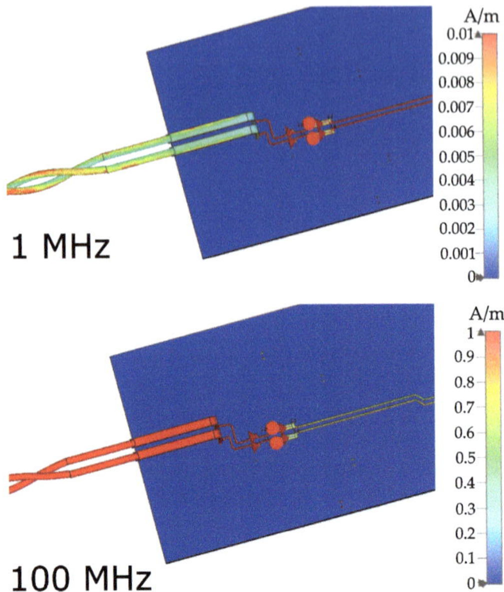

Fig. 13.15 Surface current in the vicinity of the common mode choke. The connector and cable plastic housings are hidden from view. Top: 1 MHz. Bottom: 100 MHz. Since the current injected at 1 MHz is lower than at 100 MHz, the color scaling in both figures is different

Above 100 MHz, I_{DM} becomes equal to, and even exceeds, the common mode current at the IC pins, as the common mode choke provides only little suppression of differential mode currents.

To better understand the impact of the common mode choke on I_{CM}, we plot in Fig. 13.15 the surface current density in the vicinity of the choke of PCB *B* for two different frequencies. The lower frequency, 1 MHz (top), is below the first resonance seen in Fig. 13.14. The upper frequency (bottom side of the plot) is at 100 MHz. At 1 MHz, the current is constant on both sides of the common mode choke. Along the wires, the surface current in the plot is lower than on the PCB traces, because the wires have a larger surface area. At 100 MHz, the surface current is high on the left section of the common mode choke. On the PCB side of the CMC, the surface current is lower because the choke blocks the current at 100 MHz: the common mode impedance of the CMC is tenfold the common mode impedance at 1 MHz.

Imbalance caused by asymmetry leads to mode conversion. In BCI tests, where the source injects a common mode signal, mode conversion becomes the key mechanism to analyze when monitoring disturbances at differential input pins of the DUT. To further investigate this effect, we intentionally introduce imbalance into the model and examine its impact on the disturbance at the differential pair of the IC.

Modifying the layout requires changes to the CAD model followed by an additional EM simulation. For a qualitative understanding, we begin with a circuit simulation, which is significantly faster and sufficient to provide insight into relevant parameters and their ranges. To introduce imbalance in the model, we modify the

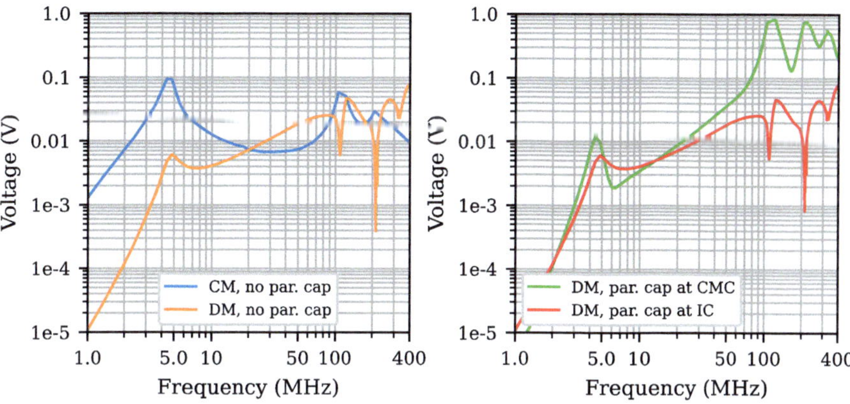

Fig. 13.16 Left: V_{CM} and V_{DM} at the IC pins of PCB *B* for the original layout. Right: V_{DM} at the IC pins when imbalance is introduced by lumped capacitances

circuit simulation by adding two capacitors to the receiving PCB *B*, shown in orange in Fig. 13.13.

One capacitor is placed at the IC pin (top right in Fig. 13.13), introducing imbalance directly at the IC. The second capacitor (bottom right in the circuit schematic) is positioned between the DC suppression capacitor and the negative branch of the CM choke. It is connected to port B_{CyN} in Fig. 13.13, linking the PCB trace to the PCB ground. To mount the second capacitor, we utilize one of the auxiliary ports defined in the EM simulation, as described in Sect. 13.2.1. Defining such auxiliary ports during the EM model setup is advantageous, as it enables further circuit-level analysis without requiring modifications to the EM simulation model.

The results are presented in Fig. 13.16. On the left-hand side, we show the voltages V_{CM} and V_{DM} at the IC pins for the layout without the additional capacitors. Similar to the currents I_{CM} and I_{DM} in Fig. 13.14, V_{CM} dominates at low frequencies (up to 20 MHz). Beyond this frequency, either V_{CM} or V_{DM} is stronger, with both being of the same order of magnitude.

On the right-hand side, we observe the impact of imbalance on V_{DM}. The red curve represents the case where the capacitance is attached directly to the IC pins. This curve is nearly identical to the orange curve in the left-hand plot, indicating that the effect of this capacitance is negligible. In contrast, the green curve shows V_{DM} when the imbalance is introduced between the DC blocking capacitors and the common mode choke. Above 20 MHz, the green curve surpasses the red curve, and beyond 100 MHz, the difference exceeds an order of magnitude.

Mode conversion has a significant impact when introduced before the common mode choke. Any differential mode current generated in that part of the layout is not filtered by the CMC. In contrast, the effect of the imbalance introduced at the IC pins remains negligible.

Finally, we validate this observation by introducing a geometric modification to the layout. Specifically, we remove part of the ground fill of PCB *B*, as shown on the

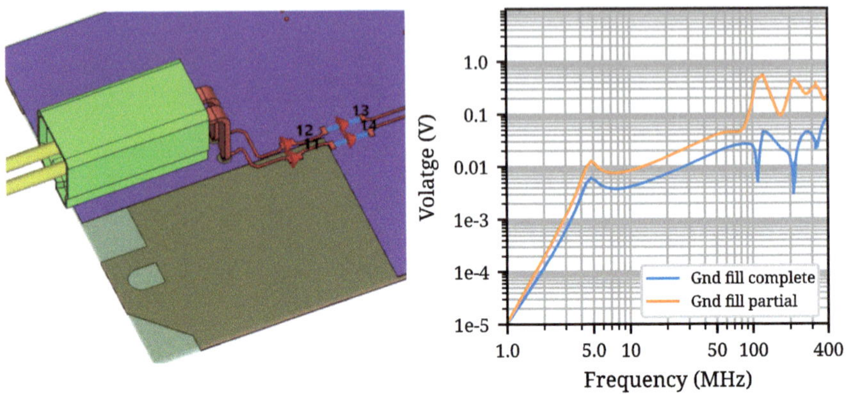

Fig. 13.17 Left: overview of the ground fill removal in a part of the second PCB layer. Right: V_{DM} at the IC pin of PCB B for the ideal reference layout and for an imbalance introduced by removing part of the ground fill

left side of Fig. 13.17. This removal alters the parasitic capacitive network that exists between the traces, the connector, and the ground. The removed surface on this simple board is illustrative and much larger than it may be on a densely populated PCB. Since the capacitance between the traces and the ground are dominated by structures in the close vicinity of the traces, the removal of a smaller piece of ground close to the trace will also be effective.

We repeat the EM simulation of the BCI test using the modified layout to evaluate its impact on the differential mode. The results are presented on the right side of Fig. 13.17. The lower blue curve represents the differential mode voltage at the IC pins of PCB B for the reference case, corresponding to the scenario where all capacitors are set to zero, meaning no imbalance is introduced. The orange curve, on the other hand, shows the differential voltage for the layout with part of the ground fill removed. At low frequencies, where capacitive coupling is low, there is little impact. Above 100 MHz, the layout with the imbalance causes V_{DM} to be up to a factor of 10 higher.

The impact of imbalance is particularly strong when the reference case is highly symmetric. Increasing an existing imbalance has much less effect. In this context, the increase in V_{DM} when comparing the layout without a ground fill to the one with the added ground fill is particularly strong.

13.2.3 Cascading of Model Sections

The two wires of the TWP are thin and positioned close to each other, resulting in a large number of mesh cells along the entire length of the cable. In the BCI model, only a small portion of the entire setup is occupied by the DUT, while the majority consists of the wires, the clamp, and the table. While optimizing the DUT requires modifications to its geometry, the other components—namely, the wires, clamp, and table—remain fixed.

In this section, we discuss how to decompose the BCI model into segments. These segments are then connected on circuit level, providing an efficient approach to accelerating BCI simulations for multiple DUTs. We compare in this section models that are composed of different segments and study their performance with respect to the non-segmented model. This model cascading is also applied in Sect. 12.6 for radiated emission models and investigated in Reddy et al. (2017) .

Each individual segment is computed using an EM simulation and then connected within a circuit simulation. Since each segment is smaller than the entire BCI setup, its modeling is faster, reducing overall computation time. If the geometry of the DUT is modified, only the DUT model needs to be recalculated. The updated EM model is then substituted in the circuit simulation, which runs within seconds only.

Additionally, the EM model for the segment containing only wires can be computed using MTL (cf. Chap. 8), enabling the consideration of wire harnesses with complex geometries.

We apply segmentation to the closed-loop setup shown in Fig. 13.2. The wire bundle is 1m long, with the injection clamp positioned at (900 ± 10) mm from the DUT and the monitoring clamp at (50 ± 10) mm. The wires are terminated at both ends by two PCBs, with PCB A on the left acting as the transmitter and PCB B on the right serving as the receiver.

We divide the model into four segments: PCB A, the injection clamp, a wire segment that includes the probes for the monitoring clamp, and PCB B. Two of these segments are illustrated in Fig. 13.18.

On the left side of the figure, the EM clamp model, introduced in Sect. 13.1.1, is shown. The clamp is loaded with the TWP wire, and the length of this segment is 22 cm. Since the clamp itself measures 6.3 cm along the wire, the wire extends approximately 8 cm on either side of the clamp. The PCB model is cut at the TWP wire model, 1 cm behind the connector.

Wherever the TWP is cut, the segment is terminated with a large metal plate of the same width as and connected to the metal plate representing the table. A

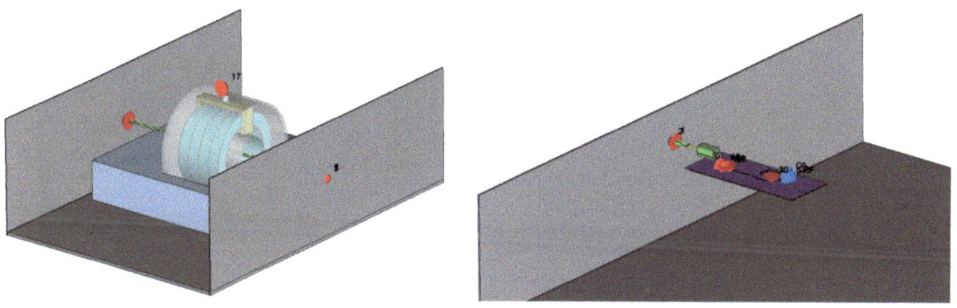

Fig. 13.18 Left: EM model of the clamp with TWP wire. Right: PCB model

lumped port connects each wire end of the TWP to the terminating plate. These ports measure the currents flowing through and the nodal (single-ended) voltages with respect to ground at each wire end. Since the metal plates have a wide geometry, their contribution to the total inductance of the system is minor. In the subsequent circuit simulation, the individual segments are cascaded by connecting the corresponding ports of adjacent segments.

There is a capacitive coupling between the wire and the metal sheet, which distorts the electric and magnetic fields in its vicinity. As the frequency increases, the coupling effect becomes stronger, making model segmentation with the metal sheet less accurate. In Sect. 12.6, we discuss an alternative approach to model segmentation. In that case, the model does not include a metallic table plate, and waveguide ports are used to cascade the individual segments.

Using metal sheets for model segmentation is more accurate at low frequencies (below 1 MHz), where capacitive coupling is minimal and waveguide ports are less accurate. At higher frequencies, however, model cascading with waveguide ports provides better results. The choice between metal sheets and waveguide ports for segmentation in a given application and frequency range can only be determined by comparing the segmented model to the complete, unsegmented model.

One segment consists of the remaining wire between the injection clamp and PCB B. There are two possible approaches to modeling this wire. The first is to treat it similarly to the clamp or the PCBs by attaching vertical metal planes and solving it with a full-wave solver. The second approach is to use MTL. The BCI test is conducted over a frequency range of 1–400 MHz. MTL (cf. Chap. 8) enables accurate and efficient cable modeling within this range. Applying MTL facilitates the computation of complex wire harness geometries and allows for the convenient generation of cascaded models for individual wire segments. This is particularly useful for accommodating various wire lengths in both open-loop and closed-loop BCI tests, including different clamp positions. In the present BCI test setup for the 100BASE-T1, this wire segment has a length of approximately 750 mm.

The segments are cascaded in a circuit simulation, as shown in Fig. 13.19. The circuit model consists of the PCB model on the left, followed by the injection clamp model, the wire segment, and, finally, the same PCB model as the termination on the right. The segments are connected via lumped ports, which link each wire end to the metal plate termination. At each port, we obtain the current flowing through the wire and the voltage between the wire end and the metal plate, which represents the table.

In Fig. 13.19, an S-parameter block generated by the MTL solver is used to represent the wire. To the right of this block, two mode converters transform the single-ended voltages at the wire ends into common and differential modes. Between these converters, we probe the common mode current I_{CM}.

Since we perform closed-loop BCI, measuring I_{CM} in the wires is essential for monitoring the maximum required current and determining the correct scaling of the clamp excitation (cf. Sect. 13.1.2). The probe is positioned between the end of the segment of PCB B and the end of the wire segment—only 1cm away from the connector instead of the standard requirement of approximately 5 cm (Fig. 13.2).

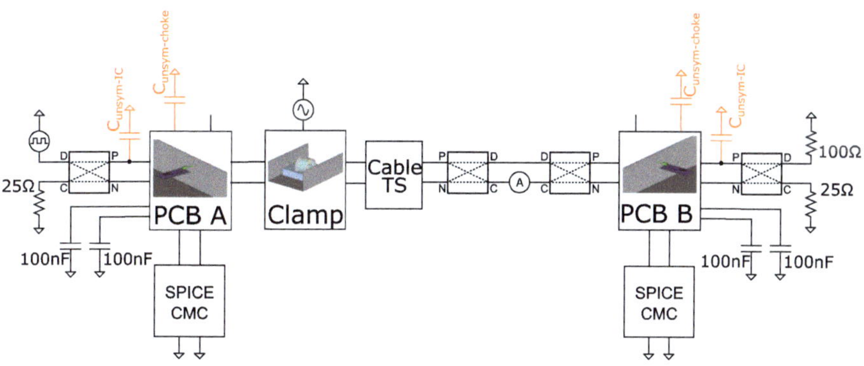

Fig. 13.19 The cascaded model as a circuit simulation. In the sketch, the cable is represented by a touchstone file as generated from MTL. Alternatively, an EM model can be placed here

However, according to MTL theory, TEM modes propagate along the wire following the 1D wave equation (Eq. 8.13a). Given that the minimum wavelength of the BCI test is 0.75 m at 400 MHz , we conclude that a shift of the clamp position by a few centimeters has only a minor impact on the monitored amplitude.

The computation time for the individual model segments is 4 minutes and 7 seconds for the PCB model, 19 minutes 7 seconds for the clamp model with the TWP as load, and about 30 seconds for the TWP model computed using MTL. The full-wave model of the wire segment requires 23 minutes and 53 seconds. The non-cascaded model (Sect. 13.2.1) requires more than 3 hours. Whenever changes in the layout of the DUT are made, the recomputation of the non-cascaded model requires another 3 hours, whereas the cascaded model requires only 4 minutes of EM computation time.

We compare the results of the non-cascaded and cascaded models in Fig. 13.20. Two cascaded models are considered: one in which the wire harness is computed using a full-wave solver and another in which the wire segment is computed using MTL. As in Sect. 13.2.2, we compare the differential voltage V_{DM} at the IC pins of PCB *B* for closed-loop BCI using a frequency-domain model. We analyze the completely symmetric case, meaning no capacitors are added, and there is no layout asymmetry. The frequency range under consideration is 1–400 MHz. Below 1 MHz, V_{DM} is negligible. The non-cascaded model is plotted in blue, the cascaded model using a full-wave solver for the wire segment is shown in orange, and the model based on MTL theory is represented in green.

The three models coincide within the thickness of the plot lines in the frequency range from 3 MHz to 50 MHz. Below 3 MHz and at resonant frequencies, deviations can exceed 6 dB, corresponding to a factor of 2. However, there is no clear trend indicating whether the full-wave or the MTL model of the wire is more accurate in these frequency ranges.

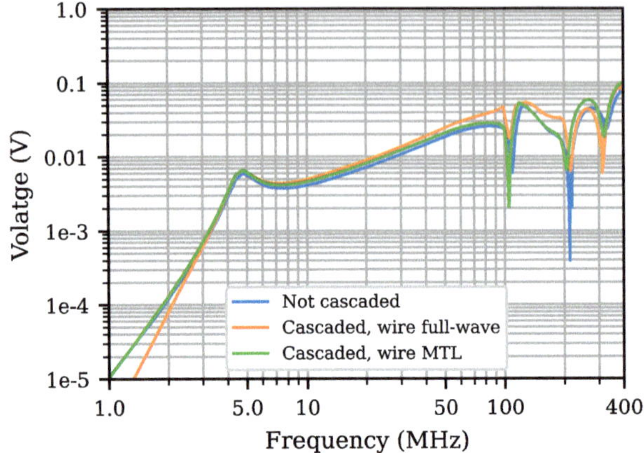

Fig. 13.20 V_{DM} at the IC pins of PCB *B* for cascaded and non-cascaded models

When multiple DUTs need to be evaluated in the BCI test, using a cascaded model offers a practical approach to reducing both modeling and computation time while maintaining an applicable level of accuracy.

13.3 Burst, Surge, and Pulses

Domains: 

Several immunity tests evaluate the effects of transients, including repetitive electrical fast transients (IEC 61000-4-4, bursts); high-energy, high-voltage pulses (IEC 61000-4-5); and ISO pulses on supply lines (ISO 7637-2). The modeling approach for all these tests is the same. Each standard defines a specific pulse waveform, which is then coupled into the DUT. The exact design of the pulse-forming network depends on its hardware implementation and varies between different suppliers.

The modeling approach presented here assumes that the interaction between the pulse source and the DUT occurs exclusively through voltages and currents, without accounting for electric or magnetic field coupling. The circuit model must generate pulse waveforms that meet the requirements of the standard. The foundation of these circuit models is the discharge networks provided in the explanatory sections of the standards (e.g., Keller et al. (2025)). These networks are adapted for implementation in circuit solvers. If needed, tuning the circuit parameters allows the pulse waveforms to be adjusted to match measurements from specific pulse generators.

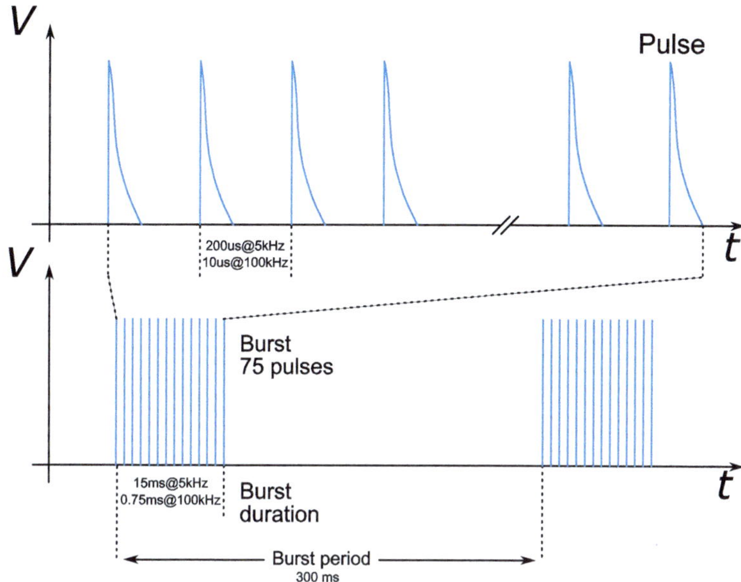

Fig. 13.21 In IEC 61000-4-4 there are repeated bursts which consist of 75 transient pulses

In this section, we develop and analyze a model for burst generation (IEC 61000-4-4) as an example. The presented approach can also be applied to derive generator models for surge and ISO pulses.

The required burst pulse waveforms are illustrated in Fig. 13.21. A burst consists of 75 pulses with a repetition frequency of either 5 kHz or 100 kHz. The burst period is 300 ms. Each pulse features a rise time of 5 ns ± 30% (from 10% to 90% of the pulse amplitude) and a pulse width of 50 ns ± 30% measured at half the pulse amplitude.

The discharge network is illustrated in Fig. 13.22. While the switch is open, the burst generator charges through an RC network consisting of the elements R_C and C_C, with a given discharge voltage V_{Ch}. When the switch closes, the network discharges toward the DUT through the elements $R_M = 50\,\Omega$ and $C_D = 10\,\text{nF}$,

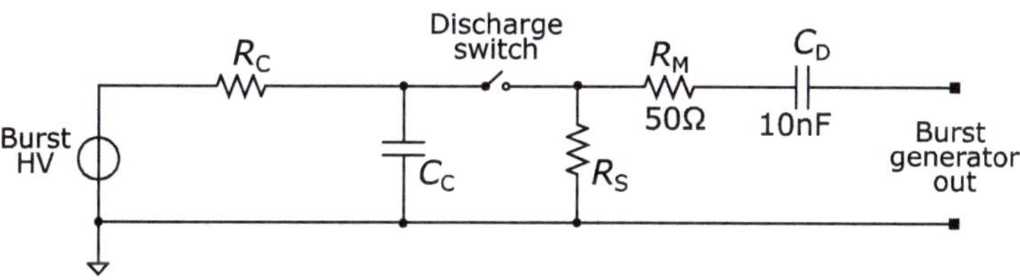

Fig. 13.22 Discharge network as described by IEC 61000-4-4. The values for R_M and C_D are given by the standard

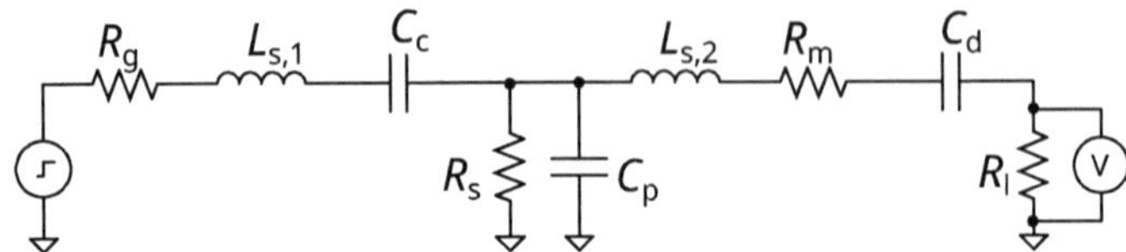

Fig. 13.23 The circuit model of the burst generator uses a step function as source. The parasitic elements can be tuned to match a measured pulse form. $R_M = 50\,\Omega$ and $C_D = 10\,\text{nF}$ are fixed by the standard

which are specified by the standard. Additionally, there is a pulse duration shaping resistor R_S connected to ground.

We convert this sketch into a SPICE simulation, following the procedure described in Cerri et al. (2001). Using the resulting circuit, we simulate a single transient pulse. For all linear DUT models, this is sufficient, as it allows us to analyze, for example, the voltages at IC pins and verify whether any voltage specifications of the IC are exceeded. In such cases, repeating the pulses does not provide additional information. Only detailed functional IC models are sensitive to the pulse repetition frequency.

We extend this circuit by adding three parasitic elements: two parasitic inductances and one parasitic capacitance. Additionally, instead of using a switch, we excite the circuit with a step waveform, resulting in the circuit shown in Fig. 13.23. In this configuration, the discharge capacitor C_C is charged by the voltage step function of a voltage generator with a source impedance of R_G. This section of the network also includes an ESL of $L_{S,1}$. Furthermore, we introduce the parasitic capacitance C_P in parallel with the resistor R_S. The network leading to the DUT is modified by incorporating the ESL $L_{S,2}$.

The value of the charging capacitor C_C is determined based on the energy W that the standard specifies for transfer into the load and the charging voltage V_{Ch}. Using

$$W = \frac{1}{2} C_C V_{Ch}^2, \tag{13.6}$$

we obtain a charging capacitor value of $C_C = 2\,\text{nF}$ for $W = 4\,\text{mJ}$ and a charging voltage of $V_{Ch} = 2\,\text{kV}$.

We analyze the effect of parasitic elements on the voltage across a $50\,\Omega$ target, represented by the load resistor R_l in the circuit shown in Fig. 13.23. As an example, we conduct a sensitivity analysis by varying the circuit parameters $L_{S,1}$, $L_{S,2}$, C_P, R_S, R_G to examine their influence on the pulse waveform. The reference scenario, listed in Table 13.1, is based on values referring to those in Cerri et al. (2001), where the resulting pulse is matched to measurements using a generator hardware and a target.

In the sensitivity analysis, all circuit parameter values are simply doubled, while the generator source impedance is increased by a factor of 10. This adjustment is

Table 13.1 Reference parameter values of the generator model

$L_{S,1}$ (nH)	$L_{S,2}$ (nH)	C_P (pF)	R_S (Ω)	R_G (Ω)
140	200	100	200	0.5

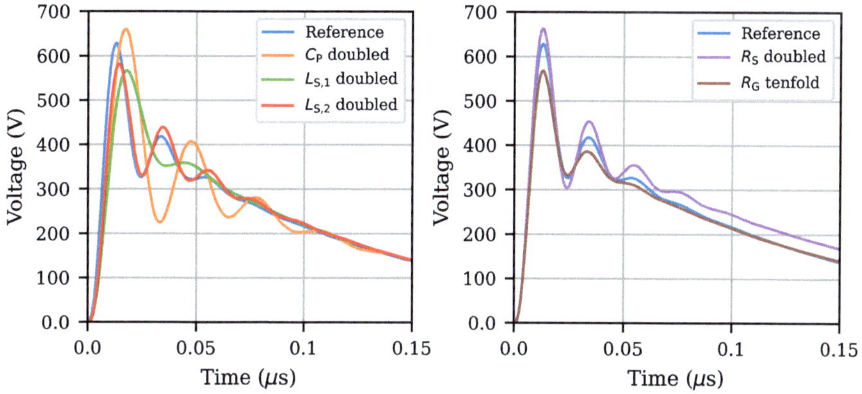

Fig. 13.24 Voltage at the load as a function of the parameters of the circuit model shown in Fig. 13.23. Left: Variation of L and C. Right: Variation of R

made solely to observe its impact on the voltage across the target, regardless of whether the chosen values are realistic.

The results are presented in Fig. 13.24, focusing on the first 150 ns of the pulse. In both plots, the reference scenario is depicted in blue. The left plot illustrates variations in the parasitic capacitance C_P and the parasitic inductances $L_{\text{S},1}$ and $L_{\text{S},2}$, while the right plot shows variations in R_S and the generator resistance R_G.

Across all pulses, we observe a sharp rise time, a peak at approximately 600 V, a damped ringing, and a long decay.

When varying the circuit parameters, we observe the following effects:

- $C_\text{P} = 200$ pF (twice the reference, left plot, orange): The rise time increases, and the peak voltage is 40 V higher. The ringing period extends from approximately 20ns to 30 ns, which corresponds to an increase by a factor of about $\sqrt{2}$. This result indicates that C_P contributes to an LC resonance, which determines the ringing frequency.
- $L_{\text{S},1} = 280$ nH (twice the reference, left plot, green): This parameter reduces the rise time, peak voltage, oscillation frequency, and oscillation amplitude.
- $L_{\text{S},2} = 400$ nH (twice the reference, left plot, red): This parameter decreases both the rise time and peak voltage.
- $R_\text{P} = 400\ \Omega$ (twice the reference, right plot, purple): This parameter affects the decay of the waveform. The 10% level of the initial pulse, corresponding to 67 V across the target impedance, is reached after 280 ns instead of 230 ns. (This is not shown in the figure as the tail is not depicted.)

- $R_G = 5\,\Omega$ (ten times the reference, right plot, brown): This parameter attenuates the oscillation. In particular, the initial peak decreases from 625 V to 570 V.

The sensitivity analysis reveals how different parameters influence the pulse voltage at the 50 Ω target. This information allows to tune the model on available measurement data. The model will then describe the pulses produced by the specific generator the measurement is taken from. It will not describe pulses from other generators, because pulses of different generators are different.

If the engineering task only requires a qualitative understanding of the pulse's impact on components of the DUT, such tuning is not necessary. Qualitatively, the reference model can be used to identify design weaknesses, assess the effectiveness of protection elements, and estimate the order of magnitude of voltages and currents at relevant pins of the DUT.

To apply the presented circuit model to a DUT, we replace the 50 Ω load resistor R_L with a model of the DUT. This model can be implemented either as a circuit model or as an EM model, which is connected to the generator as a network block.

13.4 Radiated Immunity: ALSE

Domains: 

In the RI test (ISO 11452-2), the DUT is placed on a metal table and irradiated with an antenna while in operation. The choice of antenna depends on the frequency range being tested. Typically used antenna types are biconical, log-periodic, and horn antennas. The RI test begins at 80 MHz and its range is up to several GHz.

In this section we show how to set up a model of the RI test and calibrate it according to the standard. We introduce the simplification to replace the antenna model by a plane wave. We discuss the validity of this approach and apply it to study the impact of imbalance on the differential voltage induced at an IC pin of a PCB exposed to RI.

The RI test setup according to ISO11452-2 is shown in Fig. 13.25. The distance between the table and the antenna is 0.9 m. The entire setup is enclosed within an ALSE, ensuring a free-space environment without reflections due to the presence of absorbers. The dimensions and grounding of the table, as well as the position of the DUT, wiring, terminations, and the absorbers are the same as in the radiated emission test described by the CISPR 25 standard (Sect. 12.4). In radiated emission, the DUT acts as the source and the antenna as the sink, whereas in radiated immunity, the antenna serves as the source.

In this section, we focus on a radiated immunity test using a log-periodic antenna, with its phase center aligned with the center of the longitudinal part of the wiring harness. The frequency range extends up to 1 GHz, and we consider both horizontal

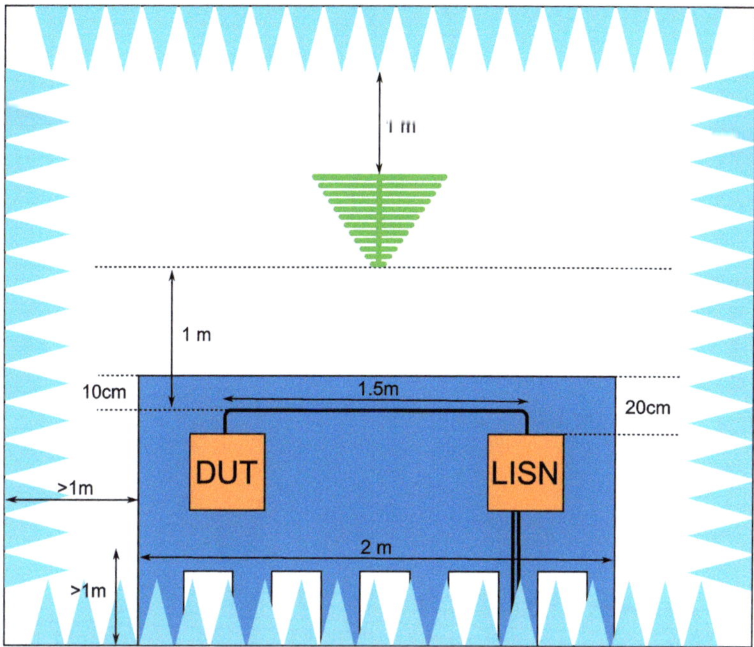

Fig. 13.25 CISPR 25 sketch of a radiated immunity setup

and vertical polarizations. The antenna used, a biconical log-periodic antenna, is the same as the one described in Sect. 12.2.2.

If no non-linear components, such as diodes, are present, the coupling path in a radiated immunity simulation is identical to that of radiated emission. If no further approximations are used, an RI setup can be used for RE, and vice versa.

Due to the size of the table and the presence of the antenna, both RI and RE tests result in computationally large models. The relevant frequency range includes both the near and the far field. The lowest frequency of RI, 80 MHz, corresponds to a wavelength of 3.75 m which is twice the length of the metal table and more than four times the distance between the antenna and the metal table. The DUT is in the near-field of the antenna. However, the entire model is also electrically large. At 1 GHz, the wavelength is 30 cm. Given that the table measures approximately 2 m, the antenna is placed beside it, and the surrounding chamber is included; the setup spans at least 10 wavelengths.

In Chap. 12, we reduce the complexity of the setup by replacing the antenna model by a field probe. In this chapter, we reduce the complexity by replacing the antenna by a plane wave, i.e., a wavefront of constant amplitude perpendicular to its direction of propagation. After introducing the calibration of the RI setup, we discuss the applicability of this plane wave approximation, before using it for an application example.

Table 13.2 Test severity levels

Test severity level	Value (V/m)
I	25
II	50
III	75
IV	100
V	Specific value agreed between users

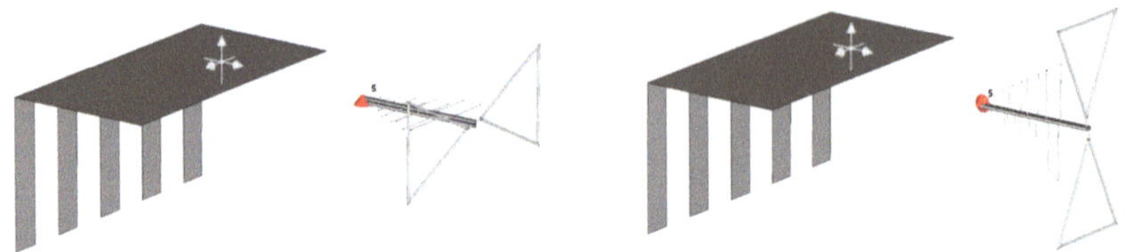

Fig. 13.26 Calibration models. Left: horizontal polarization. Right: vertical polarization

13.4.1 ALSE Model and Calibration

The standard requires to calibrate the RI setup to a predefined, constant field strength E_{cal}. This is measured at a defined probe position 15 cm above the table, in the phase center of the antenna, and 10 cm away from the table edge. This is 10 cm above the center of the cable harness, which, however, is not placed on the table during calibration. Several test severity levels are defined, as listed in Table 13.2.

In this section, we describe how to perform this calibration in a simulation. The ALSE setup for calibration is shown in Fig. 13.26 for both polarizations of the antenna. The table, on which the DUT is later placed, has dimensions of 2000×1000 mm. It is grounded using five ground straps, each measuring 200 mm in width and 1000 mm in length. The table is made of PEC material. We model the absorbing walls of the chamber using PML open boundaries and represent the metallic floor as an electric boundary. The biconical log-periodic antenna is positioned 0.9 m from the table edge, with the tip of the antenna located 10 cm above the table surface.

The standard requires to achieve a field strength according to the test levels of Table 13.2 at the probe position 15 cm above the table in the center of the wire harness, but without the harness being in place. We place a field probe at the specified position and monitor the electric field across the entire frequency range. The probe records all three components of the electric field separately, as indicated by the three arrows in the plots shown in Fig. 13.26.

e_x component Parallel to the table surface, aligning with the wire and the direction of the horizontally polarized electric field

e_y component Points along the axis of the antenna

e_z component Perpendicular to the table surface, corresponding to the direction of the vertically polarized electric field

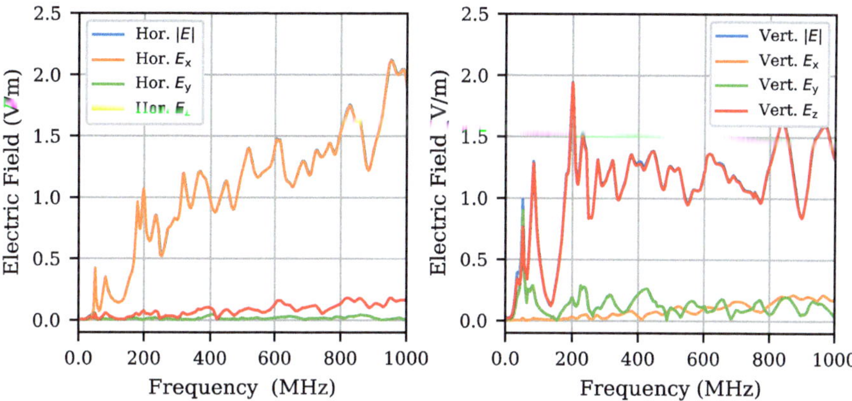

Fig. 13.27 Field strength at calibration point for an excitation of 1 V. Left: horizontal polarization. Right: vertical polarization. $\mathbf{e}_x$: parallel to table surface and in line with the wire. $\mathbf{e}_y$: in direction of antenna. $\mathbf{e}_z$: perpendicular to table surface. On the left-hand side, the blue and orange lines are overlaying

Due to the 2 m wide table with 1 m ground straps and the maximum frequency of 1GHz, the model response exhibits multiple resonances. Each EM model—one for horizontal and one for vertical polarization—contains only a single port. We efficiently compute the frequency range from DC to 1 GHz using the FIT solver. We prefer to use the FIT solver, because the DUT is simple and thus the cell lumping of the TLM solver does not improve the simulation time significantly (cf. Sect. 4.10).

We use a transient co-simulation. The EM model and the circuit are computed simultaneously. At each time step, voltages and currents are exchanged between the EM model and the circuit elements at the designated ports. A detailed explanation of this simulation approach can be found in Sect. 14.3. The simulation is excited at the circuit level with a Gaussian pulse of 1 V amplitude. The simulation time is 52 minutes and 9 seconds.

The resulting electric field is shown in Fig. 13.27. For the horizontal orientation of the antenna (left-hand side), the dominant field component is $E_{\mathrm{x,1V}}$. The components $E_{\mathrm{y,1V}}$ and $E_{\mathrm{z,1V}}$ are negligible, resulting in $E_{\mathrm{1V}} = |E_{\mathrm{x,1V}}|$. For the vertical orientation (right-hand side), the dominant component is $|E_{\mathrm{z,1V}}|$. A resonance occurs at 50 MHz, involving the $E_{\mathrm{y,1V}}$ and $E_{\mathrm{z,1V}}$ components. Analyzing the field lines at this frequency reveals that the antenna wings, the table edge, and the chamber floor resonate with each other. Figure 13.27 illustrates that when using a Gaussian pulse with an amplitude of 1 V as excitation, the resulting field strength E_{1V} exhibits resonances and increases with frequency.

To calibrate a simulation that includes a DUT, we introduce the scaling factor γ_{RI}:

$$\gamma_{\mathrm{RI}} = \frac{E_{\mathrm{cal}}}{E_{\mathrm{1V}}}. \tag{13.7}$$

In the following, we consider only linear DUTs. The calibration procedure is as follows: We use the same Gaussian pulse with an amplitude of 1 V to excite the DUT. In such a simulation, all observed voltages and currents correspond to a field strength of E_{1V} at the probe position, as shown in Fig. 13.27. Since the system is linear, we scale the voltages or currents of interest using the factor γ_{RI}. This scaling ensures that the adjusted voltages and currents correspond to a field strength of E_{cal} at the monitoring probe, as required by the standard.

13.4.2 DUT Model

- Antenna or plane wave
- Voltages at PCB pins
- Full-wave, S-parameters
- FIT method
- AC
- 0–1 GHz

The DUT is a simple design, similar to the one used in Sect. 12.2.1. It consists of two identical PCBs, each measuring 4 × 8 cm. Each PCB has two layers, with the lower layer serving as the ground plane.[3]

Unlike the model used for radiated emission, this setup features a differential pair: the upper layer consists of two parallel traces, each 50 mm long and separated by 0.2 mm. A connector with two pins links the wires to the traces, with a center-to-center pin spacing of 0.4 mm, matching the trace separation.

The two boards are positioned 50 mm above the metal table and connected by a cable with parallel wires, measuring 1500 mm in length. As specified by the standard, the cables are placed 100 mm from the table edge. We do not model the bend in the wires described in the standard.

In the RI test, the sink is any sensitive electronic component on the DUT. As in the application example of the BCI test (Sect. 13.2), we do not incorporate a functional model of an IC as the sink. Instead, we apply passive terminations at the location of the input pins of the IC and analyze the differential voltage at this position.

The EM model consists of a total of five ports: two on each PCB and a fifth port for antenna excitation. The complete model is shown in Fig. 13.28.

We evaluate the model up to a frequency of 1 GHz. Due to the 1.5 m cable length and the mismatched terminations, resonances occur along the wire, and the

[3] The possibilities to simulate RI for more complex DUTs are as follows: 1. stock up on computer power; 2. replace the antenna by a plane wave, as discussed in Sect. 13.4.3; or 3. do both.

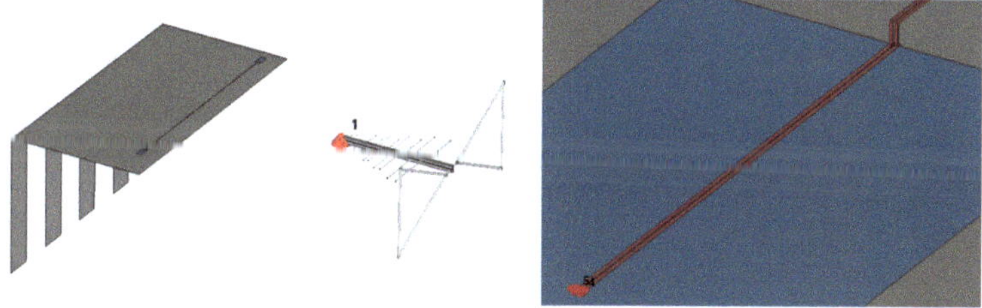

Fig. 13.28 EM model of radiated immmunity. The PCB model is very simple and consists of two traces only

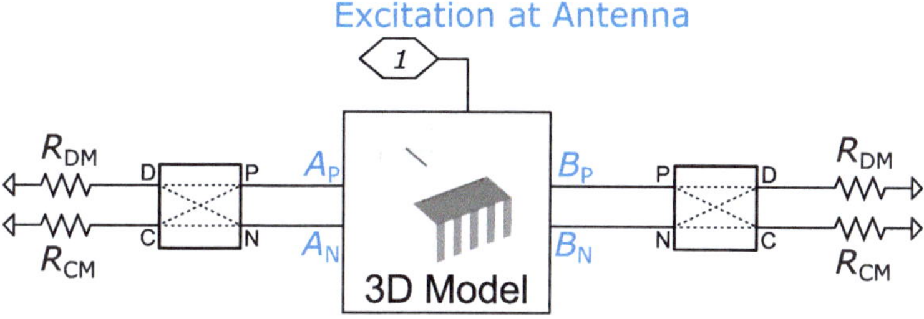

Fig. 13.29 Combined circuit-EM model of the DUT

computational domain is large. We compute over a wide frequency range using the time-domain FIT solver with GPU (cf. Sect. 4.9). We use a co-simulation as described in Sect. 13.4.1.

The circuit model of the DUT is shown in Fig. 13.29. It includes a single excitation port, which applies a Gaussian pulse with an amplitude of 1 V to the antenna port of the EM model. On either side of the model, there are ports corresponding to the differential pair on PCB A and PCB B. Mode converters are used to monitor the differential mode voltage V_{DM} and the common mode voltage V_{CM} on each PCB. The differential mode termination is set to $R_{DM} = 100\,\Omega$, while the common mode termination is set to $R_{CM} = 25\,\Omega$.

The main advantage of the co-simulation is that only a single EM simulation with the time-domain solver is needed to compute these voltages. However, if circuit elements are modified, the entire EM simulation must be re-run. Co-simulation is recommended when only a few circuit simulations are required. However, if multiple circuit-level simulations need to be performed, it is more efficient to use a sequentially coupled simulation in which the EM model calculates the full S-parameter matrix in the first step and the circuit simulation is performed in the second step. This approach is discussed in detail in Chaps. 9 and 10. In the present example, the sequentially coupled simulation requires five runs of the EM model with the time-domain FIT solver, one for each port. The subsequent circuit model runs within seconds.

The co-simulation of this model requires 2 hours, 38 minutes, and 43 seconds. The long wires and the detailed structure of the antenna necessitate a fine mesh. The wires also lead to extended signal propagation times, and since energy dissipates slowly within the model, a long pulse duration must be simulated.

13.4.3 Model Simplification Using Plane Wave Excitation

The presence of the antenna and the table inside the ALSE, combined with the upper frequency limit of 1 GHz, makes the computational domain electrically large. In this section we investigate a modeling approach to reduce the size of the computational domain.

A drastic approach to reducing the size of the domain and the simulation time involves converting the radiated model into a conducted model. In RI, the irradiating antenna generates an electric field that interacts with the DUT and the wiring harness. Unless operating at multiple GHz frequencies, where the wavelength is only a few centimeters, the cross section of the wire harness remains small compared to the wavelength. As a result, all wires within the harness experience a similar field strength across a given cross section. The RI test primarily induces a common mode voltage rather than differential mode voltages between individual wires in the harness. Qualitatively, RI follows the same coupling mechanism as the BCI test described in Sect. 13.1.

The simplest approximation of the RI test involves injecting a common mode voltage directly into the wire harness. In this approach, the antenna model is omitted, and the finite metal table is replaced with a perfectly conducting ground plane. This simplification reduces the RI test to its core coupling mechanism into the wire harness, which is identical to that of BCI. However, this approach is rough, as it simplifies the field coupling of the entire setup on the table to a single common mode injection point along the wire.

In this section, we introduce a less drastic simplification that uses a radiating interference source. Instead of modeling the antenna, we replace the source in the RI setup with a plane wave featuring either horizontal or vertical linear polarization. This modification reduces the computational volume of the model from Sect. 13.4.2 to 40% of its original size. To assess the applicability of this approach, we compare the differential mode voltages V_{DM} induced at the ports of PCB B of the original model, which is irradiated by the biconical log-periodic antenna, and the simplified model using the plane wave.

The model is shown in Fig. 13.30. The table with the DUT is illuminated by a plane wave with either horizontal or vertical polarization, with a field strength of $1\,\frac{\mathrm{V}}{\mathrm{m}}$. The model does not use an electric boundary condition to represent the floor of the ALSE. This occurs because the electric boundary condition enforces a zero tangential field, $E_{\mathrm{t}} = 0\,\frac{\mathrm{V}}{\mathrm{m}}$, which contradicts the imposed plane wave boundary condition with an amplitude of $1\,\frac{\mathrm{V}}{\mathrm{m}}$ at the edge where they coincide. To resolve this conflict, we model the floor as a finite plate made of PEC material.

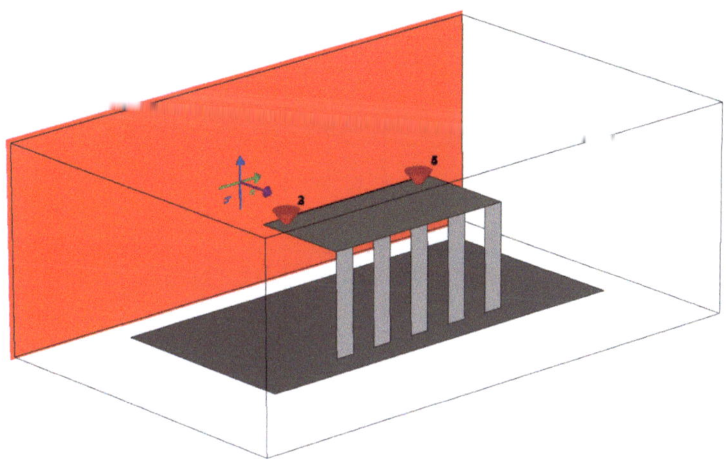

Fig. 13.30 Simplified model RI model with plane wave excitation

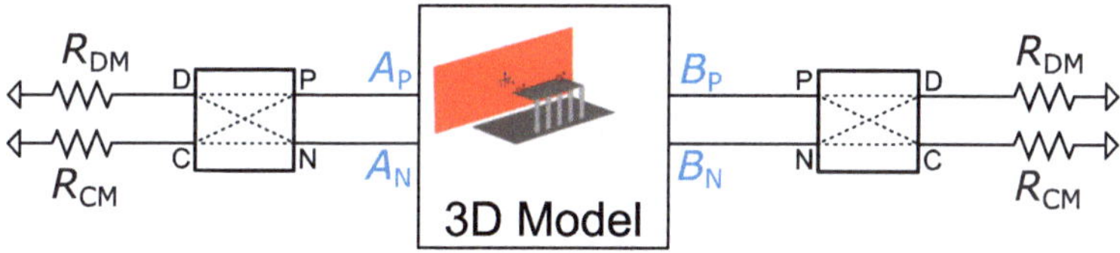

Fig. 13.31 Simplified circuit model using a plane wave excitation

At a distance of 200 mm below this floor, we impose an open boundary condition. While this issue does not arise for a vertically polarized plane wave, we use in the following the same finite-floor approach to ensure consistency in model geometry for both polarizations.

The EM model consists of four ports—one at each end of a trace—with the plane wave serving as the excitation source. The corresponding circuit model is shown in Fig. 13.31. We use the same terminations as in Fig. 13.29, with $R_{DM} = 100\,\Omega$ and $R_{CM} = 25\,\Omega$.

As described in Sect. 13.4.2, we perform a transient co-simulation. The simulation takes 1 hour, 40 minutes, and 50 seconds, resulting in a reduction to 60% of the time of the model that includes the antenna. Replacing the antenna by a plane wave reduces the model size, but the small times steps and long simulation times due to the pulse propagation on the wires still prevail.

Before computing any voltage at the DUT, we calibrate the simulation with the plane wave excitation. To achieve this, we construct a calibration model using an empty table and place a field probe at the same position as shown in Fig. 13.26. The probe measures the absolute value of the electric field for both polarizations of the plane wave. We also use a co-simulation. This takes only a few minutes, because without antenna and wires, there are no detailed structures in the model.

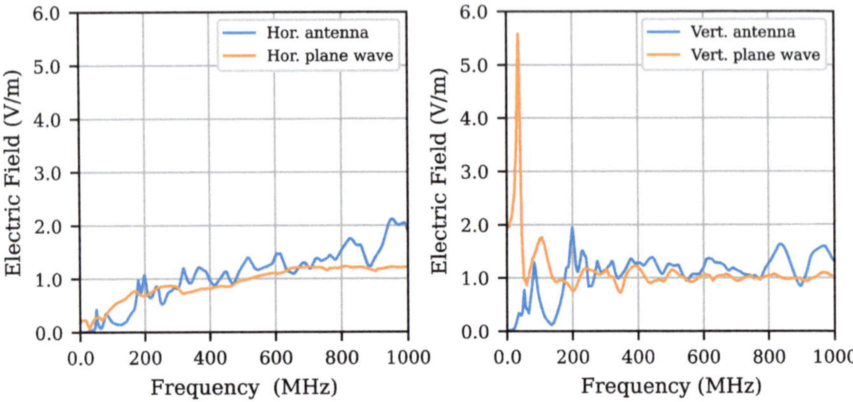

Fig. 13.32 Magnitude of the electric field at the calibration probe for a 1 V excitation with antenna and with plane wave. Left: horizontal polarization. Right: vertical polarization

The results are plotted in Fig. 13.32, along with the electric field curves obtained from the antenna excitation (as shown in Fig. 13.27). For horizontal polarization, the electric field strength monitored at the probe position for the plane wave excitation (orange curve) follows the same general trend as that generated by the antenna. However, the plane wave excitation results in a smoother response. The maximum electric field in the plot reaches $2\,\frac{\mathrm{V}}{\mathrm{m}}$, which is almost twice the excitation field strength reached by the plane wave excitation. This amplification occurs due to constructive interference and is only visible for the antenna excitation at 950 MHz. It is absent for the plane wave excitation.

For vertical polarization, the electric field remains relatively flat and, above 200 MHz, is approximately $1\,\frac{\mathrm{V}}{\mathrm{m}}$. However, the plane wave excitation induces a strong resonance at 50 MHz, with an amplitude exceeding $5\,\frac{\mathrm{V}}{\mathrm{m}}$. This resonance is also present in the antenna excitation but reaches only $1\,\frac{\mathrm{V}}{\mathrm{m}}$. The investigation of the electric field shows that at 50 MHz for the plane wave excitation, the ground straps resonate with the table and the floor of the ALSE.

Finally, we examine the differential mode voltage V_{DM} at the $100\,\Omega$ resistor on PCB B. We compare two circuit simulations: one excited by the antenna (Fig. 13.29) and the other by the plane wave (Fig. 13.31). Both simulations use a calibrated field strength of $E_{\mathrm{cal}} = 100\,\frac{\mathrm{V}}{\mathrm{m}}$ for horizontal and vertical polarization.

The simulations excited by the antenna are calibrated using the field values obtained from the respective calibration setups: either the setups including the antenna (blue curves in Fig. 13.27) or the ones using the plane wave (orange curves in the same figure).

The results are shown in Fig. 13.33. The left-hand side presents V_{DM} generated by the horizontally polarized source, while the right-hand side shows V_{DM} from the vertically polarized source. The blue curves represent V_{DM} obtained from antenna excitation, while the orange curves correspond to plane wave excitation. Both

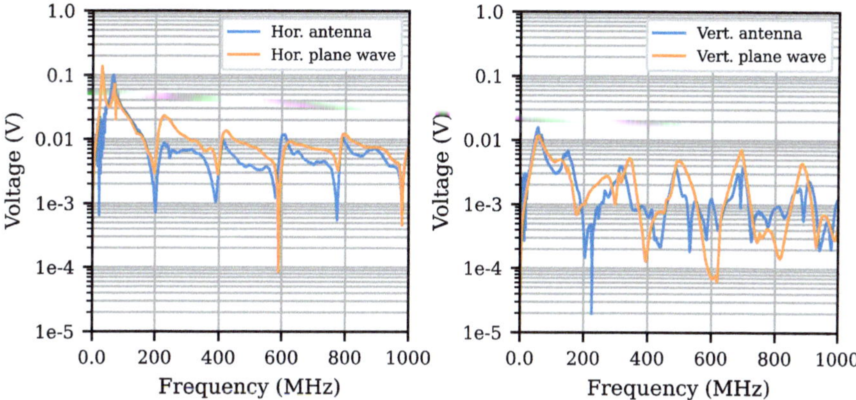

Fig. 13.33 Differential mode voltage for both antenna and plane wave as excitation source. Left: horizontal polarization. Right: vertical polarization

curves follow a similar overall trend, with deviations on the order of a factor of 2 (approximately 8 dB).

We conclude that if the inclusion of an antenna model is computationally expensive or unavailable, the plane wave excitation provides a reasonable approximation.

13.4.4 Application Example: Impact of Imbalance on RI

We analyze the impact of imbalance on the differential voltage V_{DM} received at the ports of one of the PCBs in the DUT design described in Sect. 13.4.2. Imbalance is introduced through an unsymmetry in the layout, similar to the example discussed in Sect. 13.2.2 about BCI.

The analysis is made with the RI model with plane wave excitation, as described in Sect. 13.4.3. The excitation is horizontally polarized, and the calibration field strength is set to $E_{cal} = 100\,\frac{\mathrm{V}}{\mathrm{m}}$.

The EM model described in Sect. 13.4.2 is highly symmetric. The wires and traces are parallel to the table and the PCB ground, and they have equal lengths. Since the incoming electromagnetic field is nearly uniform across the wire cross section, we assume that it predominantly excites a much larger common mode voltage V_{CM} compared to the differential mode voltage V_{DM}.

In a symmetric geometry, no mode conversion occurs, resulting in a significantly lower differential mode voltage V_{DM} at the ends of the traces compared to the common mode voltage V_{CM}.

However, introducing asymmetry into the setup increases imbalance, giving rise to mode conversion and leading to a higher V_{DM}. This effect is also demonstrated in Sect. 13.2.2 for the BCI test, where a capacitive imbalance is introduced by removing the ground fill on the PCB.

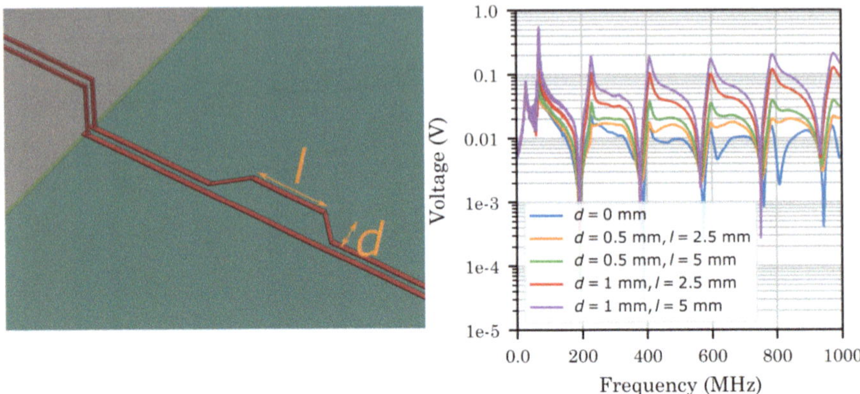

Fig. 13.34 Left: sketch of the layout unsymmetry. Right: the differential mode voltage V_{DM} increases with increasing unsymmetry

We introduce an unsymmetry into the layout of PCB B, as shown in Fig. 13.34. In the reference case, the differential pair consists of two parallel traces. To create an imbalance, we modify a segment of the traces by increasing their separation. Over a length of l, we move the traces apart by a distance of d. We consider four cases, varying d between 0.5 mm and 1 mm, and l between 2.5 mm and 5 mm.

The resulting differential mode voltages V_{DM} are shown in Fig. 13.34. The reference case ($d = 0$) exhibits the lowest V_{DM}. As d and l increase, the mode conversion strengthens, leading to a higher V_{DM}. Among these parameters, d has a greater impact than l: doubling l results in approximately a 50% increase in V_{DM}, whereas doubling d causes V_{DM} to rise by more than a factor of 2.

As in the BCI case study (Sect. 13.2.2), we analyze a highly symmetric configuration. Consequently, even a small imbalance has a noticeable effect. In more complex hardware designs with multiple layers and numerous nets, controlling imbalance becomes more challenging. However, reducing imbalance at critical areas where mode conversion occurs can significantly enhance a DUT's immunity.

13.5 TEM Cell and Stripline

Domains: 

For the simulation of a DUT inside a transverse electromagnetic mode (TEM) cell (ISO 11452-3) or a stripline (ISO 11452-5), we can replicate all geometrical details of an available hardware in an EM simulation model. This time-consuming process is the preferred method when investigating the behavior of the DUT within a specific hardware measurement setup.

The principle of both measurement procedures is to expose the DUT to a homogeneous electric field with a specified field strength. In an experiment this requires careful design. In contrast, the generation of a homogeneous field in a simulation is straightforward. If the purpose of the simulation is simply to test a DUT in a homogeneous field, there is no need to replicate a specific hardware setup.

In the following, we consider such a simplified model of a μTEM cell. Rather than constructing the cell according to predefined dimensions and including the required septum inside its housing, we create a simple box that generates a homogeneous electric field to which the DUT is exposed.

This procedure shown here can also be used to create models of other TEM cell geometries or of striplines.

13.5.1 *Simplified Model of a μTEM Cell*

A μTEM cell is a small TEM cell used for testing an IC. The IC is mounted on a PCB to enable its operation, and this PCB is placed at the bottom of the cell. The cell has an impedance of 50 Ω.

To replicate this setup, we construct a box, in which a homogeneous TEM wave with an impedance of 50 Ω is excited. This requires two identical, parallel plates—one at the top and one at the bottom. These plates have a specific width, length, and spacing between them, as illustrated in Fig. 13.35. The metal plates on the top and bottom impose electric boundary conditions, meaning the electric field is perpendicular to the plates. On the left and right sides, we impose magnetic boundary conditions, ensuring that the magnetic field remains parallel to the plates and perpendicular to the magnetic boundaries.

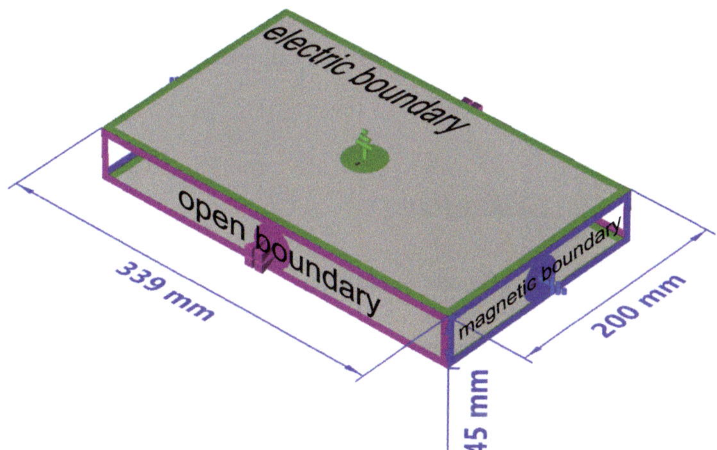

Fig. 13.35 Simple model of a TEM cell using two identical parallel plates

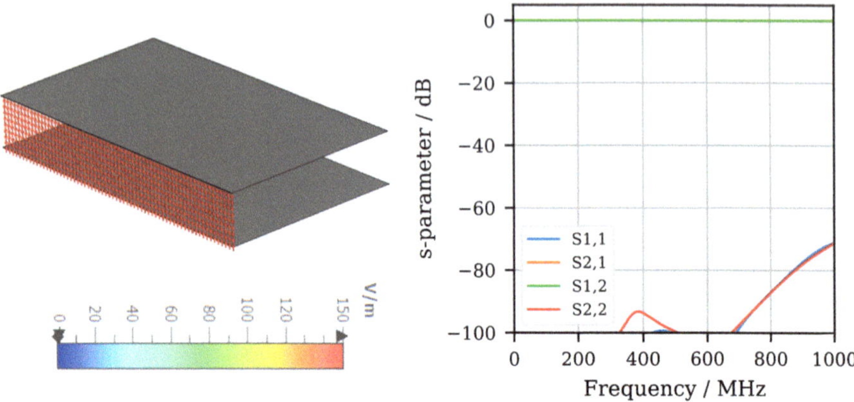

Fig. 13.36 Left: the model exhibits a homogeneous electric field. Right: the reflection less than −50 dB

We excite this structure using a waveguide port. The waveguide port computes the modes of the structure. We use the fundamental mode, which propagates from one side of the cell to the opposite, with the respective electric and magnetic boundary conditions parallel to the propagation axis. This corresponds to a TEM wave with a specific line impedance (cf. Sect. 8.1.1). The line impedance depends on the width of the metal plates and their separation distance. To achieve a line impedance of 50 Ω, we set the width to 339 mm and the height to 45 mm.

The homogeneous electric field is illustrated on the left side of Fig. 13.36. Both ends of the cell are terminated with a 50 Ω waveguide port. The transmission through the cell is nearly perfect, as demonstrated by the S-parameters on the right-hand side of the figure. In the frequency range from DC to 3 GHz, the reflections S_{11}, S_{22} remain below −50 dB. We compute the EM model of the structure using a full-wave FEM solver. The result is a 2-port circuit model, where each port represents one of the two waveguide ports. The computation time is several seconds only.

13.5.2 TEM Cell Calibration

The field inside the TEM cell must be calibrated. The standard specifies a required electric field strength E_{cal}. Since the cell behaves like a parallel-plate capacitor, the field strength is related to the excitation voltage applied between the two plates as

$$E_{\text{cal}} = \frac{V_{\text{cal}}}{d} \tag{13.8}$$

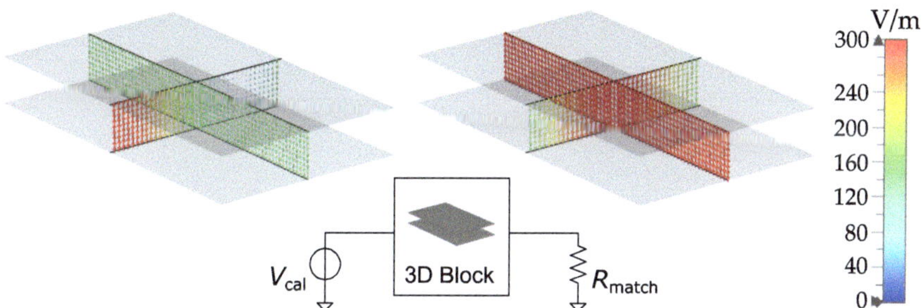

Fig. 13.37 Phase animation of the homogeneous electric field inside the μTEM call, calibrated to $E_{\text{cal}} = 300\,\frac{\text{V}}{\text{m}}$. Left: 0° phase. Right: 55° phase

where d is the distance between the plates. To achieve the required field strength E_{cal}, we excite the cell with a voltage of $V_{\text{cal}} = E_{\text{cal}} \cdot 45\,\text{mm}$, since the height of the cell is $d = 45\,\text{mm}$.

The circuit model and the resulting electric field are shown in Fig. 13.37. The circuit model is displayed at the center of the figure. We excite the circuit with a voltage of V_{cal} and terminate the other end with a matched load $R_{\text{match}} = 50\,\Omega$. In this example, we target a field strength of $E_{\text{cal}} = 300\,\frac{\text{V}}{\text{m}}$, which requires an excitation voltage of $V_{\text{cal}} = 300\,\frac{\text{V}}{\text{m}} \cdot 45\,\text{mm} = 13.5\,\text{V}$.

The electric field inside the cell is shown for two cross sections: one along the propagation axis of the TEM wave and one perpendicular to it. The plot on the left corresponds to a phase of 0°, while the plot on the right corresponds to a phase of 60°. The maximum field strength is $E = 300\,\frac{\text{V}}{\text{m}}$, as calibrated. Due to the finite length of the cell, we observe the propagating wave: at a phase of 0°, the field maximum appears at one end of the cell, whereas at 60°, it shifts to the center.

13.5.3 IC Package in a μTEM Cell

The package geometry of an IC affects its immunity inside a μTEM cell. The electric field inside the cell induces a voltage at the IC pins, which can lead to operational failures.

In this section, we monitor and analyze the differential voltage V_{DM} at the input of a pair of IC pins inside the μTEM model described in Sect. 13.5.1. We use a simplified, abstract package model and investigate resonances of V_{DM} that occur in the frequency range between 1 GHz and 3 GHz. Furthermore, we examine the root cause of these resonances as a function of the IC input impedance and the bond wire geometry within the package.

The geometric model of the package is shown in Fig. 13.38. The package itself is depicted on the top. It is modeled by the ground ring of the IC, which is square with a side length of 10 mm. At each corner of the ground ring, a ground bond

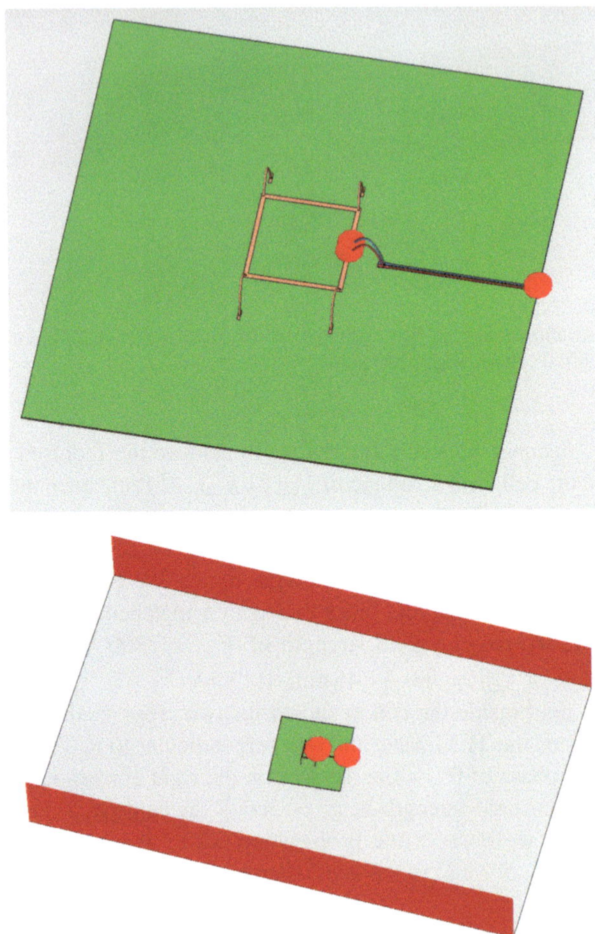

Fig. 13.38 Simplified model of an IC package in the μTEM cell. We model a differential pin pair, and the ground pins connected to the IC lead frame. Top: close-up of the simplified IC lead frame on the PCB, bottom: PCB in the μTEM cell

connects the IC to the PCB ground pads, which are located 2 mm below. The PCB is also square, with a side length of 50 mm. It consists of two layers: the lower layer contains a continuous ground plane, which is mounted on and touches the bottom of the TEM cell. The four ground pads are connected to this ground plane using vias. On one side of the IC, there is a bond pair. One side of the pair is connected to the package's ground ring via ports, while the other side is connected to two parallel traces, each 15 mm in length. All bond wires have a length of 7 mm. The traces extend to the edge of the board, where they are terminated with ports connected to the board's ground plane. The traces and the ground plane are separated by a layer of FR4, which has a thickness of 0.2 mm and a relative permittivity of $\varepsilon_r = 4.2$.

We do not model the substrate of the IC or its mounting on the board. Instead, we assume a vacuum between the ground ring and the PCB, which has a relative permittivity of $\varepsilon_r = 1$. As a result, the parasitic coupling between the ground ring and the PCB ground is lower than it would be if the IC substrate were included in the model, as the substrate typically has a higher ε_r. This capacitive coupling can be approximated as $C = \varepsilon_r \varepsilon_0 \frac{A}{d}$ where A is the surface area of the ground ring and d is its distance from the PCB ground plane (cf. Sect. 3.3 for details on this estimation). Considering the permittivity ε_r of the substrate and glue is meaningful only if the values of A and d are also known, as capacitive coupling depends on all three parameters. These values are only available if the IC layout is known. If this information is unavailable, the capacitance remains an unknown parameter. Its impact can be analyzed by assuming different substrate materials and evaluating their effect on V_{DM}.

On the bottom of Fig. 13.38, the package model is shown mounted inside the TEM cell. The model is aligned such that the magnetic field is parallel to the traces with a tilt of 4°. This slight tilt prevents artifacts in the simulation results caused by perfect symmetry, which do not occur in practical laboratory measurements (cf. Sect. 8.3.2). The EM model includes four ports and is solved in the frequency domain over a range from 100 MHz to 3 GHz. Including mesh adaption, the model computes in 1 minute, 46 seconds.

We expose this model to the homogeneous electric field of the TEM cell by combining the circuit model of the TEM cell with the EM simulation, as shown in Fig. 13.39. We investigate the root cause of any resonances in the differential voltage V_{DM} at the bond wires using a voltage probe. Since we are not interested in the absolute field strength, we excite the waveguide port at one end with an arbitrary voltage of 1 V. The other waveguide port is terminated with 50 Ω. The input capacitance of the IC is modeled as $C_{in} = 6$ pF at both bonds. Additionally, we assume a high-ohmic termination of $R_{board} = 10$ kΩ for the PCB traces. This models a digital input, where only a minimal current flows through the traces.

Next, we perform four simulations in which we vary two parameters:

1. We simulate three different values of the input capacitance C_{in}: 3, 6, and 9 pF.

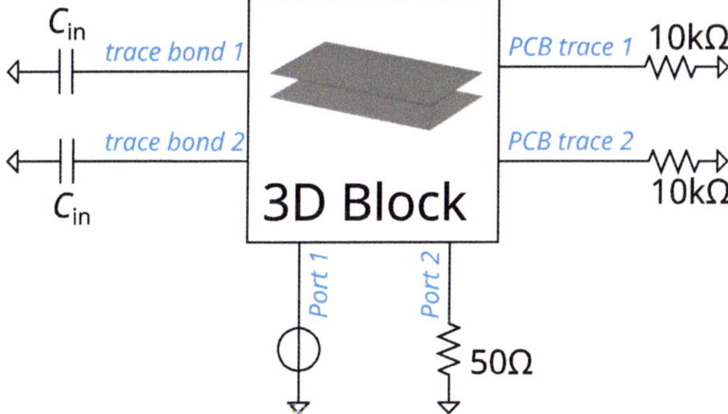

Fig. 13.39 Circuit model of the TEM cell

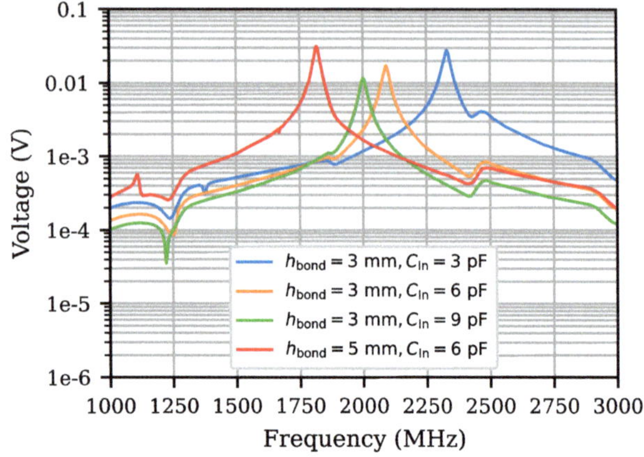

Fig. 13.40 Differential mode voltage V_{DM} at the IC pins

2. For $C_{in} = 6\,\text{pF}$, we vary the height of the two bonds of the differential input, h_{bond}, from 3 mm to 5 mm.

While the variation of C_{in} requires only a circuit simulation, modifying the bond height necessitates an additional EM simulation.

The results are shown in Fig. 13.40, covering the frequency range from 1 GHz to 3 GHz. The reference configuration, with $C_{in} = 6\,\text{pF}$ and $h_{bond} = 3\,\text{mm}$, is represented by the orange curve. In this case, V_{DM} exhibits a resonance at 2090 MHz, which may lead to a failure in the immunity test at this frequency. When C_{in} is reduced by a factor of 2 (blue curve), the resonance shifts to 2330 MHz. Since the resonance frequency of an LC circuit follows $f_r = \frac{1}{2\pi\sqrt{LC}}$, we expect f_r to increase as C_{in} decreases. The simulation confirms that C_{in} influences f_r. However, C_{in} is not the only contributing capacitance. If it were, f_r would increase by a factor of $\sqrt{2}$, reaching approximately 2800 MHz instead of 2330 MHz. Conversely, when C_{in} is increased from 6 pF to 9 pF, f_r decreases by approximately 90 MHz (green curve).

Increasing the height of the bond wire also leads to a reduction in f_r, shifting it to 1820 MHz (red curve). A higher bond wire height increases its inductance, which in turn lowers f_r.

The simulation results indicate that, for this package model, both C_{in} and the bond wire inductance contribute to a resonance of V_{DM} within the frequency range of 1–3 GHz.

Identifying this root cause enables the development of countermeasures if the IC fails the immunity test in the TEM cell at these frequencies. While shielding the IC is an effective but costly solution, an alternative approach is to reduce the bond wire height. This shifts f_r to higher frequencies, where either no immunity requirements apply or the excitation of V_{DM} is insufficient to cause a failure in the immunity measurement.

Further Reading

Bulk Current Injection

B. Adamczyk, *In Compliance Magazine* (2018). https://incompliancemag.com/bulk-current-injection-bci-substitution-method-and-closed-loop-method-with-power-limitation/

C. Austermann, S. Frei, Immunity of CAN, CAN FD and automotive Ethernet 100/1000BASE-T1 to crosstalk from power electronic systems. IEEE Trans. Electromagn. Compat. **64**(6), 2283–2291 (2022). https://doi.org/10.1109/TEMC.2022.3206334

J. Benz, J. Hansen, S. Frei, Simulation and measurement of narrowband susceptibilities of digital automotive sensors, in *2019 International Symposium on Electromagnetic Compatibility (EMC Europe)*, Barcelona (2019), pp. 1088–1093

J. Benz et al., Analysis of immunity failures and optimization measures in automotive sensors, in *2019 International Symposium on Electromagnetic Compatibility (EMC Europe)*, Barcelona (2019), pp. 1014–1019

S. Caniggia, F. Maradei, SPICE-like models for the analysis of the conducted and radiated immunity of shielded cables. IEEE Trans. Electromagn. Compat. **46**(4), 606–616 (2004). https://doi.org/10.1109/TEMC.2004.837841

P. DeRoy et al., Bulk current injection assessment of automotive remote keyless entry systems, in *2017 IEEE International Symposium on Electromagnetic Compatibility & Signal/Power Integrity (EMCSI)* (2017), pp. 660–664. https://doi.org/10.1109/ISEMC.2017.8077950

Fischer Custom Communications, *Bulk current injection probes, F130A-1*. https://www.fischercc.com/products/f-130a-1/

F. Grassi et al., Assessment of the bulk current injection test procedure based on the substitution method. *Asia-Pacific Symposium on Electromagnetic Compatibility (APEMC)* (2013)

S. Miropolsky, A. Sapadinsky, S. Frei, A generalized accurate modelling method for automotive bulk current injection (BCI) test setups up to 1 GHz. *2013 9th International Workshop on Electromagnetic Compatibility of Integrated Circuits (EMC Compo)* (2013), pp. 63–68. https://doi.org/10.1109/EMCCompo.2013.6735174

B.P. Nayak et al., Circuit models for bulk current injection (BCI) clamps with multiple cables. *2018 IEEE International Symposium on Electromagnetic Compatibility and 2018 IEEE Asia-Pacific Symposium on Electromagnetic Compatibility (EMC/APEMC)* (2018), pp. 1160–1163. https://doi.org/10.1109/ISEMC.2018.8393970

B.P. Nayak et al., Model-based system-level EMI/EMC simulation for BCI pass-fail prediction. IEEE Lett. Electromagn. Compat. Pract. Appl. **2**(2), 28–33 (2020). https://doi.org/10.1109/LEMCPA.2020.2979227

I. Oganezova et al., A new and easy approach to create BCI models, in *2014 IEEE International Symposium on Electromagnetic Compatibility (EMC)* (2014), pp. 91–96. https://doi.org/10.1109/ISEMC.2014.6898949

Open Alliance, *Automotive Ethernet specifications* (2024). https://opensig.org/automotive-ethernet-specifications/

S. Pignari, F.G. Canavero, Theoretical assessment of bulk current injection versus radiation. IEEE Trans. Electromagn. Compat. **38**(3), 469–477 (1996). https://doi.org/10.1109/15.536077

S.A.H. Razavi, S. Frei, Characterization of current transformers for impedance measurements in automotive immunity test setups, in *2016 International Symposium on Electromagnetic Compatibility – EMC EUROPE* (2016), pp. 411–416. https://doi.org/10.1109/EMCEurope.2016.7739238

V.S. Reddy, P. Kralicek, J. Hansen. A novel segmentation approach for modeling of radiated emission and immunity test setups. IEEE Trans. Electromagn. Compat. **59**(6) (2017).

G. Spadacini et al., Bulk current injection as an alternative radiated susceptibility test enforcing a statistically quantified overtesting margin. IEEE Trans. Electromagn. Compat. **60**(5), 1270–1278 (2018). https://doi.org/10.1109/TEMC.2018.2810074

TDK Product Center, *ACT1210L-201-2P-TL00*. https://product.tdk.com/en/search/emc/emc/cmf_cmc/info?part_no=ACT1210L-201-2P-TL00

N. Toscani et al., Circuit and electromagnetic modeling of bulk current injection test setups involving complex wiring harnesses. IEEE Trans. Electromagn. Compat. **60**(6), 1752–1760 (2018). https://doi.org/10.1109/TEMC.2018.2794823

Burst, Surge, and Pulses

S. Caniggia, E. Dudenhoeffer, F. Maradei, Full-wave investigation of EFT injection clamp calibration setup, in *2010 IEEE International Symposium on Electromagnetic Compatibility* (2010), pp. 602–607. https://doi.org/10.1109/ISEMC.2010.5711345

G. Cerri, R. De Leo, V.M. Primiani, Electrical fast-transient test: conducted and radiated disturbance determination by a complete source modeling. IEEE Trans. Electromagn. Compat. **43**(1) (2001).

R. Keller, S. Landolt, I. de Mendizábal, *Academy of EMC*. https://www.academyofemc.com/ (visited on 2025)

J. Zhang et al., Modeling of the immunity of ICs to EFTs, in *2010 IEEE International Symposium on Electromagnetic Compatibility* (2010), pp. 484–489. https://doi.org/10.1109/ISEMC.2010.5711323

J. Zhang et al., Modeling injection of electrical fast transients into power and IO pins of ICs. IEEE Trans. Electromagn. Compat. **56**(6), 1576–1584 (2014). https://doi.org/10.1109/TEMC.2014.2332499

Radiated Immunity

P. Alotto et al., Dual-PEEC modeling of a two-port TEM cell for VHF applications. IEEE Trans. Magn. **47**(5), 1486–1489 (2011). https://doi.org/10.1109/TMAG.2010.2088381

S. Caniggia, F. Maradei, SPICE-like models for the analysis of the conducted and radiated immunity of shielded cables. IEEE Trans. Electromagn. Compat. **46**(4), 606–616 (2004). https://doi.org/10.1109/TEMC.2004.837841

S. Deng et al., Mode suppressed TEM cell design for high frequency IC measurements, in *2007 IEEE International Symposium on Electromagnetic Compatibility* (2007), pp. 1–6. https://doi.org/10.1109/ISEMC.2007.13

W. Fang et al., Orientation effect of the electromagnetic field coupling for device in a TEM cell. IEEE Trans. Microw. Theory Techn. **70**(2), 980–991 (2022). https://doi.org/10.1109/TMTT.2021.3128931

F. Grassi, G. Spadacini, S.A. Pignari, The concept of weak imbalance and its role in the emissions and immunity of differential lines. IEEE Trans. Electromagn. Compat. **55**(6), 1346–1349 (2013). https://doi.org/10.1109/TEMC.2013.2261302

F. Grassi et al., Radiated immunity test involving crosstalk and enforcing equivalence with field-to-wire coupling. IEEE Trans. Electromagn. Compat. **58**(1), 66–74 (2016). https://doi.org/10.1109/TEMC.2015.2503599

K. Hu et al., Unbalanced currents in integrated circuits and their effect on TEM cell emissions. IEEE Trans. Electromagn. Compat. **53**(3), 600–610 (2011). https://doi.org/10.1109/TEMC.2011.2118214

T. Jobava et al., Full wave simulation of a vertically polarized transverse electromagnetic mode cell, in *2019 XXIVth International Seminar/Workshop on Direct and Inverse Problems of*

Electromagnetic and Acoustic Wave Theory (DIPED) (2019), pp. 45–48. https://doi.org/10.1109/DIPED.2019.8882608

M. Koohestani et al., Perturbations of electric and magnetic fields due to the presence of materials in TEM cells. IEEE Trans. Electromagn. Compat. **62**(4), 997–1006 (2020). https://doi.org/10.1109/TEMC.2019.2928215

C. Shi et al., Using termination effect to characterize electric and magnetic field coupling between TEM cell and microstrip line. IEEE Trans. Electromagn. Compat. **57**(6), 1338–1344 (2015). https://doi.org/10.1109/TEMC.2015.2459063

G. Spadacini, S.A. Pignari, Numerical assessment of radiated susceptibility of twisted-wire pairs with random non-uniform twisting. IEEE Trans. Electromagn. Compat. **55**(5), 956–964 (2013). https://doi.org/10.1109/TEMC.2012.2235446

G.-J. Stockman, H. Rogier, D. Vande Ginste, Efficient modeling of interactions between radiating devices with arbitrary relative positions and orientations. IEEE Trans. Electromagn. Compat. **56**(6), 1313–1321 (2014). https://doi.org/10.1109/TEMC.2014.2345127

G.-J. Stockman et al., Full-wave modeling of interacting multiport devices with arbitrary relative positions and orientations for efficient EMI assessment. IEEE Trans. Electromagn. Compat. **58**(4), 1322–1330 (2016). https://doi.org/10.1109/TEMC.2016.2576564

TDK, *ACT1210L-201-2P-TL00. TDK Corporation* (2024). https://product.tdk.com/de/search/emc/emc/cmf_cmc/info?part_no=ACT1210L-201-2P-TL00

S. Xia et al., A segmentation approach for predicting plane wave coupling to PCB structures. IEEE Trans. Electromagn. Compat. **66**(2), 483–492 (2024). https://doi.org/10.1109/TEMC.2024.3349911

Chapter 14
Electrostatic Discharge (ESD)

ESD is a natural phenomenon that can impact electronic systems. It may cause temporary disturbances or permanent damage, during manufacturing, assembly, or handling by humans. A characteristic of ESD events is their short transients, with rise times as fast or even faster as 1 ns. The exact shape of the transient depends on the specific scenario, such as a discharge from a human into an IC, from a human into a metal object, or the charging of a semiconductor during manufacturing. Additionally, environmental conditions—such as air humidity or the speed at which the charged device approaches the target—also influence the pulse shape (Pommerenke and Aidam 1996).

In this chapter, we focus on ESD testing according to the standard IEC 61000-4-2, which describes the discharge of a human into a metal object held in their hand. We begin by describing the setup of the standard. Then, we discuss the properties of the ESD generator, which generates the standardized discharge current waveform, and present two ESD generator models. The first is a SPICE model designed to reproduce the current waveform. The second is an EM model, which also accounts for the interaction of the generated electromagnetic field with the DUT.

We introduce transient co-simulation as a suitable method for computing the response of electronic systems to ESD. As an application example, we investigate contact discharge on a USB pin. We describe how the two failure criteria, hard failure and soft failure, are applied in ESD simulation. Finally, we briefly discuss the topic of air discharge.

J. Hansen, A. Barchanski, *EMC Simulation of Electronic Systems*,
https://doi.org/10.1007/978-3-032-09237-3_14

14.1 Experimental Setup

Domains: 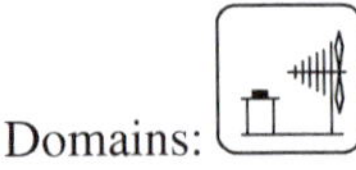

In the IEC 61000-4-2 standard, the DUT is separated from the metallic horizontal coupling plate by a 0.5 mm insulating layer (Fig. 14.1). The discharge can be applied directly to the DUT or indirectly to either the horizontal or another vertical coupling plate. The vertical coupling plate is positioned close to the DUT on the insulating layer. The floor beneath the table serves as the metallic reference plate. The horizontal and vertical coupling plates, as well as the ESD pulse generator, are galvanically connected to this plate using two 470 kΩ resistors in series.

The bottom part of Fig. 14.2 shows the discharge current pulse defined in IEC 61000-4-2 for contact mode ESD. To measure this current, the generator is discharged into a metal plate. The current is measured using a low-ohmic resistor (typically 2 Ω) placed between the generator and the metal plate.

There are four discharge voltages, each with a specified current waveform (Table 14.1). The amplitude of the first discharge current peak, its rise time, and the current levels at 30 ns and 60 ns are defined, allowing for some uncertainty. At the

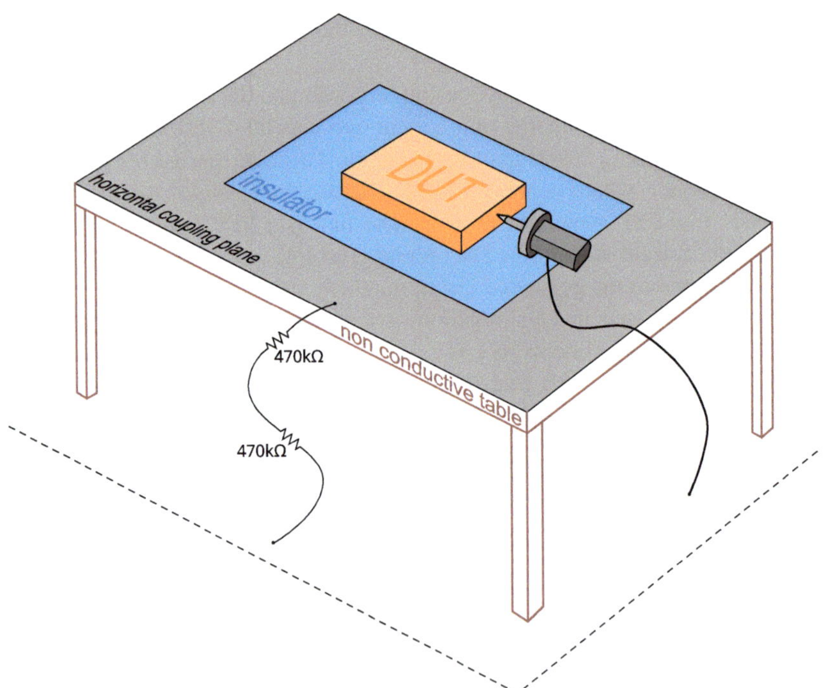

Fig. 14.1 Setup of IEC 61000-4-2

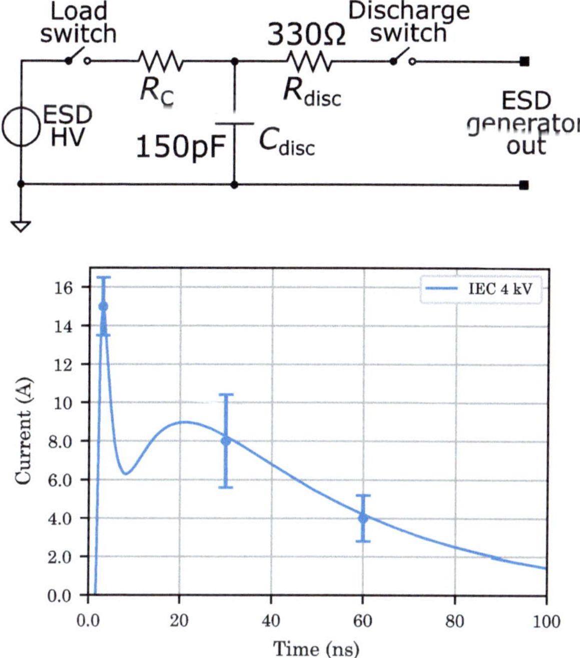

Fig. 14.2 Top: schematic of the discharge network. Bottom: current pulse form in a low-ohmic target at a discharge voltage of 4 kV as proposed by IEC 61000-4-2

Table 14.1 Discharge current waveform parameters according to IEC 61000-4-2 ed. 2 (Javor 2014)

Test level	Discharge voltage	Discharge current waveform parameter			
		First discharge current peak (±15%)	Rise time t_r with discharge switch	Current at 30 ns (±30%)	Current at 60 ns (±30%)
1	2 kV	7.5 A	0.6 to 1 ns	4 A	2 A
2	4 kV	15 A	0.6 to 1 ns	8 A	4 A
3	6 kV	22.5 A	0.6 to 1 ns	12 A	6 A
4	8 kV	30 A	0.6 to 1 ns	16 A	8 A

bottom of Fig. 14.2, we show the current waveform for Test level 2, with a discharge voltage of 4 kV.

It features a short transient pulse with a peak current of 15 A (±15%) and a rise time of $t_r = 0.6\,\text{ns} - 1\,\text{ns}$. The waveform consists of an initial peak followed by a broader second peak. At 30 ns, the second peak must produce a current of 8 A (±30%) in the resistor, and at 60 ns, it must produce 4 A (±30%). The current scales linearly with the applied voltage test levels.

ESD generators produce the pulse using an RC network. In IEC 61000-4-2, the specified values are $R_{disc} = 330\,\Omega$ and $C_{disc} = 150\,\text{pF}$. A simple schematic of the discharge network is shown on the top of Fig. 14.2. When the load switch is closed and the discharge switch is open, the 150 pF capacitor charges to the discharge voltage V_{DC}. Then, the load switch is opened and the discharge switch is closed, allowing the stored energy in the capacitor to discharge through the $R_{disc} = 330\,\Omega$ resistor. The resistor R_C is high-ohmic and in the order of MΩ.

14.2 ESD Generator Models

Domains: 

A generator model can serve two purposes: (1) generating a current waveform that meets the standard's requirements or (2) modeling the current waveform of a specific hardware pulse generator.

We focus on the first aspect. In Sect. 14.2.1, we describe a SPICE model of an ESD generator, derived from the functional schematic shown in Fig. 14.2. The model generates the current pulse required by the standard (bottom part of Fig. 14.2). We use its parameters as reference. As with the burst generator discussed in Sect. 13.3, we perform a sensitivity analysis of the reference SPICE model parameters. This analysis identifies which parameters can be adjusted to match measured data from available hardware. However, if the goal is solely to analyze the response of an electronic system to the standardized ESD pulse, replicating a specific generator is not necessary.

If the DUT is affected only by the generated current, a SPICE model of the generator is sufficient. However, if both the current and the electric or magnetic fields have an effect, a more detailed EM model is required. Such a model is described in Sect. 14.2.2. This EM model generates electric and magnetic fields which interact with the DUT. This field differs from those produced by any specific generator that is available in hardware. While it is possible to model a particular hardware device to describe the electric and magnetic fields emitted during discharge, this process is complex and not demonstrated here (Qing et al. 2008).

14.2.1 Circuit Model of the Pulse Generator

Figure 14.3 shows the circuit model of the generator used in this chapter. The model is based on Speckbacher et al. (2022). Its topology follows the model from (Li et al. 2015, Figure 17), where the authors simulate a TESEQ NSG 438. However, the circuit values have been optimized to best match the discharge current specified by IEC 61000-4-2.

The circuit model is an implementation of the functional schematic shown in Fig. 14.2. It uses only a single switch. The capacitor C_{disc} is charged through SPICE

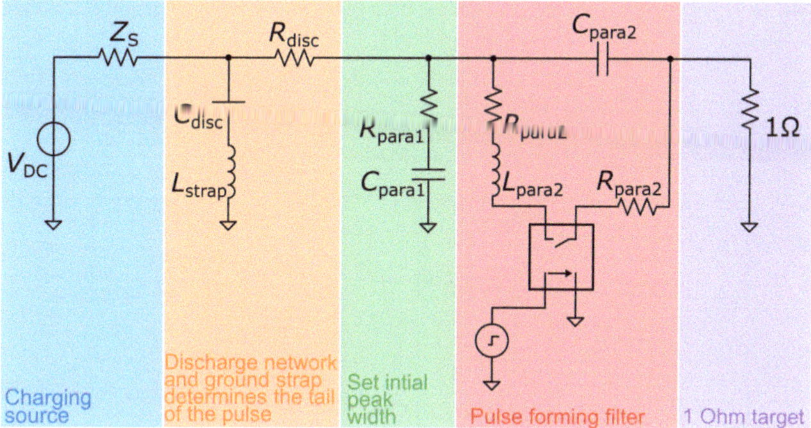

Fig. 14.3 Circuit model of an ESD generator that produces the current specified by the standard

initial conditions by the voltage source V_{DC}. The resistor $Z_S = 1\,\mathrm{M}\Omega$ is set high-ohmic to prevent the storage capacitor from recharging during the short duration of the discharge event.

In series with C_{disc} is the generator's ground strap. For commercial generators, this strap is typically about 2 m long. We model the ground strap using an inductor L_{strap} in the μH range. The value of this inductance depends on the loop area spanned by the ground strap and can be adjusted to match the required current pulse shape. In parallel with the energy storage capacitor, the generator includes the discharge resistor R_{disc}.

The generator model includes a switch that is triggered by a rectangular voltage pulse. When the pulse is applied, the switch closes, initiating the discharge of C_{disc}. In series with the switch is the pulse-forming filter, which consists of two identical resistors, R_{para2}, an inductor L_{para2}, and a parallel capacitor C_{para2}. To the left of this filter, an RC combination, R_{para1}, C_{para1}, shapes the initial height of the pulse. On the right-hand side of the model, the target resistor is connected to the metal plane where the generator discharges. The value of this resistor should be sufficiently low to avoid affecting the pulse shape. Here, we choose 1 Ω.

The sensitivities of these values with respect to the discharge current were first investigated in Wang et al. (2003) for a commercial generator that was modified to validate numerical modeling techniques. Using the model shown in Fig. 14.3, we reproduce this sensitivity analysis through a parameter sweep. We compare the results of this sweep with the nominal curve defined by the IEC 61000-4-2 standard, using a source voltage of 4 kV. The reference model employs the values listed in Table 14.2, which are selected to best match the current waveform specified in the standard.

We perform a parameter sweep, individually varying each value in Table 14.2 by 30%. Since we run a pure circuit model without any EM computation, each run takes only about 1 second. The results are shown in Fig. 14.4. On the left-hand side,

Table 14.2 Reference parameter values of the ESD generator model (according to Speckbacher et al. 2022)

R_{disc} (Ω)	C_{disc} (pF)	L_{strap} (μH)	C_{para1} (pF)	C_{para2} (pF)	R_{para1} (Ω)	R_{para2} (Ω)	L_{Para1} (nH)
280	140	2.4	12	15	68	36	47

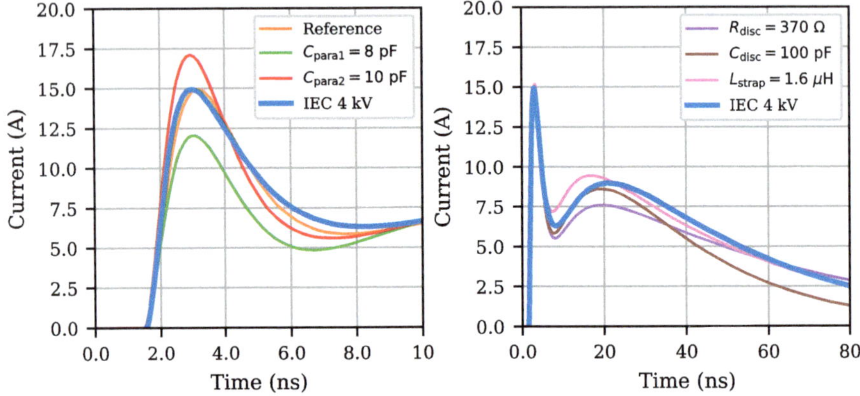

Fig. 14.4 The sensitivity analysis of the circuit

we analyze the parameters that influence the initial peak and the pulse-forming filter. For clarity, the plot is limited to the parameters that have the greatest impact on the amplitude and rise time of the first peak.

We plot four curves:

- Orange curve: Reference case using the values from Table 14.2. The rise time is 1.6 ns, reaching 15 A.
- Green curve: C_{para1} reduced by one-third, from 12 pF to 8 pF. The rise time is 1.5 ns, reaching 12 A.
- Red curve: C_{para2} reduced by one-third, from 15 pF to 10 pF. The rise time is 1.4 ns, reaching 17 A.
- Thick blue curve: The current pulse shape defined by the standard. It closely matches the reference case, with a maximum deviation of 0.6 V at 7 ns.

On the right-hand side of Fig. 14.4, we plot a time range up to 90 ns, focusing on the parameters of the discharge network and the ground strap. These affect the peak of the second pulse and the tail of the current waveform. The results are as follows:

- Thick blue curve: The current pulse shape defined by the standard. In this plot, it coincides almost exactly with the reference case (using the values from Table 14.2), so the reference case is omitted for clarity.
- Purple curve: R_{disc} increased to 370 Ω. The amplitude of the second peak decreases from 9 A to 7.5 A.

- Brown curve: C_{disc} reduced from 140 pF to 100 pF. The pulse tail decays faster, with the current at 60 ns dropping to 2.7 A instead of 4.3 A.
- Pink curve: L_{strap} reduced to 1.6 μH. The second peak reaches its maximum 4 ns earlier.

A parameter sweep, varying each value individually, can be performed efficiently. If measurement data from a physical generator is available, the model parameters can be adjusted accordingly. However, using the reference generator with the parameters given in Table 14.1 already ensures compliance with the standard requirements.

For the simulation of contact discharge on an electronic system, the system model is connected to the circuit model of the generator. The system model can be either a circuit model or an EM model. In the latter case, the EM model of the DUT must include ports to connect with the circuit model of the generator in a way that the current return path is physically modeled.

14.2.2 EM Model of the Pulse Generator

The circuit model described in Sect. 14.2.1 generates the discharge current of the generator based on the attached load (cf. Yang and Pommerenke 2018). However, if the DUT is affected not only by the current but also by the transient electric and magnetic fields of the generator, the SPICE model alone is insufficient, as it does not account for these fields.

In this section, we introduce a geometric model of an ESD generator. When the generator discharges, this model accounts for both the injected current into the DUT and the electric field that builds up between the generator and the DUT. We describe the generator model in detail, explaining how lumped elements, geometrical features, and material properties influence the generated currents and the field distribution when the generator discharges into a metallic plate.

A sketch of the geometric model of an ESD generator is shown in Fig. 14.5. The generator features interchangeable tips for air or contact discharge (orange). The discharge module houses the relay that initiates the discharge of the charged generator. On the right side, the discharge resistor R_{disc} (light blue) and capacitor C_{disc} (purple) can be replaced, and the polarity of the pulse can be adjusted (light green). A trigger mechanism is used to start the discharge event (orange). The discharge return cable, which is 2 m long, is connected to the generator's body (not shown in the figure).

The EM generator model must fulfill two criteria: (1) it must generate a current waveform that complies with the standard, and (2) it must produce an electromagnetic field that couples the generator body to the DUT in the same way as a real generator. However, there are no specific requirements for this electromagnetic field. Each commercially available generator model has its own unique geometry

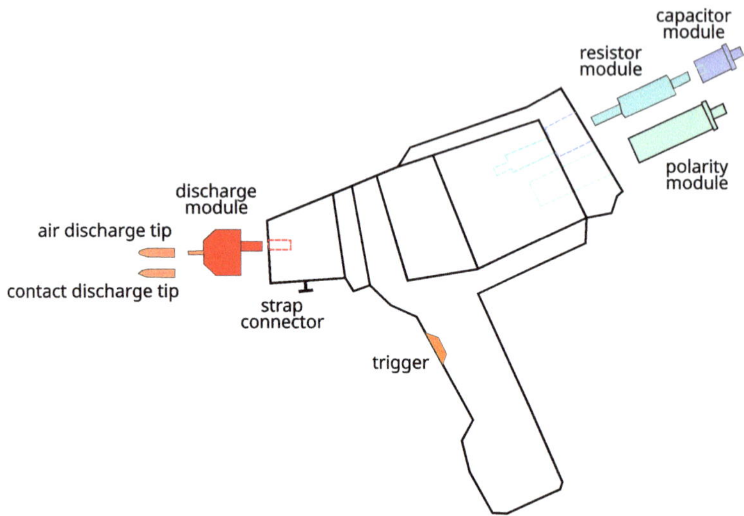

Fig. 14.5 Block diagram of an ESD generator model describing the function and positions of the individual units

and discharge characteristics, leading to variations in field distribution and coupling behavior.

There are two approaches to modeling the fields. The first is to accurately replicate the geometry of a specific generator. This requires dismantling the generator, determining its geometry and material parameters, and validating the constructed EM generator model against measurements of the tip current and the fields generated during the discharge event. The result is an EM model specific to that particular generator. If a different generator is used, the modeling process must be repeated. Such modeling approaches have been reported and validated in the literature for various types of generators.

We present an alternative approach. The primary field interaction between the generator and the DUT is driven by the high electric and magnetic fields near the DUT, particularly those originating from the discharge module and around the generator's tip. To capture this interaction, we introduce a generic generator model with a simplified geometry, featuring a cylindrical representation of the discharge module.

The size and conductivity of the discharge module determine the amplitude of the first peak. By ensuring that the current peak is accurately modeled and that the discharge module's size is a reasonable representation of a real generator, we assume that the generated field serves as a suitable approximation of the fields produced by more detailed generator models. The geometry of this representative model can be adjusted to study its impact on the discharge pulse and the resulting field. Due to the model's simplicity and lack of geometric details, the fields remain a good representation up to a frequency of 500 MHz, which corresponds to the resolution

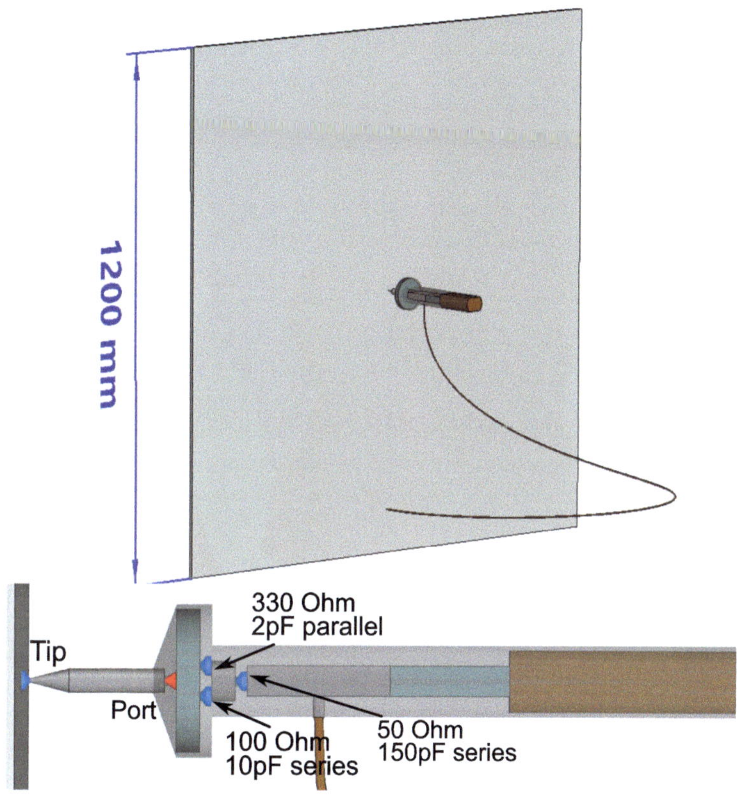

Fig. 14.6 Geometrical model of an ESD generator. Top: Generator with ground strap on discharge plate, bottom: Cross sectional view with inner details

of the rise time of the first peak. Any finer details of this current rise cannot be represented by this generator model.

The generator model described in the following is based on the one introduced by Caniggia and Maradei (2005). Various adaptations of this model exist and have been applied in different publications. This model is included in the model library of CST and is shown in Fig. 14.6.

It has a cylindrical shape with an overall length of 30 cm. The tip, which is 6 cm long, is made of PEC material. Above the tip, there is a cylindrical disk with a diameter of 70 mm (light blue). This disk is composed of a highly lossy material with a conductance of $\sigma_c = 1\,\frac{\mathrm{S}}{\mathrm{m}}$. Above the disk, there is a cylindrical aluminum block with a height of 10 mm, followed by another aluminum body with a height of 70 mm. Both have a conductance of $\sigma_c = 3.56 \cdot 10^7\,\frac{\mathrm{S}}{\mathrm{m}}$. The ground strap is attached to the upper block.

Above this, there is a lossy block, with a height of 50 mm (light blue in the figure), with a conductance of $\sigma_c = 1\,\frac{\mathrm{S}}{\mathrm{m}}$. At the top, a copper body of nearly 100 mm height (brown) is present, with a conductance of $\sigma_c = 5.8 \cdot 10^7\frac{\mathrm{S}}{\mathrm{m}}$. All parts except the tip are enclosed within a polycarbonate cover with a relative permittivity of

$\varepsilon_r = 2.9$. The high-conductance shapes are modeled as lossy metal, and the low-conductance shape as volumetric material.

The discharge network consists of a 150 pF capacitor placed between the upper and middle cylindrical aluminum blocks. This capacitor has a 150 Ω resistor in series. The 330 Ω resistor of the discharge network is mounted between the middle cylindrical block and the disk, with a parallel capacitance of 2 pF. Additionally, another 100 Ω–10 pF series element is connected in parallel. The excitation source is placed between the cylindrical disk and the discharge tip. In a time-domain simulation, it ramps up the voltage from 0 V to 430 V within 3 ns to yield a 4 kV discharge, with an impedance of 25 Ω.

Unlike the SPICE model described in Sect. 14.2.1, the EM generator is not discharged but rather charged. This setup is required as the co-simulation models require to start with a zero field. Simulating the charging of the capacitor and discharging it after the charging process has finished would increase the simulation time, without being necessary: since the current in the tip is driven by a charge difference, it makes no difference whether the excitation source is modeled as a charge or a discharge.

The values of the lumped circuit elements influence the second peak of the pulse and the tail of the current waveform. These values are selected to ensure that the waveform meets the specifications defined by the standard. Resistors and the material with $\sigma_c = 1\,\frac{\mathrm{S}}{\mathrm{m}}$ are included to dampen oscillations that would otherwise appear in the discharge waveform. The excitation voltage waveform and the source impedance are chosen to produce the required current for a given discharge voltage.

Since the generator model is linear, doubling the discharge voltage results in a proportional doubling of the discharge current, as required by the standard (cf. Table 14.1). To accommodate different discharge voltages, the predefined time-domain voltage used to excite the source is scaled accordingly.

The height of the initial peak is determined by the size and conductivity of the cylindrical disk adjacent to the discharge tip, where the excitation port is connected. The rapidly rising excitation voltage charges this disk. Since $\mathrm{I} = \dot{Q} = C\dot{V}$, and the voltage is defined by the excitation waveform, a larger capacitance—resulting from either an increased disk size or higher conductivity—leads to a higher current. The charges accumulated on the disk generate electric field lines between the disk and the DUT.

The purpose of the EM generator model is to simulate the electric and magnetic field interaction between the generator and the DUT. The shape and conductivity of the disk, combined with the presence of the metal tip, have the greatest influence on the field distribution due to their proximity to the DUT. In the presented model, the disk size is chosen to closely match the geometry of the discharge module in commercial generators, while the conductivity is adjusted to produce the required amplitude of the discharge current waveform. With these two parameters set, the resulting field strength of the first pulse is considered to be modeled realistically.

For calibration, the generator is discharged through a low-ohmic resistor into a metal plate, to which the ground strap is attached. The simulation is performed using

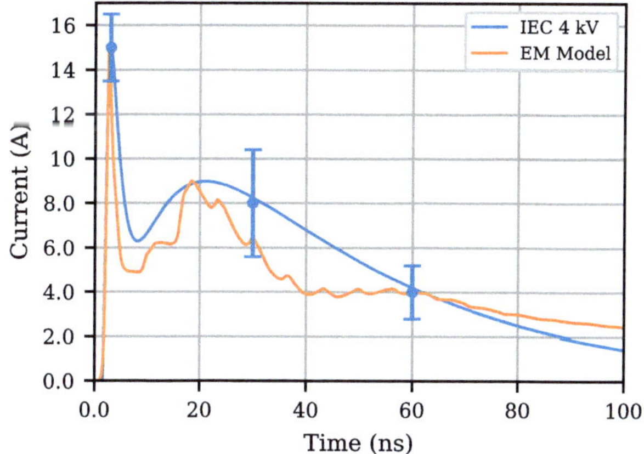

Fig. 14.7 Discharge current of the EM generator model compared to the IEC 61000-4-2 pulse

the TLM solver and a co-simulation (see Sect. 14.3). For a signal duration of 90 ns, a simulation time of 4 minutes and 36 seconds is required.

The corresponding discharge current is shown in Fig. 14.7. The orange curve represents the current generated by the model, while the blue curve represents the standardized current pulse. During the rise time of the first pulse, within 3 ns, the two pulses match within 1 A. The curve simulated by the EM model is within the specifications but, compared to the circuit model, exhibits some ringing.

The pulse from the EM generator decays more rapidly, reaching its first minimum at 5.5 ns and the peak of the second pulse at approximately 19 ns, with an amplitude of 9 A. In contrast, the standardized pulse decays to 6 A at 8 ns and reaches its second peak at around 21 ns, also at 9 A. However, these specific points are not explicitly defined by the standard. According to the standard, the current must be between approximately 6 A and 10 A at 30 ns and between 3 A and 5 A at 60 ns. The EM generator model produces 6.4 A and 4 A at these respective time instances. Like the SPICE model, the EM generator model meets the required specifications.

The key feature of the EM model is its ability to generate electric and magnetic fields and simulate their coupling with the DUT. The magnetic field between the generator and the table is shown in Fig. 14.8. The field is recorded at the time $t = 1$ ns, i.e., at the peak of the discharge pulse. The field is concentrated around the discharge pin, between the discharge module and the table. This field concentration highlights the impact of the local geometry. In the picture, the target is a flat sheet. If the generator tip is applied, e.g., within a connector housing, the field distribution is affected.

To test an electronic system, we combine the CAD models of both the system and the generator to create a comprehensive EM model. For contact discharge, the tip of the generator is positioned at the designated discharge location. An application of this approach is described in Sect. 14.4. Air discharge is briefly addressed in Sect. 14.7.

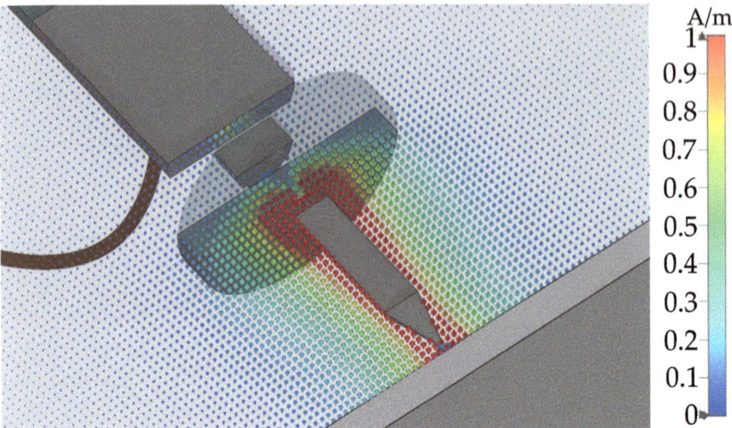

Fig. 14.8 Magnetic field in a cross section around the generator tip and the metal target at $t = 1$ ns

By integrating the generator and the DUT into a single model, they can interact electromagnetically, rather than solely through currents and voltages, as it is the case with the SPICE model described in Sect. 14.2.1.

14.3 Transient Co-simulation

Domains: 

In Sect. 6.7, we introduce a coupled EM and circuit simulation method that uses S-parameters to transfer the results of the EM simulation to the circuit. This approach can be applied to simulate ESD effects in a transient circuit simulation using vector-fitted S-parameters (cf. Sect. 10.3 for details on vector fitting). With this method, we can calculate voltages and currents at all ports within the computational domain on the circuit level.

However, the combined results methodology introduced in Sect. 6.7.5 is not suitable for ESD simulations, as it assumes steady-state behavior, whereas ESD is inherently a transient phenomenon. To calculate the transient field distribution in the EM model, a transient co-simulation is required. This simulation type is also applied in Sect. 13.4.2 for modeling radiated immunity.

In the following, we describe the transient co-simulation of the circuit with an explicit full-wave time-domain method, such as FIT or TLM. For quasi-static simulations, implicit methods for coupling the circuit to an EM solver exist but are not discussed in this book. Such quasi-static methods are predominantly used in motor simulations and are unsuitable for ESD due to the fast rise time of an ESD pulse.

In the transient co-simulation, the EM model and the circuit are calculated simultaneously. After calculating one time step in the EM model, a voltage is derived from the electric field using Eq. (4.12), and a current is derived from the magnetic field using Eq. (4.13) at all ports. These voltages and currents are then used to perform a time step on the circuit, and the circuit-level voltages and currents are transferred back to the EM model, where the next time step is calculated. This simultaneous time-stepping approach enables the calculation of the transient behavior of the fields, including the effects of all components defined in the circuit.

We briefly introduce transient co-simulation in Sect. 8.6.2 for coupling the TLM solver to the MTL. In this section, we apply it to ESD simulations. This method can also be used for other applications, such as visualizing current flow during a switching event.

The visualization of transient EM fields using transient co-simulation is a unique capability of this method. However, it has limitations. As the transient EM solvers in CST must start simulations with zero fields (Sect. 4.9.2), this restriction also applies to transient co-simulations (Wang et al. (2003) implemented a more general method). Initial conditions on the circuit cannot be used, and all sources must start at zero voltage or current. Additionally, the time step used in the co-simulation must satisfy the stability criterion of the EM simulation, which prevents the use of adaptive time-stepping methods typically employed in transient circuit simulations (Sect. 3.8.1).

Due to these time-stepping properties, the signal duration that can be simulated using transient co-simulation within a reasonable time frame is limited to a few microseconds. A transient co-simulation for signals in the millisecond range could require days of simulation time and is therefore not used in other chapters. For ESD simulations, where the signal duration is on the order of tens of nanoseconds, this method is highly efficient.

Since the circuit and EM model are simulated simultaneously, any change to a circuit element requires re-running the complete simulation. This is in contrast to the coupled, sequential simulations used in other chapters.

14.4 Application Example: Contact Discharge to a USB Pin

- EM model of ESD generator
- Current through and voltage across diodes on schematic
- Transient co-simulation
- TLM method
- Transient

0–3 GHz

We simulate a contact discharge to a USB signal trace on the PCB also used in Chaps. 11 and 12. The simulation predicts the voltage and current levels inside the IC during an ESD event and allows us to estimate the effects of individual components in the USB channel on an ESD discharge event. By performing the study step by step, we gain insights into the behavior of individual components that cannot be easily obtained through physical ESD testing.

To begin, we introduce a generic model of the ESD protection inside the IC, which can be used when an accurate IBIS model of the IC is unavailable. Following the introduction of the IC model, we describe how this model is integrated into an EM simulation that includes the PCB, the ESD generator, and the coupling plane.

In the next section we simulate the IEC 61000-4-2 standard including the full-size horizontal coupling plane and the correct distance to the table. We compare the full-size model to a reduced-size model that is computationally more efficient. This reduced model is then used to examine the effects of various sections of the USB channel. Specifically, we study:

- The effect of the power delivery network (PDN) and the effect of a connection from the PCB to the horizontal coupling plane
- The effect of a connector breakout
- The effect of the common mode choke and TVS protection diodes

14.4.1 Model of the ESD Protection Inside an IC

In Sect. 2.3.5, we introduce IBIS models, which are used to model the behavior of an IC. An IBIS model can include the internal ESD protection of an IC by describing the ESD protection diodes using $V - I$ curves. This is depicted in Fig. 2.11. Unfortunately, the quality of this data varies among IBIS models provided by vendors. While some IBIS models accurately describe the diode behavior, others may omit this information, rendering them unsuitable for ESD simulation. For instance, snapback protection might not be included, and changes to the ESD protection or power clamps during power-on and power-off conditions may not be represented. To accurately simulate ESD robustness, a correct IC model with defined voltage and current limits is essential.

To address this limitation, we develop a generic model of the ESD protection inside an IC. The goal of this model is to illustrate the ESD simulation methodology in the subsequent simulations. The structure of the generic model is identical to that of the IBIS model (Fig. 2.11) but excludes the logic and switching elements, focusing solely on modeling the ESD protection.

The generic ESD protection circuit inside an IC is depicted on the left-hand side of Fig. 14.9. A diode is placed from the signal trace to VCC, and another diode is placed from GND to the signal trace. The diode models are derived from a standard

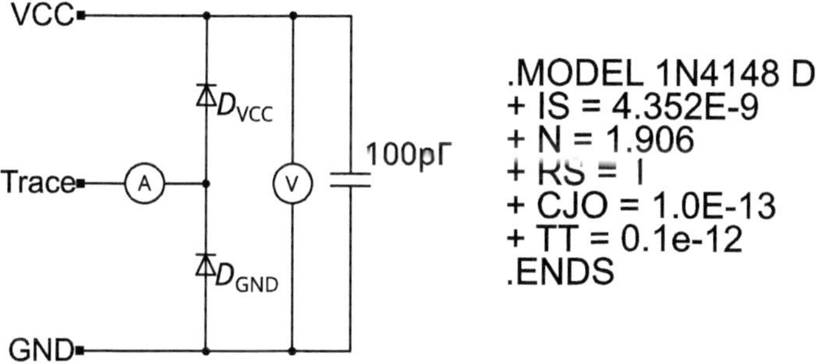

Fig. 14.9 Left: generic ESD protection circuit of an IC. Right: SPICE parameters of the protection diodes

1N4148 diode, but the SPICE parameters have been modified to better represent typical protection circuits. The series resistance of the diode is set to 1 Ω, the parallel capacitance to 0.1 pF, and the transition time to 0.1 ps, as reverse recovery is not modeled. The parameters of the SPICE model are shown on the right-hand side of Fig. 14.9.

A 100 pF capacitor is placed between VCC and GND to represent the capacitance between the reference and the power supply inside the IC. The resistive behavior of the IC's power distribution network (PDN), internal power clamps, or different power domains is not included in this model; nevertheless, it is still suitable to model a large number of ICs. In the schematic of the protection circuit, the locations for voltage and current measurements are indicated. These represent the voltages and currents observed in the subsequent simulations (Speckbacher et al. 2022; Xu et al. 2023).

14.4.2 Model Setup

The details of the simulation model at the PCB level are shown in Fig. 14.10, with the corresponding schematic plotted in Fig. 14.11. In the top section of Fig. 14.10, the surroundings of the connector are depicted. The ESD generator tip connects to the USB_P trace (red color) at the connector.

The pins of a USB-C connector are so small that the generator tip would connect to both pins if placed at one of them. To prevent this in the simulation model, the length of the pin for the USB_N trace is reduced, ensuring no connection to the generator tip. The impact of a more realistic connector breakout is discussed in Sect. 14.5.2. For now, we use the modified connector pins in the simulation model.

Behind the connector, two ports to the GND layer are placed. These ports are connected to TVS diodes in Sect. 14.5.3. At this stage, they are connected to high-ohmic resistors in the schematic model (ports $D6_2$ and $D6_1$). Behind the protection diodes, four ports are defined in the EM model to connect the common mode choke,

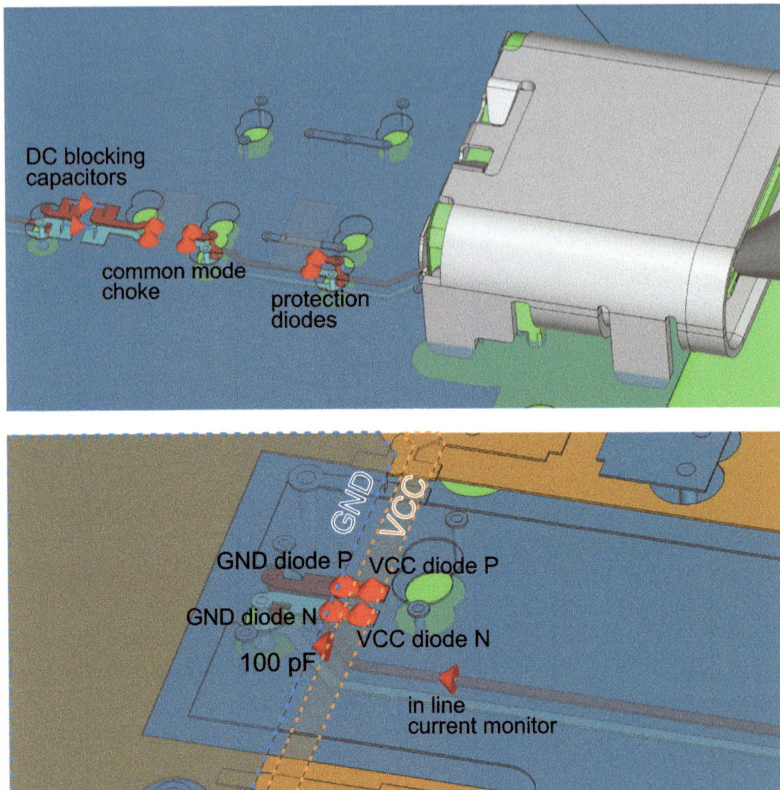

Fig. 14.10 Setup of the ESD simulation model on the PCB. Top: connector section. Bottom: IC section

as described in Sect. 14.5.3. For now, a direct connection for both traces, linking one side of the choke pads to the other, is included in the schematic (ports $CM1_X$). Following the common mode choke ports, two additional ports are used to integrate the DC blocking capacitors in the EM model. The corresponding circuit models of the capacitors are located at ports C25 and C26. This port configuration is consistent with the setup used in Sect. 11.1.2 (Fig. 11.2).

The implementation of the internal ESD protection circuit in the IC, described in the previous section, requires a port setup in the EM model that has not been used in earlier examples. The port setup at the IC location is depicted in the bottom section of Fig. 14.10. The GND layer is shown in blue, and the VCC layer in orange. Right before the IC, an in-line port is placed on the USB_P trace, where the current flowing toward the IC is monitored. As the USB traces are routed in a stripline configuration, the top layer of the GND net is visualized as a transparent shape.

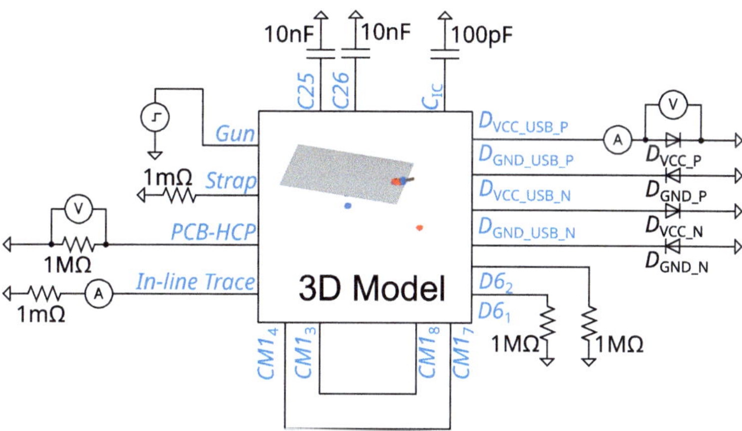

Fig. 14.11 Schematic for the simulation of an ESD discharge to a USB pin. The current and voltage measurements inside the IC are defined at all four diodes but depicted only for the USB_P VCC diode to avoid clutter on the schematic

To implement the protection circuit inside the IC, two PEC shapes are added to the model. These PEC shapes are also visualized as transparent to avoid obscuring the port definitions. Their outlines are indicated by dashed lines. The shape connecting to the GND diodes matches the footprint of the IC on the PCB and is marked by blue dashed lines with a blue GND label. This shape is connected to the GND pins of the IC through small PEC leads. The shape connecting to the VCC net is represented by dashed orange lines and modeled as a strip around the IC footprint.

At this stage of the model, the VCC shape is not connected to the PDN of the PCB. The effect of the PDN is studied in Sect. 14.5.1. For now, only the 100 pF capacitor from Fig. 14.9 is connected between the GND and VCC shapes in the schematic (Port C_{IC}). At each USB trace, one port connects the trace to the VCC shape, and another port connects the trace to the GND shape. These ports are connected to the internal IC ESD protection diode models in the schematic (ports D_{VCC} and D_{VCC}). The diodes in the ESD protection of the IC need to be oriented in reverse direction (antiparallel) in each trace, as both ports in the EM model are defined to point from the trace to the respective PEC shape.

14.5 Reduction of the Model Size

In this application example, we use the EM ESD generator model described in Sect. 14.2.2. According to the IEC 61000-4-2 standard discussed in Sect. 14.1, the DUT is placed on an insulator located on a horizontal coupling plane (HCP) on a

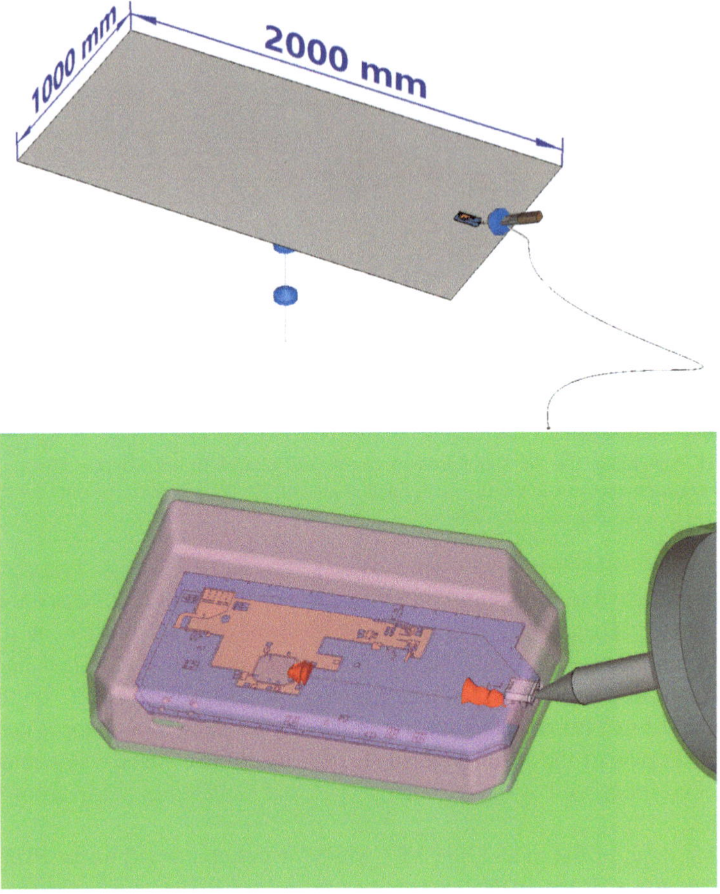

Fig. 14.12 Simulation model of the IEC 61000-4-2 with a full-size coupling plane. Top: overview of the model. Bottom: zoom in on the PCB inside a housing

nonconductive table. The coupling plane has a size of 1×2 m, and the table height is 90 cm.

We set up a model that includes the full size of the coupling plane and the correct distance to the floor. An overview of the model is shown in Fig. 14.12. The floor below the table is modeled as a PEC boundary. The coupling plane is connected to the floor through two 470 kΩ resistors. The strap from the ESD generator is also connected to the floor via a port, allowing the strap current to be monitored. Additionally, a thin wire with a port (port PCB-HCP in Fig. 14.11) is placed between the PCB and the coupling plane. This port measures the voltage between the PCB and the HCP. This full-size model leads to a large computational domain, resulting in extended simulation times.

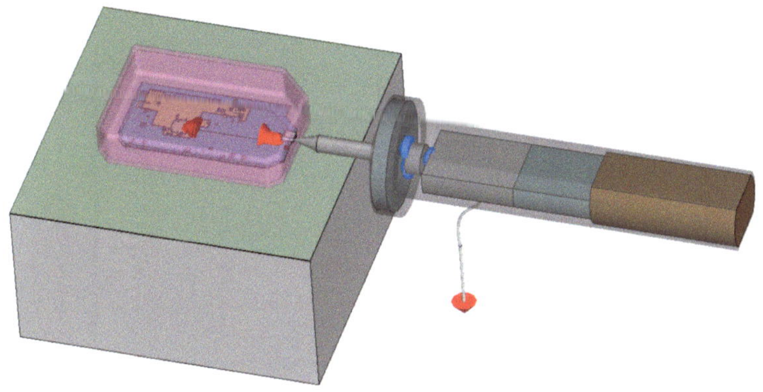

Fig. 14.13 ESD simulation model with a reduced size of the computational domain

The EM model, along with the corresponding schematic, is simulated using a transient co-simulation. The selected EM solver is the TLM solver. Due to the electrically large model (caused by the long strap of the ESD generator) and the need to resolve small details on the PCB within a large computational domain, the octree mesh of the TLM solver provides the only feasible discretization. The cell lumping of the octree mesh reduces the initial mesh count from 781 million to 39 million. For a signal duration of 50 ns, the simulation requires approximately 8 hours. To reduce this simulation time, we discuss a size reduction to the scenario in the following.

A reduced-size alternative to the complete model is depicted in Fig. 14.13. In this model, the coupling plane below the PCB is directly connected to the PEC boundary, which also connects to a shortened strap of the ESD generator. In this setup, the strap of the generator is connected to the coupling plate instead of the floor, as in the full-size model. To account for the real length of the strap, the 1 mΩ resistor at the port Strap in Fig. 14.11 is replaced by a 2.4 μH inductor. To model the losses in the strap, a 300 Ω resistor is placed in parallel with the inductor (Luo et al. 2023). The reduced-size model requires only 47 minutes to simulate a 50 ns signal duration.

In the following, we discuss the differences between the full-size and reduced-size models. In Fig. 14.14, we plot the current monitored by the in-line port for both simulation models. The first peak, occurring at 3.2 ns, differs by 5% between the two models. The amplitude of the peak is 19.6 A for the full model and 18.7 A for the reduced model. Since this peak has a much higher amplitude compared to the subsequent current, it represents the highest probability of damage to the IC. Unlike a discharge to a calibration target, there is no second peak at 20 ns, which is typical for discharges to a floating target. After the first peak, some ringing of the current is observed, with slightly higher amplitude in the full model. However, this ringing is not relevant to the likelihood of damage to the IC.

On the left-hand side of Fig. 14.15, we plot the voltage between the coupling plane and the GND net of the PCB. The voltage between the GND net of the PCB

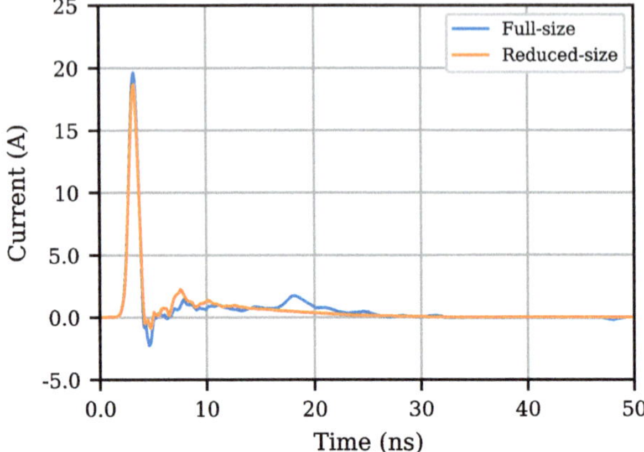

Fig. 14.14 Comparison of the results of the full-size model and the reduced-size model: current on the USB_P trace monitored by the in-line port close to the IC pin

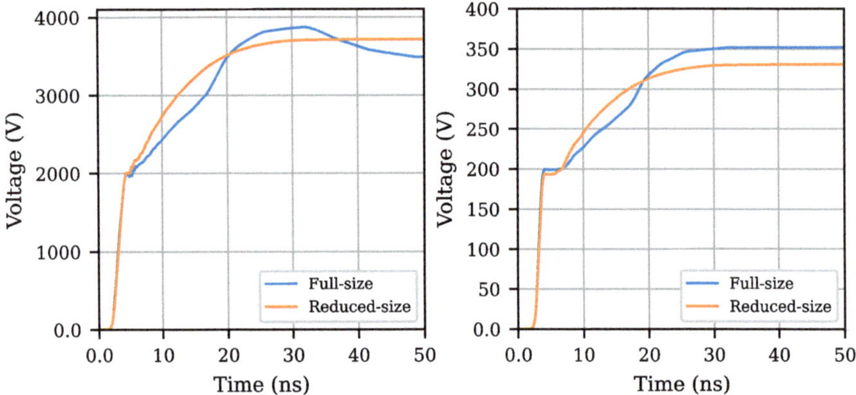

Fig. 14.15 Comparison of the results from the full-size model and the reduced-size model. Left: voltage between the PCB and the horizontal coupling plane. Right: voltage between the VCC and GND nets inside the IC

and the coupling plane rises to 370 V after 20 ns. The ESD generator discharges a voltage of 4000 V between the coupling plane and the USB trace. After the current propagates along the trace, entering the ESD protection inside the IC and through the D_{VCC} diode, the voltage divides between the GND and VCC nets, and the GND net and the coupling plane.

On the right-hand side of Fig. 14.15, the voltage between the VCC and GND nets, measured at the 100 pF capacitor, is plotted. The sum of the voltage between the VCC and GND nets and the voltage between the GND net and the coupling plane equals the voltage applied by the generator.

The calculated voltage between the VCC and GND nets reaches over 300 V after 20 ns. This high value is due to the simplifications in the IC model, which omits the switching transistors and does not impose a voltage limit on the 100 pF capacitor. Nevertheless, this simplified IC model is effective for comparing different designs and studying various testing scenarios.

When comparing the full-size and reduced-size models, the reduced model underestimates the current in the USB trace and the voltage between the VCC and GND nets by approximately 5%. However, the computational time is reduced by a factor of 8. Due to the significant reduction in simulation time, we continue our study using the reduced model.

14.5.1 *Effect of the Power Delivery Network and Connection to the Coupling Plane*

In the model from the previous section, the PDN on the PCB is not included in the simulation. Only a 100 pF capacitor represents the PDN inside the IC. The decoupling capacitors on the PCB, which are placed between the VCC and GND nets, are expected to influence the current through the IC. We study this effect in the following.

To include the PCB's PDN in the simulation model, the VCC shape around the IC footprint is connected to the VCC net of the PCB (see the bottom part of Fig. 14.10). This connection is established using small PEC leads, in the same manner as the connection between the GND shape and the GND net on the PCB.

The simulation results, including the PCB's PDN, are compared to the model without the PCB's PDN in Fig. 14.16. Connecting the PDN significantly reduces the voltage between the GND and VCC nets inside the IC compared to the model where only the 100 pF capacitor is present. The PDN is connected in parallel to the

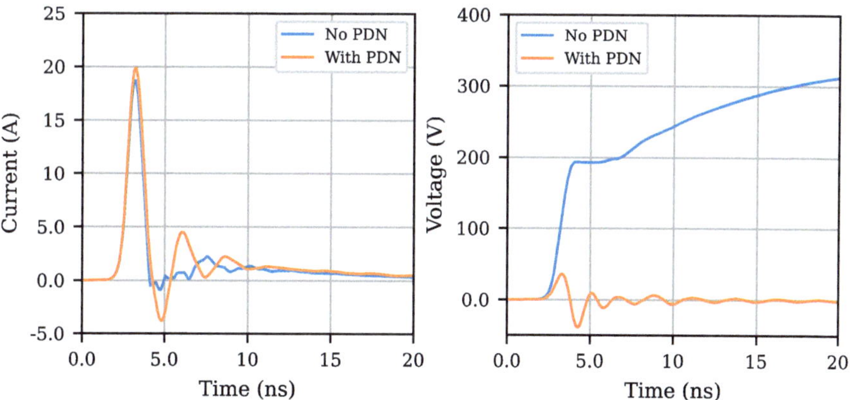

Fig. 14.16 Comparison of models with and without the PDN of the PCB. Left: current on the USBP trace monitored by the in-line port. Right: voltage between the VCC and GND nets inside the IC

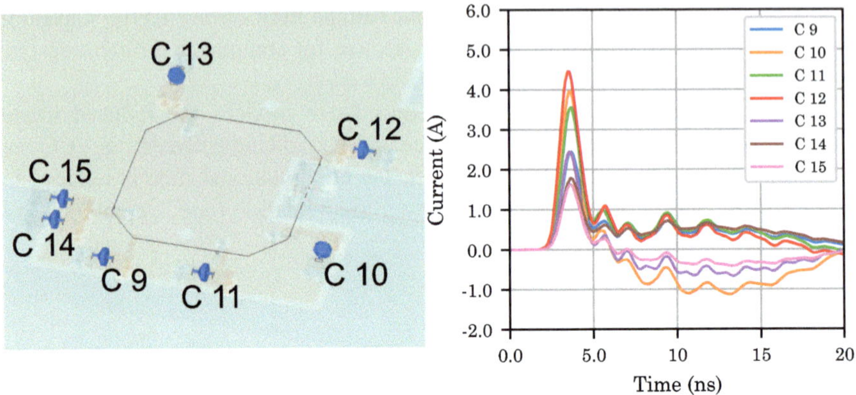

Fig. 14.17 Left: location of decoupling capacitors close to the IC. Right: current through the decoupling capacitors

100 pF capacitor. With the larger capacitance values of the PDN capacitors, a lower impedance path is established between VCC and GND. This added low-impedance path also affects the trace current, increasing the peak current at 3.2 ns from 18.7 A to 19.8 A and causing increased ringing after the peak. This ringing of the connected PDN is also visible in the voltage results.

The current through the decoupling capacitors near the IC is plotted in Fig. 14.17. The capacitors closest to the USB trace exhibit higher current compared to those further away. Additionally, the individual capacitance ratings influence the distribution of current among the decoupling capacitors.

The connection to the PCB PDN is kept from now on in all following simulations.

In all models simulated so far, the PCB is not directly connected to the coupling plane. While a wire is placed between the PCB ground and the coupling plane, a high resistance is included in the wire on the schematic, allowing for the measurement of the voltage between the PCB's GND net and the coupling plane. To connect the GND net of the PCB to the coupling plane, this resistor can be set to 1 mΩ (resistor at port PCB-HCP in Fig. 14.11). In the following, we compare the models with and without a direct connection of the PCB's GND net to the coupling plane (Fig. 14.18).

We observe that connecting the GND net of the PCB to the coupling plane significantly impacts the current entering the IC. The peak current increases from 19.8 A to 22 A. After the peak, the current stabilizes at approximately 5 A and decays only slowly. The waveform of the current is similar to the waveform observed during the calibration of the ESD generator.

The connection of the GND net to the coupling plane has only a minor effect on the voltage between the VCC and GND nets inside the IC.

From the results of this section, we conclude that adding the PDN to the simulation model significantly affects the voltage measured inside the IC, whereas connecting the PCB's GND layer to the reference plane leads to an increase of the current through the IC.

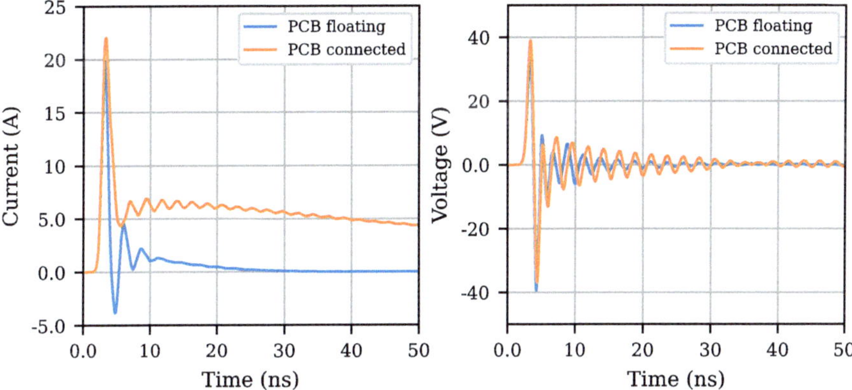

Fig. 14.18 Comparison of models with and without a connection of the GND net of the PCB to the coupling plane. Left: current on the USB_P trace monitored by the in-line port. Right: voltage between the VCC and GND nets inside the IC

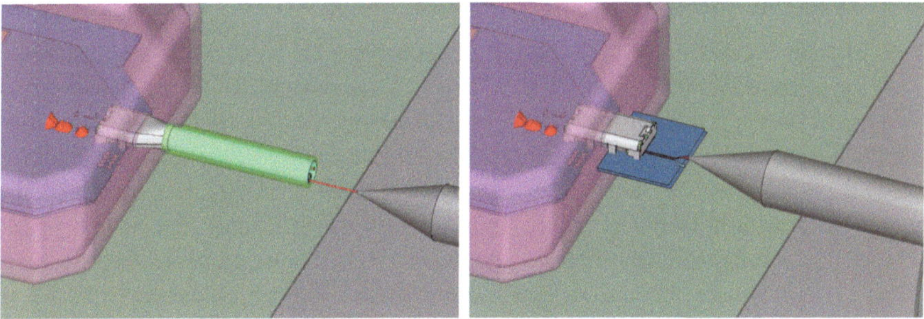

Fig. 14.19 Two possible breakout scenarios. Left: breakout realized over a cable section. Right: breakout using a small PCB

14.5.2 *Effect of Connector Breakout*

In the models discussed up to this point, the generator tip is placed directly at the connector. This requires reducing the length of the connector pin for the USB_N trace, as the discharge is applied only to the USB_P trace. However, in physical testing, such modifications are not possible. Instead, physical tests use a breakout structure to connect the generator tip to the target net. In this section, we study the effect of two breakout structures, as shown in Fig. 14.19.

The first breakout structure uses a short cable section, with one of the wires of the TWP separated and connected to the generator tip. This model employs the same connector-to-cable transition as the RE study in Chap. 12, with details depicted in Fig. 12.26.

The second breakout model is realized using a small PCB with a connector mounted on it. For this model, the existing connector is mirrored, and the interior

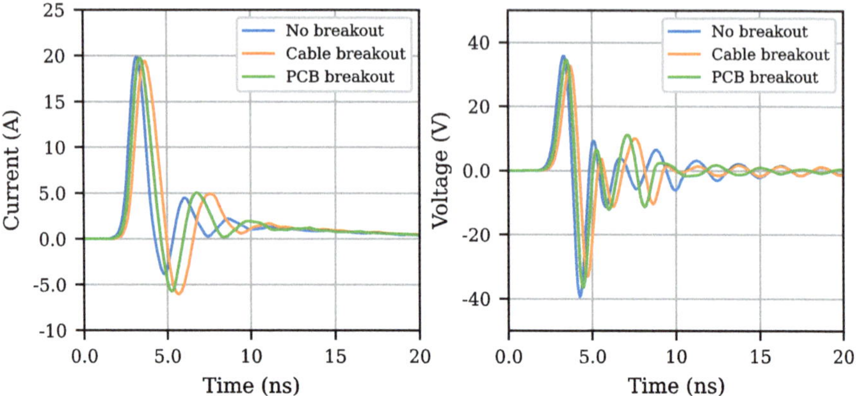

Fig. 14.20 Effect of the breakout structure on the current and voltage in the ESD simulation. Left: current on the USB_P trace monitored by the in-line port. Right: voltage between the VCC and GND nets inside the IC

pins of both connectors are connected to each other. A small PCB is constructed with the stack-up matching the PCB in the DUT. It consists of two GND layers and two traces on the top layer that connect to the connector pins. These simple models can be quickly built using the CAD modeling functionality of the simulation tool.

The current entering the IC for the two breakout scenarios, compared to the model without a breakout structure, is plotted on the left-hand side of Fig. 14.20. The peak current is slightly reduced and delayed when a breakout structure is used. The peak current decreases from 19.8 A in the model without a breakout to 19.7 A in the model with the PCB breakout, and 19.4 A in the model with the cable breakout. Additionally, the ringing of the current after the peak is increased when a breakout structure is used. These changes are more pronounced in the cable-based breakout due to the increased length of the current path, which is greater in the cable breakout model compared to the PCB-based breakout.

A similar behavior is observed in the voltage results. The peak voltage is slightly reduced, and the ringing increases when a breakout structure is added. Simulating the breakout structure enables engineers to evaluate its effect and, if necessary, optimize the design of the breakout.

14.5.3 Effect of Common Mode Choke and Protection Diode

In this section, we study the effect of the common mode choke in the USB signal path and of TVS diodes placed near the connector. Up to this point, the DC blocking capacitors are the only components on the PCB included in the simulation model. In this first study, we add a common mode choke model to the signal path. The ports for the common mode choke are already present in the model, making the inclusion of

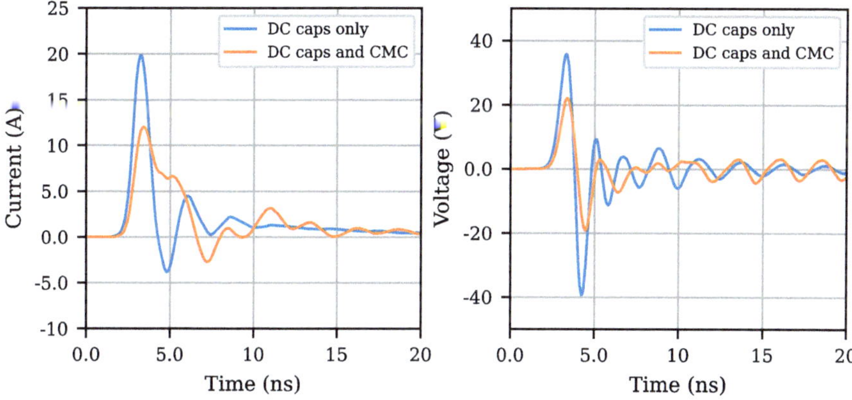

Fig. 14.21 Comparison of the simulation results for a model with only the DC blocking capacitors and a model with an added common mode choke. Left: current on the USBP trace monitored by the in-line port. Right: voltage between the VCC and GND nets inside the IC

the SPICE model straightforward. A more detailed overview of the layout is given in Fig. 11.2. We use the same model of the common mode choke[1] as in the SI Chap. 11 (DLM0NSN900HY2 from (Murata)).

Adding the common mode choke to the signal path increases the inductance of the line and significantly decreases the peak current entering the IC. A comparison of the current through the IC, with and without the common mode choke, is plotted on the left-hand side of Fig. 14.21. The peak current is reduced by nearly half, though the peak is also wider.

The addition of the common mode choke also reduces the voltage inside the IC, as shown on the right-hand side of Fig. 14.21. This reduction demonstrates the effect of the common mode choke in the USB signal channel on both current and voltage peaks during an ESD event.

In the final study of this application example, we add TVS diodes to the signal path. The part number of the diode is PUSB3FR4 (Nexperia) from NXP, and it is the same diode model used in the signal integrity and radiated emission chapters. The SPICE model of the diode is downloaded from the vendor's homepage and used as provided.

The current in the USB trace and the voltage inside the IC after adding the TVS diodes near the connector are plotted in Fig. 14.22. With the addition of the TVS diodes, the peak in the current entering the IC is no longer present, and the current remains below 0.6 A for the entire simulation duration. The voltage between the VCC and GND nets is reduced to less than 1 V, as shown on the right-hand side of Fig. 14.22.

[1] In the ESD simulations presented in the following, the modeling of the saturation of the choke's ferrite material is not included.

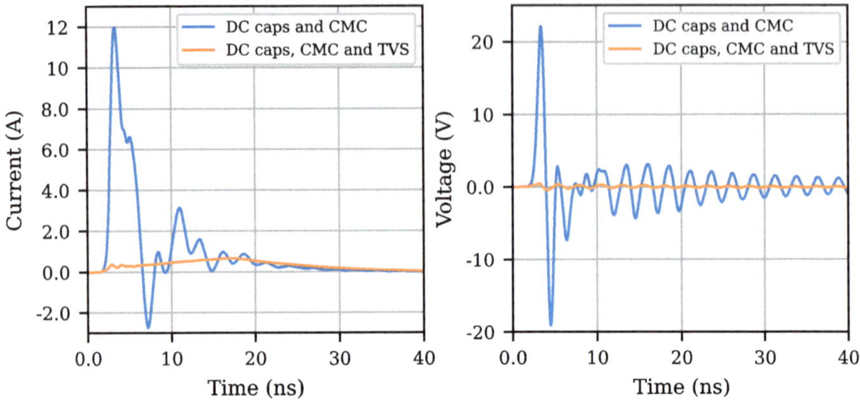

Fig. 14.22 Effect of adding a TVS diode to the USB signal path. Left: current on the USB_P trace monitored by the in-line port. Right: voltage between the VCC and GND nets inside the IC

As expected, the addition of the TVS diodes significantly reduces both the current and the voltage inside the IC, demonstrating its effectiveness in mitigating ESD events.

14.6 Failure Criteria

Domains: 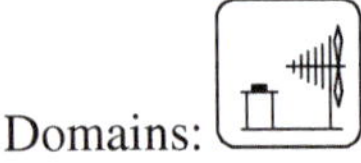

In the application example presented above, we discuss how different model setups influence the current and the voltage through the IC. The evaluation whether a given calculated current or voltage profile leads to a failure of the IC is a wide field of research and is discussed briefly in this section.

14.6.1 Hard Failure

The direct application of the discharge of the ESD generator to a pin of the connector is used for testing hard failure of the IC. In a hard failure the IC is irreversibly damaged. The mechanisms of the failure inside the IC can include metallization failure, gate oxide failure, latch-up, and thermal failure.

Wunsch and Bell (1968) have developed a thermal model that calculates a limit line above which a thermal damage is expected. The equation for modeling the

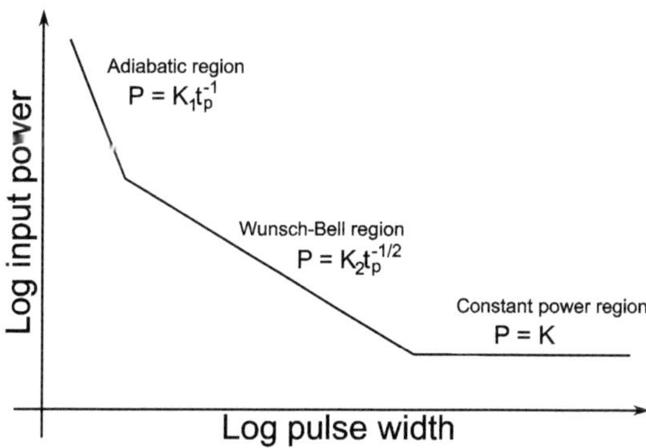

Fig. 14.23 Typical form of the thermal limit as estimated by Eq. (14.1)

thermal limit is

$$\frac{P}{A} = K_1 t_{\mathrm{p}}^{-1} + K_2 t_{\mathrm{p}}^{-\frac{1}{2}} + K \tag{14.1}$$

where P is the failure power, A the semiconductor junction area, and t_{p} the duration of a rectangular power pulse acting on the junction. The parameters K_1, K_2, and K are linked to the geometry and conductivity of the semiconductor junction. The shape of the thermal limit, as obtained from Eq. (14.1), is plotted in Fig. 14.23.

The difficulty in applying this equation lies in the fact that the parameters K_1, K_2, and K are typically unknown to the simulation engineer. These parameters can be determined through measurements, as described in Li et al. (2015). It is up to the simulation engineer to evaluate whether the investment in such measurements is justified.

The integral of the curve shown in Fig. 14.23 is an energy. In order to examine semiconductor hard failure in simulation, the energy $W_{SC}(t)$ dissipated in the semiconductor is monitored as a function of the time and compared to the area under the Wunsch-Bell curve,

$$A_{\mathrm{WB}}(t) = \int_0^t P_{\mathrm{fail}}(t')\mathrm{d}t'. \tag{14.2}$$

Hard failure occurs at the time t at which

$$W_{\mathrm{SC}}(t) > A_{\mathrm{WB}}(t). \tag{14.3}$$

The Wunsch-Bell curve is valid for rectangular pulses such as those generated from Transmission Line Pulse measurements (see the literature at the end of the chapter for further reading), so its application is typically limited to cases in which detailed information about the pulse shape is known.

14.6.2 *Soft Failure*

In a soft failure event, the discharge of the ESD generator causes a reversible malfunction of the DUT. The system-level simulation described in this chapter is not capable of directly detecting such malfunctions. However, the same simulation flow used in the BCI simulation in Sect. 13.2.2 can be applied. In this approach, the signal traces expected to lead to the malfunction are driven with their intended signals. The ESD discharge event is then simulated, and its effect on these signals is analyzed. For this type of simulation, it is essential for the simulation engineer to identify in advance which signals in the DUT are likely to contribute to the observed failure.

The safe operation area of the semiconductor provides a useful guideline for this analysis. During an ESD simulation, the currents through and voltages at all pins of the IC are monitored. If any of these exceed a certain threshold, there is a risk of soft failure. Countermeasures can be designed, and layout space can be reserved for ESD protection elements. Whether these elements are ultimately assembled or not can be determined later when hardware is available. This approach minimizes the risk of late-stage ESD failures by proactively allocating space for potential countermeasures. Simulation helps to identify the locations where these measures are likely needed.

14.7 Air Discharge

Domains: 

The IEC 61000-4-2 standard includes air discharge tests. The reproducibility of air discharge tests is lower than that of contact discharge tests. However, air discharge events are the dominating mechanisms in products used in the field. Physically precise modeling of the sparks that occur during air discharge events requires simulation methods capable of modeling the plasma in the spark. Such plasma modeling significantly increases the complexity of the simulation, and the authors have no experience in these methods. Therefore, this section describes a simplified approach to modeling air discharge events. With this approach, plasma effects such as corona discharge cannot be modeled.

In the first step, the position at which the spark occurs is estimated using an electrostatic simulation (cf. Sect. 5.4). For this simulation, the generator is placed at the discharge position, and the tip of the generator is set to a fixed potential value, such as 4 kV. The calculated electric field is evaluated using streamline plots (Fig 14.24). These streamlines help identify the path along which the field values are the highest. The electric field along these streamlines is then integrated in a post-processing step to obtain the absolute value of the electric field along each streamline. If this value exceeds the arcing threshold, the position of the streamline is identified as the point at which the arcing occurs.

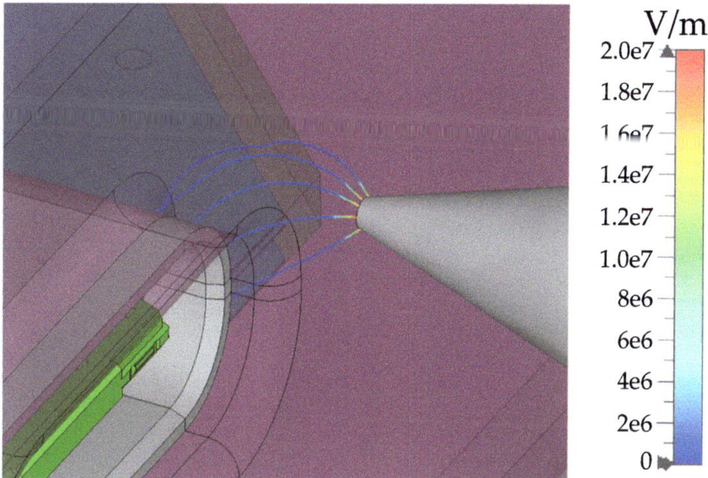

Fig. 14.24 Streamline plot of the electric field calculated using the electrostatic solver. A potential of 4 kV is applied to the tip of the ESD generator

Once the position of the arc is identified, the ESD generator is modeled using a full-wave solver as described in Sect. 14.2.2. A port is placed from the tip of the generator to the estimated arc position where it connects to the DUT. The non-linear behavior of the arc is modeled using a circuit component in SPICE format. The SPICE model is based on the Rompe-Weizel equation, which describes the resistance of the arc R_{arc} as:

$$R_{\text{arc}}(t) = \frac{d}{\sqrt{2 \cdot a \cdot \int_0^t I^2 dt}}. \tag{14.4}$$

In Eq. (14.4), d is the arc length, I is the current in the arc, and a is an empirical constant with the value $a = 1.5 \cdot 10^{-4} \, \frac{\text{m}^2}{\text{V}^2 \cdot \text{s}}$. Using this modeling approach, the remaining part of the simulation can be set up in the same way as the simulation for contact discharge.

Further Reading

ESD Generator and Device Models

S. Caniggia, F. Maradei, Circuital and numerical modeling of ESD coupling to shielded cables, in *EMC Europe 2004, International Symposium on Electromagnetic Compatibility* (2004), pp. 1–6. https://doi.org/10.23919/EMC.2004.10805922

S. Caniggia, F. Maradei, Circuital and numerical modeling of electrostatic discharge generators, in *Fourtieth IAS Annual Meeting. Conference Record of the 2005 Industry Applications Conference, 2005*, vol. 2 (2005), pp. 1119–1123. https://doi.org/10.1109/IAS.2005.1518497

S. Caniggia, F. Maradei, Circuit and numerical modeling of electrostatic discharge generators. IEEE Trans. Ind. Appl. **42**(6), 1350–1357 (2006). https://doi.org/10.1109/TIA.2006.882686

Y. Cao et al., ESD simulation with Wunsch-Bell based behavior modeling methodology, in *EOS/ESD Symposium Proceedings* (2011), pp. 1–10.

V. Javor, New function for representing IEC 61000-4-2 standard electrostatic discharge current. Facta Univ. Ser. Electr. Energ. **27**, 509–520 (2014). https://doi.org/10.2298/FUEE1404509J

T. Li et al., System-level modeling for transient electrostatic discharge simulation. IEEE Trans. Electromagn. Compat. **57**(6), 1298–1308 (2015). https://doi.org/10.1109/TEMC.2015.2443844

D.Z. Li et al., Transient electromagnetic co-simulation of electrostatic air discharge, in *2017 39th Electrical Overstress/Electrostatic Discharge Symposium (EOS/ESD)* (2017), pp. 1–8. https://doi.org/10.23919/EOSESD.2017.8073436

G.-X. Luo et al., Human metal ESD: design of a full-wave model and an improved compact generator. IEEE Trans. Electromagn. Compat. **65**(3), 634–642 (2023). https://doi.org/10.1109/TEMC.2023.3257240

D. Pommerenke, M. Aidam, ESD: waveform calculation, field and current of human and simulator ESD. J. Electrostat. **38**(1), 33–51 (1996). https://doi.org/10.1016/S0304-3886(96)00028-9

C. Qing et al., Advanced full wave ESD generator model for system level coupling simulation, in *2008 IEEE International Symposium on Electromagnetic Compatibility* (2008), pp. 1–6. https://doi.org/10.1109/ISEMC.2008.4652137

L. Speckbacher et al., Modular measurement system for system-efficient ESD design on system and component level. IEEE Trans. Electromagn. Compat. **64**(6), 1812–1819 (2022). https://doi.org/10.1109/TEMC.2022.3214438

K. Wang et al., Numerical modeling of electrostatic discharge generators. IEEE Trans. Electromagn. Compat. **45**(2), 258–271 (2003). https://doi.org/10.1109/TEMC.2003.810817

J. Yousaf et al., Efficient circuit and EM model of electrostatic discharge generator, in *2017 IEEE International Symposium on Electromagnetic Compatibility & Signal/Power Integrity (EMCSI)* (2017), pp. 164–168. https://doi.org/10.1109/ISEMC.2017.8077860

S. Yang, D.J. Pommerenke, Effect of different load impedances on ESD generators and ESD generator SPICE models. IEEE Trans. Electromagn. Compat. **60**(6), 1726–1733 (2018). https://doi.org/10.1109/TEMC.2017.2785739

System Simulation

B. Arndt et al., Virtual ESD testing of automotive electronic systems, in *2010 Asia-Pacific International Symposium on Electromagnetic Compatibility* (2010), pp. 683–686. https://doi.org/10.1109/APEMC.2010.5475598

S. Caniggia, F. Maradei, Numerical prediction and measurement of ESD radiated fields by free-space field sensors. IEEE Trans. Electromagn. Compat. **49**(3), 494–503 (2007). https://doi.org/10.1109/TEMC.2007.902196

S. Caniggia, F. Maradei, Analytical and numerical simulation models for calculating EMI into circuits due to ESD radiated fields, in *2011 IEEE International Symposium on Electromagnetic Compatibility* (2011), pp. 602–606. https://doi.org/10.1109/ISEMC.2011.6038382

G. Fellner et al., Quantification of ESD pulses caused by collision of objects, in *2022 International Symposium on Electromagnetic Compatibility - EMC Europe* (2022), pp. 273–278. https://doi.org/10.1109/EMCEurope51680.2022.9901123

G. Fellner, A. Pak, D. Pommerenke, Detection and localization of CDM like ESD using a novel sensor derived from leaky-coax, in *2023 International Symposium on Electromagnetic Compatibility – EMC Europe* (2023), pp. 1–6. https://doi.org/10.1109/EMCEurope57790.2023.10274257

J. Koo et al., A nonlinear microcontroller power distribution network model for the characterization of immunity to electrical fast transients. IEEE Trans. Electromagn. Compat. **51**(3), 611–619 (2009). https://doi.org/10.1109/TEMC.2009.2023670

J.-C. Lee et al., An improved experimental setup for electrostatic discharge (ESD) measurements based on transmission line pulsing technique. IEEE Trans. Instrum. Meas. **50**(6), 1808–1814 (2001). https://doi.org/10.1109/19.982985

D.Z. Li et al., Full-wave simulation of system-level disruption during secondary ESD events in a smartphone. IEEE Trans. Electromagn. Compat. **61**(1), 40–47 (2019). https://doi.org/10.1109/TEMC.2018.2815024

D. Liu et al., Full-wave simulation of an electrostatic discharge generator discharging in air-discharge mode into a product. IEEE Trans. Electromagn. Compat. **53**(1), 28–37 (2011). https://doi.org/10.1109/TEMC.2010.2087025

L. Pertoll, A. Pak, D. Pommerenke, Automatic creation of TVS SPICE models for ESD system level simulations, in *2023 International Symposium on Electromagnetic Compatibility – EMC Europe* (2023), pp. 1–6. https://doi.org/10.1109/EMCEurope57790.2023.10274181

J.S. Meiguni et al., Transient analysis of ESD protection circuits for high-speed ICs. IEEE Trans. Electromagn. Compat. **63**(5), 1312–1321 (2021). https://doi.org/10.1109/TEMC.2021.3071644

K. Muhonen, E. Grund, R. Ashton, High-speed TLP and ESD characterization of ICs, in *2021 IEEE BiCMOS and Compound Semiconductor Integrated Circuits and Technology Symposium (BCICTS)* (2021), pp. 1–6. https://doi.org/10.1109/BCICTS50416.2021.9682498

Murata, *Common Mode Choke Coil DLM0NSN900HY2*. https://www.murata.com/products/productdetail?partno=DLM0NSN900HY2#

Nexperia, *ESD Application Handbook*. https://www.nexperia.com/dam/jcr:34d76aab-f203-49de-8693-2168d2117f98/Nexperia_ESD_Application_Handbook_Automotive_Edition.pdf

Nexperia, *PUSB3AB4. ESD Protection for Ultra High-Speed Interfaces*. https://assets.nexperia.com/documents/data-sheet/PUSB3AB4.pdf

D. Pommerenke, J. Fan, J. Drewniak, Simulation challenges in system level electrostatic discharge modeling, in *2016 IEEE/ACES International Conference on Wireless Information Technology and Systems (ICWITS) and Applied Computational Electromagnetics (ACES)* (2016), pp. 1–2. https://doi.org/10.1109/ROPACES.2016.7465439

A. Reithofer et al., The distribution of discharge amplitudes of randomly colliding charged spheres, in *2023 International Symposium on Electromagnetic Compatibility – EMC Europe* (2023), pp. 1–6. https://doi.org/10.1109/EMCEurope57790.2023.10274325

G. Shen et al., ESD immunity prediction of D flip-flop in the ISO 10605 standard using a behavioral modeling methodology. IEEE Trans. Electromagn. Compat. **57**(4), 651–659 (2015). https://doi.org/10.1109/TEMC.2015.2418715

H. Wolf, H. Gieser, W. Wilkening, Analyzing the switching behavior of ESD-protection transistors by very fast transmission line pulsing. J. Electrostat. **49**(3), 111–127 (2000). https://doi.org/10.1016/S0304-3886(00)00010-3

D.C. Wunsch, R.R. Bell, Determination of threshold failure levels of semiconductor diodes and transistors due to pulse voltages. IEEE Trans. Nucl. Sci. **15**(6), 244–259 (1968). https://doi.org/10.1109/TNS.1968.4325054

Y. Xu et al., Improved SEED modeling of an ESD discharge to a USB cable. IEEE Trans. Electromagn. Compat. **65**(3), 625–633 (2023). https://doi.org/10.1109/TEMC.2022.3232616

X. Yan et al., A physics-based model for snapback-type ESD protection devices. IEEE Trans. Electromagn. Compat. **65**(5), 1273–1281 (2023). https://doi.org/10.1109/TEMC.2023.3303493

S.C. Yener, S. Frei, S. Scheier, Behavioural model based simulation of the ESD-soft-failure-robustness of micro-controller inputs, in *2016 International Symposium on Electromagnetic Compatibility – EMC EUROPE* (2016), pp. 541–545. https://doi.org/10.1109/EMCEurope.2016.7739293

J. Zhang et al., Modelling electromagnetic field coupling from an ESD gun to an IC, in *2011 IEEE International Symposium on Electromagnetic Compatibility* (2011), pp. 553–558. https://doi.org/10.1109/ISEMC.2011.6038373

J. Zhou et al., Transient response of ESD protection devices for a high-speed I/O interface. IEEE Trans. Electromagn. Compat. **64**(4), 907–914 (2022). https://doi.org/10.1109/TEMC.2022.3168855

Chapter 15
Platform RF Interference

Platform RF interference, a common cause of RF desense, is becoming an increasingly significant concern in modern electronic systems. While it is not typically classified as a traditional EMC issue—since it primarily involves electromagnetic couplings within an electronic system—it shares similarities with emission-related problems. For this reason, we have included it in this book. RF interference is a wide topic which we treat on an introductory level in this book by restricting the discussion on an example of a noise source to antenna coupling.

Connected electronic devices, such as tablets, laptops, smartphones, and the extensive ecosystem of Internet-of-Things (IoT) products, integrate multiple RF systems. For instance, a smartphone supports connectivity for WiFi at 2.4 GHz and 5 GHz, Bluetooth, GPS, GSM, NFC, and multiple LTE bands, all coexisting within a compact platform. Additionally, modern high-speed data buses, such as USB 3.0 and DDR buses, operate at high clock frequencies, with harmonics extending well into the RF spectrum. These harmonics and data signals can couple with antennas, causing RF system desensitization (desense) and degrading communication quality. Desense reduces a receiver's sensitivity to low-level signals, ultimately diminishing the device's range and data rate.

In this chapter, we introduce basic concepts for modeling RF interference. We present transmitter and receiver models and describe the coupling mechanisms between them using an S-parameter matrix. After establishing the methodology, we apply it to a practical example, examining the coupling between a flexible PCB and a dual-band WiFi antenna.

RF interference is not a new phenomenon; it has been studied extensively since at least the 1980s, primarily in the aerospace and marine industries. In these domains, RF-based communication is safety-critical, requiring the reliable operation of all RF systems onboard. In this book, we adapt the methods developed over decades in these industries to address RF interference challenges in general electronic systems.

J. Hansen, A. Barchanski, *EMC Simulation of Electronic Systems*,
https://doi.org/10.1007/978-3-032-09237-3_15

15.1 Modeling of RF Interference

Domains: 

We conduct a simple RF interference study in Sect. 6.6. In this chapter we describe the setup of an RF interference in detail, by defining transmitter and receiver models as well as defining an EMI margin for the measure of interference. The core concept in analyzing RF interference is to calculate the power coupled from an interferer into a (radio) receiver. The interferer could be another RF system that is not the intended transmitter, a digital signal, or noise emitted from a power electronic system. By comparing the received power to the sensitivity profile of the receiver, we can quantify the potential RF interference. The coupling between the source and the receiver is modeled using S-parameters.

15.1.1 RF Transmitter Models

An RF transmitter spectrum represents the transmitted power (for narrowband signals) or power density (for broadband noise) as a function of frequency. A simple transmitter spectrum is illustrated on the left-hand side of Fig. 15.1. The transmitter profile can be modeled either parametrically or imported from an external source, such as measurement data. The shape of this profile is directly influenced by the modulation scheme used by the transmitter. On the right-hand side of Fig. 15.1, we have plotted the Tx spectrum for the sixth 2.4 GHz WiFi channel using orthogonal frequency-division multiplexing (OFDM) modulation.

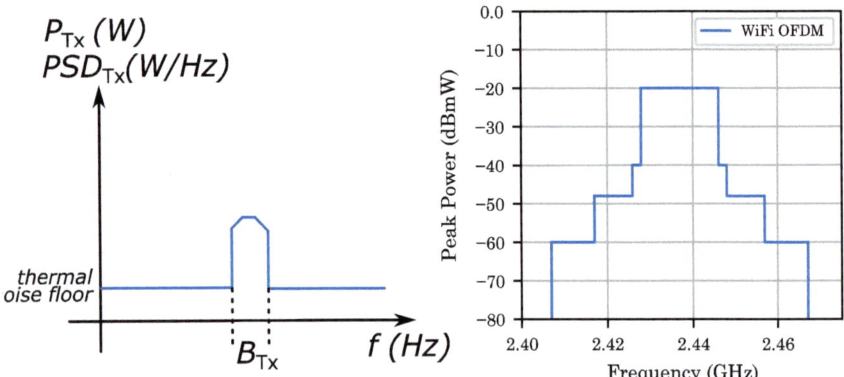

Fig. 15.1 Left: Sketch of the power spectrum of a simplified transmitter model. Right: Power spectrum of the sixth 2.4 GHz WiFi channel

An RF transmitter operates on a specific channel but may have multiple channels to choose from. While the spectral shape of the transmitter is assumed to be consistent across channels, the center frequency may vary between them. To model this, we define the number of channels, their start/stop frequencies, and the channel spacing. The transmitter typically also emits power at higher harmonics of the channel frequency. These harmonics are modeled using formulas that describe the taper profile over the frequency range. Additionally, spurious emissions can be accounted for by defining isolated spurs along with their respective power levels.

Beyond the defined power profile, the transmitter emits constant broadband background noise due to thermal effects. This noise is proportional to Boltzmann's constant, the temperature, and the transmitter bandwidth. With these parameters, we can comprehensively model an RF transmitter spectrum.

15.1.2 Digital Noise Sources

For the modeling of digital noise sources, we distinguish between two types of emissions from the digital system: noise due to clock sources and noise due to random data signals. Each source type requires a different approach.

When the noise source is a clock signal, the power of the noise source is calculated using analytical formulas. These formulas take as input parameters the voltage level, duty cycle, rise/fall times, clock period, and driver impedance. The resulting spectrum consists of narrowband peaks, as shown on the left-hand side of Fig. 15.2. The clock spectrum in this figure is calculated using the following parameters: peak voltage of 0.6 V, clock frequency of 80 MHz, duty cycle of 0.5, rise and fall times of 250 ps, and a driver impedance of 10 Ω. This spectrum is used later in this chapter as part of the application example.

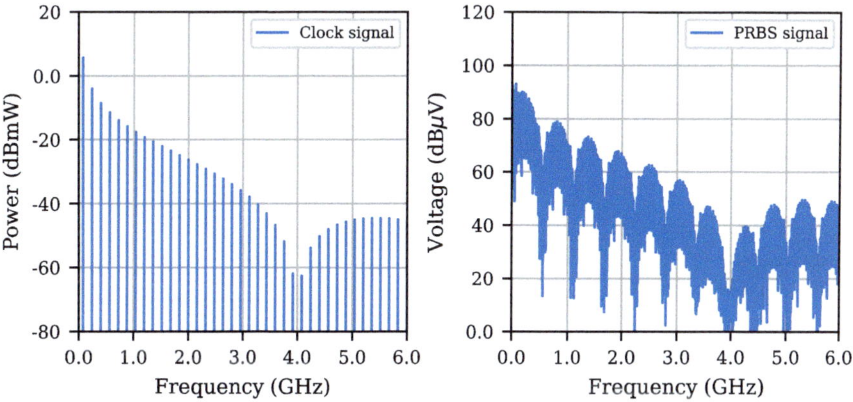

Fig. 15.2 Transmitter spectra. Left: Power spectrum of a clock signal. Right: Voltage spectrum of a PRBS signal

For more complex waveforms, such as PRBS signals, the source spectrum is calculated upfront using a transient circuit simulation. This simulation can utilize IBIS models or ideal voltage/current sources. Once the source spectrum is obtained, it is applied in the RF interference calculation, similar to the clock signal spectrum. This method bridges the gap between time-domain simulations, often used in the field of SI, and frequency-domain analysis for RF interference. A typical PRBS spectrum is shown on the right-hand side of Fig. 15.2. This spectrum, also used in the application example, is obtained from a PRBS simulation with a peak voltage of 0.6 V, signal frequency of 280 MHz, rise and fall times of 250 ps, and a driver impedance of 10 Ω. Similarly, source spectra for power electronics systems can be calculated.

15.1.3 Receiver Models

The receiver sensitivity profile is defined in terms of power density, similar to the transmitter spectrum but in the inverse sense. Within the receiver band, the sensitivity defines the minimum power required for proper operation, which varies depending on the type of radio. Outside the receiver band, a saturation value is defined (Fig. 15.3).

The total power arriving at the receiver can be decomposed into three components:

$$P_{\text{received}} = P_{\text{in}} + N_{\text{in}} + \sum_i I_{i,\text{in}} \tag{15.1}$$

where:

P_{in} The input power from the link transmitter, representing the intended signal power.

N_{in} The input (thermal) noise.

$\sum_i I_{i,\text{in}}$ The power received from one or more interferers.

A receiver is modeled by an amplifier input stage and a detector. The amplifier input stage is characterized by its gain and noise figure, which are used to calculate

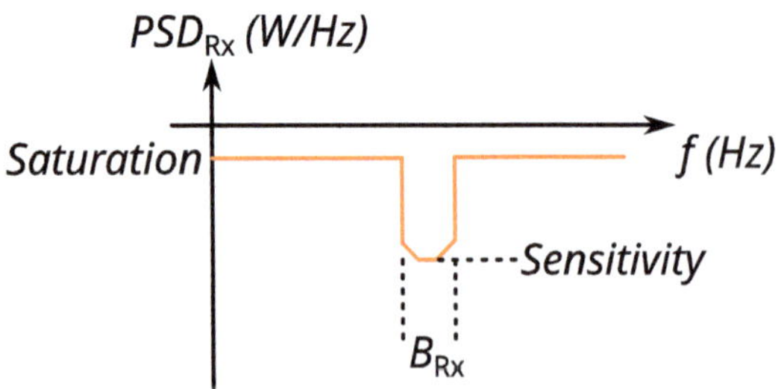

Fig. 15.3 Sketch of the power spectrum of a simplified receiver model

the power at the detector, P_{detector} from the power arriving at the receiver. The power at the detector is the sum of

$$P_{\text{detector}} = P_{\text{detector}} + N_{\text{detector}} + \sum_i I_{i,\text{detector}} \tag{15.2}$$

Here:

P_{detector} The amplified signal power at the detector.
N_{detector} The noise power at the detector, consisting of the amplified noise from the receiver input and additional noise generated by the amplifier.
$\sum_i I_{i,\text{detector}}$ The interference power at the detector, representing the amplified interference power present at the receiver input.

The sensitivity of the receiver, P_{sens}, is defined as the minimum input signal power required to maintain a minimum receiver quality in the absence of interference. This value is typically provided in the data sheets of the individual radio systems.

For interference analysis, the susceptibility of the receiver is defined as the maximum allowed interference level, $\sum_i I_{i,\text{detector,max}}$, at the input of the receiver for a given received power of the link signal. This reflects the fact that when the link signal power is high, the effect of interference is less significant.

The receiver susceptibility is expressed as

$$\text{Susceptibility} = (P_{\text{in}} - P_{\text{sens}}) \cdot \frac{N_{\text{detector}} + \sum_i I_{i,\text{detector,max}}}{P_{\text{detector}}} \tag{15.3}$$

If the power from the link transmitter is unknown, as in the examples discussed in this section, we assume $P_{\text{in}} = 2 \cdot P_{\text{sens}}$.

15.1.4 EMI Margin

With the transmitter and receiver models defined, we can calculate a measure of the expected interference: the so-called EMI margin. Assuming the coupling between the systems is described by the S-parameters of the transmit-receive channel, the power arriving at the receiver, under the assumption that the receiver and transmitter are matched, is given by

$$P_{\text{received}} = P_{\text{Tx}} \cdot \left|S_{\text{Rx,Tx}}\right|^2 . \tag{15.4}$$

If the (unintended) power arriving at the receiver exceeds the receiver saturation level, an interference event is always reported, regardless of whether the transmitted signal falls inside the receiver band.

Within the receiver band, we define the point margin as the ratio of the power at the receiver to the receiver's susceptibility at every frequency point:

$$M_{\text{Point}} = \frac{P_{\text{received}}}{\text{Susceptibility}} \tag{15.5}$$

In the logarithmic (dB) scale, this ratio becomes a difference. For narrowband transmitters with only one peak within the receiver's bandwidth, the point margin is sufficient to describe the interference event.

For cases where multiple narrowband peaks or broadband noise exist within the receiver's band, the total accepted power must first be calculated. The accepted power is determined by summing the power contributions from individual narrowband peaks at different frequencies:

$$P_{\text{accepted}} = \left(\sum_i \sqrt{P_i}\right)^2 \tag{15.6}$$

Additionally, broadband noise is integrated over the receiver's bandwidth. In these scenarios, the in-band margin is defined as

$$M_{\text{InBand}} = \frac{P_{\text{accepted}}}{\text{Susceptibility}} \tag{15.7}$$

In the case of a single narrowband peak within the receiver's band, the in-band margin equals the point margin.

The EMI margin determines whether interference is expected between the interferer and the receiver:

Negative EMI margin No interference is expected.

Positive EMI margin Interference is likely, resulting in a degradation of the receiver's performance.

The magnitude of the margin quantifies the severity of the interference.

This is a simplified view of the RF interference calculation, which we use in the following application example. It handles only one-on-one events, involving a single receiver and a single transmitter. For interference scenarios involving multiple transmitters, a combination of the transmitters must be considered. For radios with channel hopping, such as Bluetooth, a statistical analysis is required. High-power transmitters used in the marine and aerospace industries may also necessitate the inclusion of non-linear effects, which create intermodulation products in the RF systems. These topics are outside the scope of this book, as our focus is on the simulation of electronic systems rather than high-power RF systems.

In the aerospace industry, RF interference is often referred to as co-site interference. Several commercial tools are available for its calculation, including Andro E3 Expert, Ansys EMIT, and Altair's co-site interference tool. For the following example, we use the Interference Task in CST to calculate the EMI margins. Commercial tools for RF interference simulation often provide libraries of RF systems, which significantly reduce the time required to define transmitters and receivers. These tools can handle scenarios with a large number of systems and present a simplified view of interference events in the form of a matrix.

For simpler scenarios, such as the one in the following application example, the received power can also be calculated using circuit simulation with an AC task. The use of specialized RF interference tools is not mandatory, but it simplifies the setup of calculations and provides an overview of different margins and interference events. In this example, we use the Interference Task to demonstrate the potential outputs of these tools and provide the reader with insights into their capabilities.

15.2 Application Example: Coupling of a Flex PCB into a WiFi Antenna

- Digital PRBS signal as an IBIS model, clock signal
- Received power at antenna, simple receiver model
- Full-wave, S-parameters
- FEM method
- Interference analysis based on AC
- 0–6 GHz

The coupling of signals routed on flexible PCBs into RF systems in highly integrated electronic systems can result in the desensitization (desense) of the receiver. In the application example presented in this section, we calculate the coupling from an LVDS signal routed over a flexible PCB into a WiFi antenna.

Both the flexible PCB and the antenna are simulated using a full-wave solver to calculate the S-parameter coupling from the LVDS channel into the WiFi antenna. The calculated coupling is then used in the Interference Task to estimate the EMI margin. Since the EMI margin provides an indication of the likelihood of desensitization of the WiFi channel, we explore various design improvements for the flexible PCB to mitigate the interference.

To reduce simulation time, design optimizations are initially performed on the standalone flexible PCB. Once an improved design is identified, it is tested again in the complete model to evaluate its performance in the integrated system.

The flex PCB connects two rigid PCBs: a larger rigid PCB containing the WiFi system and a smaller rigid PCB modeling a PCB on which an IC is mounted. An overview of the simulation model is shown in Fig. 15.4. In electronic systems, flexible PCBs are commonly used to connect larger rigid PCBs, as they require

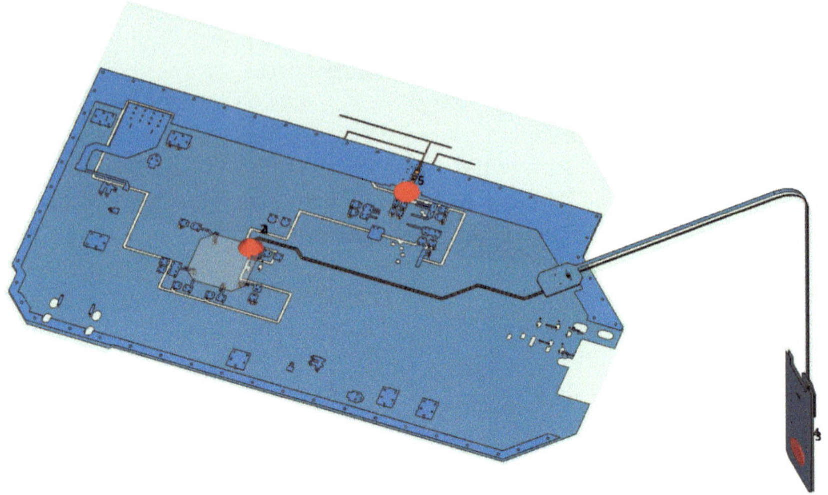

Fig. 15.4 Simulation model for the calculation of RF interference from an LVDS signal routed on a flexible PCB into a WiFi antenna

less space than traditional connectors and cables. Additionally, flex PCBs can accommodate a large number of signals and be designed with controlled impedance.

In this example, the digital board used throughout the book is modified to connect to a flex PCB. For demonstration purposes, only a single differential pair is routed across all three PCBs, although an LVDS channel typically consists of multiple differential pairs. On the main board, the differential pair is routed on an inner layer. In Fig. 15.4, the top GND layer is hidden to visualize the inner routing. The flex PCB is 10 cm long and bent at a 90° angle at its center. At the other end of the flex PCB, it connects to a small four-layer PCB, where the differential pair terminates into an IC.

The connectors on both ends of the flex PCB were designed and validated in separate simulations. At the ends of the differential pair, single-ended ports to the reference are defined. Mode converters are used in the schematic for mixed-mode conversion. The S-parameter calculation for the EM model takes approximately 1 hour and 57 minutes using the FEM solver.

The WiFi antenna in this example is a dual-band PIFA antenna driven by a WiFi IC located on the top side of the larger PCB. The antenna was designed using Antenna Magus and optimized for mounting on this specific PCB. The PCB includes component placeholders for a matching circuit between the antenna and the IC. However, as the antenna's performance is satisfactory without a matching circuit, it is not used; the planned location of the series matching element is shorted. The S-parameter of the antenna is shown in Fig. 15.5. In the 2.4 GHz WiFi band, the return loss is below −6 dB, and in the 5 GHz band, it remains below −6 dB from 5 GHz to 5.6 GHz. Since this example only considers the upper channels of the 5 GHz band, the antenna design is sufficient.

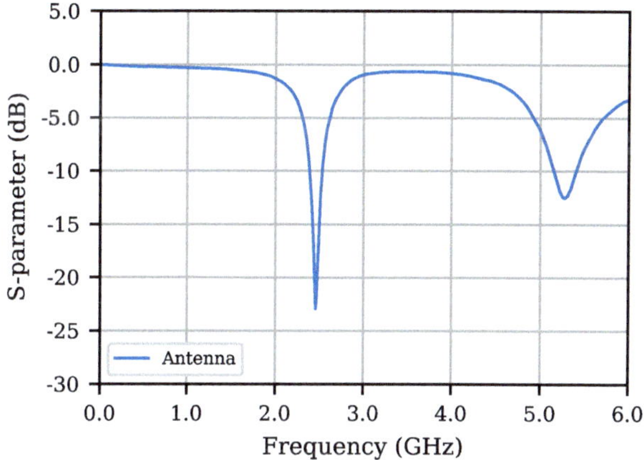

Fig. 15.5 S-parameter of the dual-band WiFi antenna

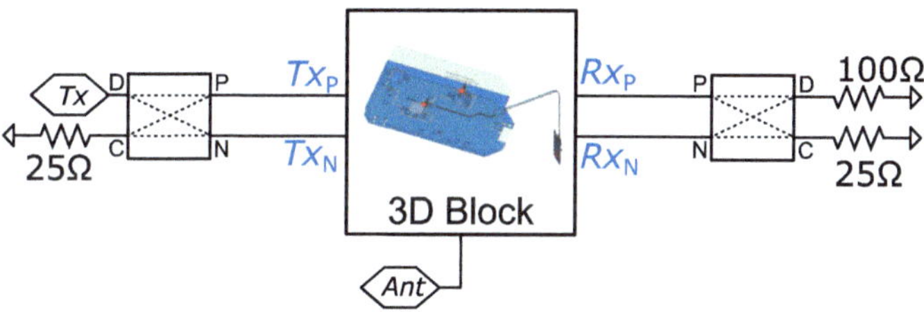

Fig. 15.6 Schematic for the reduction of the five-port matrix into a two-port matrix

The EM simulation model uses four single-ended ports to terminate the differential LVDS channel. Along with the antenna port, this results in a 5×5 S-parameter matrix. Since the Interference Task requires the definition of receivers and transmitters, as described earlier, it is not feasible to use simple terminations and mode converters for calculating the EMI margin. Therefore, we reduce the number of ports before performing the interference calculation, where the LVDS channel is excited by a differential signal. This is achieved using a schematic with mode converters and terminations, which reduces the S-parameter matrix to two ports.

The schematic for the port reduction is shown in Fig. 15.6. A port is placed at the differential pin of the mode converter connected to the transmitter side, which is located on the digital PCB. The receiver side, on the smaller PCB, as well as the common-mode pin on the transmitter side, is terminated with resistors. A second port is connected to the antenna port of the EM model. The resulting two-port S-parameter matrix is calculated from this schematic and passed to the Interference Task.

In the Interference Task, a WiFi radio is defined at the antenna port. Since the antenna only receives in this analysis, only the receiver properties are required. In the 2.4 GHz band, 13 channels are defined with a width and spacing of 20 MHz each.

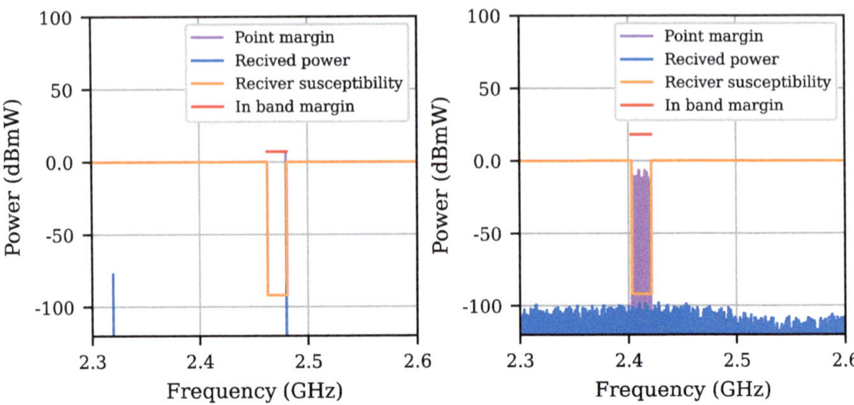

Fig. 15.7 EMI margins and power from two different signals. Left: Clock signal. Right: PRBS signal

In the 5 GHz band, only the upper part of the available channels is considered. Four channels are defined with a bandwidth and spacing of 40 MHz, starting at 5.15 GHz.

We analyze the EMI margin for two driving signals: an 80 MHz clock signal and a PRBS signal at 280 MHz. The spectra of both signals are shown in Fig. 15.2.

The results of the interference analysis are shown in Fig. 15.7. We first analyze the results for the clock signal, which are plotted on the left-hand side of Fig. 15.7. The received power is represented in blue. Two peaks are observed at the harmonics of the clock signal: one at 2.32 GHz and another at 2.48 GHz.

The peak at 2.32 GHz lies outside the receiver's band and is significantly below the receiver's saturation value. Therefore, it does not result in an interference event. In contrast, the peak at 2.48 GHz falls into the 13th channel of the WiFi receiver, with a received power of −84.9 dBm. Since the receiver sensitivity is set to −92 dBm, an interference event is detected. The point margin is calculated as 7.1 dB, which is equal to the in-band margin because only this single peak falls within the receiver's band.

The results for the PRBS signal are plotted on the right-hand side of Fig. 15.7. The imported spectrum from the transient PRBS simulation is treated as a series of narrowband sources in the Interference Task. Since multiple peaks are located within the receiver's band, their contributions are summed according to Eq. 15.6. The resulting in-band margin is calculated to be 18 dB, indicating that the interference event from the PRBS signal is more severe than the interference event caused by the clock signal.

Additionally, the PRBS signal causes interference events across all channels of the 2.4 GHz WiFi band, whereas the clock signal results in an interference event in only one channel. In the plot of the PRBS interference event, only the first channel of the 2.4 GHz WiFi band is depicted for clarity.

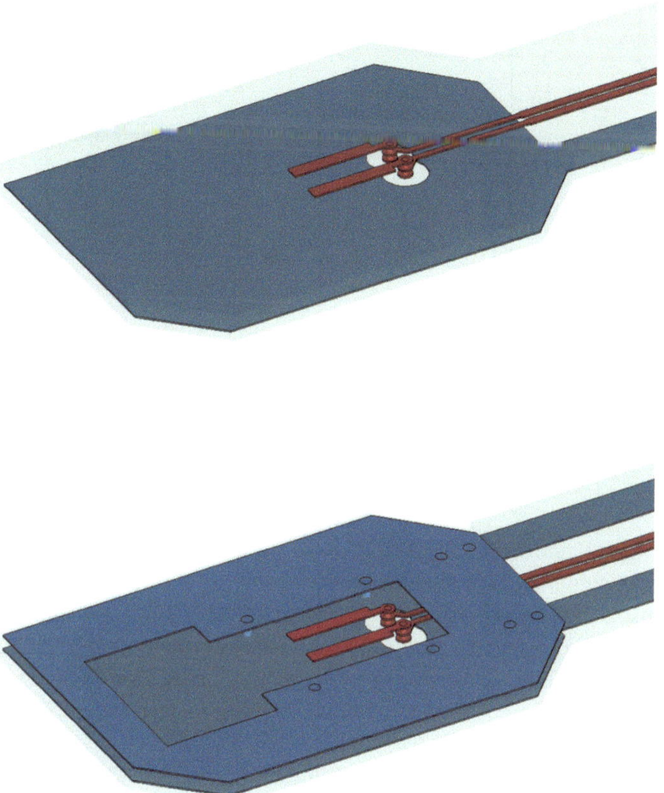

Fig. 15.8 Design of the flexible PCB. Top: Original design. Bottom: Symmetric design

15.2.1 Mitigation of RF Interference

As the interference analysis reports a positive EMI margin for the flex example, we discuss possible improvements to the flex PCB aimed at reducing interference. We suspect that the flex PCB is the primary source of interference, as the routing on both rigid PCBs uses a stripline configuration and is shielded by ground layers for most of the channel length.

In the original design, the reference layer on the flex PCB is located only on one side of the differential pair. Such an imbalanced design can lead to differential-to-common-mode conversion. To address this, we improve the design by adding a second reference trace along the differential pair. Additionally, around the connector, the reference is originally located only on the top layer; to balance this, we copy the copper fill to the bottom layer. Both designs are shown in Fig. 15.8.

We apply the RF interference simulation to the improved design and observe that the in-band margin in the 2.4 GHz band increases from 7.1 dB to 10.9 dB for

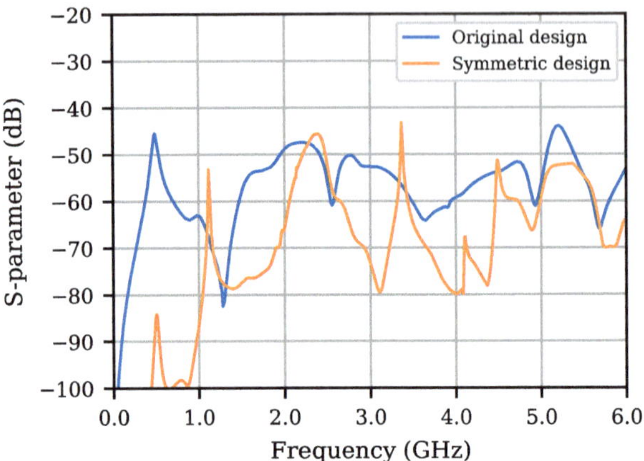

Fig. 15.9 Coupling from the LVDS channel into the WiFi antenna

the clock signal, and from 18 dB to 22.3 dB for the PRBS signal. Surprisingly, the symmetric design increases the probability of interference rather than reducing it.

This result is unexpected, as we anticipate the symmetric layout to reduce interference. To identify the root cause, we analyze the coupling from the LVDS channel to the antenna across the entire frequency range. For this purpose, we take a step back and, instead of relying on the EMI margin, examine the S-parameters of the corresponding EM model.

The S-parameter from the LVDS channel to the WiFi antenna is plotted in Fig. 15.9. Below 1 GHz, the coupling in the symmetric design is more than 30 dB lower than in the original design, indicating that the symmetric design has significantly improved the coupling in this frequency range, where the coupling is dominated by the common mode. However, at 1.1 GHz, 2.4 GHz, and 3.6 GHz, we observe resonances in the symmetric design that lead to an increase in the EMI margin. Since only the flex PCB has been modified while the rest of the model remains unchanged, we suspect that the resonance originates from the flex PCB itself.

To investigate the flex PCB in detail, we create a standalone simulation model. This approach not only isolates the problem but also results in a simulation model with reduced runtime. In this model, the four ports driving the differential trace are retained but are placed at the connector pins of the flex.

As the coupling to the antenna is proportional to the electromagnetic field around the flex, we place a field probe at a distance of 5 cm from the flex. A single probe is sufficient for this analysis, as we are primarily comparing the field strength between the different designs on a relative basis. The field probe serves as an indicator of the coupling level. This simplified model requires only 1 minute and 30 seconds of simulation time.

We drive the standalone flex PCB model with a 1 V broadband differential signal and observe that the resonances are also visible in the electric field at the probe, as

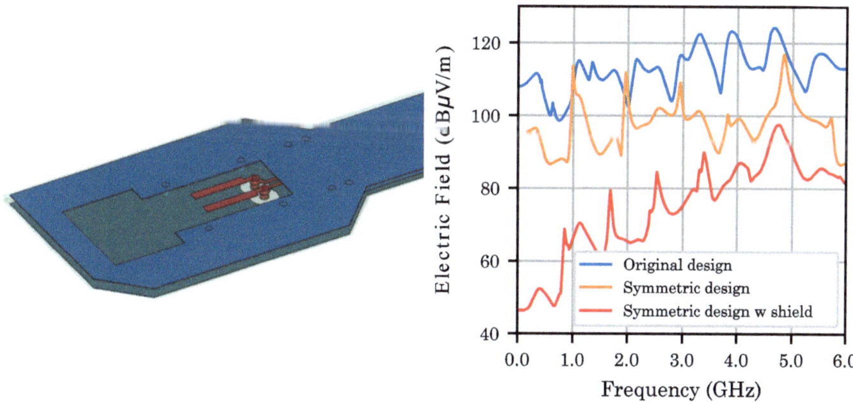

Fig. 15.10 Left: Modified symmetric design including a shielding layer on top and bottom. Right: Electric field at the probe in the standalone flex PCB model for three different designs

shown on the right-hand side of Fig. 15.10. Compared to the model that includes the rigid PCBs, the resonances are now shifted to 1 GHz, 2 GHz, and 3 GHz. However, we suspect that these are the same resonances observed in the S-parameters in Fig. 15.9.

To understand the origin of these resonances, we plot the electric field at the peaks in Fig. 15.11. From the field distribution, we observe that length resonances occur at these frequencies. Upon closely examining the design, we identify the mechanism responsible for these resonances: the reference layer on the flex PCB forms a slot antenna over which the differential trace pair is routed. The differential pair couples to this slot, causing it to resonate. This resonance is absent in the original design because the reference layer is located only on one side of the differential pair. This example illustrates that at frequencies where resonances become significant, merely reducing differential-to-common-mode conversion might not suffice. Field plots are crucial for revealing the coupling mechanisms.

To improve the design and validate our hypothesis, we apply a simple modification to the symmetric flex PCB. On both the top and bottom layers, we extend the reference net to fully cover the differential pair along its entire length, as shown on the left-hand side of Fig. 15.10. Depending on the manufacturing process of the flex PCB and the intended bending radius, a continuous copper layer over the entire length of the flex might not be feasible. However, in simulation, this modification allows us to test whether it reduces the coupling. If the modification proves effective, a more refined layout, such as a hatched reference, can be implemented.

In the standalone model, adding the shielding layer significantly reduces the electric field at the probe, as seen on the right-hand side of Fig. 15.10. While the resonances are still present, the magnitude of the electric field is reduced across the entire frequency band. With this modification, we return to the model that includes the rigid PCBs and perform an RF interference analysis of the shielded flex design.

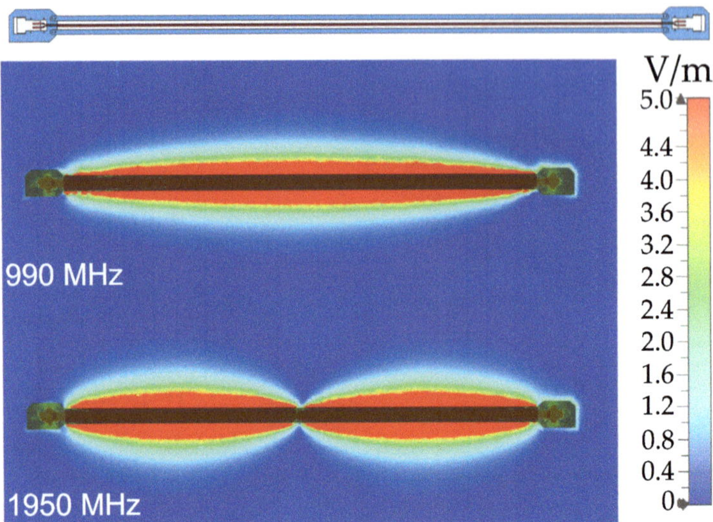

Fig. 15.11 Top: Overview of the symmetric flex PCB. The top layers close to the connector are hidden. Center: Electric field around the flex PCB at 990 MHz. Bottom: Electric field around the flex PCB at 1950 MHz. Both are driven by a differential, broadband 1 V signal

Table 15.1 EMI margin obtained from the RF interference simulation for three different flex PCB designs

	2.4 GHz		5 GHz	
	Clock	PRBS	Clock	PRBS
Original Design	7.1 dB	18.0 dB	3.4 dB	17.3 dB
Symmetric Design	10.9 dB	22.3 dB	−5.1 dB	11.0 dB
Symmetric Shielded Design	−8.3 dB	0.4 dB	−10.8 dB	5.2 dB

The calculated EMI margin for the three designs is summarized in Table 15.1. With the shielded layout, interference in the 2.4 GHz band is almost completely eliminated. The EMI margin for the PRBS signal in this band is still positive but reduced to only 0.4 dB, representing an 18 dB improvement compared to the original design. This margin is tolerable. In the 5 GHz band, the improvement is approximately 12 dB. As in the 2.4 GHz band, the PRBS signal results in a positive EMI margin, amounting to 5.2 dB. It is up to the designer to determine whether this level of interference is acceptable or if further modifications are necessary.

In the simulations conducted in this section, we calculate the EMI margin as a measure of potential RF interference. Since the first design modification does not lead to improvement and resonances are observed, we analyze a simplified model containing only the flex PCB to understand the coupling mechanism. After identifying the cause of the coupling, we return to the complete model including all PCBs.

This strategy of simplifying models and reducing complexity is highly effective for simulating complex electronic products. Running the full model, which can take several hours to compute, is not necessary at every step of the design process. Particularly when testing design modifications, qualitative comparisons using simplified models are a valuable approach.

Further Reading

RF Interference Modeling

E.X. Alban et al., Statistical modeling of the interference noise generated by computing platforms. IEEE Trans. Electromagn. Compat. **54**(3), 574–584 (2012). https://doi.org/10.1109/TEMC.2011.2164255

J. Bommer, B. Li, F. German, Design-by-desense: a multi-fidelity approach to interference-free design for electronic devices, in *2017 IEEE International Symposium on Electromagnetic Compatibility & Signal/Power Integrity (EMCSI)* (2017), pp. 447–455. https://doi.org/10.1109/ISEMC.2017.8077912

W.G. Duff, EMC figure of merit for receivers, in *1969 IEEE Electromagnetic Compatibility Symposium Record* (1969), pp. 257–262. https://doi.org/10.1109/TEMC.1969.4307212

S. Grivet-Talocia et al., Fast evaluation of electromagnetic interference between antenna and PCB traces for compact mobile devices, in *2008 IEEE International Symposium on Electromagnetic Compatibility* (2008), pp. 1–5. https://doi.org/10.1109/ISEMC.2008.4652050

Q. Huang et al., Physics-based dipole moment source reconstruction for RFI on a practical cellphone. IEEE Trans. Electromagn. Compat. **59**(6), 1693–1700 (2017). https://doi.org/10.1109/TEMC.2017.2705641

Q. Huang et al., Radio frequency interference estimation using transfer function based dipole moment model, in *2018 IEEE International Symposium on Electromagnetic Compatibility and 2018 IEEE Asia-Pacific Symposium on Electromagnetic Compatibility (EMC/APEMC)* (2018), pp. 115–120. https://doi.org/10.1109/ISEMC.2018.8393750

J. Joo et al., Modeling of RF interference caused by solid-state drive noise, in *2022 Asia-Pacific International Symposium on Electromagnetic Compatibility (APEMC)* (2022), pp. 685–687. https://doi.org/10.1109/APEMC53576.2022.9888351

D. Kostka et al., Modeling of RF desense for mobile electronics, in *2017 IEEE International Symposium on Electromagnetic Compatibility & Signal/Power Integrity (EMCSI)* (2017), pp. 474–480. https://doi.org/10.1109/ISEMC.2017.8077917

S. Lee et al., Analytical intra-system EMI model using dipole moments and reciprocity, in *2018 IEEE International Symposium on Electromagnetic Compatibility and 2018 IEEE Asia-Pacific Symposium on Electromagnetic Compatibility (EMC/APEMC)* (2018), pp. 1169–1173. https://doi.org/10.1109/ISEMC.2018.8393972

L. Li et al., Radio-frequency interference estimation by reciprocity theorem with noise source characterized by Huygens's equivalent model, in *2016 IEEE International Symposium on Electromagnetic Compatibility (EMC)* (2016), pp. 358–363. https://doi.org/10.1109/ISEMC.2016.7571673

J. Pan et al., Radio-frequency interference estimation using equivalent dipole-moment models and decomposition method based on reciprocity. IEEE Trans. Electromagn. Compat. **58**(1), 75–84 (2016). https://doi.org/10.1109/TEMC.2015.2496210

G. Ramsubbaraj, L. Cheng, K. Rao, RF desense risk prediction using EM simulation, in *2023 IEEE Symposium on Electromagnetic Compatibility & Signal/Power Integrity (EMC+SIPI)* (2023), pp. 65–69. https://doi.org/10.1109/EMCSIPI50001.2023.10241649

K. Slattery, H. Skinner, *Platform Interference in Wireless Systems: Models, Measurement, and Mitigation* (Newnes/Elsevier, Amsterdam, 2008)

H. Wang et al., Estimating radio-frequency interference to an antenna due to near-field coupling using decomposition method based on reciprocity. IEEE Trans. Electromagn. Compat. **55**(6), 1125–1131. https://doi.org/10.1109/TEMC.2013.2248090

Chapter 16
Machine Learning

The application of machine learning (ML) in EMC simulation is still in its early stages. Many activities are underway, and it remains difficult to predict which of them will mature into widely adopted simulation procedures. In this chapter, we present one particularly promising approach to leveraging ML for EMC design. We acknowledge that a few years from now, this chapter may look quite different, as improved methods may emerge or research priorities may shift. It is not the goal of this chapter to provide a comprehensive overview of all existing methods or their respective applications. Potential future applications of ML are outlined in Sect. 18.3.

EMC simulation supports a wide range of engineering decisions using physically motivated simulation models. This approach is bottom-up, as the model is built for a specific system under test. Parameter studies are used to explain the system's behavior under various conditions. ML significantly extends these opportunities. An ML model is trained over a defined range of parameters and can generalize beyond a single system to represent a broad class of systems. Instead of describing just one specific configuration, the ML model encompasses a multitude of possible designs. Because the ML model evaluates results within milliseconds, it can be seen as a container of ready-to-evaluate system designs. This shifts the design approach from bottom-up to top-down: rather than analyzing variations of a specific design, the objective becomes identifying the optimal design among many candidates.

This concept is illustrated in Fig. 16.1. The design space is defined by N design parameters, which may include, for example, nominal or parasitic values of components, or the dimensions of a mechanical structure. Within this design space, various functions can be defined—schematically represented by the colored box at the center of the figure—to evaluate design properties. These functions serve to identify optimal designs based on specified requirements. For instance, a company may offer a high-performance product at a higher cost, as well as a budget-oriented variant with reduced performance. Each of these offerings corresponds to a distinct set of requirements, which in turn maps to a specific region within the design space.

J. Hansen, A. Barchanski, *EMC Simulation of Electronic Systems*,
https://doi.org/10.1007/978-3-032-09237-3_16

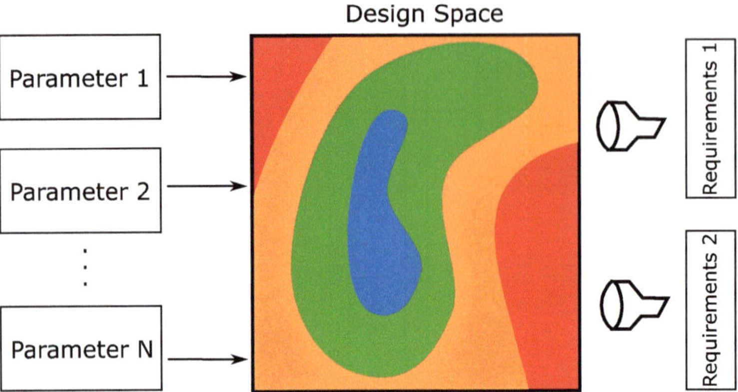

Fig. 16.1 An ML operates on a design space. Its evaluation is so fast that it is a container of solutions. Requirements are imposed to find optimal solutions in this space

Some functions defined over the design space may be simple or approximate. For example, properties such as the volume and cost of circuit components can be obtained from databases or estimated using analytical formulas. In contrast, evaluating performance aspects related to EMC—such as conducted emissions—requires the use of detailed physical models, as discussed throughout this book. ML enables the training of behavioral models based on a set of training data sampled from the design space. The resulting model is essentially a complex analytical expression. Unlike formulas derived directly from physical laws—such as those in Sect. 3.3—which are limited to cases where analytical solutions exist, ML allows the creation of parameterized, equation-based models for a much broader range of cases. This includes models derived from any type of physical or even measurement-based data.

In its current state, any ML model remains inferior to the judgment of an experienced engineer. Engineers draw on years of hands-on experience across numerous designs to assess which solutions are sufficiently promising to pass laboratory evaluation. However, an ML model can screen millions of designs per day. When this screening is performed in an unsupervised, brute-force manner, the vast majority of designs are not feasible. Yet, the sheer speed of this process makes it possible to identify the few viable options—those that an experienced engineer might intuitively recognize. In this way, the ML model acts as a support tool in the design process. Its value becomes particularly evident when little is known about the correlations between different functions in the design space. When applied to new products or devices incorporating novel technologies, ML can serve as an accelerator, helping to build engineering intuition more quickly than is possible through incremental lab work or traditional simulation methods.

The potential of ML in EMC simulation is not yet explored. Its practical success depends on factors such as the size of the parameter ranges within the design space and the amount of training data required to build ML models of engineering-grade quality. There is limited value in models that cover too narrow a parameter space, require excessive training time, or lack the accuracy needed to support design decisions. Nonetheless, the growing number of published models is a highly promising sign. Many potential use cases—ranging from multi-objective optimization (MOO) to model calibration and risk analysis—are still in their early stages and offer ample opportunities for high-impact research. Some of these topics are outlined in Sects. 18.2 and 18.3.

In this chapter, we focus on one such use case: the MOO of conducted emissions from the traction inverter of an electric vehicle. Before presenting this application, we introduce the generation of a machine learning model and the formulation of MOO using simplified examples. The specific ML algorithm used is detailed in the advanced Sect. 16.4.

16.1 Generation of Machine Learning Models

Domains: 

There are many machine learning algorithms available, such as neural networks (NN) with various topologies, polynomial chaos expansion (PCE), and Gaussian process regression (GPR). All of these perform a nonparametric approximation of an unknown function defined over multiple input parameters. Nonparametric in this context means that no specific functional form—such as polynomial and exponential—is assumed for the unknown function. Machine learning algorithms are designed to handle a wide variety of data types, which accounts for their broad applicability.

When an ML model is trained using data from a physical model, the physical model must be evaluated multiple times. The ML model is then constructed from this simulation data. Once the training data has been collected, model generation takes minutes to a few hours. Since the number of training samples is in the range of hundreds to thousands, the most time-consuming step is the generation of this training data. Therefore, algorithms that reach the target accuracy with a minimal number of training samples are desirable.

The performance of a machine learning algorithm can only be assessed by applying it to actual data. In the context of EMC simulations, many variants of NNs have been reported to be data-hungry. However, there are cases where NNs with specific architectures achieve good results with surprisingly few training samples. PCE approximates the target function by interpolating the data using orthogonal polynomials. This method can converge rapidly to a good approximation if the problem structure aligns well with the chosen polynomials. If not, PCE may fail to produce a usable model, regardless of the number of training samples. In this chapter, we use GPR, which has proven effective in modeling various EMC setups with an acceptable or remarkably low number of training samples. The method is robust: once a sufficient number of samples are provided, GPR yields a model of acceptable accuracy. Its main limitation is that Gaussian processes inherently produce smooth functions. As a result, they perform poorly at resonance frequencies with high quality factor, where the system response exhibits sharp peaks.

We can apply GPR to models with up to 20 design parameters. These models can be either circuit models or EM models. In circuit models, parameters such as nominal and parasitic component values can be varied. In EM models, material properties or geometric dimensions can be adjusted. In coupled EM-circuit models, both types of parameters may be varied simultaneously. The broader the selected parameter ranges, the more general the resulting ML model becomes, and the more training data is required to construct it. However, it is not possible to predict in advance how much training data will be needed for a model with a given number of parameters and specific parameter ranges.

It may seem appealing to generate ML models as a function of all design parameters in a simulation model. However, the number of such parameters can far exceed the roughly 20 that GPR can handle effectively. As the number of design parameters increases, so does the required training effort. While artificial intelligence techniques can manage hundreds of parameters, they often require hundreds of thousands of training samples. This number is prohibitive for the training with complex simulation models.

ML is a powerful tool to support engineering decisions, but it must not replace engineering intuition. No EMC engineer is expected to master the design of a product involving more than 20 simultaneously interacting parameters. As with any simulation approach, the precise formulation of the question the ML model is intended to answer is essential. Well-posed, technically relevant engineering problems typically involve between 5 and 15 design parameters, making GPR an excellent method for solving them.

16.1.1 Gaussian Process Regression (GPR)

GPR is a robust method for generating a behavioral model that is fast to evaluate. Such a model is also referred to as a surrogate model. GPR operates under the assumption that the data can be described as a realization of a Gaussian process. In this section, we introduce the intuition behind GPR using scalar output data and a single design parameter. The general case, involving K design parameters and vector-valued outputs in the frequency domain with M samples, along with the derivation of the corresponding predictor equations, is presented in Sect. 16.4.

The foundation of GPR is the Gaussian process, where the random variable Y is distributed as

$$Y(x) \sim \mathcal{N}(\mu(x), \sigma(x)^2) \tag{16.1}$$

at each point x. The variable x can represent, for example, time. In the context of this work, it denotes a design variable such as the nominal value of a filter capacitor, the parasitic inductance of a component, or the magnetic permeability of a ferrite material. The outcome of the random variable Y depends on the design variable x, and its distribution varies accordingly.

A Gaussian process is characterized by

$$E[Y(x)] = \mu(x) \qquad \text{mean value,} \tag{16.2a}$$

$$E\left[(Y(x) - E[Y(x)])^2\right] = \sigma(x)^2 \qquad \text{variance,} \tag{16.2b}$$

$$\operatorname{cov}\left(Y(x'), Y(x'')\right) = E\left[\left(Y(x') - E[Y(x')]\right)\left(Y(x'') - E[Y(x'')]\right)\right] \qquad \text{covariance.} \tag{16.2c}$$

The effect of the covariance on the behavior of a Gaussian process is illustrated in Fig. 16.2. In both plots, five realizations of a Gaussian process Y are shown, with $E[Y(x)] = \mu = 0$ and $\sigma = 1$.

The covariance of the process is described by the function:

$$E\left[\left(Y(x')Y(x'')\right)^2\right] = \sigma^2 \exp\left(-\theta_1 |x' - x''|^{\theta_2}\right) \tag{16.3}$$

The parameter θ_1 corresponds to the inverse of the correlation length: the larger its value, the faster the correlation between x' and x'' decays. The parameter θ_2 defines the distance metric used. In this example, we use the Euclidean distance with $\theta_2 = 2$.

The curves on the left-hand side of Fig. 16.2 are generated with $\theta_1 = 10$, while those on the right use $\theta_1 = \frac{1}{2}$. The correlation length in the right-hand plot is thus 20 times greater, resulting in significantly slower variation of the process over x.

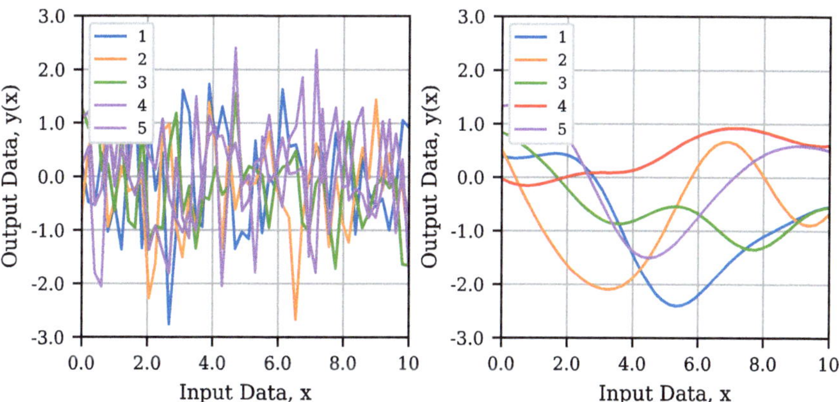

Fig. 16.2 Five realizations of a Gaussian process. Left, $\theta_1 = 10$; right, $\theta_1 = \frac{1}{2}$

In this figure, the data $y(x)$ is generated from a known Gaussian process. In GPR, we proceed in the opposite direction: given observed data, we estimate the optimal values of the parameters $\hat{\theta}_1$ and $\hat{\theta}_2$ so that a Gaussian process best fits the data. The parameters θ_1 and θ_2 are referred to as hyperparameters. The correlation function in Eq. 16.3, also known as the kernel, describes an exponential decay. While this kernel is commonly used, alternative types of kernels are also available and can be selected based on the characteristics of the underlying data.

In GPR, both the mean μ and the variance σ^2 are functions of the hyperparameters. We assume that the observed realizations—i.e., the training data $y_1(x_1), \ldots, y_N(x_N)$—provide a representative sample of the underlying function to be predicted. The values of μ and σ^2 are determined through maximum likelihood estimation. This method assumes that the observed training data is the most probable outcome generated by the corresponding Gaussian process. The estimation yields the optimal hyperparameters $\hat{\theta}_1$ and $\hat{\theta}_2$, along with the corresponding estimates for the mean and variance, denoted by $\hat{\mu}(\hat{\theta}_1, \hat{\theta}_2)$ and $\hat{\sigma}^2(\hat{\theta}_1, \hat{\theta}_2)$. Once the Gaussian process parameters have been estimated, the probability density function of any $Y(x)$ at any point x is fully defined. The most probable value of $Y(x)$ is given by $\hat{\mu}(x)$, which serves as the predictor in the GPR model.

We examine the proposed procedure using a set of N data points $y_n(x_n)$, for $n = 1, \ldots, N$. The example is based on a case study provided in the toolbox described in Marelli (2014), which has been widely used to generate surrogate models for EMC applications. Implementation details are available in the toolbox documentation.

The training data is generated using the function $y(x) = x \cdot \sin x$, which is featured as the Kriging example in the metamodeling section of the toolbox (Lataniotis et al. 2022). This function has a one-dimensional input parameter x and a scalar output $y(x)$. We analyze how the accuracy of the GPR model prediction evolves as a function of the number of training samples.

The results are shown in the two plots of Fig. 16.3. Each plot contains three curves and a shaded area: the analytical function $y(x) = x \cdot \sin x$ is shown in blue, the training samples are marked as orange dots, and the estimated mean $\hat{\mu}(x)$ from the GPR model is shown in green. The shaded area corresponds to the 95% confidence interval, i.e., the range within which 95% of all sample realizations lie in case $y(x)$ is Gaussian distributed. The limits of this interval correspond to $2\hat{\sigma}$, because for a Gaussian distribution, about 95% of all samples are located within 2σ around the mean.

In the left plot, with $N = 6$ training samples, the mean $\hat{\mu}(x)$ is a good estimate of the true curve for the first three samples. With increasing oscillation of $x \cdot \sin x$, the GPR does not describe the peaks of the true curve. The 95% interval exceeds the oscillation amplitude by far. In the right plot, where $N = 8$ samples are used, the prediction quality is significantly better. The mean $\hat{\mu}(x)$ coincides with the analytical curve by almost the thickness of the line for the first seven samples. Between the 7th and the 8th samples, the oscillation becomes too strong and the peak at about $x = 14$ is not captured. At the peak, the true curve is outside the 95% confidence interval. To the left of the first training sample, the confidence interval

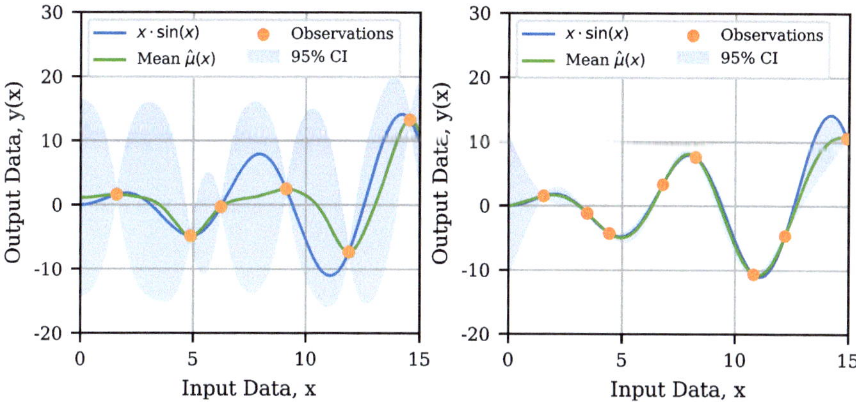

Fig. 16.3 Prediction of the function $x \cdot \sin x$ with $N = 6$ and $N = 8$ training points (example taken from Marelli 2014)

becomes large because the GPR extrapolates and there is no information about the behavior of the data at x close to 0.

A GPR model assumes no prior knowledge about the underlying data. Any physical insight, such as the fact that the values $y(x)$ may be solutions to Maxwell's equations or Kirchhoff's current law, is disregarded. On the one hand, the robustness of GPR, along with its ability to produce accurate models with relatively few training samples, is remarkable. On the other hand, it is reasonable to ask whether incorporating physical knowledge into GPR or other machine learning methods could reduce the amount of training data required or improve the accuracy of the resulting models. This question is further outlined in Sect. 18.3.

16.1.2 Sampling Plans and Model Training

A GPR model is nonparametric, meaning that it does not assume a predefined functional form. Instead, the model information is derived from training on selected samples of the parameter space using a physical model. Since the goal is to capture as much relevant information as possible with as few training samples as necessary, the selection of the training dataset is critical.

For practical engineering problems, the parameter space typically has a dimensionality between 5 and 20. The most straightforward and intuitive sampling strategy is full-factorial sampling, where the parameter space is uniformly covered with k samples along each of the K dimensions (see Fig. 16.4, left, for an example with $K = 2$ and $k = 8$, resulting in $N = k^K = 64$). This approach is thorough and minimizes the risk of missing important regions of the space. However, it quickly becomes infeasible as dimensionality increases. For instance, sampling each dimension with $k = 5$ points yields $k^2 = 25$ samples for $K = 2$, $k^5 = 3125$ samples

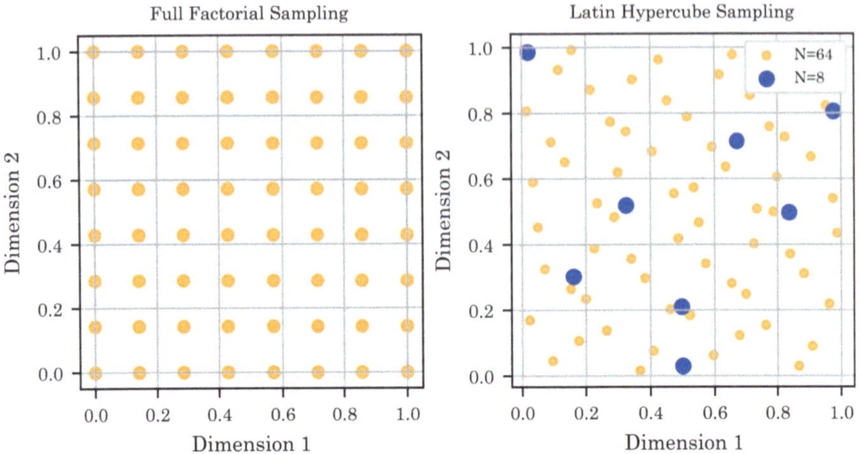

Fig. 16.4 Sampling plans. Left: Full-factorial sampling with $N = 64$. Right: Latin hypercube sampling with $N = 8$ and $N = 64$

for $K = 5$ and $k^{15} = 453$ million samples for $K = 15$. Assuming a simulation time of 10 minutes per sample, the total training time becomes approximately 4 hours for $K = 2$, over 3 weeks for $K = 5$, and more than 8,600 years for $K = 15$. This exponential growth in computational effort with increasing dimensionality is referred to as the curse of dimensionality.

Practical engineering problems do not require full-factorial sampling. Engineering intuition tells that changing 2 out of 15 parameters does not drastically alter a model's behavior with respect to all of the remaining 13. In other words, we assume a degree of independence or weak coupling between parameters. If it were known that all parameters were fully independent, then only k samples would be needed, regardless of the number of dimensions K. In that case, we could sample along the diagonal of the parameter space, ensuring that each dimension is still represented with k distinct values, while dramatically reducing the number of required simulations.

A priori, we do not know which parameters in a given engineering problem are independent of one another, nor over which ranges this independence might hold. Diagonal sampling leaves most of the parameter space unexplored and is therefore inadequate. Sampling strategies commonly used in design-of-experiments and machine learning offer a practical compromise. These approaches aim to distribute the N training samples such that they are as uniformly spread as possible in each dimension. In addition, the distribution should be space-filling, meaning the samples are placed to cover the parameter space as evenly as possible, with each point maximally distant from the others. While diagonal sampling does ensure uniformity along each dimension, it does not satisfy the space-filling criterion.

An example of a space-filling design, Latin hypercube sampling, is shown on the right-hand side of Fig. 16.4 for $K = 2$ and two sampling sizes, $N = 8$ and $N = 64$.

In both cases, the samples are distributed such that in each dimension, the points are randomly but uniformly spread along the respective axes. Whether $N = 8$ samples are sufficient, or a larger number are required, depends on the complexity of the function being modeled. With increasing number of dimensions K, Latin hypercube sampling provides good coverage of the parameter space without the exponential growth in sample size seen in full-factorial sampling. Latin hypercube sampling is a generalization of the Latin square. Beyond Latin hypercube sampling, various other space-filling schemes exist, each with different strategies for distributing samples as evenly as possible across the design space.

Achieving a homogeneous distribution of samples in the parameter space requires the definition of a distance metric and the solution of an optimization problem. For large K, uniformly covering the space—particularly its corners—becomes increasingly challenging.

By employing a space-filling design, the growth in sample count associated with full-factorial sampling is less than exponential. The authors of Loeppky et al. (2009) recommend $10K$ as size of the initial sampling set, i.e., a size linear in K. However, the actual number of training samples N required to meet a desired target accuracy must still be determined individually for each problem, as it depends on the complexity and sensitivity of the underlying model.

Once the training sample set has been generated, the physical model is evaluated at each training point. In Sect. 16.4, we derive Eq. 16.33, which predicts the output vector at an unseen point in the parameter space. This prediction relies on a maximum-likelihood estimation of the hyperparameters $\hat{\theta}_1$ and $\hat{\theta}_2$, based on the available training data.

The accuracy of the resulting predictor equation must be validated. This is typically done using a separate subset of the parameter space reserved for testing, commonly around 20% of the total samples. These testing points are excluded from the training set because a GPR model, by construction, predicts the training points exactly and thus cannot provide a meaningful error estimate at those locations.

To evaluate the accuracy of the machine learning model, the predicted outputs from the GPR model are compared with the results of the physical model at all testing points. To determine when to stop the training process, an appropriate error metric and an acceptance criterion must be defined, both tailored to the specific problem at hand. Error criteria relevant to our application are discussed in Sect. 16.3.

If the error criterion is not satisfied, additional training samples are required. Adaptive learning offers strategies to identify the most informative new sample points based on the current model performance. There are two strategies: local exploitation, which focuses on refining the model in regions of interest, such as near an optimal design. Additionally, there is global exploration, which aims to improve the model's accuracy across the entire design space.

The initially selected range for each design parameter also influences model behavior along that dimension. If the model is particularly sensitive to changes within a given range, more training samples are needed along that dimension. High sensitivity correlates with a short correlation length in the Gaussian process, reflected by large values of the hyperparameters θ_1 and θ_2.

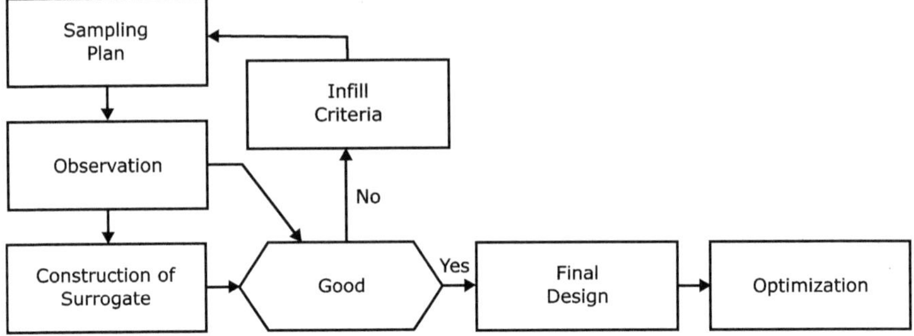

Fig. 16.5 Generation of a machine learning model, including adaptive learning

The model generation process, including adaptive training, is summarized in Fig. 16.5. There is no guarantee that a machine learning model can be successfully trained for every problem. However, numerous successful applications have been reported across a wide variety of physical models and machine learning algorithms.

16.2 Multi-Objective Optimization

Domains: 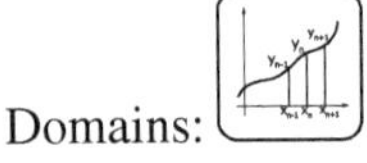

Designing a competitive electronic system can be stated as an MOO problem. Cost must be minimized while simultaneously meeting performance targets and ensuring electromagnetic compliance. These objectives are often in conflict with one another. Only a trade-off between these competing requirements can be achieved.

A sound final design decision can only be made if the full range of possible trade-offs is understood. MOO provides a framework to compute these trade-offs. In the following, we explore the concept of the Pareto front, the set of optimal trade-off solutions, and illustrate it using a simple example with two design variables and two objectives.

16.2.1 The Pareto Front

The Pareto front represents the set of optimal designs that capture all trade-offs between the considered design goals. In optimization terminology, these goals are referred to as objectives. In this section, we explain how a Pareto front is constructed. Figure 16.6 (left) illustrates three typical and conflicting objectives encountered in electronic system design. The primary goal is often system perfor-

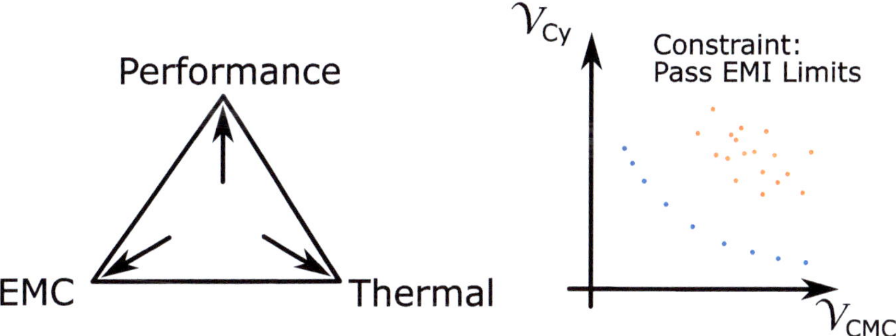

Fig. 16.6 Conflicting requirements in electronic system design

mance. In the case of a power converter, for example, one relevant performance metric is a smooth output current, which is an objective that can be improved by increasing the switching frequency. However, a higher switching frequency leads to increased switching losses and higher thermal dissipation. It also results in stronger electromagnetic emissions, thus raising the need for more extensive EMI mitigation.

The EMI effort can be reduced by slowing down the rise and fall times of the switching transistors. However, this increases power losses and places greater demands on thermal management. Conversely, minimizing power loss favors fast switching transitions and lower switching frequencies—both of which worsen voltage ripple and exacerbate EMI. The interaction between these conflicting objectives is complex. For instance, the EMI generated by fast rise and fall times can be mitigated through the use of capacitors that act at the high-frequency side of the spectrum introduced by rapid switching transitions (cf. Fig. 2.5). In contrast, the switching frequency influences the EMI spectrum across a much broader frequency range.

Some engineers, through years of experience and deep system knowledge, are able to make excellent design decisions in such trade-off scenarios. When this level of insight is not yet available, its development can be supported by performing multiple simulation runs. This process is made practical by the high computational efficiency of surrogate models, which enable rapid exploitation of the design space.

Using a surrogate model, a large number of design parameter combinations can be evaluated efficiently. If these combinations are selected at random, many resulting designs are unfeasible. However, some of them may be as good as a design can be, representing optimal trade-offs.

This situation is illustrated on the right side of Fig. 16.6 using the example of the volumes $\mathcal{V}_{\mathrm{Cy}}$ and $\mathcal{V}_{\mathrm{CMC}}$ required for C_{y} capacitors and the common mode choke of a filter. As shown in Sect. 9.5, both C_{y} capacitors and the common mode choke are effective for reducing the emission of a filter. Given that minimizing the total volume of these components is a key design objective, we investigate the individual volumes $\mathcal{V}_{\mathrm{Cy}}$ and $\mathcal{V}_{\mathrm{CMC}}$. Our goal is to minimize both simultaneously, which leads to an MOO problem with two objectives.

In the figure, the blue dots represent the optimal samples on the Pareto front. Any decrease in $\mathcal{V}_{\mathrm{Cy}}$ requires an increase in $\mathcal{V}_{\mathrm{CMC}}$. On the upper side of the Pareto front, there are arbitrarily many samples displayed as yellow dots. These are suboptimal, because samples with either or both lower $\mathcal{V}_{\mathrm{Cy}}$, $\mathcal{V}_{\mathrm{CMC}}$ exist.

Solving this optimization problem is meaningful only if the resulting filter design complies with the specified EMI limits. Therefore, in addition to the two objectives, we can introduce a constraint: only those combinations of C_{y} and common mode choke volumes that yield an EMI-compliant filter are accepted. This transforms the problem into a constrained multi-objective optimization task.

The volumes $\mathcal{V}_{\mathrm{Cy}}$ and $\mathcal{V}_{\mathrm{CMC}}$ depend on several design variables, such as their electrical values (in Farads and Henrys, respectively) and material properties. We denote these variables collectively as $\mathbf{x}$ and express the volumes as functions of them. The resulting two-objective optimization problem with one constraint is formulated as follows:

$$\min_{\mathbf{x}} \left(\mathcal{V}_{\mathrm{Cy}}(\mathbf{x}), \mathcal{V}_{\mathrm{CMC}}(\mathbf{x})\right) \tag{16.4}$$

subject to: the design satisfies the EMI limits.

The goal is to simultaneously minimize both $\mathcal{V}_{\mathrm{Cy}}$ and $\mathcal{V}_{\mathrm{CMC}}$ by selecting suitable values for the design variables $\mathbf{x}$, while ensuring that the resulting design meets the required EMI limits.

The resulting set of all feasible solutions is shown on the right-hand side of Fig. 16.6. This set includes many valid filter designs. For example, selecting large values for both the C_{y} capacitors and the common mode choke results in a filter that satisfies the EMI requirements. However, such solutions use more volume than necessary and are therefore suboptimal. These designs are represented by the orange markers in the plot. The designs of interest are indicated by the blue markers. For any blue point, the volumes $\mathcal{V}_{\mathrm{Cy}}$ and $\mathcal{V}_{\mathrm{CMC}}$ are Pareto optimal, which means that a reduction in one volume inevitably results in an increase in the other. These blue points form the Pareto front, representing the set of optimal trade-offs between the two objectives.

Equation 16.4 represents just one of many possible formulations for a filter design objective. For instance, one might focus solely on minimizing the total volume $\mathcal{V}_{\mathrm{Cy}} + \mathcal{V}_{\mathrm{CMC}}$. This defines a valid, single-objective optimization problem, which is faster to solve due to its reduced complexity. However, this formulation can obscure important trade-offs. For example, common mode chokes are typically more expensive than capacitors. By separating the problem into two objectives—minimizing $\mathcal{V}_{\mathrm{Cy}}$ and $\mathcal{V}_{\mathrm{CMC}}$ independently—we obtain more detailed insights. From the resulting Pareto front, one can then estimate additional factors such as component cost.

The optimal design strongly depends on how the problem is formulated. It is common to solve several variants of the optimization problem and then evaluate the resulting optima using engineering judgment. The comparative study of these

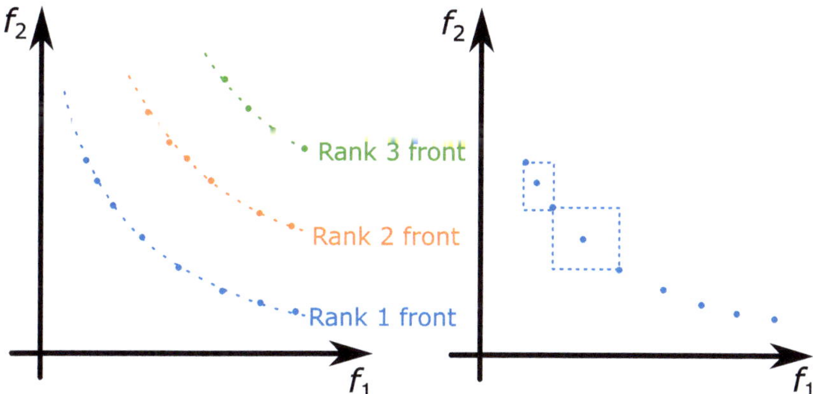

Fig. 16.7 Search strategies to determine the Pareto front

problem formulations is already a valuable step in the design process, as it reveals the trade-offs and sensitivities inherent to the system.

There are several algorithms available to solve MOO problems. One widely used approach is evolutionary optimization. It begins with an initial population of sampled designs, evaluates the objectives, and checks whether the constraints are satisfied. In the example shown on the right of Fig. 16.6, this involves selecting several combinations of C_y capacitors and common mode chokes, evaluating their volumes, and determining whether the design complies with the imposed EMI limits. A new generation of design candidates is then produced using an evolutionary mechanism, such as particle swarm optimization. For each new candidate, the objectives and constraints are reevaluated.

The selection of preferred samples for the next generation is guided by two key criteria, ranking and crowding, as illustrated in Fig. 16.7 for two arbitrary objectives f_1 and f_2. On the left-hand side of the figure, the ranking criterion favors samples that lie on Pareto fronts of lower rank (closer to optimal). On the right-hand side, the crowding criterion prefers samples that are farther from existing points, promoting a diverse and well-distributed set of solutions. As the number of objectives increases, the computational effort required to generate and visualize the Pareto front increases. Therefore, careful selection of the optimization objectives is essential. Exploring and comparing different formulations is an important step in the design process.

16.2.2 An MOO Example Using an Analytical Model

We illustrate the MOO procedure and its analysis using a very simple analytical model. In this case, no ML model is needed, as the system is defined by explicit

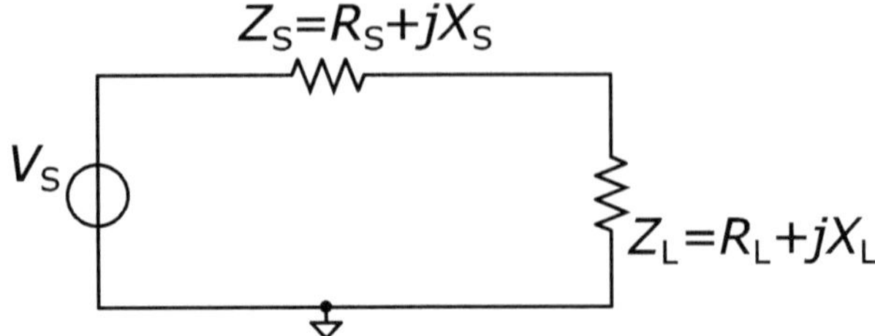

Fig. 16.8 Simple circuit to study the trade-off between efficiency and power maximization

formulas. The Pareto front can also be derived analytically, allowing us to study it in terms of both the objectives and the underlying design parameters.

We consider the circuit shown in Fig. 16.8. A voltage source with voltage V_S and source impedance $Z_S = R_S + jX_S$ drives a load impedance $Z_L = R_L + jX_L$. Our goal is to determine suitable values for the load parameters R_L and X_L such that both the power delivered to the load and the overall efficiency of the circuit are optimized.

The current through the circuit is

$$I = \frac{V_S}{Z_{tot}} = \frac{V_S}{Z_S + Z_L} = \frac{V_S}{(R_S + R_L) + j(X_S + X_L)}, \tag{16.5}$$

and the power in the load

$$P_L = R_L |I|^2 = V_S^2 \frac{R_L}{(R_S + R_L)^2 + (X_S + X_L)^2}. \tag{16.6}$$

The total generated power is

$$P_G = \Re\{V_S I\} = V_S^2 \frac{R_S + R_L}{(R_S + R_L)^2 + (X_S + X_L)^2}. \tag{16.7}$$

The efficiency η_C of this circuit is the ratio between power in the load and the total power:

$$\eta_C = \frac{P_L}{P_G} = \frac{R_L}{R_S + R_L}. \tag{16.8}$$

We have the unconstrained two-objective optimization problem:

$$\max_{\mathbf{x}} \left(P_L(\mathbf{x}), \eta(\mathbf{x})\right) \tag{16.9}$$

$$\mathbf{x} = [R_L,\ X_L]^T.$$

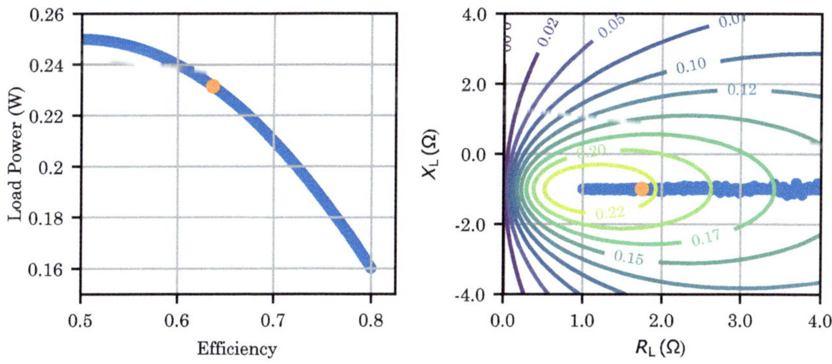

Fig. 16.9 Pareto front of efficiency η_C and load power P_L. Left: as a function of η_C, P_L Right: in the design parameter space R_L, X_L

The optimal solutions for each objective, disregarding the other, are known. In order to maximize the output power, we require $Z_L^* = Z_S$, i.e., $R_L = R_S$, $X_L = -X_S$ (matched load impedance). In order to maximize the efficiency, we need $R_L \gg R_S$. The Pareto front shows all trade-offs which include and are in between these corner solutions.

The example is a rare case where we can compute the Pareto front analytically. With two the Eqs. 16.7 and 16.8 and the two objectives η_C, P_L, we solve for P_L and obtain:

$$P_L = \frac{V_S^2}{R_S}\eta_C(1 - \eta_C). \tag{16.10}$$

A plot of the Pareto front is shown in Fig. 16.9 on the left. It is evaluated for $V_S = 1\,\text{V}$, $R_S = 1\,\Omega$, $X_S = 1\,\Omega$.

The maximum load power of $P_L = \frac{1}{4}\,\text{W}$ is achieved at $R_L = 1\,\Omega$ and $X_L = -1\,\Omega$, i.e., when the load is impedance-matched to the source. At this operating point, the efficiency is $\eta_C = \frac{1}{2}$.

As the efficiency η_C increases beyond this point, the load power P_L decreases. Conversely, when η_C decreases below 0.5, P_L also drops. These latter operating points are not on the Pareto front, because they are dominated by points with higher efficiency and equal or higher load power. Eventually, the maximum efficiency $\eta_C = 1$ is approached, but this occurs when no power is delivered to the load, i.e., $P_L = 0$.

On the right-hand side of Fig. 16.9, the Pareto front is shown as a function of the design parameters R_L and X_L. The front begins at $R_L = 1\,\Omega$, $X_L = -1\,\Omega$, corresponding to the point of maximum load power. Since $X_L = -1\,\Omega$ maximizes P_L and does not influence the efficiency η_C, it remains constant across the Pareto front. As a result, the Pareto front in the design parameter space appears as a horizontal line at $X_L = -1\,\Omega$, with R_L increasing toward infinity. This plot becomes particularly important once a suitable trade-off point (P_L, η_C) on the Pareto front is selected.

The orange dot on the Pareto front, for example, at $\eta_C = 0.63$ and $P_L = 0.23$, corresponds to the orange point in the design space on the right-hand side. To build a circuit with the efficiency and the load power of the orange point, we need to use approximately $R_L = 1.8\,\Omega$ and $X_L = -1\,\Omega$.

The selection of the orange point is ultimately left to the engineer. In this example, the trade-off is between efficiency and maximum output power, and the final decision would depend on the specific requirements of the product being developed. For each selected point on the Pareto front shown on the left side of Fig. 16.9, the corresponding design parameters can be read from the right side of the figure. With these parameters determined, the circuit shown in Fig. 16.8 is fully defined.

16.3 Application Example: MOO of a Traction Inverter

- Fitted transistor models
- LISNs and EMI receiver
- RLC models
- SPICE solver
- Transient
- 0–108 MHz

We demonstrate the solution of an MOO problem using an ML model, applied to the example of a traction inverter in an electric vehicle. The objective of the MOO process is to design an EMI filter that ensures that the inverter's common mode voltage is compliant with conducted emission limits defined by CISPR 25 Class 3. We focus on identifying filter designs that offer low component volume, while also meeting a range of functional requirements—such as specific rise and fall times, and switching frequency constraints. In addition, the filter components possess mechanical characteristics, including parasitic capacitances and inductances, which must be taken into account during the optimization.

We begin by describing the inverter model, which is similar to the motor drive analyzed in Chap. 10. This model is used to train a surrogate model based on wide-band GPR, using a set of design parameters as outlined in Sect. 16.1. Subsequently, we define a set of objectives and a constraint and solve the corresponding MOO problem following the procedure described in Sect. 16.2.

The optimization strategies and the model introduced in Sect. 16.3 constitute a simplified variant of those presented in Gsoels et al. (2025). In this publication, a model is used that is in parts experimentally validated, a more complex filter is applied to qualify the design for CISPR 25 class 4, and both common and differential model are considered.

16.3.1 The Inverter Model and Its Design Parameters

The circuit model of the traction inverter is shown in Fig. 16.10. It consists of a B6 bridge, a machine model, and an EMI filter with a C-L-C topology. At the center of the filter is a common mode choke, flanked on each side by a C_x capacitor and a pair of C_y capacitors.

The model follows the same topology as the inverter discussed in Sect. 10.1. However, since a traction inverter typically operates at power levels of several tens of kilowatts or more, the values of the circuit parameters differ accordingly. We give a rough overview of the model; the details are given in Gsoels et al. (2025).

The inverter is supplied by a DC voltage of 400 V, and the modulation index is chosen such that the peak current through the electric motor reaches approximately 100 A. It runs with a switching frequency of f_{sw}.

The silicon carbide semiconductor switches are modeled based on a data sheet of a commercially available product. The rise and fall time can be adjusted using the gate driver resistance R_{driver}. The power modules include parasitic capacitances C_{phase} to the inverter's cooler, which are significant in high-power systems. Additionally, the electrical motor exhibits parasitic capacitances C_{CM} between the stator and the housing. All capacitors and inductors in the filter and interconnects are modeled as RLC elements, as described in Sect. 2.4.2.

The EMI filter has a C-L-C structure with both C_x and C_y capacitors. The inductor is a nanocrystalline core. It is represented by an impedance model based on a data sheet. The model is made such that its impedance scales linearly with the length l_{CMC} of the core. The C_y capacitors are modeled by their nominal capacitance C_{y1}, C_{y2} on the battery and the inverter side, and by the respective parasitic inductances L_{Cy1}, L_{Cy2}. The inverter also has a nanocrystalline core on its three AC wires. This core is not changed in the MOO and the surrogate model is not trained on any of its parameters.

The system setup follows the CISPR 25. The inverter has a housing, to which the filter capacitances C_{y1}, C_{y2}, as well as the parasitic capacitances C_{phase} of the power modules and C_{CM} of the electric motor are referenced. The housing is mounted 50 mm above a metallic ground plane, to which it is connected with a ground strap. Shielded high-voltage cables connect the inverter's positive and negative DC terminals to the two LISNs. With this setup, the conducted emission can be computed over frequency by applying an EMI receiver in the post-processing stage, using the average detector (cf. Sect. 2.5.4).

At this stage, no specific circuit parameter values are fixed. To train the machine learning model, we consider defined parameter ranges rather than fixed values. These ranges are discussed in detail in Sect. 16.3.2.

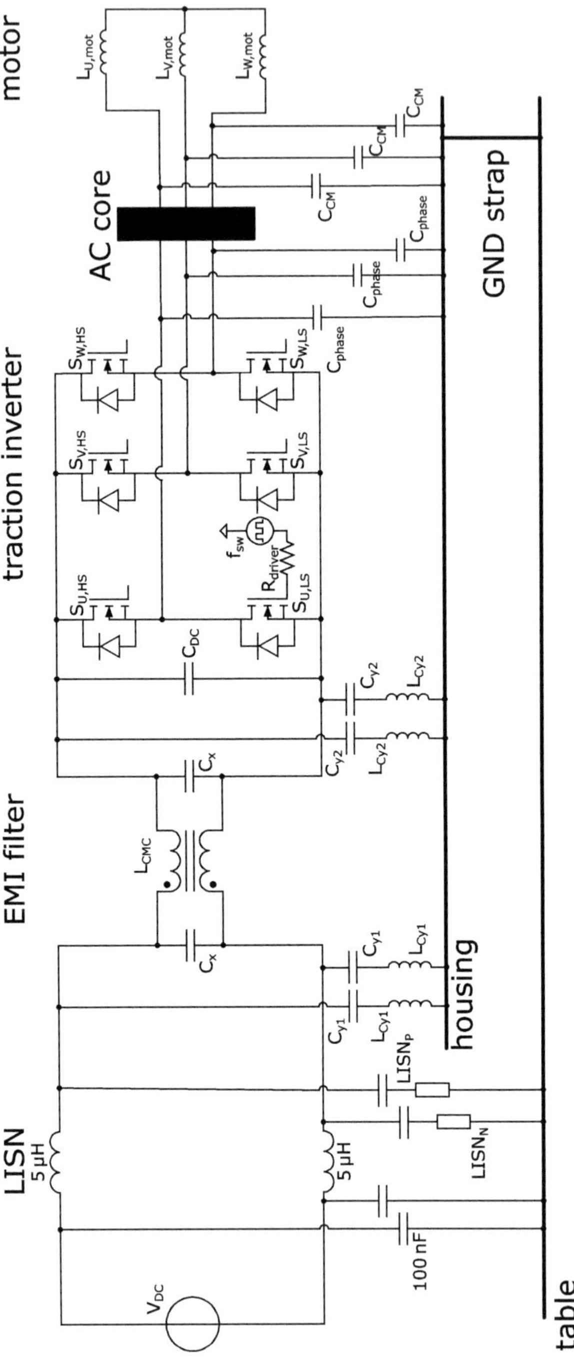

Fig. 16.10 Circuit model of the traction inverter

16.3.2 GPR Model Generation and MOO

To set up the model training, we first define the optimization task in detail. The objective is to identify an EMI filter design that minimizes physical size while satisfying several requirements: a measure of the inverter's efficiency, an indicator of its mechanical design characteristics, and compliance with the specified EMI limits. In this section, we restrict the analysis to the common mode emission V_{CM} measured at the LISNs. The treatment of differential mode emission V_{DM} follows analogously and can be addressed using the same framework.

We describe these requirements in the following way:

EMI Filter The EMI filter is characterized by the length of the common mode choke, l_{CMC}, and by the nominal values C_{y1} and C_{y2} of the C_{y} capacitors on the input and output sides of the filter (cf. Fig. 16.10). For training the machine learning model, the choke length l_{CMC} is varied within the range from 10 mm to 120 mm. The C_{y} capacitors on both sides of the choke, C_{y1} and C_{y2}, are also varied independently within the range of 55 nF–880 nF. Since only V_{CM} is considered, we neglect any C_{x} capacitor.

Efficiency A simplified efficiency indicator is used, which estimates switching losses as a function of the rise and fall time $t_{\mathrm{r,f}}$, itself determined by the gate resistance R_{driver}. Lower values of R_{driver} result in shorter $t_{\mathrm{r,f}}$ and hence lower switching losses. We vary R_{driver} from 6 Ω to 18 Ω, and the inverter's switching frequency is varied between 6 kHz and 16 kHz.

Mechanical Design The parasitics of the mechanical design impact V_{CM} (cf. Sects. 9.5 and 10.6). We parameterize the following parasitic elements: the parasitic inductances L_{Cyi} of all C_{y} capacitors with values between 20 nH and 40 nH. Further, the parasitic capacitance of the electric motor, C_{CM}, with a range between 1.5 nF and 6 nF. The phase capacitance C_{phase} describes the capacitance between the power module and the cooler of the inverter. Its range is from 0.5 nF – 2.0 nF. Among all feasible solutions, we favor those with large parasitic values, because they are simpler to design.

EMI Limit We assume CISPR 25 class 3 as limit:[1] chosen designs must fulfill this limit. We restrict the analysis to the average detector.

Table 16.1 summarizes the parameter ranges used in this study. These values reflect the characteristics of automotive traction inverters operating in several tens of kilowatts. In addition, we define a reference design, which serves as a baseline for later comparisons. The parameters of the reference design are chosen near the center of each specified range and reflect representative values for a typical automotive traction inverter. The reference design is used later for comparison in case studies.

It would be possible to train the model using V_{CM} data in the time domain. However, time-domain waveforms containing sharp transients, fast rise or fall times, or high-frequency oscillations require a large number of samples to be accurately

[1] This limit is also applied in Sect. 10.6.

Table 16.1 Design parameter bounds for the system model, and reference design

	f_{sw} (kHz)	R_{driver} (Ω)	l_{CMC} (mm)	C_{y1} (nF)	C_{y2} (nF)	L_{Cy1} (nH)	L_{Cy2} (nH)	C_{CM} (nF)	C_{phase} (nF)
Lower bound	6	6	10	55	55	10	10	1.5	0.5
Upper bound	16	18	120	880	880	40	40	6	2
Reference	10	12	40	220	220	20	20	3	1

represented. Even small variations—such as a shift in oscillation frequency—can lead to significant changes in the waveform, making the training process data-intensive. In contrast, the corresponding frequency-domain representation is less complex and requires fewer training samples to capture the essential characteristics of the waveform. For this reason, frequency-domain training is more efficient for such signals. The training of the differential mode voltage follows the same approach (cf. Gsoels et al. 2025).

According to Table 16.1, the design parameter vector $\mathbf{x}$ of the inverter is

$$\mathbf{x} = \left[f_{sw}, R_{driver}, l_{CMC}, C_{y1}, C_{y2}, L_{Cy1}, L_{Cy2}, C_{CM}, C_{phase}\right]^T . \tag{16.11}$$

The design space consists of $K = 9$ dimensions. The training samples are distributed using Latin hypercube sampling with maximin criterion (Forrester et al. 2008) to ensure good coverage of the parameter space. We train the surrogate model using the wideband GPR algorithm described in Sect. 16.4. Model evaluations are performed in parallel using 96 instances on a computing cluster, using circuit simulation based on LTSpice. The training process is terminated after 13 computation rounds, resulting in a training dataset of $N = 1248$ samples.

To determine whether the number of training samples is sufficient, an appropriate error criterion is required. A common choice is the normalized root-mean-square error (NRMSE), averaged over a set of training and testing samples. Since the V_{CM} spectra of power inverters typically exhibit resonances, a more suitable metric is provided by feature-selective validation (FSV) (Orlandi et al. 2006). Unlike point-wise error measures, FSV not only assesses the spectral distance between predicted and reference values across frequency. It also accounts for features such as resonant peaks even if they do not occur at exactly the same frequencies. The testing is performed with a set of 125 samples again chosen from a random Latin hypercube, which corresponds to 10% of the number of training samples. For the given training dataset, the model achieves a "very good" rating according to the feature-selective validation method.

Next, we define the optimization task using the nine available model parameters. In words, it can be described as follows:

Filter volume Minimize the length of the common mode choke l_{CMC} and select the smallest practical values for the C_y capacitors. Note that this assumption is a simplification, as the physical form factor of a capacitor does not scale linearly with its capacitance.

Efficiency Study variants with different efficiency. Efficiency requires fast rise and fall times, i.e., a low R_{driver}. Lower switching frequency, f_{sw}, also reduces switching loss and the emission. High f_{sw} is favorable though for low ripple and a small DC link capacitor.

Mechanical design Designs with moderate values of parasitic inductances and capacitances are preferred, as lower values may be difficult or even impractical to realize in a physical system.

EMI Pass the class 3 EMI limits.

We can pose the last part of the task, passing the EMI limits, as a constraint. In a constrained optimization problem, as indicated in Fig. 16.6, we discard all samples $\mathbf{x}$ that violate the limits. We risk discarding samples that are almost optimal, but which just fail the limits by a few dB. Using an evolutionary algorithm, the offspring of such samples could just pass the limits.

As suggested by Gsoels et al. (2025), rather than enforcing limit violations as hard constraints, a penalty term $P(\mathbf{x})$ is incorporated into the objective function. This term penalizes a sample $\mathbf{x}$ with a function that exponentially grows with the violation. Violations may occur at different frequencies. In $P(\mathbf{x})$, we choose that frequency where the difference d between the amplitude at this frequency and the limit line is largest. The penalty term is then $P(\mathbf{x}) = \exp(d(\mathbf{x})/5)$ in case there is any violation in the spectrum. In case the samples pass the limits, we set $P(\mathbf{x}) = 0$.

Based on these considerations, we pose the following multi-objective optimization problem:

$$\text{obj}_l = \min_{\mathbf{x}} \left(l_{\text{CMC}} \mid R_{\text{driver,i}}, C_{\text{phase,j}} \right) + P(\mathbf{x}) \tag{16.12a}$$

$$\text{obj}_C = \min_{\mathbf{x}} \left(C_{\text{Cy1}} + C_{\text{Cy2}} | R_{\text{driver,i}}, C_{\text{phase,j}} \right) + P(\mathbf{x}) \tag{16.12b}$$

$$R_{\text{driver,i}} \in \{6, 11, 16\}\ \Omega \tag{16.12c}$$

$$C_{\text{phase,j}} \in \{0.5, 1.3, 2.0\}\ \text{nF}. \tag{16.12d}$$

The obj_l in Eq. 16.12a minimizes the length l_{CMC} of the choke, which is a key factor to minimize the volume of the filter. The obj_C in Eq. 16.12b minimizes the combined size of the C_{y} capacitors. This results in a two-objective optimization problem, which is both straightforward for solvers to handle and intuitive to analyze using a 2D Pareto front. The setup enables an explicit trade-off between the length of the common mode choke and the capacitance values of the C_{y} capacitors, both of which impact filter volume and performance. To both objectives, the penalty term is added, which makes the corresponding samples the more unlikely to survive the more they violate the class 3 limits.

Equation 16.12c and 16.12d defines fixed values for the gate driver resistance R_{driver} and the parasitic capacitance C_{phase} between the power module and the cooler. Each parameter is assigned three discrete values within its respective range. This setup allows us to investigate the influence of R_{driver} and C_{phase} independently by evaluating their impact across all combinations.

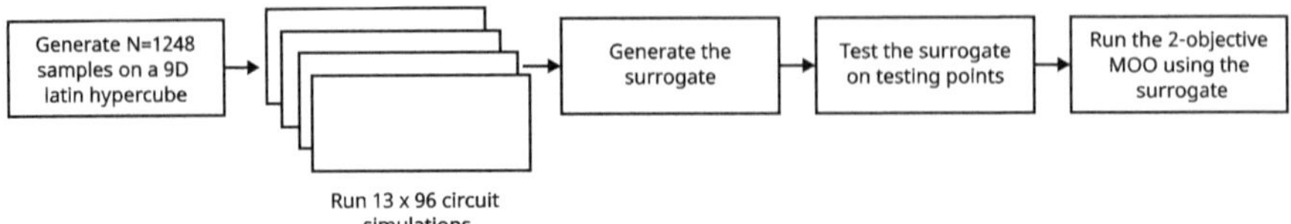

Fig. 16.11 Workflow of surrogate model generation and subsequent MOO

In total, we perform nine simulation runs, one for each unique combination of R_{driver} and C_{phase}, resulting in nine distinct Pareto fronts. The remaining four parameters of the design space, f_{sw}, L_{Cy1}, L_{Cy2}, and C_{CM}, are left without any condition and are freely varied.

Alternative problem formulations are also possible. One option is to define a one-dimensional minimization problem by combining the contributions of the choke length and the C_{y} capacitor values into a single objective. This approach requires assigning appropriate weights to the different terms. Simply summing the choke length in millimeters and the capacitance in nanofarads to form a joint objective is arbitrary and lacks physical justification. Another approach is to define a three-objective optimization problem, treating l_{CMC}, C_{y1}, and C_{y2} as separate objectives. This results in a Pareto front in three dimensions, allowing for more detailed trade-off analysis between all three design contributions but increasing the computational effort. Additionally, we could define f_{sw} to be a fixed value such as R_{driver} and C_{phase}. This would result in 3^3 simulation runs and the same number of Pareto fronts.

All formulations are valid. In this work, we opt for the two-objective setup, as it allows us to study the contributions of the choke and the capacitors individually. This formulation provides a good balance between insight and computational efficiency. Compared to the three-objective case, the two-dimensional Pareto front is easier to visualize and interpret.

To solve the MOO problem, we use the multi-objective optimizer NSGA-II (Deb et al. 2002) to solve the optimization problem. The total circuit simulation time for generating the training data is 10,326 s on 96 cores, with a total of 256 GB RAM. The time for a single circuit simulation is approximately 750 s.

Once the training data is available, the surrogate model is generated in 1045 s using a GPU. After training, one evaluation of the surrogate model takes approximately 2.5 ms. The model evaluation time reduces by a factor of 30,000 and enables rapid MOO and efficient exploitation of the design space. To generate a Pareto front, several 100,000 up to 1 million designs are screened. A sketch of the model generation and the optimization is shown in Fig. 16.11. Some 100 optimal samples are required to achieve convergence of the Pareto front.

16.3.3 Model Evaluation and Optimal Designs

In this section, we describe how to extract optimal designs from the solution set of the optimization problem defined by Eqs. 16.12a–16.12d. The design space spans nine dimensions, making direct interpretation challenging. To facilitate analysis, we group related parameters and investigate their interactions to better navigate the space of potentially optimal solutions. We introduce several types of plots to visualize and interpret the results, helping to identify trade-offs and correlations among the design variables. From the large number of solutions that satisfy the optimization criteria, we select a subset of designs based on engineering intuition.

Since we have formulated a two-dimensional optimization problem, the resulting Pareto fronts can be visualized in a 2D scatter plot, as shown in Fig. 16.12. The x-axis represents the first objective obj_l, the minimization of the choke length l_{CMC}. The y-axis represents the second objective obj_C, the minimization of the total C_{y} capacitance, expressed as $\sum_i C_{\text{yi}}$. Because we evaluate 3×3 discrete combinations of the parameters $\left\{R_{\text{driver}}, C_{\text{phase}}\right\}$, the plot displays a total of nine distinct Pareto fronts, one for each pair of objectives.

All Pareto fronts exhibit a bent or convex shape, showing that either the choke length l_{CMC} or the total capacitance $\sum_i C_{\text{y,i}}$ can be minimized, but not both simultaneously. To achieve a compact choke, larger C_{y} capacitor values are required, and conversely, reducing the capacitor sizes necessitates a longer choke. Among the nine Pareto fronts, five cluster closely in the region where both objectives attain relatively low values. These configurations are feasible. The remaining four Pareto fronts are shifted toward high values on both axes, with objective values exceeding one. These samples fail the CISPR 25 class 3 emission limits. The elevated objective values in these fronts result from the penalty term $P(\mathbf{x})$ that is added when the emission constraint is violated. This term is chosen to exceed the objectives except when the limits are only slightly violated (see paragraph above Eqs. 16.12a–16.12d). The penalty applies to all samples of the front, because no sample passes the limits. These infeasible fronts correspond to the three cases with largest $C_{\text{phase}} = 2\,\text{nF}$, and one additional case with $R_{\text{driver}} = 6.6\,\Omega$ and $C_{\text{phase}} = 1.3\,\text{nF}$. The latter combines the second-largest parasitic phase line capacitance with the smallest gate resistance.

The plot reveals that the selected filter topology and parameter ranges do not permit compliance with the class 3 EMI limits under the following conditions:

- when C_{phase} takes its maximum value, because the common mode current is largest,
- when C_{phase} takes its second-largest value and the rise/fall time is fastest (the smallest R_{driver}), i.e., the emission spectrum of the transistors is largest.

The results suggest that, within the tested range, the impact of C_{phase} on conducted emissions is more critical than that of R_{driver}. Even samples with the lowest R_{driver}, and therefore the fastest switching transitions, can still meet the emission limits, provided that C_{phase} remains within a moderate range.

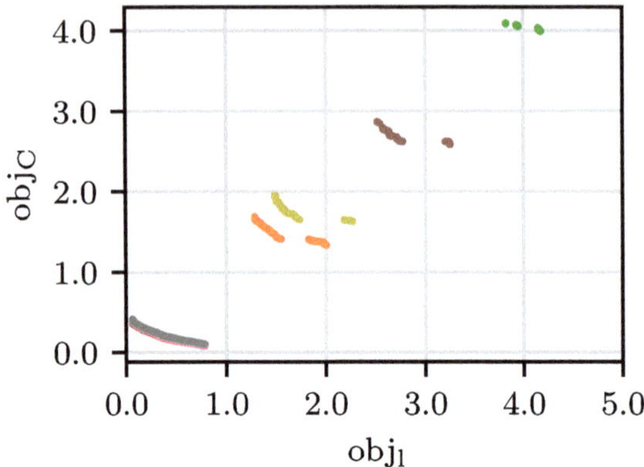

Fig. 16.12 The Pareto fronts for the inverter and nine fixed values for R_{driver} and C_{phase}

The Pareto fronts that satisfy the emission constraints yield similar objective values. To explore these solutions in more detail, we select one representative case: the sample with the lowest gate driver resistance, $R_{\text{driver}} = 6.6\,\Omega$ corresponding to the fastest rise and fall times, and with the lowest parasitic phase capacitance, $C_{\text{phase}} = 0.5\,\text{nF}$. This configuration represents a design with minimal switching losses and minimal capacitive coupling between the power module and the inverter cooler. By fixing R_{driver} and C_{phase}, the original 9D design space is reduced to seven dimensions, allowing for more focused analysis of the remaining parameters.

We analyze the correlations between the design parameters and objectives using a scatter plot matrix shown in Fig. 16.13. The matrix contains 11×11 entries: each row and column corresponds to one of the nine design parameters, along with the two optimization objectives (occupying the last two rows and columns). The diagonal of the matrix displays histograms representing the distribution of each design variable or objective across all points on the inspected Pareto front. The off-diagonal entries show pairwise scatter plots, which help identify correlations and trade-offs between variables.

We analyze the following values (i, j) of the ith row and the jth columns:

Matrix element (1,1) This represents the distribution of the switching frequency f_{sw}. Among all samples on the Pareto front, lower switching frequencies are

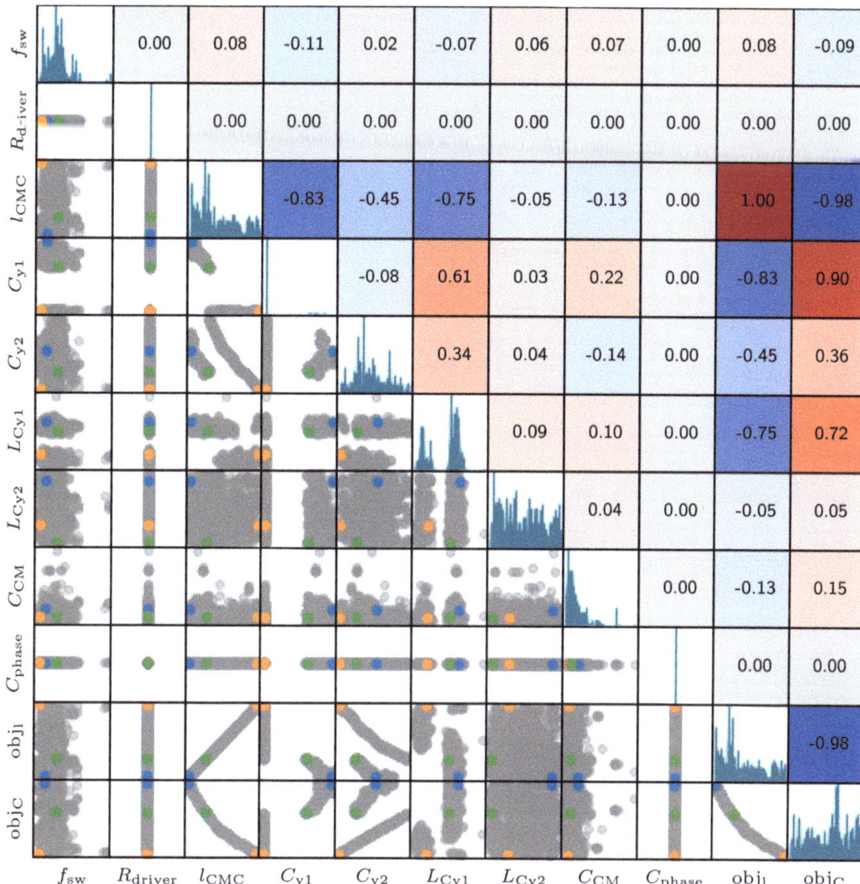

Fig. 16.13 The scatter plot matrix displays the correlations between all design variables and objectives

dominant. Slower switching generates less electromagnetic emission, making it easier to meet emission constraints with smaller filter components, resulting in lower objective values.

Matrix element (2,2) This shows the distribution of the gate driver resistance R_{driver}. Since R_{driver} is fixed to 6.6 Ω for this analysis, the distribution appears as a Dirac pulse.

Matrix element (6,6) The distribution of the parasitic inductance L_{Cy1} reveals clustering in two distinct value ranges. This suggests that only certain inductance values are compatible with Pareto-optimal designs, potentially due to interactions with circuit resonance conditions.

Matrix element (7,7) The distribution of L_{Cy2} is nearly uniform, indicating that this parameter has relatively little influence on the optimal designs within the investigated range. The parameter L_{Cy} is only one part of the inductive loop that also consists of the commutation cell and the housing.

Matrix element (9,9) The distribution of the parasitic phase capacitance C_{phase} also appears as a Dirac pulse, as it is fixed to $C_{phase} = 0.5\,nF$ in the selected scenario.

In the lower triangle of the matrix, we see at each position (i, j) the joint distribution of the parameters i and j. In the following, we analyze some of them exemplarily:

Matrix element (9,2) This is a point, because the values of the pair R_{driver}, C_{phase} are fixed.

Matrix elements (:,2), (9,:) All distributions in the second column and the ninth row are lines, because R_{driver} and C_{phase} are fixed.

Matrix element (10,3) This is a diagonal line, because the first objective equals the value of l_{CMC}.

Matrix element (11,3) This resembles the Pareto front, because with increasing l_{CMC}, the C_y capacitors are smaller.

Matrix elements (4,3), (5,3) They show more complex correlation. With increasing l_{CMC}, the value of C_{y1} at first becomes lower, until it drops to its minimum value. Simultaneously, the values of C_{y2} decrease. But when C_{y1} drops to its minimum, C_{y2} takes over and must be increased. The sum $C_{y1} + C_{y2}$ still decreases, as seen in the matrix element $(11, 3)$.

Matrix elements (7,:), (:,7) $L_{Cy,2}$ can assume any value for the chosen parameter range, for all design parameters

The scatter plot matrix is a powerful tool for visualizing correlations between design variables and objectives within a single comprehensive plot. Identifying correlations—or the absence thereof—between pairs of variables is essential during the design process. If one design parameter needs to be adjusted to a more favorable value (e.g., due to cost, availability, or manufacturability), the scatter plot matrix helps determine which other parameters are likely to be affected. It reveals whether a trade-off must be made or whether the adjustment can be implemented independently without impacting the rest of the design.

To further support interpretation, the upper triangle of the scatter plot matrix displays the estimated correlation coefficients, calculated from the joint distributions. Each matrix element (i, j) in the upper triangle corresponds to the two-dimensional scatter plot shown in the mirrored element (j, i) on the opposite side of the diagonal. The correlation strength is visually encoded using color: dark red indicates strong positive correlation, dark blue indicates strong negative correlation, and gray tones represent weak or no correlation. This visual cue helps quickly identify which parameter pairs are tightly linked and which can be varied independently.

We look at the correlation coefficients:

2nd row, 9th column They are zero, because R_{driver} and C_{phase} are fixed.

Matrix element (3,10) The correlation coefficient between l_{CMC} and objective 1 is one, because they are equal.

Matrix element (4,11) The correlation coefficient between C_{y1} and the second objective is 0.90, because it increases with increasing capacity (see also the joint distribution in the matrix element $(11, 4)$).

5th row, 11th column The correlation coefficient between C_{y2} and the second objective is 0.36 because this objective increases with increasing C_{y2}. Due to interaction between C_{y1} and C_{y2}, there are two clusters of C_{y1}. These clusters are also visible in the marginal distribution of C_{y1} in the scatter plot matrix element $(4, 4)$.

10th row, 11th column The correlation between the objectives for l_{CMC} and the capacitors is -0.98, i.e., we can choose either a small choke or low values of the capacitors, but not both.

The values of the correlation coefficient serve as an indicator of correlation, but not a proof. In particular, when the data is clustered or non-linearly distributed, the correlation coefficient may be low, even if a meaningful structure or dependency exists between the variables.

We use the scatter plot matrix to identify pairs of design variables whose dependencies suggest closer examination. Each combination of design variables results in a unique emission spectrum. When one variable is changed, the spectrum may change significantly in specific frequency ranges but remain largely unaffected in others. To explore these sensitivities, we focus on the four design variables C_{y1}, C_{y2}, l_{CMC}, L_{Cy1}. These variables exhibit noticeable correlations with the optimization objectives, as indicated by their correlation coefficients in the scatter plot matrix: -0.83, -0.45, -0.75, and 0.61, respectively.

In Fig. 16.14 we plot four spectra, C_{y1} (upper left), C_{y2} (upper right), l_{CMC}, (lower left), and L_{Cy1} (lower right). Each plot has three curves. The green curve corresponds to the reference design given in the last row of Table 16.1. The orange curve corresponds to the case, where the one design parameter indicated in each figure is increased to the upper bound of its design range. The blue curve corresponds to the lower bound. In all subplots, the emission limit lines for the average detector class 3 are included. The results show that the reference design does not comply with the class 3 emission limits. Improvement of the values for either C_{y1}, C_{y2}, or l_{CMC} is required to make the design pass.

Each subplot in Fig. 16.14 highlights the frequency range in which the corresponding design parameter influences the emission spectrum. These observations are valid for the specific values chosen for the other eight design parameters in the reference design. The C_y capacitors affect the spectrum below 1 MHz, where they contribute to the suppression of low-frequency common mode noise. The common mode choke influences the same low-frequency region but also shows an effect above 20 MHz, where it helps attenuate higher-frequency components. The parasitic series inductance L_{Cy1} of C_{y1} affects the emission spectrum in the high-frequency range above 20 MHz.

The values of l_{CMC} and the C_y capacitors are anticorrelated in the context of meeting the class 3 emission limits. Since both components act in the same frequency range, increasing one allows for a reduction in the other. The sensitivity

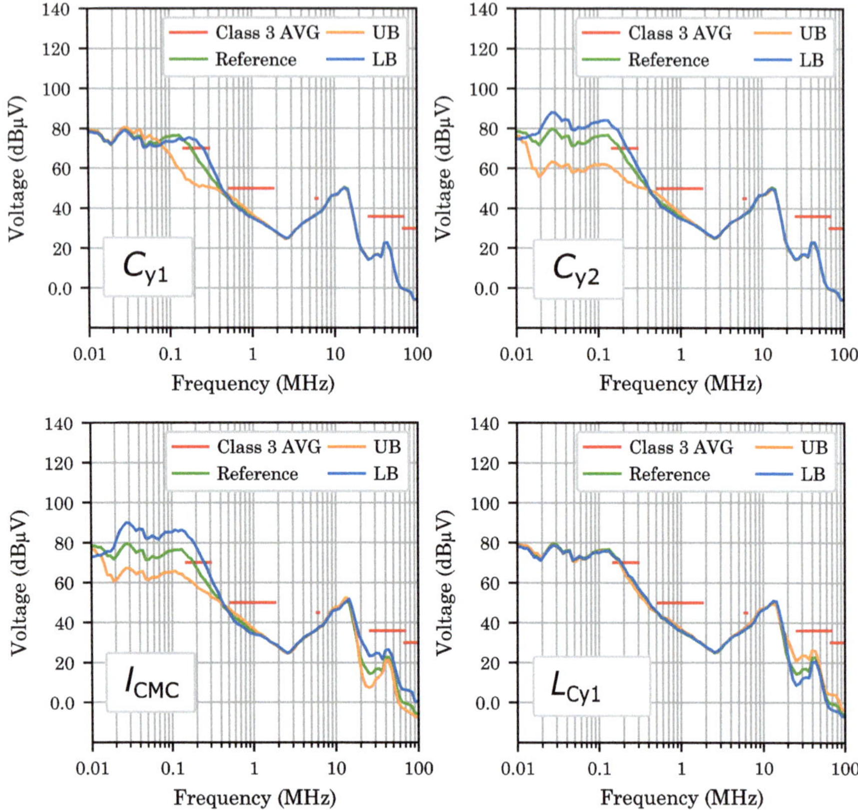

Fig. 16.14 The sensitivities highlight the behavior of individual design samples over frequency

plots for l_{CMC} and L_{Cy1} (Fig. 16.14) show that, for the reference design, a large choke or a small parasitic inductance reduces emissions above 20 MHz. The scatter plot matrix element (6, 3), which represents the joint distribution of l_{CMC} and L_{Cy1}, reveals a more complex picture when considering the full parameter space. While some combinations follow the expected trend—small parasitic inductance with large choke or vice versa—other valid configurations exist. For example, medium-sized values of L_{Cy1} satisfy the emission limits when paired with relatively small chokes, and conversely, low values of L_{Cy1} may appear alongside large chokes. This complexity is especially pronounced at higher frequencies with a position of resonances relative to the emission limit lines. A small shift in design parameters may move a resonance peak into or out of a restricted band, affecting compliance. A robust understanding of these relationships is best achieved through a combination of scatter plot matrix analysis and targeted best—medium—worst-case evaluations, as illustrated in Fig. 16.14.

Finally, we select and investigate three designs that pass the class 3 limits. We trade the length of the choke against the values of the two C_{y} capacitors.

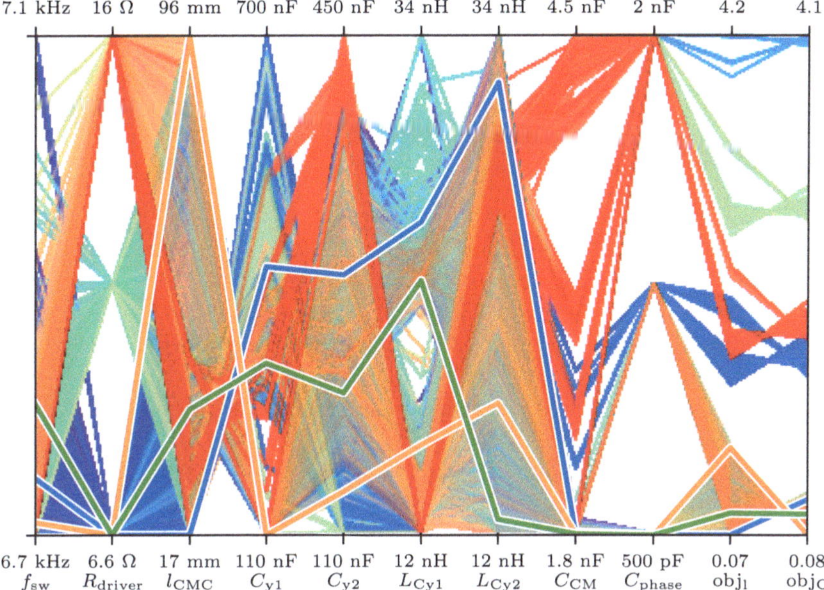

Fig. 16.15 The parallel coordinate plot displays all solutions over all design parameters

In particular, we choose the three designs that have (1) the minimum possible min (l_{CMC}), or (2) the minimum sum of the capacitor values, min $\left(\sum_i C_{y1,i}\right)$, and (3) a trade-off solution between them. These three solutions are highlighted in the scatter plot matrix by color. The design with the minimum l_{CMC} is colored blue, the one with the minimum capacitors is colored orange, and the compromise is green. In the scatter plot matrix, we find the blue point at the minimum of obj_l and l_{CMC}, and the orange point at the opposite end. The green point is the trade-off solution which favors both reasonable values for choke length and size of the capacitors. It is chosen to minimize the maximum of the value of each (l_{CMC}, $\sum_i C_{yi}$) pair, which results in pairs in which both values are both small and close to each other. The three colored points appear in all 2D histograms of Fig. 16.13 and indicate the value of the respective design parameter of this sample.

Finally, we introduce the parallel coordinate plot as another means to represent multidimensional data. The parallel coordinate plot visualizes the values of each design parameter across all samples, with respect to their minimum and maximum values. This type of plot is well-suited for examining how close a given design is to the boundaries of each parameter's range. The parallel coordinate plot for the investigated samples is shown in Fig. 16.15. The x-axis lists the nine design parameters followed by the two objective functions. The y-axis represents the scale from the minimum to the maximum value of each parameter or objective. The actual values corresponding to the minimum and maximum bounds are annotated at the bottom and top of each vertical axis.

The designs are color-coded according to R_{driver}. All of the thin red lines belong to samples with $R_{\text{driver}} = 16\,\Omega$. In the plot, all red lines pass through this point. All thin green lines belong to samples with $R_{\text{driver}} = 11\,\Omega$, and they pass through this point in the parallel coordinate plot. All thin blue curves belong to $R_{\text{driver}} = 6.6\,\Omega$. This color coding helps distinguish the influence of R_{driver} on the distribution of the remaining parameters and objectives. In the parallel coordinate plot, we find the following correlations:

- The red plot lines tend to have small l_{CMC}. Large R_{driver} corresponds to long rise/fall time, which helps to reduce the length of the choke.
- The red lines need rather large values for C_{y2} and small values of L_{Cy1}.
- The red lines allow larger values for motor and phase capacitance, C_{CM} and C_{phase}.
- The thin green lines have faster rise/fall times. The corresponding designs cover a larger range of l_{CMC}, because faster switching increases the emission.
- The permissible values for C_{y2} cover a larger range for the thin green lines, and more L_{Cy1} can be large.
- The thin blue lines, which represent $R_{\text{driver}} = 6.6\,\Omega$ and fast rise/fall times require a large range of l_{CMC}, and also for C_{y2}.
- The values of the switching frequency f_{sw} tend to cluster toward the lower end of the design range. This behavior results from the fact that no objective or constraint has been explicitly assigned to this variable. Since lower switching frequencies generally lead to reduced electromagnetic emissions, the objectives—minimizing choke length and C_{y} capacitance—are more easily achieved at low f_{sw}. As a result, high switching frequencies are absent from the Pareto-optimal set. If, for engineering or application-specific reasons, higher switching frequencies must be considered, then f_{sw} should be fixed, similar to how R_{driver} and C_{phase} are fixed in this study to ensure targeted analysis within a constrained subspace.

The information conveyed by the parallel coordinate plot in Fig. 16.15 is similar to that of the scatter plot matrix in Fig. 16.13, with the added advantage that color coding can be used to highlight relations across parameters and to distinguish different subsets of designs.

In addition to displaying all samples in thin lines, the plot also highlights the three selected samples with minimum choke length (thick blue line), minimum capacitor size (thick orange line), and the trade-off sample (thick green line). For all three samples, R_{driver} and C_{phase} are selected to be minimal, because they are part of the Pareto front selected from Fig. 16.12. All three curves pass the point $R_{\text{driver}} = 6.6\,\Omega$ and $C_{\text{phase}} = 0.5\,\text{nF}$ in the parallel coordinate plot.

For the three selected designs, we observe the following:

Orange line Imposing the smallest possible C_{y} capacitor values results in the need for the largest common mode choke. In this configuration, C_{y1} reaches its minimum, while C_{y2} is slightly above its lower bound. The switching frequency f_{sw} for this design is relatively low, supporting reduced emissions in the absence of large filtering capacitors.

Blue line Minimizing the choke length leads to large C_y capacitors. In this case, the parasitic inductances L_{Cy1} and L_{Cy2} are allowed to be relatively high. The switching frequency f_{sw} is higher than in the orange design.

Green line This is one compromise (out of many) between moderate choke size and capacitor values. The switching frequency that corresponds to this design is largest, but still in the lower third of the design parameter range. The parasitic inductance L_{Cy1} is moderate. The parasitic inductance L_{Cy2} is tight.

The design parameter values for the three selected samples are summarized in Table 16.2, which quantifies the qualitative observations made so far. From this table, we observe that the permissible lengths of the common mode choke vary significantly from 17 mm to 96 mm, a difference of nearly 8 cm. The design with the smallest choke length requires approximately 80% less choke length than the design with the smallest C_y capacitors. To compensate, the value of C_{y1} must be increased by a factor of four, and C_{y2} must be doubled. The trade-off design, which balances choke length and capacitor size, features a choke that is twice as long as in the minimum case, while C_{y1} is increased by a factor of three and C_{y2} by 50%.

All three designs operate at a low switching frequency between 6 kHz and 7 kHz, and their equivalent series inductance values range between 13 nH and 32 nH. In addition to the fixed minimum value of C_{phase}, all selected designs also assume a close-to-minimum motor capacitance, $C_{CM} = 1.8$ nF. The optimizer consistently selects this minimum because it contributes to minimizing the defined objectives. Since no penalty is associated with doing so, there is no incentive to increase C_{CM}. If designs with larger machine capacitance are desired, e.g., for more realistic or conservative configurations, this parameter must also be fixed in the optimization setup.

The designs shown in Table 16.2 are selected based on the two defined objectives, which directly involve three design parameters: l_{CMC}, C_{y1}, C_{y2}. To ensure a meaningful optimization outcome, the parameters R_{driver} and C_{phase} are fixed in advance. This prevents the optimizer from trivially pushing them to their respective corner points in the design space.

In the absence of constraints, the optimizer would select the maximum possible value for R_{driver} (resulting in slow rise and fall times) and the minimum value for C_{phase} (minimizing parasitic coupling), as both lead to reduced electromagnetic emission. Fixing these parameters allows the optimization to focus on the trade-offs between choke length and capacitor sizing, rather than converging to a degenerate solution driven by those other effects.

Table 16.2 List of selected best designs

	f_{sw} (kHz)	R_{driver} (Ω)	l_{CMC} (mm)	C_{y1} (nF)	C_{y2} (nF)	L_{Cy1} (nH)	L_{Cy2} (nH)	C_{CM} (nF)	C_{phase} (nF)
$\min(l_{CMC})$	6.72	6.6	17	429	285	25	32	1.8	0.5
$\min(\sum_i C_{yi})$	6.68	6.6	96	110	139	16	18	1.8	0.5
trade-off	6.79	6.6	37	314	206	23	13	1.8	0.5

Having evaluated five out of the nine design parameters, four of them still need to be assessed. These are f_{sw}, L_{Cy1}, L_{Cy2}, C_{CM}. Among them, the optimizer consistently drives f_{sw} and C_{CM} toward their minimum values. This behavior arises because neither parameter is explicitly constrained or included in the objective functions, and reducing them helps minimize emissions. Similarly, the parasitic inductances of the C_{y} capacitors, L_{Cy1} and L_{Cy2}, are not fixed but tend to be minimized by the optimizer. Lower values for these inductances improve high-frequency filtering performance, which contributes to satisfying the emission constraints without increasing the other objective values.

The designs presented in Table 16.2 represent three out of many possible solutions. Since it is not feasible to investigate all design parameters simultaneously, we must fix those considered most critical in order to narrow down the design space. However, alternative optimal designs may exist. For example, there may be configurations that allow for significantly larger parasitic inductances in the C_{y} capacitors without negatively affecting the key objectives l_{CMC} and $C_{\text{y1}}+C_{\text{y2}}$. Local exploration may result in even more optimal designs.

The designs listed in Table 16.2 satisfy the constraint of passing class 3 conducted emission limits according to CISPR 25 with average detector. However, this result is derived from the trained surrogate model, which serves as an approximation of the underlying circuit simulation model from which it is generated. Therefore, validation using the original circuit simulations must confirm that these designs truly comply with the emission requirements.

The corresponding emission spectra, computed with LTSpice for the three selected designs, are shown in Fig. 16.16. The blue, red, and orange curves represent the emission profiles of the designs with (1) the minimum possible l_{CMC}, (2) the minimum total C_{y} capacitance, and (3) the trade-off solution between the two, respectively. All three designs are confirmed to meet the class 3 emission limits, validating the surrogate model's predictions.

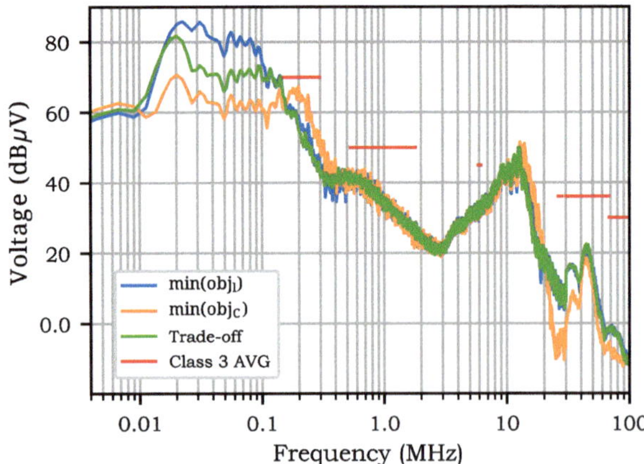

Fig. 16.16 Evaluation of the designs of Table 16.2 in the circuit model

16.4 Advanced: Gaussian Process Regression

Domains: 

In the following we explain the details of the modeling procedure that is used to train the surrogate model used in Sect. 16.3. We rely on the publications Rezende et al. (2024) and Gsoels et al. (2025).[2]

The intuitive idea of GPR is explained in Sect. 16.1.1 on an example which has a 1D input x, and a scalar output $Y(x)$. The case treated here is a generalization such that:

1. the generated model operates over a K dimensional input parameter space, which describes the design parameters of the underlying physical simulation model,
2. it predicts frequency-domain responses, i.e., vectors over M frequency samples.

As a consequence, we operate with:

1. an input matrix instead of a vector of input samples
2. an output matrix instead of a 1D curve
3. a matrix-variate Gaussian distribution instead of a vector-valued

We stack the K design parameters into the K-dimensional column vector $\mathbf{x}$. We assume N training vectors and stack them into the matrix $\mathbf{X} = [\mathbf{x}_1, \mathbf{x}_2, \ldots, \mathbf{x}_N]$. This matrix

$$\mathbf{X} = [\mathbf{x}_1, \mathbf{x}_2, \ldots, \mathbf{x}_N] = \begin{bmatrix} x_{11} & \cdots & x_{1N} \\ \vdots & \ddots & \vdots \\ x_{K1} & \cdots & x_{KN} \end{bmatrix} \tag{16.13}$$

is the input matrix of the Gaussian process, with which the model is trained.

We use each training vector $\mathbf{x}_n$ to evaluate the simulation model on M frequency points. The result is the frequency response of the model at this realization of input parameters. This model can be a SPICE model, an EM model, or even measurement data. We stack these samples in a vector $\mathbf{y}_n$ of length M, and the output vectors that refer to the N training samples into the matrix:

$$\mathbf{Y} = [\mathbf{y}_1, \mathbf{y}_2, \ldots, \mathbf{y}_N] = \begin{bmatrix} y_{11} & \cdots & y_{1N} \\ \vdots & \ddots & \vdots \\ y_{M1} & \cdots & y_{MN} \end{bmatrix}. \tag{16.14}$$

[2] In these publications, the outlined concept is denoted as wideband Kriging. Kriging and GPR have been developed from two different communities, spatial statistics and machine learning, but result in the same concept. Since for EMC simulation the concept matters but not its origin, we do not discuss the different semantics which are linked to each community.

This is the output matrix of the Gaussian process. In the presented formulation each entry y_{mn} is scalar, an extension to complex variables is possible.

We define a Gaussian process on these random variables. In order to consider the covariance between all $M \times N$ random variables, we need to stack the vectors $\mathbf{y}_1, \ldots, \mathbf{y}_N$ into one large vector $\text{vec}(\mathbf{Y}) = \left[\mathbf{y}_1^T, \ldots, \mathbf{y}_N^T\right]^T$ and obtain the vector-variate Gaussian process:

$$\text{vec}(\mathbf{Y}) \sim \mathcal{N}\left(\text{vec}\left(\boldsymbol{\mu}_1, \ldots, \boldsymbol{\mu}_N\right), \boldsymbol{\Sigma}_{MN \times MN}\right) \tag{16.15}$$

where $\boldsymbol{\mu}_i = E[\mathbf{y}_i]$ is the vector of mean values of $\mathbf{y}_i$.

The covariance matrix $\boldsymbol{\Sigma}_{MN \times MN}$ describes the correlation between the observed magnitudes at any frequency point m of training sample n and any other frequency point m' of the training sample n'. The size of this matrix is $(M \cdot N) \times (M \cdot N)$. In case of several hundred training data points and a similar amount of frequency samples, this is in the order of some $10^4 \times 10^4$.

The authors in Rezende et al. (2024) suggest reducing the size of Σ by assuming a correlation structure between its entries. They propose to apply a correlation that allows to decompose $\boldsymbol{\Sigma}_{MN \times MN}$ into the form:

$$\boldsymbol{\Sigma}_{MN \times MN} = \boldsymbol{\Sigma}_M \otimes \boldsymbol{\Sigma}_N \tag{16.16}$$

where $\otimes$ denotes the Kronecker product.

This assumption allows to derive the continuous matrix-variate distribution of the random matrix $\boldsymbol{Y}$ (Glanz 2013; Gupta and Nagar 2018) of the form:

$$\boldsymbol{Y} \sim \mathcal{MN}(\mathbf{M}, \boldsymbol{\Sigma}_M, \boldsymbol{\Sigma}_N). \tag{16.17}$$

with mean value matrix $\mathbf{M} = E\left[\mathbf{y}_1, \ldots, \mathbf{y}_N\right]$.

The probability density of this distribution can be given in closed form as

$$p_{\boldsymbol{Y}}(\mathbf{Y}) = \frac{\exp\left(-\frac{1}{2}\text{tr}\left(\boldsymbol{\Sigma}_M^{-1}(\mathbf{Y}-\mathbf{M})\,\boldsymbol{\Sigma}_N^{-1}(\mathbf{Y}-\mathbf{M})^T\right)\right)}{\sqrt{(2\pi)^{MN}\det(\boldsymbol{\Sigma}_M)^N \det(\boldsymbol{\Sigma}_N)^M}}. \tag{16.18}$$

The authors further simplify the matrix $\boldsymbol{\Sigma}_M$ to be diagonal, i.e.,

$$\boldsymbol{\Sigma}_M = \text{diag}\left(\sigma_1^2, \ldots, \sigma_M^2\right). \tag{16.19}$$

This assumption means that the correlation between computed magnitudes of training samples n and n' are, up to a scaling factor σ_m, not a function of the frequency m at which they are observed. Further, samples at different frequencies m and m' are independent from each other.

The covariance $\mathbf{\Sigma}_N$ between the samples in the observed training data space is

$$\mathbf{\Sigma}_N = \begin{bmatrix} \operatorname{cov}(y_{m1}, y_{m1}) & \cdots & \operatorname{cov}(y_{m1}, y_{mN}) \\ \vdots & \ddots & \vdots \\ \operatorname{cov}(y_{mN}, y_{m1}) & \cdots & \operatorname{cov}(y_{mN}, y_{mN}) \end{bmatrix}, \tag{16.20}$$

This covariance is chosen to be not a function of the frequency sample m.

We use an exponential function to describe the correlation between the samples. In Eq. 16.3 we argue that with larger distance of the two samples x', x'', the lower is their correlation. In K dimensions this generalizes to

$$\kappa\left(\mathbf{x}', \mathbf{x}''\right) = \exp\left(-\sum_{k=1}^{K} \theta_{\mathrm{k}}^{(1)} \left|x_k' - x_k''\right|^{\theta_{\mathrm{k}}^{(2)}}\right), \tag{16.21}$$

in which $\mathbf{x}'$, $\mathbf{x}''$ are arbitrary vectors in the design parameter space. We sum over the distances $\left|x_k' - x_k''\right|$ of each dimension k. We allow variable correlation lengths in each k, expressed by the hyperparameters $\theta_{\mathrm{k}}^{(1)}$, $\theta_{\mathrm{k}}^{(2)}$:

Large $\theta_{\mathrm{k}}^{(1)}$, $\theta_{\mathrm{k}}^{(2)}$ make the kernel decay fast in the dimension k, there is little correlation. Small $\theta_{\mathrm{k}}^{(1)}$, $\theta_{\mathrm{k}}^{(2)}$ describe long-range correlation. A parameter $\theta_{\mathrm{k}}^{(2)} = 2$ corresponds to the Euclidian distance.

This kernel function $\kappa\left(\mathbf{x}', \mathbf{x}''\right)$ fulfills the properties of a correlation matrix. It is symmetric and has a unit diagonal. In the Gaussian process, we now set:

$$\mathbf{\Sigma}_N = \sigma^2 \mathbf{R}_{\mathrm{xx}} \tag{16.22}$$

with σ serving as a yet unknown scaling factor. With Eq. 16.22, we relate the correlation between the magnitudes of the physical model to the underlying design space parameters:

$$\mathbf{R}_{\mathrm{xx}} = \begin{bmatrix} \kappa(\mathbf{x}_1, \mathbf{x}_1) & \cdots & \kappa(\mathbf{x}_1, \mathbf{x}_N) \\ \vdots & \ddots & \vdots \\ \kappa(\mathbf{x}_N, \mathbf{x}_1) & \cdots & \kappa(\mathbf{x}_N, \mathbf{x}_N) \end{bmatrix}. \tag{16.23}$$

We re-sort the covariance matrix $\mathbf{\Sigma}$ introduced in Eq. 16.16 as

$$\mathbf{\Sigma}_M \otimes \mathbf{\Sigma}_N = \boldsymbol{\sigma}_M^2 \otimes \mathbf{R}_{\mathrm{xx}}, \tag{16.24}$$

with $\boldsymbol{\sigma}_M^2$ being defined as

$$\boldsymbol{\sigma}_M^2 = \sigma^2 \mathbf{\Sigma}_M = \sigma^2 \operatorname{diag}\left(\sigma_1^2, \ldots, \sigma_M^2\right). \tag{16.25}$$

Finally we assume, as in Rezende et al. (2024), that the mean matrix **M** has identical mean values across each row:

$$\mathbf{M} = \boldsymbol{\mu} \cdot \mathbf{1}^T = \begin{bmatrix} \mu_1 & \cdots & \mu_1 \\ \vdots & \ddots & \vdots \\ \mu_M & \cdots & \mu_M \end{bmatrix}, \tag{16.26}$$

with $\mathbf{1} = [1, \ldots, 1]^T$ being a column vector of N ones, and $\boldsymbol{\mu} = [\mu_1, \ldots, \mu_M]^T$. While the correlation between the computed samples is identical for all M frequencies, the mean and the variance of the process are individual for each m.

We can estimate the mean value vector $\boldsymbol{\mu}$ of the Gaussian process and the correlation matrix $\mathbf{R}_{\text{XX}}$ using the available training data matrix $\mathbf{Y}$. We use maximum likelihood estimation, which assumes that $\mathbf{Y}$ is a typical representative of the sampling space. We stack the hyperparameters into a column vector: $\boldsymbol{\theta} = \left[\theta_1^{(1)}, \ldots, \theta_{\text{M}}^{(2)}\right]^T$ and obtain the likelihood from the probability density Eq. 16.19 as

$$\mathcal{L}(\boldsymbol{\theta}|\mathbf{Y}) = \frac{\exp\left(-\frac{1}{2}\text{tr}\left(\boldsymbol{\sigma}_M^{-2}(\mathbf{Y}-\mathbf{M})\,\mathbf{R}_{\text{XX}}^{-1}(\mathbf{Y}-\mathbf{M})^T\right)\right)}{\sqrt{(2\pi)^{NM}\det\left(\boldsymbol{\sigma}_M^2\right)^N\det\left(\mathbf{R}_{\text{XX}}\right)^M}}. \tag{16.27}$$

where we use Eqs. 16.22 and 16.24 to introduce $\mathbf{R}_{\text{XX}}$ and $\boldsymbol{\sigma}_M^2$ into Eq. 16.18.

In Eq. 16.27, $\mathbf{Y}$ is computed from the physical model. All variables $\mathbf{M}$, $\mathbf{R}_{\text{XX}}$, $\boldsymbol{\sigma}$ are functions of the hyperparameters $\boldsymbol{\theta}$. The optimal, estimated, set of hyperparameters $\hat{\boldsymbol{\theta}}$ is obtained using maximum likelihood estimation and solving the optimization problem:

$$\hat{\boldsymbol{\theta}} = \text{argmax}_{\boldsymbol{\theta}}\,\ell(\boldsymbol{\theta}|\mathbf{Y}). \tag{16.28}$$

Optimizers such as Nelder-Mead, gradient-based optimizer, or stochastic methods such as particle swarm optimization have successfully solved Eq. 16.28.

Based on $\hat{\boldsymbol{\theta}}$, we obtain the maximum-likelihood estimator for the mean value $\hat{\boldsymbol{\mu}}(\hat{\boldsymbol{\theta}})$:

$$\hat{\boldsymbol{\mu}}\left(\hat{\boldsymbol{\theta}}\right) = \frac{\mathbf{Y}\mathbf{R}_{\text{XX}}^{-1}\mathbf{1}}{\mathbf{1}^T\mathbf{R}_{\text{XX}}^{-1}\mathbf{1}}. \tag{16.29}$$

This vector of length M contains the mean values of the Gaussian process for each frequency sample.

Similarly, we obtain the variance as the maximum of the log-likelihood with respect to the matrix $\boldsymbol{\sigma}_M^2$:

$$\sigma_M^2 = \sigma_M^2\left(\hat{\theta}\right) = \operatorname{diag}\left(\frac{(\mathbf{Y}-\mathbf{M})\,\mathbf{R}_{\mathrm{XX}}^{-1}\,(\mathbf{Y}-\mathbf{M})^T}{n}\right). \tag{16.30}$$

This matrix contains the M variances, one for each frequency sample.

Equations 16.29 and 16.30 characterize the Gaussian process at every training sample $\mathbf{y}_n(\mathbf{x}_n), n = 1, \ldots, N$. In order to interpolate at yet untrained design parameters $\mathbf{x}_*$, we use that at $\mathbf{x}_*$, the variable $\mathbf{y}_*(\mathbf{x}_*)$ is also the realization of a matrix-variate Gaussian process. We estimate the mean $\boldsymbol{\mu}(\mathbf{y}_*)$ and the variance at these points based on the trained samples $\mathbf{Y}$ and with the aid of the correlation function $\mathbf{R}_{\mathrm{XX}}$ and the estimated mean value $\hat{\boldsymbol{\mu}}$.

To make new predictions $\hat{\mathbf{y}}_*$ for a design parameter vector $\mathbf{x}_*$ that is not yet trained, its correlation to already trained points is considered. This correlation is expressed by the vector:

$$\mathbf{r}_{\mathrm{X}}(\mathbf{x}_*, \hat{\boldsymbol{\theta}}) = \begin{bmatrix} \kappa\,(\mathbf{x}_1, \mathbf{x}_*) \\ \vdots \\ \kappa\,(\mathbf{x}_N, \mathbf{x}_*) \end{bmatrix} \tag{16.31}$$

which itself is a function of the optimal hyperparameters $\hat{\boldsymbol{\theta}}$.

In order to use this correlation, we create an augmented correlation matrix $\tilde{\mathbf{R}}_{\mathrm{XX}}$ as

$$\tilde{\mathbf{R}}_{\mathrm{XX}}(\hat{\boldsymbol{\theta}}) = \begin{bmatrix} \mathbf{R}_{\mathrm{XX}} & \mathbf{r}_{\mathrm{X}} \\ \mathbf{r}_{\mathrm{X}}^T & 1 \end{bmatrix} \tag{16.32}$$

We use the augmented correlation matrix $\tilde{\mathbf{R}}_{\mathrm{XX}}$, the augmented observation matrix $\tilde{\mathbf{Y}} = \left[\mathbf{Y}, \hat{\mathbf{y}}_*\right]$, and apply the partitioned inverse method for $\tilde{\mathbf{R}}_{\mathrm{XX}}^{-1}$. This results in an augmented expression which is inserted into the log-likelihood function 16.27. Maximizing this function with respect to $\hat{\mathbf{y}}$ allows to derive an estimate for the mean $\boldsymbol{\mu}(\mathbf{x}_*)$ of the Gaussian process at the point $\mathbf{x}_*$ in the design parameter space:

$$\hat{\mathbf{y}}\,(\mathbf{x}_*) = \hat{\boldsymbol{\mu}}(\boldsymbol{\theta}) + \left(\mathbf{Y} - \hat{\boldsymbol{\mu}}(\boldsymbol{\theta})\mathbf{1}^T\right)\mathbf{R}_{\mathrm{XX}}^{-1}\mathbf{r}_{\mathrm{X}}. \tag{16.33}$$

This estimated mean is a function of the available training data $\mathbf{Y}$, the mean value vector $\boldsymbol{\mu}(\hat{\boldsymbol{\theta}})$, and the correlations among the training vectors themselves and the training vectors and the yet untrained location in the design parameter space.

In a similar way, we can approximate an estimator for the variance matrix: $\hat{\boldsymbol{\sigma}}\,(\mathbf{x}_*)$,

$$\hat{\boldsymbol{\sigma}}_M^2(\mathbf{x}_*) \approx \operatorname{diag}\left(\hat{\boldsymbol{\sigma}}_M^2 \cdot \left(1 - \mathbf{r}_{\mathrm{X}}^T\mathbf{R}_{\mathrm{XX}}^{-1}\mathbf{r}_{\mathrm{X}}\right)\right). \tag{16.34}$$

Equations 16.33 and 16.34 completely characterize the probability distribution of any $\mathbf{y}(\mathbf{x})$ as a function of the training data. When the training sample $\mathbf{x}_*$ coincides with a testing point, the term $\mathbf{r}_{\mathrm{x}}^T \mathbf{R}_{\mathrm{xx}}^{-1} \mathbf{r}_{\mathrm{x}}$ equals one, because in this case, the correlation between the training and the testing sample is one. Equation 16.34, i.e., the variance of the distribution at this point, is 0 (Forrester et al. 2008). At the training samples, the prediction of the GPR model is exact, as already seen in Fig. 16.3.

The assumptions made by Rezende et al. (2024) regarding the structure of the correlation and mean value matrices are empirical. In several application cases, these assumptions have enabled the construction of accurate surrogate models across various design spaces using a reasonable number of training samples. The method is considered robust; however, like any GPR approach, it struggles to accurately interpolate sharp resonances due to the inherent smoothness of Gaussian processes.

Numerous alternative correlation models and kernel functions beyond the one presented here can be explored. For example, the complexity of the correlation matrix can be increased by introducing off-diagonal terms in $\mathbf{\Sigma}_M$, thereby capturing additional dependencies between frequency samples. This, however, increases the number of parameters required to describe the Gaussian process. There is no guarantee that with a given number of training samples, and the still identical number of hyperparameters $\theta_{\mathrm{k}}^{(1)}$, $\theta_{\mathrm{k}}^{(2)}$, this increase improves the prediction quality of the trained model.

Further Reading

Machine Learning and Applications

Akash et al., Multi-objective EMI Optimisation using a Metamodel based SiC/GaN converter and NSGA II, in *IEEE Electrical Design of Advanced Packaging and Systems (EDAPS) Symposium*. Mauritius & online (2023)

M.A. Alvarez, L. Rosasco, N.D. Lawrence, *Kernels for Vector-Valued Functions: A Review* (2012). arXiv: 1106.6251 [stat.ML]. https://arxiv.org/abs/1106.6251

N. Ambasana et al. S-Parameter and frequency identification method for ANN based eye-height/width prediction. IEEE Trans. Compon. Packag. Manuf. Technol. **7**(5), 698–709 (2017)

A. Bingler, S. Bilicz, M. Csörnyei, Global sensitivity analysis using a kriging metamodel for EM design problems with functional outputs. IEEE Trans. Magn. **58**(9), 1–4 (2022). https://doi.org/10.1109/TMAG.2022.3167105

A. Bingler, S. Bilicz, M. Csörnyei, Polynomial chaos kriging metamodel for automotive EMC simulations, in *2022 International Symposium on Electromagnetic Compatibility—EMC Europe* (2022), pp. 873–878. https://doi.org/10.1109/EMCEurope51680.2022.9901150

A. Bingler et al., Surrogate model-based parameter identification for 3-D modeling of common mode chokes. IEEE Trans. Magn. **60**(3), pp. 1–4 (2024). https://doi.org/10.1109/TMAG.2023.3322036

Z. Chen, B. Wang, A.N. Gorban, Multivariate Gaussian and Student-t process regression for multi-output prediction. Neural Comput. Appl. **32**(8), 3005–3028 (2020). ISSN: 1433-3058. https://doi.org/10.1007/s00521-019-04687-8

A. Forrester, A. Sóbester, A. Keane, *Engineering Design via Surrogate Modeling* (Wiley, Chichster, 2008)

F. Garbuglia et al., Modeling electrically long interconnects using physics-informed delayed gaussian processes. IEEE Trans. Electromagn. Compat. **65**(6), 1715–1723 (2023). https://doi.org/10.1109/TEMC.2023.3317917

H. Glanz, L. Carvalho. *An Expectation-Maximization Algorithm for the Matrix Normal Distribution* (2013). arXiv: 1309.6609 [stat.ME]. https://arxiv.org/abs/1309.6609

P. Gsoels et al., CM-EMI prediction for 3-Level T-Type traction inverters using Wideband-Kriging, in *International Symposium on Electromagnetic Compatibility, EMC Europe*. Bruges, Belgium (2024)

P.D. Gsoels et al., Achieving EMC compliance in traction inverters via multi-objective optimization, in *IEEE Open Journal of the Industrial Electronics Society* (2025). in preparation

A.K. Gupta, D.K. Nagar, *Matrix Variate Distributions* (Chapman and Hall/CRC, New York, 2018). https://doi.org/10.1201/9780203749289. https://www.taylorfrancis.com/books/mono/10.1201/9780203749289/matrix-variate-distributions-nagar-gupta

J. Henninger, J. Hansen, D. Pommerenke, Measurement procedure to generate a GaN power transistor surrogate model at GHz frequencies, in *International Symposium on Electromagnetic Compatibility, EMC Europe*. Bruges, Belgium (2024)

C. Lataniotis et al., UQLab *user manual—Kriging (Gaussian process modeling)*. Chair of Risk, Safety and Uncertainty Quantification (ETH Zurich, Switzerland, 2022)

T. Leitgeb, J. Hansen, S. Ranftl, A Warped Gaussian process regression which obeys Kirchhoff's current law, in *25th International Conference on the Computation of Electromagnetic Fields, COMPUMAG*. to be submitted. Naples, Italy (2025)

J.L. Loeppky, J. Sacks, W.J. Welch, Choosing the sample size of a computer experiment: a practical guide. Technometrics **51**(4), 366–376 (2009). ISSN: 0040-1706, 1537-2723. http://www.jstor.org/stable/40586647 (visited on 08/19/2024)

P. Manfredi et al., Stochastic transmission line analysis via polynomial chaos methods: an overview. IEEE Electromagn. Compat. Mag. **6**(3), 77–84 (2017). https://doi.org/10.1109/MEMC.0.8093844

S. Marelli, B. Sudret, UQLab: a framework for uncertainty quantification in MATLAB, in *2nd International Conference on Vulnerability and Risk Analysis and Management* (University of Liverpool, United Kingdom, 2014), pp. 2554–2563. https://doi.org/10.1061/9780784413609.257

A. Orlandi et al., Feature selective validation (FSV) for validation of computational electromagnetics (CEM). Part II- assessment of FSV performance. IEEE Trans. Electromagn. Compat. **48**(3), 460–467 (2006). https://doi.org/10.1109/TEMC.2006.879360

K. Patra et al., Surrogate modeling for predicting shielded cable emissions. IEEE Trans. Electromagn. Compat. **65**(1), 249–256 (2023). https://doi.org/10.1109/TEMC.2022.3225631

C.E. Rasmussen et al., *Gaussian Processes for Machine Learning* (The MIT Press, New York, 2006). https://gaussianprocess.org/gpml/

R.S. Rezende et al., Multi-output variable-fidelity bayesian optimization of a common mode choke, in *Joint IEEE International Symposium on Electromagnetic Compatibility, Signal & Power Integrity, and EMC Europe*. Virtual Conference (2021)

R.S. Rezende et al., Wideband gaussian process modeling of a high-voltage EMI filter, in *20th International IGTE Symposium 2022 on Computational Methods in Electromagnetics and Multiphysics* (Graz, Austria, 2022)

R.S. Rezende et al., Wideband Kriging for multiobjective optimization of a high-voltage EMI filter. IEEE Trans. Electromagn. Compat. **66**(4), 1116–1124 (2024). https://doi.org/10.1109/TEMC.2024.3390556

R.S. Rezende et al., A Wideband Gaussian process as surrogate model of non-smooth functions in EMC, in *21st International IGTE Symposium 2024 on Computational Methods in Electromagnetics and Multiphysics*. Graz, Austria (2024)
K. Scharff, H.-D. Brüns, C. Schuster, Efficient crosstalk analysis of differential links on printed circuit boards up to 100 GHz. IEEE Trans. Electromagn. Compat. **61**(6), 1849–1859 (2019). https://doi.org/10.1109/TEMC.2019.2894536
J. Wang, An intuitive tutorial to Gaussian process regression. Comput. Sci. Eng. **25**(4), 4–11 (2023). https://doi.org/10.1109/MCSE.2023.3342149
Ö.F. Yildiz, H.-D. Brüns, C. Schuster, Variance-based iterative model order reduction of equivalent circuits for EMC analysis. IEEE Trans. Electromagn. Compat. **61**(1), 128–139 (2019). https://doi.org/10.1109/TEMC.2018.2845676

Multi-Objective Optimization

J.S. Arora, *Introduction to Optimum Design*, 3rd edn., pp. 96, 713, 717, 727, 737 (Academic Press, New York, 2011). ISBN: 978-0123813756
D.O. Boillat, F. Krismer, J.W. Kolar, Design space analysis and $\rho - \eta$ Pareto optimization of *LC* output filters for switch-mode AC power sources, in IEEE Trans. Power Electron. **30**(12), 6906–6923 (2015). https://doi.org/10.1109/TPEL.2015.2393151
A. Bréard, R. Moulla, C. Vollaire, Metamodel of power electronic converters using learning SVR method coupling with wavelet compression. IEEE Trans. Electromagn. Compat. **58**(2), 588–598 (2016). https://doi.org/10.1109/TEMC.2016.2519684
K. Deb et al., A fast and elitist multiobjective genetic algorithm: NSGA-II". IEEE Trans. Evol. Comput. **6**(2), 182–197 (2002). https://doi.org/10.1109/4235.996017
R. Fletcher, *Practical Methods of Optimization*. Second. 450 (Wiley, New York, 1988). ISBN: 978-0471915478
P. Gsöls, J. Romero-Lopera, J. Hansen, Multi-objective optimization for surrogate model based traction inverter systems, in *21st International IGTE Symposium 2024 on Computational Methods in Electromagnetics and Multiphysics*. Graz, Austria (2024)
K.M. Miettinen, *Nonlinear Multiobjective Optimization* (Kluwer's Interational Series, New York, 1999)
K. Niewiadomski et al., Sensitivity analysis of parasitics in power electronic circuit through Sobol' indices, in *2021 Asia-Pacific International Symposium on Electromagnetic Compatibility (APEMC)* (2021), pp. 1–4. https://doi.org/10.1109/APEMC49932.2021.9597088
N. Rashidi et al., Multi-objective design and optimization of power electronics converters with uncertainty quantification-Part I: parametric uncertainty. IEEE Trans. Power Electron. **36**(2), 1463–1474 (2021). https://doi.org/10.1109/TPEL.2020.3005456

Chapter 17
Computing the Real World

The simulation models presented in this book are a compromise between a realistic application and the simplicity required to not overwhelm the reader with unnecessary details. In this chapter, we let the experts speak, whose simulation work contributes to design decisions in electronic industry. They tell about their favorite modeling strategies, and why and how simulation in their companies contributes to the design of electronic systems.

17.1 Analog Devices

17.1.1 About the Author: Patrick DeRoy

Patrick DeRoy is a Sr. Manager/Principal Engineer at Analog Devices, Inc. (ADI) within the automotive business unit, automotive audio, and networking team. Patrick leads the A2B Applications Engineering Team in the USA, supporting customers on released transceiver products and supporting IC design for next-generation A2B transceiver developments. Additionally, he provides EM simulation support and EMC/SI/PI consulting to other product lines within the automotive BU (such as DSP, E2B, GMSL) as a core team member of ADI's Automotive EMC Council. Prior to ADI, Patrick began his career at CST of America, Inc., and has been using EM simulation tools for more than 15 years. Patrick is a Producer of SAE J2962-3 (Communication Transceivers Qualification Requirements—Ethernet) and a Member of SAE EMC Standards, CISPR-D USTAG, and ISO TC22 SC32 WG3 USTAG. Patrick is a Senior Member, IEEE, and active within IEEE EMC Society as a volunteer within TC-9 (Computational EM) and TC-10 (Signal and Power Integrity). Patrick has served in various leadership roles within the EMC Society (Young Professionals (YP) Representative, Boston Chapter Chair, and as a Member

J. Hansen, A. Barchanski, *EMC Simulation of Electronic Systems*,
https://doi.org/10.1007/978-3-032-09237-3_17

of the Board of Governors ('25-'27)). Patrick earned his B.S. ECE and M.S. EE degrees from the University of Massachusetts (UMass) Amherst.

This work is made by Patrick in his individual capacity and not as a representative of his employer, ADI, or any organization to which he belongs, and all views and opinions are his own.

17.1.2 EMC Simulation Perspective from the IC Level

17.1.2.1 Introduction

Before diving into the fascinating world of EMC simulation and sharing some of my experiences with the reader, I must first thank the authors for the invitation to contribute to this book. It's an honor and pleasure to do so, and I hope my perspective can be helpful and inspirational for the readers and engineers working on EMC, SI, and PI simulations now and in the future. I'd also be remiss if I didn't thank and acknowledge those who have helped me over the years, as their collaboration, insights, and lessons learned have enabled me to be successful in my modeling and simulation endeavors. Discussions with many colleagues at both Computer Simulation Technology (CST) and Analog Devices, Inc. (ADI), many customers, researchers, colleagues within IEEE EMC Society (especially those in TC-9/TC-10) have all shaped how I think and approach modeling and simulation problems, as well as how I aim to teach and mentor the engineers on my team. There are too many to thank and mention, but I do want to acknowledge my ADI colleagues Abhishek Ramanujan and Pete Sealey who have been my most engaging collaborators and enablers in recent years.

Computational EM software tools today are more capable and easier to use than ever before. It's important to share ideas and experiences, and do not hesitate to ask for help. No matter how capable the tools are, or will become, working together with others will accelerate and amplify success. We are all still learning, and with that in mind, the reader should be encouraged to make the extra effort to collaborate with others on simulation.

This contribution provides a glimpse into the why, what, and how with respect to EMC simulation. It will then dive deeper into a few specific examples related to our work over recent years within the automotive audio and networking team at Analog Devices, Inc (ADI).

17.1.2.2 Why EMC Simulation?

The fundamental why is rather simple: if EM simulation tools were not used for EMC analysis, it would be impossible to fulfill the commitment to customers— to design innovative, robust ICs along with reference designs that enable them to achieve EMC compliance with relative ease. Could the IC be designed without per forming EMC simulation and rely on measurements alone to verify performance? Anyone who has endured the pain of going through an additional, unplanned tapeout to fix an EMC issue that was only identified through EMC measurements (or worse, after the product is in the field) knows that it is simply too important to rely on measurement alone. The reason for EMC modeling and simulation is the same reason for all simulations: to gain confidence in a design that not only functions as intended but also ensures customer success. EMC is not simply a system-level problem to be solved by customers. Many times, when there are EMC issues, solutions are only implemented at the system level (e.g., PCB filtering, shielding, etc.) and not necessarily by making a change at the IC (for instance by programming a different operating mode to improve EMC). Sometimes there is simply no choice, or perhaps the solution outside of the IC is sufficient and cost-effective. But more often, ICs are being designed with EMC compliance in mind. This can be accomplished through careful consideration of the physical interconnects (such as die layout, packaging technology, package pinout, etc.) and/or through the digital features it is capable of, providing customers with flexibility when addressing EMC issues in their designs (e.g., programmable transmit levels for transceivers or spread spectrum clocking to reduce emissions).

17.1.2.3 The Importance of Measurement and Simulation

The following sections will illustrate some of the challenges with modeling and simulation. Details on measurement and simulation will also help explain the why. The goal is to ensure the reader understands that simulation and measurements should be done hand in hand. They are complementary, not competing. One should not be done exclusively without the other, and they often enable one another. And remember that errors can happen just as often in measurement as they can within simulation tools.

Many readers may wonder why not just measure? Indeed, it is sometimes much faster, or necessary (depending on how much time is available), to create a measurement-based model. Many vector network analyzer (VNA) measurements have been performed on cable assemblies and magnetic components to obtain S-parameters and that data has been used to make recommendations (either to IC design teams or to customers). Or those S-parameters have been used in system-level simulations (often converting S-parameters to SPICE models for IC-level simulations). The experienced engineer will realize the importance of applying extreme caution in both measurements and simulations, as small details matter (especially for unshielded situations). When performing a measurement, if done properly, the result is representative of reality. Those subtle imperfections of the cable and connectors and small imbalances will inherently be captured. One problem is that it may not be apparent what those small details are that impact the measured results. This is where simulation can provide much more insight. It may take more time and effort, but simulation allows for many possible variables and to visualize the fields with relative ease compared to measurements. The overall point is that measurements and simulations are complementary to one another. Certain things cannot be measured easily (or it may be impractical to measure), while others cannot possibly be simulated easily (or within a reasonable time frame). The absolute best engineering teams are making use of both, taking advantage of their strengths and using one to inform the other. This is the ultimate goal in EMC modeling and simulation efforts.

17.1.2.4 What and How to Simulate

Through work at ADI, there has been the opportunity to gain a unique perspective on EMC simulation. This involves collaborating with the IC design team on new chip developments, supporting released products, and working directly with customers. Some perspectives and lessons learned from efforts in automotive networking transceiver development and support are shared here. This includes simulations for the design of the IC itself, but also for the individual elements of the full channel from chip to chip, including PCB interconnects, passive components (especially magnetic components such as dataline common mode chokes), connectors, and cabling. After all, when the signal path includes electrically long lengths of unshielded cabling in a modern vehicle, you must contend with the wide spectrum of EMC and system ESD requirements.

The specific usage of EMC simulation depends highly on the type of IC and its application. In the following examples, the focus is on automotive networking transceivers that operate at speeds between 10 and 100 Mbps over unshielded twisted pair (UTP) cabling. Some of these transceivers also support power delivery over the same two wires. In this case, it should be fairly obvious that a significant amount of time has been spent on simulating and understanding the EM effects of the external environment (everything outside of the IC), more so than the IC itself. However, there have been plenty of IC-/package-level simulations performed to influence the design of such transceivers, and to enable customers to perform system-level simulations themselves. The following examples will answer the what and the how together for each example.

17.1.3 Simulation Methodology

17.1.3.1 Limitations of SPICE Circuit Simulation

It is common knowledge the industry that SPICE simulation is used for the IC design workflow (SP**I**C**E** = simulation program with **integrated circuit** emphasis, after all). SPICE is used primarily with the aim of simulating the active devices' behaviors and modeling the on-chip interconnects, as well as parasitics. For a typical IC (excluding RF designs for the moment), interconnect resistances and capacitances are accounted for. However, inductance may often be ignored (or approximated at best) with Maxwell's equations sometimes forgotten completely by designers. Of course, this has its advantages in speed, allowing many simulations to be performed very quickly. For certain ICs, the packaging may typically have an order of magnitude higher inductance relative to the on-chip interconnects, so this is often another reason for ignoring inductance on the die. However, this order of magnitude difference is not always the case, and extracted inductances also need to be generated. For those situations, and for simulating the external environment (everything outside of the die, including its packaging), today's IC designs require EM simulation (alongside SPICE) to accurately model the IC behavior. And this is purely from a functionality perspective, without considering EMC analyses.

3D EM modeling is used across many IC designs to ensure signal integrity (SI) and power integrity (PI) for high-speed digital ICs, to provide sufficient isolation of sensitive analog circuits from digital core and I/O switching noise in mixed-signal ICs, and to achieve highly accurate RF designs. In addition to the IC, there are many situations that need to simulate the PCB and beyond. For example, in an automotive transceiver design where the channel routes over meters of unshielded cabling within the vehicle, it's essential to model the external environment because most of the signal path is outside the IC. There is also the need to consider the EM effects happening in the external environment, and how they will impact the IC itself. The goal is to not only provide an IC but to offer customers robust and relevant PCB reference designs that they can use. Considering all of this, there are plenty of reasons for EM simulation whether it's used standalone to inform IC design decisions, or in tandem with SPICE simulation in a coupled simulation workflow.

17.1.3.2 EM Simulations

EM simulations of the external environment are often converted into SPICE models (e.g., using vector fitting) and brought into the IC design framework. This could be with simpler, faster parasitic extraction tools, 2D or 2.5D EM field solvers, or more accurate and complex (relatively slower) 3D full-wave methods that generate S-parameters. (There is a tradeoff between speed and accuracy to be considered.) It should be clear for the transceiver example described above, or even for high-speed digital ICs communicating on a large PCB, that providing accurate EM models of the external environment is crucial for the most accurate IC design. Conversely, EM simulations of the external environment can be performed within the EM simulation tool, with sources and/or terminations defined according to device-level (SPICE) simulations. Both approaches have their advantages and disadvantages, and engineering judgment is necessary to determine the best method based on the specific goals of the simulations.

For many ICs, 3D EM simulation is not needed for die level simulation (and maybe even for the packaging, depending on the situation). Does this mean the laws of physics don't apply for an IC? Not at all. There is a tradeoff between speed and accuracy, and the relative importance of capturing die level EM effects compared to the external environment must be considered on a case-by-case basis. From a historical perspective, it has been rare that EMC analyses and coupling effects were simulated at the die level. Whether this will increase significantly in the future or not is hard to predict. It is recommended for any IC or electronic device (whether it's a power converter, a communication transceiver, or a GPU) to always consider its EM fields. Whether choosing to simulate them or not, EM fields are fundamental, and they are everywhere—including on the IC and especially within its packaging. At the die level, specialized EM modeling methods have been used for many years. Of course, with specialized methods there are approximations, and these may not be sufficient for the types of low-level coupling analyses sometimes required. Is 3D EM simulation required for all IC design flows, all the time? Absolutely not. However, it's worth noting that for some of the cutting-edge, higher-speed, and/or higher-frequency IC designs (including 3D ICs), there is indeed a need for 3D EM simulation at both the die level and for modeling the external environment, especially in recent years. With that said, 3D EM is not typically used for entire ICs at once, due to the enormous complexity involved.

It is clear that EM simulation is critical, even for engineers designing ICs. Despite this, the complexity of modern ICs makes it extremely difficult and challenging to integrate EM simulation into the IC design flow. Not everything can or should be simulated given practical limitations with respect to time, compute resources, or personnel with the skill. Even when incorporating 3D EM extracted or measured models into the IC design framework, there are practical limitations on how much detail of the external environment can be captured while maintaining reasonable simulation times. The emergence of high-performance computing (HPC)/Cloud and more recently AI/ML is pushing these boundaries further, but that's a discussion for another time.

The importance of EM simulation for ensuring the intended functionality of the ICs is clear, but why is it particularly useful for EMC?

17.1.3.3 Simulation of EMC: Component vs. System vs. Test Level

EMC simulation does not exclusively mean simulations of specific EMC test setups. Although possible with today's software tools, hardware acceleration, and a knowledgeable user, it is less common to go through the effort of a detailed CISPR-25 or ISO 11452-x virtual bench test setup complete with the battery, line impedance stabilization networks (LISNs), DUT PCB with a representative IC model, cabling, load simulator, antennas, or current probes within the 3D EM environment. However, as the saying goes, there is a time and a place for it. In recent years, the focus has been more on modeling and measuring device-level equivalent tests, such as IC emissions and immunity, which are designed from standards like IEC 61967 and IEC 62132. These tests are designed to correlate with bench-level tests and ultimately with vehicle-level performance. The advantage of these virtual tests is that they have fewer variables and less complexity and allow for faster simulations. Still, there are many gaps to be filled between the different levels of testing, and the challenge of translating vehicle-level EMC requirements down to IC design targets remains. 3D EM simulation tools can play a crucial role in bridging these gaps, as shown in an example below.

What else can EMC simulation mean, then, if not simulating the EMC test itself? When the root causes of the emissions and/or immunity issues that may occur for a certain product within those test setups are understood, and especially during the design phase of an IC, work smarter (not harder) by addressing those root causes through simulation. For instance, performing channel simulations to minimize mode conversion in UTP-based communication transceivers can lead to lower emissions and improved immunity. If not yet aware of the root causes, simulation is a great way to explore them. However, even in this case, simulating within the context of a complete EMC test setup model isn't always practical. That level of detail is best left for the final design verification steps. With that in mind, let's look at a few examples.

17.1.4 Application Examples

17.1.4.1 Magnetics (Dataline CMC and Bus Power Inductors)

One of the most crucial components in the signal path for UTP-based communication transceivers is the common mode choke (CMC) or common mode filter (CMF). Although ADI does not design such components, the engineers have spent considerable time and effort trying to model them and simulate their effects (including with 3D EM simulations). The primary role of the CMC is to attenuate common mode noise that may couple to the cable. However, it must also have sufficiently low common to differential mode conversion (and vice versa). This ensures that excessive common mode noise coupled to the line does not convert into differential noise that the transceiver IC pins can detect. Similarly, it should prevent too much differential signal from converting into common mode noise on the line, which helps minimize emissions. Very critically, the CMC model must be accurate enough to capture these effects. Otherwise, the efforts to simulate realistic noise levels would be insufficient.

Common mode chokes are fundamentally designed for a certain (common mode) inductance value, for example, 200 μH. As with all components, there is a tolerance to be accounted for. While SPICE simulations can be run to gauge the impact of the common mode inductance variation, what about for mode conversion? This is not something that is specified by the vendor. They typically share measured results for mode conversion, but be careful with this data. The provided curve is often for a single component with a specific inductance value (likely close to the typical value), tested on a particular PCB with its unique layer stackup and coil winding parasitics. The winding construction, which can vary due to manufacturing tolerances, also plays a crucial role.

Vendors have been challenged to provide measurements for 20–30 samples, showing how much the mode conversion parameters can vary across multiple samples. It may be surprising to learn that this variation can be significant. Using one sample over another could easily mean the difference between passing and failing EMC tests.

SPICE models for CMCs are sometimes provided, but more often S-parameters are offered instead. S-parameters have their advantages, but they can be challenging because the data is typically for a single component and represents a black-box model. Exploring the impact of tolerances through SPICE or 3D simulations can be helpful, but both methods have their challenges. SPICE models may not accurately capture 3D field effects that are present in the actual PCB. 3D models are extremely hard to attain, and difficult to create from scratch (even when information is shared by the vendor to assist). The same challenge exists with respect to power inductors. Some transceiver products support bus-powered modules, using the same two-wire cable for both data communication and power delivery. This means that the low-pass filter, particularly its inductors, is as critical for EMC performance as the CMC. The inductors must be well matched (balanced), because any imbalances will increase mode conversion. Additionally, their nonideal effects, such as near-field coupling with nearby components or mutual coupling between closely spaced inductors, need to be carefully considered (Fig. 17.1). The same simulation challenges exist for these components. It is possible to create models using a combination of vendor information and several assumptions. While these models are not perfect, they provide useful insight that can influence the IC design, the PCB reference design, and customer recommendations.

While there can be some success in modeling the magnetic components in 3D, it can take a significant amount of time and effort to develop the models. A call to action for all magnetic component vendors is to enable customers with 3D models,

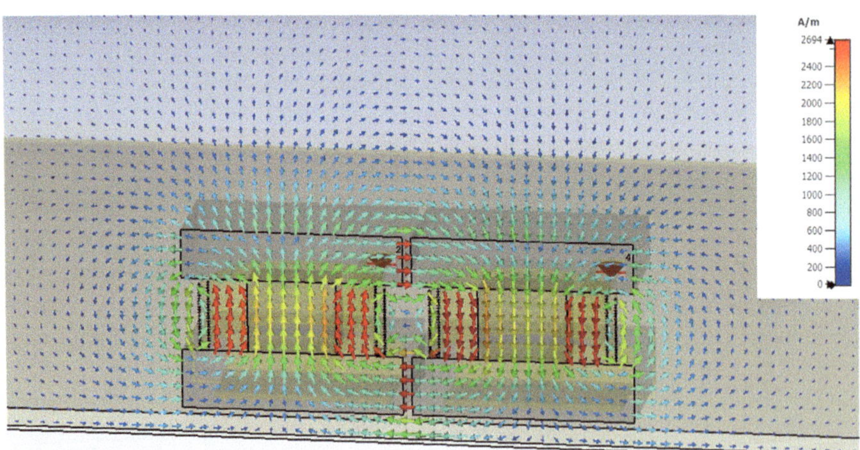

Fig. 17.1 Magnetic field at 1 MHz for two closely spaced inductors (unintended coupling). The impact on differential and common mode impedances can be derived from this 3D simulation model, using the calculated S-parameters

even if those need to be encrypted or protected models in which certain physical features are not visible. This is a win-win situation as the vendor can model their component more accurately, learn from the simulations, and optimize the part to improve its performance. At the same time, these types of near-field and EMC modeling efforts can be made much easier and much faster for customers. At the very least, enabling customers with the appropriate measurement data (or supporting them in their own independent measurements) to create the required models is highly recommended.

17.1.4.2 (Unshielded) Cabling and Connectors

Almost the entire communication channel for these transceivers is over the cabling, including connectors, which are especially critical when considering in-line connectors. Channel models are essential for fundamental signal integrity simulations, but they are equally important for analyzing crosstalk issues and EMC performance. For these types of communication transceivers, maintaining electrical balance is one of the most crucial factors to EMC success. In the past, cable and connector models have proven too ideal from a mode conversion standpoint, leading to unrealistically low noise levels (Fig. 17.2). To make these models more realistic, some asymmetries must be introduced. This could include slightly different distances of each wire to the common reference, slightly different wire lengths, or offsets from the center of the twisting. Connectors and the transitions from twisted to untwisted sections are typically the weakest links of the channel. While these transitions often dominate the effects, this is not always the case. Therefore, it is important that the cable portion of the model is not overly idealized.

Many simulations make use of 2D (e.g., multiconductor transmission line or MTL) solvers for modeling cables. These can often capture the physical effects sufficiently with the advantage of being faster than 3D solvers. When modeling the full transition from the PCB through the connector (especially if it is a right angle or multi-pin connector) and then through the untwisted portion to the fully twisted section, a 3D simulation is the best choice to accurately capture the physics, especially those related to common mode. The time and compute resources required for modeling just the transition are reasonable compared to attempting to model both the transition region and a long length of cabling in a single 3D model. This transition can then be effectively cascaded with a longer length of cabling modeled using the 2D solver, and the results have proven to be useful and reasonably accurate. As with any cascading approach, be sure to carefully define the ports in the individual models to ensure that the relevant modes are captured.

Considerable time has been spent modeling and illustrating the impact of the untwisted portions to convince customers and harness manufacturing teams of the importance for transceivers operating at these frequencies and beyond. However, this often leads to a classic tradeoff between cost and EMC performance. Any improvements made at the IC level to ease the requirements on cables and connectors, while keeping cost at a minimum, offer significant benefits to customers.

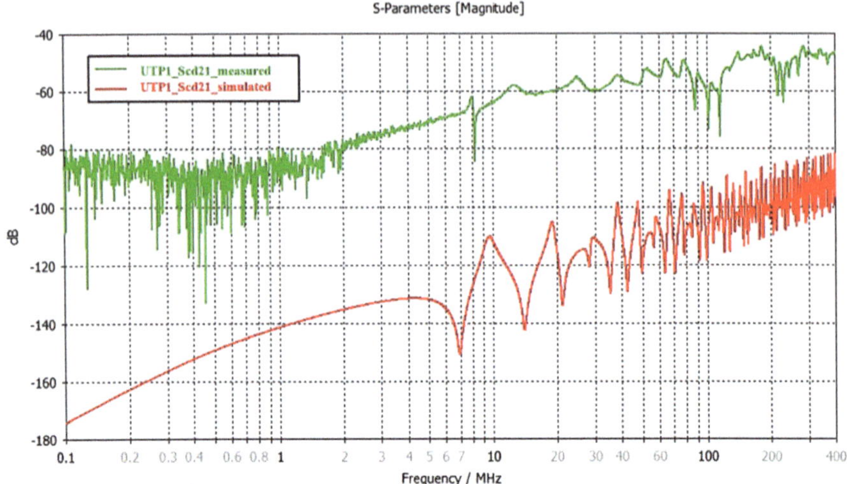

Fig. 17.2 UTP cable measured vs. simulated Scd21 (differential to common mode conversion, from port 1 to port 2). If cable twisting is perfect along its entire length, or if the connector/untwist imbalances are not modeled, results are far too ideal

The above discussion primarily addresses unshielded cables and connectors, but for shielded cases, there are equally intriguing effects to explore in which simulations can add value. These include shield transfer impedance, shield termination impact, and the degradation of shielding effectiveness over time due to thermal and mechanical stresses. EMC simulations will play a significant role in the development of much higher-speed communication links in the future, typically using coax or shielded twisted pair (STP).

17.1.4.3 Simulations to Inform and Influence Standards Development

As alluded to above, there are device-level (IC EMC) standards, bench (or module)-level standards, and vehicle-level standards. Many OEMs deviate from the international standards and have their own preferences particularly in regard to emissions limits or the ingress field levels (or target injected current levels in bulk current injection (BCI)). Sometimes, these tests are specific to only one OEM, causing significant headaches for Tier1 suppliers and IC designers, especially considering how quickly the standards are evolving.

Another challenge is the gap between device-level and vehicle-level tests. This was evident in a recent experience where a transceiver passed the device-level emissions test but failed at both the bench and vehicle levels. To investigate the root cause of this, simulations were done, modeling the device-level test fixture (IC Stripline), along with the IC packaging on the test PCB.

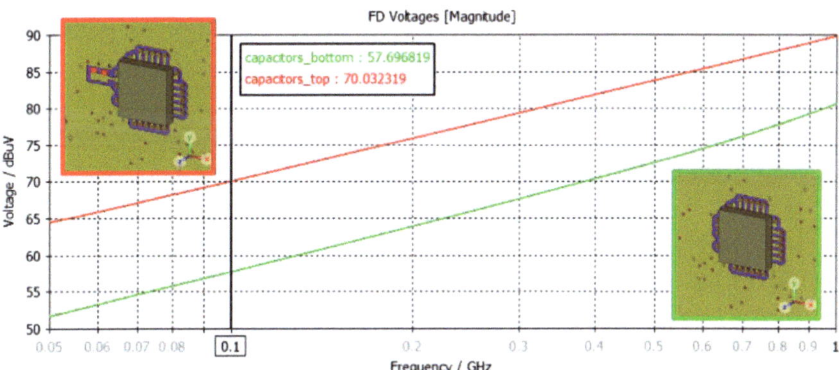

Fig. 17.3 Result comparison for two IC Stripline emissions simulations, with decoupling capacitors top vs. bottom side placement. 1 Amp current source excitation on die (DVDD-DVSS), resulting voltage (dBuV) at IC Stripline test fixture port. 12 dB higher emissions for the case of capacitors placed on the top side along with the IC

The device-level standard defines a certain PCB size that fits within the fixture, aiming to measure the emissions from the IC, by routing power and signal nets internal to the PCB or on the back side of the PCB. The front side, where the IC is mounted, faces the Stripline, while any other components, such as decoupling capacitors, are on the back side, away from the Stripline (Fig. 17.3). This setup helps isolate the IC's emissions, but does it accurately reflect the system-level implementation if the PCB design is different? Often, it does not, as the current loops that radiate change in dimension and orientation relative to the measurement antenna.

Simulations can demonstrate that if the PCB design in the device-level test more closely resembled the real application (customer PCB), the emissions levels would be much higher in the device-level test. This would better correlate with the emissions failure observed at the system level.

This is just one example of the gaps between different testing levels. Another significant gap is assessing the worst-case coupled pin voltages due to RF immunity tests at device level (direct power injection (DPI)), bench level and vehicle level (BCI and radiated immunity (RI)). Ideally, IC-level immunity tests would closely match the worst-case levels seen in bench- or vehicle-level tests, but there are unfortunately still some gaps due to a combination of test methodology and underlying physics.

To elaborate, cable length and grounding-related resonances can easily result in 2–3x the target injected current levels in a BCI (substitution method) test at certain frequencies, which may not align to the worst-case DPI ingress. Simulation can play a key role in optimizing device-level EMC testing procedures to better reflect system-level behavior. Additionally, simulations can help improve and develop future EMC standards at all levels. Investigating these topics through simulation is part of the initiative to translate system-level EMC requirements into IC design targets.

17.1.5 Conclusion

The times of designing an IC in isolation, without considering EMC, are long gone. Engineers need to be executing an EMC-aware IC design flow to ensure customers can meet their EMC requirements. Failing to do so can lead to significant risks including delays in schedules, increased project costs, and being too late to market.

Being EMC-aware is impossible without EMC simulation. EM simulation tools are invaluable for deepening understanding of the fundamental physics that occur within designs, at the IC level and beyond. Over the years, the models have challenged understanding and prompted new ways of thinking. They have imparted valuable lessons, providing new rules of thumb and best practices that can be applied to future projects, leading to better efficiency. The value of these tools cannot be overstated.

17.2 BMW AG

17.2.1 About the Author: Sergey Kochetov

Dr. Sergey Kochetov is an experienced researcher and engineer, who began his academic career at the St. Petersburg State Polytechnic University. There, he completed his M.S.E. in 1996 and his Ph.D. study in 2000, respectively. He gained initial experience in electromagnetic simulation already in 1992 during his work as a student assistant at the Department of Theoretical Electrical Engineering and stay on this way until today. His thirst for knowledge led him to the École polytechnique fédérale de Lausanne in 1997 and later to the University of Magdeburg, where he deepened his knowledge and completed his habilitation in 2008 in the field of theoretical electrical engineering with a specialization in EMC modeling.

These academic achievements paved the way for his industrial career at Robert Bosch GmbH, where from 2008 to 2012 he applied his expertise in practical EMC simulation on the component level. In parallel, Dr. Kochetov shared his knowledge with new generations of engineers as a lecturer at the University of Stuttgart.

Since 2012, he has been contributing his expertise in EMC architecture and system design for the entire vehicle with BMW AG in Munich, where he applies his experience in EMC simulation to solve practical problems in the automotive development.

Dr. Kochetov's recognized scientific work includes more than 50 conference and journal publications, some of which have been awarded, underscoring his reputation as an expert in his field. Throughout his career, Dr. Kochetov has successfully bridged the gap between industrial applications and academic research, enabling him to make contributions that are both practically relevant and theoretically sound.

17.2.2 EMC Simulation in the Automotive Development

The development of automobiles involves a detailed and collaborative process that moves a car from an initial abstract concept to the beginning of mass production, known as the Start of Production (SOP). This process is crucial to ensure that all the main features and functions of the new vehicle meet the quality standards expected by customers and set by the car manufacturer, also known as the Original Equipment Manufacturer (OEM). It also ensures compliance with safety regulations and legal requirements for the product.

An important feature of a vehicle is its electromagnetic compatibility (EMC). This refers to the car's ability to meet various legal requirements and national standards (e.g., UN ECE-R10 and GB/T 18387), as well as the quality requirements of customers. These standards are concerned with how much electromagnetic interference the vehicle and its parts emit and how they are affected by external electromagnetic fields. OEMs usually set the EMC customer quality standards by considering their business goals and international guidelines (e.g., CISPR 25, ICNIRP, etc.).

In the following, we make a deep dive into the application of EMC simulation within the automotive industry, examining the contexts in which it is employed and identifying the typical solution methods, domains, and techniques. This discussion intentionally omits specific references to software vendors to maintain an unbiased perspective.

The automotive development process typically follows to the V-cycle model, which delineates the shared responsibilities between OEMs and its Tier 1 suppliers. Within this framework, EMC simulation appears as an important method for analysis and virtual testing across various stages of development. We will explore the general timeline of successive development for multiple vehicle generations in the relation to EMC simulation. The development cycle is generally bifurcated into two phases, the introduction of a new vehicle generation and the mid-generation refresh, often referred to as a facelift. An illustrative example provided in Fig. 17.4 demonstrates how the development of the "gen2" generation benefits from synergies with both the preceding "gen1" and the forthcoming "gen3," using models of chassis and components from these generations. This approach underscores the iterative nature of automotive design and shows the benefits of EMC simulation.

The development process for a new vehicle starts with the Start of Development (SoD), which kicks off the early phase. This phase focuses on creating concepts and outlining the specifications for components and the vehicle's overall package. The EMC simulation is a valuable estimation method during this stage, it helps to estimate how components will behave once integrated into the vehicle. The EMC

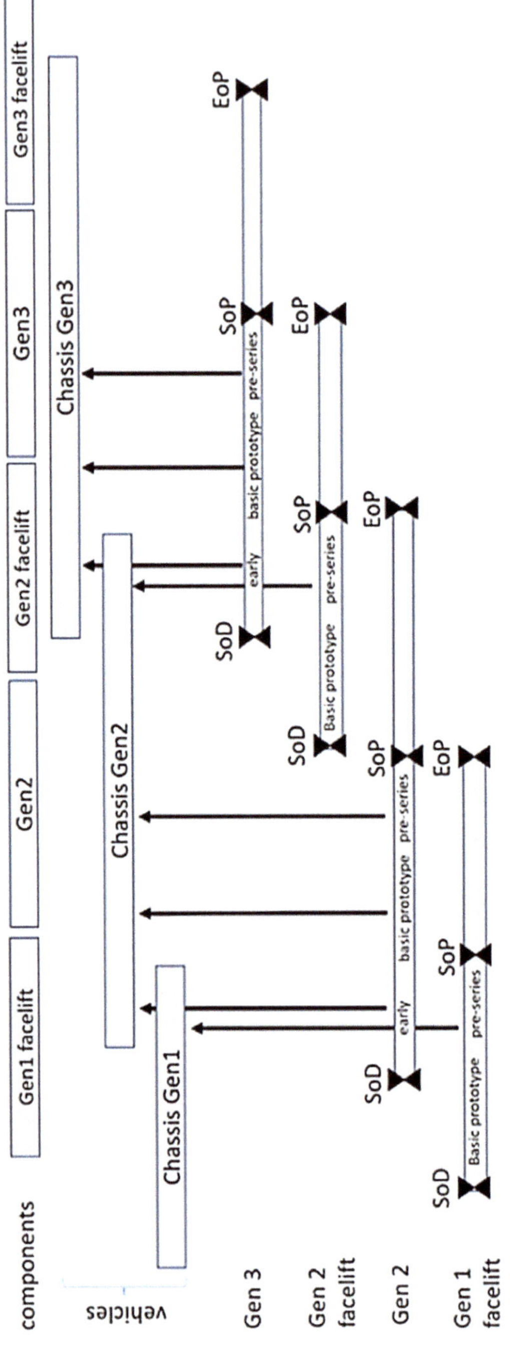

Fig. 17.4 An abstract life cycle of an EMC model in the automotive development

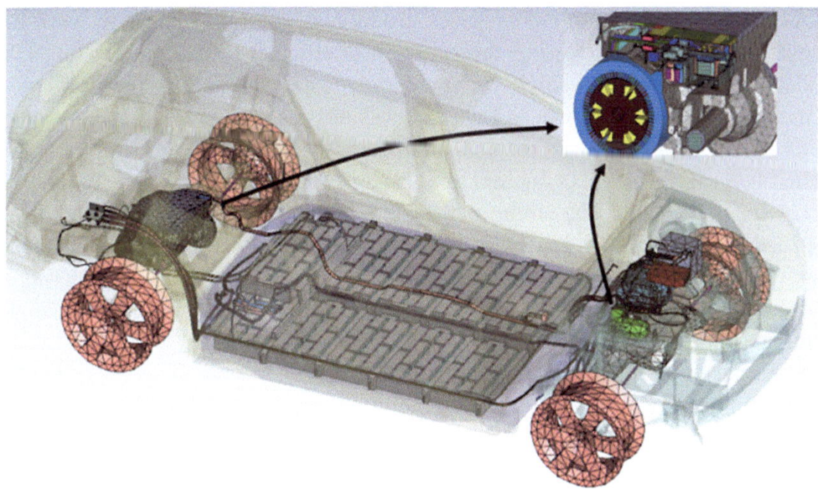

Fig. 17.5 A full-wave 3D EMC model of the HV power train system in a BEV vehicle

component requirements are related to component-level setups that are designed to check their compliance with EMC norms. However, once these components are assembled into the vehicle's package, they might behave differently and produce unexpected results. This discrepancy is where EMC simulation is most beneficial. It allows engineers to scale vehicle-level requirements down to the component level, considering how components will interact within the full vehicle package. For a new generation or concept in the early phase, it's practical to use EMC simulation with early prototype models of the chassis or even with a modified version of an older chassis. These models should be coupled with validated component models from the previous generation. For instance, the EMC model of a battery electric vehicle (BEV), as shown in Fig. 17.5, was used precisely for this purpose. This model incorporates a validated 3D full-wave model of a high-integrated powertrain unit, which includes the inverter, gearbox, and electric drive from the earlier generation. The new generation of vehicles introduces significant changes to the board network, chassis, and powertrain system. By virtually integrating components from the older model into the prototype chassis and cable package of the new vehicle, engineers can conduct comparative simulations. These simulations are valuable to see advantages and disadvantages of the new package and facilitate early-stage optimization of the EMC concept.

Following the early phase, the V-cycle progresses to the development of the basic vehicle prototypes and their components, and then to the pre-series vehicle prototypes. At this point, the chassis model of the new generation is refined and consistent, which allows us to validate the EMC models through EMC tests. These models are particularly useful for diagnosing the root causes of EMC issues and finding solutions to these problems. This phase of development culminates with the Start of Production (SoP), but the life of the EMC model does not end there.

As the new generation enters series production, the OEM starts to work on a facelift for the vehicle. During this stage, it is efficient to use the existing EMC models of the pre-series vehicle for comparative analysis. That is, the EMC model of the electric vehicle shown in Fig. 17.5 was also used to predict changes during its facelift. The vehicle went through an update, where the inverter technology was shifted from IGBT semiconductors to SiC semiconductor technology. To adapt the model for this change, the circuit part of the 3D model for the IGBT module was replaced with the circuits for SiCs, while the 3D model itself remained unchanged. This simulation provided a relative comparison of the emissions, considering the higher switching speeds of the SiC technology in the very early project phase with minimal costs.

The final stage in the life cycle of an EMC model is when it serves as the base for building of the model of the next generation, thus completing the development loop. This cyclical approach ensures that each generation benefits from the insights and data gathered from the previous one, fostering continuous improvement and innovation in vehicle design and EMC performance.

A vehicle is a sophisticated object that includes the body in white, various mechanical parts of the chassis with intricate electromagnetic properties, numerous kilometers of cable harness, about 100 electronic control units (ECUs), and hundreds of sensors and actuators operating in different modes. Simulating the electromagnetic compatibility (EMC) of all these components together at the vehicle level is not realistic. Instead, it is feasible to simulate individual or multiple systems integrated into the vehicle with a representative part of the harness, which is referred to as the "vehicle-level EMC model." The typical examples of such models are "the electric powertrain system in a vehicle," "the radio antenna system in a vehicle," and a model of the electromagnetic interference between these systems in a full vehicle configuration.

Moreover, the mentioned above vehicle-level models remain sophisticated and may be calculated only by means of two paradigms: the hybridization of numerical methods and the hierarchical partitioning of models

In the following, we will explore established techniques for hybridizing computational methods in electromagnetics. These techniques are often associated with commercial software packages that integrate these methods into their solution workflows as proven hybrid combinations. The widespread hybrid solvers for automotive applications use a full-wave 3D differential method, i.e., FDTD, FIT, and TLM in the time domain, or FEM in the frequency domain. It is intended for modeling the chassis, antennas, electrically large components, and test setups. The cable harness, on the other hand, is typically modeled using transmission-line theory due to its elongated structure and the need for analyzing signal propagation along its length.

Circuit-level models are integrated into a network simulator coupled to a 3D solver in order to take into account the components that significantly affect high-frequency (HF) behavior. Certain critical components may also be granted their own 3D models, which are then incorporated into the overall vehicle 3D model. When it comes to printed circuit boards (PCBs), which can be exceedingly complex, direct full-wave modeling may be impractical. Therefore, the tool chain is supplemented with a PCB solver that employs a 2.5D method, such as transmission-line theory or the partial element equivalent circuit (PEEC) method. All the mentioned methods need to be coupled, which is typically achieved using a system or circuit solver that can handle the interaction between different simulation environments.

The second widespread type of hybrid solver utilizes an integral equation electromagnetic method instead of the differential full-wave solver. The most common approach here is the Method of Moments (MoM), and a less common but still relevant technique is the PEEC method.

Comparing the abovementioned tool chains is not particularly meaningful as both are highly developed and can be applied to a broad range of problems. However, there are certain tendencies in their application. Differential methods, such as FDTD, FIT, TLM, and FEM, are better suited for setups with relatively small volumes and denser constructions. These methods solve Maxwell's equations in differential form, typically using a grid that discretizes the problem space, and they are well-adapted to handle complex material properties and intricate geometries. On the other hand, integral methods, like MoM and PEEC, may be preferable for scenarios involving large volumes, often on the scale of ten times the wavelength or more, and are particularly effective for modeling antennas. Integral methods solve Maxwell's equations in integral form, which can be more efficient for problems where the region of interest is surrounded by a large empty space. However, the true measure of efficiency in electromagnetic simulation does not rely solely on the choice of method but also significantly on the strategy employed in constructing the model, particularly in terms of hierarchical partitioning.

This approach involves decomposing the overall problem into smaller, more manageable pieces, which can be solved individually and then recombined to form the complete solution. By intelligently segmenting the model and applying the most appropriate method to each segment, simulation efficiency and accuracy can be greatly improved, regardless of whether differential or integral methods are used.

The paradigm of the hierarchical partitioning means decomposing the overall problem into smaller, more manageable pieces, which can be solved individually and then coupled in the complete solution. By intelligently partitioning the model and using the most appropriate method to each part, simulation efficiency and accuracy can be greatly improved, regardless of whether differential or integral methods are used. In the case of a vehicle EMC model, we should use three levels of partitioning:

- vehicle and system level,
- component level (e.g., ECU, sensor, actuator),
- sub-component level (e.g., PCBs, ICs).

The EMC models on the sub-component level are developed based on measurements, EDA and CAD data. They must be validated and may be quite sophisticated, but this complexity may be reduced by model order reduction (MOR) methods, i.e., vector fitting. Thus, the simplified sub-component models flow into the next hierarchical level (component models). This makes the component models feasibly complex. The component models must also be validated, typically through a series of tests on a component test bench, to ensure they reflect real-world behavior. Once validated, they are similarly condensed using MOR methods. In the vehicle model, attention is concentrated on the chassis, cable harness, and environmental setup. The complexities of the remaining model are encapsulated within the MOR circuits, serving as proxies for the actual sub-components and components. A key advantage of this approach is the seamless integration of models developed by tier 1 suppliers. This standardization allows for efficient collaboration across the automotive industry, where the assembly of different suppliers' components into a unified vehicle-level EMC model is commonplace.

The next discussion topic pertains to the data flow that facilitates EMC simulations at the vehicle level. Here, we focus exclusively on data specific to vehicle-level simulations. A vehicle is characterized by CAD data, which encapsulate comprehensive constructive details, including materials, fine structural features like screws and clinchers, and history lists of all modifications. The most of this information is superfluous for EMC simulation purposes, and its huge volume of data can be enormous.

Despite the ability to import these extensive CAD datasets into any commercial tool for use as a geometric foundation for EMC models, this approach is quite effort intensive. A more widespread method uses synergies with the mechanical crash simulation field.

OEMs have applied the mechanical crash simulation since for a long time, they mesh chassis from CAD data through a standardized process. By adopting these existing meshes, EMC simulations can bypass the original CAD data, which accelerate and simplify the setup process. This simplification is illustrated in Fig. 17.5, where the discrete geometric model of the chassis, imported as a mesh, is integrated alongside analytical CAD objects (such as stp or igs files), which are then meshed within the simulator (e.g., antenna structure in Fig. 17.6). It's important to note that the comprehensive CAD data, while unrelated to the electronics domain, consist of purely mechanical elements that nonetheless have a significant impact on the electromagnetic behavior of the vehicle.

The second component of vehicle data is the onboard network, which includes the cable harness and various components. These datasets are often created with specialized system design tools that operate with proprietary formats. To facilitate data exchange for the vehicle harness, the German Association of the Automotive Industry (VDA) has introduced a specialized format known as KBL. System development tools are capable to export the complete onboard network system diagram in the KBL format.

The KBL format stores comprehensive geometrical details of the harness setup, detailed information on every wire within the cables, including data sheets, circuit connectivity details, as well as information about connectors and shielding. These KBL files can be imported into an EMC simulator, where the representative sections of the harness are translated into transmission line models and incorporated into the

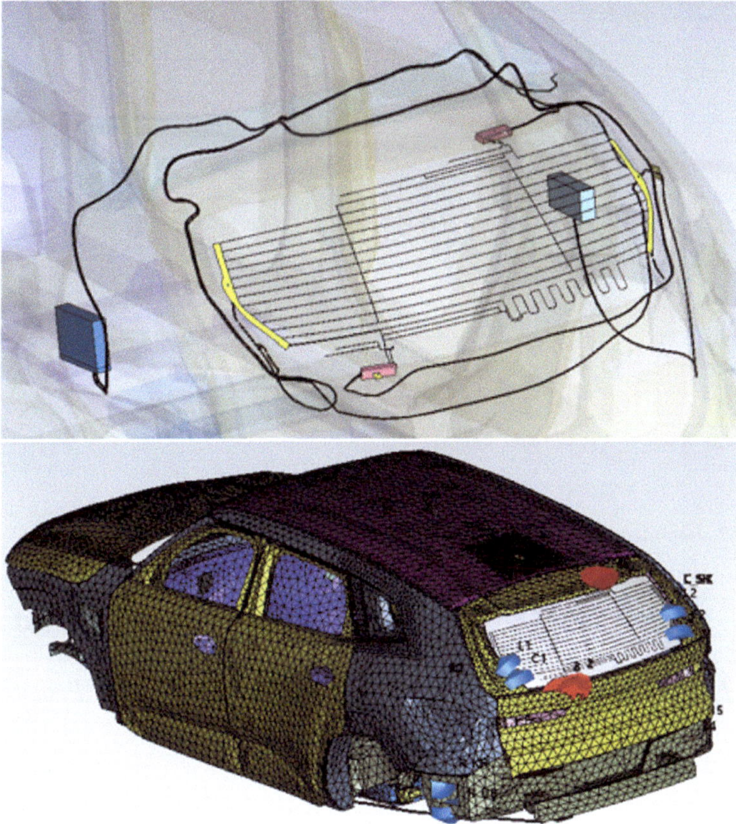

Fig. 17.6 EMC model of a radio antenna system integrated in the vehicle model

simulation. Some tools are even equipped to perform this conversion automatically, as illustrated in Fig. 17.7

In conclusion, an EMC model of a vehicle is composed of a 3D full-wave model that includes the chassis, antennas, various interconnection structures, and the surrounding environment. This model is interconnected with an EMC system model, which integrates the transmission-line model of the cable harness and couples it to the full-wave model. Additionally, this system encompasses a lot of circuit or behavioral models of various components. Such a comprehensive level of modeling is actively employed in automotive development, yielding results of significant value.

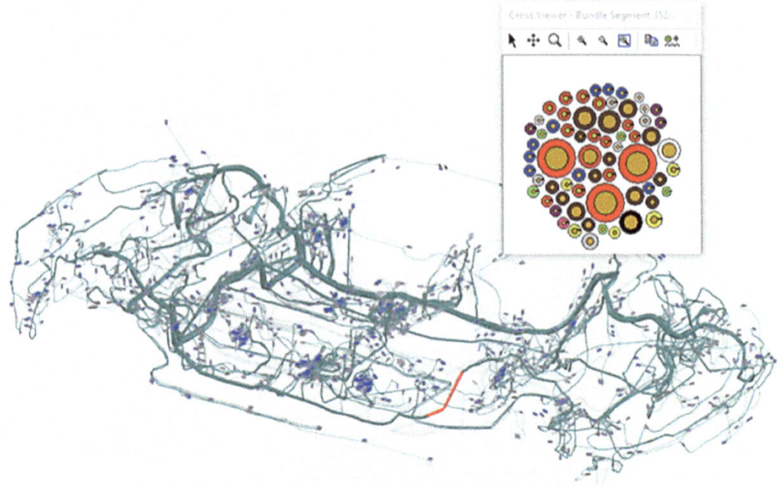

Fig. 17.7 Imported cable harness of a vehicle and a cross section of a cable

17.3 Bosch Global Software Technologies Private Limited

17.3.1 About the Authors: Sayantan Dhar and Bibhu Prasad Nayak

Sayantan Dhar was born in Calcutta, in 1983. He received the B.E. degree from Visvesvaraya Technological University, Belgaum, in 2006 and the M.E.Tel.E. and Ph.D. degrees from Jadavpur University (JU), Kolkata, in 2010 and 2015, respectively. In 2013, he joined Heritage Institute of Technology as an Assistant Professor. Since 2019 he is associated with Bosch Global Software Technologies, Bengaluru, India, where he is currently working as a Technical Lead. He has authored or coauthored more than 20 technical papers. His recent research interests include system simulation, electromagnetic interference/EMC, machine learning in EMC, and antenna design/analysis.

Bibhu Prasad Nayak was born in Cuttack, India, in 1980. He received the B.E. degree in Electronics and Telecommunication Engineering from the University College of Engineering, Sambalpur University, Orissa, India, in 2002. He earned the M.Tech. degree in Microwave and Photonics from IIT Kanpur, India, in 2005, and completed his Ph.D. in Electrical Communication Engineering at the Indian Institute of Science, Bengaluru, India, in 2021. From 2005 to 2007, he worked with the Indian Space Research Organization (ISRO) in Bengaluru. From 2007, he was been engaged in electromagnetic compatibility (EMC) simulations for automotive products at Bosch Global Software Technology, Bengaluru, India. His current research interests include 2.5D methods, hybridization of computational techniques for system simulation, electromagnetic interference/EMC, and the application of machine learning in EMC. Since 2025, he works as Senior Postdoc at the Institute of Electronics, Graz University of Technology.

17.3.2 EMC Simulation at an Automotive Parts Supplier

The automotive industry compared to the IC industry has experienced a silent but remarkable transformation over the years. The integration of electronic components into vehicles has increased at an exponential rate since the 1990s, leading to a shift toward a more electronic-driven automotive landscape. This transformation has been largely driven by advancements in automotive electronics, which have played a pivotal role in reshaping the industry. With the advent of hybrid electric vehicles (HEVs), plug-in hybrid electric vehicles (PHEVs), and battery electric vehicles (BEVs), the number of electronic components ranging from sensors to electronic control units (ECUs) and vehicle computing units (VCUs) has been steadily on the rise. These electronic devices operate in an environment with high switching currents and voltages, compounded by the lengthy cables that supply power and facilitate critical communication between various modules of the vehicle. This creates a system that can generate and/or be vulnerable to electromagnetic interference (EMI).

Typically any module or the entire vehicle undergoes rigorous tests to be declared compliant with EMC regulations. Despite thorough efforts to design a compliant product, numerous variables often make it challenging to perfect the design on the first attempt. It is crucial to remember that achieving a fully functional product requires the collaboration of multiple teams, including hardware, software, mechanical, reliability, and EMC specialists, among others. As a result, finding an optimal solution that satisfies all functional teams is inherently challenging. Generally, products are refined through multiple design iterations. Each iteration addresses the shortcomings of the previous design and aims to meet the target in the shortest design cycle possible. However, testing each product after a design cycle and addressing its deficiencies requires significant time and financial resources. Given the intense competition, a crucial factor is the speed at which you can bring your product to market. Simulations play a key role in this process. They offer valuable insights into potential failure mechanisms and can predict the behavior of components, thereby reducing the need to develop and test multiple prototypes.

System-level simulation is increasingly crucial in addressing EMC issues in the automotive industry. The complexity and diverse geometric scales involved in automotive simulation present significant challenges, ranging from fraction of wavelengths to multiple wavelengths. The time and memory requirements for solving these problems using complete 3D full-wave accuracy become prohibitive. Therefore, efficient simulations can only be performed successfully by decoupling the simulation domain hierarchically and employing suitable simulation techniques for each domain. A typical EMC test setup includes a device under test (DUT), cable harness, termination, and measuring probes such as current clamps, antennas, and line impedance stabilization networks (LISNs). The corresponding modeling environments involve full-wave 3D resolution at the DUT level, transmission-line methods at the cable harness, and circuit formulations for the passive components and termination. However, obtaining accurate electrical parameters, such as physical dimensions and frequency-dependent material properties of these devices from manufacturers, can be challenging.

In such scenarios, measurement-based behavioral models become a feasible and alternate way of modeling. Accurate modeling of non-linear components, such as integrated circuits (ICs), bipolar junction transistors (BJTs), and diodes, is crucial for emission and immunity measurements. The sources of EM noise in emission are mainly the high-speed clocks of communication ICs or the switching noise from the converter ICs. Immunity failures can also be attributed to the malfunction of the ICs under the exposure of EM noise, conducted either through the supply pins or exposure to the high EM field. This emphasizes the need for EMC behavioral models for ICs, along with other subsystem models, for the EMC prediction of systems. To simulate such a complex electronic system with design constraints for EMI/EMC, a robust simulation framework aided by judicious selection of solver technology is needed. The complexity, diversity, and multi-scale nature of the system require the use of hybrid solvers, where a harmonious integration of solver technology is carried out to exploit optimal speed of solution at tolerable levels of accuracy. While volume-based methods such as finite-difference time domain (FDTD) and finite-element methods (FEM) are more suited to model-guided wave-based EMI/EMC tests, surface-based formulations such as Method of Moments (MoM) are ideal techniques to address open-ended radiation problems. Therefore, a combination of such techniques may be hybridized to effectively solve a combination of EMI/EMC diagnostic scenarios. In addition, realistic practical electronic systems house a plethora of active and passive components, terminations, sources, and port parameter models, which may require circuit theory-based simulation techniques and robust circuit solvers. This comprehensive approach is essential for ensuring the EMC compliance and reliability of automotive electronic systems.

With the increasing complexity of electronic control units (ECUs), there is a growing trend toward utilizing machine learning-based simulation approaches in electromagnetic compatibility (EMC) testing. One such potential application is the use of machine learning to study radiated emission from shielded cables. As ECUs become more intricate and interconnected, traditional simulation methods may struggle to accurately capture the complex interactions and behaviors within the system. Machine learning offers a promising alternative by leveraging advanced algorithms to analyze and predict EMC phenomena with greater precision and efficiency. In the context of radiated emission from shielded cables, machine learning can be employed to model the intricate electromagnetic interactions between the cable, its shielding, and the surrounding environment. By training the model on a diverse set of data, it can learn to accurately predict the radiated emissions under various operating conditions, leading to more reliable and comprehensive EMC assessments. Furthermore, machine learning-based simulation approaches can also enable the identification of potential EMC issues earlier in the design phase, allowing for proactive mitigation strategies and ultimately reducing the time and cost associated with EMC compliance testing. Overall, the integration of machine learning into EMC testing represents a significant advancement in addressing the challenges posed by the increasing complexity of ECUs, offering a more sophisticated and data-driven approach to ensuring electromagnetic compatibility in modern electronic systems.

Traditional simulations for shielded cables often rely on 2D or full 3D simulations, but these methods may fall short in accurately modeling the complexities of a real harness. Shielded cables encompass a multitude of parameters that can significantly impact radiated emissions (RE). Gaining a comprehensive understanding of these parameters is crucial for effectively selecting a cable that minimizes emissions. The intricate nature of shielded cables, including factors such as conductor layout, shielding effectiveness, cable geometry, and the presence of connectors, makes it challenging for traditional simulations to capture the full spectrum of interactions and their impact on RE. As a result, there is a growing recognition of the limitations of conventional simulation methods in providing a holistic understanding of cable behavior in real-world scenarios. By delving into the diverse parameters that influence RE in shielded cables, engineers can gain valuable insights into the complex interplay of electromagnetic fields, material properties, and geometric configurations. This deeper understanding facilitates informed decision-making when selecting cables, enabling the identification of options that offer the lowest emissions for a given application. Moreover, the utilization of advanced modeling techniques, such as machine learning-based approaches, holds promise in effectively capturing the intricate relationships between the myriad parameters and their collective influence on RE. By leveraging these cutting-edge methods, engineers can enhance their ability to predict and optimize cable performance in terms of electromagnetic compatibility. In summary, a comprehensive grasp of the multifaceted parameters governing RE in shielded cables is pivotal for making informed choices in cable selection, and the integration of advanced simulation techniques can offer a more nuanced and accurate representation of cable behavior in EMC scenarios (see Chap. 16, Patra et al. 2023).

17.4 Meta Reality Labs

17.4.1 About the Author: Deepak Pai

Deepak Pai earned his master's degree from the University of Texas at Dallas, Richardson, TX, USA, in 2014. Currently, he serves as an integral part of the wireless at Meta Reality labs in Sunnyvale, CA, USA, focusing on Desense design in the Smart device product line up. He has over 10 years of experience in the consumer electronics industry—prior to joining Meta Reality labs, he held significant roles as a Wireless RF engineer at Amazon Lab 126 and an antenna engineer at Starkey

Labs. Throughout his career, Deepak has made significant research contributions and is recognized as an author of multiple IEEE conference papers. Furthermore, his innovation and problem-solving skills have led to the granting of multiple US Patents in the fields of EM, desense, and antenna design. Deepak's research interests encompass a wide range of topics, including RFI, near-field EM, computational optimization algorithms, design of experiments, noise-aware antenna design and miniaturization, NF-FF transformation, and EMC simulation strategies.

17.4.2 Need for Predictive Analysis

Modern hardware product development demands unprecedented speed, driven by compressed timelines and market pressures that necessitate elegant, efficient designs. This universal imperative spans industries, compelling teams to deliver superior solutions that optimize performance, cost, and time-to-market parameters.

The consumer electronics sector faces a distinctive challenge: executing rapid go/no-go decisions through accelerated prototyping and strategic failure analysis. This methodology enables real-time refinement of design strategies, aesthetic language, and product roadmaps. Success hinges on developing sophisticated electrical systems and wireless architectures that satisfy stringent electromagnetic compatibility and wireless performance criteria. As contemporary development cycles preclude traditional iterate-and-test approaches, electromagnetic simulations have become fundamental to the design process.

This evolution mandates advanced predictive analysis, precise correlation, rigorous validation, and accelerated design iterations. Multi-fidelity simulation analysis enables comprehensive risk assessment across board, system, and device levels. Through sophisticated simulation methodologies and optimized workflows, teams can systematically minimize risk factors while providing decisive guidance on requirement feasibility. This framework accelerates development while establishing precise technical parameters, enabling designers to achieve targeted objectives with maximum efficiency.

Design workflows now intrinsically incorporate simulation analysis across multiple dimensions: from comparative A/B evaluations and targeted scenario testing to comprehensive end-to-end analysis utilizing maximum-fidelity models for PCB, mechanical, and electrical components. While confidence levels vary across simulation types, these workflows continuously evolve through enhanced understanding of specific aggressors and unintentional radiators, facilitating increasingly refined system optimization.

17.4.3 Products, Sensors, and Interfaces

Consumer electronics manufacturers and designers pursue the creation of feature-rich experiences that integrate seamlessly into customers' daily routines, delivering comprehensive solutions across entertainment, security, and convenience applica-

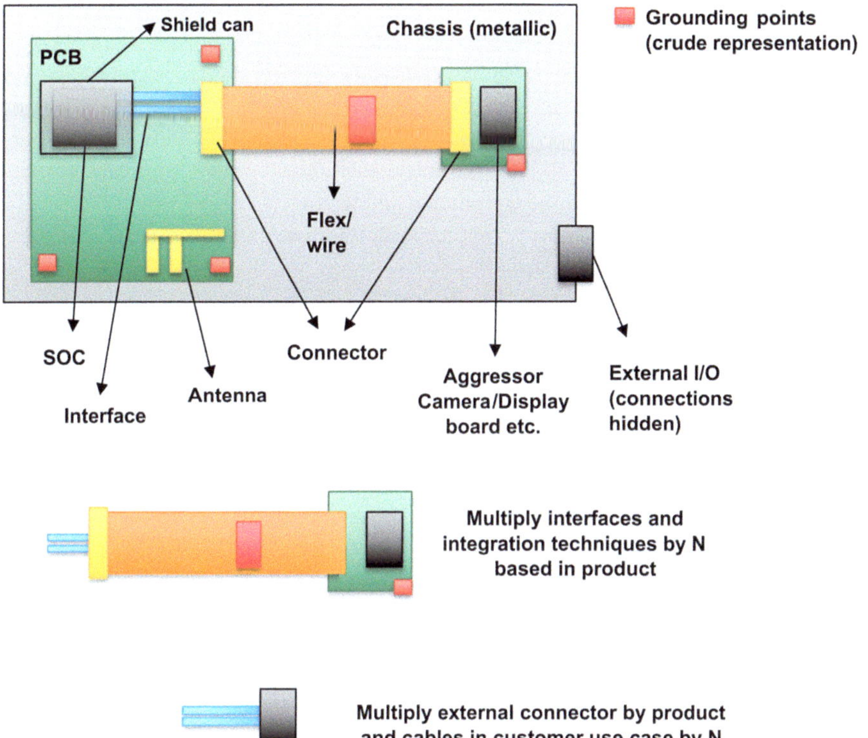

Fig. 17.8 Simplified example of a generic smart device and interconnects

tions. This commitment to enhanced functionality has driven the proliferation of sophisticated input and output sensors throughout modern devices.

The resulting system architecture encompasses an intricate network of synchronized components: internal and customer-facing sensors, high-speed connectors, and output mechanisms operating in parallel. While smart devices share core architectural elements—including System-on-Chip (SoC), memory subsystems, and I/O interfaces—specialized products incorporate distinct sensor arrays optimized for specific use cases. These arrays may feature high-resolution cameras, advanced display technologies, and precision motor systems, all orchestrated to deliver a cohesive user experience.

The high-speed operation of these sensors, whether functioning independently or in conjunction with long cable runs, can generate EMI (electromagnetic interference) and RFI (RF interference) affecting the devices' wireless systems. Although EMI/EMC evaluation typically occurs at the product level, detailed analysis of individual sensors and their system integration is crucial. This integration encompasses mechanical considerations, electrical requirements, and interface specifications to ensure optimal performance (Fig. 17.8).

Interface and sensor performance optimization requires a dual-pronged analytical approach leveraging both theoretical knowledge and empirical data from previous generations. The first approach involves comprehensive "what-if" scenario analysis to validate existing designs against new constraints and requirements. The second approach establishes baseline performance metrics for novel interfaces through systematic risk assessment and correlation matrix development, validated through iterative hardware prototype measurements and testing.

Design optimization scenarios require systematic evaluation through structured workflow analysis, exploring variables such as:

- Ground point modifications (e.g., reducing connection points from 6 to 4, relocating grounds by 2 mm)
- Metallic surface adjustments (e.g., extending shield length by 0.5 mm, increasing trace width by 0.1 mm)
- Interface signal modifications (e.g., boosting drive strength by 3dB, adjusting termination resistance by 10%)

Each analysis scenario employs tailored workflows that align with product development milestones. Engineering judgment determines appropriate confidence levels based on design criticality and available validation windows. For instance, early-stage architectural decisions may require higher confidence levels with extended analysis periods, while late-stage optimizations might accept moderate confidence levels to meet production schedules and supplement with measurement datasets.

17.4.4 EMC Modeling and Simulation Analysis

A critical consideration in simulation analysis, beyond the timing factors discussed earlier, is the available computational resources. While a comprehensive, high-fidelity simulation represents the ideal scenario, organizational constraints often limit such capabilities. Though technically feasible, executing full-scale simulations may not yield a positive return on investment within required time frames. Therefore, engineers must conduct strategic trade-off analyses, carefully evaluating interfaces and aggressors to determine appropriate analytical depth for each simulation level.

In EMC analysis, every problem can be fundamentally decomposed into three key components: the emission source, the coupling path (including unintentional radiating elements), and the receptor (whether a victim antenna or compliance measurement antenna). This systematic breakdown enables engineers to strategically focus their efforts on the elements where design control offers maximum impact, leading to optimized solutions and more efficient problem resolution.

When evaluating simulation results, the focus should extend beyond simple dB value comparisons against compliance requirements. The optimal approach is to develop comprehensive emission profiles of the interface or problem area, recognizing that measurement data remains the authoritative reference. While simulation analysis provides valuable insights, it serves as a complementary tool to physical measurements. Engineers can construct detailed emission profiles through various methods, including:

- Implementation of Huygens' surfaces or volumes
- Analysis of unintentional radiators using antenna parameters (gain, NF, and FF patterns)
- Strategic placement of probes within EM simulator tools for field strength assessment

These methodologies enable robust correlation between simulation and measurement data, supporting effective product design decisions. Some tools offer calculators to use the data outputs and scale results to an EMC plot. Once a perspective has been established for analysis, specific scenarios can be implemented to mitigate unwanted emissions at each fundamental level.

17.4.4.1 Source Analysis

Source modeling, though inherently complex, benefits from standardized excitation models available in modern simulation tools for common interfaces. Engineers can leverage circuit and 3D co-simulation capabilities to evaluate source emission profiles using simplified CAD structures that incorporate critical metallic elements and grounding points. This early analysis serves multiple purposes. Through this approach, engineers can identify critical grounding point impacts and enable proactive filtering strategy development. The analysis also highlights opportunities for source manipulation while facilitating early-stage design trade-off evaluation. While this approach may not provide definitive emission solutions, it establishes a robust framework for strategic design decisions during initial development phases.

Strategic port placement at critical interface points, such as connectors and key transitions, enables incremental enhancement of model complexity and fidelity. This methodical approach allows engineers to progressively refine their analysis from interface origin to completion, implementing targeted solutions at both mechanical integration and component levels as needed. The systematic evaluation of multiple interface points facilitates comprehensive optimization opportunities throughout the design hierarchy.

17.4.4.2 Unintentional Radiators

When source analysis proves insufficient due to design complexity or modeling limitations, engineers can employ an alternative strategy using standardized excitation sources combined with high-fidelity modeling. This approach focuses on comprehensive analysis of unintentional radiator interfaces and their associated elements, including PCB and flex stackup, contoured heat sinks, and shield cans.

This methodology provides a distinct analytical perspective while accommodating system uncertainties through appropriate source approximations. The approach excels as a comparative analysis tool, enabling engineers to evaluate the relative impact of various design choices on emission characteristics. Such comparative insights guide product design decisions until additional validation through measurement data becomes available.

17.4.4.3 Comprehensive End-to-End Analysis

End-to-end analysis represents the most sophisticated simulation approach, typically reserved for complex designs where development timelines permit extensive evaluation or where product delivery carries significant implications. This comprehensive methodology integrates source analysis and reconstruction with detailed modeling to create highly accurate representations of physical devices.

The approach encompasses several advanced capabilities:

- Integration of source analysis with complete model details
- Extensive parameter evaluation and output analysis
- Tolerance analysis of emission-critical components
- Circuit and 3D electromagnetic co-simulation with full design details

This thorough analysis methodology achieves maximum predictive accuracy and can be further refined through correlation with measurement data from specific subassemblies and interfaces.

17.4.4.4 Strategic Implementation

Simulation tools and methodologies represent essential instruments in the RF/EMC engineering design process. Their strategic and selective application can dramatically improve design cycles and outcomes. Success in implementation hinges on engineering expertise to identify optimal simulation opportunities and understand inherent limitations. Professional judgment remains the cornerstone for managing dynamic design changes and adapting to real-world engineering challenges.

Chapter 18
Open Problems and Research Questions

In this book we highlight and detail the many cases where simulation supports EMC design. In the chapters we also describe the limits, beyond which simulation cannot yet contribute. In this final chapter, we give a list of open problems of EMC simulation. Some of them may be straightforward to solve, others are challenging and may require years of research. Some are long known and need a brief sketch only; others are involved and need detailed background knowledge. In particular for the latter we pose research questions, which shall give an orientation on how to solve these problems. The present list is based on experience and personal favors. It is by no means complete, but it can serve, e.g., PhD students, as a guideline to identify interesting research topics.

18.1 Component Model Availability

Throughout this book, we use and demonstrate how an EMC simulation model of an electronic system is created using coupled EM and circuit models. We use circuit models of components obtained from vendors' webpages. As discussed in Chap. 2, the availability and quality of circuit models from vendors varies significantly. The availability of models suitable for EMC simulation is a long-prevailing issue. This is in parts a matter of research, but also a matter of standardization and guarantee of quality of a defined set of properties.

Robust Transistor Models for EMI Assessment While some transistor vendors provide robust and accurate simulation models; others either do not offer models at all or supply models that result in long simulation times due to the way the transistor is represented in the SPICE netlist. Available models typically describe functional behavior, as needed for tasks such as switching loss extraction. Parasitic elements required to model the transistor as a source of emission are often added

J. Hansen, A. Barchanski, *EMC Simulation of Electronic Systems*,
https://doi.org/10.1007/978-3-032-09237-3_18

separately. For EMC simulations, however, higher accuracy at elevated frequencies is necessary.

Current switched power transistor models operate in the time domain. To analyze emissions, their time-domain response must be transformed into the frequency domain. However, as rise and fall times become faster, high-frequency emissions increase (cf. Fig. 2.5), making these models more difficult to use. For example, predicting emissions at 1 GHz from a power converter switching at 100 kHz requires accurately capturing the amplitude of the 10,000th harmonic. Time-domain models are not designed for this purpose, and such simulations require extremely small time steps, leading to long simulation durations. It remains an open research question how reliable frequency-domain models can be developed to predict emissions from switching transistors in the range of several hundred MHz to beyond.

Availability of IBIS Models The IBIS syntax was developed to facilitate the electronic exchange of modeling data between semiconductor vendors and simulation engineers. Although the IBIS Open Forum began its work decades ago, the availability of models that can be directly downloaded from vendor websites remains limited. We encourage vendors to invest in the development of IBIS models and to ensure these models are easily accessible to simulation engineers.

Passive Components While vendor-provided models for capacitors are relatively common and widely available, equivalent models for inductors are less prevalent. As a result, simulation engineers often have to rely on data sheets and reverse-engineer the inductor using the methodology described in Sect. 9.6. A potential improvement for the future would be the availability of validated EM models of inductors provided directly by vendors, as we demonstrate that relying on measured data alone can lead to discrepancies if not applied correctly (see Sect. 9.4).

Magnetic Materials In Sect. 9.6, we describe the reverse-engineering of apparent magnetic material properties based on measurements and the Debye model (Eq. 4.11). In Sect. 9.7, this approach is extended by incorporating a DC model of the BH-hysteresis curve, combining a parameterized BH-curve with Debye modeling based on rough estimates from literature. However, for the frequency range relevant to EMC simulation, measurement data is generally unavailable. Moreover, there is no existing model that accurately describes magnetic material behavior across a broad range of magnetization levels and frequencies. A model that links magnetic permeability to intrinsic material properties is also lacking. These challenges have remained unresolved for a long time.

Cable Shield Properties Efficient modeling of cable shields requires transfer impedance data, as discussed in Sect. 8.5.2. Unfortunately, such data is rarely included in data sheets, which typically provide only basic information about the shield construction. Cable manufacturers could significantly improve the situation by offering more detailed and accurate data to designers and simulation engineers.

Electrical Motor Modeling for EMC Section 10.4 discusses the modeling of electrical motors and highlights the limitations of available model types, none of which are fully satisfactory. These models are typically restricted in frequency range—being accurate either at low frequencies (below the first resonance) or at high frequencies (with poor representation of the DC operating point). Reliable prediction with such frequency-limited models remains a significant challenge. Additionally, incorporating load conditions into an EMC simulation is nontrivial: time-domain simulations of motors are magnetoquasi-static, while frequency-domain simulations require fixing a single operating point (usually at 0 A).

Motor modeling for EMC presents two main challenges: (1) the large variation in geometric scales—from the coating of hairpin windings to the full size of the motor—demands substantial computational resources and high-quality CAD data. This raises the question of whether advanced modeling strategies such as multi-fidelity approaches could be applied, or whether a brute-force simulation is feasible. (2) Electric motors contain magnetic materials and laminated steel sheets. All issues discussed in the previous section on magnetic materials also apply here. It remains an open question whether an effective model can be developed that accurately describes motor behavior across the full frequency range relevant for EMC simulations.

Model Standards Models allow to digitally distribute product data between engineers in a much more detailed way than data sheets do. In this book we show the large variety of models, and how the various model types refer to various problem types. Standardized interfaces to couple simulation models are available (e.g., the Functional Mock-up Interface, FMI[1]) They are too generic to be applied between EMC simulation engineers. The exchange of EMC simulation models requires, beyond the specification of the interface, more information. For instance, the model purpose must be determined to allow the existence of models of various complexity and fidelity. Model-based test scenarios are required to allow the efficient application of these models, with the test scenario adapted to the purpose it shall fulfill. The exchange of models between industry partners in the supply chain would significantly ease the effort spent on testing.

18.2 Complex Systems

It is often the subtleness of the electromagnetic coupling paths and the availability of models that prohibits the examination of ever more complex electronic systems for their electromagnetic compatibility. There are many cases for which either economic necessity requires or the availability of novel methods allows to cross the present-day limits in the complexity of EMC simulation. Some examples are given below.

EMC Qualification by Simulation Measurements with respect to given limits still set the ultimate criterion for EMC qualification, i.e., the assessment whether

[1] https://fmi-standard.org/.

a product does or does not qualify to be sold in the market. Simulations could be an efficient alternative to measurements. But most ready-to-sell electronic systems are yet too complex to be qualified by simulation in a way sufficiently practical to compete with established measurement procedures. In particular, the standardized test setups are both complex and, in part, difficult to reproduce.

Within several companies, simulation models are already used to test for changes in an already qualified electronic system. Validated simulation models are used to predict the changes in emissions or immunity if individual components are replaced. We propose to apply simulation for EMC qualification. To reach this goal, we need to answer at least the following questions:

- How can realistic qualification scenarios look like that are accessible to the currently achievable level of modeling and simulation and simultaneously guarantee the EMC of the tested product?
- Are there any submodules of electronic systems, which can be qualified separately?

Since EMC qualification relies on the cooperation between system integrators and suppliers, this topic requires the implementation of model standards, an issue that is raised in Sect. 18.1.

Automated Simulation Model Creation Although almost the entire design process of electronic systems is carried out using CAE software, the simulation models presented in this book are primarily created manually.

The main reason for this is the highly heterogeneous landscape of CAE tools. The circuit schematic of an electronic system is typically created using an ECAD tool. When simulation models are stored alongside circuit design data, they often originate from a variety of sources and are provided in different formats, such as SPICE, Touchstone, and IBIS. Acquiring these circuit models is often a manual process. While some component libraries are available, they frequently do not cover all elements of a design. Additionally, schematics used for simulations that rely on S-parameter models differ from those used in the design process, requiring manual adaptation.

Mechanical components, such as enclosures or connectors, are designed using mechanical CAD tools. While connector models can often be integrated into the ECAD environment, positioning the PCB within an enclosure is typically done in the mechanical CAD tool—often using simplified placeholders that omit critical PCB trace details.

Cable harnesses are designed using yet another class of tools. These focus primarily on mechanical properties such as length, thickness, and bending radius. However, the electrical characteristics required for simulation—such as the transfer impedance of cable shielding—are often not included in the design data. While information mapping the ends of cable harnesses to connector pins may be available, it is typically geared toward manufacturing and may not be easily converted into a usable simulation model.

To the best of the authors' knowledge, no software tool currently exists that can handle all of the above input data and automatically generate an EMC simulation model. Despite the fragmented nature of the design tool landscape, we encourage software vendors to pursue automation in this area. Automating simulation model creation would lead to significant time savings and enable simulation engineers to perform analyses earlier in the design process.

More Automatized Vector Fitting In Chap. 10, we introduce vector fitting as a method for converting frequency-domain S-parameter data into passive and causal macromodels. The advanced section on vector fitting (Sect. 10.9) discusses error control of the system response.

Vector fitting remains an active area of research. In particular, its application to models with a large number of ports, or to models where the matrix entries to be fitted exhibit high impedance contrast, still offers room for improvement.

An enhanced vector fitting methodology could automate the selection of the reference impedance prior to the vector fitting step (cf. Sect. 10.9) by incorporating system-level information during macromodel generation. This automatic selection would save simulation engineers time, as manual testing of the system-level accuracy of the fitted model would become unnecessary.

Simulation of Systems with Multiple Emission Sources In this book we focus on electronic systems which have one, or at most several similar, emission sources. But electronic systems, such as computers, cell phones, or vehicles, are comprised of a large variety of subsystems, which each bring their own emission sources and specific properties such as emission frequency ranges, source models, or relevant coupling mechanisms. Abundant research exists of the simulation of these individual parts, but only little on the assembly of such systems out of many (say dozens) of individual subsystems.

There are scenarios in which emission is caused by not only dozens but hundreds or more sources. Even if the contribution of each individual source may be negligible, a vast number of them can produce significant emissions. The electrification of our environment with LEDs to illuminate buildings or streets easily exceeds this limit to add up thousands of individual sources. In this case, statistical approaches rather than the combination of deterministic models by prove a suitable approach for emission modeling.

We see in this respect at least the following questions:

- Which design decisions in the EMC of complex systems can be made on which level of model simplifications (e.g., decomposition of EM models to individual subsets which are connected on circuit level)?
- How many models of subsystems with acceptable quality can be combined to reproduce the EMC of the entire system so that further design decisions can be made?
- Which quantities of interest can be observed at which point of the system to derive conclusions about the system's EMC performance?

- Do universal scaling laws exist that allow to extrapolate the total emitted power computed for several sources to any arbitrary number of sources?

System-Efficient ESD Design (SEED) The SEED concept allows to simulate the ESD current path in order to investigate the interaction between the IC and its protection elements. With ever shrinking structure sizes within the IC, coupling paths, sensitivities, and failure mechanisms change and require further model improvements. Any design must ensure that the discharge pulse goes along the expected path. The simulation-based approach needs further development of:

- models of the IC robustness in power-on and power-off conditions, and other conditions that are relevant,
- models for the expected ESD type,
- models for the transient voltage suppressor (TVS) diodes
- as well as models for the interconnects.

Model Calibration/Bayesian Inference The higher the frequency, the more details have impact on the emission or immunity behavior of an electronic system. In a simulation model, this is reflected by an increasing number of model parameters that must be determined. Collecting the parameter values, e.g., for the conducted emission simulation of a power converter, is time-consuming in a way that prohibits the generation of accurate models in an industrially meaningful time frame. Models are validated based on available measurements of subsystems, and parameters are tuned based on experience and intuition until a sufficient match between measurement and simulation is achieved. This process is time-consuming, and there is no guarantee that the found parameter combination is valid when the model is used for further engineering tasks.

Model calibration, also known as Bayesian inference and, e.g., successfully applied in civil engineering, provides a rigorous method for parameter estimation. Model calibration applies Bayesian statistics to infer the most likely set of input parameters that a model should have to match an observed outcome, e.g., of a measurement.

It requires the generation of an, e.g., ML-trained surrogate model over chosen ranges of the uncertain parameters. Exploiting the computational speed of such a model, it is possible to statistically derive the probability of sets of input parameters to match a given observation.

It is open research to clarify if model calibration improves the prediction quality of EMC simulation models. In particular, we need to understand:

- How many parameters can we calibrate, and over which ranges?
- Up to which frequency does model calibration improve simulation models, and to what level?
- Which measurements are suitable, and how rich must the information derived from these measurement be to best support model calibration?

Multiphysics Simulation The coupling of multiple physical domains—such as electromagnetic and thermal simulations—based on CAD data has been established for more than a decade and is now supported by many commercial software tools. The software industry has promoted joint optimization workflows across domains; however, within the context of EMC, such multiphysics approaches have yet to achieve widespread practical impact.

A key reason may be the domain-specific nature of required modeling details. Each simulation type emphasizes different physical features. For example, EM analysis of a filter requires an accurate model of the mechanical connections for mounted capacitors but does not depend on the representation of cooling fins. Conversely, a thermal simulation needs to capture the cooling fins but can omit the capacitor model. A combined simulation model thus inherits the complexity of both domains, often without providing direct benefits for each. This added complexity reduces the attractiveness of multiphysics modeling when simulations are performed separately.

Moreover, in larger organizations where simulation engineers from different disciplines are available, these domains typically fall under separate departments. This division can hinder collaboration, knowledge exchange, and the integration of domain-specific modeling strategies.

Multiphysics considerations are more common at the level of small components, such as transistors, where thermal behavior is frequently included in the modeling.

However, integrating data from different physical domains into EM system-level models remains uncommon. Two central questions must be addressed:

- Is the physical interaction between domains sufficiently weak that the inclusion of coupled simulations in the design process offers limited added value?
- If not, how can a joint design process—e.g., combining thermal and EMC simulation—be effectively implemented? This question is closely related to the discussion on automated simulation model generation in this chapter.

18.3 Machine Learning and Artificial Intelligence

Machine learning (ML) and artificial intelligence (AI) are, despite the attention these topics have attracted in recent years, still at their infancy in EMC simulation. There is huge potential, but as so often, the path from scientific curiosity to industrial benefit is long. We list some topics that are being discussed in the scientific community and point out that the presented picture may be outdated within only a few years after the publication of this book.

From ML to AI: Managing Higher Complexity The training of machine learning models can break down when too many parameters are involved. Even massive training does not improve the model prediction quality any further. The reported bound for Gaussian Process Regression, introduced in Chap. 16, is less than 20 parameters.

A multitude of engineering problems should be solvable within this restriction. Any electronic system the EMC of which depends in a complex manner on a set of more than 20 intertwined and inseparable parameters may be too complex to be realized in a competitive design.

However, at some point ML models will be established as a convenient method to produce insight in the EMC properties of electronic systems, and in the relation between the parameters that determine it. The hunger will come to characterize more complex systems, and to add more parameters, e.g., from other domains, such as performance, thermal design, and environmental impact. Methods to tackle high-dimensional problems are available, for instance in image processing or large-language models. AI methods have already been tested on EMC problems, such as the routing on PCBs, or the placement of decoupling capacitors. High-dimensional AI models are successfully applied in scenarios where gathering of an enormous amount of training data is possible. The concept of transfer learning teaches how to leverage the results of already existing surrogate models to generate novel models.

We can ask the following research question: how can we extend the number of parameters to train, or likewise the parameter range on which EMC models can operate. Equivalently, which quality can a model achieve without the need for massively more training data? The rich research in the ML and AI community provides certainly answers.

Physics-Informed Learning The machine learning models presented in Chap. 16 are purely data driven. They do not make use of any physical law. This is counterintuitive, because in the last 150 years, we have learned to solve Maxwell's equations in so many powerful ways. So obvious the combination of physical models with machine learning appears, so tricky is its implementation. There are two strategies. Physics-informed models train on data and penalize model evaluations which are far from a physical response. During the training process, nonphysical responses tend to be discarded, as a function of the strength of the penalization term. In the resulting model, nonphysical answers are less likely, but possible, as a matter of the intensity of the training.

Physics-induced models incorporate the laws of physics into the model itself. Nonphysical responses are impossible. A prominent example is a physically constrained Gaussian process. These can be derived for any linear differential equation or linear constraint of the physical system under investigation. Physically constrained Gaussian processes use that fact that the sum of Gaussian processes is again a Gaussian process and derive, based on the linear constraint of the physical system, a correlation matrix that inherently produces only physically consistent solutions.

The goal of applying physical laws to learning is to reduce the number of training samples or improve surrogate model accuracy. The intuitive charm of physical knowledge is the better understanding of the problem, which reduces the massive trial and error of the training process. For abstract examples such as a plane wave in a two-dimensional domain, this relation turns out true.

Unfortunately, many more complex problems, and EMC models are among them, show the opposite trend. Physics-informed models require more training data, because the selection process is harder and nonphysical solutions are discarded. Physics-induced learning increases the complexity of the surrogate to be trained. The optimization process becomes much more time-consuming, and the runtime of the resulting surrogate increases significantly. Small errors in the estimates of the model coefficients can have deteriorating impact on the model results, because physical laws are inconsistently applied.

At this state, the research task is to dismantle the apparent contradiction that adding information to a model increases the effort rather than to streamline it. The application range must be identified. Whenever massive data is available, a machine that learns 24/7 converges to a good model without the need to feed further insight. If data is rare, we have a hint that the underlying physical laws are hard to solve. The research questions to be addressed here need to target to which level physical information is introduced and how this is done. With too little physics, we cannot expect any reduction in training effort. With too complex laws, either we restrict ourselves to very simple scenarios or we end up in equation systems which are very time-consuming to solve.

Risk-Based EMC Design The capability to run simulation models millions of times opens up new application areas and the solution of problems unsolved today. Design for EMC targets to produce electronic systems which work reliably and safely in human hands. The infinite variety of scenarios in which electronic systems operate has led to agreements on standardized test setups as physically sensible representatives. If successfully tested in these scenarios, the system is considered electromagnetically compliant.

Decades of successful operation of myriads of electronic systems in even more various scenarios have proven the success of this approach. However, at times there are failures in the field. The examination of these failures can only be made after the incident. Overdesign based on experienced-based safety margins is a common strategy to reduce the probability of failure. But with ever-new emerging technology, the diversity of scenarios to build the experience on shrinks, and estimates become uncertain.

Risk-based EMC design quantifies the risk of failure (many references are listed at the end of this chapter). It investigates the multitude of scenarios in the field, instead of hooking a pass-or-fail criterion to individual, well-engineered reference scenarios. Incidences with very low probability of failure are studied, because the risk of failure must be small.

The study of rare events is a well-known field of statistics. A good model of all possible scenarios is as essential as the study of a large amount of them: if the probability of failure shall be as low as 10^{-6}, we need to study at least ten times the number of cases to reliably tell the transition from safe to noncompliant states.

Machine learning models allow to run more than 10 million cases a day, enabling to assess events that occur with a probability of 10^{-6}. Based on physical models, these models can be made to represent realistic scenarios. Test runs of electromagnetic compliance can be performed based on numerical data in the same way as automatic driving scenarios are partly produced by massive computer simulations.

The modeling of the complex coupling paths in multi-source electromagnetic environments may turn out complex. The risk-based EMC based on surrogates of physical models requires to answer (at least) the following research questions:

- Can we extend the modeling of individual electronic systems to realistic assemblies of them?
- Can surrogate models represent a set of scenarios which is sufficiently rich to describe a useful range of realistic scenarios?
- Are they sufficiently accurate to include the study of rare events?

These are exciting and challenging research questions, which require to significantly extend our present knowledge about machine learning and artificial intelligence in EMC research.

Further Reading

Complex Systems

I. Echeverria et al., Common mode noise propagation and effects in a four-wheel drive electric vehicle. IEEE Trans. Electromagn. Compat. **60**(1), 132–139 (2018). https://doi.org/10.1109/TEMC.2017.2701885

K. Ghosh et al., Growth of radiated emission in multi-modular systems. IEEE Trans. Electromagn. Compat. **62**(2), 612–616 (2020). https://doi.org/10.1109/TEMC.2019.2907455

J. Hein, J. Hippeli, T.F. Eibert, Efficient EMC parameter analysis for the verification of complex automotive simulation models by the utilization of design of experiments. IEEE Trans. Electromagn. Compat. **60**(6), 1965–1973 (2018). https://doi.org/10.1109/TEMC.2017.2784879

D. Liu et al., Accurate evaluation of field interactions between cable harness and vehicle body by a multiple scattering method. IEEE Trans. Electromagn. Compat. **59**(2), 383–393 (2017). https://doi.org/10.1109/TEMC.2016.2611006

D.G. Lopez, M. Ignatenko, D.S. Filipovic, Eigenmode prediction of high RF exposure frequency region inside vehicles. IEEE Trans. Electromagn. Compat. **59**(1), 43–47 (2017). https://doi.org/10.1109/TEMC.2016.2593023

F. de Paulis, C. Olivieri, System level cumulative EM radiation prediction from stacked machines in data centers. IEEE Trans. Instrum. Meas. **74**, 1–12 (2025). https://doi.org/10.1109/TIM.2025.3551022

R.S. Rezende et al., Multi-output variable-fidelity Bayesian optimization of a common mode choke, in *2021 IEEE International Joint EMC/SI/PI and EMC Europe Symposium* (2021), pp. 1093–1098. https://doi.org/10.1109/EMC/SI/PI/EMCEurope52599.2021.9559196

I. Wu et al., Radiated noise measurement from multiple LED lights using reverberation chamber, in *2023 International Symposium on Electromagnetic Compatibility—EMC Europe* (2023), pp. 1–4. https://doi.org/10.1109/EMCEurope57790.2023.10274397

W. Zhang et al., System-level EMI of an artificial router system with multiple radiators: prediction and validation. IEEE Trans. Electromagn. Compat. **62**(4), 1601–1610 (2020). https://doi.org/10.1109/TEMC.2020.3006517

Machine Learning and Artificial Intelligence

G. Andrieu, N. Ticaud, F. Lescoat, On the risk to declare EMC compliant a faulty EUT during radiated susceptibility tests in reverberation chambers. IEEE Trans. Electromagn. Compat. **62**(3), 645–653 (2020). https://doi.org/10.1109/TEMC.2019.2919247

T. Bdour, C. Guiffaut, A. Reineix, Use of adaptive kriging metamodeling in reliability analysis of radiated susceptibility in coaxial shielded cables. IEEE Trans. Electromagn. Compat. **58**(1), 95–102 (2016). https://doi.org/10.1109/TEMC.2015.2501899

M. Das et al., Time-efficient EMI risk evaluation method in a hospital environment, in *2021 IEEE International Joint EMC/SI/PI and EMC Europe Symposium* (2021), pp. 857–861. https://doi.org/10.1109/EMC/SI/PI/EMCEurope52599.2021.9559356

H. De Gersem et al., Magnetic field simulation with data-driven material modeling. IEEE Trans. Magn. **56**(8), 1–6 (2020). https://doi.org/10.1109/TMAG.2020.3002092

L. Devaraj et al., Risk assessment approach for EM resilience in complex systems using Bayesian networks, in *2021 IEEE International Joint EMC/SI/PI and EMC Europe Symposium* (2021), pp. 760–764. https://doi.org/10.1109/EMC/SI/PI/EMCEurope52599.2021.9559243

E. Ecik et al., A statistically evaluated decision tree approach for SI-compliant PCB design, in *2024 International Symposium on Electromagnetic Compatibility—EMC Europe* (2024), pp. 140–145. https://doi.org/10.1109/EMCEurope59828.2024.10722564

A. Galetzka, D. Loukrezis, H. De Gersem, Data-driven solvers for strongly nonlinear material response. Int. J. Numer. Methods Eng. **122**(6), 1538–1562 (2021). https://doi.org/10.1002/nme.6589. https://onlinelibrary.wiley.com/doi/pdf/10.1002/nme.6589. https://onlinelibrary.wiley.com/doi/abs/10.1002/nme.6589

F. Garbuglia, D. Deschrijver, T. Dhaene, On the role of Bayesian learning for electronic design automation: a survey. IEEE Electromagn. Compat. Mag. **12**(4), 77–84 (2023). https://doi.org/10.1109/MEMC.2023.10466425

E. Garcia, Electromagnetic compatibility uncertainty, risk, and margin management. IEEE Trans. Electro-Magnetic Compatibility **52**(1), 3–10 (2010). https://doi.org/10.1109/TEMC.2009.2035108

V. Gkatsi et al., Accurate terminology and uniform methodology for risk-based EMC engineering. IEEE Letters on Electro-Magnetic Compatibility Practice Appl. **7**(1), 5–13 (2025). https://doi.org/10.1109/LEMCPA.2024.3506855

S. Guglani et al., Artificial neural networks with fast transfer learning for statistical signal integrity analysis of MWCNT and MLGNR interconnect networks. IEEE Trans. Electro-magnetic Compatibility **66**(3), 939–948 (2024). https://doi.org/10.1109/TEMC.2024.3360279

T. Houret et al., Controlled stratification based on kriging surrogate model: an algorithm for determining extreme quantiles in electromagnetic compatibility risk analysis. IEEE Access **8**, 3837–3847 (2020). https://doi.org/10.1109/ACCESS.2019.2961851

C. Jidling et al., *Linearly constrained Gaussian processes*. NeurIPS 2017 (2017). arXiv: 1703.00787 [stat.ML]. https://arxiv.org/abs/1703.00787

G.E. Karniadakis et al., Physics-informed machine learning. Nat. Rev. Phys. **3**(6), 422–440 (2021). ISSN: 2522-5820. https://doi.org/10.1038/s42254-021-00314-5

K. Kontolati et al., A survey of unsupervised learning methods for high-dimensional uncertainty quantification in black-box-type problems, J. Comput. Phys. **464**, 111313 (2022). ISSN: 0021-9991. https://doi.org/10.1016/j.jcp.2022.111313. https://www.sciencedirect.com/science/article/pii/S0021999122003758.

M. Lange-Hegermann, Algorithmic linearly constrained gaussian processes, in *Advances in Neural Information Processing Systems*, ed. by S. Bengio et al., vol. 31. (Curran Associates, Inc., New York, 2018). https://proceedings.neurips.cc/paper_files/paper/2018/file/68b1fbe7f16e4ae3024973f12f3cb313-Paper.pdf

D. Loukrezis et al., High-dimensional uncertainty quantification for an electrothermal field problem using stochastic collocation on sparse grids and tensor train decompositions. Int. J. Numer. Modell. (2017). https://doi.org/10.1002/jnm.2222

D. Loukrezis, A. Galetzka, H. De Gersem, Robust adaptive least squares polynomial chaos expansions in high-frequency applications. Int. J. Numer. Modell. Electron. Networks Devices Fields **33**(6), e2725 (2020). https://doi.org/10.1002/jnm.2725. https://onlinelibrary.wiley.com/doi/pdf/10.1002/jnm.2725. https://onlinelibrary.wiley.com/doi/abs/10.1002/jnm.2725

J.B. Nagel, B. Sudret, A unified framework for multilevel uncertainty quantification in Bayesian inverse problems. Probab. Eng. Mech. **43**, 68–84 (2016)

J.B. Nagel, J. Rieckermann, B. Sudret, Principal component analysis and sparse polynomial chaos expansions for global sensitivity analysis and model calibration: Application to urban drainage simulation. Reliab. Eng. Syst. Saf. **195** (2020). ISSN: 0951-8320. https://doi.org/10.1016/j.ress.2019.106737. https://www.sciencedirect.com/science/article/pii/S0951832019301747

P. Offermann et al., Uncertainty quantification and sensitivity analysis in electrical machines with stochastically varying machine parameters. IEEE Trans. Magn. **51**(3), 1–4 (2015). https://doi.org/10.1109/TMAG.2014.2354511

M. Raissi, P. Perdikaris, G.E. Karniadakis, *Physics Informed Deep Learning (Part I): Data-driven Solutions of Nonlinear Partial Differential Equations*. arXiv preprint, arXiv:1711.10561. 2017. https://arxiv.org/abs/1711.10561

S. Ranftl, W. von der Linden, Bayesian surrogate analysis and uncertainty propagation. Physical Sciences Forum **3**(1), 6 (2021). ISSN: 2673-9984. https://doi.org/10.3390/psf2021003006. https://www.mdpi.com/2673-9984/3/1/6.

A.R. Ruddle, A.J.M. Martin, Adapting automotive EMC to meet the needs of the 21st century. IEEE Electromagn. Compat. Mag. **8**(3), 75–85 (2019). https://doi.org/10.1109/MEMC.2019.8878241

F. Sabath, A systematic approach for electromagnetic interference risk management. IEEE Electromagn. Compat. Mag. **6**(4), 99–106 (2017). https://doi.org/10.1109/MEMC.0.8272296

N.G. Shoaee et al., Enhanced reinforcement learning methods for optimization of power delivery networks on PCB, in *2024 International Symposium on Electromagnetic Compatibility—EMC Europe* (2024), pp. 157–161. https://doi.org/10.1109/EMCEurope59828.2024.10722737

V. Yaghoubi et al., Sparse polynomial chaos expansions of frequency response functions using stochastic frequency transformation. Probab. Eng. Mech. **48** (2016). https://doi.org/10.1016/j.probengmech.2017.04.003. https://ethz.ch/content/dam/ethz/special-interest/baug/ibk/risk-safety-and-uncertainty-dam/publications/reports/RSUQ-2016-006.pdf

Appendix: Notation

Notation of Electrical Symbols

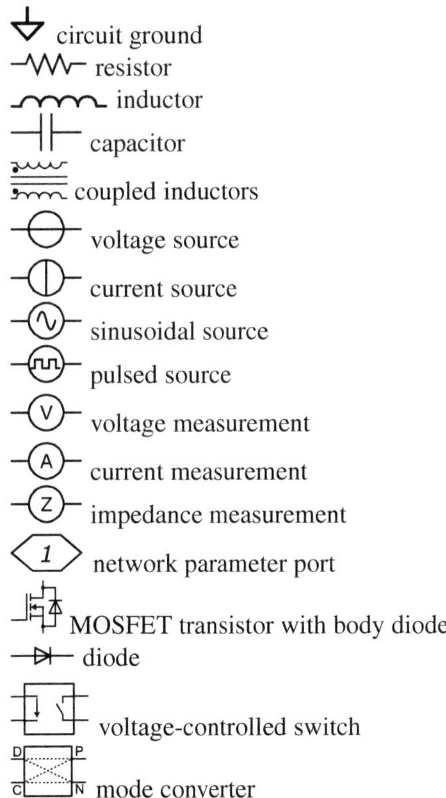

J. Hansen, A. Barchanski, *EMC Simulation of Electronic Systems*,
https://doi.org/10.1007/978-3-032-09237-3

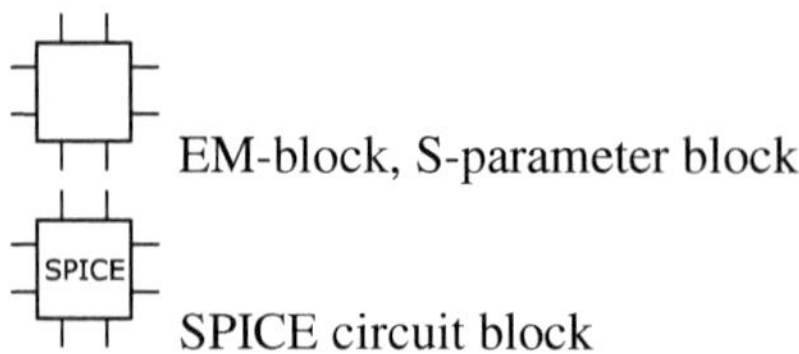
EM-block, S-parameter block
SPICE
SPICE circuit block

Nomenclature

Electronics and Electromagnetics

A	Cross section (area) of an object
f	Frequency
k, k_{coup}	Mutual inductive coupling coefficient
l	Length of an object
n	Natural number, e.g., number of ports of a network
nD	n-dimensional, e.g., 2D and 3D
Q	Electric charge
S	Surface of an object
T	Time after which a periodic signal repeats itself
t	Time
AF	Antenna factor
$\bar{A}$	Amplitude of a signal in time domain
δ_{sk}	Skin depth
η	Thickness of an object
γ, γ_i	Propagation constant (of a transmission line mode i)
$\hat{\mathbf{e}}_i$	Unit vector in direction of coordinate i
$\hat{\mathbf{n}}$	Unit normal vector to a surface
λ	Wavelength
$\mathbf{A}$	The magnetic vector potential
$\mathbf{a}$	Incident power wave vector with entries b_i, related to an S-parameter matrix $\mathbf{S}$
$\mathbf{B}$	Magnetic flux density
$\mathbf{b}$	Reflected power wave vector with entries b_i, related to an S-parameter matrix $\mathbf{S}$
$\mathbf{b}_s$	Vector containing the sources (right-hand side) in FEM or FIT
$\mathbf{C}$	Capacitance matrix
$\mathbf{C}'$	Per unit length capacitance matrix
$\mathbf{D}$	Electric flux density

J. Hansen, A. Barchanski, *EMC Simulation of Electronic Systems*,
https://doi.org/10.1007/978-3-032-09237-3

$\mathbf{E}$	Electric field
$\mathbf{E}_{tr}$	Transverse electric field, i.e., perpendicular to the direction of propagation
$\mathbf{E}_t$	Electric field tangential to a surface
$\mathbf{G}$	Conductance matrix
$\mathbf{G}'$	Per unit length conductance matrix
$\mathbf{H}$	Magnetic field
$\mathbf{H}_{tr}$	Transverse magnetic field, i.e., perpendicular to the direction of propagation
$\mathbf{H}_t$	Magnetic field tangential to a surface
$\mathbf{I}$	Current (vector)
$\mathbf{J}$	Electric current density
$\mathbf{L}$	Inductance matrix
$\mathbf{L}'$	Per unit length inductance matrix
$\mathbf{M}$	Magnetic current density.
$\mathbf{r}$	Position in 3D-space
$\mathbf{R}'$	Per unit length resistance matrix
$\mathbf{S}$, $S_{i,j}$	Scattering parameter (S-parameter) matrix, S-parameter between ports i, j
$\mathbf{T}_I$	Similarity transform between modal and nodal currents on a transmission line
$\mathbf{T}_V$	Similarity transform between modal and nodal voltages of a transmission line
$\mathbf{V}$	Voltage (vector)
$\mathbf{x}$	Vector of unknowns in the FEM or FIT (e.g., discretized electric and magnetic fields)
$\mathbf{Y}$	Admittance matrix.
$\mathbf{Z}$	Impedance matrix
$\mathcal{A}$	System matrix describing the FEM or FIT matrices
$\mathcal{H}$	Transfer function that relates the output of a system with its input
μ_∞	Permeability at infinite frequency in the Debye model
μ_s	Static permeability in the Debye model
ω	Angular frequency
ϕ	The scalar electric potential
ρ_c	Electrical resistivity, $\rho_c = \frac{1}{\sigma c}$
ρ_V, ρ_S	Electric charge density (volume/surface)
σ_c, σ_i	Electrical conductivity (of material i)
τ	Duty cycle, $\tau = \frac{t_{on}}{T}$
τ_r	Magnetic relaxation time in the Debye model
Δf_{RBW}	Resolution bandwidth of the EMI receiver
Φ	Magnetic flux
φ	Phase of a complex number of signal
a_i	Amplitude of the incident power wave at port i
B_s	Saturation flux density

b_i	Amplitude of the reflected power wave at port i
C, C_i	Capacitance (of element i)
C_DS	Drain source capacitance of a transistor
C_D	Diode reverse capacitance
C_x	Capacitor which is mounted between a positive and a negative line. In a layout, this may look like an x
C_y	Capacitor which is mounted from a line to the PCB ground. In a layout, this may look like a y
d, D, s	Distance between two objects
E_i	Electric field strength of the field component i
E	Absolute value of the electric field
f_relax	Magnetic relaxation frequency in the Debye model
f_r, f_i	Resonance frequency (ith resonance)
f_sw, $f_\mathrm{c,i}$	Clock frequency (of source i)
f_tr	Frequency above which the spectrum of a trapezoidal signal with rise and fall time t_r decays with -40 dB/dec
G, G_i	Conductance (of element i) of a transmission line
I, I_i	Current (scalar, through element, or edge i)
I_CM	Common-mode current of two reference currents I_P, I_N, $I_\mathrm{CM} = I_\mathrm{P} + I_\mathrm{N}$
I_DM	Differential-mode current of two reference currents I_P, I_N, $I_\mathrm{DM} = \frac{1}{2}(I_\mathrm{P} - I_\mathrm{N})$
I_DS	Drain-source current through a transistor
I_L	Load current, e.g., of a power converter
I_P, I_N	Currents flowing in the positive and negative supply line
L, L_i	Inductance (of element i)
l_λ	Electrical length of a component
l_λ	Electrical length of an object
L_L	Load inductance
n_w	Number of wires in a wire harness
P_fwd	Forward power, square of the absolute value of the incident power wave
R, R_i	Resistance (of element i)
r, r_i	Radius of a circle or wire, or distance $\lvert\mathbf{r}\rvert$ to a chosen origin
R_L	Load resistance
R_on	On-resistance of a transistor
t_f	Fall time of a signal
t_on	Time during which a digital signal is HI
$t_\mathrm{r,f}$	Signal with equal rise and fall time, $t_\mathrm{r} = t_\mathrm{f}$
t_r	Rise time of a signal
V, V_i, V_n	Voltage (scalar, over element i, or at node n)
V_CM	Common-mode voltage of two reference voltages V_P, V_N, $V_\mathrm{CM} = \frac{1}{2}(V_\mathrm{P} + V_\mathrm{N})$
V_DM	Differential-mode voltage of two reference voltages V_P, V_N, $V_\mathrm{DM} = (V_\mathrm{P} - V_\mathrm{N})$

V_{DC}	DC voltage, e.g., from a battery
V_{DS}	Drain-source voltage across a transistor
V_{in}	Input voltage
$V_{LISN,N}$	Voltage at the 50 Ω measurement output of the LISN in the negative supply path
$V_{LISN,P}$	Voltage at the 50 Ω measurement output of the LISN in the positive supply path
V_{out}	Output voltage
$V_P,\ V_N$	Voltages in the positive and the negative supply line with respect to a reference conductor such as a metallic table plate
V_S	Voltage of an ideal voltage source (with source impedance)
$w,\ h$	Width and height of a structure
$W,\ W_i$	Energy (of type i, e.g., the energy of the electric field W_e)
X, X_i	Reactance (of element i)
$Z,\ Z_i$	Impedance (of element i) in Ω
Z_0	Characteristic impedance of a transmission line
Z_T	Transfer impedance of a shielded wire or connector or of a current clamp
E	$E = \|\mathbf{E}\|$ Magnitude of the electric field
Machine Learning	
K	Dimension of the design space of a Gaussian process
M	Number of frequency samples simultaneously predicted by the Gaussian Process
N	Number of training samples of a Gaussian process
x	Input parameter of a Gaussian process
$\boldsymbol{\kappa}$	Kernel function/covariance matrix of a machine learning model
κ_{ij}	i, jth entry of the Kernel function $\boldsymbol{\kappa}$ of a machine learning model
$\boldsymbol{\mu}$	Mean value vector of a Gaussian process
$\boldsymbol{\Sigma}$	Covariance matrix of a vector or matrix variate Gaussian process
$Y,\ y$	Scalar random variable and its realization
$\mathbf{M}$	Mean value matrix of a matrix variate Gaussian process
$\mathbf{R}_{xx}$	Correlation matrix of random vectors $\mathbf{x}$
$\mathbf{X}$	Matrix of training samples of a machine learning model
$\mathbf{x}_i$	ith design parameter vector (out of N training samples) to train a machine learning model
$\mathbf{Y}$	Training sample matrix of a machine learning model, N samples with p frequencies each
$\mathbf{y}_i$	ith out of N training sample vectors of a machine learning model with p frequencies each
$\mathcal{MN}$	Matrix variate normal distribution
$\mathcal{N}$	Normal distribution
$\mathcal{V}_X$	Volume of the object X
μ	Mean value of a normal distribution

σ	Standard deviation of a Gaussian process
θ_1, θ_2	Vector/ith entry of the hyperparameters of a machine learning model
$P(.)$	Penalty term used in multi-objective optimization
x_{ij}	ij-th entry of the matrix of training samples $\mathbf{X}$ of a machine learning model
y_{ij}	ij-th entry of the training sample matrix $\mathbf{Y}$ of a machine learning model

Mathematical Notation

j	The imaginary unit $\sqrt{-1}$
$\bar{A}$	Amplitude of a sinusoidal signal
$\det \mathbf{X}$	Determinant of the matrix $\mathbf{X}$
$\frac{\mathrm{d}}{\mathrm{d}x}$	Total derivative
$\frac{\partial}{\partial x}$	Partial derivative
$\hat{x}$	Estimate of the value of x using a statistical estimator
$\iiint \mathrm{d}V$	Volume integral
$\Im$	Imaginary part of a complex number
$\mathbf{1}$	Column vector having all values as 1
$\mathbf{I}_N$	Identity matrix of size $N \times N$
$\mathbf{X}^T$	Transpose of the matrix $\mathbf{X}$
$\mathcal{J}$	Jacobian matrix
$\mathrm{diag}(\alpha_1, \ldots, \alpha_N)$	Diagonal matrix with diagonal elements α_i, $i = 1, \ldots, N$
$\mathrm{tr}\,\mathbf{X}$	Trace of the matrix $\mathbf{X}$
$\nabla f(x)$	Gradient of the scalar function $f(x)$
$\nabla\cdot$	Divergence
$\nabla\times$	Curl
$\oiint_S \mathrm{d}\mathbf{S}$	Closed surface integral over the surface S
$\oint_l \mathrm{d}\mathbf{l}$	Closed contour integral over the curve l
$\otimes$	Kronecker product of two matrices
$\Re$	Real part of a complex number
$\times$	Cross product between two vectors
$\underline{\mathbf{A}}$	Phasor of the vector $\mathbf{A}$, $\underline{\mathbf{A}} = \bar{\mathbf{A}} \exp j(\omega t + \varphi)$
$\underline{A}$	Phasor of the scalar A, $\underline{A} = \bar{A} \exp j(\omega t + \varphi)$
$\vert.\vert$	Absolute value / magnitude
$\vert.\vert$	Absolute value of a complex number
$E[.]$	Expectation value of a random variable

Mathematical and Physical Constants

μ	Complex permeability $\mu = \mu_0\mu_\mathrm{r}$
μ_0	Permeability of vacuum, $\mu_0 \approx 4\pi \cdot 10^{-7}\,\frac{H}{m}$
μ_r	Complex relative permeability, $\mu_\mathrm{r} = \mu_\mathrm{r}' - j\mu_\mathrm{r}''$
π	The number π
ε	Complex permittivity $\varepsilon = \varepsilon_0\varepsilon_\mathrm{r}$
ε_0	Permittivity of vacuum, $\varepsilon_0 = 8.854\ldots \cdot 10^{-12}\,\frac{F}{m}$
ε_r	Complex relative permittivity, $\varepsilon_\mathrm{r} = \varepsilon_\mathrm{r}' - j\varepsilon_\mathrm{r}''$
c_0	Speed of light in vacuum, $c_0 = 2.99792\ldots \cdot 10^8\,\frac{m}{s}$

Acronyms

100BASE-T1 100BASE-T1 automotive ethernet standard.

AC Alternate Current.
AI Artificial intelligence.
ALSE Absorber-lined shielded enclosure.
AV Average detector of an EMI receiver.

BCI Bulk Current Injection.

CAD Computer aided design.
CAE Computer aided engineering.
CATIA Computer-Aided Three-Dimensional Interactive Application, a CAD system made by the French company Dassault Systèmes.
CB Citizens Band radio (26 — 28 MHz).
CE Conducted emission.
CFL Courant–Friedrichs–Lewy condition.
CISPR 25 The CISPR 25 automotive EMC standard for conducted and radiated emission.
CM Common Mode.
CMC Common mode choke.
CPU Central processing unit.
CST The software CST Studio Suite from Dassault Systè mes.

DAB Digital Audio Broadcasting radio band (170 — 230 MHz).
DC Direct Current.
DFT Discrete Fourier Transform.
DM Differential Mode.
DPT Double pulse test.
DUT Device under test.

J. Hansen, A. Barchanski, *EMC Simulation of Electronic Systems*,
https://doi.org/10.1007/978-3-032-09237-3

ECU Electronic control unit.
EDA Electronic design automation.
EM Electromagnetic. When used as EM simulation, we refer to geometry based simulation methods that solve Maxwell's Equations on a grid..
EMC Electromagnetic Compatibility.
EMI Electromagnetic Interference.
ESD Electrostatic discharge.
ESL Equivalent Series Inductance.
ESR Equivalent Series Resistance.

FCC Federal Communications Commission.
FDFD Finite difference frequency domain method.
FDTD Finite difference time domain method.
FEM Finite Element Method.
FFT Fast Fourier Transform.
FIT Finite Integration Method.
FM Frequency Modulation radio band (76 — 108 MHz).
FMM Fast multipole method.
FR4 Glass-reinforced epoxy laminate material, used as a substrate material of PCBs.

GaN Gallium nitride.
GND Ground node (circuit) or ground net (layout).
GPR Gaussian process regression.
GPU Graphics processing unit.

HF High Frequency (as opposed to DC).

IBIS I/O Buffer Information Specification.
IC Integrated Circuit.
ICNIRP International Committee for Non Ionizing Radiation Protection.
IDEM A commercial implementation of the vector fitting algorithm (IdEM, Dassault Systemes).
IEC International Electrotechnical Commission.
ISO International Organization for Standardization.
ITU International Telecommunication Union.

KBL Kabelbaumliste (Harness Description List), XML-data format for the description of wire harnesses.
KCL Kirchhoff's current law.

LISN Line Impedance Stabilization Network.
LTE Long term evolution standard for wireless broadband communication.
LTI Linear time invariant system.
LVDS Low voltage digital signaling.

LW Long Wave broacast band (0.15 — 0.30 MHz).

ML Machine learning.
MLFMM Multilevel fast multipole method.
MoM Method of Moments.
MOO Multi-objective optimization.
MOSFET Metal-oxide-semiconductor field-effect transistor.
MQS Magnetoquasistatic.
MTL Theory for Multiconductor Transmission Lines.
MW Medium Wave broadcast band (0.53 — 1.8 MHz).

NEC Numerical Electromagnetics Code, a popular antenna modeling computer code.
NN Neural Network.

PAM3 Pulse Amplitude Modulation with three amplitude levels.
PBA Perfect boundary approximation.
PCB Printed circuit board.
PCE Polynomial Chaos Expansion.
PDN Power delivery network.
PEC Perfect electric conductor.
PEEC Partial Element Equivalent Circuit.
PI Power integrity.
PK Peak detector of an EMI receiver.
PMC Perfect magnetic conductor.
PML Perfectly matched layer boundary condition.
PRBS Pseudo Random Binary Sequence.
PWL Piece-wise linear.
PWM Pulse width modulation.

QP Quasi-peak detector of an EMI receiver.

RAM Random-Access Memory.
RBW Receiver bandwidth.
RCS Radar cross section.
RE Radiated emission.
RF Radio Frequency, roughly between the upper limit of audio frequencies and the lower limit of the infrared spectrum.
RI Radiated immunity.
RLC As in RLC models: models that consist only of resistors (R), Inductors (L), and capacitors (C).
RLCG Multiconductor transmission line model composed of resistors (R), Inductors (L), capacitors (C), and Conductance (G).
Rx Receiver.

SD Secure Digital, used in SD card.
SDIO Secure Digital Input Output.
SET Surface Equivalent Theorem.
SI Signal integrity.
SIBC Surface impedance boundary condition.
SiC Silicon carbide.
SMA connector type (SubMiniature version A).
SMD Surface-mounted device.
SMPS Switched Mode Power Supply.
SOIC Small outline integrated circuit, a surface-mounted IC package.
SPICE SPICE simulation. (Simulation Program with Integrated Circuit Emphasis).
SSN Simultaneous switching noise.
STEP STEP format: Standard for the Exchange of Product model data.
STFT Short-time Fourier Transform.
STWP Shielded twisted pair wire.
SW Short Wave broadcast band (5.9 — 6.2 MHz).

TDR Time domain reflectometry.
TE Transverse electric mode.
TEM Transverse electromagnetic mode.
TLM Transmission Line Matrix Method.
TM Transverse magnetic mode.
TST Thin sheet technique.
TVS Transient voltage suppression (used e.g. in TVS diode, to protect circuits from fast transients such as occuring in ESD).
TV I TV Band I (41 — 88 MHz).
TWP Twisted pair wire.
Tx Transmitter.

USB Universal Serial Bus.

VCC Positive supply node (circuit) or supply network (layout).
VCSW Voltage controlled switch in a SPICE simulator.
VDD Positive supply node (circuit) or supply network (layout).
VHF Very high frequency, wavelength between 1 m and 10 m.
VHF I VHF Band I (30 — 54 MHz).
VHF II VHF Band II (68 — 87 MHz).

Index

J. Hansen, A. Barchanski, *EMC Simulation of Electronic Systems*,
https://doi.org/10.1007/978-3-032-09237-3